ADDICTIVE SUBSTANCES AND NEUROLOGICAL DISEASE

ADDICTIVE SUBSTANCES AND NEUROLOGICAL DISEASE

ALCOHOL, TOBACCO, CAFFEINE, AND DRUGS OF ABUSE IN EVERYDAY LIFESTYLES

Edited by

RONALD ROSS WATSON

University of Arizona, Arizona Health Sciences Center
Tucson, AZ, USA

SHERMA ZIBADI

Department of Pathology, University of South Florida Medical School
Tampa, FL, USA

Academic Press is an imprint of Elsevier
125 London Wall, London EC2Y 5AS, United Kingdom
525 B Street, Suite 1800, San Diego, CA 92101-4495, United States
50 Hampshire Street, 5th Floor, Cambridge, MA 02139, United States
The Boulevard, Langford Lane, Kidlington, Oxford OX5 1GB, United Kingdom

Notices
Knowledge and best practice in this field are constantly changing. As new research and experience broaden our understanding, changes in research methods, professional practices, or medical treatment may become necessary.

Practitioners and researchers must always rely on their own experience and knowledge in evaluating and using any information, methods, compounds, or experiments described herein. In using such information or methods they should be mindful of their own safety and the safety of others, including parties for whom they have a professional responsibility.

To the fullest extent of the law, neither the Publisher nor the authors, contributors, or editors, assume any liability for any injury and/or damage to persons or property as a matter of products liability, negligence or otherwise, or from any use or operation of any methods, products, instructions, or ideas contained in the material herein.

Library of Congress Cataloging-in-Publication Data
A catalog record for this book is available from the Library of Congress

British Library Cataloguing-in-Publication Data
A catalogue record for this book is available from the British Library

ISBN: 978-0-12-805373-7

For information on all Academic Press publications visit our website at https://www.elsevier.com/books-and-journals

Publisher: Mara Conner
Acquisition Editor: April Farr
Editorial Project Manager: Timothy Bennett
Production Project Manager: Edward Taylor
Designer: Matthew Limbert

Typeset by TNQ Books and Journals

Contents

7. Zebrafish Models of Alcohol Addiction

S. TRAN, R. GERLAI

8. Effect of Alcohol on the Regulation of α-Synuclein in the Human Brain

P. JANECZEK, J.M. LEWOHL

9. Consumption of Ethanol and Tissue Changes in the Central Nervous System

L.O. BITTENCOURT, F.B. TEIXEIRA, K.L. VIEIRA, D.P. SANTOS, C.S.F. MAIA, R.R. LIMA

10. Ethanol Consumption and Cerebellar Disorders

A.C.A. DE OLIVEIRA, B. PUTY, L.K.R. LEÃO, R.M. FERNANDES, C.S.F. MAIA, R.R. LIMA

11. Gene Expression in CNS Regions of Genetic Rat Models of Alcohol Abuse

W.J. MCBRIDE

12. Role of TLR4 in the Ethanol-Induced Modulation of the Autophagy Pathway in the Brain

A. PLA, M. PASCUAL, C. GUERRI

13. Ghrelinergic Signaling in Ethanol Reward

L.J. ZALLAR, H.M. BAUMGARTNER, E.E. GARLING, S. ABTAHI, R. PASTOR, P.J. CURRIE

List of Contributors

S. Abtahi Reed College, Portland, OR, United States

M. Balconi Catholic University of the Sacred Heart, Milan, Italy

V. Batra Louisiana State University Health Sciences Center, Shreveport, LA, United States

H.M. Baumgartner Reed College, Portland, OR, United States

L.O. Bittencourt Federal University of Pará, Belém, Brazil

E.I. Bon Grodno State Medical University, Grodno, Belarus

G. Brolese Federal University of Rio Grande do Sul, Porto Alegre, Brazil

S.J. Brooks University of Cape Town, Cape Town, South Africa

A.N. Bukiya The University of Tennessee Health Science Center, Memphis, TN, United States

R. Camarini University of São Paulo, São Paulo, Brazil

S.C. Cartágenes Federal University of Pará, Belém, Brazil

M.K. Cavalcanti Galdino Federal University of Paraiba, João Pessoa, Brazil

J. Clifford Virginia Commonwealth University, Richmond, VA, United States

C.L. Crutcher, II Louisiana State University Health Science Center, New Orleans, LA, United States

P.J. Currie Reed College, Portland, OR, United States

E. Fontes de Andrade, Jr. Federal University of Pará, Belém, Brazil

J.A. da Silva Federal University of Pelotas, Pelotas, Brazil

A.C.A. de Oliveira Federal University of Pará, Belém, Brazil

K. Dodson Villanova University, Villanova, PA, United States

A.M. Dopico The University of Tennessee Health Science Center, Memphis, TN, United States

N.A. dos Santos Federal University of Paraiba, João Pessoa, Brazil

L. Fattore Institute of Neuroscience-Cagliari, National Research Council, Cagliari, Italy

L.M.P. Fernandes Federal University of Pará, Belém, Brazil

R.M. Fernandes Federal University of Pará, Belém, Brazil

R. Finocchiaro Catholic University of the Sacred Heart, Milan, Italy

E.E. Garling Reed College, Portland, OR, United States

R. Gerlai University of Toronto, Toronto, ON, Canada; University of Toronto Mississauga, Mississauga, ON, Canada

C.-A. Gonçalves Federal University of Rio Grande do Sul, Porto Alegre, Brazil

E. González-Reimers Universidad de La Laguna, Tenerife, Canary Islands, Spain

C. Guerri Príncipe Felipe Research Center, Valencia, Spain

Z.H. Gursky University of Delaware, Newark, DE, United States

S.R. Hauser Indiana University School of Medicine, Indianapolis, IN, United States

J. Ipser University of Cape Town, Cape Town, South Africa

T. Isomura Reset Behavior Research Group, Nagoya, Japan

P. Janeczek Griffith University, Gold Coast Campus, QLD, Australia

J. Kaiser Goethe University, Frankfurt am Main, Germany

M. Kano Shin-Nakagawa Hospital, Yokohama, Japan

A.Y. Klintsova University of Delaware, Newark, DE, United States

G. Langi Virginia Commonwealth University, Richmond, VA, United States

L.K.R. Leão Federal University of Pará, Belém, Brazil

J.M. Lewohl Griffith University, Gold Coast Campus, QLD, Australia

R.R. Lima Federal University of Pará, Belém, Brazil

L.H. Lobo Torres Federal University of Alfenas, Alfenas, Brazil

F. Lopes Federal University of Rio Grande do Sul, Porto Alegre, Brazil

P. Lunardi Federal University of Rio Grande do Sul, Porto Alegre, Brazil

C.S.F. Maia Federal University of Pará, Belém, Brazil

T. Marcourakis University of São Paulo, São Paulo, Brazil

M.T. Marin São Paulo State University (UNESP), Araraquara, Brazil

M.C. Martín-González Universidad de La Laguna, Tenerife, Canary Islands, Spain

W.J. McBride Indiana University School of Medicine, Indianapolis, IN, United States

A.S. Melo Federal University of Pará, Belém, Brazil

D.J. Meyerhoff University of California San Francisco, San Francisco, CA, United States

M.C. Monteiro Federal University of Pará, Belém, Brazil

G. Morais-Silva São Paulo State University (UNESP), Araraquara, Brazil

T. Murai Kyoto University, Kyoto, Japan

M.J. Naumer Goethe University, Frankfurt am Main, Germany

M. Pascual Príncipe Felipe Research Center, Valencia, Spain

R. Pastor Universitat Jaume I, Castellón, Spain

K.M. Phedina Grodno State Medical University, Grodno, Belarus

B.G. Pinheiro Federal University of Pará, Belém, Brazil

A. Pla Príncipe Felipe Research Center, Valencia, Spain

R.D. Prediger Federal University of Santa Catarina, Florianópolis, Brazil

E. Prom-Wormley Virginia Commonwealth University, Richmond, VA, United States

B. Puty Federal University of Pará, Belém, Brazil

G. Quintero-Platt Universidad de La Laguna, Tenerife, Canary Islands, Spain

J. Real Virginia Commonwealth University, Richmond, VA, United States

Z.A. Rodd Indiana University School of Medicine, Indianapolis, IN, United States

L. Romero-Acevedo Universidad de La Laguna, Tenerife, Canary Islands, Spain

B.D. Sachs Villanova University, Villanova, PA, United States

P. Sahota HSTMV Hospital, University of Missouri, Columbia, MO, United States

F. Santolaria-Fernández Universidad de La Laguna, Tenerife, Canary Islands, Spain

D.P. Santos Federal University of Pará, Belém, Brazil

M.A. Schrager Stetson University, DeLand, FL, United States

R. Sharma HSTMV Hospital, University of Missouri, Columbia, MO, United States

D.J. Stein University of Cape Town, Cape Town, South Africa; MRC Unit on Anxiety & Stress Disorders, Cape Town, South Africa

R.C. Tamborelli Garcia Federal University of São Paulo (UNIFESP), Diadema, Brazil

F.B. Teixeira Federal University of Pará, Belém, Brazil

G.C. Tender Louisiana State University Health Science Center, New Orleans, LA, United States

M.M. Thakkar HSTMV Hospital, University of Missouri, Columbia, MO, United States

S. Tran University of Toronto, Toronto, ON, Canada

J. Veith Louisiana State University Health Science Center, New Orleans, LA, United States

K.L. Vieira Federal University of Pará, Belém, Brazil

J.A. Wilden Willis-Knighton Health Systems, Shreveport, LA, United States

Y. Yalachkov University Hospital Frankfurt, Frankfurt am Main, Germany; Goethe University, Frankfurt am Main, Germany

L.J. Zallar Reed College, Portland, OR, United States

M.T. Zanda University of Cagliari, Cagliari, Italy

W.B. Zheng The Second Affiliated Hospital, Medical College of Shantou University, Shantou, China

S.M. Zimatkin Grodno State Medical University, Grodno, Belarus

Preface

Bioactive foods such as alcohol, tobacco, and caffeine modulate nerves producing neurological diseases and their complications. The effects of such foods produce diseases with neurological function and structure changes that are reviewed by researchers writing *concise*, definitive, and focused chapters. Reflecting the widespread use and abuse of alcohol, a significant portion of the book relates to ethanol and related neurological actions.

ALCOHOL AND NEUROLOGICAL DYSFUNCTION INCLUDING NUTRITIONAL THERAPY

This section primarily focuses on the role of alcohol (ethanol) in the diet and brain function and damage. The mechanisms and symptoms of the wide variety of brain and other neurological systems as caused, accentuated, or modified by addictive foods and alcoholic beverages are discussed. Zheng reviews structural and metabolite changes using noninvasive brain imaging. Brolese and his group describe alcohol's effects on neuroglial changes in neurochemistry and the resulting changes in activity and actions. Zimatkin et al. describe dietary alcohol and histaminergic neurons. Crutcher et al. review excess and debilitating alcohol consumption, as well as the basic and clinical events of intoxication in spinal cord.

Alcohol impacts cellular processes, ranging from involvement in preventing DNA damage to influence in intracellular signaling. Neurological damage due to alcohol affects the functions of the eye and ear due to acute consumption. Cavalcanti Galdino's group reviewed the effects of neurological changes involved. Tran and Gerlai describe the effects of alcohol addiction by developing a novel model, zebrafish. Pinheiro et al. describe a variety of neurological syndromes induced by alcohol. Its actions on brain and neurological disorders causing biochemical, structural, and electrical abnormalities with adverse symptoms are described. Janeczek and Lewohl review specific modulators as changed by alcohol in the human brain, specifically alpha-synuclein. Potential therapy using deep brain stimulation is described for human brains suffering from alcohol-addictive behaviors. This is followed by Bittencourt et al. describing alcohol consumption and related changes in neurological tissues. As the chapter by de Oliveira et al. shows, alcohol abuse is significantly involved in various cerebellar disorders. McBride discusses the role of gene expression and inhibition in neurons of rats in alcohol models. TLR4 actions in alcohol modulation of autophagy pathway are critical to brain functions. Marin and Morais-Silva describe mechanism from lipid changes to specific protein receptor binding. Sachs and Dodson review the effects of serotonin deficiency and alcoholism and alcohol intake. Finally, Gonzalez-Reimers and collaborators review effects of antioxidant vitamin on brain dysfunction in alcohol abusers.

TOBACCO AND NICOTINE IN NEUROMODULATION AND BEHAVIORAL HEALTH

Bioactive drugs of abuse as nonnutrient dietary materials play key roles in accentuating neurological diseases, and possible toxicities and/or lack of function. Yalachkov and Kaiser extend their interest to the reward and habit functions of the brain for a variety of additive materials as an example of how they can function. Alcohol's neurotoxicity directly induces behavior disorders as reviewed by Fernandes et al. Deep brain stimulation is described as a therapeutic technique for treating drug addiction behaviors by Hauser et al. Brooks et al. review the various actions of nicotine on brain functions focusing on fMRI studies with examples. Torres and Garcia evaluate the materials after tobacco combustion in neurotoxicity of the developing brain. Isomura, Murai, and Kano use their experience and the literature to develop new explanations for neuronal functions and psychological changes in nicotine addiction subjects. Frequently heavy smokers also consume lots of alcohol, as if there were a relationship and regulation between the two drugs. McBride reviews the CNS sites and their role in both alcohol and nicotine abuse.

Sharma et al. also discuss alcohol–nicotine co-abuse focusing on the basal forebrain.

DRUGS OF ABUSE AND MECHANISMS AND MODELS OF ADDICTION RELEVANT TO DRUGS OF ABUSE

Several less traditional but frequently used drugs modify nerves and their functions with less damage than alcohol and nicotine. For example, about 85% of adults use caffeine. Bukiya and Dopico look at caffeine's role as a modulator of the brain. Besides, the additional role of caffeine in modulating alcohol's neuromodulation is focused upon. The Cartagenes et al. chapter on ketamine adds neurotoxicity and causes neurobehavioral disorders.

Zanda and Fattore review additional, novel psychoactive substances beyond nicotine as future mental health modulators contributing to behavioral changes. Finocchiaro and Balconi evaluate data showing that hemispheric imbalance affects addiction and may explain addiction in some patients.

Acknowledgments

The work of Dr. Watson's editorial assistant, Bethany L. Stevens, in communicating with authors and working on the manuscripts was critical to the successful completion of the book. The help of Kristi L. Anderson is also very much appreciated. Support for Ms. Stevens' and Dr. Watson's work was graciously provided by Natural Health Research Institute, www.naturalhealthresearch.org—an independent, nonprofit organization that promotes science-based research on natural health and wellness. The institute is committed to informing about scientific evidence on the usefulness and cost-effectiveness of diet, supplements, and a healthy lifestyle to improve health and wellness and to reduce disease. The work of Mari Stoddard, the librarian of Arizona Health Science Library, was vital and very helpful in identifying key researchers who contributed to this book.

P A R T I

ALCOHOL AND NEUROLOGICAL DYSFUNCTION

C H A P T E R

1

Acute Ethanol-Induced Changes in Microstructural and Metabolite Concentrations on the Brain: Noninvasive Functional Brain Imaging

W.B. Zheng

The Second Affiliated Hospital, Medical College of Shantou University, Shantou, China

INTRODUCTION

Alcoholism is a major health issue that afflicts people all over the world. In addition, researchers have raised concerns about health and the social consequences of excessive drinking (Das, Balakrishnan, & Vasudevan, 2006). The consumption of alcohol and the subsequent production of its oxidative metabolites have many direct and indirect effects both on the developing and the developed nervous system, and have acute and chronic complications (Alderazi & Brett, 2007). Alcohol impairs cognitive function and is associated with a variety of behavioral changes resulting in deficits in perceptual and emotional function. Alcohol consumption has immediate effects on multiple cognitive–motor processing domains and leads to damage of multiple attentional abilities (Mongrain & Standing, 1989). Given the high prevalence of alcohol abuse, and the current limited and inefficient treatment options, the need for a better understanding of the effects of alcohol is clear (Nielsen & Nielsen, 2015).

Alcohol and Neurological Dysfunction

Historically, it is well documented that acute alcohol intoxication could result in changes in regional brain function, as assessed by changes in glucose metabolism or cerebral blood flow (Volkow et al., 1988, 1990), cognitive performance (Lau, Pihl, & Peterson, 1995; Lindman, Sjoholm, & Lang, 2000; Reynolds, Richards, & de Wit, 2006), motor function (Lemon, 1993), and behavior (Kong, Zheng, Lian, & Zhang, 2012; Zheng, Kong, Chen, Zhang, & Zheng, 2015; Mongrain & Standing, 1989).

Laboratory studies, which typically examine blood alcohol concentrations (BACs) in the range of 0.03–0.08%, indicate that, at these doses, the influence of alcohol on memory depends on the cognitive functions required by the particular experimental task (e.g., Bisby, Leitz, Morgan, & Curran, 2010; Söderlund, Parker, Schwartz, & Tulving, 2005).

Acute alcohol intake has pronounced effects on brain function in a general way. Until recently, the effect of alcohol on neural mechanisms had not been fully elucidated. These effects include neurotoxicity of the ethanol molecule itself, and the consequences of nutritional deficiencies or liver dysfunction, each of which can lead to the possibility of alcohol-induced neuro-inflammation. A number of measures have been used to investigate these issues. For example, George used the P300 event-related potentials (ERPs) component to assess the relationship of alcoholism with frontal lobe damage, indicating that the frontal lesion subject group of the study had significant P300 amplitude reduction. There was a similar trend for the alcohol-dependent group, but not for the subcortical group when compared to the control subjects (George, Potts, Kothman, Martin, & Mukundan, 2004).

The misuse of alcohol can affect the adult central and peripheral nervous systems, and direct effects are the result of the toxic and intoxicating effects of alcohol. It

Addictive Substances and Neurological Disease
http://dx.doi.org/10.1016/B978-0-12-805373-7.00001-3

is well accepted that alcohol intoxication results in changes in physical and mental impairments including thought, judgment, coordination, concentration, and reasoning. Moreover, alcohol can lead to confusion, ataxia, and loss of social inhibition, and in some circumstances, aggression, putting a person at risk of traumatic injury.

Neuroimaging in Acute Ethanol Consumption

A number of measures have been used to examine the effects of acute ethanol consumption on the brain. Previously, magnetic resonance (MR) imaging has been applied to the study of human chronic alcoholism. Early observations utilizing structural MR imaging studies reported that white matter (WM) subjacent to the cortex and the pons suffers structural volume in those with uncomplicated alcoholism. Specific brain regions affected by chronic alcohol exposure as determined on structural MR imaging include the cortical gray matter (GM) and WM, the thalamus (Bellis et al., 2005), and particularly the prefrontal areas in older alcoholic individuals (Cardenas, Studholme, & Gazdzinski, 2007).

Traditionally, neuroimaging has provided noninvasive anatomic views of the brain, but the applications of neuroimaging have now expanded to include the measurement of neurochemical concentrations and specific proteins, such as subtypes of neurotransmitter receptors or transporters, rates of metabolic pathways and blood flow, and the detection of functional or pharmacological changes in the brain and evaluations of connections among brain regions (Niciu & Mason, 2014). In vivo modern neuroimaging modalities, including diffusion tensor imaging (DTI), magnetic resonance spectroscopy (MRS), and functional magnetic resonance imaging (fMRI), provide important information regarding coexistent structural and functional brain damage, are powerful approaches to the study of brain function, and may provide novel opportunities for investigating the impact of alcohol on the nervous system. The effects of chronic alcohol exposure on the brain and its neurochemistry can be assessed through MRS (Kroenke et al., 2013). In addition, as the microstructural scale can also be assessed through DTI, neuroimaging has dramatically improved researchers' ability to understand the neuropathology of alcoholism (Nagel & Kroen, 2008). Neuroimaging data delineating alcohol effects on brain function are, however, scant. Therefore, the use of these techniques will no doubt provide important information in the near future regarding the mechanisms related to alcohol-induced brain function impairments. This chapter reviews what we have done, to date, regarding the specific imaging features of acute alcohol-induced brain dysfunction, including microstructural changes on the brain and metabolic product concentrations using these noninvasive functional brain imaging techniques.

DTI- and DKI-Detected Acute Ethanol-Induced Changes on Microstructures in the Brain

DTI has identified reduced diffusion anisotropy within the frontal WM of chronic alcoholics (Harris et al., 2008) in the genu of the corpus callosum, the centrum semiovale (Rosenbloom, Sullivan, & Pfefferbaum, 2003), and the frontal and superior sites, such as frontal forceps, internal and external capsules, and the fornix (Pfefferbaum, Rosenbloom, Rohlfing, & Sullivan, 2009), which is interpreted as a manifestation of alcohol-related WM damage. DTI has also revealed evidence for microstructural disruption of WM in alcoholic men and women, even in regions appearing normal on conventional volume imaging (Yeh, Simpson, Durazzo, Gazdzinski, & Meyerhoff, 2009). These findings are consistent with the impaired attention and emotion processing seen with WM fiber disruption (Schulte, Müller-Oehring, Sullivan, & Pfefferbaum, 2012). Individuals with alcohol use disorders underwent DTI, revealing reduced axial diffusivity in the bilateral frontal and temporal WM and lower fractional anisotropy (FA) in bilateral parietal regions, and exhibited abnormalities in subcortical areas associated with sensory processing and memory (Monnig, Tonigan, Yeo, Thoma, & McCrady, 2013). In summary, DTI measures enable examination of the effect of chronic excessive alcohol consumption on the microstructural integrity of major fiber bundles in vivo.

Ethanol is also known to increase the fluidity of cell membranes, thereby causing a change in ion permeability and cell membrane function. It was shown that acute exposure to alcohol induced cellular edema in neonatal rat primary astrocyte cultures and was also associated with a dose-dependent increase in astrocyte volume (Aschner, Mutkus, & Allen, 2001), but the acute effect ethanol exerts on the human brain has not yet been addressed by longitudinal DTI experiments. In the only DTI study of acute alcoholism in humans, researchers demonstrated that ADC values of the frontal lobe, thalamus, and middle cerebellar peduncle were significantly reduced, reaching a minimum value in 1 or 2 h. In contrast, BrAC (BAC) was significantly increased to reach a peak at 0.5 h in both doses and decreased gradually. In addition, FA values (indicating an increase in diffusion along a specific axis or trajectory) of the frontal lobe were significantly increased. Researchers concluded that DTI can detect alcohol-induced cytotoxic brain edemas that are not detectable by conventional magnetic resonance imaging (MRI), and that the frontal lobe, thalamus, and middle cerebellar peduncle are among regions most sensitive to the effects of acute alcohol consumption (Kong et al., 2012).

DTI measurements are based on the assumption of a Gaussian displacement probability distribution of water molecules due to water self-diffusion, such as water in isotropic liquid media (Hui, Cheung, Chan, & Wu, 2010). However, the diffusion of water molecules in most biological tissues, especially brain tissues, is restricted by barriers, such as cellular membranes, which cause diffusion to deviate substantially from a Gaussian form (Filli et al., 2014), making DTI a limited indicator of complexity. Therefore, new imaging methods based on non-Gaussian diffusion models, such as diffusion kurtosis imaging (DKI), may assess microstructural complexity more accurately than DTI, especially in GM (Hori et al., 2012; Jensen & Helpern, 2010; Jensen, Helpern, Ramani, Lu, & Kaczynski, 2005). Moreover, DKI datasets generally include DTI datasets as part of the total measurements (Stokum et al., 2015).

A new 7.0 T MRI study uses DKI to test the microstructural changes in the brains of rats after acute alcohol intoxication. In this study, compared with DTI, DKI can provide a more comprehensive evaluation of EtOH-related brain changes at varying time points, and not only in anisotropic WM, but also in GM. DKI possesses sufficient sensitivity for tracking pathophysiological changes at various stages associated with acute alcohol intoxication and may provide additional information that may be missed by conventional DTI parameters. In addition, the thalamus may be especially vulnerable to effects of acute alcohol intoxication (Chen, Zeng, Kong, & Zheng, 2016).

MRS-Detected Acute Ethanol-Induced Changes on Metabolites in the Brain

Particularly, acute alcohol consumption has marked effects on brain metabolism, by decreasing glucose metabolism throughout the human brain while also causing functional and morphological changes in cells. Alcohol affects multiple neurotransmitter systems in brain and brain functions that depend on a delicate balance between excitation and inhibitory neurotransmission (Mukherjee, Das, Vaidyanathan, & Vasudevan, 2008). Substantial evidence now indicates that alcohol selectively alters the activity of specific complexes of proteins embedded in the membranes of cells that bind neurotransmitters such as gamma-aminobutyric acid (GABA), glutamate, serotonin, acetylcholine, and glycine (White, 2003).

In vivo ^{1}H MRS provides unique information about brain function metabolism. The concentrations of N-acetyl aspartate (NAA), total creatine phosphate phosphocreatine (CrPCr), choline-containing compounds (Cho), myo-inositol (myo-Ins), glutamate (Glu), glutamine (Gln), and glucose (Glc) can be determined using a steady-state ^{1}H MR spectra. It has been repeatedly demonstrated that MRS is the most direct MR-based technique for studying alcohol in the brain. This approach has been used to characterize alcohol pharmacodynamics in animals (Adalsteinsson, Sullivan, Mayer, & Pfefferbaum, 2006; Lee et al., 2012), humans (Biller, Bartsch, Homola, Solymosi, & Bendszus, 2009), and nonhuman primates (Kroenke et al., 2013).

Previous MRS studies have shown the chronic effects of EtOH on regional brain structure and brain metabolism (Ende et al., 2005; Meyerhoff et al., 2004). The results of Braunová may indicate that myo-Ins loss, reflecting a disorder in astrocytes, might be one of the first changes associated with alcoholism, which could be detected in the brain using in vivo ^{1}H MRS (Braunová et al., 2000). Some MRS studies have shown that individuals who have chronically consumed alcohol demonstrate persistent N-acetyl aspartate decreases in the frontal lobe, the thalamus, and the cerebellum, while others have found improvement in the levels of N-acetyl aspartate and choline. Moreover, alcohol interferes with the conversion of thiamine to its metabolically active form, thereby interrupting the production of glucose-derived neurotransmitters. These metabolic deficits can contribute to neuronal and WM damage (Zahr, Kaufman, & Harper, 2011). Most investigators report reduced levels of NAA and Cho in alcoholics compared with those levels in healthy subjects and alcoholics before and after abstinence, respectively. A study examining ^{1}H MRS and high resolution magic angle spinning nuclear magnetic resonance (HR-MAS NMR) spectroscopy in a rat model for long-term alcohol exposure using the liquid diet technique, to assess neurochemical changes in the frontal cortex region, suggested that the reduced myo-Ins concentrations and increased Cho concentrations might be utilized as key markers in chronic alcohol intoxication, providing useful neurochemical information about human chronic alcoholism-related brain damage (Lee et al., 2013). Biller longitudinally characterized cerebral metabolism changes in 15 healthy individuals by^{1}H-MRS subsequent to the ingestion of a standard beverage, and discovered that supratentorial creatine, choline, inositol, and aspartate levels decreased after ethanol administration, whereas glucose levels increased (Biller et al., 2009).

Subsequent research provided additional evidence suggesting a link between ethanol and GABA receptors (Follesa et al., 2015; Silveri, 2014). Follesa measured hippocampal gamma-aminobutyric acid type A (GABAAR) expression at different time points during and after voluntary EtOH consumption following forced EtOH vapor exposure in mice by comparing molecular data with concomitant EtOH concentrations (BECs). The findings indicate that hippocampal GABAAR delta subunit expression changes transiently over the course of a chronic intermittent ethanol (CIE) exposure model associated with voluntary intake, in response to ethanol-mediated disturbance of GABAergic neurotransmission (Follesa et al., 2015). Flory extended ethanol MRS techniques to nonhuman primate subjects, and demonstrated that brain ^{1}H MRS following intravenous infusion of ethanol can be used to perform quantitative measurements of ethanol MRS signal intensity in GM and WM within 12 monkeys (Flory, O'Malley, Grant, Park, & Kroenke, 2010). Furthermore, Kroenke used in vivo MRS to measure the GM and WM ethanol methyl ^{1}H MRS intensity in 18 adult male primates throughout the course of a chronic drinking experiment, and found that chronic exposure to ethanol is associated with brain changes that result in differential increases in ethanol MRS intensity in GM and WM (Kroenke et al., 2013).

Increased Tau levels have been reported in the frontal cortex and nucleus accumbens under acute and chronic ethanol administration in animals (Lee et al., 2012). Acute exposure of ethanol perturbs the level of neurometabolites and decreases the excitatory and inhibitory activity differentially across the regions of the brain (Tiwari, Veeraiah, Subramaniam, & Patel, 2014). ^{1}H MRS has been used to study acute ethanol effects before and during a 1-h intravenous alcohol administration in humans. This study found that ethanol acutely reduced cortical GABA and NAA levels. Reductions in GABA levels are consistent with the facilitation of GABA receptor function by ethanol. The gradual decline in NAA levels suggests the inhibition of neural or metabolic activity in the brain. The time course of ethanol in the brain is similar to its time course in the breath, but was faster than that seen in the venous blood (Gomez et al., 2012). In rat brains, MRS findings of acute metabolite changes after intravenous, intragastric, and intraperitoneal ethanol application revealed stable as well as reduced Cho and stable Cr concentrations (Adalsteinsson et al., 2006).

Diffusion Weighted Imaging- and MRS-Detected Acute Ethanol-Induced Changes in the Brain

Liu et al. designed a study to evaluate brain edema and the metabolism of the rat brain tissue acutely exposed to ethanol, and to assess the capability of these techniques in revealing brain metabolic changes. They examined the rat brain tissue in vivo, by means of diffusion weighted imaging (DWI) together with ^{1}H MRS techniques at a 7.0 T MR, and found that apparent diffusion coefficient (ADC) values in the frontal lobe are lower than other regions at 3 h post exposure. EtOH levels also significantly affected choline, taurine, and glutamate concentrations in the frontal lobes, and EtOH/tCr (tCr: creatine and phosphocreatine) correlates well with these metabolite levels.

The reduction of ADC values in different brain areas reflects the process of cytotoxic edema in vivo. The characterization of metabolic frontal lobe changes and the correlation between metabolic concentrations provide a better understanding of the biological mechanisms in neurotoxic effects of EtOH on the brain. These data provide further evidence that the frontal lobe is more vulnerable to the effects of acute alcohol consumption. Moreover, the correlation between metabolite concentrations and ADC help to understand the development of the ethanol-induced brain cytotoxic edema (Liu et al., 2014).

Resting-State fMRI Study of Alcohol Effects

Another advantage of in vivo MR tools is the ability to conduct behavioral experiments during imaging to determine brain structure–function relationships. Resting-state fMRI (rf-MRI) techniques were applied to demonstrate abnormalities in various neuropsychiatric disorders (Garrity et al., 2007; Zhang et al.,2010). The BOLD signal has been confirmed to indirectly reflect neural activity. The default mode network (DMN) has first been observed as a task-negative network, showing increased metabolic demand during the "baseline" activity and has therefore been hypothesized to reflect intrinsic default brain processes (Raichle & Snyder, 2007).

It may be helpful to further understand about the abnormalities of brain activity in participants in a resting state while under the acute effect of alcohol, as the absence of demanding cognitive activities and instructions might mitigate subjects' differences in motivation or cognitive abilities when comparing brain activity across groups. Numerous studies indicate that acute alcohol intoxication attenuates activity in fronto-parietal areas during high-conflict and error trials, most prominently in the anterior cingulate cortex (ACC), suggesting that cognitive control functions are vulnerable to acute alcohol intoxication (Anderson et al., 2011; Kovacevic et al., 2012; Marinkovic, Rickenbacher, Azma, & Artsy, 2012). Alcohol intoxication may impair top–down regulative functions by attenuating the anterior cingulate cortex (ACC) activity, resulting in behavioral disinhibition and decreased self-control (Marinkovic, Rickenbacher, Azma, Artsy, & Lee, 2013). However, all of these studies examined the brain activation during the fMRI tasks. In contrast, an rf-MRI study investigated the acute effects of alcohol on the human brain by detecting the functional connectivity of DMN, and using ALFF and ReHo it showed that an rf-MRI could detect selectively vulnerable brain regions including the superior frontal gyrus, cerebellum, hippocampal gyrus, basal ganglia, and internal capsule which were affected by alcohol. These different brain regions, which are related to memory, motor control, cognitive ability, and spatial functions might provide a neural basis for alcohol's effects on behavioral performance (Zheng et al., 2015).

Alcohol Intoxication and Traumatic Brain Injury

Decades of research have established that alcohol use results in the deterioration of judgment, alertness, attention, and a loss of fine motor coordination, as well as a slowing in reaction times and a diminishing of sensory perceptions (Ferguson, 2012). There is no question that alcohol impairs the ability to drive safely. Driving while under the influence of alcohol is a major public health issue whose neural basis is not well understood. Alcohol intoxication is a significant risk factor for TBI, and TBI should be appreciated as a heterogeneous, dynamic pathophysiological process that occurs at the moment of impact and continues over time with sequelae potentially seen many years after the initial event (Currie et al., 2016).

Today the treatment of traumatic brain edema remains a therapeutic challenge, and diagnosis is still largely symptomatic in nature. All treatment modalities presently used are focused on decreasing intracranial pressure. For example, steroids are postulated to seal the endothelial lining, thus, lessening vasogenic brain edema formation. The prevalence of cytotoxic edema formation, however, might explain the limited efficacy of steroids to treat traumatic brain edema (Unterberg, Stover, Kress, & Kiening, 2004). Clinical trials targeting cytotoxic and vasogenic mechanisms of edema formation may benefit from using DWI and FLAIR MRI as a means to differentiate the predominant edema type after TBI (Hudak et al., 2014). Ethanol is known to increase the fluidity of cell membranes, thus causing a change in ion permeability and cell membrane function. Acute EtOH intoxication has been shown to increase the permeability of the blood–brain barrier (BBB) in the injured area following cerebral stab wounds, but the extent of the BBB disruption and brain edema that follows TBI is obviously intensified under conditions of acute EtOH intoxication (Yamakamiet et al., 1995). Brain edema is a critical event in the pathophysiology of TBI, and ethanol adversely affects morbidity and mortality after TBI by accelerating brain edema (Katada et al., 2009). Increased brain edema has been described in TBI rats receiving higher doses of alcohol compared to TBI rats exposed without alcohol (Opreanu, Kuhn, & Basson, 2010).

To characterize the effect of acute EtOH intoxication on the brain following TBI, Kong et al. used DTI and evaluated aquaporin-4 (AQP4) expression changes in rat brain stems following acute alcohol intoxication with diffuse axonal injury (DAI). The results showed changes in ADC and FA values in DAI with acute alcoholism indicating that ethanol can aggravate brain edema and the severity of axonal injury. The correlations between ADC values and the brainstem AQP4 expression at different time points suggest that AQP4 expression follows an adaptative profile to the severity of brain edema (Kong, Lian, Zheng, Liu, & Zhang, 2013).

CONCLUSION

Taken together, the in vivo modern neuroimaging studies reviewed here reflect several themes on the effects of alcohol that have been investigated. Noninvasive functional brain imaging such as DWI, DTI, DKI, MRS, and rf-MRI can detect abnormalities that may be due to alcohol intoxication, while conventional MRI scans are unable to detect such abnormalities and thus make the scan results look "normal." The selectively vulnerable brain regions including the frontal lobe, cerebellum, and thalamus were affected by alcohol. This series of studies suggests that acute ethanol intake can cause cytotoxic brain edema and exacerbate brain edema after acute alcoholism, and lead to detectable brain metabolic abnormalities. The effect of acute ethanol administration on the severity of axonal injury was also detected by these functional brain imaging.

Acknowledgments

This work was supported by Natural Science Foundation of Guangdong Province, China (grant No. S2012010008974, 2014A030313481), and was sponsored by Characteristic Innovation Project of Ordinary University of Guangdong Province, China (No. 922-38040223, No. 923-38040404).

References

Adalsteinsson, E., Sullivan, E. V., Mayer, D., & Pfefferbaum, A. (2006). In vivo quantification of ethanol kinetics in rat brain. *Neuropsychopharmacology, 31*, 2683–2691.

Alderazi, Y., & Brett, F. (2007). Alcohol and the nervous system. *Current Diagnostic Pathology, 13*, 203–209.

Anderson, B. M., Stevens, M. C., Meda, S. A., Jordan, K., Calhoun, V. D., & Pearlson, G. D. (2011). Functional imaging of cognitive control during acute alcohol intoxication. *Alcoholism: Clinical and Experimental Research, 35*, 156–165.

Aschner, M., Mutkus, L., & Allen, J. W. (2001). Aspartate and glutamate transport in acutely and chronically ethanol exposed neonatal rat primary astrocyte cultures. *NeuroToxicology, 22*, 601–605.

Bellis, D., Narasimhan, A., Thatcher, D. L., Keshavan, M. S., Soloff, P., & Clark, D. B. (2005). Prefrontal cortex, thalamus, and cerebellar volumes in adolescents and young adults with adolescent-onset alcohol use disorders and comorbid mental disorders. *Alcoholism: Clinical and Experimental Research, 29*, 1590–1600.

Biller, A., Bartsch, A. J., Homola, G., Solymosi, L., & Bendszus, M. (2009). The effect of ethanol on human brain metabolites longitudinally characterized by proton MR spectroscopy. *Journal of Cerebral Blood Flow & Metabolism, 29*(5), 891–902.

Bisby, J. A., Leitz, J. R., Morgan, C. J. A., & Curran, H. V. (2010). Decreases in recollective experience following acute alcohol: A dose–response study. *Psychopharmacology, 208*, 67–74. http://dx.doi.org/10.1007/s00213-009-1709-y.

Braunová, Z., Kasparová, S., Mlynárik, V., Mierisová, S., Liptaj, T., Tkác, I., & Gvozdjáková, A. (2000). Metabolic changes in rat brain after prolonged ethanol consumption measured by 1H and 31P MRS experiments. *Cellular and Molecular Neurobiology, 20*(6), 703–715.

Cardenas, V. A., Studholme, C., & Gazdzinski, S. (2007). Deformation-based morphometry of brain changes in alcohol dependence and abstinence. *NeuroImage, 34*, 879–887.

Chen, X. R., Zeng, J. Y., Kong, L. M., & Zheng, W. B. (February 2016). Diffusion kurtosis imaging detects acute microstructural changes in white and gray matter after alcohol intoxication in rat. In *IEEE international conference on biological sciences and technology.*

Currie, S., Saleem, N., Straiton, J. A., Macmullen-Price, J., Warren, D. J., & Craven, I. J. (2016). Imaging assessment of traumatic brain injury. *Postgraduate Medical Journal, 92*(1083), 41–50.

Das, S. K., Balakrishnan, V., & Vasudevan, D. M. (2006). Alcohol: Its health and social impact in India. *The National Medical Journal of India, 19*(2), 94–99.

Ende, G., Welzel, H., Walter, S., Weber-Fahr, W., Diehl, A., Hermann, D., … Mann, K. (2005). Monitoring the effects of chronic alcohol consumption and abstinence on brain metabolism: A longitudinal proton magnetic resonance spectroscopy study. *Biological Psychiatry, 58*(12), 974–980.

Ferguson, S. A. (2012). Alcohol-impaired driving in the United States: Contributors to the problem and effective countermeasures. *Traffic Injury Prevention, 13*(5), 427–441.

Filli, L., Wurnig, M., Nanz, D., Luechinger, R., Kenkel, D., & Boss, A. (2014). Whole-body diffusion kurtosis imaging: Initial experience on non-Gaussian diffusion in various organs. *Investigative Radiology, 49*(12), 773–778.

Flory, G. S., O'Malley, J., Grant, K. A., Park, B., & Kroenke, C. D. (2010). Quantification of ethanol methyl (1)H magnetic resonance signal intensity following intravenous ethanol administration in primate brain. *Methods, 50*(3), 189–198.

Follesa, P., Floris, G., Asuni, G. P., Ibba, A., Tocco, M. G., Zicca, L., … Gorini, G. (2015). Chronic intermittent ethanol regulates hippocampal GABA(A) receptor delta subunit gene expression. *Frontiers in Cellular Neuroscience, 9*, 445.

Garrity, A. G., Pearlson, G. D., McKiernan, K., Lloyd, D., Kiehl, K. A., & Calhoun, V. D. (2007). Aberrant "default mode" functional connectivity in schizophrenia. *The American Journal of Psychiatry, 164*(3), 450–457.

George, M. R., Potts, G., Kothman, D., Martin, L., & Mukundan, C. R. (2004). Frontal deficits in alcoholism: An ERP study. *Brain and Cognition, 54*(3), 245–247.

Gomez, R., Behar, K. L., Watzl, J., Weinzimer, S. A., Gulanski, B., Sanacora, G., … Mason, G. F. (2012). Intravenous ethanol infusion decreases human cortical γ-aminobutyric acid and N-acetylaspartate as measured with proton magnetic resonance spectroscopy at 4 tesla. *Biological Psychiatry, 71*(3), 239–242.

Harris, G. J., Jaffin, S. K., Hodge, S. M., Kennedy, D., Caviness, V. S., Marinkovic, K., … Oscar-Berman, M. (2008). Frontal white matter and cingulum diffusion tensor imaging deficits in alcoholism. *Alcoholism: Clinical and Experimental Research, 32*(6), 1001–1013.

Hori, M., Fukunaga, I., Masutani, Y., Taoka, T., Kamagata, K., Suzuki, Y., & Aoki, S. (2012). Visualizing non-Gaussian diffusion: Clinical application of q-space imaging and diffusional kurtosis imaging of the brain and spine. *Magnetic Resonance in Medical Sciences, 11*(4), 221–233.

Hudak, A. M., Peng, L., Marquez de la Plata, C., Thottakara, J., Moore, C., Harper, C., … Diaz-Arrastia, R. (2014). Cytotoxic and vasogenic cerebral oedema in traumatic brain injury: Assessment with FLAIR and DWI imaging. *Brain Injury, 28*(12), 1602–1609.

Hui, E. S., Cheung, M. M., Chan, K. C., & Wu, E. X. (2010). B-value dependence of DTI quantitation and sensitivity in detecting neural tissue changes. *NeuroImage, 49*(3), 2366–2374.

Jensen, J. H., & Helpern, J. A. (2010). MRI quantification of non-Gaussian water diffusion by kurtosis analysis. *NMR in Biomedicine, 23*(7), 698–710.

Jensen, J. H., Helpern, J. A., Ramani, A., Lu, H., & Kaczynski, K. (2005). Diffusional kurtosis imaging: The quantification of non-Gaussian water diffusion by means of magnetic resonance imaging. *Magnetic Resonance in Medicine, 53*(6), 1432–1440.

Katada, R., Nishitani, Y., Honmou, O., Okazaki, S., Houkin, K., & Matsumoto, H. (2009). Prior ethanol injection promotes brain edema after traumatic brain injury. *Journal of Neurotrauma, 26*(11), 2015–2025.

Kong, L. M., Lian, G. P., Zheng, W. B., Liu, H. M., & Zhang, H. D. (2013). Effect of alcohol on diffuse axonal injury in rat brainstem: Diffusion tensor imaging and aquaporin-4 expression study. *Biomed Research International, 2013*, 798261.

Kong, L. M., Zheng, W. B., Lian, G. P., & Zhang, H. D. (2012). Acute effects of alcohol on the human brain: Diffusion tensor imaging study. *American Journal of Neuroradiology, 33*(5), 928–934.

Kovacevic, S., Azma, S., Irimia, A., Sherfey, J., Halgren, E., & Marinkovic, K. (2012). Theta oscillations are sensitive to both early and late conflict processing stages: Effects of alcohol intoxication. *PLoS One, 7*, e43957.

Kroenke, C. D., Flory, G. S., Park, B., Shaw, J., Rau, A. R., & Grant, K. A. (2013). Chronic ethanol (EtOH) consumption differentially alters gray and white matter EtOH methyl 1H magnetic resonance intensity in the primate brain. *Alcoholism: Clinical and Experimental Research, 37*(8), 1325–1332.

Lau, M. A., Pihl, R. O., & Peterson, J. B. (1995). Provocation, acute alcohol intoxication, cognitive performance, and aggression. *Journal of Abnormal Psychology, 104*, 150–155.

Lee, D. W., Kim, S. Y., Lee, T., Nam, Y. K., Ju, A., Woo, D. C., … Choe, B. Y. (2012). Ex vivo detection for chronic ethanol consumption-induced neurochemical changes in rats. *Brain Research, 1429*, 134–144.

Lee, D. W., Kim, S. Y., Kim, J. H., Lee, T., Yoo, C., Nam, Y. K., … Choe, B. Y. (2013). Quantitative assessment of neurochemical changes in a rat model of long-term alcohol consumption as detected by in vivo and ex vivo proton nuclear magnetic resonance spectroscopy. *Neurochemistry International, 62*(4), 502–509.

Lemon, J. (1993). Alcoholic hangover and performance: a review. *Drug and Alcohol Review, 12*, 299–314.

Lindman, R. E., Sjoholm, B. A., & Lang, A. R. (2000). Expectations of alcohol-induced positive affect: A cross-cultural comparison. *Journal of Studies on Alcohol, 61*, 681–687.

Liu, H. M., Zheng, W. B., Yan, G., Liu, B. G., Kong, L. M., Ding, Y., … Zhang, G. (2014). Acute ethanol-induced changes in edema and metabolite concentrations in rat brain. *Biomed Research International, 2014*, 351903.

Marinkovic, K., Rickenbacher, E., Azma, S., & Artsy, E. (2012). Acute alcohol intoxication impairs top–down regulation of stroop incongruity as revealed by blood oxygen level-dependent functional magnetic resonance imaging. *Human Brain Mapping, 33*, 319–333.

Marinkovic, K., Rickenbacher, E., Azma, S., Artsy, E., & Lee, A. K. (2013). Effects of acute alcohol intoxication on saccadic conflict and error processing. *Psychopharmacology, 230*(3), 487–497.

Meyerhoff, D. J., Blumenfeld, R., Truran, D., Lindgren, J., Flenniken, D., Cardenas, V., ... Weiner, M. W. (2004). Effects of heavy drinking, binge drinking, and family history of alcoholism on regional brain metabolites. *Alcoholism: Clinical and Experimental Research, 28*(4), 650–661.

Mongrain, S., & Standing, L. (1989). Impairment of cognition, risk taking, and self-perception by alcohol. *Perceptual and Motor Skills, 69*(1), 199–210.

Monnig, M. A., Tonigan, J. S., Yeo, R. A., Thoma, R. J., & McCrady, B. S. (2013). White matter volume in alcohol use disorders: a meta-analysis. *Addiction Biology, 18*(3), 581–592.

Mukherjee, S., Das, S. K., Vaidyanathan, K., & Vasudevan, D. M. (2008). Consequences of alcohol consumption on neurotransmitters—an overview. *Current Neurovascular Research, 5*(4), 266–272.

Nagel, B. J., & Kroen, C. D. (2008). The use of magnetic resonance spectroscopy and magnetic resonance imaging in alcohol research. *Alcohol Research & Health, 31*(3), 243–246.

Niciu, M. J., & Mason, G. F. (2014). Neuroimaging in alcohol and drug dependence. *Current Behavioral Neuroscience Reports, 1*(1), 45–54.

Nielsen, A. S., & Nielsen, B. (2015). Implementation of a clinical pathway may improve alcohol treatment outcome. *Addiction Science & Clinical Practice, 10*, 7.

Opreanu, R. C., Kuhn, D., & Basson, M. D. (2010). Influence of alcohol on mortality in traumatic brain injury. *Journal of the American College of Surgeons, 210*(6), 997–1007.

Pfefferbaum, A., Rosenbloom, M., Rohlfing, T., & Sullivan, E. V. (2009). Degradation of association and projection white matter systems in alcoholism detected with quantitative fiber tracking. *Biological Psychiatry, 65*(8), 680–690.

Raichle, M. E., & Snyder, A. Z. (2007). A default mode of brain function: A brief history of an evolving idea. *NeuroImage, 37*(4), 1083–1090.

Reynolds, B., Richards, J. B., & de Wit, H. (2006). Acute-alcohol effects on the Experiential Discounting Task (EDT) and a question-based measure of delay discounting. *Pharmacology Biochemistry and Behavior, 83*, 194–202.

Rosenbloom, M., Sullivan, E. V., & Pfefferbaum, A. (2003). Using magnetic resonance imaging and diffusion tensor imaging to assess brain damage in alcoholics. *Alcohol Research & Health, 272*, 146–152.

Schulte, T., Müller-Oehring, E., Sullivan, E., & Pfefferbaum, A. (2012). White matter fiber compromise contributes differentially to attention and emotion processing impairment in alcoholism, HIV-infection, and their comorbidity. *Neuropsychologia, 50*(12), 2812–2822.

Silveri, M. M. (2014). GABAergic contributions to alcohol responsivity during adolescence: Insights from preclinical and clinical studies. *Pharmacology & Therapeutics, 143*(2), 197–216.

Söderlund, H., Parker, E. S., Schwartz, B. L., & Tulving, E. (2005). Memory encoding and retrieval on the ascending and descending limbs of the blood alcohol concentration curve. *Psychopharmacology, 182*, 305–317. http://dx.doi.org/10.1007/s00213-005-0096-2.

Stokum, J. A., Sours, C., Zhuo, J., Kane, R., Shanmuganathan, K., & Gullapalli, R. P. (2015). A longitudinal evaluation of diffusion kurtosis imaging in patients with mild traumatic brain injury. *Brain Injury, 29*(1), 47–57.

Tiwari, V., Veeraiah, P., Subramaniam, V., & Patel, A. B. (2014). Differential effects of ethanol on regional glutamatergic and GABAergic neurotransmitter pathways in mouse brain. *Journal of Neurochemistry, 128*(5), 628–640.

Unterberg, A. W., Stover, J., Kress, B., & Kiening, K. L. (2004). Edema and brain trauma. *Neuroscience, 129*(4), 1021–1029.

Volkow, N. D., Hitzemann, R., Wolf, A. P., Logan, J., Fowler, J. S., Christman, D., ... Hirschowitz, J. (1990). Acute effects of ethanol on regional brain glucose metabolism and transport. *Psychiatry Research, 35*, 39–48.

Volkow, N. D., Mullani, N., Gould, L., Adler, S. S., Guynn, R. W., Overall, J. E., & Dewey, S. (1988). Effects of acute alcohol intoxication on cerebral blood flow measured with PET. *Psychiatry Research, 24*, 201–209.

White, A. M. (2003). What happened? Alcohol, memory blackouts, and the brain. *Alcohol Research & Health, 27*(2), 186–196.

Yamakami, I., Vink, R., Faden, A. I., Gennarelli, T. A., Lenkinski, R., & McIntosh, T. K. (1995). Effects of acute ethanol intoxication on experimental brain injury in the rat: Neurobehavioral and phosphorus-31 nuclear magnetic resonance spectroscopy studies. *Journal of Neurosurgery, 82*, 813–821.

Yeh, P. H., Simpson, K., Durazzo, T. C., Gazdzinski, S., & Meyerhoff, D. J. (2009). Tract-based spatial statistics (TBSS) of diffusion tensor imaging data in alcohol dependence: Abnormalities of the motivational neurocircuitry. *Psychiatry Research, 173*(1), 22–30.

Zahr, N. M., Kaufman, K. L., & Harper, C. G. (2011). Clinical and pathological features of alcohol-related brain damage. *Nature Reviews Neurology, 7*(5), 284–294.

Zhang, Z., Lu, G., Zhong, Y., Tan, Q., Liao, W., Wang, Z., ... Liu, Y. (2010). Altered spontaneous neuronal activity of the default-mode network in mesial temporal lobe epilepsy. *Brain Research, 1323*, 152–160.

Zheng, H. Y., Kong, L. M., Chen, L. M., Zhang, H. D., & Zheng, W. B. (2015). Acute effects of alcohol on the human brain: A resting-state fMRI study. *Biomed Research International, 2015*, 947529.

CHAPTER

2

Prenatal Alcohol Exposure and Neuroglial Changes in Neurochemistry and Behavior in Animal Models

G. Brolese, P. Lunardi, F. Lopes, C.-A. Gonçalves

Federal University of Rio Grande do Sul, Porto Alegre, Brazil

ETHANOL EXPOSURE AND BRAIN DEVELOPMENT

Drugs such as alcohol have the ability to change our state of consciousness, which explains our behavioral change after few drinks. In general, we can feel anything from euphoric, talkative, and sociable to dizzy, sick, sad, and sleepy. Alcohol pharmacology and its effects on the brain are still relatively poorly characterized, and the mechanism of its action is considered unspecific, because ethanol can have long-lasting effects on different kinds of neurons and glial cells. However, it is already well accepted in the literature that ethanol can act not only in exacerbating the inhibitory GABAergic synapses, but also in reducing the excitatory glutamatergic synapses. This effect could explain why ethanol is considered a "depressor of the central nervous system," CNS, increasing excitation and decreasing inhibition, and why symptoms such as alcohol intoxication, amnesia, and ataxia appear. Alcohol reward effects are mediated especially by the mesolimbic dopaminergic reward system. These effects are mediated not only by the gamma-aminobutyric acid (GABA) and glutamate systems, but also by direct and indirect pathways by opioid and cannabinoid synapses.

Prenatal ethanol exposure (PEE) has been linked to widespread impairments in brain structure and function. Many countries do not have a clear recommendation regarding alcohol abstinence during gestation. Prenatal alcohol consumption has been shown to be associated with a higher incidence of juvenile behavioral and cognitive problems, including deficits in cognitive behavior. Following PEE, general CNS disorganization is observed, with errors in neuronal migration, glial differentiation, and microcephaly, as well as abnormalities of the brain stem, cerebellum, basal ganglia, hippocampus, corpus callosum, pituitary gland, and optic nerve (Jones & Smith, 1973). In fact, the developing hippocampus is one of the areas of the brain that is most vulnerable to effects of ethanol. However, the range of alcohol-related disorders depends on differences in the duration, timing, and pattern of ethanol exposure.

The most severe effect of ethanol exposure during pregnancy is fetal alcohol syndrome (FAS), and the more consistent characteristics of FAS are (1) facial abnormalities with short palpebral fissures, thin vermilion border of the upper lip, and epicanthal folds; (2) mental retardation varying in degree from mild to severe; (3) small weight and height at birth that persist in the postnatal period; and (4) abnormalities in the cardiovascular and skeletal systems (Wilhelm & Guizzetti, 2016). It is clear that binge drinking during PEE can produce significant and severe brain damage and cognitive behavioral dysfunction. Damage becomes apparent in juveniles, including a reduction in general intellectual functioning and academic skills, as well as deficits in learning, spatial memory and reasoning, reaction time, balance, and other cognitive and motor skills. These deficits are pervasive and can persist throughout the person's life.

Additionally, the negative consequences of PEE are not limited to high levels of alcohol; moderate (or social) prenatal ethanol consumption has been shown to be associated with a higher incidence of behavioral and cognitive problems in childhood and/or adolescence, including alterations in the neurophysiology of the CNS and in the glutamate system, oxidative stress, and glial proteins (Brolese et al., 2014, 2015). Although there are no agreed guidelines in the literature defining the exact amounts for

Addictive Substances and Neurological Disease
http://dx.doi.org/10.1016/B978-0-12-805373-7.00002-5

low, moderate, or high ethanol doses, in animal models, commonly used low to moderate doses result in a blood ethanol concentration (BAC) of between 80 and 150 mg/dL; binge drinking and/or FAS usually reach BAC > 200 mg/dL (Patten, Fontaine, & Christie, 2014).

Ethanol exposure during brain development, even in moderate doses, induces a variety of disruptions in normal neuronal development patterns. Abnormalities in glial development have been also suspected to contribute to the adverse effects of ethanol on the developing brain. Significant differentiation of the CNS occurs during gestational days (GDs) 11–21 in rats and is highlighted by a burst of neurogenesis and population of the cerebral cortex and hippocampus by migrating neurons (Guerri, 1998). The newborn cells differentiate into neurons and glia and start to mature, forming axonal and dendritic processes. Ethanol exposure during this time causes decreased neurogenesis and disrupted radial glia, as well as reduced migration and survival of neurons. Additionally, a key maturation called the brain growth spurt, which occurs during the third trimester of gestation in humans, takes place postnatally in rats; during this period a substantial increase in brain size, dendritic arborization, and synaptogenesis, which corresponds to proliferation of astrocytes, oligodendrocyte precursors, and initiation of myelination, occurs from late gestation up to postnatal day (PND) 9 in rats. This time corresponds to a period of major development of glial and myelin structures and suggests a potential effect of ethanol on glial cells (see review by Wilhelm & Guizzetti, 2016).

During these particular periods, changes including morphological and functional alterations affect the balance of synaptic plasticity and alter the protective and supportive functions of glial cells. Alcohol exposure at different stages of development can harm different populations of neurons, glial cells, and neurotransmitter systems. During early stages of cell differentiation, ethanol can alter synapse formation, some types of neurons and glial cells. In other words, cells die when alcohol exposure either prevents them from migrating properly or induces a delayed cell death that occurs after migration, even though exposure occurred before migration started. Moreover, multiple mechanisms may operate simultaneously to produce abnormal cell development or cell death. To better understand the mechanisms and ethanol actions during brain development, different kinds of animal models have been used.

Animal Models of Prenatal Ethanol Exposure

The effects of alcohol are detrimental throughout the developing nervous system, and therefore heavy alcohol exposure can be harmful to the fetus at any stage of gestation; moderate doses that cause lower BACs can also have long-lasting effects. For a greater understanding of alcohol effects on brain development and behavioral effects of PEE, several animal models have been developed. At the structural level, rodent models of fetal alcohol spectrum disorder (FASD) exhibit similar brain alterations as those seen in humans. There are several animal models that can be used to study the structural and functional deficits caused by PEE. Regarding the variability of experimental protocols using rodents to study the effects of alcohol during the prenatal period, there are three major methods of ethanol administration commonly used: ingestion (through liquid diet, water, or near beer, or intubation), injection, or inhalation (see Fig. 2.1).

Ingestion

Dietary. The liquid diet model developed by Lieber and DeCarli is one of the most commonly utilized routes of delivery in rodent models of ethanol exposure during pregnancy and was one of the first models to be developed. In this model the food provided to pregnant dams in their cages is a liquid diet in which a percentage of the calories (usually ~35%, which equals 6.61% v/v) is derived from ethanol. The diet has all the nutrients specifically required for pregnancy and is offered throughout gestation as the only source of nutrition. The rats consume on average 12 g ethanol/kg/day (and up to 18 g/kg/day) (Gil-Mohapel, Boehme, Kainer, & Christie, 2010). Pair-fed control groups are used to manage and compare the weight gain by the dams. This group receives an isocaloric diet, with maltose dextrin substituting the ethanol calories. The delivery of the diet usually begins on GD1 of pregnancy and over 3, or sometimes more, days the ethanol is added to the diet gradually (i.e., one-third final ethanol concentration on GD1, two-thirds of final ethanol concentration on GD2, and final ethanol concentration on GD3 and for the remainder of the pregnancy). The BAC produced by the liquid diet depends on the ethanol concentration chosen; it usually fluctuates between 80 and 180 mg/dL in rats.

Voluntary drinking. There are other kinds of oral administration similar to the liquid diet model; ethanol can also be administered through the drinking water. In this case the animal keeps receiving the laboratory food (rat chow), and the ethanol is added to the water bottle. This method is stressless; however, the female rats have to be trained to voluntarily consume a saccharin-sweetened 10% ethanol solution prior to pregnancy. The control group receives saccharin-sweetened water only. Using this paradigm, rodents consume an average of 14 g ethanol/kg/day and the BAC achieved is 120 mg/dL (Choi, Allan, & Cunningham, 2005). Another voluntary drinking model is the "beer model," as an alternative to oral self-administration of ethanol without stress. In this

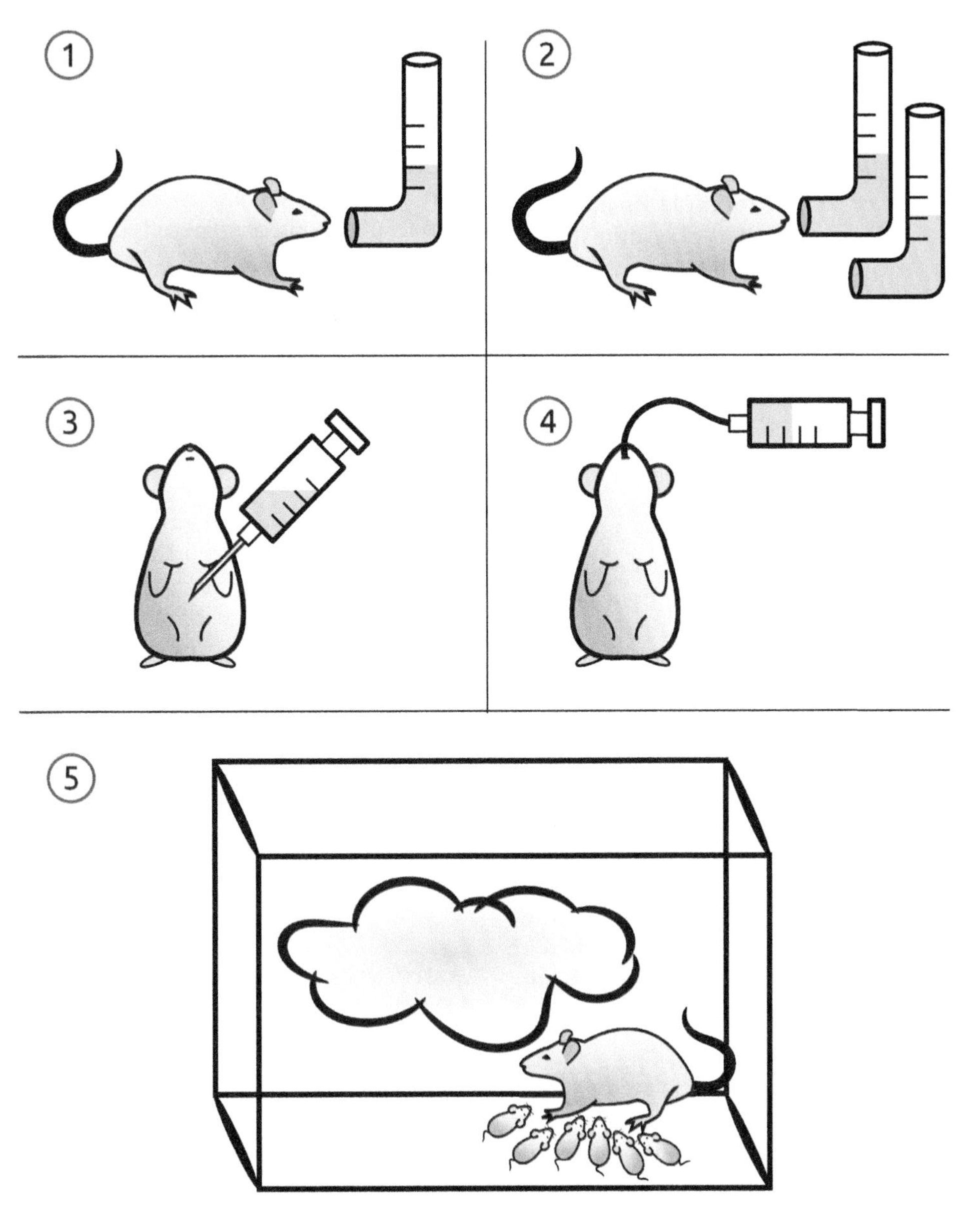

FIGURE 2.1 Animal models of ethanol exposure. Different kinds of animal models are used to administer ethanol to the rodents. (1) liquid diet; (2) voluntary drinking; (3) injection; (4) intubation (gavage); and (5) chamber with vaporized ethanol.

method the control group receives near beer and the treated group receives near beer with an added ethanol dose. Near beer makes the solution more palatable than plain water, and it is easier to make rats drink higher doses during the whole pregnancy. It is important to choose a near beer that guarantees 0.0% of ethanol to ensure the correct dose chosen for the study. Using this model the BAC can reach between 80 and 150 mg/dL (Brolese et al., 2014; Samson, Denning, & Chappelle, 1996).

One of the advantages of the dietary or voluntary drinking models is that much less stress (i.e., less handling) is involved for the dams. Additionally, the techniques are simple; there is less risk of fatality with the dose, and it is less labor-intensive when compared to other methods. However, there are some disadvantages; with these methods it is not possible to be precise with the dose and timing of ethanol administration, and this can lead to variability in the BAC achieved. Also, administration during the extrauterine time, equivalent to the third trimester of pregnancy, could be continued on a liquid diet or voluntary drinking during the suckling period, but it is not possible to control how much ethanol crosses into the breast milk. Thus, dams exposed to ethanol drinking during the suckling period may not engage in appropriate maternal behavior. Nevertheless, this model has been very well used to expose dams to moderate doses and still has validity and significant legitimacy, mimicking the human condition.

Injection

When the study needs to control the exact dose administered and the acute ethanol effects during development, subcutaneous (s.c.) (Ikonomidou et al., 2000) or intraperitoneal (i.p.) (De Licona et al., 2009) ethanol injection, either acutely or across multiple days during gestation, can be useful. This method of administration allows a rapid increase in BAC, reaching an average between 180 and 200 mg/dL, with limited handling-induced stress. However, this method of administration far from resembles ethanol consumption in human beings and may not accurately replicate several important aspects of human PEE.

Intragastric Intubation (Gavage)

Ethanol can also be delivered directly to the stomach of the pregnant dam using an intubation method. In this method, ethanol can be diluted in saline or water and administered with a syringe that is attached to a curved steel gavage needle and inserted down the esophagus to the entrance to the stomach. For the control group, an isocaloric control liquid solution (such as maltose dextrin or sucrose) can also be administered by gavage. The BAC reached by this method is one of the highest, greater than 200 mg/dL. Depending on the dose, it is common to divide administration into two doses in a 24-h period. The main advantage of this model is the precise control over the administered dose and the BAC achieved. However, it is a very invasive and stressful method, with higher risk of mortality and comprising further behavior studies. Also, researchers need to undergo specific training before doing this kind of procedure.

Inhalation

The inhalation model allows pregnant dams and neonatal pups, or even the dam and her litter to be placed in an inhalation chamber filled with vaporized ethanol for several hours (Karanian et al., 1986). The advantages of this method are the rapid increase in BAC without the stress of intubation or injection. It is also much less labor-intensive than other methods, especially because multiple animals can be placed in the chamber at the same time. However, the administration does not mimic the route of intake in human beings, and the vaporized ethanol can cause irritation to the upper respiratory tract of the rodent. If this method is used to expose pups, they will have to be removed from their mothers for determined periods of time that may result in stress and bad nutrition associated with the separation, which can have lifelong effects on the litter's behavior (Marais, van Rensburg, van Zyl, Stein, & Daniels, 2008).

The fact that the brain passes through several developmental stages and is particularly vulnerable to ethanol means that the threshold dose of alcohol capable of producing detrimental effects is likely to vary according to the alcohol dose and the stage of development. Ethanol detrimentally affects the glial cells of the developing brain, and disruption of the function of these cells leads to neuronal deficits, which in turn has an impact on cognition and behavior. In the next section we describe the impact of ethanol on glial cells during brain development.

ETHANOL, GLIA, AND NEURODEVELOPMENT

Since 1980s, new and exciting roles for astroglial cells in brain development have been described. Astrocytes belong to one of the three major classes of glial cells and constitute one of the most common cell types in the brain. Both astrocytes and their precursors are essential for neuronal development, survival, and function, and ultimately the development of the proper brain architecture and connectivity (Vangipuram & Lyman, 2010).

Mounting evidence, which has reported alterations in both neurons and astrocytes by ethanol during brain development, has also assumed that the effects of fetal alcohol are due, at least in part, to effects on astrocytes, affecting their ability to modulate neuronal development and function (Guerri, 2001; Rubert, Miñana, Pascual, & Guerri, 2006).

Astrocytes As Ethanol Targets During Brain Development

Astrocyte proliferation coincides with birth. In rodents, which have a gestation period of 3 weeks, the peak of neuronal formation occurs at 2 weeks during the embryonic stage. Then, there is astrocyte peak at birth or 2 days postnatally. In humans, the timeframe follows a similar course. As we are born, astrocytes will be placed next to neuronal networks and will help to build up this crucial supportive microenvironment during CNS development.

Abnormalities in neuronal development by ethanol have been extensively investigated since the end of the 1970s, when it became clearly established that ethanol is an important teratogenic substance and its consumption during pregnancy induces harmful effects on the developing fetus (Clarren & Smith, 1978). For a long period, the majority of in vivo studies in rodents, both prenatal and postnatal ethanol exposure models, have focused on neuronal susceptibility and the consequent long-lasting behavioral abnormalities (see recent reviews by Sadrian, Wilson, & Saito, 2013; Wilson, Peterson, Basavaraj, & Saito, 2011). Notwithstanding this, brain structural analysis of FASD patients has shown specific alterations in the size of white-matter areas and corpus callosum, indicating that ethanol also affects glial cells and myelin development (Riley et al., 1995; Swayze et al., 1997). Moreover, Guerri and Renau-piqueras (1997) and Guerri (1998) have pointed out important alterations in astrocyte functions caused by ethanol exposure, suggesting that astroglial damage would likely have profound effects on many developmental processes in the CNS (see also Guerri, 2001; Phillips, 1992).

Studies of CNS development have already characterized the role of radial glia on proper neuronal migration and astrocyte maturation. For instance, radial glia begin to express GFAP (glial fibrillary acidic protein), an intermediate filament and astrocyte-specific cytoskeleton marker, as they become astrocytes during brain development (Bayrakthar, Fuentealba, Alvarez-Buylla, & Rowitch, 2014; De Juan Romero & Borrell, 2015). Ethanol exposure during brain development significantly alters the expression of GFAP, and these observations strongly suggest an alteration in astrocyte profile (such as proliferation or activity) commonly observed in response to several forms of injury (Dalçik et al., 2009; Franke, Kittner, Berger, Wirkner, & Schramek, 1997). Specifically, it has been shown that ethanol exposure during this critical time window significantly delays the onset of GFAP expression and decreases the expression of GFAP protein and mRNA levels both in the fetal brain and in radial glia cultures, consistently followed by a reduction in the number of astrocytes (Miller & Robertson, 1993; Nash, Krishnamoorthy, Jenkins, & Csete, 2012; Rubert et al., 2006; Taléns-Visconti et al., 2011; Vallés, Pitarch, Renau-Piqueras, & Guerri, 1997; Vemuri & Chetty, 2005). Other studies have demonstrated that high levels of alcohol during brief postnatal exposure may cause a transient increase in GFAP that may result from an astrogliosis reaction (Goodlett, Leo, O'Callaghan, Mahoney, & West, 1993; Sofroniew, 2005). Therefore, the effect of neonatal exposure to alcohol on GFAP levels is not very conclusive. The overall differences in astrocytes with respect to GFAP may depend on timing of exposure relative to the stage of astrocyte maturation, on ethanol levels, route of administration, length of treatment, and alcohol withdrawal status (Brolese et al., 2014; Eckardt et al., 1998).

Ethanol-induced damage to the cytoskeleton may have functional consequences, including changes in the intra- and intercellular transport of proteins. Results from several in vitro and in vivo studies have indicated that early ethanol exposure depresses the proliferative activity of both astrocytes and other glial cells by interfering with the stimulatory effect of trophic factors (Luo & Miller, 1999). The mechanisms vary from a decrease in neuritogenic extracellular matrix proteins (Guizzetti, Moore, Giordano, VanDeMark, & Costa, 2010) to an increase of neurite outgrowth inhibitors, both in vitro and after neonatal ethanol exposure (Zhang et al., 2014).

In particular, some studies have demonstrated a reduction in S100B expression and release in the midline raphe of prenatal alcohol-exposed rats and mice (Eriksen, Gillespie, & Druse, 2000; Tajuddin, Orrico, Eriksen, & Druse, 2003; Zhou, Sari, Zhang, Goodlett, & Li, 2001), which may underlie the reduced number of serotonergic neurons as well as the migration and development observed in FASD animal models, suggesting an important role of S100B as a trophic factor for serotonergic neuron development (Tajuddin & Druse, 1999; Zhou et al., 2001). Importantly, S100B is a calcium-binding protein mainly expressed and secreted by astrocytes (Donato et al., 2013). Although S100B secretion mechanisms are not fully described, S100B has been shown to act as a stimulator of cell proliferation and migration and an inhibitor of apoptosis and differentiation in physiological conditions (Arcuri, Bianchi, Brozzi, & Donato, 2005; Tubaro, Arcuri, Giambanco, & Donato, 2011) as well as to directly interact with cytoskeletal proteins, including GFAP (Tramontina, Souza, Frizzo, & Gonçalves, 2006). Moreover, S100B induces GFAP expression, thereby increasing astrocyte branching (Evrard et al., 2006; Reeves et al., 1994), modulates synaptic plasticity, and has some important impacts on neuronal and astroglial metabolic pathways (see review by Donato, 2003). Results from our group have reported an increase in S100B levels in offspring cerebrospinal fluid following prenatal ethanol treatment (Brolese et al., 2014). An increase in S100B level may be considered a marker of impairment, but may also reflect an attempt to save and protect cell regions that have been damaged by alcohol, as a response to astrocyte toxicity (Kleindienst & Ross Bullock, 2006).

In this scenario, it is worth highlighting the major relevance of the capability of astrocytes and other glial cells to provide CNS formation and maturity. Overall, the vast majority of studies published so far have indicated a supportive and regulatory process drives from astrocytes to neuronal development in order to constrain ethanol toxicity. However, in addition to their supportive role, astrocytes may have multiple influences on postnatal synaptic development and functioning due to their dynamic enwrapping of synapses.

Ethanol-Induced Changes to Astrocytes Alter Synaptic Function and Maturation

Astrocytes can dramatically change the synapse's appearance through different signals during brain development (see review by Allen, 2013). They can transform synapses especially due to their capability to induce postnatal differentiation of several neuronal features (Allen & Barres, 2005; Slezak & Pfrieger, 2003). Additionally, there is increasing evidence that different phases of brain development depend on neuron–glia interactions, including key postnatal events like neuronal synapse formation and maturation in the developing brain. Accordingly, it has been demonstrated that astrocytes may regulate the developmental appearance of inhibitory and excitatory currents in the hippocampus (Allen et al., 2012; Hughes, Elmariah, & Balice-Gordon, 2010; Jones et al., 2011). Moreover, several factors, including glia-derived tumor necrosis factor alpha (TNF-α) (Beattie et al., 2002), activity-dependent neurotrophic factor (ADNF) (Blondel et al., 2000), and other membrane-delimited signals (Nägler, Mauch, & Pfrieger, 2001), could also control synaptic maturation. In fact, neurons can form synapses without glia, but may require glia-derived cholesterol to form numerous and efficient synapses (Pfrieger, 2002). Notably, evidence has supported the theory that some of the deleterious effects of ethanol in the developing brain may be due to the disruption of cholesterol homeostasis in astrocytes (Guizzetti & Costa, 2007; Pfrieger, 2002).

Apart from their synapse formation and maturation, it is worth noting that the concept that astrocytes may exert an active and rapid feedback control of neuronal activity is of great functional importance for understanding how the brain could work under both normal and pathological or toxic conditions. Given the importance of glutamate as the main neuroexcitatory amino acid in the CNS, its concentration, both intra- and extracellularly, is tightly regulated by membrane transporters. It has already been reported that astrocytes mainly express two isoforms of plasma membrane glutamate transporters: L-glutamate/L-aspartate transporter (GLAST), also referred to as excitatory amino acid transporter 1 (EAAT-1), and glial L-glutamate transporter 1 (GLT-1), also called excitatory amino acid transporter 2 (EAAT-2) (see review by Danbolt, 2001). Both astrocyte transporters are responsible for the long-term maintenance of low extracellular concentrations of glutamate, maintaining synaptic signaling and, consequently, protecting neurons from excitotoxicity (reviewed by Tanaka, 2000). Important aspects of astrocytic glutamate transporter in alcohol use disorder (AUD) were reviewed, highlighting the regulating glutamate homeostasis; for example, EAAT-2 regulation as a potential therapeutical target for recovery from AUD (Ayers-Ringler, Jia, Qiu, & Choi, 2016). On the other hand, ethanol exposure during development has shown more extended effects. For instance, data from our group have suggested a significant decrease in hippocampus glutamate uptake followed by moderate pre- and postnatal ethanol exposure (Brolese et al., 2015). Concurrently, the glutathione (GSH) level was also reduced. Indeed, GSH is the main nonenzymatic endogenous antioxidant of glial cells, and has a pivotal role in cellular homeostasis (Dringen, 2000; Dringen & Hirrlinger, 2003) in order to counteract oxidative stress-induced neurotoxicity. Therefore, these findings have corroborated numerous studies that have reported that molecular mechanisms of ethanol neurotoxicity are due to reactive oxygen species (ROS) production and consequent decrease of antioxidant defenses (Kumral et al., 2005; Patten et al., 2013; reviewed by Brocardo, Gil-Mohapel, & Christie, 2011).

In conclusion, besides the relevance of studies that have aimed to explore the direct neuronal impairment resulting from ethanol exposure during development, mounting evidence has been reported of an important astroglia vulnerability to ethanol-induced effects both pre- and postnatally. These findings may represent ethanol-induced indirect effects on neuronal development, synapse formation, and signaling. Moreover, they may indicate that ethanol effects on proper astroglial development will play a relevant role in the developmental neurotoxicity of ethanol (Fig. 2.2).

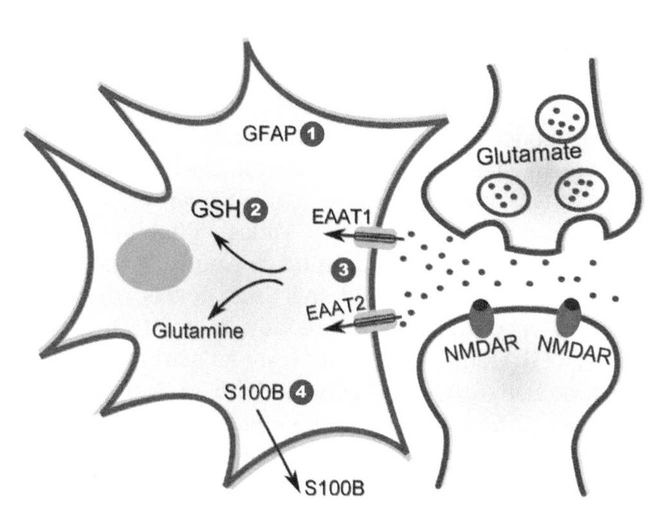

FIGURE 2.2 Illustration of glutamate synaptic signaling at the tripartite synapse. Extracellular glutamate is captured by astrocytes by specific glial transporters, EAAT-1 and EAAT-2. In the astrocytes, glutamate can be converted into glutamine or can also be used for glutathione (GSH) synthesis, one of the main antioxidant defenses. The astrocytes have specific markers, such as S100B protein, that can be secreted and used as a trophic factor, promoting differentiation and neuronal growth, and GFAP, a cytoskeletal protein from mature astrocytes. The main ethanol changes in astrocyte functions are represented by (1) GFAP expression alteration; (2) a significant decrease of glutamate uptake function; (3) a decrease in GSH levels; and (4) an increase of S100B release. *Adapted from Brolese, G., Lunardi, P., Broetto, N., Engelke, D.S., Lírio, F., Batassini, C.,...Gonçalves, C.A. (2014). Moderate prenatal alcohol exposure alters behavior and neuroglial parameters in adolescent rats.* Behavioural Brain Research, *269, 175–184. http://dx.doi.org/10.1016/j.bbr.2014.04.023; Brolese, G., Lunardi, P., de Souza, D. F., Lopes, F. M., Leite, M. C., & Gonçalves, C. A. (2015). Pre- and postnatal exposure to moderate levels of ethanol can have long-lasting effects on hippocampal glutamate uptake in adolescent offspring.* PLoS One, *10(5), e0127845. http://dx.doi.org/10.1371/journal.pone.0127845.*

COGNITIVE AND BEHAVIORAL EFFECTS OF PRE- AND POSTNATAL ETHANOL EXPOSURE

Drugs of abuse, such as alcohol, cocaine, heroin, and others that act on the CNS, alter the functioning of the brain, which causes cognition and behavioral consequences. Although each of them have their particular mechanism of action, all drugs of abuse share in common the fact that they all increase the activity of dopamine in the reward pathway of the brain, which crosses the limbic region and connects with the cortex. It means that the emotional center and cognition parts of the brain are altered, affecting intentional behaviors. Therefore, cognitive impairments and emotional disorders, including anxiety and mood disorders, can be caused or triggered by continued drug consumption.

Alcohol, for example, has already been shown to interfere with many molecular, neurochemical, and cellular events occurring during the normal development of the brain, noting that the earlier the consumption, the more significant the impairments. Certain areas of the brain and cell populations are more vulnerable than others, but it is known that the prefrontal cortex, hippocampus, and cerebellum are especially susceptible to alcohol and have been associated with behavioral deficits (Guerri, 2002). The prefrontal cortex, considered to be the executive center of the brain, is involved in goal-directed behaviors, decision making, planning, emotional restraint, impulsivity, focused attention, and consolidation of memories. In turn, the hippocampus is considered the memory center, and the cerebellum is involved in balance and coordination of movement.

There are many experimental difficulties and ethical concerns in studying cognitive and behavioral effects of alcohol exposure on human brains. In this sense, animal models can be useful once they provide a rigorous means to precisely control environmental context and drug exposure, as well as assessing behavioral and cognitive performance prior to and after drug administration. While animal models can never reproduce the complex social, psychological, and environmental reasons why people abuse drugs, to date, they have provided a valuable means to investigate the association between brain, cognition, and behavior (see review by Belin-Rauscent, Fouyssac, Bonci, & Belin, 2016). Especially in alcohol research, animal models allow neural manipulations and thus, establish the causal influences of neural region and cellular and molecular substrates of alcohol intake on behavior and cognition.

Animal model studies have shown cognitive and behavioral impairments in offspring produced by ethanol exposure during the brain growth spurt, which is both prenatal and early postnatal in rats. Several studies have documented negative effects of high concentrations of ethanol on the developing brain that directly impact on cognitive functions, showed by impairment in many learning and memory tasks (see review in Berman & Hannigan, 2000). Similarly, risk-taking behavior and depressant effects were found in prenatally ethanol-exposed offspring in both high and moderate concentrations, but the effects were dose dependent, which means that higher doses produced higher losses (Carneiro et al., 2005). On the other hand, a systematic review of studies in humans on the effects of low to moderate levels of prenatal alcohol consumption (up to 10.4 UK units or 83 g/week) found no convincing evidence of adverse effects of prenatal alcohol exposure at these levels of exposure (Henderson, Gray, & Brocklehurst, 2007). In turn, some animal model studies found long-lasting behavioral consequences of moderate ethanol exposure during brain development. Adolescent male offspring of dams exposed to moderate concentrations of ethanol (10% w/v) presented more pronounced risk-taking behavior (Brolese et al., 2014), and adult dams exposed to 5% ethanol (v/v) showed abnormalities in social behavior such as a decrease in social interaction (Hamilton et al., 2010).

In short, it is well established in the literature and widely accepted that high doses of alcohol intake during pregnancy affect the developing brain and have long-lasting detrimental effects on cognition and behavior, such as those outcomes of FASD described earlier in this chapter. However, for moderate consumption is underappreciated, and more studies are needed to determine which structures in the CNS are affected and whether it imparts some level of cognitive and behavioral impairment.

Behavioral Tasks After Pre- and Postnatal Alcohol Exposure

Regarding effects of moderate alcohol exposure during the pre- and postnatal period on behavioral parameters, there are some specific tasks in animal models that have been successfully used in benchmark studies. Some of these main tasks and their results are briefly described in the following.

Radial Arm Maze

The Radial Arm Maze (RAM) was designed by Olton and Samuelson (1976) to measure spatial learning and memory in rats. It is an apparatus consisting of eight horizontal equidistantly spaced arms (usually 57 × 11 cm) radiating from a small circular central platform (30 cm in diameter) elevated (70 cm) off the floor. At the entrance of each arm there is an opaque door and at the end of each arm a small food cup is placed, which is not visible from the central platform. Experimental subjects are placed on the central platform from which they have to collect the hidden baits placed at the end of the arms. In the standard version of the RAM, animals are habituated to the environment by exploring the maze for 15 min per day for 3 days. After that they are trained one session per day for 8 consecutive days. Each session lasts 10 min or until all eight arms have been entered or 2 min have passed since the animal's last arm entrance. In order to analyze the animal's performance, the following are considered: the number of errors (i.e., entering an arm that has been visited previously) in each session and the total number of errors across eight sessions; the number of correct choices in the first eight arm entries of each session; the number of adjacent arm entries in each session; total time to complete the session divided by the total number of arm entries; and the number of sessions to reach the criterion of one error or no errors, averaged over 4 consecutive days of training (Olton & Samuelson, 1976). Over the years, variations of the RAM have developed, and all of them have confirmed that it is a consolidated paradigm for the evaluation of learning and memory.

Studies investigating effects of moderate alcohol exposure during the pre- and postnatal period found impairments in these two cognitive processes, revealed by worse performance of experimental groups compared to control groups in the RAM task. The first study to report deficits in RAM performance showed that both groups exposed to moderate (3.3% v/v) or high (6.7% v/v) concentrations of ethanol required twice as many trials in the task as their respective pair-fed controls, suggesting that in utero administration of moderate doses of ethanol affects spatial memory capacity (Reyes, Wolfe, & Savage, 1989). Rats exposed to moderate doses of ethanol in different stages of their fetal life showed that the second quarter of the gestation period was more sensitive to harmful effects of alcohol on the areas of the brain (hippocampus) involved in learning and memory; behavior of the animals exposed to alcohol during the first and the second quarter of the gestation period demonstrated that only the latter were weak in solving the RAM task (Salami, Aghanouri, Rashidi, & Keshavarz, 2004).

Elevated Plus Maze

The Elevated Plus Maze (EPM) was described by Handley and Mithani (1984) and validated by Pellow and colleagues in 1995 to measure anxiety in rats. This apparatus consists of a four-armed maze, two open arms crossing in the middle two arms enclosed by walls, elevated off the floor (Pellow, Chopin, File, & Briley, 1985). In the standard version of the EPM, experimental subjects are placed at the junction of the four arms of the maze, facing an open arm, and entries/duration in each arm are recorded and observed for 5 min. The assessment of anxiety behavior of rodents is described by using the ratio of time spent in the open arms to the time spent in the closed arms, so that an increase in open arm activity (duration and/or entries) reflects anti-anxiety behavior. It has been the most extensively used task to assess anxiety in animal models because of its face, construct, and predictive validity. Besides measuring spontaneous anxiety-like behavior levels, it has also been used to assess anxiolytic and anxiogenic effects of pharmacological agents, drugs of abuse, and hormones, as well as to define brain regions and mechanisms underlying anxiety-related behavior (Walf & Frye, 2006).

Regarding effects of moderate alcohol exposure during the pre- and postnatal period, studies using the EPM task usually find significant increases in the number of entrances to the open arms and in the time of permanence in the open arms in prenatal ethanol-exposed offspring, as compared to controls, indicating anxiolytic effects (Carneiro et al., 2005). In other words, alcohol being a depressor of the CNS, a benzodiazepine-like effect is expected and, in some cases, it can be interpreted as a more pronounced risk-taking behavior which, in turn, is also observed as a consequence of the mechanism of action of alcohol in the prefrontal cortex.

Plus Maze Discriminative Avoidance Task

The Plus Maze Discriminative Avoidance Task (PMDAT) was developed by Silva and colleagues in 1997 to measure learning and memory in mice and rats (Silva & Frussa-Filho, 2000). This paradigm uses the same EPM apparatus, including two aversive stimuli, and a 100 W lamp placed over and 80 dB noise equipment placed under one of the closed arms (see Fig. 2.3). Experimental subjects are conditioned over a period of 10 min (training session) to choose between two enclosed arms (an aversive and a nonaversive arm), while avoiding the open arms of the apparatus, so that every time the animal enters the aversive enclosed arm, the illumination and noise turns on, turning off when the animal leaves that arm. In the test session (usually performed 24 h after the training session) the animal is placed in the center of the apparatus, but only for 3 min, without receiving the aversive stimuli when entering the aversive enclosed arm. In both sessions (training and test), the time spent in each arm is recorded as a learning/memory parameter. Long-term memory evaluation also compares the time spent in the conditioned arm (aversive arm) and unconditioned arms (nonaversive+open arms) (Brolese et al., 2014). Besides being a useful model for measuring learning/memory, the PMDAT also evaluates the time spent in the open arms of the apparatus, which can

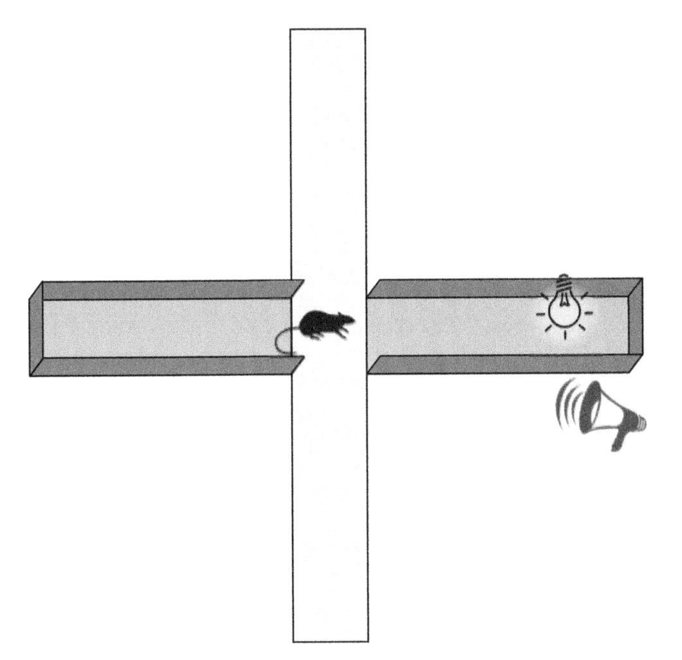

FIGURE 2.3 Illustrative model of Plus Maze Discriminative Avoidance Task.

provide simultaneous information about the anxiety level of the same animals (Silva & Frussa-Filho, 2000).

Considering the advantage of it being a task for learning/memory–anxiety interactions, the PMDAT has been an effective behavioral tool to evaluate the diversity of ethanol effects, including effects of moderate alcohol exposure during the pre- and postnatal period. Studies have found alterations in behavioral parameters, especially those related to anxiety, such as a more pronounced risk-taking behavior. Since there is a close relationship between learning/memory and anxiety parameters, it is supposed that the two cognitive processes could be impaired when the emotional process is altered (Brolese et al., 2014; Silva & Frussa-Filho, 2000).

Final Considerations

It is well established in the literature that prenatal alcohol exposure affects the brain, impacting cognitive and behavioral parameters. Taken together, studies confirm the general idea that high doses of alcohol exposure during the pre- and postnatal period appear to be a leading cause of mental retardation and congenital malformation in humans, and the most affected cognitive processes are learning and memory. In this chapter we pointed out putative causal links between glial dysfunction and brain structural and functional abnormalities, as well as behavioral impairments.

Although many countries do not have a clear recommendation regarding alcohol abstinence during gestation, studies with animal models have found that offspring prenatally exposed to moderate doses of ethanol also show pronounced risk-taking behavior that could also interfere in decision making in risky situations. Indeed, it is better to avoid alcohol exposure during gestation, since even moderate doses may affect different parameters in the CNS. Moreover, it is important to include astrocytes as targets in the therapeutic strategy against ethanol exposure.

References

Allen, N. J. (2013). Role of glia in developmental synapse formation. *Current Opinion in Neurobiology, 23*(6), 1027–1033. http://dx.doi.org/10.1016/j.conb.2013.06.004.

Allen, N. J., & Barres, B. A. (2005). Signaling between glia and neurons: Focus on synaptic plasticity. *Current Opinion in Neurobiology, 15*(5), 542–548. http://dx.doi.org/10.1016/j.conb.2005.08.006.

Allen, N. J., Bennett, M. L., Foo, L. C., Wang, G. X., Chakraborty, C., Smith, S. J., & Barres, B. A. (May 27, 2012). Astrocyte glypicans 4 and 6 promote formation of excitatory synapses via GluA1 AMPA receptors. *Nature, 86*(7403), 410–414. http://dx.doi.org/10.1038/nature11059.

Arcuri, C., Bianchi, R., Brozzi, F., & Donato, R. (2005). S100B increases proliferation in PC12 neuronal cells and reduces their responsiveness to nerve growth factor via Akt activation. *The Journal of Biological Chemistry, 280*(6), 4402–4414. http://dx.doi.org/10.1074/jbc.M406440200 [pii]10.1074/jbc.M406440200.

Ayers-Ringler, J. R., Jia, Y. F., Qiu, Y. Y., & Choi, D. S. (2016). Role of astrocytic glutamate transporter in alcohol use disorder. *World Journal of Psychiatry, 6*(1), 31. http://dx.doi.org/10.5498/wjp.v6.i1.31.

Bayraktar, O. A., Fuentealba, L. C., Alvarez-Buylla, A., & Rowitch, D. H. (November 20, 2014). Astrocyte development and heterogeneity. *Cold Spring Harbor Perspectives in Biology, 7*(1). a020362. http://dx.doi.org/10.1101/cshperspect.a020362

Beattie, E. C., Stellwagen, D., Morishita, W., Bresnahan, J. C., Ha, B. K., Von Zastrow, M., … Malenka, R. C. (2002). Control of synaptic strength by glial TNFalpha. *Science (New York, N.Y.), 295*(5563), 2282–2285. http://dx.doi.org/10.1126/science.1067859.

Belin-Rauscent, A., Fouyssac, M., Bonci, A., & Belin, D. (2016). How preclinical models evolved to resemble the diagnostic criteria of drug addiction. *Biological Psychiatry, 79*(1), 39–46. http://dx.doi.org/10.1016/j.biopsych.2015.01.004.

Berman, R. F., & Hannigan, J. H. (2000). Effects of prenatal alcohol exposure on the hippocampus: Spatial behavior, electrophysiology, and neuroanatomy. *Hippocampus, 10*(1), 94–110. http://dx.doi.org/10.1002/(SICI)1098-1063(2000)10:1%3c94::AID-HIPO11%3e3.0.CO;2-T.

Blondel, O., Collin, C., McCarran, W. J., Zhu, S., Zamostiano, R., Gozes, I., … McKay, R. D. (2000). A glia-derived signal regulating neuronal differentiation. *The Journal of Neuroscience, 20*(21), 8012–8020. Retrieved from http://www.ncbi.nlm.nih.gov/pubmed/11050122.

Brocardo, P. S., Gil-Mohapel, J., & Christie, B. R. (2011). The role of oxidative stress in fetal alcohol spectrum disorders. *Brain Research Reviews, 67*(1–2), 209–225. http://dx.doi.org/10.1016/j.brainresrev.2011.02.001.

Brolese, G., Lunardi, P., Broetto, N., Engelke, D. S., Lírio, F., Batassini, C., … Gonçalves, C. A. (2014). Moderate prenatal alcohol exposure alters behavior and neuroglial parameters in adolescent rats. *Behavioural Brain Research, 269*, 175–184. http://dx.doi.org/10.1016/j.bbr.2014.04.023.

Brolese, G., Lunardi, P., de Souza, D. F., Lopes, F. M., Leite, M. C., & Gonçalves, C. A. (2015). Pre- and postnatal exposure to moderate levels of ethanol can have long-lasting effects on hippocampal glutamate uptake in adolescent offspring. *PLoS One, 10*(5), e0127845. http://dx.doi.org/10.1371/journal.pone.0127845.

Carneiro, L. M. V., Diógenes, J. P. L., Vasconcelos, S. M. M., Aragão, G. F., Noronha, E. C., Gomes, P. B., & Viana, G. S. B. (2005). Behavioral and neurochemical effects on rat offspring after prenatal exposure to ethanol. *Neurotoxicology and Teratology, 27*(4), 585–592. http://dx.doi.org/10.1016/j.ntt.2005.06.006.

Choi, I. Y., Allan, A. M., & Cunningham, L. A. (2005). Moderate fetal alcohol exposure impairs the neurogenic response to an enriched environment in adult mice. *Alcoholism, Clinical and Experimental Research, 29*(11), 2053–2062. Retrieved from http://www.ncbi.nlm.nih.gov/pubmed/16340464.

Clarren, S. K., & Smith, D. W. (1978). The fetal alcohol syndrome. *The New England Journal of Medicine, 298*(19), 1063–1067. http://dx.doi.org/10.1056/NEJM197805112981906.

Dalçik, H., Yardimoglu, M., Filiz, S., Gonca, S., Dalçik, C., & Erden, B. F. (2009). Chronic ethanol-induced glial fibrillary acidic protein (GFAP) immunoreactivity: An immunocytochemical observation in various regions of adult rat brain. *The International Journal of Neuroscience, 119*(9), 1303–1318. Retrieved from http://www.ncbi.nlm.nih.gov/pubmed/19922358.

Danbolt, N. C. (2001). Glutamate uptake. *Progress in Neurobiology, 65*(1), 1–105. Retrieved from http://www.ncbi.nlm.nih.gov/pubmed/11369436.

De Juan Romero, C., & Borrell, V. (2015). Coevolution of radial glial cells and the cerebral cortex. *Glia, 63*(8), 1303–1319. http://dx.doi.org/10.1002/glia.22827.

De Licona, H. K., Karacay, B., Mahoney, J., McDonald, E., Luang, T., & Bonthius, D. J. (2009). A single exposure to alcohol during brain development induces microencephaly and neuronal losses in genetically susceptible mice, but not in wild type mice. *Neurotoxicology, 30*(3), 459–470. http://dx.doi.org/10.1016/j.neuro.2009.01.010.

Donato, R. (2003). Intracellular and extracellular roles of S100 proteins. *Microscopy Research and Technique, 60*(6), 540–551. http://dx.doi.org/10.1002/jemt.10296.

Donato, R., Cannon, B. R., Sorci, G., Riuzzi, F., Hsu, K., Weber, D. J., & Geczy, C. L. (2013). Functions of S100 proteins. *Current Molecular Medicine, 13*(1), 24–57. http://dx.doi.org/10.2174/1566524011307010024.

Dringen, R. (2000). Metabolism and functions of glutathione in brain. *Progress in Neurobiology, 62*.

Dringen, R., & Hirrlinger, J. (2003). Glutathione pathways in the brain. *Biological Chemistry, 384*(4), 505–516. http://dx.doi.org/10.1515/BC.2003.059.

Eckardt, M. J., File, S. E., Gessa, G. L., Grant, K. A., Guerri, C., Hoffman, P. L., … Tabakoff, B. (1998). Effects of moderate alcohol consumption on the central nervous system. *Alcoholism, Clinical and Experimental Research, 22*(5), 998–1040. Retrieved from http://www.ncbi.nlm.nih.gov/pubmed/9726269.

Eriksen, J. L., Gillespie, R. A., & Druse, M. J. (2000). Effects of in utero ethanol exposure and maternal treatment with a 5-HT(1A) agonist on S100B-containing glial cells. *Brain Research. Developmental Brain Research, 121*(2), 133–143. Retrieved from http://www.ncbi.nlm.nih.gov/pubmed/10876026.

Evrard, S. G., Duhalde-Vega, M., Tagliaferro, P., Mirochnic, S., Caltana, L. R., & Brusco, A. (2006). A low chronic ethanol exposure induces morphological changes in the adolescent rat brain that are not fully recovered even after a long abstinence: An immunohistochemical study. *Experimental Neurology, 200*(2), 438–459. http://dx.doi.org/10.1016/j.expneurol.2006.03.001.

Franke, H., Kittner, H., Berger, P., Wirkner, K., & Schramek, J. (1997). The reaction of astrocytes and neurons in the hippocampus of adult rats during chronic ethanol treatment and correlations to behavioral impairments. *Alcohol, 14*(5), 445–454.

Gil-Mohapel, J., Boehme, F., Kainer, L., & Christie, B. R. (2010). Hippocampal cell loss and neurogenesis after fetal alcohol exposure: Insights from different rodent models. *Brain Research Reviews, 64*(2), 283–303. http://dx.doi.org/10.1016/j.brainresrev.2010.04.011.

Goodlett, C. R., Leo, J. T., O'Callaghan, J. P., Mahoney, J. C., & West, J. R. (1993). Transient cortical astrogliosis induced by alcohol exposure during the neonatal brain growth spurt in rats. *Brain Research. Developmental Brain Research, 72*(1), 85–97. Retrieved from http://www.ncbi.nlm.nih.gov/pubmed/8453767.

Guerri, C. (1998). Neuroanatomical and neurophysiological mechanisms involved in central nervous system dysfunctions induced by prenatal alcohol exposure. *Alcoholism, Clinical and Experimental Research, 22*(2), 304–312. Retrieved from http://www.ncbi.nlm.nih.gov/pubmed/9581633.

Guerri, C. (2001). Glia and fetal alcohol syndrome. *Neurotoxicology, 22*(5), 593–599. http://dx.doi.org/10.1016/S0161-813X(01)00037-7.

Guerri, C. (2002). Mechanisms involved in central nervous system dysfunctions induced by prenatal ethanol exposure. *Neurotoxicity Research, 4*(4), 327–335. http://dx.doi.org/10.1080/1029842021000010884.

Guerri, C., & Renau-piqueras, J. (1997). Alcohol, astroglia and brain development. *Molecular Neurobiology, 15*(1), 65–81.

Guizzetti, M., & Costa, L. G. (2007). Cholesterol homeostasis in the developing brain: A possible new target for ethanol. *Human & Experimental Toxicology, 26*(4), 355–360. http://dx.doi.org/10.1177/0960327107078412.

Guizzetti, M., Moore, N. H., Giordano, G., VanDeMark, K. L., & Costa, L. G. (2010). Ethanol inhibits neuritogenesis induced by astrocyte muscarinic receptors. *Glia, 58*(12), 1395–1406. http://dx.doi.org/10.1002/glia.21015.

Hamilton, D. A., Akers, K. G., Rice, J. P., Johnson, T. E., Candelaria-Cook, F. T., Maes, L. I., … Savage, D. D. (2010). Prenatal exposure to moderate levels of ethanol alters social behavior in adult rats: Relationship to structural plasticity and immediate early gene expression in frontal cortex. *Behavioural Brain Research, 207*(2), 290–304. http://dx.doi.org/10.1016/j.bbr.2009.10.012.

Handley, S. L., & Mithani, S. (1984). Effects of alpha-adrenoceptor agonists and antagonists in a maze-exploration model of "fear"-motivated behaviour. *Naunyn-Schmiedeberg's Archives of Pharmacology, 327*(1), 1–5. Retrieved from http://www.ncbi.nlm.nih.gov/pubmed/6149466.

Henderson, J., Gray, R., & Brocklehurst, P. (2007). Systematic review of effects of low-moderate prenatal alcohol exposure on pregnancy outcome. *BJOG: An International Journal of Obstetrics and Gynaecology, 114*(3), 243–252. http://dx.doi.org/10.1111/j.1471-0528.2006.01163.x.

Hughes, E. G., Elmariah, S. B., & Balice-Gordon, R. J. (2010). Astrocyte secreted proteins selectively increase hippocampal GABAergic axon length, branching, and synaptogenesis. *Molecular and Cellular Neurosciences, 43*(1), 136–145. http://dx.doi.org/10.1016/j.mcn.2009.10.004.

Ikonomidou, C., Bittigau, P., Ishimaru, M. J., Wozniak, D. F., Koch, C., Genz, K., … Olney, J. W. (2000). Ethanol-induced apoptotic neurodegeneration and fetal alcohol syndrome. *Science (New York, N.Y.), 287*(5455), 1056–1060. Retrieved from http://www.ncbi.nlm.nih.gov/pubmed/10669420.

Jones, E. V., Bernardinelli, Y., Tse, Y. C., Chierzi, S., Wong, T. P., & Murai, K. K. (2011). Astrocytes control glutamate receptor levels at developing synapses through SPARC-beta-integrin interactions. *The Journal of Neuroscience, 31*(11), 4154–4165. http://dx.doi.org/10.1523/JNEUROSCI.4757-10.2011.

Jones, K. L., & Smith, D. W. (1973). Recognition of the fetal alcohol syndrome in early infancy. *Lancet (London, England), 302*(7836), 999–1001. Retrieved from http://www.ncbi.nlm.nih.gov/pubmed/4127281.

Karanian, J., Yergey, J., Lister, R., D'Souza, N., Linnoila, M., & Salem, N. (1986). Characterization of an automated apparatus for precise control of inhalation chamber ethanol vapor and blood ethanol concentrations. *Alcoholism, Clinical and Experimental Research,*

10(4), 443–447. Retrieved from http://www.ncbi.nlm.nih.gov/pubmed/3530024.

Kleindienst, A., & Ross Bullock, M. (2006). A critical analysis of the role of the neurotrophic protein S100B in acute brain injury. *Journal of Neurotrauma, 23*(8), 1185–1200. http://dx.doi.org/10.1089/neu.2006.23.1185.

Kumral, A., Tugyan, K., Gonenc, S., Genc, K., Genc, S., Sonmez, U., & Ozkan, H. (2005). Protective effects of erythropoietin against ethanol-induced apoptotic neurodegenaration and oxidative stress in the developing C57BL/6 mouse brain. *Brain Research. Developmental Brain Research, 160*, 146–156. http://dx.doi.org/10.1016/j.devbrainres.2005.08.006. PMID:16236368.

Luo, J., & Miller, M. W. (1999). Platelet-derived growth factor-mediated signal transduction underlying astrocyte proliferation: Site of ethanol action. *The Journal of Neuroscience, 19*(22), 10014–10025. Retrieved from http://www.ncbi.nlm.nih.gov/pubmed/10559409.

Marais, L., van Rensburg, S. J., van Zyl, J. M., Stein, D. J., & Daniels, W. M. U. (2008). Maternal separation of rat pups increases the risk of developing depressive-like behavior after subsequent chronic stress by altering corticosterone and neurotrophin levels in the hippocampus. *Neuroscience Research, 61*(1), 106–112. http://dx.doi.org/10.1016/j.neures.2008.01.011.

Miller, M. W., & Robertson, S. (1993). Prenatal exposure to ethanol alters the postnatal development and transformation of radial glia to astrocytes in the cortex. *The Journal of Comparative Neurology, 337*(2), 253–266. http://dx.doi.org/10.1002/cne.903370206.

Nägler, K., Mauch, D. H., & Pfrieger, F. W. (2001). Glia-derived signals induce synapse formation in neurones of the rat central nervous system. *The Journal of Physiology, 533*(Pt 3), 665–679. Retrieved from http://www.pubmedcentral.nih.gov/articlerender.fcgi?artid=2278670&tool=pmcentrez&rendertype=abstract.

Nash, R., Krishnamoorthy, M., Jenkins, A., & Csete, M. (2012). Human embryonic stem cell model of ethanol-mediated early developmental toxicity. *Experimental Neurology, 234*(1), 127–135. http://dx.doi.org/10.1016/j.expneurol.2011.12.022.

Olton, D. S., & Samuelson, R. J. (1976). Remembrance of places passed: Spatial memory in rats. *Journal of Experimental Psychology. Animal Behavior Processes, 2*(2), 97–116. http://dx.doi.org/10.1037/0097-7403.2.2.97.

Patten, A. R., Brocardo, P. S., Sakiyama, C., Wortman, R. C., Noonan, A., Gil-Mohapel, J., & Christie, B. R. (2013). Impairments in hippocampal synaptic plasticity following prenatal ethanol exposure are dependent on glutathione levels. *Hippocampus, 23*(12), 1463–1475. http://dx.doi.org/10.1002/hipo.22199.

Patten, A. R., Fontaine, C. J., & Christie, B. R. (2014). A comparison of the different animal models of fetal alcohol spectrum disorders and their use in studying complex behaviors. *Frontiers in Pediatrics, 2*(93). http://dx.doi.org/10.3389/fped.2014.00093.

Pellow, S., Chopin, P., File, S. E., & Briley, M. (1985). Validation of open: closed arm entries in an elevated plus-maze as a measure of anxiety in the rat. *Journal of Neuroscience Methods, 14*(3), 149–167. Retrieved from http://www.ncbi.nlm.nih.gov/pubmed/2864480.

Pfrieger, F. W. (2002). Role of glia in synapse development. *Current Opinion in Neurobiology, 12*(5), 486–490. Retrieved from http://www.ncbi.nlm.nih.gov/pubmed/12367626.

Phillips, D. E. (1992). Effects of alcohol on the development of glial cells and myelin. In R. R. Watson (Ed.), *Alcohol and neurobiology: Brain development and hormone regulation* (pp. 83–108). Boca Raton, FL: CRC Press.

Reeves, R. H., Yao, J., Crowley, M. R., Buck, S., Zhang, X., Yarowsky, P., & Hilt, D. C. (1994). Astrocytosis and axonal proliferation in the hippocampus of S100b transgenic mice. *Proceedings of the National Academy of Sciences of the United States of America, 91*(12), 5359–5363. Retrieved from http://www.pubmedcentral.nih.gov/articlerender.fcgi?artid=43994&tool=pmcentrez&rendertype=abstract.

Reyes, E., Wolfe, J., & Savage, D. D. (1989). The effects of prenatal alcohol exposure on radial arm maze performance in adult rats. *Physiology & Behavior, 46*(1), 45–48. Retrieved from http://www.ncbi.nlm.nih.gov/pubmed/2813555.

Riley, E. P., Mattson, S. N., Sowell, E. R., Jernigan, T. L., Sobel, D. F., & Jones, K. L. (1995). Abnormalities of the corpus callosum in children prenatally exposed to alcohol. *Alcoholism, Clinical and Experimental Research, 19*(5), 1198–1202. Retrieved from http://www.ncbi.nlm.nih.gov/pubmed/8561290.

Rubert, G., Miñana, R., Pascual, M., & Guerri, C. (2006). Ethanol exposure during embryogenesis decreases the radial glial progenitor-pool and affects the generation of neurons and astrocytes. *Journal of Neuroscience Research, 84*(3), 483–496. http://dx.doi.org/10.1002/jnr.20963.

Sadrian, B., Wilson, D. A., & Saito, M. (2013). Long-lasting neural circuit dysfunction following developmental ethanol exposure. *Brain Sciences, 3*(2), 704–727. http://dx.doi.org/10.3390/brainsci3020704.

Salami, M., Aghanouri, Z., Rashidi, A. A., & Keshavarz, M. (2004). Prenatal alcohol exposure and dysfunction of hippocampal formation in cognition. *Iranian Journal of Reproductive Medicine, 2*(2), 43–50.

Samson, H. H., Denning, C., & Chappelle, A. M. (1996). The use of nonalcoholic beer as the vehicle for ethanol consumption in rats. *Alcohol (Fayetteville, NY), 13*(4), 365–368. Retrieved from http://www.ncbi.nlm.nih.gov/pubmed/8836325.

Silva, R. H., & Frussa-Filho, R. (2000). The plus-maze discriminative avoidance task: A new model to study memory-anxiety interactions. Effects of chlordiazepoxide and caffeine. *Journal of Neuroscience Methods, 102*(2), 117–125. Retrieved from http://www.ncbi.nlm.nih.gov/pubmed/11040408.

Slezak, M., & Pfrieger, F. W. (October 2003). New roles for astrocytes: Regulation of CNS synaptogenesis. *Trends in Neurosciences, 26*(10), 531–535. http://dx.doi.org/10.1016/j.tins.2003.08.005.

Sofroniew, M. V. (2005). Reactive astrocytes in neural repair and protection. *The Neuroscientist: A Review Journal Bringing Neurobiology, Neurology and Psychiatry, 11*(5), 400–407. http://dx.doi.org/10.1177/1073858405278321.

Swayze, V. W., Johnson, V. P., Hanson, J. W., Piven, J., Sato, Y., Giedd, J. N., & Andreasen, N. C. (1997). Magnetic resonance imaging of brain anomalies in fetal alcohol syndrome. *Pediatrics, 99*(2), 232–240. Retrieved from http://www.ncbi.nlm.nih.gov/pubmed/9024452.

Tajuddin, N. F., & Druse, M. J. (1999). In utero ethanol exposure decreased the density of serotonin neurons. Maternal ipsapirone treatment exerted a protective effect. *Brain Research. Developmental Brain Research, 117*(1), 91–97.

Tajuddin, N. F., Orrico, L. A., Eriksen, J. L., & Druse, M. J. (2003). Effects of ethanol and ipsapirone on the development of midline raphe glial cells and astrocytes. *Alcohol, 29*(3), 157–164. http://dx.doi.org/10.1016/S0741-8329(03)00024-7.

Taléns-Visconti, R., Sanchez-Vera, I., Kostic, J., Perez-Arago, M. A., Erceg, S., Stojkovic, M., & Guerri, C. (2011). Neural differentiation from human embryonic stem cells as a tool to study early brain development and the neuroteratogenic effects of ethanol. *Stem Cells and Development, 20*(2), 327–339. http://dx.doi.org/10.1089/scd.2010.0037.

Tanaka, K. (2000). Functions of glutamate transporters in the brain. *Neuroscience Research, 37*(1), 15–19. Retrieved from http://www.ncbi.nlm.nih.gov/pubmed/10802340.

Tramontina, A.C., Souza, D.F., Frizzo, J.K., & Gonçalves, C.A. (2006). High Glutamate Decreases S100B Secretion by a Mechanism Dependent on the Glutamate Transporter. 815–820. http://dx.doi.org/10.1007/s11064-006-9085-z.

Tubaro, C., Arcuri, C., Giambanco, I., & Donato, R. (2011). S100B in myoblasts regulates the transition from activation to quiescence and from quiescence to activation and reduces apoptosis. *Biochimica et Biophysica Acta, 1813*(5), 1092–1104. http://dx.doi.org/10.1016/j.bbamcr.2010.11.015.

Vallés, S., Pitarch, J., Renau-Piqueras, J., & Guerri, C. (1997). Ethanol exposure affects glial fibrillary acidic protein gene expression and transcription during rat brain development. *Journal of Neurochemistry, 69*, 2484–2493..

Vangipuram, S. D., & Lyman, W. D. (2010). Ethanol alters cell fate of fetal human brain-derived stem and progenitor cells. *Alcoholism: Clinical and Experimental Research, 34*, 1574–1583.

Vemuri, M. C., & Chetty, C. S. (2005). Alcohol impairs astrogliogenesis by stem cells in rodent neurospheres. *Neurochemistry International, 47*(1), 129–135.

Walf, A. A., & Frye, C. A. (2006). A review and update of mechanisms of estrogen in the hippocampus and amygdala for anxiety and depression behavior. *Neuropsychopharmacology, 31*(6), 1097–1111. http://dx.doi.org/10.1038/sj.npp.1301067.

Wilhelm, C. J., & Guizzetti, M. (2016). Fetal alcohol spectrum disorders: An overview from the glia perspective. *Frontiers in Integrative Neuroscience, 9*, 65. http://dx.doi.org/10.3389/fnint.2015.00065.

Wilson, D. A., Peterson, J., Basavaraj, B. S., & Saito, M. (2011). Local and regional network function in behaviorally relevant cortical circuits of adult mice following postnatal alcohol exposure. *Alcoholism, Clinical and Experimental Research, 35*(11), 1974–1984. http://dx.doi.org/10.1111/j.1530-0277.2011.01549.x.

Zhang, X., Bhattacharyya, S., Kusumo, H., Goodlett, C. R., Tobacman, J. K., & Guizzetti, M. (2014). Arylsulfatase B modulates neurite outgrowth via astrocyte chondroitin-4-sulfate: Dysregulation by ethanol. *Glia, 62*(2), 259–271. http://dx.doi.org/10.1002/glia.22604.

Zhou, F. C., Sari, Y., Zhang, J. K., Goodlett, C. R., & Li, T. (2001). Prenatal alcohol exposure retards the migration and development of serotonin neurons in fetal C57BL mice. *Brain Research. Developmental Brain Research, 126*(2), 147–155. Retrieved from http://www.ncbi.nlm.nih.gov/pubmed/11248348.

Further Reading

Göritz, C., Mauch, D. H., Nägler, K., & Pfrieger, F. W. (2002). Role of glia-derived cholesterol in synaptogenesis: New revelations in the synapse-glia affair. *Journal of Physiology, Paris, 96*(3–4), 257–263. Retrieved from http://www.ncbi.nlm.nih.gov/pubmed/12445904.

Lieber, C. S., & DeCarli, L. M. (1989). Liquid diet technique of ethanol administration: 1989 Update. *Alcohol and Alcoholism (Oxford, Oxfordshire), 24*(3), 197–211. Retrieved from http://www.ncbi.nlm.nih.gov/pubmed/2667528.

Namba, T., Mochizuki, H., Onodera, M., Mizuno, Y., Namiki, H., & Seki, T. (2005). The fate of neural progenitor cells expressing astrocytic and radial glial markers in the postnatal rat dentate gyrus. *The European Journal of Neuroscience, 22*(8), 1928–1941. http://dx.doi.org/10.1111/j.1460-9568.2005.04396.x.

Nimmerjahn, A. (2009). Astrocytes going live: Advances and challenges. *Journal of Physiology, 587*(Pt 8), 1639–1647. http://dx.doi.org/jphysiol.2008.167171 [pii]10.1113/jphysiol.2008.167171.

Singh, S. P., Ehmann, S., & Snyder, A. K. (1994). Ethanol and glucose-deprivation neurotoxicity in cortical cell cultures. *Metabolism: Clinical and Experimental, 43*(9), 1108–1113. Retrieved from http://www.ncbi.nlm.nih.gov/pubmed/7916117.

Sonnewald, U., Westergaard, N., & Schousboe, A. (1997). Glutamate transport and metabolism in astrocytes. *Glia, 21*(1), 56–63. Retrieved from http://www.ncbi.nlm.nih.gov/pubmed/9298847.

Stipursky, J., Romão, L., Tortelli, V., Neto, V. M., & Gomes, F. C. A. (2011). Neuron-glia signaling: Implications for astrocyte differentiation and synapse formation. *Life Sciences, 89*(15–16), 524–531. http://dx.doi.org/10.1016/j.lfs.2011.04.005.

CHAPTER

3

Alcohol on Histaminergic Neurons of Brain

S.M. Zimatkin, K.M. Phedina

Grodno State Medical University, Grodno, Belarus

INTRODUCTION

Histamine is a transmitter in the nervous system and a signaling molecule in the gut, skin, and immune system (Haas, Sergeeva, & Selbach, 2008). Histaminergic neurons in mammalian brain are located exclusively in the tuberomamillary nucleus of the posterior hypothalamus, forming five cluster groups (E1–E5) (Zimatkin, Kuznetsova, & Strik, 2006).

According to Zimatkin et al. (2006) the total bilateral volume of the histaminergic nuclei of the rat brain amounts to 0.5 mm^3: the E2, E4, E3, E5, and E1 nuclei occupy 40%, 35%, 13%, 9%, and 3% of the total volume, respectively. The distribution density of neurons in the histaminergic nuclei of the rat hypothalamus decreases in the order E1 > E3 > E2 (the "compact" nuclei) > E4 (an "intermediate" density nucleus) > E5 (the "diffuse part"). The mean number of histaminergic neurons in the rat hypothalamus is 37,200 ± 2800; the E2, E3, E4, E1, and E5 nuclei contain 54%, 23%, 6%, 7%, and 0.32% of these cells, respectively. The E1–E3 nuclei are dominated by small- and intermediate-sized neurons, round in shape, and the E5 nucleus is dominated by intermediate- and large-sized neurons, fusiform in shape. A subpopulation of giant histaminergic neurons is detected in the E4–E5 nuclei.

Histaminergic neurons send their axons throughout the central nervous system, regulating other transmitter systems. Active solely during waking, they maintain wakefulness and attention. Three of the four known histamine receptors bind to glutamate NMDA receptors and serve multiple functions in the brain, particularly control of excitability and plasticity. H1 and H2 receptor-mediated actions are mostly excitatory; H3 receptors (H3Rs) act as inhibitory auto- and heteroreceptors (Haas et al., 2008). Histaminergic system participates in regulation of many functions, systems, and reactions of the body: neuroendocrine and cardiovascular, brain blood flow, sleep and awakening, hibernation, feeding and drinking behavior, memory, cognition, and learning (reviews of Blandina, Munari, Provensi, & Passani, 2012; Haas & Panula, 2003; Haas et al., 2008). This system stimulates serotonin, norepinephrine, and dopamine transmission in the brain (Flik, Folgering, Cremers, Westerink, & Dremencov, 2015). Brain histamine also functions as a suppressor in bioprotection against various noxious and unfavorable stimuli of convulsion, drug sensitization, denervation supersensitivity, ischemic lesions, and stress susceptibility (Yanai & Tashiro, 2007). It participates in some pathological states and diseases, including addictions (reviews of Blandina et al., 2012; Haas & Panula, 2003; Haas et al., 2008).

Histamine is synthesized from L-histidine by the enzyme histidine decarboxylase. No high-affinity and selective uptake system for histamine has been found in the central nervous system. Histamine is removed by histamine-N-methyltransferase by conversion to tele-methylhistamine, which is further metabolized to N-tele-methylimidazole acetic acid by the sequential actions of type B monoamine oxidase and aldehyde dehydrogenase. In peripheral tissues about 30–40% of total histamine is oxidized by diamine oxidase and aldehyde oxidase to imidazole acetic acid, a full agonist of type A GABA receptors. In brain this can take place when the histamine-N-methyltransferase is inhibited (Prell, Morrishow, Duoyon, & Lee, 1997).

Thus, the histamine and ethanol metabolic pathways in the brain have the common enzyme—aldehyde dehydrogenase, therefore the highly active ethanol metabolite, acetaldehyde can interfere with histamine degradation by competing with histamine metabolite N-tele-methylimidazole acetaldehyde for this enzyme (Ambroziak & Pietruszko, 1987). Therefore, this may represent a metabolic basis for the alcohol–histamine interaction in the brain (Zimatkin & Anichtchik, 1999).

Addictive Substances and Neurological Disease
http://dx.doi.org/10.1016/B978-0-12-805373-7.00003-7

THE EFFECTS OF ALCOHOL ON HISTAMINE LEVELS IN THE BRAIN

Studies investigating the effects of alcohol on histamine levels in the brain obtained different results depending on the animal species tested, the brain region studied, and the dose of alcohol administered (Zimatkin & Anichtchik, 1999). Although alcohol has been shown to increase histamine concentration in many brain regions such as the hypothalamus, the thalamus, and the cortex in rats (Prell, Bielkiewicz, & Mazurkiewicz-Kwilecki, 1982; Rawat, 1980; Subramanian, Mitznegg, & Estler, 1978; Subramanian, Schinzel, Mitznegg, & Estler, 1980), Subramanian et al. (1980) did not detect any changes in the striatum. Moreover, high doses of alcohol administered directly into the stomach decreased histamine concentration in the hypothalamus of rats (Prell et al., 1982) and in the hypothalamus, cerebral cortex, and whole brain of guinea pigs (Nowak & Maslinski, 1984). Following alcohol administration, the activity of the first histamine metabolizing enzyme, histamine N-methyltransferase failed to change in any brain region (Prell & Mazurkiewicz-Kwilecki, 1981; Subramanian et al., 1980). Ethanol increases the steady-state N-tele-methylhistamine levels in the mouse hypothalamus, probably by inhibiting the elimination of this metabolite in the brain (Itoh, Nishibori, Oishi, & Saeki, 1985).

When the whole mouse brain was considered, Papanicolaou and Fennessy (1980) found that administration of low doses of ethanol (0.175 g/kg, intraperitoneally, i.p.) enhanced brain histamine levels, whereas higher doses (1.75 g/kg, i.p.) decreased histamine concentrations. Interestingly, the increase of brain histamine induced by the low alcohol dose paralleled with behavioral hyperactivity and hyperthermia. In contrast, the higher doses that reduced histamine caused sedation and a decrease in body temperature.

The aforementioned investigations showed that alcohol and histamine interact in the body. Further studies suggest that the brain histaminergic system and especially the H3 histamine receptors participate in the regulation of alcohol consumption and alcohol-related behavior. Thus, lowering the brain histamine levels significantly increases ethanol sensitivity of alcohol tolerant rats. In keeping with these data, a neuronal histamine N-methyltransferase polymorphism has been linked to alcoholism in humans (Reuter et al., 2007).

BRAIN HISTAMINERGIC SYSTEM IN HUMAN ALCOHOLICS

However, limited data are available about brain histaminergic system in human alcoholics. Histamine levels in cortical gray matter were higher in type 1 alcoholics (late onset, often females, low degree of association with violence) than in normal control brains (Alakarppa et al., 2002). The levels of the first metabolite of histamine, tele-methylhistamine, were significantly increased in type 2 alcoholics (early onset, often males, high degree of association with violence) indicating increased histamine release and turnover. The concentrations of histamine and tele-methylhistamine in white matter were much lower than in gray matter. This may mean that histamine synthesis and/or metabolism are primarily altered in alcoholics, or that the possibly associated liver pathology with increased blood histidine lies behind the abnormal findings. Thus, these results show that histamine may play some role in alcoholism and alcoholism-associated aggressive behavior in man.

This finding is supported by a previous study, in which behavioral tests were carried out on mutant mice lacking H1 receptors. The mutant mice were less aggressive than the wild-type mice (Yanai et al., 1998). In the same study, it was found that the turnover rate of 5-hydroxytryptamine (5-HT) was significantly increased in H1 receptor null mice. Low 5-HTergic activity is associated with aggressive behavior (Davidson, Putnam, & Larson, 2000). It is possible that histamine participates in controlling the release of 5-HT through H1 receptors.

THE ROLE OF H3 RECEPTORS IN MODULATION OF ALCOHOL STIMULATION AND REWARD

Growing evidence supports a role of the central histaminergic system in a modulatory influence on drug addiction in general and alcohol-use disorders in particular through histamine H3 receptor (H3R).

H3R is one of the four receptors (H1−H4) mediating the effects of neuronal histamine. H3R is highly expressed in the brain and differs from other histamine receptor types in its exceptionally multifunctional role. In the beginning of the 1980s, H3R was characterized as a typical autoreceptor, which regulates the release and synthesis of histamine (Arrang, Garbarg, & Schwartz, 1983, 1987). In 1999 it was cloned (Lovenberg et al., 1999), but later it was found that H3R can act also as a heteroreceptor, which regulates the release of many other neurotransmitters, such as dopamine, acetylcholine, noradrenaline, and GABA (Haas & Panula, 2003; Schlicker, Malinowska, Kathmann, & Gothert, 1994). These properties of H3Rs initiated a broad interest within the pharmaceutical industry to develop specific H3R ligands for the treatment of, for example, sleep–wake disorders and obesity (histamine dependent), and attention and cognitive deficits (noradrenaline and acetylcholine dependent). H3R cooperate with dopamine D2 receptors in the regulation of striatal gene expression (Pillot et al., 2002). Related interactions of histamine with dopamine, other amines, GABA, and glutamate (Selbach, Brown, & Haas, 1997; Selbach, Stehle, & Haas, 2007) may be relevant for both learning and memory, as well as addiction and compulsion. The high expression level of H3R in the mesolimbic system and the ability of H3Rs to affect brain dopaminergic functions via regulation of dopamine release and interaction with postsynaptic dopaminergic receptors raise a question whether H3Rs could be involved in brain reward processes underlying the development of addictive behaviors.

Brain histamine levels and signaling via H3Rs play an important role in modulation of alcohol stimulation and reward in rodents. For example, high alcohol preference and low alcohol sensitivity correlate with brain histamine and H3R-mediated processes (Lintunen et al., 2001, 2002; Nuutinen, Karlstedt, Aitta-Aho, Korpi, & Panula, 2010). AA (Alko, alcohol) line rats with high inborn preference to alcohol have higher contents of histamine and it's primarily metabolite, methylhistamine, as well as increased density of histaminergic fibers in brain regions as compared to refused ethanol ANA (Alko, nonalcohol) rats. H3 histamine receptor ligands modulate the voluntary alcohol consumption in AA rats (Lintunen et al., 2001). H3Rs are abundant in cortico-mesolimbic areas of the brain in rodents and humans (Anichtchik, Peitsaro, Rinne, Kalimo, & Panula, 2001; Pillot et al., 2002; Pollard, Moreau, Arrang, & Schwartz, 1993). These are the same neuroanatomical areas on which alcohol and other drugs of abuse exert their reinforcing and subjective effects (George et al., 1998).

Interestingly, H3R antagonists/inverse agonists stimulate acetylcholine transmission in different brain areas, facilitate memory in animal models and can reverse learning deficits induced by drugs such as scopolamine, dizocilpine and alcohol (review of Alleva, Tirelli, & Brabant, 2013).

Studies on both rats and mice indicate that histamine H3R antagonists (they activate brain histaminergic neurons) decrease alcohol drinking in several models, such as operant alcohol administration and drinking in the dark paradigm, inhibit ethanol-evoked stimulation of locomotor activity and potentiate an ethanol reward. Alcohol-induced place preference is also affected by these drugs (Nuutinen et al., 2011; Panula & Nuutinen, 2011). For instance, Nuutinen et al. (2011) suggested that histamine H3R antagonists can diminish motivational aspects of alcohol dependence. They studied the role of H3Rs in alcohol-related behaviors using H3R knockout mice and ligands. H3R knockout mice consumed less alcohol than wild-type littermates in a two-bottle free-choice test and in a "drinking in the dark" model. H3R antagonist ciproxifan suppressed and H3R agonist immepip increased alcohol drinking in C57BL/6J mice. Impairment in reward mechanisms in H3R knockout mice was confirmed by the lack of alcohol-evoked conditioned place preference. Plasma alcohol concentrations of H3R knockout and wild-type mice were similar. There were no marked differences in brain biogenic amine levels in H3R knockout mice compared with the control animals after alcohol drinking. The findings of this study provide evidence for the important role of H3R in alcohol-related behaviors, especially in alcohol drinking and alcohol reward.

Galici et al. (2011) have shown that subcutaneous administration of a novel, selective, and brain penetrant H3R antagonist (JNJ-39220675) dose dependently reduced both alcohol intake and preference in alcohol-preferring rats—a genetic animal model of high alcohol preference. JNJ-39220675 also reduced alcohol preference in the same strain of rats following a 3-day alcohol deprivation. The compound significantly and dose dependently reduced alcohol self-administration without changing saccharin self-administration in alcohol nondependent rats. Galici et al. measured the effects of JNJ-39220675 on alcohol-induced dopamine release in the nucleus accumbens in freely moving rats as alcohol is believed to have its rewarding effects mediated, at least in part, by stimulation of dopamine in the nucleus accumbens (Imperato & Di Chiara, 1986). The compound did not change alcohol-induced dopamine release in nucleus accumbens. Furthermore, it did not change the ataxic effects of alcohol, nor alcohol elimination rate.

Bahi, Sadek, Schwed, Walter, and Stark (2013) have found that following administration of the H3R antagonist ST1283 (2.5, 5, and 10 mg/kg, i.p.), there was a significant dose-dependent decrease in alcohol consumption and preference. More interestingly, systemic administration of ST1283 inhibited ethanol-induced place preference and ethanol-enhanced locomotion. This inhibition was blocked when mice were pretreated with the selective H3R agonist R-(alpha)-methylhistamine (10 mg/kg). Thus, ST1283 may decrease voluntary ethanol consumption and ethanol-induced place preference by altering its reinforcing effects.

It can be concluded that the brain histaminergic system has an inhibitory role in alcohol reward. Increasing neuronal histamine released via H3R blockade could therefore be a novel way of treating alcohol dependence (Vanhanen et al., 2013). Thus, H3R antagonists are promising candidates for use in human alcoholic patients in the future (Nuutinen et al., 2011; Panula & Nuutinen, 2011). Data continue to appear showing the involvement of brain histaminergic system in the mechanisms of action of alcohol on the brain and the pathogenesis of alcoholism (Bahi, Sadek, Nurulain,

Lazewska, & Kiec-Kononowicz, 2015; Morais-Silva, Ferreira-Santos, & Marin, 2016; Nuutinen et al., 2015).

HISTOLOGICAL CHANGES IN HISTAMINERGIC NEURONS FOLLOWING ALCOHOL ADMINISTRATION

The action of alcohol on histaminergic neurons using microscopic methods has been studied only in our laboratory. We have examined histological, histochemical, and ultrastructural changes developing in brain histaminergic neurons following acute (single i.p. doses of 1 or 4 g/kg as 20% ethanol solution, decapitation at 1 and 6 h after administration), subacute (20% ethanol solution at a dose of 4 g/kg/day for 7 days, decapitation at 24 h after the last dose) and chronic (20% ethanol solution as the sole source of liquid, 3.5 g/kg/day for 6 months) exposure to alcohol (Zimatkin, Fedina, & Kuznetsova, 2013; Zimatkin & Phedina, 2015a, 2015b; Zimatkin, Phedina, & Anichtchik, 2016).

Ethanol has unequal and sometimes multidirectional effects on histaminergic neurons of rat brain structure depending on the dose, time after administration, and duration of exposure to alcohol. Thus, acute (single) administration at a low dose of ethanol (1 g/kg) does not cause changes in the degree of neurons chromatophilia (the intensity of staining of neurons cytoplasm by basic dies), large dose (4 g/kg) causes a slight increase in the amount of cells-shadows (very pale remnants of died neurons with no visible nucleus) (Zimatkin et al., 2013), subchronic (repeated) causes a sharp increase in the number of cells-shadows and hypochromic (pale-staining) neurons (Zimatkin & Phedina, 2015b), chronic alcohol consumption leads to an increase of the amount of cells-shadows and hyperchromic (intense-staining) neurons, that indicates the toxic effect of ethanol to histaminergic neurons (Zimatkin & Phedina, 2015a).

Acute, subacute, and chronic alcohol intoxication leads to perikarya rounding of histaminergic neurons. Nuclei of such cells also become more roundish and spherical after single ethanol administration in narcotic dose and its chronic consumption. The administration at a low dose of ethanol causes a decrease in size of the neurons perikarya and their nuclei. Under the influence of narcotic ethanol dose the average sizes of the cells are not significantly changed. At 24 h following 7-day ethanol administration the histaminergic neurons bodies became smaller, and after 6-month chronic alcohol consumption they became larger. After single ethanol administration in high dose and chronic alcohol consumption the observed increase in sphericity and rounding of histaminergic neurons perikarya and nuclei may be associated with a disturbance of the electrolyte balance, and an alteration of the cytoskeleton of neurons, induced by ethanol. After 24 h following the last administration of the seventh high ethanol dose the cells decrease, may be the way of adaptation of neurons to the predictable alcohol-induced swelling of their bodies (Zimatkin et al., 2013; Zimatkin & Phedina, 2015a, 2015b; Zimatkin et al., 2016).

HISTOCHEMICAL CHANGES IN HISTAMINERGIC NEURONS FOLLOWING ALCOHOL ADMINISTRATION

Under the influence of alcohol there is a significant reorganization of the metabolism in brain histaminergic neurons. It was found that alcohol administration (except low dose) leads to the activity increase of acid phosphatase and lactate dehydrogenase. The significant increases in acid phosphatase activity reflect enhancement of autophagic processes directed to eliminating damaged membranes and organelles of histaminergic neurons in conditions of ethanol toxicity. The increase in activity of lactate dehydrogenase is evidence of increases in the late stages of glycolysis occurring in anaerobic conditions and required for the compensatory maintenance of neuron viability (Zimatkin et al., 2013; Zimatkin & Phedina, 2015a, 2015b; Zimatkin et al., 2016).

Under the acute and chronic exposure to alcohol the activity of the key enzyme of histamine metabolism in brain histaminergic neurons—monoamine oxidase type B—significantly increases (Zimatkin et al., 2013; Zimatkin & Phedina, 2015a, 2015b; Zimatkin et al., 2016). It indicates the intensification of oxidative diamination of the inactive first metabolite of histamine, tele-methylhistamine, and formation of the active second metabolite N-tele-methylimidazole acetaldehyde, competing with the acetaldehyde formed in the course of ethanol oxidation in the brain (Ambroziak & Pietruszko, 1987). This may be the metabolic basis for the alcohol—histamine interaction in the brain (Zimatkin & Anichtchik, 1999). Our previous investigations demonstrated that histamine H1 receptor antagonists that pass the blood—brain barrier, increase the ethanol metabolism in rats, but decrease tolerance to hypnotic effect of ethanol, because they increase the sensitivity of the brain to ethanol (Zimatkin, Liopo, & Zakharov, 1997).

Low ethanol dose leads to the increase of NADPH dehydrogenase activity. It indicates the intensifying extramitochondrial energy metabolism processes. After single and repeated ethanol administration in narcotic dose the enzyme activity decreases but remains invariable at chronic alcohol intoxication.

Sevenfold ethanol administration and its chronic consumption oppress the activity of succinate dehydrogenase that may reflect the Krebs cycle deceleration in mitochondria. In 6 h after alcohol administration at a dose of 4 g/kg the enzyme activity increases.

The activity of glucose-6-phosphate dehydrogenase decreases after chronic ethanol consumption and single (4 g/kg, 1 h after injection) alcohol administration. It indicates the deceleration of pentose phosphate pathway.

Narcotic ethanol dose (6 h after administration) and chronic alcohol intoxication lead to increase in the activity of NADH dehydrogenase, which participate in transport of electrons. Sevenfold ethanol influence inhibits the enzyme activity. Thus, the histochemical changes in histaminergic neurons may reflect both the disturbances of energy metabolism and metabolic adaptation of neurons to alcohol, taking into consideration the alcohol-induced hypoxia (Zimatkin et al., 2013; Zimatkin & Phedina, 2015a, 2015b; Zimatkin et al., 2016).

c-Fos is a marker of the functional activity of neurons, which plays an important role in neuronal adaptation and plasticity of the brain (Dragunow & Faull, 1989; Herrera & Robertson, 1996; Kovacs, 2008). Immunohistochemical staining of c-Fos showed that while control animals display only faint c-Fos staining in histaminergic neurons cytoplasm, both low and high single doses of ethanol increase it. Notably, higher dose of ethanol causes bigger increase in c-Fos staining. It may indicate an increase of functional neuronal activity following ethanol administration at both doses. When co-stained with mitochondrial marker ATP-synthase-β, we were able to demonstrate that low dose of ethanol causes relative increase in this staining in cytoplasmic mitochondria, while higher dose of ethanol leads to the relative decrease of ATP-synthase-β-positive staining, showing possible fragmentation of mitochondria. Low dose of ethanol increased the expression of ATP-synthase-β both in neuron bodies and surrounding neuropile reflecting activation of mitochondria (Zimatkin et al., 2016).

ALCOHOL EFFECTS INTO A HISTAMINERGIC NEURONS ULTRASTRUCTURE

Administration of ethanol to animals produces a variety of ultrastructural changes in brain histaminergic neurons, some being common to all administration regimes and others being dependent on dose, time after administration, and duration of exposure to alcohol. It can be suggested that repair processes (intracellular regeneration) in histaminergic neurons following ethanol administration occur in parallel with destruction processes. Thus, regardless of dose and duration of exposure to alcohol, ethanol led to hypertrophy and displacement of nucleoli toward the nuclear envelope, accumulation of electron-dense granular material (possibly ribosomal subunits) between the nucleolus and the nuclear envelope, bulging it into the cytoplasm, with increases in the extent of folding of the nuclear envelope, widening of the perinuclear space, increases in the number of nuclear pores, and even local degradation of the nuclear envelope. Thus, histaminergic neurons following alcohol administration exhibit the structural signs of hyperactivity, the intensive functioning probably related to their adaptation to alcohol (Zimatkin et al., 2013; Zimatkin & Phedina, 2015a, 2015b, 2015c).

Twenty-four hours after the 7-day ethanol administration some nuclei contained unusual vacuoles of polygonal shape surrounded by single biological membrane (Zimatkin & Phedina, 2015b, 2015c). The formation of nuclear vacuoles has previously been noted in other types of neurons as they develop destructive processes (Manina, 1971). From our point of view, this may be due to widening of the perinuclear space with shunting of the internal nuclear membrane.

One hour after administration of a low dose and 24 h after the 7-day ethanol administration of the narcotic dose of alcohol, the cytoplasm of some histaminergic neurons showed round or oval shaped bodies of about 1–2 μm in size, consisting of osmophilic granules aggregations. They well resembled the nucleoli and may be called "nucleolus-like bodies" (Zimatkin & Phedina, 2015b, 2015c). The similar "nucleolus-like bodies" were described earlier in cytoplasm of neurons and glial cells of hypothalamus in other experimental conditions (Kawabata, 1965). In some publications they are called processing bodies or stress granules (Decker & Parker, 2012; Thomas, Loschi, Desbats, & Boccaccio, 2011). They contain the RNA-binding proteins and mRNA, regulate the transcription, involve in mRNA degradation or storage, and contribute to cell survival.

A further ultrastructural phenomenon seen particularly frequent in the cytoplasm of histaminergic neurons after single or repeated administration of alcohol at high doses was closure of stacks of Golgi complex cisterns to form rings, as well as the appearance of unusual layered structures and myelin-like figures in the neuron cytoplasm. The appearance of lamellar, myelin-like figures in histaminergic neurons seen in alcohol intoxication is a sign of pressurized functioning of cells and even their exhaustion (Manina, 1971) and correlates at the light microscope level with an increase in the number of hyperchromic neurons and cells-shadows on exposure to ethanol (Zimatkin et al., 2013; Zimatkin & Phedina, 2015b, 2015c).

High alcohol doses and chronic consumption of alcohol lead to widening of the endoplasmic reticulum channels and a significant increase in the number of free ribosomes. Hypertrophy always occurred, and mitochondrial hyperplasia was sometimes present, accompanied by mitochondrial swelling and fragmentation and destruction of cristae. Transverse stretching of mitochondria was sometimes seen. Hypertrophied mitochondria sometimes collected together and became tightly pressed to the nucleus, endoplasmic reticulum, or Golgi complex, presumably supporting the increased energy needs of the nucleus and organelles (Manina, 1971). There are increases in the number and area of mitochondria in all experimental treatments with alcohol (Zimatkin et al., 2013; Zimatkin & Phedina, 2015a, 2015b, 2015c).

Alcohol leads to lysosomal hypertrophy and hyperplasia. It evidently reflects increases in autophagy processes directed to removing damaged neuron membranes and organelles in conditions of recent toxic exposure to alcohol and adaptation to it (Zimatkin et al., 2013; Zimatkin & Phedina, 2015a, 2015b, 2015c).

The high sensitivity of histaminergic neurons to alcohol can be theoretically predicted because of high activity of ethanol oxidizing enzyme catalase and low activity of aldehyde dehydrogenase, providing the conditions for toxic acetaldehyde accumulation in brain aminergic neurons (Zimatkin & Lindros, 1996).

Thus, all structural and metabolic disturbances have revealed that in brain histaminergic neurons of rats after exposure to alcohol are nonspecific, as they can be observed in other types of neurons under other experimental conditions. They may reflect both alcohol-induced metabolic disturbances and an adaptation of neurons to ethanol.

In conclusion, the literature data indicate an important role of brain histaminergic system on the effect of alcohol on brain, as well as mechanisms of inclination and resistance to ethanol, which are key links in alcoholism pathogenesis. At the same time alcohol has significant influence on the histamine content and metabolism in the brain and causes various morphological and functional disorders of brain histaminergic neurons, as well as adaptation to structural and metabolic changes, depending on the dose, time after administration, and duration of exposure to ethanol. These changes are directed at restoring and maintaining of neurons function and are serving as signs of the protective reactions of body required for the maintenance of homeostasis in conditions of exposure to the toxic agent. It is further supporting participation of central histaminergic system in modulating addictive or toxic effects of ethanol.

References

Alakarppa, K., Tupala, E., Mantere, T., Sarkioja, T., Rasanen, P., Tarhanen, J., ... Tuomisto, L. (2002). Effect of alcohol abuse on human brain histamine and tele-methylhistamine. *Inflammation Research, 51*(Suppl. 1), S40–S41.

Alleva, L., Tirelli, E., & Brabant, C. (2013). Therapeutic potential of histaminergic compounds in the treatment of addiction and drug-related cognitive disorders. *Behavioural Brain Research, 237*, 357–368.

Ambroziak, W., & Pietruszko, R. (1987). Human aldehyde dehydrogenase: Metabolism of putrescine and histamine. *Alcoholism: Clinical and Experimental Research, 11*(6), 528–532.

Anichtchik, O. V., Peitsaro, N., Rinne, J. O., Kalimo, H., & Panula, P. (2001). Distribution and modulation of histamine H3 receptors in basal ganglia and frontal cortex of healthy controls and patients with Parkinson's disease. *Neurobiology of Disease, 8*(4), 707–716.

Arrang, J. M., Garbarg, M., & Schwartz, J. C. (1983). Auto-inhibition of brain histamine release mediated by a novel class (H3) of histamine receptor. *Nature, 302*(5911), 832–837.

Arrang, J. M., Garbarg, M., & Schwartz, J. C. (1987). Autoinhibition of histamine synthesis mediated by presynaptic H3-receptors. *Neuroscience, 23*(1), 149–157.

Bahi, A., Sadek, B., Nurulain, S. M., Lazewska, D., & Kiec-Kononowicz, K. (2015). The novel non-imidazole histamine H3 receptor antagonist DL77 reduces voluntary alcohol intake and ethanol-induced conditioned place preference in mice. *Physiology & Behavior, 151*, 189–197.

Bahi, A., Sadek, B., Schwed, S. J., Walter, M., & Stark, H. (2013). Influence of the novel histamine H3 receptor antagonist ST1283 on voluntary alcohol consumption and ethanol-induced place preference in mice. *Psychopharmacology, 228*(1), 85–95.

Blandina, P., Munari, L., Provensi, G., & Passani, M. B. (2012). Histamine neurons in the tuberomamillary nucleus: A whole center or distinct subpopulations. *Frontiers in Systems Neuroscience, 6*(Article 33), 1–6.

Davidson, R. J., Putnam, K. M., & Larson, C. L. (2000). Dysfunction in the neural circuitry of emotion regulation – a possible prelude to violence. *Science, 289*(5479), 591–594.

Decker, C. J., & Parker, R. (2012). P-bodies and stress granules: Possible roles in the control of translation and mRNA degradation. *Cold Spring Harbor Perspectives in Biology, 4*(9), a012286.

Dragunow, M., & Faull, R. (1989). The use of c-Fos as a metabolic marker in neuronal pathway tracing. *Journal of Neuroscience Methods, 29*(3), 261–265.

Flik, G., Folgering, J. H., Cremers, T. I., Westerink, B. H., & Dremencov, E. (2015). Interaction between brain histamine and serotonin, norepinephrine, and dopamine systems: In vivo microdialysis and electrophysiology study. *Journal of Molecular Neuroscience, 56*(2), 320–328.

Galici, R., Rezvani, A. H., Aluisio, L., Lord, B., Levin, E. D., Fraser, I., ... Bonaventure, P. (2011). JNJ-39220675, a novel selective histamine H3 receptor antagonist, reduces the abuse-related effects of alcohol in rats. *Psychopharmacology, 214*(4), 829–841.

George, F. K., Amanda, J. R., Gery, S., Loren, H. P., Charles, J. H., Petri, H., ... Friedbert, W. (1998). Neurocircuitry targets in ethanol reward and dependence. *Alcoholism: Clinical and Experimental Research, 22*, 3–9.

Haas, H., & Panula, P. (2003). The role of histamine and the tuberomamillary nucleus in the nervous system. *Nature Reviews Neuroscience, 4*(2), 121–130.

Haas, H., Sergeeva, O., & Selbach, O. (2008). Histamine in the nervous system. *Physiological Reviews, 88*(3), 1183–1241.

Herrera, D. J., & Robertson, H. A. (1996). Activation of c-Fos in the brain. *Progress in Neurobiology, 50*(2–3), 83–107.

Imperato, A., & Di Chiara, G. (1986). Preferential stimulation of dopamine release in the nucleus accumbens of freely moving rats by ethanol. *Journal of Pharmacology and Experimental Therapeutics, 239*(1), 219–228.

Itoh, Y., Nishibori, M., Oishi, R., & Saeki, K. (1985). Changes in histamine metabolism in the mouse hypothalamus induced by acute administration of ethanol. *Journal of Neurochemistry, 45*(6), 1880–1885.

Kawabata, I. (1965). Electron microscopy of the rat hypothalamic neuro-secretory system. II. Nucleolus-like inclusion bodies in the cytoplasm of neurosecretory cells. *Archivum Histologicum Japonicum, 26*(2), 101–113.

Kovacs, K. J. (2008). Measurement of immediate-early gene activation-c-Fos and beyond. *Journal of Neuroendocrinology, 20*(6), 665–672.

Lintunen, M., Hyytia, P., Sallmen, T., Karlstedt, K., Tuomisto, L., Leurs, R., … Panula, P. (2001). Increased brain histamine in an alcohol-preferring rat line, and modulation of ethanol consumption by H3 receptor mechanisms. *FASEB Journal, 15*(6), 1074–1076.

Lintunen, M., Raatesalmi, K., Sallmen, T., Anichtchik, O., Karlstedt, K., Kaslin, J., … Panula, P. (2002). Low brain histamine content affects ethanol-induced motor impairment. *Neurobiology of Disease, 9*(1), 94–105.

Lovenberg, T. W., Roland, B. L., Wilson, S. J., Jiang, X., Pyati, J., Huvar, A., … Erlander, M. G. (1999). Cloning and functional expression of the human histamine H3 receptor. *Molecular Pharmacology, 55*(6), 1101–1107.

Manina, A. A. (1971). *Ultrastructural changes and repair processes in the central nervous system in response to various treatments.* Leningrad: Meditsina.

Morais-Silva, G., Ferreira-Santos, M., & Marin, M. T. (2016). Conessine, an H3 receptor antagonist, alters behavioral and neurochemical effects of ethanol in mice. *Behavioural Brain Research, 305*, 100–107.

Nowak, J., & Maslinski, C. (1984). Ethanol-induced changes of histamine content in guinea-pig brain. *Polish Journal of Pharmacology and Pharmacy, 36*(6), 647–651.

Nuutinen, S., Karlstedt, K., Aitta-Aho, T., Korpi, E. R., & Panula, P. (2010). Histamine and H3 receptor-dependent mechanisms regulate ethanol stimulation and conditioned place preference in mice. *Psychopharmacology, 208*(1), 75–86.

Nuutinen, S., Lintunen, M., Vanhanen, J., Ojala, T., Rozov, S., & Panula, P. (2011). Evidence for the role of histamine H3 receptor in alcohol consumption and alcohol reward in mice. *Neuropsychopharmacology, 36*(10), 2030–2340.

Nuutinen, S., Maki, T., Rozov, S., Backstrom, P., Hyytia, P., Piepponen, P., & Panula, P. (2015). Histamine H3 receptor antagonist decreases cue-induced alcohol reinstatement in mice. *Neuropharmacology, 30*, 1–8.

Panula, P., & Nuutinen, S. (2011). Histamine and H3 receptor in alcohol-related behaviors. *Journal of Pharmacology and Experimental Therapeutics, 336*(1), 9–16.

Papanicolaou, J., & Fennessy, M. (1980). The acute effect of ethanol on behaviour, body temperature, and brain histamine in mice. *Psychopharmacology, 72*(1), 73–77.

Pillot, C., Heron, A., Cochois, V., Tardivel-Lacombe, J., Ligneau, X., Schwartz, J. C., & Arrang, J. M. (2002). A detailed mapping of the histamine H3 receptor and its gene transcripts in rat brain. *Neuroscience, 114*(1), 173–193.

Pollard, H., Moreau, J., Arrang, J. M., & Schwartz, J. C. (1993). A detailed autoradiographic mapping of histamine H3 receptors in rat brain areas. *Neuroscience, 52*(1), 169–189.

Prell, G., Bielkiewicz, B., & Mazurkiewicz-Kwilecki, I. (1982). Rat brain histamine concentration, synthesis and metabolism: Effect of acute ethanol administration. *Progress in Neuro-psychopharmacology & Biological Psychiatry, 6*(4–6), 427–432.

Prell, G. D., & Mazurkiewicz-Kwilecki, I. M. (1981). The effects of ethanol, acetaldehyde, morphine and naloxone on histamine methyltransferase activity. *Progress in Neuro-psychopharmacology, 5*(5–6), 581–584.

Prell, G. D., Morrishow, A. M., Duoyon, E., & Lee, W. S. (1997). Inhibitors of histamine methylation in brain promote formation of imidazoleacetic acid, which interacts with GABA receptors. *Journal of Neurochemistry, 68*(1), 142–151.

Rawat, A. (1980). Development of histaminergic pathways in brain as influenced by maternal alcoholism. *Research Communications in Chemical Pathology and Pharmacology, 27*(1), 91–103.

Reuter, M., Jeste, N., Klein, T., Hennig, J., Goldman, D., Enoch, M. A., & Oroszi, G. (2007). Association of THR105Ile, a functional polymorphism of histamine N-methyltransferase (HNMT), with alcoholism in German Caucasians. *Drug and Alcohol Dependence, 87*(1), 69–75.

Schlicker, E., Malinowska, B., Kathmann, M., & Gothert, M. (1994). Modulation of neurotransmitter release via histamine H3 heteroreceptors. *Fundamental & Clinical Pharmacology, 8*(2), 128–137.

Selbach, O., Brown, R. E., & Haas, H. L. (1997). Long-term increase of hippocampal excitability by histamine and cyclic AMP. *Neuropharmacology, 36*(11–12), 1539–1548.

Selbach, O., Stehle, J., & Haas, H. L. (2007). Hippocampal long-term synaptic plasticity is controlled by histamine, hypocretins (orexins) and clock genes. *Society for Neuroscience Abstracts, 928*, 13.

Subramanian, N., Mitznegg, P., & Estler, C. (1978). Ethanol-induced alterations in histamine content and release in the rat hypothalamus. *Naunyn-Schmiedeberg's Archives of Pharmacology, 302*(1), 119–121.

Subramanian, N., Schinzel, W., Mitznegg, P., & Estler, C. (1980). Influence of ethanol on histamine metabolism and release in the rat brain. II. Regions of the histaminergic pathway. *Pharmacology, 20*(1), 42–45.

Thomas, M. G., Loschi, M., Desbats, M. A., & Boccaccio, G. L. (2011). RNA granules: The good, the bad and the ugly. *Cellular Signalling, 23*(2), 324–334.

Vanhanen, J., Nuutinen, S., Lintunen, M., Maki, T., Ramo, J., Karlstedt, K., & Panula, P. (2013). Histamine is required for H(3) receptor-mediated alcohol reward inhibition, but not for alcohol consumption or stimulation. *British Journal of Pharmacology, 170*(1), 177–187.

Yanai, K., Son, L. Z., Endou, M., Sakurai, E., Nakagawasai, O., Tadano, T., … Watanabe, T. (1998). Behavioural characterization and amounts of brain monoamines and their metabolites in mice lacking histamine H1 receptors. *Neuroscience, 87*(2), 479–487.

Yanai, K., & Tashiro, M. (2007). The physiological and pathophysiological roles of neuronal histamine: An insight from human positron emission tomography studies. *Pharmacology & Therapeutics, 113*(1), 1–15.

Zimatkin, S. M., & Anichtchik, O. V. (1999). Alcohol-histamine interactions. *Alcohol and Alcoholism, 34*(2), 141–147.

Zimatkin, S. M., Fedina, E. M., & Kuznetsova, V. B. (2013). Histaminergic neurons in the rat brain after acute exposure to alcohol. *Neuroscience and Behavioral Physiology, 43*(6), 691–696.

Zimatkin, S. M., Kuznetsova, V. B., & Strik, O. N. (2006). Spatial organization and morphometric characteristics of histaminergic neurons in the rat brain. *Neuroscience and Behavioral Physiology, 36*(5), 467–471.

Zimatkin, S. M., & Lindros, K. O. (1996). Distribution of catalase in rat brain: Aminergic neurons as possible targets for ethanol effects. *Alcohol and Alcoholism, 31*(2), 167–174.

Zimatkin, S. M., Liopo, A. V., & Zakharov, O. Y. (1997). Effects of histamine H1 receptor antagonists on alcohol related behavior and ethanol metabolism. *Alcohol and Alcoholism, 32*(3), 361.

Zimatkin, S. M., & Phedina, E. M. (2015a). Influence of chronic alcohol consumption on histaminergic neurons of the rat brain. *Alcohol and Alcoholism, 50*(1), 51–55.

Zimatkin, S. M., & Phedina, E. M. (2015b). Seven-day ethanol administration influence on the rat brain histaminergic neurons. *Alcohol, 49*(6), 589–595.

Zimatkin, S. M., & Phedina, E. M. (2015c). Ultrastructural changes in cerebral histaminergic neurons on exposure to alcohol. *Neuroscience and Behavioral Physiology, 45*, 873–877.

Zimatkin, S. M., Phedina, E. M., & Anichtchik, O. (2016). Rat brain histaminergic neurons following single ethanol exposure: Comparison of low and high dose effects. *Journal of GrSMU, 53*(1), 55–59.

CHAPTER

4

Antenatal Alcohol and Histological Brain Disturbances

S.M. Zimatkin, E.I. Bon

Grodno State Medical University, Grodno, Belarus

INTRODUCTION

Alcohol consumption during pregnancy induces specific disorders in offspring that are combined under the term fetal alcohol syndrome (FAS), which is a part of fetal alcohol spectrum disorders (FASDs) (Jones, 1973; Lemoine, 2012; Riley, Infante, & Warren, 2011). It is well known that the nervous system, especially the brain, is particularly sensitive to prenatal alcohol exposure. It induces severe and various mental and behavioral disorders in children and adults; violation of the intellectual sphere; impairment of attention, memory, and other cognitive functions; disturbances of the emotional sphere; impaired social interaction; sensomotor activity; preference to alcohol both in human and experimental animals (Brocardo, Budni, Pavesi, & Franco, 2012; Kodituwakku, 2009; Shea, Hewitt, & Olmstead, 2012). Magnetic resonance imaging (MRI) after prenatal alcohol exposure revealed the significant changes of neurovisual parameters of a brain, including thickening of the brain and cerebellum cortex, diffuse white matter disorders, increase in the volume of the ventricles, anomalies of the corpus callosum, significant reduction of subcortical nuclei. The severity and the prevalence of them in the brain structures vary considerably depending on the dose and timing of alcohol exposure and the time of the study (De Guio et al., 2014; Parnell et al., 2009; Yang et al., 2012).

The aim of this chapter is to analyze histological, cellular, and molecular biological data as a basis of these various mental and behavioral disorders.

NEUROHISTOLOGY

Ethanol has a negative impact on the processes of the neural tube formation of the rat embryo. Pregnant animals were given a liquid diet in which 20% or 25% of calories came from ethanol on 7 or 8 days of embryonic development. In 60% of embryos at 13 days and 20% on day 15 the violation was detected in the formation of the diencephalic vesicles and later completion of the neural tube formation (Zhou, Sari, Powrozek, Goodlett, & Li, 2003). In pregnant rats water was replaced by 15% ethanol for the entire period of gestation. Macroscopic examination of the brain of fetuses and newborns showed hyperemia of the meninges, leptomeningeal heterotopias and microcephaly in 6 cases of 24 (25%) (Chikhladze, Ramishvili, Tsagareli, & Et Kikalishvili, 2011).

With alcohol exposure on the 14th and 18th day of intrauterine development in fetuses of mice were revealed the following anomalies: absence of olfactory bulbs, defects of medial septa and cerebral cortex, and connection of lateral ventricles, and decrease in the thickness of ventricular wall (Schambra, Lauder, & Petrusz, 1990). The effects of moderate doses of alcohol during prenatal development leads to stunted growth of the brain, reducing its weight in the newborn rat pups, the severity of which varies in animals of different lines (Wainright & Gagnon, 1985; Woodson & Ritchey, 1979).

One of the key features of experimental FAS is fetal microcephaly. The cerebral cortex is particularly sensitive to prenatal ethanol exposure. Its total mass is reduced; it becomes thinner and contains fewer neurons

Addictive Substances and Neurological Disease
http://dx.doi.org/10.1016/B978-0-12-805373-7.00004-9

and glia. Ethanol causes a thinning of the sensorimotor cortex in rats that were subjected to its influence in utero (Miller & Nowakowski, 1991). Ethanol causes a thinning of the sensorimotor cortex in rats that were subjected to its influence in utero (Minciacchi, Granato, Santarelli, & Sbriccoli, 1993). These data contradict the results of the MRI showing thickening of the cortex in most parts (with the exception of the cingulate gyrus).

On the 18th day of embryogenesis the reduced weight of brain, amount of cells, and content of DNA and RNA were detected. Ethanol in rodents induces apoptosis of neurons, neurodegenerative changes, and long-term behavioral abnormalities (Wilson, Peterson, Basavaraj, & Saito, 2011). Antenatal alcohol exposure in rats did not affect the number of neurons or glial cells in medial prefrontal cortex C, but changed the branching of dendrites along the longitudinal axis (Lawrence, Otero, & Kelly, 2012). Prenatal exposure to alcohol causes a decrease in the number and size of pyramidal neurons in the cortex of the brain in animals, a decrease in protein content, and maturation of cytoplasm. Morphologically, in the cortex of the fetus under the influence of alcohol there is massive destruction of neurons and their mitochondria, involutional changes in the dendrites, and impaired proliferation of glia (Chikhladze et al., 2011).

In systematic research carried out by Popova the postnatal changes of the ultrastructure of sensorimotor cortex in offspring of rats consumed moderate alcohol doses during pregnancy. Three categories of ultrastructural changes were found: long-term delayed maturation of cortical structures, their destructive changes, and signs of reparative processes, having its own dynamics in postnatal ontogenesis (Popova, 2010).

In another study the impact of ethanol on the reproduction of cells in two proliferative zones of the neocortex, the ventricular zone (VZ), and subventricular zone (SZ) was examined. During 5–21 days of pregnancy, the rats were fed a liquid diet containing 6.7% of ethanol. Pregnant rats were given injections of bromosuccinimide (BrdU). After immunohistochemical processing, the ratio of cells labeled with BrdU in each proliferative zone was determined. In VZ under the influence of ethanol the total duration of the cell cycle was increased, but the duration of S-phase and growth phase remained the same. In contrast, SZ was the slow growth phase, and the duration of a full cycle and S-phase was not changed (Miller & Nowakowski, 1991).

Orderly migration of neuroblasts is required for the development of layered structures such as the cerebral cortex. Rats were fed a liquid diet containing 6.6% of ethanol from 6 to 21 days of pregnancy. In embryos at 21 days of fetal development, the migration was delayed 4–6 days, and often these neurons had completed the migration to ectopic locations. Ethanol significantly reduced the rates of migration, and cells in the post-mitotic phase remained in proliferative zones. Thus, the migration of young neurons was deeply altered by prenatal exposure to ethanol. Such delays can lead to an asynchronous development of the cortex and disorders of the central nervous system (CNS) (Miller, 1993).

To clarify the molecular mechanisms of violations of ethanol for migration of neurons, a transcription factor, Pax6, which is responsible for the process of radial migration of neuroblasts and glioblasts in the developing cortex was studied. A month before mating, during pregnancy and lactation, female rats were fed a liquid diet with the addition of a 5.9% solution of ethanol. The cerebral cortex of rats was investigated immunohistochemically for the detection of vimentin, nestin, S-100b, and Pax6 on 12 days of embryogenesis and on day 3 after birth. A delay in the migration of neurons, a decreased number of neuroblasts, and the decrease in the level of vimentin, nestin, S-100b, and Pax6 were found (Aronne, 2011).

Prenatal alcoholization significantly slows down and break the processes of myelinization of all fiber tracts of the brain (Rosman & Malone, 1976).

On the 7th day after birth, the mice were administered ethanol (20%, 2.5 g/kg per injection). After 3 months, it was revealed that the mice that were exposed to ethanol developed a destruction of cells with the highest levels in the neocortex, intermediate levels in the dorsal hippocampus, and relatively low levels in the olfactory bulb (Wilson et al., 2011).

Pregnant macaques at different stages of pregnancy (from 105 to 155 days) were given access to alcohol for 8 h daily. Then the fetuses were extracted using cesarean section, and their brains were fixed by perfusion. The brain observed a 60-fold increase in apoptosis compared with the control. It is expected that many of the neuropathological changes and long-term neuropsychiatric disorders FASD can be explained by the apoptogenic effects of alcohol on the fetal brain (Farber, Creeley, & Olney, 2010). In another experiment macaque were given ethanol once a week (1.8 g/kg body weight) from 2 to 19 weeks of pregnancy. The fetus subjected to prenatal exposure to ethanol revealed microcephaly, absence of olfactory bulbs, optic nerves and chiasm, lateral extension of the ventricle of the brain, parietal–occipital hernia, and dysplasia of the cerebellum (Siebert, Astley, & Clarren, 1991). In the third study, ethanol was administered to pregnant macaques orally once weekly in doses of 0, 0.3, 0.6, 1.2, 1.8, 2.5, or 4.1 g/kg after 1 week of pregnancy, or at doses of 2.5, 3.3, or 4.1 g/kg after 5 weeks. The average concentration of ethanol in plasma of the mother ranged from 24 ± 6 mg/DL at a dose of 0.3 g/kg to 549 ± 71 mg/DL at 4.1 g/kg. Microphthalmia and reduction in the number of retinal ganglion cells were observed in 3 of 26 animals exposed to ethanol (Clarren et al., 1990).

The ***hippocampus*** is one of the areas of the brain most vulnerable to the effects of ethanol. Studies using morphometric techniques have shown that prenatal exposure to ethanol disrupts the development of the dentate gyrus (Miki, Harris, Wilce, Takeuchi, & Bedi, 2003; Miki et al., 2008). On 4th and 9th days after birth the rat pups were given ethanol in milk formula, the total dose was 5.25 g/kg per day. On 50th and 80th days after birth the brains were fixed by perfusion, and in hippocampal slices marker of proliferation, BrdU, and marker of mature neurons, NeuN, were examined immunohistochemically. In the dentate gyrus there was a significant reduction in the number of mature neurons as well as a decrease in the number of new neurons that were formed between 30 and 50 days (Miki et al., 2008).

Ethanol intragastric intubation throughout 7–21 gestation days at a daily dose of 6 g/kg was performed. Isocaloric intubation and intact control groups were included. Unbiased stereological estimation of hippocampal volume, the total number of pyramidal and granular cells, and doublecortin expressing neurons were carried out for postnatal days (PDs) 1, 10, 30, and 60. Its effect on hippocampal morphology was limited to a marginally lower number of granular cells in dentate gyrus on PD30 (Elibol-Can et al., 2014).

Prenatal alcohol intoxication was shown to induce the disturbances in proliferative activity of granular layer cells in the hippocampal dentate gyrus, neuro- and glioblasts migration, enhancement of free NO and lipoperoxide production and cell death. This resulted in the changes in the number of neurons in cortical and subcortical structures of the rat brain limbic system and in FAS formation (Svanidze, Museridze, Didimova, & Sanikidze, 2012).

Ethanol was given to rats during periods equivalent to the first, second, and third trimester of pregnancy in humans. When the rats reached adulthood the stereological estimation of the total number of pyramidal and granular cells in CA1 and CA3 regions of the hippocampus and the dentate gyrus was carried out. The results suggest that the CA1 region is very susceptible to the effects of ethanol in the early neonatal period, and area CA3 and dentate gyrus are more resistant to the effects of ethanol in all of the periods of development of the hippocampus (Tran & Kelly, 2003).

In the ***brain cortex*** antenatal alcohol exposure disrupts the cell cycle (Tran & Kelly, 2003), development, differentiation, and migration of neurons of the motor cortex of rats (Alvares and Stone, 1988; Miller, 1986).

To find out to what extent the developing visual system is vulnerable to apoptogenic effects of ethanol, rats and mice received subcutaneous injections of ethanol after birth, in single or two doses in 1 day with intervals of 2 h. Electron microscopical study was conducted 4–24 h after administration of ethanol. Ganglionic cells in the retina, neurons of the lateral geniculate body, the midbrain and visual cortex showed high sensitivity to apoptogenic action of ethanol. Peak sensitivity to ganglion cells is 1–4 days after birth, but for other neurons, from 4 to 7 days. The concentration of alcohol in the blood of about 120 mg/DL was sufficient to activate cell death in the optic neurons (Dursun et al., 2011).

Using an acute (single day) model of moderate (3 g/kg) to severe (5 g/kg) alcohol exposure in postnatal day (P) 7 or 8 to mice, it was found that alcohol-induced neuroapoptosis in the neocortex is closely correlated in space and time with the appearance of activated *microglia* near dead cells. Although microglia rapidly mobilized to contact and engulf late-stage apoptotic neurons, apoptotic bodies temporarily accumulated in neocortex, suggesting that in severe cases of alcohol toxicity the neurodegeneration rate exceeds the clearance capacity of endogenous microglia. Nevertheless, most dead cells were cleared and microglia began to deactivate within 1–2 days of the initial insult. Coincident with microglial activation and deactivation, there was a transient increase in expression of pro-inflammatory factors, tumor necrosis factor alpha (TNFα) and interleukin 1 beta (IL-1β), after severe (5 g/kg), but not moderate (3 g/kg) ethanol levels. Alcohol-induced microglial activation and pro-inflammatory factor expression were largely abolished in BAX-null mice lacking neuroapoptosis, indicating that microglial activation is primarily triggered by apoptosis rather than the alcohol (Ahlers, Karaçay, Fuller, Bonthius, & Dailey, 2015).

The three major classes of glial cells, astrocytes, oligodendrocytes, and microglia, as well as their precursors are affected by ethanol during brain development. Alterations in glial cell functions by ethanol dramatically affect neuronal development, survival, and function and ultimately impair the development of the proper brain architecture and connectivity. For instance, ethanol inhibits astrocyte-mediated neuritogenesis and oligodendrocyte development, survival and myelination; furthermore, ethanol induces microglia activation and oxidative stress leading to the exacerbation of ethanol-induced neuronal cell death (Guizzetti, Zhang, Goeke, & Gavin, 2014).

The vulnerability of Purkinje cells to ethanol depends on the degree of the maturation of neurons—mature Purkinje cells being more vulnerable. Quantitative analysis showed that the most pronounced reduction in the density of Purkinje cells of the cerebellum observed in pups whose mothers consumed ethanol during pregnancy compared to lactation (Lewandowska et al., 2012). Studies conducted on rats and sheep confirmed that prenatal exposure of ethanol during the first trimester of pregnancy affects Purkinje cells in a nonselective manner, whereas exposure during the third trimester selectively affects postmitotic Purkinje cells in specific regions of the cerebellar vermis with a vulnerable period of differentiation and synaptogenesis (Sawant et al., 2013).

The rats were given a liquid diet containing 20% ethanol for 2 weeks before mating and until the completion of breastfeeding. The brain samples were taken on the 27th day after birth. The results showed a significant reduction in the size of Purkinje cells and width of molecular and granular layers of the cerebellum (Ghimire, Saxena, Rai, & Dhungel, 2009; Lee, Rowe, Eskue, West, & Maier, 2008). B6D2F1 mice were fed by liquid diet containing 25% ethanol from 12 to 17 days after fertilization. During this time half of them underwent two 1-h period of stress daily. As a result, the offspring of the second group showed developmental defects of the cerebellar tracts (Ward & Wainwright, 1991).

In utero exposure of the fetal brain to alcohol on a single occasion triggers widespread acute apoptotic death of neurons and oligodendrocytes throughout white matter regions of the developing brain. The rate of oligodendrocyte apoptosis in alcohol-exposed brain was 12.7 times higher than the natural oligodendrocyte apoptosis rate. Oligodendrocytes become sensitive to the apoptogenic action of alcohol when they are just beginning to generate constituents of myelin in their cytoplasm, and they remain vulnerable throughout later stages of myelination (Creeley, Dikranian, Johnson, Farber, & Olney, 2013).

Thus, prenatal or early postnatal exposure to ethanol induces a significant and varied histological changes in many structures of the brain, depending on the dose and timing of ethanol exposure.

CELLULAR AND MOLECULAR DISORDERS

Estimates of total neuron numbers in brain cortex showed a trend level reduction of about 8%, due mainly to reduced cortical volume, but unchanged neuron density. However, counts of calretinin (CR) and parvalbumin (PV) subtypes of GABAergic neurons showed a striking >30% reduction of neuron number. Similar ethanol effects were found in male and female mice, and in C57BL/6By and BALB/cJ mouse strains. This may contribute to the lasting cognitive and behavioral deficits in FASD (Smiley et al., 2015).

Antenatal alcohol exposure affects micro-RNA that controls the expression of genes (Miranda, 2012). Ethanol disturbs the communication of regulatory micro-RNAs, which are important for maturation of nerve cells (Balaraman, Winzer-Serhan, & Miranda, 2012).

Antenatal ethanol influences neural stem cells (NSCs), both glial cells precursors (GCPs) and nerve cell precursors (NCPs). These human cells were isolated in the second trimester for a positive selection using magnetic beads labeled with antibodies to CD133 (NSC), A2B5 (GCP), or polysialylated neuronal cell adhesion molecule (PSA-NCAM; NCP) and exposed to the effect of a solution containing 0 or 100 mm ethanol for 120 h. In the control, differentiated GCP, and GCP and NCP form neurosphere, which is significantly smaller than that in the medium with ethanol (Vangipuram & Lyman, 2010).

One of the possible mechanisms for the formation of FAS is the effect of alcohol on the system of excitatory and inhibitory neurotransmitters. Activating one of the subtypes of glutamate receptors NMDA (N-methyl-D-aspartate) plays a key role in the formation of synapses. Alcohol by blocking NMDA receptors promotes apoptotic neurodegeneration in the anterior brain, which could be the reason for the decrease of brain mass, various mental disorders, and also reduce learning ability in the offspring. Hyperactivity of NMDA receptors leads to excitotoxic cell death (Tsai & Coyle, 1998).

Immunohistochemical study of the neurotrophin receptor p75 (p75 NTR) in the sensorimotor cortex of 10- and 20-day-old rat pups exposed to ethanol during the first postnatal week of life was performed. In both age groups the number of p75 NTR immunoreactive neurons was higher in experimental animals compared to that of the control group (Moore, Madorsky, Paiva, & Barrow Heaton, 2004).

The developing cerebellum is particularly vulnerable to the effects of prenatal alcohol exposure. Ethanol inhibits axonal growth of neurons of the granular layer of the cerebellar cortex. Axonal growth normally stimulated by Netrin-1, neurotrophic factor glial cell (GDNF), and adhesion L1, through the activation of SFK (Src kinase), a signaling protein Cas, and excretion of extracellular regulated kinases 1 and 2 (ERK1/2). Neurotrophic factor (brain-derived neurotrophic factor, BDNF) stimulates the growth of axons and the activation of ERK1/2 without first activating SFK or Cas. Clinically relevant concentrations of ethanol inhibited the growth of axons and the activation of the SFK–Cas–ERK1/2 route, but it does not disrupt BDNF-induced axon growth or the activation of ERK1/2. These results show that SFK, but not ERK1/2, is the main target for ethanol-induced suppression of axons growth (Chen & Charness, 2012).

The effects of moderate maternal exposure to ethanol (10% v/v in the drinking water) throughout gestation or gestation and lactation was examined, on crucial 21-day-old offspring of Wistar rat, for brain parameters, such as the activities of acetylcholinesterase (AChE) and two adenosine triphosphatases (Na(+),K(+)-ATPase and Mg(2+)-ATPase), in major offspring CNS regions (frontal cortex, hippocampus, hypothalamus, cerebellum and pons). It has revealed a CNS region-specific susceptibility of the examined crucial neurochemical parameters to the ethanol exposure schemes (Stolakis et al., 2015).

Ample evidence shows that ethanol causes the "classic" FAS face (short palpebral fissures, elongated upper lip, deficient philtrum) because it suppresses prechordal plate outgrowth, thereby reducing neuroectoderm and neural crest induction and causing holoprosencephaly. Prenatal alcohol exposure (PAE) at premigratory stages elicits a different facial appearance, indicating FASD may represent a spectrum of facial outcomes. PAE at this premigratory period initiates a calcium-transient increase that activates calcium/calmodulin-dependent protein kinase II (CaMKII) and destabilizes transcriptionally active β-catenin, thereby initiating apoptosis within neural crest populations. Contributing to neural crest vulnerability are their low antioxidant responses. Ethanol-treated neural crests produce reactive oxygen species, and free radical scavengers attenuate their production and prevent apoptosis. Ethanol also significantly impairs neural crest migration, causing cytoskeletal rearrangements that destabilize focal adhesion formation; their directional migratory capacity is also lost. Genetic factors further modify vulnerability to ethanol-induced craniofacial dysmorphology and genes important for neural crest development, including shh signaling, PDFGA, vangl2, and ribosomal biogenesis. Because facial and brain development are mechanistically and functionally linked, research into effects of ethanol on neural crest also inform our understanding of CNS pathologies (Smith, Garic, Flentke, & Berres, 2014).

Using animal models of FASD, it has been recently discovered that ethanol induces neuroimmune activation in the developing brain. The resulting microglial activation, production of proinflammatory molecules, and alteration in the expression of developmental genes are postulated to alter neuronal survival and function and also lead to long-term neuropathological and cognitive defects. It has also been discovered that microglial loss occurs, reducing microglia's ability to protect neurons and contribute to neuronal development. Interestingly, the behavioral consequences of microglial depletion and neuroimmune activation in the fetal brain are particularly relevant to FASD (Drew & Kane, 2014).

Upon fetal alcohol exposure, heat shock factor 2 (HSF2) is essential for the triggering of HSF1 activation, which is accompanied by distinctive posttranslational modifications, and HSF2 steers the formation of atypical alcohol-specific HSF1−HSF2 heterocomplexes. This perturbs the in vivo binding of HSF2 to heat shock elements (HSEs) in genes that control neuronal migration in normal conditions, such as p35 or the MAPs (microtubule-associated proteins, such as Dclk1 and Dcx), and alters their expression. In the absence of HSF2, migration defects as well as alterations in gene expression are reduced. Thus, HSF2, as a sensor for alcohol stress in the fetal brain, acts as a mediator of the neuronal migration defects associated with FASD (Fatimy et al., 2014).

The C57BL/6J mouse was used to assess the dynamics of genomic alterations following binge alcohol exposure. Ethanol-exposed fetal (short-term effect) and adult (long-term effect) brains of C57BL/6J mice were assessed for gene expression, and miRNA changes using Affymetrix mouse arrays; 48 and 68 differentially expressed genes in short- and long-term groups, respectively, were identified. No gene was common between the two groups. Short-term (immediate) genes were involved in cellular compromise and apoptosis, which represent toxic effects of ethanol. Long-term genes were involved in various cellular functions, including epigenetics. Using quantitative RT-PCR, the downregulation of long-term genes: Camk1g, Ccdc6, Egr3, Hspa5, and Xbp1, was confirmed. miRNA arrays identified 20 differentially expressed miRNAs, one of which (miR-302c) was confirmed. miR-302c was involved in an inverse relationship with Ccdc6. The data support the critical role of apoptosis in FASD, and the potential involvement of miRNAs in the adaptation of gene expression following prenatal ethanol exposure. The ultimate molecular footprint involves inflammatory disease, neurological disease, and skeletal and muscular disorders as major alterations in FASD. At the cellular level, these processes represent abnormalities in redox, stress, and inflammation, with potential underpinnings to anxiety (Mantha, Laufer, & Singh, 2014).

Using immunohistochemical methods three calcium-binding proteins (CaBPs) were studied: calbindin D28k, calretinin, and parvalbumin in the cerebellum in 10-day rat pups subjected to prenatal exposure to ethanol. The number of cells containing calbindin D28k and parvalbumin decreased in all experimental groups, while the content calretinin increased in gusset neurons (Wierzba-Bobrowicz, Lewandowska, Stepień, & Szpak, 2011).

Prenatal exposure to ethanol interferes with cellular growth and differentiation of astroglia, and reduces the level of gliofibrillar acidic protein (marker of astrocytes) and expression of its gene (Guerri, Pascual, & Renau-Piqueras, 2001).

Ethanol decreases the viability of neurons and disrupts their functions in two main ways: (1) inhibits insulin signals, which is required for viability, metabolism, synapse formation, and acetylcholine production; (2) functions as a neurotoxin, causing oxidative stress,

DNA damage, and mitochondrial dysfunction. Ethanol inhibits insulin signals indirectly at the level of the insulin receptor, disrupting binding to the receptor and increasing the activity of the phosphatase. As a result, insulin activates PI3K-Akt, which mediates the mobility, energy metabolism, and plasticity of neurons. Thus, chronic intrauterine exposure to ethanol leads to a state of insulin resistance in the CNS (de la Monte & Wands, 2010).

Structural, metabolic, and functional changes of many identified parts of the brains are associated with specific mental and behavioral disorders. For example, behavioral disorders in rats following prenatal alcohol exposure, correlate with the reduced accumulation of C14-deoxyglucose in the brain structures related to the limbic and motor systems, while the rate of glucose utilization in structures of the hypothalamic–pituitary axis was increased (Williams, 1991).

POSSIBLE MECHANISMS FOR THE DEVELOPMENT OF BEHAVIORAL AND MENTAL DISORDERS AFTER PRENATAL ALCOHOL EXPOSURE

In the process of ontogenesis, alcohol causes many defects in molecular, neurochemical, and cellular processes that occur during normal brain development, disturbing the functions of the glia, changing the regulation of gene expression and the molecules involved in intercellular interactions, and increasing the formation of free radicals. Alcohol acts on specific proteins of the membrane receptors, for example, GABA-A ion channels (Ca^{2+} channels L-type) and signaling pathways (e.g., PKA and PKC signaling). These effects may underlie a wide spectrum of behavioral disorders induced by ethanol (Alfonso-Loeches & Guerri, 2011). This is accompanied by disorder of the brain structures at the ultramicroscopic, microscopic, and even anatomical level, and is determined by a variety of in vivo changes of neurovisual parameters as well as histological, histochemical, and ultrastructural disorders of the brains in children with FAS and experimental animals exposed to prenatal alcohol abuse. In particular, we detected a stoppage of growth and progressive shrinking of neurons in the cerebral cortex of rats subjected to prenatal exposure to alcohol (Zimatkin & Bon, 2015a,b).

Ethanol freely passes through the placental barrier into the fetal blood and across the blood–brain barrier into the brain. We can assume that the high sensitivity of the fetal brain to alcohol is due to several metabolic reasons: (1) increased activity in the brain of the ethanol oxidizing enzymes (particularly catalase) that may lead to increased local production of its toxic metabolite acetaldehyde (AA); (2) low activity in crucial brain structures of the AA-oxidizing enzyme, aldehyde dehydrogenase (ALDH) (Zimatkin & Lis, 1990). Both of them can lead to increased accumulation of the toxic AA in the fetal brain. In addition, the low ALDH activity in barrier structures of the fetal brain facilitates the penetration of AA from the blood into the brain, under conditions of increased formation from ethanol in peripheral tissues of the fetus, the placenta, and in the maternal liver, especially in late period of pregnancy. This creates conditions for the accumulation and manifestation of AA toxic action in the fetal brain (Zimatkin & Bon, 2014).

In conclusion, prenatal alcohol exposure leads to diverse, severe, and long-term (often irreversible) disorders in the offspring at the molecular, cellular, tissue, and organ level, which disrupts the organization and function of the brain.

References

Ahlers, K. E., Karaçay, B., Fuller, L., Bonthius, D. J., & Dailey, M. E. (2015). Transient activation of microglia following acute alcohol exposure in developing mouse neocortex is primarily driven by BAX-dependent neurodegeneration. *Glia, 63*, 1694–1713.

Alfonso-Loeches, S., & Guerri, C. (2011). Molecular and behavioral aspects of the actions of alcohol on the adult and developing brain. *Critical Reviews in Clinical Laboratory Sciences, 48*, 19–47.

Alvares, M. R., & Stone, D. I. (1988). Hypoploidy and hyperplasia in the developing brain exposed to alcohol in utero. *Teratology,* (3), 233–238.

Aronne, M. P. (2011). Effects of prenatal ethanol exposure on rat brain radial glia and neuroblast migration. *Experimental Neurology, 229*, 364–371.

Balaraman, S., Winzer-Serhan, U. H., & Miranda, R. C. (2012). Opposing actions of ethanol and nicotine on microRNAs are mediated by nicotinic acetylcholine receptors in fetal cerebral cortical-derived neural progenitor cells. *Alcoholism: Clinical and Experimental Research, 36*, 1669–1677.

Brocardo, P. S., Budni, J., Pavesi, E., & Franco, J. (2012). Anxiety- and depression-like behaviors are accompanied by an increase in oxidative stress in a rat model of fetal alcohol spectrum disorders: Protective effects of voluntary physical exercise. *Neuropharmacology, 62*, 16–18.

Chen, S., & Charness, M. E. (2012). Ethanol disrupts axon outgrowth stimulated by netrin-1, GDNF, and L1 by blocking their convergent activation of Src family kinase signaling. *Journal of Neurochemistry, 123*, 602–612.

Chikhladze, R. T., Ramishvili, N. S., Tsagareli, Z. G., & Et Kikalishvili, N. O. (2011). The spectrum of hemispheral cortex lesions in intrauterine alcoholic intoxication. *Georgian Medical News, 192*, 81–87.

Clarren, S. K., Astley, S. J., Bowden, D. M., Lai, H., Milam, A. H., Rudeen, P. K., & Shoemaker, W. J. (1990). Neuroanatomic and neurochemical abnormalities in nonhuman primate infants exposed to weekly doses of ethanol during gestation. *Alcoholism: Clinical and Experimental Research, 14*, 674–683.

Creeley, C. E., Dikranian, K. T., Johnson, S. A., Farber, N. B., & Olney, J. W. (2013). Alcohol-induced apoptosis of oligodendrocytes in the fetal macaque brain. *Acta Neuropathologica Communications, 12*, 23–31.

De Guio, F., Mangin, J. F., Rivière, D., Perrot, M., Molteno, C. D., Jacobson, S. W., ... Jacobson, J. L. (2014). A study of cortical morphology in children with fetal alcohol spectrum disorders. *Human Brain Mapping, 35*, 2285–2296.

Drew, P. D., & Kane, C. J. (2014). Fetal alcohol spectrum disorders and neuroimmune changes. *International Review of Neurobiology, 118*, 41–80.

Dursun, I., Jakubowska-Doğru, E., van der List, D., Liets, L. C., Coombs, J. L., & Berman, R. F. (2011). Effects of early postnatal exposure to ethanol on retinal ganglion cell morphology and numbers of neurons in the dorsolateral geniculate in mice. *Alcoholism: Clinical and Experimental Research, 35*, 2063–2074.

Elibol-Can, B., Dursun, I., Telkes, I., Kilic, E., Canan, S., & Jakubowska-Dogru, E. (2014). Examination of age-dependent effects of fetal ethanol exposure on behavior, hippocampal cell counts, and doublecortin immunoreactivity in rats. *Developmental Neurobiology, 74*, 498–513.

Farber, N. B., Creeley, C. E., & Olney, J. W. (2010). Alcohol-induced neuroapoptosis in the fetal macaque brain. *Neurobiology of Disease, 40*, 200–206.

Fatimy, R., Miozzo, F., Le Mouël, A., Abane, R., Schwendimann, L., Sabéran-Djoneidi, D., … Mezger, V. (2014). Heat shock factor 2 is a stress-responsive mediator of neuronal migration defects in models of fetal alcohol syndrome. *EMBO Molecular Medicine, 6*, 1043–1061.

Ghimire, S. R., Saxena, A. K., Rai, D., & Dhungel, S. (2009). Effect of maternal alcohol consumption on cerebellum of rat pups: A histological study. *Nepal Medical College Journal, 11*, 268–271.

Guerri, C., Pascual, M., & Renau-Piqueras, J. (2001). Glia and fetal alcohol syndrome. *Neurotoxicology, 22*, 593–599.

Guizzetti, M., Zhang, X., Goeke, C., & Gavin, D. P. (2014). Glia and neurodevelopment: Focus on fetal alcohol spectrum disorders. *Frontiers in Pediatrics, 2*, 123–130.

Jones, K. L. (1973). Pattern of malformation in offspring of chronic alcoholic mothers. *Lancet, 1*, 1267–1271.

Kodituwakku, P. W. (2009). Neurocognitive profile in children with fetal alcohol spectrum disorders. *Developmental Disabilities Research Reviews, 15*, 218–224.

Lawrence, R. C., Otero, N. K., & Kelly, S. J. (2012). Selective effects of perinatal ethanol exposure in medial prefrontal cortex and nucleus accumbens. *Neurotoxicology and Teratology, 34*, 128–135.

Lee, Y., Rowe, J., Eskue, K., West, J. R., & Maier, S. E. (2008). Alcohol exposure on postnatal day 5 induces Purkinje cell loss and evidence of Purkinje cell degradation in lobule I of rat cerebellum. *Alcohol, 42*, 295–302.

Lemoine, P. (2012). The history of alcoholic fetopathies. *Journal of Population Therapeutics and Clinical Pharmacology, 19*, 224–226.

Lewandowska, E., Stepień, T., Wierzba-Bobrowicz, T., Felczak, P., Szpak, G. M., & Pasennik, E. (2012). Alcohol-induced changes in the developing cerebellum. Ultrastructural and quantitative analysis of neurons in the cerebellar cortex. *Folia Neuropathologica, 50*, 397–406.

Mantha, K., Laufer, B. I., & Singh, S. M. (2014). Molecular changes during neurodevelopment following second-trimester binge ethanol exposure in a mouse model of fetal alcohol spectrum disorder: From immediate effects to long-term adaptation. *Developmental Neuroscience, 36*, 29–43.

Miki, T. S., Harris, J., Wilce, P. A., Takeuchi, Y., & Bedi, K. S. (2003). Effects of alcohol exposure during early life on neuron numbers in the rat hippocampus. I. Hilus neurons and granule cells. *Hippocampus, 13*, 388–398.

Miki, T., Yokoyama, T., Sumitani, K., Kusaka, T., Warita, K., Matsumoto, Y., … Takeuchi, Y. (2008). Ethanol neurotoxicity and dentate gyrus development. *Congenital Anomalies, 48*, 110–117.

Miller, M. W. (1986). Effects of alcohol on the generation and migration of cerebral cortical neurons. *Science, 233*, 1308–1311.

Miller, M. W. (1993). Migration of cortical neurons is altered by gestational exposure to ethanol. *Alcoholism: Clinical and Experimental Research, 17*, 304–314.

Miller, M. W., & Nowakowski, R. S. (1991). Effect of prenatal exposure to ethanol on the cell cycle kinetics and growth fraction in the proliferative zones of fetal rat cerebral cortex. *Alcoholism: Clinical and Experimental Research, 15*, 229–232.

Minciacchi, D., Granato, A., Santarelli, M., & Sbriccoli, A. (1993). Modifications of thalamo-cortical circuitry in rats prenatally exposed to ethanol. *Neuroreport, 4*, 415–418.

Miranda, R. C. (2012). MicroRNA and fetal brain development: Implication for ethanol teratology during the second trimester period of neurogenesis. *Frontiers in Genetics, 3*, 70–77.

de la Monte, S. M., & Wands, J. R. (2010). Role of central nervous system insulin resistance in fetal alcohol spectrum disorders. *Journal of Population Therapeutics and Clinical Pharmacology, 17*, 390–404.

Moore, D. B., Madorsky, I., Paiva, M., & Barrow Heaton, M. (2004). Ethanol exposure alters neurotrophin receptor expression in the rat central nervous system: Effects of neonatal exposure. *Journal of Neurobiology, 60*, 114–126.

Parnell, S. E., O'Leary-Moore, S. K., Godin, E. A., Dehart, D. B., Johnson, B. W., Allan Johnson, G., … Sulik, K. K. (2009). Magnetic resonance microscopy defines ethanol-induced brain abnormalities in prenatal mice: Effects of acute insult on gestational day 8. *Alcoholism: Clinical and Experimental Research, 33*, 1001–1011.

Popova, E. N. (2010). *Brain ultrastructure, alcohol and offspring*. Moscow: Scientific World (in Russian).

Riley, E. P., Infante, M. A., & Warren, K. R. (2011). Fetal alcohol spectrum disorders: An overview. *Neuropsychology Review, 21*, 73–80.

Rosman, N. P., & Malone, M. J. (1976). An experimental study of fetal alcohol syndrome. *Neurology, 26*, 365–366.

Sawant, O. B., Lunde, E. R., Washburn, S. E., Chen, W. J., Goodlett, C. R., & Cudd, T. A. (2013). Different patterns of regional Purkinje cell loss in the cerebellar vermis as a function of the timing of prenatal ethanol exposure in an ovine model. *Neurotoxicology and Teratology, 35*, 7–13.

Schambra, U. B., Lauder, J. M., & Petrusz, P. (1990). Development of neurotransmitter systems in the mouse embryo following acute ethanol exposure: A histological and immunocytochemical study. *Development Neuroscience, 8*, 507–522.

Shea, K. M., Hewitt, A. J., & Olmstead, M. C. (2012). Maternal ethanol consumption. *Behavior Pharmacology, 23*, 105–112.

Siebert, J. R., Astley, S. J., & Clarren, S. K. (1991). Holoprosencephaly in a fetal macaque (*Macaca nemestrina*) following weekly exposure to ethanol. *Teratology, 44*, 29–36.

Smiley, J. F., Saito, M., Bleiwas, C., Masiello, K., Ardekani, B., Guilfoyle, D. N., … Vadasz, C. (2015). Selective reduction of cerebral cortex GABA neurons in a late gestation model of fetal alcohol spectrum disorder. *Alcohol, 49*, 571–580.

Smith, S. M., Garic, A., Flentke, G. R., & Berres, M. E. (2014). Neural crest development in fetal alcohol syndrome. *Birth Defects Research. Part C, Embryo Today: Reviews, 102*, 210–220.

Stolakis, V., Liapi, C., Zarros, A., Kalopita, K., Memtsas, V., Botis, J., … Tsakiris, S. (2015). Exposure to ethanol during neurodevelopment modifies crucial offspring rat brain enzyme activities in a region-specific manner. *Metabolic Brain Disease, 30*, 1467–1477.

Svanidze, I. K., Museridze, D. P., Didimova, E. V., & Sanikidze, T. V. (2012). Disorders of neurogenesis of cortical and subcortical structures in rat brain limbic system during fetal alcohol syndrome formation. *Morfologiia, 141*, 18–22.

Tran, T. D., & Kelly, S. J. (2003). Critical periods for ethanol-induced cell loss in the hippocampal formation. *Neurotoxicology and Teratology, 25*, 519–528.

Tsai, G., & Coyle, J. T. (1998). The role of glutamatergic neurotransmission in the pathophysiology of alcoholism. *Annual Review of Medicine, 49*, 173–184.

Vangipuram, S. D., & Lyman, W. D. (2010). Ethanol alters cell fate of fetal human brain-derived stem and progenitor cells. *Alcohol: Clinical Experimental Research, 34*, 1574–1583.

Wainright, P., & Gagnon, M. (1985). Moderate prenatal ethanol exposure interacts with strain in affecting brain development in BALB/c and C57BL/6 mice. *Experimental Neurology, 88*, 84–94.
Ward, G. R., & Wainwright, P. E. (1991). Effects of prenatal stress and ethanol on cerebellar fiber tract maturation in B6D2F2 mice: An image analysis study. *Neurotoxicology, 12*, 665–676.
Wierzba-Bobrowicz, T., Lewandowska, E., Stepień, T., & Szpak, G. M. (2011). Differential expression of calbindin D28k, calretinin and parvalbumin in the cerebellum of pups of ethanol-treated female rats. *Folia Neuropathologica, 49*, 47–55.
Williams, S. (1991). Alcohol's possible covert role: Brain dysfunction, paraphilias, and sexually aggressive behaviors. *Sexual Abuse, 11*, 147–158.
Wilson, D. A., Peterson, J., Basavaraj, B. S., & Saito, M. (2011). Local and regional network function in behaviorally relevant cortical circuits of adult mice following postnatal alcohol exposure. *Alcoholism: Clinical and Experimental Research, 35*, 1974–1984.
Woodson, P., & Ritchey, S. (1979). Effects of maternal alcohol consumption of fetal brain cell number and cell size. *Nutrition Reports International, 20*, 225–228.
Yang, Y., Roussotte, F., Kan, E., Sulik, K. K., Mattson, S. N., Riley, E. P., ... Sowell, E. R. (2012). Abnormal cortical thickness alterations in fetal alcohol spectrum disorders and their relationships with facial dysmorphology. *Cerebral Cortex, 22*, 1170–1179.
Zhou, F. C., Sari, Y., Powrozek, T., Goodlett, C. R., & Li, T. K. (2003). Moderate alcohol exposure compromises neural tube midline development in prenatal brain. *Brain Research. Developmental Brain Research, 144*, 43–55.
Zimatkin, S. M., & Bon, E. I. (2014). *Fetal alcohol syndrome*. Minsk: Novoe Znanie (in Russian).
Zimatkin, S. M., & Bon, E. I. (2015a). Dynamics of histological changes in the parietal cortex of rats, exposed antenatal alcohol exposure. *Novosti medico-biologicheskish nauk, 2*, 146–151 (in Russian).
Zimatkin, S. M., & Bon, E. I. (2015b). Involution of neurons in the cerebral cortex of the offspring rats, consuming alcohol during pregnancy. *Vesci nacionalnoy acadrmii nauk Belarusi, 3*, 124–127 (in Russian).
Zimatkin, S. M., & Lis, R. E. (1990). The activity of aldehyde dehydrogenase in the developing brain. *Archiv Anatomii, Gistologii i Embriologii, 5*, 27–33 (in Russian).

CHAPTER

5

Alcohol Intoxication and Traumatic Spinal Cord Injury: Basic and Clinical Science

C.L. Crutcher, II, J. Veith, G.C. Tender

Louisiana State University Health Science Center, New Orleans, LA, United States

INTRODUCTION

Traumatic spinal cord injury (SCI) victims often suffer permanent neurological damage. Depending on the anatomical level of injury, SCI patients may be left paraplegic or quadriplegic. The incidence of traumatic SCI in the United States is 54 cases per million annually, with about 17,000 new cases per year (National Spinal Cord Injury Statistical Center, 2015). The incidence of traumatic SCI worldwide is 10.4 to 83 cases per million annually (Wyndaele & Wyndaele, 2006). The most common causes of SCI, in descending order of frequency, are: traffic accidents, work, sports and recreation, falls, and violence (Devivo, 2012). Traumatic SCI can be separated into two components: the primary injury and the secondary injury. The primary injury encompasses the inciting mechanism, such as physical spinal cord compression, and the secondary injury involves downstream pathological processes, such as ischemia and altered cellular metabolism.

Alcohol intoxication is a major risk factor for traumatic SCI and is present in up to 34% of cases (Levy et al., 2004). The lifetime prevalence of alcohol abuse disorders in the United States is 17.8% (Hasin, Stinson, Ogburn, & Grant, 2007). Alcohol use is the third leading cause of mortality from modifiable risky behaviors in the United States (Mokdad, Marks, Stroup, & Gerberding, 2004). Many SCI patients have higher rates of preinjury alcohol abuse than the general population (Kolakowsky-Hayner et al., 1999). Alcohol intoxicated patients often present with more severe injuries in the trauma population (Waller, Hill, Maio, & Blow, 2003). Basic science and clinical science research has investigated alcohol intoxication and its influence on traumatic SCI. This chapter is a review of the pertinent literature regarding pathophysiology of SCI, alcohol intoxication and its effects on the pathophysiology of SCI, and alcohol intoxication and its impact on clinical SCI patients.

PATHOPHYSIOLOGY OF TRAUMATIC SPINAL CORD INJURY

An acute SCI can be divided into two components, the primary and secondary injury. The primary injury typically consists of compression, contusion, distraction, or transection of the neural tissue, blood vessels, or supporting tissues as a direct result of a traumatic event. Common causes of primary injury include herniated disks, fractures and/or dislocations of the spine, and bullets or other sharp penetrating objects. The secondary injury involves edema, hemorrhage, ischemia, inflammation, and injury to neuronal cells and their membranes. Over 25 secondary mechanisms of SCI have been described (Oyinbo, 2011). See Table 5.1 for a list of major secondary injury mechanisms.

Secondary injury to the spinal cord involves a complex cascade of vascular, cellular, and inflammatory events.

VASCULAR INJURY

Immediately following an SCI, there are changes in the local and systemic blood flow. Experimental SCI models demonstrate that there is a local reduction in vascularity and blood flow at the site of trauma, contributing to spinal cord ischemia (Figley, Khosravi, Legasto, Tseng, & Fehlings, 2014; Yeo et al., 1984; Yeo, Payne, Hinwood, & Kidman, 1975). Within 30 min of trauma, local spinal cord microvasculature experiences

Addictive Substances and Neurological Disease
http://dx.doi.org/10.1016/B978-0-12-805373-7.00005-0

TABLE 5.1 Major Secondary Mechanisms of Spinal Cord Injury

Major Secondary Injury Mechanisms	
• Vascular compromise	Ischemia/reperfusion injury
	Hemorrhage
	Neurogenic shock
	Vasospasm
• Membrane dysfunction	Increased membrane permeability
	Fluid accumulation and edema
	Altered ionic homeostasis
	Lipid peroxidation
• Cellular dysfunction	Apoptosis/programmed cell death
	Astroglial scar formation
	Cavitation
	Chromatolysis
	Altered myelination demyelination
	Altered energy utilization and production
	Oligodendrocyte apoptosis
• Other	Injury due to excitotoxic neurotransmitters
	Excessive cytokine release and immune response
	Inflammation
	Increased oxidative stress
	Cytokine production
	Free radical formation

vasospasm that results in reduced blood flow (Dohrmann & Allen, 1975). Systemically, the mean arterial pressure temporarily increases, followed by an extended period of hypoperfusion. The heart rate and peripheral vascular resistance also decrease (Guha & Tator, 1988). The combination of these effects results in spinal cord hypoperfusion, which leads to further spinal cord ischemia.

Direct contusion of the spinal cord leads to development of intraparenchymal hemorrhage (Choo et al., 2007). Congestion of the microvasculature of the gray matter with erythrocytes, followed by endothelial injury, further contribute to hemorrhage formation within the spinal cord (Dohrmann, Wagner, & Bucy, 1971). The spinal cord tissue adjacent to these sites of hemorrhage demonstrates increased susceptibility to ischemic damage further contributing to the initial SCI (Tator & Fehlings, 1991). In addition to the direct injury to the spinal cord tissue caused by the contusion, intraparenchymal hemorrhage acts as a catalyst for blood–brain–spinal cord barrier (BBSCB) breakdown, plasma membrane breakdown, and free radical production (Losey, Young, Krimholtz, Bordet, & Anthony, 2014; Sadrzadeh, Anderson, Panter, Hallaway, & Eaton, 1987).

Anderson examined the effects of alcohol on traumatic SCI using ferrets. The animals were intoxicated to a level of 100 mg/dL. The study demonstrated that intoxicated animals had larger amounts of iron accumulation at the injury site and in adjacent tissue when compared to the non-intoxicated controls. Iron accumulation at the site on injury was suggested to represent either hemorrhage or vascular congestion (Anderson, 1986). Other studies have shown that intoxicated animals have much larger areas of hemorrhage and surrounding edema when compared to non-intoxicated subjects (Flamm et al., 1977). Free iron derived from hemoglobin potentiates secondary SCI by inhibiting sodium–potassium ATPase (Na^+/K^+ ATPase) activity and catalyzes peroxidation of central nervous system lipids (Sadrzadeh et al., 1987). Therefore, the initial injury that causes hemorrhage initiates a cyclical cascade of secondary injury mechanisms that are self-propagating and are exacerbated by alcohol.

MEMBRANE DYSFUNCTION AND EDEMA FORMATION

The BBSCB is an anatomical and physiological barrier preventing the passage of certain substances from the systemic blood circulation into the extracellular fluid space of the brain and spinal cord. The BBSCB is semi-permeable and highly selective. The BBSCB is composed of astrocyte-induced tight junctions of the basement membrane around capillaries. The BBSCB prevents passage of large and hydrophilic molecules, large proteins, and other toxic substances into the central nervous system. Certain substances such as glucose or specific proteins may be transported selectively across the BBSCB by simple diffusion, simple diffusion through an aqueous channel, facilitated diffusion, or active transport through protein channels.

Examinations of spinal cords after experimental transection demonstrate increased blood–brain barrier breakdown due to increased transendothelial transport of proteins (Noble & Wrathall, 1987). This increased leakage of intravascular content contributes to spinal cord edema and worsens functional outcome (Leonard, Thornton, & Vink, 2015). Furthermore, direct spinal cord contusion leads to both temporary and permanent increases in cellular membrane permeability that is proportional to the degree of contusion (Simon, Sharif, Tan, & LaPlaca, 2009).

The loss of cellular membrane continuity leads to a loss of ionic homeostasis and edema formation due to a leaky plasma membrane and ion channel dysfunction.

The cells of the central nervous system utilize several different ion channels to maintain homeostasis and ionic equilibrium. Sodium and calcium concentrations are two main ionic components of merit in the discussion regarding SCI. The sodium–potassium channel (Na^+/K^+ ATPase) and the sodium–calcium exchanger (NCX), among others, are two principal ion exchangers that maintain osmotic and ionic homeostasis. The Na^+/K^+ ATPase normally functions to exchange three extracellular potassium ions for two intracellular sodium ions. Similarly, the NCX removes one intracellular calcium ion in exchange for three extracellular sodium ions. NCX is one of the most important methods for removing intracellular calcium and maintaining calcium homeostasis (DiPolo & Beauge, 2006).

Sodium–potassium channel malfunction after SCI is evident as early as 5 min post contusion in the experimental SCI model (Clendenon, Allen, Gordon, & Bingham, 1978). Experimental studies show that increased intracellular sodium concentrations contribute to cytotoxic edema formation (Lemke & Faden, 1990), intracellular pH alterations (Agrawal & Fehlings, 1996), and calcium homeostasis derangement (Blaustein & Lederer, 1999; Tomes & Agrawal, 2002). Agrawal and Fehlings (1996) have demonstrated that increased intracellular sodium concentrations post SCI correlate with worsened traumatic axonal injury and that prevention of intracellular acidosis by blocking the sodium–hydrogen exchanger had neuroprotective effects. This suggests that increased intracellular acidosis has a negative impact in traumatic SCI.

Experimental animals pretreated with alcohol have more extensive white matter edema, and evidence of tissue loss when compared to non-intoxicated controls (Brodner, Van Gilder, & Collins, 1981). Although there are no experimental studies that examine alcohol intoxication and its impact on Na^+/K^+ ATPase and NCX in the setting of traumatic SCI, alcohol has been studied in nontraumatic models. Alcohol treated animals have altered Na^+/K^+ ATPase activity resulting in intracellular sodium accumulation (Blachley, Johnson, & Knochel, 1985). Additionally, studies have shown that ethanol exposed animals have higher levels of lactate (Halt, Swanson, & Faden, 1992) and oxygen consumption than nonalcohol treated animals (Blachley et al., 1985). It is possible that alcohol exposure puts the neural tissue in an altered physiological state that exacerbates the secondary mechanisms after SCI.

Calcium influx is evident within 45 min of experimental SCI (Happel et al., 1981). In the setting of experimental SCI, the increased sodium concentration causes the NCX to reverse direction and contribute to calcium influx rather than efflux (DiPolo & Beauge, 2006; Li, Jiang, & Stys, 2000). Alcohol exposure to microsomes and synaptosomes from murine neural tissue causes a release of calcium from intracellular storage and an increase in resting calcium concentration (Daniell, Brass, & Harris, 1987; Daniell & Harris, 1989). Increased intracellular calcium concentration post SCI leads to mitochondrial dysfunction, free radical and reactive oxygen species (ROS) formation, and phospholipase oxygenation (Young, 1992). Mitochondrial dysfunction and ROS formation occurs within the first hour after SCI, in conjunction with calcium influx (Azbill, Mu, Bruce-Keller, Mattson, & Springer, 1997). Excess intracellular calcium also promotes activation of proteases with resulting membrane breakdown and initiation of cellular apoptosis, also known as programmed cell death (Ray, Matzelle, Wilford, Hogan, & Banik, 2001; Shields, Schaecher, Hogan, & Banik, 2000). Alcohol exposure increases the resting calcium concentration in neural tissue potentially exacerbating these deleterious effects of excess calcium in traumatic SCI.

INFLAMMATION AND IMMUNE-MEDIATED RESPONSE TO SPINAL CORD INJURY

Inflammation and immune-mediated responses to SCI have been identified as key elements in secondary SCI. After traumatic SCI, the site is invaded by leukocytes, neutrophils, and monocytes (Bao et al., 2009). These cells contribute to the release of pro-inflammatory cytokines and free radical production. In experimental rat studies, pro-inflammatory cytokines, tumor necrosis factor alpha (TNF-α), interleukin-1 beta (IL-1β), and interleukin-6 (IL-6) expression, have been shown to increase after SCI (Yang et al., 2005). Increased production of cytokines after SCI contributes to apoptosis and increased production of free radical species (Wang, Kong, Qi, Ye, & Song, 2005; Yune et al., 2003).

Exogenous cytokine administration during in vitro studies has shown that early exposure, within 1 day of injury, leads to increased recruitment and activation of macrophages and microglial cells. The size of tissue loss appears greater in the cytokine treated spinal cords compared to controls, but the difference is not statistical. The same study demonstrated that late exposure, 4 days after injury, had increased macrophage recruitment and decreased microglial activation. The animals that were exposed to exogenous cytokines 4 days after injury had reduced tissue loss at 7 days when compared to controls. This suggests a possible protective effect with late exposure to cytokines (Klusman & Schwab, 1997).

In human subjects, traumatic SCI patients have higher neutrophil and monocyte free radical production, higher free radical production from leukocytes, higher concentration of oxidative enzymes, and pro-inflammatory factors, transcription factor nuclear factor kappa B (NF-κB) and cyclooxygenase-2 (COX-2) than their non–spinal cord injured and able bodied controls. There is also evidence that the heightened inflammatory state in SCI patient increases oxidative stress and leads to worsening lipid peroxidation (Bao et al., 2009).

COX 2 is responsible for the enzymatic conversion of arachidonic acid into breakdown products such as thromboxanes, prostacyclins, and prostaglandins. Resnick et al. suggest that COX 2 production leads to worse functional outcome. The authors studied COX 2 mRNA and protein expression after spinal cord contusion injury. They found that the study group that was administered selective COX 2 inhibitors had better functional outcomes when compared to the nontreated group (Resnick, Graham, Dixon, & Marion, 1998).

Research regarding alcohol intoxication and inflammation after traumatic SCI is lacking. Recent research in traumatic brain injury has revealed that acute alcohol intoxication delays the resolution of inflammation and inflammatory markers IL-1β, IL-6, and TNF-α after traumatic injury (Teng & Molina, 2014). Alcohol intoxication may have a similar effect in spinal cord tissue. Alcohol treated subjects have higher levels of free fatty acid and thromboxane levels than that their sober counterparts (Halt et al., 1992). The higher concentration of thromboxane represents a heightened inflammatory state as well as contributing to spinal cord hypoperfusion (Tempel & Martin, 1992). Prolonged exposure to pro-inflammatory molecules can lead to increased cytokine production, greater tissue loss, and greater free radical damage.

FREE RADICAL FORMATION AND LIPID PEROXIDATION

Free radicals are molecules with an unpaired electron that are highly reactive to oxygen molecules, DNA, and certain enzymes. ROS are a specific type of free radicals that has the unpaired electron on the oxygen atom. Increased iron concentration after experimental SCI can act as a catalyst for the formation of some ROS (Liu, Liu, Sun, Alcock, & Wen, 2003) and lipid peroxidation (Zhang, Scherch, & Hall, 1996). Free radicals can cause damage to plasma membranes in the form of lipid peroxidation. Lipid peroxidation of fatty acids can generate more free radicals leading to further damage and decreased neuron viability (Porter, Caldwell, & Mills, 1995; Zhang et al., 1996).

Free radical formation contributes to increased oxidative stress, lipid damage, nucleic acid damage, mitochondrial damage, and initiation of apoptosis (Xu et al., 2005). Increased calcium concentrations, as discussed earlier, catalyze lipid peroxidation leading to further membrane damage (Braughler, Duncan, & Chase, 1985). After experimental contusion injury to the spinal cord, there is evidence of dose-related membrane phospholipid hydrolysis (Demediuk, Daly, & Faden, 1989a) and ROS production (Luo, Li, Robinson, & Shi, 2002). Additionally, macrophages, as discussed earlier, are recruited to the site of injury as early as 6 h after contusion and produce the ROS, nitric oxide (Satake et al., 2000).

After experimental spinal cord contusion, examination of injured tissue reveals decreased levels of phospholipids and cholesterol, and increased in free fatty acids (FFA) (Demediuk et al., 1989a). Arachidonic acid is a key FFA released after spinal cord trauma. Increased arachidonic acid exposure increases intracellular calcium concentrations, increases oxidative stress, and decreases cell viability (Toborek et al., 1999). Alcohol treated subjects also have higher levels of free fatty acid and thromboxane levels than that of their sober counterparts (Halt et al., 1992). This offers further support that alcohol intoxication contributes to increased lipid peroxidation and plasma membrane damage.

EXCITATORY NEUROTRANSMITTERS

Immediately after experimental SCI there is an increase in release of excitatory neurotransmitters from injured neurons (Demediuk, Daly, & Faden, 1989b; Farooque, Hillered, Holtz, & Olsson, 1996; Panter, Yum, & Faden, 1990). Glutamate is one of the most abundant excitatory neurotransmitters in the spinal cord. Glutamate stimulates N-methyl-D-Aspartate (NMDA), kinate, and α-amino-3-hydroxy-5-methyl-4-isoxazole propionic acid (AMPA) receptors. The net result of glutaminergic stimulation of NMDA receptors is a net influx of calcium ions. Experimentally, excessive concentrations of glutamate similar to levels experienced during traumatic SCI has been shown to cause damage to local neurons (Liu, Xu, Pan, & McAdoo, 1999), to cause damage to supporting cells of the central nervous system (Xu, Hughes, Ye, Hulsebosch, & McAdoo, 2004), and to cause functional impairments in rats (Xu, Hughes, Zhang, Cain, & McAdoo, 2005). NMDA and AMPA-mediated calcium influx lead to mitochondria depolarization and cytotoxicity and increased cell death (Sen, Joshi, Joshi, & Joshi, 2008; Urushitani et al., 2001). The devastating effects of glutamate excito-toxicity are also linked to apoptosis (Xu, Liu, Hughes, & McAdoo, 2008).

Alcohol treated rats have significantly less aspartate and glutamate in their neural tissue after SCI when compared to saline infused rats (Halt et al., 1992). The decreased concentration of excitatory amino acids within the neural tissue suggest that the cells release more excitatory amino acids upon injury. As stated foregoing, increased exposure to these excitatory neurotransmitters leads to increased intracellular calcium, which contributes to cellular apoptosis, mitochondrial dysfunction, free radical formation, and ultimately decreased cell viability.

ALCOHOL INTOXICATION AND THE EFFECTS ON PRECLINICAL SPINAL CORD INJURY

The effect of alcohol intoxication on the severity of traumatic SCI has been a point of interest within the scientific community. There are numerous preclinical studies demonstrating that alcohol intoxication leads to worse traumatic SCI when compared to non-intoxicated controls. This section of the chapter focuses on alcohol intoxication and its detrimental effects on traumatic SCI outcomes.

ALCOHOL AND FUNCTIONAL OUTCOMES

Experimental traumatic SCI outcomes are routinely evaluated by measuring compound action potentials, measuring somatosensory evoked potentials, testing reflexes, or evaluating voluntary limb movement. Alcohol intoxicated animals have significantly diminished neurological recovery after contusion SCI and have a significantly higher mortality than non-intoxicated study animals (Halt et al., 1992). Four to six weeks from injury, alcohol treated animals are significantly less likely to be able to support their own weight and walk without difficulty when compared to sober study subjects (Flamm et al., 1977; Halt et al., 1992). During surgery, monitored evoked potentials are permanently lost more often in the alcohol intoxicated animals, and those intoxicated animal are more often left permanently paralyzed when compared to non-intoxicated animals (Flamm et al., 1977). Flamm et al. (1977) suggest that the immediate loss of evoked potentials results from a synergistic effect of alcohol and trauma where damages to membrane-bound Na^+/K^+ ATPase impair axon impulse transmission capabilities.

ALCOHOL INTOXICATION AND INJURY SEVERITY

In general trauma, alcohol intoxicated patients present to the hospital with more severe injuries than sober patients (Waller et al., 2003). In a small study, intoxicated patients who presented with positive blood alcohol were more likely present with more severe neurological injury or quadraparesis (Forchheimer, Cunningham, Gater, & Maio, 2005). Patient with cervical SCI are more likely to have consumed alcohol compared to patients with other levels of SCI (Garrison et al., 2004). Patient with cervical SCI are more likely to present with paralysis of all four extremities than thoracic or lumbar injuries due to the anatomy of the spinal cord. More recent research has shown that after controlling for injury severity, alcohol intoxication had no effect on 1-year mortality, or degree of functional impairment at 6 weeks, 6 months, or 1 year after SCI in patients who survive to hospital admission (Furlan & Fehlings, 2013).

Although injury severity is unaffected by degree of alcohol intoxication, intoxicated patients are more likely to suffer from in-hospital medical complications such as pneumonia, deep vein thrombosis and pulmonary embolism, urinary tract infections, and skin complications (Crutcher, Ugiliweneza, Hodes, Kong, & Boakye, 2014).

ALCOHOL USE AND SPINAL CORD INJURY PATIENTS

Up to 96% of patients with SCI report preinjury alcohol use, with 57% of users identifying as heavy users (Kolakowsky-Hayner et al., 1999). Acute SCI patients often require intense physical and occupational rehabilitation to recover some degree of functional independence. Patients with significant histories of drinking problems have lower functional independence upon rehabilitation admission and discharge than patients without a significant drinking history (Bombardier, Stroud, Esselman, & Rimmele, 2004).

CONCLUSION

Traumatic SCI often has devastating and permanent neurological consequences. There are about 17,000 cases of traumatic SCI every year. Up to 96% of SCI patients report preinjury alcohol use. Alcohol intoxication has been shown to have deleterious effects on experimental SCI severity. Alcohol intoxication and exposure primes spinal cord cells for injury and promotes a milieu to enhance and propagate many secondary injury mechanisms. Alcohol intoxication worsens the degree of vascular insult, inflammation and immune-mediated response, excitatory neurotransmitter damage, free radical damage, edema formation, and membrane dysfunction.

In human SCI, alcohol intoxication has no impact on the severity on injury.

Intoxicated patients, however, suffer more complications and have lower functional independence than sober patients.

References

Agrawal, S. K., & Fehlings, M. G. (1996). Mechanisms of secondary injury to spinal cord axons in vitro: Role of Na+, Na(+)-K(+)-ATPase, the Na(+)-H+ exchanger, and the Na(+)-Ca2+ exchanger. *Journal of Neuroscience, 16*(2), 545–552.

Anderson, T. E. (1986). Effects of acute alcohol intoxication on spinal cord vascular injury. *Central Nervous System Trauma, 3*(3), 183–192.

Azbill, R. D., Mu, X., Bruce-Keller, A. J., Mattson, M. P., & Springer, J. E. (1997). Impaired mitochondrial function, oxidative stress and altered antioxidant enzyme activities following traumatic spinal cord injury. *Brain Research, 765*(2), 283–290.

Bao, F., Bailey, C. S., Gurr, K. R., Bailey, S. I., Rosas-Arellano, M. P., Dekaban, G. A., & Weaver, L. C. (2009). Increased oxidative activity in human blood neutrophils and monocytes after spinal cord injury. *Experimental Neurology, 215*(2), 308–316. http://dx.doi.org/10.1016/j.expneurol.2008.10.022.

Blachley, J. D., Johnson, J. H., & Knochel, J. P. (1985). The harmful effects of ethanol on ion transport and cellular respiration. *The American Journal of the Medical Sciences, 289*(1), 22–26.

Blaustein, M. P., & Lederer, W. J. (1999). Sodium/calcium exchange: Its physiological implications. *Physiological Reviews, 79*(3), 763–854.

Bombardier, C. H., Stroud, M. W., Esselman, P. C., & Rimmele, C. T. (2004). Do preinjury alcohol problems predict poorer rehabilitation progress in persons with spinal cord injury? *Archives of Physical Medicine and Rehabilitation, 85*(9), 1488–1492.

Braughler, J. M., Duncan, L. A., & Chase, R. L. (1985). Interaction of lipid peroxidation and calcium in the pathogenesis of neuronal injury. *Central Nervous System Trauma, 2*(4), 269–283.

Brodner, R. A., Van Gilder, J. C., & Collins, W. F., Jr. (1981). Experimental spinal cord trauma: Potentiation by alcohol. *Journal of Trauma, 21*(2), 124–129.

Choo, A. M., Liu, J., Lam, C. K., Dvorak, M., Tetzlaff, W., & Oxland, T. R. (2007). Contusion, dislocation, and distraction: Primary hemorrhage and membrane permeability in distinct mechanisms of spinal cord injury. *Journal of Neurosurgery. Spine, 6*(3), 255–266. http://dx.doi.org/10.3171/spi.2007.6.3.255.

Clendenon, N. R., Allen, N., Gordon, W. A., & Bingham, W. G., Jr. (1978). Inhibition of Na+-K+-activated ATPase activity following experimental spinal cord trauma. *Journal of Neurosurgery, 49*(4), 563–568. http://dx.doi.org/10.3171/jns.1978.49.4.0563.

Crutcher, C. L., 2nd, Ugiliweneza, B., Hodes, J. E., Kong, M., & Boakye, M. (2014). Alcohol intoxication and its effects on traumatic spinal cord injury outcomes. *Journal of Neurotrauma, 31*(9), 798–802. http://dx.doi.org/10.1089/neu.2014.3329.

Daniell, L. C., Brass, E. P., & Harris, R. A. (1987). Effect of ethanol on intracellular ionized calcium concentrations in synaptosomes and hepatocytes. *Molecular Pharmacology, 32*(6), 831–837.

Daniell, L. C., & Harris, R. A. (1989). Ethanol and inositol 1,4,5-trisphosphate release calcium from separate stores of brain microsomes. *The Journal of Pharmacology and Experimental Therapeutics, 250*(3), 875–881.

Demediuk, P., Daly, M. P., & Faden, A. I. (1989a). Changes in free fatty acids, phospholipids, and cholesterol following impact injury to the rat spinal cord. *Journal of Neuroscience Research, 23*(1), 95–106. http://dx.doi.org/10.1002/jnr.490230113.

Demediuk, P., Daly, M. P., & Faden, A. I. (1989b). Effect of impact trauma on neurotransmitter and nonneurotransmitter amino acids in rat spinal cord. *Journal of Neurochemistry, 52*(5), 1529–1536.

Devivo, M. J. (2012). Epidemiology of traumatic spinal cord injury: Trends and future implications. *Spinal Cord, 50*(5), 365–372. http://dx.doi.org/10.1038/sc.2011.178.

DiPolo, R., & Beauge, L. (2006). Sodium/calcium exchanger: Influence of metabolic regulation on ion carrier interactions. *Physiological Reviews, 86*(1), 155–203. http://dx.doi.org/10.1152/physrev.00018.2005.

Dohrmann, G. J., & Allen, W. E. (1975). Microcirculation of traumatized spinal cord. A correlation of microangiography and blood flow patterns in transitory and permanent paraplegia. *Journal of Trauma, 15*(11), 1003–1013.

Dohrmann, G. J., Wagner, F. C., Jr., & Bucy, P. C. (1971). The microvasculature in transitory traumatic paraplegia. An electron microscopic study in the monkey. *Journal of Neurosurgery, 35*(3), 263–271.

Farooque, M., Hillered, L., Holtz, A., & Olsson, Y. (1996). Changes of extracellular levels of amino acids after graded compression trauma to the spinal cord: An experimental study in the rat using microdialysis. *Journal of Neurotrauma, 13*(9), 537–548. http://dx.doi.org/10.1089/neu.1996.13.537.

Figley, S. A., Khosravi, R., Legasto, J. M., Tseng, Y. F., & Fehlings, M. G. (2014). Characterization of vascular disruption and blood–spinal cord barrier permeability following traumatic spinal cord injury. *Journal of Neurotrauma, 31*(6), 541–552. http://dx.doi.org/10.1089/neu.2013.3034.

Flamm, E. S., Demopoulos, H. B., Seligman, M. L., Tomasula, J. J., De Crescito, V., & Ransohoff, J. (1977). Ethanol potentiation of central nervous system trauma. *Journal of Neurosurgery, 46*(3), 328–335. http://dx.doi.org/10.3171/jns.1977.46.3.0328.

Forchheimer, M., Cunningham, R. M., Gater, D. R., Jr., & Maio, R. F. (2005). The relationship of blood alcohol concentration to impairment severity in spinal cord injury. *The Journal of Spinal Cord Medicine, 28*(4), 303–307.

Furlan, J. C., & Fehlings, M. G. (2013). Blood alcohol concentration as a determinant of outcomes after traumatic spinal cord injury. *European Journal of Neurology, 20*(7), 1101–1106. http://dx.doi.org/10.1111/ene.12145.

Garrison, A., Clifford, K., Gleason, S. F., Tun, C. G., Brown, R., & Garshick, E. (2004). Alcohol use associated with cervical spinal cord injury. *The Journal of Spinal Cord Medicine, 27*(2), 111–115.

Guha, A., & Tator, C. H. (1988). Acute cardiovascular effects of experimental spinal cord injury. *Journal of Trauma, 28*(4), 481–490.

Halt, P. S., Swanson, R. A., & Faden, A. I. (1992). Alcohol exacerbates behavioral and neurochemical effects of rat spinal cord trauma. *Archives of Neurology, 49*(11), 1178–1184.

Happel, R. D., Smith, K. P., Banik, N. L., Powers, J. M., Hogan, E. L., & Balentine, J. D. (1981). Ca2+-accumulation in experimental spinal cord trauma. *Brain Research, 211*(2), 476–479.

Hasin, D. S., Stinson, F. S., Ogburn, E., & Grant, B. F. (2007). Prevalence, correlates, disability, and comorbidity of DSM-IV alcohol abuse and dependence in the United States: Results from the National Epidemiologic Survey on Alcohol and Related Conditions. *Archives of General Psychiatry, 64*(7), 830–842. http://dx.doi.org/10.1001/archpsyc.64.7.830.

Klusman, I., & Schwab, M. E. (1997). Effects of pro-inflammatory cytokines in experimental spinal cord injury. *Brain Research, 762*(1–2), 173–184.

Kolakowsky-Hayner, S. A., Gourley, E. V., 3rd, Kreutzer, J. S., Marwitz, J. H., Cifu, D. X., & McKinley, W. O. (1999). Pre-injury substance abuse among persons with brain injury and persons with spinal cord injury. *Brain Injury, 13*(8), 571–581.

Lemke, M., & Faden, A. I. (1990). Edema development and ion changes in rat spinal cord after impact trauma: Injury dose-response studies. *Journal of Neurotrauma, 7*(1), 41–54. http://dx.doi.org/10.1089/neu.1990.7.41.

Leonard, A. V., Thornton, E., & Vink, R. (2015). The relative contribution of edema and hemorrhage to raised intrathecal pressure after traumatic spinal cord injury. *Journal of Neurotrauma, 32*(6), 397–402. http://dx.doi.org/10.1089/neu.2014.3543.

Levy, D. T., Mallonee, S., Miller, T. R., Smith, G. S., Spicer, R. S., Romano, E. O., & Fisher, D. A. (2004). Alcohol involvement in burn, submersion, spinal cord, and brain injuries. *Medical Science Monitor, 10*(1), CR17–CR24.

Li, S., Jiang, Q., & Stys, P. K. (2000). Important role of reverse Na(+)-Ca(2+) exchange in spinal cord white matter injury at physiological temperature. *Journal of Neurophysiology, 84*(2), 1116–1119.

Liu, D., Liu, J., Sun, D., Alcock, N. W., & Wen, J. (2003). Spinal cord injury increases iron levels: Catalytic production of hydroxyl radicals. *Free Radical Biology & Medicine, 34*(1), 64–71.

Liu, D., Xu, G. Y., Pan, E., & McAdoo, D. J. (1999). Neurotoxicity of glutamate at the concentration released upon spinal cord injury. *Neuroscience, 93*(4), 1383–1389.

Losey, P., Young, C., Krimholtz, E., Bordet, R., & Anthony, D. C. (2014). The role of hemorrhage following spinal-cord injury. *Brain Research, 1569*, 9–18. http://dx.doi.org/10.1016/j.brainres.2014.04.033.

Luo, J., Li, N., Robinson, J. P., & Shi, R. (2002). The increase of reactive oxygen species and their inhibition in an isolated Guinea pig spinal cord compression model. *Spinal Cord, 40*(12), 656–665. http://dx.doi.org/10.1038/sj.sc.3101363.

Mokdad, A. H., Marks, J. S., Stroup, D. F., & Gerberding, J. L. (2004). Actual causes of death in the United States, 2000. *JAMA, 291*(10), 1238–1245. http://dx.doi.org/10.1001/jama.291.10.1238.

National Spinal Cord Injury Statistical Center. (2015). *Spinal Cord Injury (SCI), facts and figures at a glance (pp. 1–2, Internet)*. Birmingham, AL: University of Alabama at Birmingham. Available from https://www.nscisc.uab.edu/Public/Facts%202015.pdf.

Noble, L. J., & Wrathall, J. R. (1987). The blood–spinal cord barrier after injury: Pattern of vascular events proximal and distal to a transection in the rat. *Brain Research, 424*(1), 177–188.

Oyinbo, C. A. (2011). Secondary injury mechanisms in traumatic spinal cord injury: A nugget of this multiply cascade. *Acta Neurobiologiae Experimentalis, 71*(2), 281–299.

Panter, S. S., Yum, S. W., & Faden, A. I. (1990). Alteration in extracellular amino acids after traumatic spinal cord injury. *Annals of Neurology, 27*(1), 96–99. http://dx.doi.org/10.1002/ana.410270115.

Porter, N. A., Caldwell, S. E., & Mills, K. A. (1995). Mechanisms of free radical oxidation of unsaturated lipids. *Lipids, 30*(4), 277–290.

Ray, S. K., Matzelle, D. D., Wilford, G. G., Hogan, E. L., & Banik, N. L. (2001). Cell death in spinal cord injury (SCI) requires de novo protein synthesis. Calpain inhibitor E-64-d provides neuroprotection in SCI lesion and penumbra. *Annals of the New York Academy of Sciences, 939*, 436–449.

Resnick, D. K., Graham, S. H., Dixon, C. E., & Marion, D. W. (1998). Role of cyclooxygenase 2 in acute spinal cord injury. *Journal of Neurotrauma, 15*(12), 1005–1013. http://dx.doi.org/10.1089/neu.1998.15.1005.

Sadrzadeh, S. M., Anderson, D. K., Panter, S. S., Hallaway, P. E., & Eaton, J. W. (1987). Hemoglobin potentiates central nervous system damage. *Journal of Clinical Investigation, 79*(2), 662–664. http://dx.doi.org/10.1172/JCI112865.

Satake, K., Matsuyama, Y., Kamiya, M., Kawakami, H., Iwata, H., Adachi, K., & Kiuchi, K. (2000). Nitric oxide via macrophage iNOS induces apoptosis following traumatic spinal cord injury. *Brain Research. Molecular Brain Research, 85*(1–2), 114–122.

Sen, I., Joshi, D. C., Joshi, P. G., & Joshi, N. B. (2008). NMDA and non-NMDA receptor-mediated differential Ca2+ load and greater vulnerability of motor neurons in spinal cord cultures. *Neurochemistry International, 52*(1–2), 247–255. http://dx.doi.org/10.1016/j.neuint.2007.06.028.

Shields, D. C., Schaecher, K. E., Hogan, E. L., & Banik, N. L. (2000). Calpain activity and expression increased in activated glial and inflammatory cells in penumbra of spinal cord injury lesion. *Journal of Neuroscience Research, 61*(2), 146–150.

Simon, C. M., Sharif, S., Tan, R. P., & LaPlaca, M. C. (2009). Spinal cord contusion causes acute plasma membrane damage. *Journal of Neurotrauma, 26*(4), 563–574. http://dx.doi.org/10.1089/neu.2008.0523.

Tator, C. H., & Fehlings, M. G. (1991). Review of the secondary injury theory of acute spinal cord trauma with emphasis on vascular mechanisms. *Journal of Neurosurgery, 75*(1), 15–26. http://dx.doi.org/10.3171/jns.1991.75.1.0015.

Tempel, G. E., & Martin, H. F., 3rd. (1992). The beneficial effects of a thromboxane receptor antagonist on spinal cord perfusion following experimental cord injury. *Journal of the Neurological Sciences, 109*(2), 162–167.

Teng, S. X., & Molina, P. E. (2014). Acute alcohol intoxication prolongs neuroinflammation without exacerbating neurobehavioral dysfunction following mild traumatic brain injury. *Journal of Neurotrauma, 31*(4), 378–386. http://dx.doi.org/10.1089/neu.2013.3093.

Toborek, M., Malecki, A., Garrido, R., Mattson, M. P., Hennig, B., & Young, B. (1999). Arachidonic acid-induced oxidative injury to cultured spinal cord neurons. *Journal of Neurochemistry, 73*(2), 684–692.

Tomes, D. J., & Agrawal, S. K. (2002). Role of Na(+)-Ca(2+) exchanger after traumatic or hypoxic/ischemic injury to spinal cord white matter. *The Spine Journal, 2*(1), 35–40.

Urushitani, M., Nakamizo, T., Inoue, R., Sawada, H., Kihara, T., Honda, K., … Shimohama, S. (2001). N-methyl-D-aspartate receptor-mediated mitochondrial Ca(2+) overload in acute excitotoxic motor neuron death: A mechanism distinct from chronic neurotoxicity after Ca(2+) influx. *Journal of Neuroscience Research, 63*(5), 377–387.

Waller, P. F., Hill, E. M., Maio, R. F., & Blow, F. C. (2003). Alcohol effects on motor vehicle crash injury. *Alcoholism: Clinical and Experimental Research, 27*(4), 695–703. http://dx.doi.org/10.1097/01.ALC.0000062758.18918.7C.

Wang, X. J., Kong, K. M., Qi, W. L., Ye, W. L., & Song, P. S. (2005). Interleukin-1 beta induction of neuron apoptosis depends on p38 mitogen-activated protein kinase activity after spinal cord injury. *Acta Pharmacologica Sinica, 26*(8), 934–942. http://dx.doi.org/10.1111/j.1745-7254.2005.00152.x.

Wyndaele, M., & Wyndaele, J. J. (2006). Incidence, prevalence and epidemiology of spinal cord injury: What learns a worldwide literature survey? *Spinal Cord, 44*(9), 523–529. http://dx.doi.org/10.1038/sj.sc.3101893.

Xu, W., Chi, L., Xu, R., Ke, Y., Luo, C., Cai, J., … Liu, R. (2005). Increased production of reactive oxygen species contributes to motor neuron death in a compression mouse model of spinal cord injury. *Spinal Cord, 43*(4), 204–213. http://dx.doi.org/10.1038/sj.sc.3101674.

Xu, G. Y., Hughes, M. G., Ye, Z., Hulsebosch, C. E., & McAdoo, D. J. (2004). Concentrations of glutamate released following spinal cord injury kill oligodendrocytes in the spinal cord. *Experimental Neurology, 187*(2), 329–336. http://dx.doi.org/10.1016/j.expneurol.2004.01.029.

Xu, G. Y., Hughes, M. G., Zhang, L., Cain, L., & McAdoo, D. J. (2005). Administration of glutamate into the spinal cord at extracellular concentrations reached post-injury causes functional impairments. *Neuroscience Letters, 384*(3), 271–276. http://dx.doi.org/10.1016/j.neulet.2005.04.100.

Xu, G. Y., Liu, S., Hughes, M. G., & McAdoo, D. J. (2008). Glutamate-induced losses of oligodendrocytes and neurons and activation of caspase-3 in the rat spinal cord. *Neuroscience, 153*(4), 1034–1047. http://dx.doi.org/10.1016/j.neuroscience.2008.02.065.

Yang, L., Jones, N. R., Blumbergs, P. C., Van Den Heuvel, C., Moore, E. J., Manavis, J., … Ghabriel, M. N. (2005). Severity-dependent expression of pro-inflammatory cytokines in traumatic spinal cord injury in the rat. *Journal of Clinical Neuroscience, 12*(3), 276–284. http://dx.doi.org/10.1016/j.jocn.2004.06.011.

Yeo, J. D., Hales, J. R., Stabback, S., Bradley, S., Fawcett, A. A., & Kearns, R. (1984). Effects of a contusion injury on spinal cord blood flow in the sheep. *Spine (Phila Pa 1976), 9*(7), 676–680.

Yeo, J. D., Payne, W., Hinwood, B., & Kidman, A. D. (1975). The experimental contusion injury of the spinal cord in sheep. *Paraplegia, 12*(4), 279–298. http://dx.doi.org/10.1038/sc.1974.45.

Young, W. (1992). Role of calcium in central nervous system injuries. *Journal of Neurotrauma, 9*(Suppl. 1), S9–S25.

Yune, T. Y., Chang, M. J., Kim, S. J., Lee, Y. B., Shin, S. W., Rhim, H., … Oh, T. H. (2003). Increased production of tumor necrosis factor-alpha induces apoptosis after traumatic spinal cord injury in rats. *Journal of Neurotrauma, 20*(2), 207–219. http://dx.doi.org/10.1089/08977150360547116.

Zhang, J. R., Scherch, H. M., & Hall, E. D. (1996). Direct measurement of lipid hydroperoxides in iron-dependent spinal neuronal injury. *Journal of Neurochemistry, 66*(1), 355–361.

CHAPTER

6

Visual and Auditory Changes After Acute Alcohol Ingestion

M.K. Cavalcanti Galdino[1], *J.A. da Silva*[2], *N.A. dos Santos*[1]

[1]Federal University of Paraiba, João Pessoa, Brazil; [2]Federal University of Pelotas, Pelotas, Brazil

INTRODUCTION

Psychoactive pharmacological substances may cause structural, and functional changes in the entire organism. These alterations possibly tend to generate as consequent losses in many life aspects of individuals. Among these substances, ethanol (ethyl alcohol, or simply alcohol), one of the most commonly used substances, can negatively influence cognitive and behavioral manifestations, which are considered varied sensory. Research about the relationship between acute alcohol ingestion and visual and auditory changes have extensive applicability in issues related to everyday activities, such as orientation and mobility in transit. However, even though it is known that the sensorial–perceptual system is affected by acute alcohol ingestion, the impairment degree particularly of visual functions and the underlying neurobiological mechanisms involved are not well understood yet. In this sense, some studies have been developed in this direction with the purpose of investigating how alcohol interacts with the sensory mechanisms involved in the processing of basic visual and auditory patterns. In this chapter, the main studies and research results related to acute alcohol ingestion and its effects on visual perception and hearing developed at the Laboratory of Perception, Neuroscience, and Behavior—LPNeC/Brazil since mid-2000s are presented.

CHARACTERIZATION OF ACUTE USE OF ALCOHOL

Ethanol/alcohol is an organic chemical compound, used from its fermentation for production mainly of fuels and drinks. The preparation of each of these substances is by differentiated procedures (other more detailed information about the characterization and production of alcoholic beverages may be obtained in Cavalcanti, 2007; Silva, 2011). As a drink, alcohol has been used since 6000 BC in different contexts. The great social acceptance of its consumption may have been derived from these long years of use in several ways (Ferreira, 2010).

Although it is accepted and widely consumed, alcohol has a large number of implications for the society. The social costs linked to the use of alcoholic beverages are widely worrying, involving accidents such as automobile accidents and working accidents, hospital admissions in general clinics, and psychiatric disorders and increase in the number of episodes linked to violent behaviors. Furthermore, alcohol can facilitate the use of other psychotropic substances (Bresighello, 2005). These social consequences of the use of alcoholic beverages are due to the biochemical actuation of ethanol in the body (Planeta & Graeff, 2012). Psychic and neuropsychological consequences of alcohol can be found in the literature (Poltavski, Marino, Guido, Kulland, & Petros, 2011; Ray & Bates, 2006; Schreckenberger et al., 2004; Schweizer et al., 2006; Wetherill & Foroud, 2011).

The use of alcohol in its acute form, that is, from mild to moderate, and the frequency and amount used by people who say "social drinking," is considered by the World Health Organization (1994), as the limit alcoholic strength that causes no damage to health. However, according to Brick (2006) and Edwards, Marschall, and Cook (2005), the level of ingestion "moderate" does not have scientific or clinical–medical consonance. In addition, political and cultural issues are involved in the quantification of that threshold, which can vary from 8 to 20 g of pure ethanol per week (which on average corresponds to three daily doses of distilled beverages, two cans of beer or two

Addictive Substances and Neurological Disease
http://dx.doi.org/10.1016/B978-0-12-805373-7.00006-2

glasses of wine) (Andrade & Oliveira, 2009; Grinfeld, 2009; Klatsky, 2003).

The studies about the effects of moderate alcohol ingestion are still very controversial. Some indicate the onset of positive clinical consequences as the tensiolytic, orexigenic, coronary cerebrovascular, arterial effects, and protective agents against diabetes mellitus type II and dementias (Almeida & Barbosa Filho, 2006; Andrade & Oliveira, 2009; Clemente & Fonseca, 2002; Edwards et al., 2005; Glória, 2002; Klatsky, 2003; Meyer, Nicastri, Bordin, Nisenbaum, & Ribeiro, 2004; Planeta & Graeff, 2012). On the other hand, other studies propose occurring serious organic damage mainly concerning digestive systems (mainly liver damage), cardiac (arrhythmias and blood pressure increases), and neuropsychological impairments (nervous) (Clemente & Fonseca, 2002; Ferreira, 2010; Figueira, 2002; Glória, 2002).

All these physiological, general implications arise because beverage ingestion shows absorption and rapid distribution in all tissues and body fluids due to its easy transportation to the interior of the cell membranes that are highly permeable to ethanol (Yonamine, 2004). For any quantity of ethanol, about half is absorbed in 15 min, the maximum concentration in the blood is reached in 30 min, and the total absorption occurs in about 2 h. Variables such as the feeding or the practice of physical exercises before or during the process can delay these time values (Drummer & Odell, 2001; Leyton, Ponce, & Andreuccetti, 2009).

ALCOHOL AND VISUAL PERCEPTION

Even though alcohol causes changes in all sensory–perceptual systems of the body, it is possible to find more specific alterations, particularly related to vision. Important aspects of visual information processing such as contrast sensitivity, peripheral vision, visual acuity, and selective attention can present functional declines due to alcohol ingestion (Johnston & Timney, 2004). Although the neurophysiological studies demonstrate structural, functional, and visual changes resulting from alcohol consumption, showing evidences that alcohol has an effect potentially neurotoxic on the visual system, the results concerning the sensorial–perceptive interrelationship are still controversial.

In animals the view sense is proportionate to interaction of light with specialized receptors that are in the retina. The information encoded by visual systems runs in parallel routes of the retina to the thalamus and from there to cortex, in specialized areas in the processing of specific aspects of visual scene. These parallel routes allow the individual to perform the main visual sub-modalities, as the spatial location of bright stimuli, the identification of the objects shape, the measure of intensity, the color vision, among others. However, the identification of the form is a complex process that depends on the neurons that indicate the characteristics of the objects edges, especially the contrast and its spatial orientation, in addition to its three-dimensional characteristics. Thus, the discrimination forms allow us to differentiate and recognize the objects according to their contours (Lent, 2001).

In general, we realize the contours when there is light contrast that allows the visual system to perform distinctions of contours through neural interactions. The contrast corresponds to differences in brightness between adjacent areas (Campbell & Maffei, 1974). Other optical and psychophysical factors, such as the size of the image projected in the retina, the stimuli configuration, and the psychological aspects (e.g., motivation, attention, and emotion), also play an important role in the perception of the details, but the contrast is a necessary condition in the perception of any object, because without it a visual field would be simply a uniform luminance distribution for the observer. We also know that the objects require different levels of contrast to be detected (e.g., the perception of small objects and with details need more contrast), thus the contrast necessary to detect objects vary, as well as their intensity (Caelli, 1981).

In the studies related to vision, the contrast sensitivity function (CSF) has been used as the main tool to access the vision sensorial processing. The contrast sensitivity is an important attribute of the visual system, recognized as the minimum amount of required contrast for the human visual system to detect a stimulus or a spatial frequency. This is a type of sensory threshold. The CSF is defined as the inverse of the contrast threshold and can evaluate the ability of human space vision to detect and process stimuli in different sizes and contrasts levels (Campbell & Maffei, 1974; Colombo, Issolio, Santillán, & Aguirre, 2009). CSF is also an indicator of the functioning of hypothetical channels related to the process of analysis and visual scenes synthesis.

The contrast spectrum sensitivity curve presents a well-defined format, with the sensitivity peak in the medium frequencies and reduced sensitivity to values below or above this range. Therefore, changes in the aspect of the curve may be indicative of alterations in the mechanisms or sensory routes that handle contrast. Such changes can be due to internal or external insults (Fig. 6.1).

Usually, in studies on contrast visual perception patterns of elementary spatial frequencies for measurement are used. Such standards are described as simple sinusoidal modulations or luminance cosenoidal per space unit. They may also be described in terms of amplitude of contrast and its spatial elementary frequency, the latter being commonly called in the literature only by "spatial frequency." The spatial frequency is given by

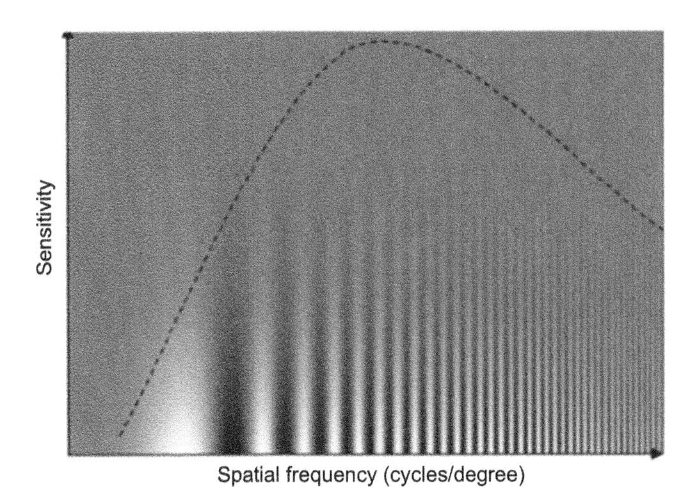

FIGURE 6.1 The sensitivity curve to the contrast. X-axis in the spatial frequencies and Y-axis the sensitivity.

the number of cycles (light and dark streaks) per unit of space, usually defined in cycles per degree of visual angle (cpd). The quantity of cycles specifies the spatial frequency of a wave. Now the contrast can be defined in relation to maximum and the minimum luminance (wave peaks and depressions, respectively), and it is illustrated mathematically by the Michelson equation (De Valois & De Valois, 1988): $C = L_{max} - L_{min}/L_{max} + L_{min}$.

The Campbell and Robson model (1968) proposes that the human visual system (HVS) is composed of a series of routes or tuned channels selectively for narrow bands of spatial frequencies. Thus, in the perception of complex visual scenes the HVS decomposes the stimuli in its most basic components of spatial frequencies, making a sort of Fourier analysis, which begins in the retina. These signals are then transmitted by specialized routes until the cortex, where synthesis, interpretation, representation, and knowledge of the objects occur. A more extensive discussion about the multiple channels model and analysis of linear systems on visual perception are given in De Valois and De Valois (1988).

The theoretical model of multiple channels assumes that the CSF is the sensitivity envelope for the total series of channels, each of them sensitive to a restricted and discreet range of the spatial frequency spectrum (Santos & Simas, 2001). The CSF allows characterizing the visual system response to spatial patterns in low, medium, and high levels of contrast level. In this sense, the CSF is a classic tool to describe the behavior of the visual system and diagnose changes resulting from abnormalities in information processing and the location of the damage in routes or areas of the nervous system during the development, in neuropsychiatric pathologies in the use or abuse of licit or illicit substances.

Some studies describe changes in cerebral blood flow (CBF) related to acute and chronic alcohol consumption (Andre, 1996; Pearson & Timney, 1998, 1999; Roquelaure et al., 1995). Roquelaure et al. (1995) measured the CSF for 30 alcoholic patients and 52 healthy volunteers, using spatial frequency of 0.1, 0.3, 0.5, 1.0, 2.6, and 6.0 cpd. Their results demonstrate a reduction in all spatial frequency of experimental group. Pearson and Timney (1998) measured the CSF to spatial frequency of 0.75, 1.5, 3.0, 5.0, 7.5, and 10 cpd at a level of blood alcohol limit of 0.077% blood alcohol concentration (BAC), in six volunteers, and observed that alcohol decreased sensitivity in the high spatial frequency. On the other hand, Andre, Tyrrell, Leibowitz, Nicholson, and Wang (1994) found no changes in the sensitivity to contrast with the use of static stimuli in 0.08% BAC levels. Results were similar to those of Quintyn, Massy, and Brausser (1999). In general terms, discrepancies between results may be arising from different methodological aspects, such as luminance level, merger alcoholic beverages, and stimuli class employed in each search.

Most of the studies that relate the CSF and alcohol intake were performed with spatial patterns of sinusoidal vertical type. It was not found until the moment studies using spatial patterns of sinusoidal concentric grid type (elementary visual stimuli of radial sinusoidal static grid type and achromatic) and elementary stimuli in polar coordinates (angular frequencies). In this sense, the use of concentric and angular spatial frequencies in studies of this nature allows an assessment of the effect of moderate alcohol ingestion in cerebral areas that had not been assessed by the CSF yet. This hypothesis can be enhanced by psychophysical (Cavalcanti-Galdino, Silva, Mendes, Santos, & Simas, 2014; Santos & Simas, 2002; Simas, Santos, & Thiers, 1997; Wilson & Wilkinson, 1998; Wilson, Wilkinson, & Asaad, 1997) and neurophysiological (Gallant, Connor, Rakshit, Lewis, & Van Essen, 1996; Heywood, Gadotti, & Cowey, 1992; Merigan, 1996; Wilkinson et al., 2000) studies that suggest that vertical sinusoidal grids, concentric and angular, are processed by distinct mechanisms and visual areas. While the stimuli of type vertical sinusoidal grid are processed by visual area V1, the frequencies concentric sinusoidal waves are processed by concentric cortical visual areas V2, V4, and TI (Gallant et al., 1996; Merigan, 1996; Wilkinson et al., 2000) and the angular frequencies are processed by V4 and by the inferotemporal cortex (Gallant et al., 1996; Simas et al., 1997) (Fig. 6.2).

ALCOHOL AND AUDITORY PERCEPTION

When compared to the number of studies directed to other systems such as visual, it is deemed to be still insufficient research related to the auditory system.

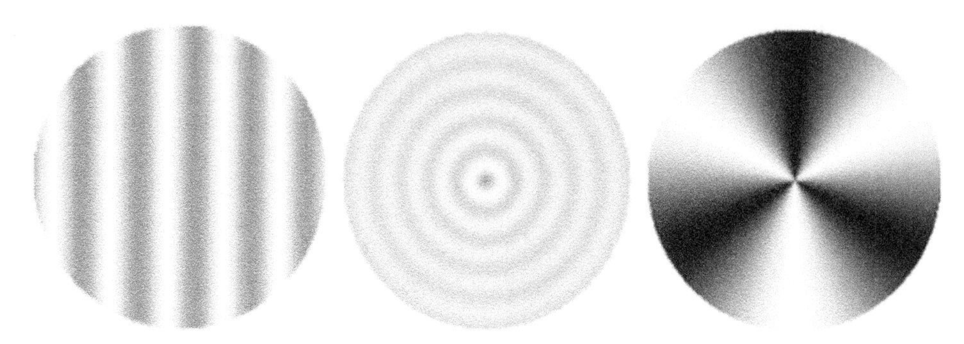

FIGURE 6.2 Examples of spatial frequency stimuli. From left to right spatial frequency sinusoidal grating 2.0 cpd, sinusoidal radial grating 2.0 cpd, and angular frequency 3 cycles/360 degrees.

The complexity and structural and functional detail of the auditory system has left it in second place by researchers, hindering the achievement of other more extensive investigations about it. The auditory perception is the final result of a process that starts with the vibration of the objects in the environment by changing the atmospheric pressure of the air and has its end with the referred cortical modulation that occurred in the encephalon (Costamilan, 2004; Mendes, 2008; Moore, 2007; Puggina, Da Silva, Gatti, Graziano, & Kimura, 2005; Samelli & Schochat, 2008).

In general, the sound enters into auditory canal through the external ear, traversing the tympanic membrane, which enables the movement of three of the middle ear ossicles (hammer, anvil, and stirrup), balancing the air input sound. These ossicles are connected to a membrane called oval window that covers a small hole of the skull, and is connected to the cochlea, in the inner ear, where it is done the first separation and interpretation of beeps (Bear, Connors, & Paradiso, 2002; Moore, 2007; Rui & Steffani, 2007).

The cochlea is an elliptical structure that has in its interior a liquid called lymph that moves by external vibrations causing action potentials in ciliated cells: kinocilia (longer) and stereocilia (shorter). When the lymph oscillates toward the kinocilia, the cells are depolarized causing the neuronal excitement increase. Now, when the oscillation happens toward the stereocilia they are hyperpolarized, causing neuronal inhibition (Bear et al., 2008; Vollrath, Kwan, & Corey, 2007).

This processed information pervades through the vestibulocochlear nerve by a series of structures until they arrive in the cortex region, which is organized tonotopically. This means that these areas have proportionate representations, such as maps and locations of specific neurons for the interpretation of the different sound frequencies (Bilecen, Scheffler, Schimid, Tschopp, & Seelig, 1998; Formisano et al., 2003; Joris, Schreiner, & Rees, 2004; Langers, Backes, & Van Dijk, 2007; Machado, 2005; Schönwiesner, von Cramon, & Rübsamen, 2002).

The cells of the thickest region of the cochlea, called basis, are driven by high-frequency sounds, while in the thinnest region of this structure, called the apex, are driven by low-frequency sounds that activate the nerve cells (Dallas, 1992; Gold, 1948; Robles & Ruggero, 2001). Other more specific information in terms of structures and the hearing processing can be obtained in Silva (2011).

The understanding of the sound stimuli, from external uptake to the more central processing in the brain has been an object of difficult control, which arose contradictions in the results of the studies found, whether they are related to the most common attributes of hearing. However, the auditory processing is an aspect of great importance for the individual's interaction with their environment, requiring greater attention.

Few studies are still found when it comes to the alcohol effects on the auditory processing. Despite this, investigations about alcohol effects on the inhibition mechanisms in the auditory system have shown to be important for the understanding of neural and behavioral hearing. The physiological mechanisms may have an inhibitory effect on afferent cells. The stimulation of these routes has indicated that these inhibitory connections of afferent mechanisms are responsible for a reduction of the excitation that is equivalent to the stimuli intensity reduction (Fex, 1962).

In addition, to reach a minimum level of neural activity in a given period of time, the relative amplitude of the neural responses is dependent on the sound frequency (Fex, 1962). Therefore, direct stimulations of specific parts of the auditory nervous system can result in lower responses of the auditory cortex after alcohol consumption (Teo & Ferguson, 1986). Due to this, studies have been conducted to assess the alcohol effects on the hearing.

Specific sound frequencies are listed individually to specific different tones that are sound attributes corresponding to direct musical notes. For this reason, discrimination tasks of musical notes are used to evaluate the perception of sound frequencies (Lecanuet,

Graniere-Deferre, Jacquet, & DeCasper, 2000; Silva et al., 2014; Silva, Cavalcanti-Galdino, Gadelha, Andrade, & Santos, 2013; Silva, Cavalcanti-Galdino, Simas, & Santos, 2015).

The physical and chemical operation of whole hearing can be altered during alcohol ingestion. Studies show neurophysiological, psychophysical, cognitive, and behavioral hearing changes due to alcohol interference (He et al., 2013; Kähkönen, Rossi, & Yamashita, 2005; Maurage et al., 2009; Maurage, Campanella, Philippot, Pham, &Joassin, 2007; Meerton, Andrews, Upile, Drenovak, & Graham, 2005; Vanneste & De Ridder, 2012). In the same way that the studies have been carried out regarding alcohol interference on visual perception, the hearing sensory-perceptual aspect studies have still been contradictory. The differences in results found by several researchers to alcohol ingestion experiments can generally be explained, to some extent, by different forms that are made with the control and measurement of alcoholic concentration (Capenter, 1962; Eggleton, 1941; Himwich, 1957; Hogan & Linfield, 1983; Levine, Kramer, & Levine, 1975; Wallgren & Barry, 1970).

Even so, some studies have shown that hearing can be impaired, for example, when it comes to the distinction among sounds of different frequencies (Bellé, Sartori, & Rossi, 2007; He et al., 2013; Pearson, Dawe, & Timney, 1999; Rossi, Bellé, & Sartori, 2010). Studies with auditory evoked potential report on hearing electrophysiological alterations, for example, Kähkönen et al. (2005) suggested the occurring damage in auditory processing of new sounds and frequency changes combining magnetoencephalography (MEG) and electroencephalography (EEG). They assessed 11 right-handed individuals through a double-blind placebo-controlled experiment (0.8 Eg/kg of ethanol or juice). According to the authors, the alcohol damages the tone processing, the frequency shifts, and the perception of new sounds originated from standard sounds in different stages of the auditory processing, similarly in both hemispheres.

In the same way, He et al. (2013) evaluated the alcohol effects on the hearing preattentive processing in four sound characteristics: frequency, intensity, location, and duration. Twelve participants were observed under the condition of alcohol ingestion (0.65 g/kg) compared to the condition without alcohol. The amplitudes of all four types of MMN wave significantly decreased after the alcohol ingestion.

Pihkanen and Kauko (1962) also analyzed the consequences of alcohol for the auditory discrimination of sound frequencies. They used three people with visual disabilities who were piano tuners. Two of the three participants showed a statistically significant decrease of their capacity to discriminate frequencies. In the same way, in studies by Pearson (1997) and Pearson et al. (1999) have investigated the effects of alcohol on hearing. They have tested the identification threshold of sound frequencies. In both studies, the alcohol ingestion to 0.08% BAC only negatively changed the perception of higher frequencies (above 1000 Hz).

RESEARCH CARRIED OUT BY THE LABORATORY OF RESEARCH IN PERCEPTION, NEUROSCIENCE, AND BEHAVIOR—LPNeC

Knowing that the intoxication by alcohol promotes physiological changes in the central nervous system (CNS), which result in sensory–motor perceptive, cognitive, and behavioral changes (Draski & Deitrich, 1996; Curtin & Fairchild, 2003; Watten & Lie, 1996), we developed, in Brazil, a series of studies about the effects on vision and hearing. We relied on the studies that showed that the alcohol effects on the CNS depend on the dose. This means that the changes are individual and subject to a variety of factors, such as absorption rates and elimination of alcohol in the body, anthropometric and physiological characteristics, hormonal problems, and hereditary, beyond the characteristics related to the type and quantity of beverage (Brick, 2006). It was manipulated the acute ingestion in university students, comparing with themselves in two different conditions. In one of the conditions, the participants were asked to drink a quantity of alcohol proportional to their body weight for which at the time of the tests they would have ethanol content in blood. In another, they only drank a placebo drink. In all studies the design was experimental, with repeated measures.

One of the first studies had an objective to characterize the sensitivity function to the contrast (i.e., CSF) for spatial frequency of 0.25, 1.0, and 4.0 cpd in the absence (Control Group—GC) and after the moderate alcohol consumption (Experimental Group—GE) for women. For that, it was used the psychophysical choice method forced by four women from 21 to 30 years, with normal or corrected visual acuity. The experimental session was initiated when the alcohol content reached the value of 0.06% BAC. The results showed significant differences between the groups in the frequency of 4.0 cpd, being the GE more sensitive to contrast than in the CG. These results suggest changes in CBF related to moderate alcohol consumption by women (Cavalcanti & Santos, 2008).

Then, a study was held to characterize the CSF to stimuli of radial sinusoidal grid frequencies of 0.25; 2.0, and 8.0 cpd in adult females without the alcohol ingestion and after moderate alcohol consumption—0.09% BAC.

Five volunteers, with age ranging from 21 to 30 years, and with normal or corrected visual acuity participated in the experiments. Experimental design was used in addition to repeated measures and the psychophysical method of forced choice. The results showed that the five participants presented injury in all frequencies tested, mainly on frequency 2 cpd ($p < .001$), indicating that the change was greater in channels that process the medium frequencies (Cavalcanti-Galdino, Mendes, Vieira, Simas, & Santos, 2011).

We measured the CSF for elementary stimuli spatial frequencies in Cartesian coordinates of 0.25, 1.25, 2.5, 4, 10, and 20 cpd, and polar coordinates 1, 2, 4, 24, 48, and 96 cycles/360 degrees in 20 volunteers with normal or corrected vision (18–31 years old, mean (M) = 21.75, standard deviation (SD) = 2.95, 10 females and 10 males). All participants were tested for visual acuity with Rasquin's chart of optotypes "E," and Ishihara's color test.

All stimuli were static, generated in grayscale, circularly symmetric, with a diameter of about 7 degrees of visual angle and presented at 300 cm distance between the monitor and the volunteer. Other details about these procedures can be found at Santos and Simas (2001, 2002). The average alcohol concentration in the experimental group was set to about 0.08%. Significant changes in contrast sensitivity were observed after alcohol intake compared with the control condition at spatial frequency of 4 cpd and 1, 24, and 48 cycles/360 degrees for angular frequency stimuli. Alcohol intake seems to affect the processing of sine wave gratings at maximum sensitivity and at the low and high frequency ends for angular frequency stimuli, both under photopic luminance conditions (Cavalcanti-Galdino et al., 2014).

In all the studies that were performed, changes were found in frequencies. The alterations were specifically reductions in contrast sensitivity in the high frequencies and increase in the low frequencies in experimental groups, in comparison with control groups. Some hypotheses may explain the results found. Virsu, Kyykka, and Vahvelainen (1974) reported the existence of a reduction of the mechanisms of neural interaction inhibitory effects after the alcohol ingestion. In the visual system, one of the main inhibitory mechanisms is the lateral inhibition process. This phenomenon implies that neighboring detectors inhibit its activities to each other (Sagi & Hochstein, 1985). In summary, when a receptor cell is strongly enabled, it causes inhibition of the response of the cells that are adjacent to receptive fields, through horizontal interconnections. This effect in the spatial frequencies dimension operates in high spatial frequencies, and thus amplifies the images contours perception.

Physiological studies in *Limulus* species (horseshoe crab) (MacNichol & Benolken, 1956; Negishi & Svaetichin, 1966) and in frogs (Backstrom, 1977) showed that alcohol reduces or completely extinguishes the lateral inhibition at the level of the retina. Psychophysical studies in humans also point in that direction (Johnston &Timney, 2004). Thus, the reduction of mechanism activity of lateral inhibition promotes sensitivity reduction at the high spatial frequencies and increase in low frequencies. This hypothesis corroborates the results found in this study.

The inhibitory mechanisms also relate to photoreceptors differently. Pearson and Timney (1999) affirm that the cones system seems to be more influenced by alcohol than the rods system, in which the inhibitory mechanisms have a less important role. In this sense, the photopic luminance level used in all studies also influenced the results presented. The condition of high luminance level (41.05 cd/m^2) possibly provides maximum cone system sensitivity, and is indicative of the responsiveness of the parvocellular route, specialized in the processing of the fine details of the objects, that is, of the mid- and high frequencies (Benedek, Benedek, Kéri, & Janáky, 2003). In this sense, the alterations found corroborate with the results of Pearson and Timney (1999) who propose that alcohol selectively acts on the system of the cones and on parvocellular route.

The results of the studies carried out with the CSF and moderate alcohol consumption point to the discussion that the visual system sensitivity to certain stimuli depends on the spatial, physical, mathematical model, and coordinate system (Cartesian or polar) that defines them. Taking into account these considerations, and the results and literature data, we can infer that the visual system uses different mechanisms or pathways for processing sine wave gratings, sinusoidal, radial, and angular frequency stimuli and that alcohol possibly affects all systems. These results involving the CSF for radial and angular stimuli after alcohol ingestion are initial pioneers in literature, differing from the results with sinusoidal grid, already widely disseminated. Although preestablished parameters are missing on the radial and angular frequencies CSF after the alcohol ingestion, data can be revealing.

These initial studies have stimulated us in magnification of sensory–perceptual data about the hearing, knowing that the ethanol, present in alcoholic beverages, ingested in its acute form, at its various levels, can also lead to a series of organic changes that can interfere in basic auditory processes. In this sense, this research was aimed at a more basic study to evaluate the influence of acute alcohol ingestion in auditory perception. The auditory perception of 40 university students was evaluated; volunteers comprised of males and females, from 18 to 30 years of age, nonmusicians, healthy, nondrug (or other toxic substance) users, except alcohol in a moderate way and acute, through discrimination test of noise frequencies corresponding to musical notes of a standard western scale.

The objective of this study was to assess whether there is a relationship between the moderate alcohol consumption and the discrimination of musical notes D, F, and A in young adults. These notes present fundamental frequencies, respectively, of 349.228, 293.665, and 440 Hz, of equal temperament, chosen because they are in central location in standard western musical scale. A software called "Psysounds," created specifically for this study, was used in which the volunteers had to choose between two musical notes, one of the notes test presented initially. The analysis showed significant differences in the frequencies of correct responses for all notes used, suggesting that the alcohol ingestion impairs musical notes discrimination of D, F, and A. The results showed that the alcohol ingestion causes perceptive damage in hearing, suggesting that the use of alcohol can be a marker of certain perceptual auditory weaknesses (Silva et al., 2015).

Another study, for complementation of these results was held for investigation of the influence of acute alcohol ingestion in auditory perception. The experiment aimed to verify the effects of acute alcohol ingestion for sound frequencies corresponding to all the musical notes of a standard western scale: C (+262 Hz), D (+294 Hz), E (+330 Hz), F (+349 Hz), G (392 Hz), A (440 Hz), and B (+491 Hz), fourth in the octave of the piano. The results obtained have confirmed the initial hypothesis that individuals answer after the alcohol ingestion would be negatively affected and that in this way, the participants would make a mistake more times in the discrimination process (identification and comparison) between the musical notes. Thus, the results suggest that there are changes in the auditory perception for musical notes as a consequence to the use of alcoholic drinks.

This study has detected that the participants are differentiated in their responses before and after the alcohol ingestion in all musical notes, being that after the use, they have committed a larger quantity of mistakes. In addition, women and men without alcohol ingestion also responded differently to all notes, with the exception of musical notes D and A, being that most of the times the men discriminated better than women. In the same way, after the alcohol ingestion, women and men are differentiated in all musical notes with the exception of F, following pattern similar to the placebo ingestion with more correct answers for the men (Silva, 2015).

ALCOHOL IN NEURAL SYSTEMS OF VISION AND HEARING

In spite of the alterations related to the alcohol consumption being idiosyncratic, generally, alcohol is a CNS depressant and interacts with the main neurotransmitters systems such as gamma-aminobutyric acid (GABA), glutamate, dopamine, serotonin, and endogenous opioids. It is difficult to assign to the neurotransmitters or specific neuromodulators any changes of visual and auditory systems after alcohol ingestion.

However, surveys show that alcohol consumption promotes a potential decrease in GABA activity, the main inhibitory neurotransmitter in the brain, present in bipolar and ganglion cells of the retina, in lateral geniculate nucleus (NGL), and in the visual cortex (Mccormick, 1989; Xiao & Ye, 2008). This reduction in the GABA activity can be one of the causes of changes in visual functions (Ogawa, Kato, & Ito, 1986).

The literature shows that many visual functions are changed after the alcohol consumption (Andre, 1996; Nicholson et al., 1995; Pearson & Timney, 1998). However, in spite of important behavioral and physiological findings, an explanatory model has not been built yet of the physiological action mechanisms of alcohol on the sensorial system (Pearson & Timney, 1998). The reduction in the GABA activity, previously reported, is one of the possible explanations for the alcohol effects on the visual functions. On the other hand, there are other hypotheses of sensory mechanisms involved and affected by alcohol.

Pearson (1997) and Pearson and Timney (1998) point out the injury in visual functions as a secondary consequence of the reduction of the oculo-motor control, which occurs after alcohol ingestion. This effect is well documented in the literature (Hill & Toffolon, 1990; Nicholson et al., 1995). The hypothesis of Virsu et al. (1974) is that alcohol acts preferentially in neural mechanisms and results in inhibitory effects, that is, affecting the lateral inhibition, a process of important function in the visual details detection. Now Hill and Toffolon (1990) propose that alcohol promotes different changes on visuals routes, with a higher injury within the magnocellular route.

Among the changes in visual perception associated with the use of alcohol, that stand out: (1) neurodegenerative alteration disorders in visual routes and retina layers integrated to functional decline of the cones and rods (Lima et al., 2005); (2) changes in cortical receptive fields of individual cells and selective hypercomplex spatial visual orientation (Medina, Krahe, & Ramoa, 2006); and (3) changes in visual processing of chromatic and achromatic stimuli (Castro, Rodrigues, Côrtes, & Silveira, 2009; Chen, Xia, Li, & Zhou, 2010; Rosenbloom, Pfefferbaum, & Sullivan, 2004; Wegner, Günthner, & Fahle, 2001).

Our results pointed to different alcohol effects in the stimuli used. As already reported the sinusoidal and radial spatial frequencies stimuli and the angular frequency are different and cannot be compared directly. However, considering that these stimuli are processed by different visual areas, the results suggest that alcohol alters the brain in a diffuse way; that is, it commits primary and superior visual areas.

In the same way, the results found that the hearing should be evaluated in terms of the neurobiological plausible mechanisms through which alcohol may act in the organism. Data obtained in this direction may not be fully accountable for possible auditory alterations found on the basis of the alcohol actions only on a single mechanism. It seems that a more complete explanation of the alcohol effects on the sensorial processing requires a model that includes more of a physiological mechanism and consequently behavioral therapy for the various individuals. Direct stimulations of medial geniculate nucleus (MGN) and auditory nerves, for example, result in responses of the auditory cortex diminished after the use of ethanol (Teo & Ferguson, 1986).

Investigations in relation to the alcohol effects on the inhibition mechanisms in the auditory system have shown to be important. The physiological mechanisms may have an inhibitory effect on afferent cells. The stimulation of these routes has indicated that these inhibitory connections of afferent mechanisms are responsible for a reduction of the excitation that is equivalent to the reduction of the stimuli intensity (Fex, 1962).

Thus, it can be inferred that the processing of the more serious sound information, as found in the lower frequencies, performed at the base of the cochlea do not change much compared to the processing of higher frequencies, since the stimuli lose the force under the effect of ethanol ingestion. Thus, it is understandable that the changes usually found in the studies were due to the use of ethanol in all musical notes that have lower frequencies.

The findings presented and discussed in this study suggest the importance to continue investigating the effects of moderate alcohol ingestion in visual and auditory processing. They also point toward the need to relate the basic research with applied research, approximating the theoretical and experimental results to social reality and to public health.

It is in this perspective that we continue with the investigations. In this sense, the future intentions are (1) to evaluate the contrast sensitivity associated with Eye-Router, a tool for routing ocular movements; (2) to include a third condition between the groups, to evaluate experimental condition, control and placebo, and (3) to compare existing results with acute alcohol ingestion to the studies that have been carried out in our laboratory with alcoholics.

References

Almeida, R. N., & Barbosa Filho, J. M. (2006). Drogas Psicotrópicas [Psychotropic drugs]. In R. N. Almeida (Ed.), *Psicofarmacologia: Fundamentos práticos [Psychopharmacology: Practical fundamentals]*. Rio de Janeiro, Brazil: Guanabara Koogan.

Andrade, A. G., & Oliveira, L. G. (2009). Principais consequências em longo prazo relacionadas ao consumo moderado de álcool [Major long-term consequences related to moderate consumption of alcohol]. In A. G. Andrade, J. C. Anthony, & C. M. Silveira (Eds.), *Álcool e suas consequências: Uma abordagem multiconceitual [Alcohol and its consequences: A multiconceitual approach]*. Barueri, São Paulo, Brazil: Minha Editora.

Andre, J. T. (1996). Visual functioning in challenging conditions: Effects of alcohol consumption, luminance, stimulus motion, and glare on contrast sensitivity. *Journal of Experimental Psychology, 2*, 250–269.

Andre, J. T., Tyrrell, R. A., Leibowitz, H. W., Nicholson, M. E., & Wang, M. (1994). Measuring and predicting the effects of alcohol consumption on contrast sensitivity for stationary and moving gratings. *Perception & Psychophysics, 53*, 261–267.

Backstrom, A. C. (1997). Effects of alcohol on ganglion cell receptive field properties and sensitivity in frog retina. *Advances in Experimental Medicine and Biology, 85*, 187–208.

Bear, M. F., Connors, B. W., & Paradiso, M. A. (2002). Os sistemas auditivo e vestibular [Auditory and vestibular systems]. In M. F. Bear, B. W. Connors, & M. A. Paradiso (Eds.), *Neurociências: Desvendando o sistema nervoso [Neuroscience: Exploring the brain]* (2nd ed.). Porto Alegre, Brazil: Artmed.

Bellé, M., Sartori, A. S., & Rossi, A. G. (2007). Alcoolismo: Efeitos no aparelho vestíbulo-coclear [Alcoholism: Effects on vestibulocochlear apparatus]. *Revista Brasileira de Otorrinolaringologia, 73*, 116–122.

Benedek, G., Benedek, K., Kéri, S., & Janáky, M. (2003). The scotopic low-frequency spatial contrast sensitivity develops in children between the ages of 5 and 14 years. *Neuroscience Letters, 345*, 161–164.

Bilecen, D., Scheffler, K., Schmid, N., Tschopp, K., & Seelig, J. (1998). Tonotopic organization of the human auditory cortex as detected by BOLD-FMRI. *Hearing Research, 126*, 19–27.

Bresighello, M. L. M. (2005). *Jovensuniversitários e álcool: Conhecimentos e atitudes [University students and alcohol: Knowledge and attitudes]* (Dissertação de Mestrado) (Master's thesis). São Carlos, São Paulo, Brazil: Universidade Federal de São Carlos [Federal University of São Carlos].

Brick, J. (2006). Standardization of alcohol calculations in research. *Alcoholism: Clinical and Experimental Research, 30*, 1276–1287.

Caelli, T. (1981). *Visual perception: Theory and practice*. Oxford: Pergamon.

Campbell, F. W., & Maffei, L. (1974). Contrast and spatial frequency. *Scientific American, 231*, 106–114.

Campbell, F. W., & Robson, F. G. (1968). Application of the Fourier analysis to the visibility of gratings. *The Journal of Physiology, 197*, 551–566.

Capenter, J. A. (1962). Effects of alcohol on some psychological processes: A critical review with special reference to automobile driving skill. *Quarterly Journal of Studies on Alcohol, 23*, 274–314.

Castro, A. J. L., Rodrigues, A. R., Côrtes, M. I. T., & Silveira, L. C. L. (2009). Impairment of color spatial vision in chronic alcoholism measured by psychophysical methods. *Psychology & Neuroscience, 2*, 179–187.

Cavalcanti, M. K. (2007). *Efeito da ingestão de etanol na percepção visual da forma em adultos [Ethanol ingestion effect on visual perception of form in adults]* (Dissertação de Mestrado) (Master's thesis). João Pessoa, Paraíba, Brazil: Universidade Federal da Paraíba [Federal University of Paraíba].

Cavalcanti-Galdino, M. K., Mendes, L. C., Vieira, J. G., Simas, M. L. B., & Santos, N. A. (2011). Percepção visual de grade senoidal radial após o consumo de álcool [Visual perception of radial sinusoidal grid after consuming alcohol]. *Psicologia USP [USP Psychology], 22*, 99–115.

Cavalcanti-Galdino, M. K., Silva, J. A., Mendes, L. C., Santos, N. A., & Simas, M. L. B. (2014). Acute effect of alcohol intake on sine-wave Cartesian and polar contrast sensitivity functions. *Brazilian Journal of Medical and Biological Research, 47*, 321–327.

Cavalcanti, M. K., & Santos, N. A. (2008). Alterações na sensibilidade ao contraste relacionadas à ingestão de álcool [Changes in the contrast sensitivity related to the alcohol ingestion]. *Psicologia: Teoria e Pesquisa [Psychology: Theory and Research], 24*, 515–518.

Chen, B., Xia, J., Li, G., & Zhou, Y. (2010). The effects of acute alcohol exposure on the response properties of neurons in visual cortex area 17 of cats. *Toxicology and Applied Pharmacology, 243*, 348–358.

Clemente, S., & Fonseca, T. (2002). Qual o benefício e toxicidade do álcool no sistema cardiovascular? [What benefit and toxicity of alcohol on the cardiovascular system?] *Revista da Faculdade de Medicina de Lisboa [Journal of the Faculty of Medicine of Lisbon], 7*, 179–183.

Colombo, E., Issolio, L., Santillán, J., & Aguirre, R. (2009). What characteristics a clinical CSF system has to have? *Optica Applicata, 39*, 415–428.

Costamilan, C. M. (2004). *Processamento auditivo em escolares: Um estudo longitudinal* [*Auditory processing in school children: A longitudinal study*] (Dissertação de Mestrado) (Master's thesis). Santa Maria, Rio Grande do Sul, Brazil: Universidade Federal de Santa Maria [Federal University of Santa Maria].

Curtin, J. J., & Fairchild, B. A. (2003). Alcohol and cognitive control: Implications for regulation of behavior during response conflict. *Journal of Abnormal and Social Psychology, 112*, 424–436.

Dallas, P. (1992). The active cochlea. *The Journal of Neuroscience, 12*, 4575–4585.

De Valois, R. L., & De Valois, K. K. (1988). *Spatial vision*. New York: Oxford University Press.

Draski, L. J., & Deitrich, R. A. (1996). Initial effects of ethanol on the central nervous system. In R. A. Deitrich, & V. G. Erwin (Eds.), *Pharmachological effects of ethanol on the nervous system*. Boca Raton, FL: CRC Press.

Drummer, O. H., & Odell, M. (2001). Pharmacokinetics, metabolism and duration of action. In O. H. Drummer, & M. Odell (Eds.), *The forensic pharmacology of drugs of abuse*. London: Arnold.

Edwards, G., Marschall, E. J., & Cook, C. C. H. (2005). *O tratamento do alcoolismo: Um guia para profissionais de saúde* [*The treatment of alcoholism: A guide for health professional*] (4th ed.). Porto Alegre, Brazil: Artmed.

Eggleton, M. G. (1941). The effect of alcohol on the central nervous system. *British Journal of Psychology, 32*, 52–61.

Ferreira, R. (2010). *Alterações no processamento da informação sensorial auditiva induzidas pela abstinência ao álcool em ratos: Importância dos mecanismos GABAérgicos e glutamatérgicos do colículo inferior* [*Changes in the processing of auditory sensory information induced by withdrawal of alcohol in rats: Importance of GABAergic and glutamatergic mechanisms of the inferior colliculus*] (Dissertação de Mestrado) (Master's thesis). Ribeirão Preto, São Paulo, Brazil: Universidade de São Paulo [University of São Paulo].

Fex, J. (1962). Auditory activity in centrifugal and centripetal cochlear fibres in cat. *Acta Physiologica Scandinavica, 55*, 1–68.

Figueira, I. (2002). Etanol e bebidas alcoólicas: Pode a atividade farmacológica do álcool explicar a diversidade de efeitos nos diferentes sistemas? [Ethanol and alcoholic beverages: Can the pharmacological activity of alcohol explain the diversity of effects on different systems?] *Revista da Faculdade de Medicina de Lisboa [Journal of the Faculty of Medicine of Lisbon], 7*, 165–171.

Formisano, E., Kim, D. S., Di Salle, F., van de Moortele, P. F., Ugurbil, K., & Goebel, R. (2003). Mirror-symmetric tonotopic maps in human primary auditory cortex. *Neuron, 40*, 859–869.

Gallant, J. L., Connor, C. E., Rakshit, S., Lewis, J. W., & Van Essen, D. C. (1996). Neural responses to polar, hyperbolic, and Cartesian gratings in area V4 of the macaque monkey. *Journal of Neurophysiology, 76*, 2718–2739.

Glória, H. (2002). A eupepsia é o único efeito benéfico gastrointestinal do álcool? [The eupepsia is the only gastrointestinal beneficial effect of alcohol?] *Revista da Faculdade de Medicina de Lisboa [Journal of the Faculty of Medicine of Lisbon], 7*, 173–178.

Gold, T. (1948). Hearing. II. The physical basis of the action of the cochlea. *Proceedings of the Royal Society of London. Series B. Biological Sciences, 135*, 492–498.

Grinfeld, H. (2009). Consumo nocivo de álcool durante a gravidez [Harmful alcohol consumption during pregnancy]. In A. G. Andrade, J. C. Anthony, & C. M. Silveira (Eds.), *Álcool e suas consequências: Uma abordagem multiconceitual [Alcohol and its consequences: A multiconceitual approach]*. Barueri, São Paulo, Brazil: Minha Editora.

He, J., Li, B., Guo, Y., Näätänen, R., Pakarinen, S., & Luo, Y. J. (2013). Effects of alcohol on auditory pre-attentive processing of four sound features: Evidence from mismatch negativity. *Psychopharmacology, 225*, 353–360.

Heywood, C. A., Gadotti, A., & Cowey, A. (1992). Cortical area V4 and its role in the perception of color. *The Journal of Neuroscience, 12*, 4056–4065.

Hill, J. C., & Toffolon, G. (1990). Effect of alcohol on sensory and sensorimotor visual functions. *Journal of Studies on Alcohol, 51*, 108–113.

Himwich, H. E. (1957). *Alcoholism*. Washington, DC: American Association for the Advancement of Science.

Hogan, R. E., & Linfield, P. B. (1983). The effects of moderate doses of ethanol on heterophoria and other aspects of binocular vision. *Ophthalmic and Physiological Optics, 3*, 21–31.

Johnston, K. D., & Timney, B. (2004). Effects of acute ethyl alcohol consumption on a psychophysical measure of lateral inhibition in human vision. *Vision Research, 48*, 1539–1544.

Joris, P. X., Schreiner, C. E., & Rees, A. (2004). Neural processing of amplitude-modulated sounds. *Physiological Reviews, 84*, 541–577.

Kähkönen, S., Rossi, E. M., & Yamashita, H. (2005). Alcohol impairs auditory processing of frequency changes and novel sounds: A combined MEG and EEG study. *Psychopharmacology, 177*, 366–372.

Klatsky, A. L. (2003). Um brinde à suasaúde? [A toast to your health?] *Scientific American Brasil, 1*, 82–89.

Langers, D. R., Backes, W. H., & Van Dijk, P. (2007). Representation of lateralization and tonotopy in primary versus secondary human auditory cortex. *NeuroImage, 34*, 264–273.

Lecanuet, J. P., Graniere-Deferre, C., Jacquet, A. Y., & DeCasper, A. J. (2000). Fetal discrimination of low-pitched musical notes. *Developmental Psychobiology, 36*, 29–39.

Lent, R. (2001). *Cem bilhões de neurônios: Conceitos fundamentais da neurociência* [*One hundred billion neurons: Basic neuroscience concepts*]. São Paulo, Brazil: Atheneu.

Levine, J. M., Kramer, G. G., & Levine, E. N. (1975). Effects of alcohol on human performance: An integration of research findings based on an abilities classification. *Journal of Applied Psychology, 60*, 285–293.

Leyton, V., Ponce, J. C., & Andreuccetti, G. (2009). Problemas específicos: Álcool e trânsito [Specific problems: Alcohol and traffic]. In A. G. Andrade, J. C. Anthony, & C. M. Silveira (Eds.), *Álcool e suas consequências: Uma abordagem multiconceitual [Alcohol and its consequences: A multiconceitual approach]*. Barueri, São Paulo, Brazil: Minha Editora.

Lima, R. F. (2005). Compreendendo os mecanismos atencionais [Understanding the attentional mechanisms]. *Ciências & Cognição [Science and Cognition], 6*, 113–122.

Machado, A. B. M. (2005). Grandes vias aferentes [Great afferent pathways]. In A. B. M. Machado (Ed.), *Neuroanatomia funcional [Functional neuroanatomy]* (2nd ed.). São Paulo, Brazil: Atheneu.

MacNichol, E. F., Jr., & Benolken, R. (1956). Blocking effect of ethyl alcohol on inhibitory synapses in the eye of *Limulus*. *Science, 124*, 681–682.

Maurage, P., Campanella, S., Philippot, P., Charest, I., Martin, S., & Timary, P. (2009). Impaired emotional facial expression decoding in alcoholism is also present for emotional prosody and body postures. *Alcohol and Alcoholism, 44*, 476–485.

Maurage, P., Campanella, S., Philippot, P., Pham, T. H., & Joassin, F. (2007). The crossmodal facilitation effect is disrupted in alcoholism: A study with emotional stimuli. *Alcohol and Alcoholism, 42*, 552–559.

Mccormick, D. A. (1989). Cholinergic and noradrenergic modulation of thalamocortical processing. *Trends in Neurosciences, 12*, 215–221.

Medina, A. E., Krahe, T. E., & Ramoa, A. S. (2006). Restoration of neuronal plasticity by a phosphodiesterase type 1 inhibitor in a model of fetal alcohol exposure. *The Journal of Neuroscience, 26*, 1057–1060.

Meerton, L. J., Andrews, P. J., Upile, T., Drenovak, M., & Graham, J. M. (2005). A prospective randomized controlled trial evaluating alcohol on loudness perception in cochlear implant users. *Clinical Otolaryngology, 30*, 328–332.

Mendes, L. C. (2008). *Caracterização da percepção visual da forma em crianças surdas e ouvintes [Characterization of visual perception of form in deaf children and hearing]* (Dissertação de Mestrado) (Master's thesis). João Pessoa, Paraíba, Brazil: Universidade Federal da Paraíba [Federal University of Paraíba].

Merigan, W. H. (1996). Basic visual capabilities and shape discrimination after lesions of extrastriate area V4 in macaques. *Visual Neuroscience, 13*, 51–60.

Meyer, M., Nicastri, S., Bordin, S. L., Nisenbaum, E. B., & Ribeiro, M. (2004). *Cuidando da pessoa com problemas relacionados com álcool e outras drogas: Guia para a família [Taking care of the person with problems related to alcohol and other drugs: A Guide for the family]*. São Paulo, Brazil: Atheneu.

Moore, B. C. J. (2007). Basic auditory processes. In E. B. Goldstein (Ed.), *Sensation and perception*. Oxford, UK: Blackwell Publishing.

Negishi, K., & Svaetichin, G. (1966). Effects of alcohols and volatile anesthetics on S-potential producing cells and on neurons. *Pflügers Archiv, 292*, 177–206.

Nicholson, D. W, Ali, A., Thornberry, N. A., Vaillancourt, J. P., Ding, C. K., Gallant, M., ... Lazebnik, Y. A. (1995). Identification and inhibition of the ICE/CED-3 protease necessary for mammalian apoptosis. *Nature, 376*, 37–43.

Ogawa, T., Kato, H., & Ito, S. (1986). Studies of inhibitory neurotransmission in visual cortex in vitro. In J. Pettigrew, K. Sanderson, & W. Levick (Eds.), *Visual neurocience* (pp. 280–289). Cambridge: Cambridge University Press.

Pearson, P. M. (1997). *The effects of ethyl alcohol on visual and auditory thresholds* (Doctoral thesis). London, Ontario: The University of Western Ontario.

Pearson, P., Dawe, L. A., & Timney, B. (1999). Frequency selective effects of alcohol on auditory detection and frequency discrimination thresholds. *Alcohol and Alcoholism, 34*, 741–749.

Pearson, P., & Timney, B. (1998). Effects of moderate blood alcohol concentrations on spatial and temporal contrast sensitivity. *Journal of Studies on Alcohol, 59*, 163–173.

Pearson, P., & Timney, B. (1999). Differential effects of alcohol on rod and cone temporal processing. *Journal of Studies on Alcohol, 60*, 879–883.

Pihkanen, T., & Kauko, O. (1962). The effects of alcohol on the perception of musical stimuli. An orientative experimental study. *Annales Medicinae Experimentalis et Biologiae Fenniae, 40*, 275–282.

Planeta, C. S., & Graeff, F. G. (2012). Abuso e dependência de substâncias psicoativas [Abuse and dependence of psychoactive substances]. In F. G. Graeff, & F. S. Guimarães (Eds.), *Fundamentos de Psicofarmacologia [Fundamentals of Psychopharmacology]*. São Paulo, Brazil: Atheneu.

Poltavski, D. V., Marino, J. M., Guido, J. M., Kulland, A., & Petros, T. V. (2011). Effects of acute alcohol intoxication on verbal memory in young men as a function of time of day. *Physiology & Behavior, 102*, 91–95.

Puggina, A. C. G., Da Silva, M. J. P., Gatti, M. F. Z., Graziano, K. U., & Kimura, M. (2005). A percepção auditiva nos pacientes em estado de coma: Uma revisão bibliográfica [The auditory perception in the patients in state of coma: A literature review]. *Acta Paulista de Enfermagem [São Paulo Acta of Nursing], 18*, 313–319.

Quintyn, J. C., Massy, M., & Brausser, G. (1999). Effects of low alcohol consumption on visual evoked potential, visual field and visual contrast sensitivity. *Acta Ophthalmologica Scandinavica, 77*, 23–26.

Ray, S., & Bates, M. E. (2006). Acute alcohol effects on repetition priming and word recognition memory with equivalent memory cues. *Brain and Cognition, 60*, 118–127.

Robles, L., & Ruggero, M. A. (2001). Mechanics of the mammalian cochlea. *Physiological Reviews, 81*, 1305–1352.

Roquelaure, Y., Gargasson, J. F. L. E., Kupper, S., Girre, C., Hispard, E., & Dally, S. (1995). Alcohol consumption and visual contrast sensitivity. *Alcohol and Alcoholism, 30*, 681–685.

Rosenbloom, M., Pfefferbaum, A., & Sullivan, E. V. (2004). Recovery of short-term memory and psychomotor speed but not postural stability with long-term sobriety in alcoholic women. *Neuropsychology, 18*, 589–597.

Rossi, A. G., Bellé, M., & Sartori, S. D. A. (2010). Avaliação audiológica básica em alcoólicos [Basic audiological evaluation in alcoholics]. *Revista de Ciências Médicas e Biológicas [Journal of Medical and Biologicalsciences], 5*, 21–28.

Rui, L. R., & Steffani, M. H. (2007). Janeiro/Fevereiro. Física: Som e audição humana. Em A. J. S. Oliveira (Coordenador geral), *XVII Simpósio Nacional de Ensino de Física. Simpósio dirigido ao encontro da Sociedade Brasileira de Física, São Luiz, Maranhão, Brasil* [January/February. Physics: Sound and human hearing. In A. J. S. Oliveira (General Coordinator), *XVII National Symposium on Physics Teaching. Symposium addressed to the meeting of the Brazilian Physics Society, São Luiz, Maranhão, Brazil*].

Sagi, D., & Hochstein, S. (1985). Lateral inhibition between spatially adjacent spatial frequency channels? *Perception & Psychophysics, 37*, 315–322.

Samelli, A. G., & Schochat, E. (2008). Processamentoauditivo, resolução temporal e teste de detecção de gap: Revisão da literatura [Auditory processing, temporal resolution and gap detection test: Literature review]. *CEFAC: Speech, Language, Hearing Sciences and Education Journal, 10*, 369–377.

Santos, N. A., & Simas, M. L. B. (2001). Percepção e processamento visual da forma: Discutindo modelos teóricos atuais [Perception and visual processing of form: Discussing current theoretical models]. *Psicologia: Reflexão e Crítica [Psychology: Research and Review], 14*, 151–160.

Santos, N. A., & Simas, M. L. B. (2002). Percepção e processamento visual da forma em humanos: Filtros de freqüências radiais de 1 e 4 cpd. *Psicologia: Reflexão e Crítica [Psychology: Research and Review], 15*, 383–391.

Schönwiesner, M., von Cramon, D. Y., & Rübsamen, R. (2002). Is it tonotopy after all? *NeuroImage, 17*, 1144–1161.

Schreckenberger, M., Amberg, R., Scheurich, A., Lochmann, M., Tichy, W., Klega, A., ... Urban, R. (2004). Acute alcohol effects on neuronal and attentional processing: Striatal reward system and inhibitory sensory interactions under acute ethanol challenge. *Neuropsychopharmacology, 29*, 1527–1537.

Schweizer, T. A., Vogel-Sprott, M., Danckert, J., Roy, E. A., Skakum, A., & Broderick, C. E. (2006). Neuropsychological profile of acute alcohol intoxication during ascending and descending blood alcohol concentrations. *Neuropsychopharmacology, 31*, 1301–1309.

Silva, J. A. (2011). *Percepção de notasmusicaisapósingestãomoderada de etanolpormulheres e homens [Perception of musical notes after moderate consumption of ethanol by women and men]* (Dissertação de Mestrado) (Master's thesis). João Pessoa, Paraíba, Brazil: Universidade Federal da Paraíba [Federal University of Paraiba].

Silva, J. A. (2015). *Ingestão aguda e crônica de etanol no funcionamento auditivo e neurocognitivo [Acute and chronic intake of ethanol in auditory and neurocognitive functioning]* (Tese de Doutorado) (Doctoral thesis).

João Pessoa, Brazil: Universidade Federal da Paraíba [Federal University of Paraiba].

Silva, J. A., Bezerra, P. C., Gadelha, M. J. N., Andrade, M. J. O., Andrade, L. M. M. S., Torro-Alves, N., & Santos, N. A. (2014). Mulheres e homens: Diferentes também na percepção de notas musicais? [Women and men: Different also the perception of musical notes?] *Psicologia: Teoria e Pesquisa [Psychology: Theory and Research], 30*, 83–87.

Silva, J. A., Cavalcanti-Galdino, M. K., Gadelha, M. J. N., Andrade, M. J. O., & Santos, N. A. (2013). Revisãosobre o processamentoneuropsicológico dos atributostonais da música no contextoocidental [Review of the neuropsychological processing of tonal attributes of music in the Western contexto]. *Avances en Psicología Latinoamericana [Advances in Latin American Psychology], 31*, 86–96.

Silva, J. A., Cavalcanti-Galdino, M. K., Simas, M. L. B., & Santos, N. A. (2015). Consequências da ingestão moderada de etanol na discriminação de notas musicais [Consequences of moderate ethanol intake in discrimination of musical notes]. *Psicologia: Reflexão e Crítica [Psychology: Research and Review], 28*, 1–10.

Simas, M. L. S., Santos, N. A., & Thiers, F. A. (1997). Contrast sensitivity to angular frequency stimuli is higher than that for sine wave gratings in the respective middle range. *Brazilian Journal of Medical and Biological Research, 30*, 633–636.

Teo, R. K., & Ferguson, D. A. (1986). The acute effects of ethanol on auditory event-related potentials. *Psychopharmacology, 90*, 179–184.

Vanneste, S., & De Ridder, D. (2012). Noninvasive and invasive neuromodulation for the treatment of tinnitus: An overview. *Neuromodulation, 15*, 350–360.

Virsu, V., Kyykka, T., & Vahvelainen, M. (1974). *Effects of alcohol on inhibition in the human visual system. II. Spatial and temporal contrast sensitivity.* Helsinki: Reports from the Institute of Psychology. University of Helsinki.

Vollrath, M. A., Kwan, K. Y., & Corey, D. P. (2007). The micromachinery of mechanotransduction in hair cells. *Annual Review of Neuroscience, 30*, 339–365.

Wallgren, H., & Barry, H., III (1970). *Actions of alcohol: Biochemical, physiological and psychological aspects.* New York: Elsevier.

Watten, R. G., & Lie, I. (1996). Visual functions and acute ingestion of alcohol. *Ophthalmic and Physiological Optics, 16*, 460–466.

Wegner, A. J., Günthner, A., & Fahle, M. (2001). Visual performance and recovery in recently detoxified alcoholics. *Alcohol and Alcoholism, 36*, 171–179.

Wetherill, L., & Foroud, T. (2011). Understanding the effects of prenatal alcohol exposure using three dimensional facial imaging. *Alcohol Research & Health, 34*, 38–41.

Wilkinson, F., James, T. W., Wilson, H. R., Gati, J. S., Menon, E. S., & Goodale, M. A. (2000). An fMRI study of the selective activation of human extrastriate form vision areas by radial and concentric gratings. *Current Biology, 10*, 1455–1458.

Wilson, H. R., & Wilkinson, F. (1998). Detection of global structure in glass patterns: Implications for form vision. *Vision Research, 38*, 2933–2947.

Wilson, H. R., Wilkinson, F., & Asaad, W. (1997). Concentric orientation summation in human form vision. *Vision Research, 37*, 2325–2330.

World Health Organization. (1994). *Lexicon of alcohol and drug terms.* Geneva: Author.

Xiao, C., & Ye, J. H. (2008). Ethanol dually modulates GABAergic synaptic transmission onto dopaminergic neurons in ventral tegmental area: Role of mu-opioid receptors. *Neuroscience, 153*, 240–248.

Yonamine, M. (2004). *A saliva como espécime biológico para monitorar o uso deálcool, anfetamina, metanfetamina, cocaína e maconha por motoristas profissionais [Saliva as biological specimen to monitor the use of alcohol, amphetamine, methamphetamine, cocaine and marijuana by professional drivers]* (Tese de Doutorado) (Doctoral thesis). São Paulo, Brazil: Universidade de São Paulo [University of São Paulo].

CHAPTER

7

Zebrafish Models of Alcohol Addiction

S. Tran[1], R. Gerlai[1,2]

[1]University of Toronto, Toronto, ON, Canada; [2]University of Toronto Mississauga, Mississauga, ON, Canada

ALCOHOL ADDICTION IN HUMANS

Alcohol (ethanol or ethyl alcohol) is one of the world's most heavily and frequently consumed drugs despite its potential for abuse. The life-time prevalence rate for alcohol dependence in the United States was estimated to be as high as 17% (Haberstick et al., 2014). Binge drinking has increased in North America over the past several years and is associated with negative health consequences (Naimi et al., 2003). About 50% of all motor vehicle mortalities are associated with alcohol impaired driving (Jewett, Shults, Banerjee, & Bergen, 2015). In addition to negative health consequences affecting a given individual, alcohol addiction also negatively impacts the society. Alcohol addiction represents a large unmet medical need and a huge economic burden with current pharmacological treatment options being limited at best (Franck & Jayaram-Lindstrom, 2013). The combined cost of alcohol addiction due to loss of productivity, crime, and health care is estimated to be as high as $250 billion annually (Sacks, Gonzales, Bouchery, Tomedi, & Brewer, 2015). Despite the economic and societal cost of alcohol addiction, the mechanisms underlying this drug's actions in the brain and the development of alcohol addiction in humans remain poorly understood.

ANIMAL MODELS OF ALCOHOL ADDICTION

Numerous animal models have been established to model and investigate the mechanisms underlying the development of alcohol addiction in humans. Among the many vertebrate models, rats and mice have been more commonly utilized for alcohol addiction research in comparison to higher-order vertebrates such as primates (Brabant, Guarnieri, & Quertemont, 2014). In contrast, invertebrate models including fruit flies and nematodes are also frequently used for alcohol addiction research (Scholtz & Mustard, 2013). A number of factors are considered when choosing an animal model for alcohol-related research. First, the complexity and translational relevance of primates are clear advantages compared to lower-order mammals. However, rodent models of alcohol addiction allow for the sophisticated complexity of mammalian vertebrates along with the reduced cost compared to primate research. In contrast, invertebrate models such as fruit flies and nematodes are phylogenetically distant, and exhibit less biological similarity and evolutionary homology, including a lower nucleotide sequence homology to corresponding human DNA. One argument for animal models employing lower-order species is that such reductionist approach may allow one to investigate evolutionarily ancient common putative mechanisms underlying the effects of alcohol on biological systems. For example, the brains of invertebrates are much simpler allowing the dissection of simple circuits and perhaps also the identification of the most important molecular mechanisms mediating alcohol-induced responses. In addition, these lower-order organisms are significantly cheaper to house as compared to traditional mammalian models (Scholtz & Mustard, 2013). Mammalian models of alcohol addiction including rodents and primates have been predominant in the literature due to their high translational relevance to humans. However, mammals are not ideal for high-throughput behavioral screens due to their large size and low number of progeny. In contrast, invertebrate models including fruit flies and nematodes are highly amenable to high-throughput behavioral screens due to their small size and high number of progenies, but these latter species are much less complex compared to mammals. Due to these limitations, zebrafish have moved to the forefront of behavioral neuroscience as a vertebrate model that strikes the right balance. It is

Addictive Substances and Neurological Disease
http://dx.doi.org/10.1016/B978-0-12-805373-7.00007-4

highly amenable to high-throughput behavioral screens, yet complex and similar enough to mammals to be considered translationally relevant.

ZEBRAFISH AS AN ANIMAL MODEL FOR BEHAVIORAL NEUROSCIENCE

The zebrafish (*Danio rerio*) is relatively a newcomer to the field of behavioral neuroscience. Zebrafish have been a favorite of geneticists and developmental biologists over the past several decades due to their rapid development and developmental transparency. Over the past decade, zebrafish have been gaining increasing attention in behavioral neuroscience due to a number of advantages over several vertebrate and invertebrate models. Zebrafish have been proposed to represent a compromise between systems complexity and practical simplicity (Gerlai, 2012). The zebrafish vertebrate brain is evolutionarily conserved and exhibits many anatomical similarities to the mammalian brain (Tropepe & Sive, 2003). Furthermore, zebrafish exhibit a rich repertoire of quantifiable behavioral responses similar to mammalian vertebrate models. In contrast to mammals, zebrafish are significantly cheaper and have a number of practical advantages. Zebrafish are highly prolific with a single female capable of spawning several hundred eggs every week. Furthermore, zebrafish are highly social allowing these animals to be housed in high-density system racks, features that make the zebrafish amenable to high-throughput behavioral screens.

ZEBRAFISH AS AN ANIMAL MODEL FOR ALCOHOL ADDICTION

The zebrafish brain is evolutionarily conserved and numerous neurotransmitter systems implicated in the development of alcohol addiction in humans are also found in zebrafish (Rico et al., 2011). Furthermore, the anatomical, pharmacological, and behavioral effects of alcohol in zebrafish are conserved from zebrafish to humans (for review, see Tran, Facciol, & Gerlai, 2016a; Tran & Gerlai, 2014). One of the major advantages of using zebrafish for alcohol addiction research is the method of alcohol administration. In vertebrate models of alcohol addiction including rodents and primates, the method of alcohol administration is often invasive or uncontrolled. For example, alcohol administration via injection or via oral gavage (i.e., gastric cannulation) allows precise control of alcohol delivery (e.g., concentration), but is stress inducing and invasive. In contrast, noninvasive alcohol administration methods including delivery of alcohol via the drinking water or through a gel diet are less controlled, and are often dependent upon motivational factors related to the experimental subject including thirst and hunger (Keane & Leonard, 1989). In contrast to these alcohol-delivery methods, alcohol can be administered to zebrafish in a controlled and less invasive manner. Zebrafish are able to absorb pharmacological compounds including alcohol from the water in which they swim. By immersing zebrafish in an alcohol bath solution, alcohol is taken up through the skin and gills without the need for invasive injections or surgeries. The internal (brain or blood) alcohol levels correlate with external alcohol concentration, and the length of alcohol immersion is also precisely controlled. Detailed time-dependent changes in brain and blood alcohol levels have been reported in zebrafish, and external bath concentrations leading to brain/blood alcohol concentrations relevant to the human clinic have been employed for zebrafish (Dlugos & Rabin, 2003; Mathur, Berberoglu, & Guo, 2011; Tran, Chatterjee, & Gerlai, 2015). In this chapter, we discuss different behavioral paradigms in zebrafish that can be used to study the effects of alcohol on brain function and also to model some aspects of human alcohol addiction.

ALCOHOL-INDUCED LOCOMOTOR ACTIVITY

Alcohol is considered a central nervous system depressant with hypnotic and sedative effects. However, alcohol is also known to produce biphasic responses depending upon dose. At low to moderate doses, alcohol has euphoric and stimulant effects as blood alcohol concentration (BAC) increases upon acute alcohol consumption. However, at higher BAC, alcohol produces sedation and motor impairing effects (Martin, Earleywine, Musty, Perrine, & Swift, 1993). Animal models are often used to examine the effects of different concentrations of alcohol on behavioral responses. For example, in rodents, a single injection of an alcohol solution can produce hyperactivity or sedation depending on the amount of alcohol delivered (Smoothy & Berry, 1985). Similarly, a single acute exposure to alcohol water bath solution can also alter locomotor activity in zebrafish depending on the concentration of alcohol and duration of exposure (Rosemberg et al., 2012; Tran, Chatterjee, et al., 2015). In the context of this chapter, acute alcohol exposure is defined as a single alcohol exposure session. Acute alcohol exposure can be used to examine alcohol intoxication in humans, and also how it may lead to the development of addiction.

In zebrafish, acute exposure to alcohol is commonly examined in the context of two commonly reported effects: locomotor-altering effects and anxiety-altering effects of alcohol. Locomotor-stimulant effect of alcohol is thought to be related to its rewarding properties and may contribute to the development of addiction (Phillips & Shen, 1996). For example, individual sensitivity to stimulant effects of alcohol has been reported to be a risk factor for the development of alcohol addiction (Holdstock, King, & de Wit, 2000). The hypothesis that alcohol-induced locomotor activity in animal models can be used to study the rewarding properties of alcohol is supported by the observation that many drugs of abuse produce dose-dependent locomotor stimulation in animals. Furthermore, similar to biphasic locomotor-stimulant effects of alcohol, the rewarding and euphoric effects are also dependent on BAC suggesting a common underlying mechanism (Phillips & Shen, 1996). In zebrafish, alcohol-induced locomotor activity is easily quantified with the use of video-tracking software. The total distance traveled is often reported as a measure of locomotor activity. It is notable that alcohol has been reported to increase locomotor activity in a time- and concentration-dependent manner in zebrafish (Rosemberg et al., 2012; Tran & Gerlai, 2013).

A number of studies have examined potential underlying mechanisms regulating alcohol-induced locomotor activity in zebrafish. Acute exposure to alcohol has been shown to increase locomotor activity (Blaser & Penalosa, 2011; Tran & Gerlai, 2013) and decrease the putative amino sulfonic acid taurine (Chatterjee, Shams, & Gerlai, 2014; Tran, Chatterjee, et al., 2015). Pretreating zebrafish with taurine has been shown to inhibit alcohol-induced locomotor activity as well as inhibit increases in brain alcohol levels (Rosemberg et al., 2012). In rodents, locomotor-stimulant effect of alcohol is often examined in the context of the dopaminergic system (Phillips & Shen, 1996). In zebrafish, a number of studies suggest that activation of the dopaminergic system may partially regulate alcohol-induced locomotor activity. Alcohol exposure has been shown to increase whole-brain levels of dopamine and its metabolite 3,4-dihydroxyphenyl acetic acid (DOPAC) (Chatterjee et al., 2014; Tran, Chatterjee, et al., 2015). Examination of upstream and downstream mechanisms regulating brain dopamine levels revealed that acute alcohol exposure increased tyrosine hydroxylase (TH) activity (rate-limiting enzyme responsible for dopamine synthesis) without altering monoamine oxidase activity (enzyme responsible for dopamine metabolism) (Chatterjee et al., 2014). Furthermore, alcohol exposure has also been shown to increase TH mRNA expression (Puttonen, Sundvik, Rozov, Chen, & Panula, 2013) and protein expression (unpublished data), suggesting that TH increases dopamine levels in response to alcohol. In support of this hypothesis, inhibition of phosphorylated TH has been shown to abolish alcohol-induced dopamine increases, and attenuate alcohol-induced locomotor activity in zebrafish (Nowicki, Tran, Chatterjee, & Gerlai, 2015). Further analysis revealed that alcohol-induced locomotor activity may be regulated by activation of dopamine D_2-like receptors (Tran, Facciol, & Gerlai, 2016b) but not D_1-like receptors (Tran, Nowicki, Muraleetharan, Chatterjee, & Gerlai, 2015). Although the question of whether alcohol-induced locomotor activity is related to its rewarding effects is still up for debate (Brabant et al., 2014), it is clear that zebrafish can be used to examine neural mechanisms regulating locomotor-stimulant effect of alcohol.

ALCOHOL-INDUCED ANXIOLYSIS

In contrast to the positive reinforcing effects of alcohol, opponents of this hypothesis argue that negative reinforcement is more important to the development of alcohol addiction. Through negative reinforcement, individuals compulsively consume alcohol to relieve a negative underlying emotional state (e.g., anxiety) (Koob, 2009). For example, neuroadaptations due to prolonged alcohol consumption may cause withdrawal symptoms following the cessation of alcohol, causing individuals to excessively consume alcohol to relieve this negative state (discussed in the following paragraphs). In addition, exposure to alcohol has been reported to reduce anxiety-like behaviors in humans and animal models, which may also contribute to the development of alcohol addiction. For example, the life-time prevalence rate of alcohol dependence is significantly higher in individuals with anxiety disorders (Vorspan, Mehtelli, Dupuy, Bloch, & Lepine, 2015).

Similar to alcohol's biphasic effect on locomotor activity, exposure to low and moderate doses of alcohol has been shown to reduce anxiety-like behavioral responses in zebrafish (Blaser & Penalosa, 2011; Egan et al., 2009). In contrast, exposure to higher concentrations of alcohol has been shown to increase anxiety-like behavioral (Tran & Gerlai, 2013) and cortisol stress responses (Tran, Chatterjee, et al., 2015). Anxiety-related behavioral paradigms have been developed to demonstrate anxiolytic effects of alcohol in zebrafish (Echevarria, Toms, & Jouandot, 2011). Pretreating zebrafish with low to moderate concentrations of alcohol has been shown to reduce anxiety-like behavioral responses in the novel tank test (Egan et al., 2009), light–dark preference task (Blaser & Penalosa, 2011), and predator exposure test (Oliveira et al., 2013; Pannia, Tran, Rampersad, & Gerlai, 2014). Similarly, alcohol exposure has also been shown to reduce stress-induced whole-body cortisol levels (Oliveira et al., 2013). Although anxiolytic effect of alcohol may contribute to the development of alcohol dependence, it is likely that this represents only a small subpopulation of individuals. For example, the efficacy of anxiolytic or antidepressant drugs, including selective serotonin reuptake inhibitors, for reducing alcohol consumption ranges from 10% to 70% depending on a number of factors including alcoholic subtypes (Naranjo & Knoke, 2001). Therefore, it is likely that other mechanisms also contribute to the transition from casual to compulsive and chronic alcohol consumption in humans.

ALCOHOL-INDUCED TOLERANCE

One of the diagnostic criteria for patients to be considered suffering from an alcohol use disorder is the development of tolerance; that is, whether the patient requires increasing number of drinks to reach a desired effect (e.g., euphoria) (American Psychiatric Association, 2013). The development of alcohol tolerance is thought to be due to neuroadaptations following chronic consumption of alcohol. In rodent models, chronic exposure to alcohol has been shown to lead to the development of tolerance to effects of alcohol (Quoilin, Didone, Tirelli, & Quertemont, 2013). The use zebrafish to study the mechanisms underlying the development alcohol tolerance is particularly unique. Zebrafish can be continuously immersed in alcohol for an extended period of time, with studies reporting no mortality after as many as 10 weeks of alcohol exposure (Damodaran, Dlugos, Wood, & Rabin, 2006; Dlugos & Rabin, 2003). Although continuous alcohol exposure in zebrafish does not resemble the most usual mode of alcohol consumption in humans, it represents a reductionist approach to investigate the mechanisms regulating alcohol tolerance without the potential confounding effect of fluctuating alcohol levels, which includes repeated withdrawal from the substance (Tran & Gerlai, 2014).

The typical protocol for continuous alcohol exposure is to directly administer alcohol to the housing tank of the fish for an extended period of time (ranging from 1 to 10 weeks) with concentrations ranging from 0.25% to 0.50% v/v (Cachat et al., 2010; Dlugos & Rabin, 2003; Gerlai, Lahav, Guo, & Rosenthal, 2000; Tran, Chatterjee, et al., 2015). Several studies have utilized a dose-escalation procedure to reduce mortality rates associated with prolonged alcohol administration. The dose-escalation procedure entails administering lower doses of alcohol and gradually increasing the concentration over a period of days to weeks to reach the desired final chronic concentration (Gerlai, Chatterjee, Pereira, Sawashima, & Krishnannair, 2009; Tran & Gerlai, 2013). Importantly, tank water containing alcohol is repeatedly changed to prevent alcohol evaporation over time and to remove organic waste and toxins that may accumulate. However, alcohol concentrations have been reported to remain relatively stable over several weeks (Gerlai et al., 2009). Continuous alcohol exposure has been shown to elicit behavioral (Gerlai et al., 2000), neurochemical (Chatterjee et al., 2014), hepatic (Tran, Nowicki, Chatterjee, & Gerlai, 2015), and endocrinological tolerance (Tran, Chatterjee, et al., 2015) to the acute effects of alcohol in zebrafish. Furthermore, continuous alcohol exposure has been shown to alter both gene and protein expression in the zebrafish brain (Damodaran et al., 2006; Pan, Kaiguo, Razak, Westwood, & Gerlai, 2011). In contrast to continuous alcohol exposure, repeated intermittent exposure to alcohol has also been shown to induce tolerance. For example, zebrafish repeatedly exposed to 1% alcohol for 60 min a day for 10 consecutive days have been shown to develop tolerance to locomotor-stimulant effect of alcohol (Tran, Nowicki, Chatterjee, et al., 2015). However, the effects of continuous alcohol exposure appear to be more robust compared to repeated intermittent exposure, likely due to the longer duration of the exposure.

ALCOHOL-INDUCED SENSITIZATION

Repeated intermittent chronic alcohol exposure may lead to the development of tolerance, but it may also lead to the development of reverse tolerance, that is, sensitization. The development of alcohol-induced sensitization is an essential component of the positive reinforcement theory of alcohol addiction in humans. Alcohol is thought to activate the reward centers of the brain, and the development of alcohol addiction may be due to the "hijacking" of this system. The incentive-sensitization theory proposes that exposure to drugs of abuse, including alcohol, hypersensitizes brain reward systems that regulate reward and motivation (Robinson & Berridge, 2008). One of the most common behavioral measures in animal studies for evaluating the development of alcohol-induced sensitization is locomotor activity. For example, mice that are repeatedly administered a stimulant dose of alcohol begin to exhibit an enhanced response to locomotor-stimulant effect of alcohol over time (Phillips, Roberts, & Lessov, 1997). Alcohol-induced sensitization has been difficult to demonstrate in zebrafish, likely due to strain, age, and alcohol exposure regimen-dependent factors (Tran & Gerlai, 2014). However, one study has shown that zebrafish exposed to 1% alcohol for 60 min a day for 8 consecutive days become sensitized to stimulant effect of alcohol, but this effect was found to be context dependent (Blaser, Koid, & Poliner, 2010). Further research is required to elucidate the factors that allow the development of alcohol-induced sensitization in zebrafish.

ALCOHOL-INDUCED WITHDRAWAL

Chronic and compulsive alcohol consumption is the hallmark of individuals diagnosed with an alcohol use disorder (National Institute on Drug Abuse, 2008). Once alcohol consumption ceases, a number of negative symptoms emerge, collectively referred to as alcohol withdrawal symptoms. It is hypothesized that neuroadaptations occur in the brain following long-term alcohol consumption, such that once an individual is withdrawn from alcohol, an underlying negative emotional state (e.g., anxiety) emerges, which reinforces continued alcohol consumption (Koob, 2009). In zebrafish, alcohol-induced withdrawal has been demonstrated following both continuous and repeated intermittent alcohol exposure regimens (Gerlai et al., 2009; Mathur & Guo, 2011; Pittman & Ichikawa, 2013; Tran, Chatterjee, et al., 2015). Zebrafish that were chronically exposed to alcohol and subsequently returned to a tank without alcohol exhibit increased anxiety-like behavioral responses including erratic movement (Tran & Gerlai, 2013), bottom dwelling (Mathur & Guo, 2011), and freezing (Pittman & Ichikawa, 2013). Withdrawal from alcohol has also been shown to increase whole-body cortisol levels (Cachat et al., 2010), as well as to alter monoamine and amino acid neurotransmitter levels in zebrafish (Chatterjee et al., 2014). Brain alcohol concentrations have been shown to be reduced by 50% following 1-min-long exposure to regular system water, confirming an immediate withdrawal from alcohol (Tran, Chatterjee, et al., 2015).

Several studies have started to examine the neural mechanisms underlying withdrawal-related behavioral responses in zebrafish. For example, withdrawal from alcohol has been shown to increase anxiety-like behavioral responses, and to alter serotonergic and glutamatergic responses in zebrafish (Chatterjee et al., 2014; Tran, Chatterjee, et al., 2015). However, pretreating zebrafish with fluoxetine (a serotonin reuptake inhibitor) or ketamine (a glutamate receptor antagonist) has been shown to abolish alcohol withdrawal–induced anxiety-like behavioral responses (Pittman & Hylton, 2015). Identifying mechanisms underlying alcohol withdrawal–induced behavioral changes may provide key insights into developing efficacious drugs capable of treating alcohol addiction.

ALCOHOL-INDUCED CONDITIONED PLACE PREFERENCE

Unlike the negative reinforcement theory underlying alcohol addiction (i.e., consuming alcohol to prevent the negative symptoms associated with withdrawal), the rewarding aspects of alcohol may cause individuals to associate certain environmental stimuli with alcohol consumption, and these stimuli, even when perceived alone, can elicit drug-seeking behavior. In humans, environmental cues commonly associated with alcohol consumption can elicit cravings and induce relapse (Snelleman, Schoenmakers, & van de Mheen, 2014). According Pavlovian conditioning theory, cues (conditioned stimuli) that are paired with the rewarding stimulus (the unconditioned stimulus), for example, alcohol, will become associated with each other following repeated pairings. Eventually, the conditioned stimulus (alcohol-paired environment) will elicit a conditioned response (craving for alcohol). In animal models, the reinforcing and motivational effects of alcohol can be examined by using conditioned place preference (CPP). In CPP, animals are exposed to two different environments, one paired with alcohol and the other not paired with this substance. Following repeated pairings, animals exhibit a preference for the alcohol-paired environment (Martin-Fardon & Weiss, 2013). Zebrafish have been successfully utilized in CPP paradigms for numerous drugs of abuse including alcohol (Collier & Echevarria, 2013; Collier, Khan, Caramillo, Mohn, & Echevarria, 2014). The development of an alcohol-induced CPP response in zebrafish has been associated with gene expression changes that are implicated in addiction in mammals (Kily et al., 2008). Furthermore, zebrafish may be uniquely well suited for high-throughput CPP since a single 60-min exposure to 1.5% alcohol has been shown to induce a robust CPP response (Mathur, Lau, & Guo, 2011). One of the limitations of alcohol-induced CPP is related to the potential memory-altering effects of alcohol. Preference for an alcohol-paired environment is dependent not only upon the motivation to obtain alcohol, but also on the subject's ability to make, consolidate, retain, and recall conditioned and unconditioned stimulus (CS–US) association, that is, to remember the environment with which alcohol was paired. This represents a confound. Alcohol rewarded place preference may be impaired by the inability of the subjects to remember the association between alcohol and where it was delivered. Briefly, the positively reinforcing aspect of alcohol and its memory impairing effects work against each other. Therefore, the use of alcohol for CPP paradigms and the results from such studies should be interpreted carefully.

VOLUNTARY ALCOHOL CONSUMPTION

The zebrafish model of alcohol addiction has been criticized due to its artificial method of alcohol administration. The most common method of alcohol administration in zebrafish is via bath immersion, whereas rodents and humans consume alcohol orally. However, this limitation has recently been addressed with the development of voluntary alcohol consumption paradigms in zebrafish. Similarly to alcohol agar gel diets employed in rodents, zebrafish can also be fed gelatin laced with alcohol. Alcohol preference in zebrafish has been shown to manifest as increased consumption of 20% alcohol–gelatin as compared to 0% alcohol–gelatin over a 2-week period, and the alcohol–gelatin-consuming fish were shown to have BAC comparable to that of human drinkers (Sterling, Karatayev, Chang, Algava, & Leibowitz, 2015). A 2016 study further extended this work, and demonstrated that embryonic alcohol exposure increased voluntary alcohol consumption in adult zebrafish (Sterling, Chang, Karatyev, Chang, & Leibowitz, 2016). These studies imply that voluntary alcohol consumption may be rewarding in zebrafish. The development and validation of voluntary alcohol consumption paradigms in zebrafish has opened up new research domains for this species. In humans, alcohol addiction has a significant genetic component (Jones, Comer, & Kranzler, 2015). To examine the underlying genetic mechanisms regulating alcohol consumption, high and low alcohol preferring rats have been selectively bred (Bell, Rodd, Lumeng, Murphy, & McBride, 2006). The development of voluntary alcohol consumption tasks will now make it possible to selectively breed zebrafish for high and low alcohol preference. The numerous alcohol-related paradigms developed for zebrafish has established this species as an effective tool for investing the mechanisms underlying the development of alcohol addiction in humans. However, similarly to many animal models, the zebrafish too has both advantages and disadvantages.

LIMITATIONS OF THE ZEBRAFISH MODEL

As discussed earlier, the controlled and less invasive manner of alcohol administration in zebrafish is an advantage of this species, but it can also be viewed as a limitation. In the clinical population, individuals consume alcohol over time in a manner that leads to fluctuating BAC, which is markedly different from how alcohol has been administered in zebrafish. Since alcohol is administered via immersion, zebrafish exhibit an initial increase in BAC that subsequently plateaus and remains stable due to the established equilibrium between the continued infusion and metabolism/secretion of alcohol. In humans, the effects of alcohol are often described in terms of the BAC curve. For example, during the ascending limb of the BAC in which alcohol concentrations increases due to alcohol consumption, individuals exhibit a number of cognitive and motor impairments. However, as alcohol is metabolized and BAC levels start to decline (descending of the BAC) cognitive ability and mood changes are different as compared to the ascending limb, despite similar BAC (Morean & Corbin, 2010).

The differential effects of alcohol on the ascending and descending limb are often examined in the context of acute tolerance. Acute tolerance is commonly defined as a decrease in responsiveness to a single alcohol exposure session. For example, individuals that develop acute tolerance often report that stimulant effects of alcohol are stronger on the ascending limb of the BAC than on the descending limb of the BAC curve, despite the same BAC (Radlow, 1994). In rodents, examination of the BAC curve and acute tolerance can be easily mimicked using oral administration or injection-based alcohol-delivery methods. In zebrafish, alcohol is usually administered through water, and BAC reaches a stable plateau after about 1 h of exposure and can be maintained elevated for at least 2 weeks by chronic exposure (Dlugos & Rabin, 2003; Tran, Chatterjee, et al., 2015). Since the alcohol immersion method does not produce a similar BAC curve as usually seen in humans, a different approach must be taken. By modifying the alcohol immersion method such that zebrafish are removed from the alcohol bath and placed in a regular tank water, BAC levels have been shown to decline similarly to the descending limb of the BAC curve in humans (Tran, Chatterjee, et al., 2015). Alternatively, alcohol can also be delivered using injection (Maximino, da Silva, Gouveia, & Herculano, 2011) or through a gel diet (Sterling et al., 2016, 2015). These alternative alcohol-delivery methods may produce a BAC curve similar to what is seen in humans. Although acute alcohol tolerance and the ascending and descending limbs of the BAC have not been examined in zebrafish, these alternative alcohol exposure methods or regimens make the examination of this question possible.

Another notable limitation, often overlooked in the analysis of the effects of alcohol on zebrafish brain function and behavior is the fact that alcohol is bacterio-toxic. Although bath immersion–based alcohol delivery is less invasive compared to many other delivery methods routinely employed with mammals, the presence of alcohol in the water poses an important problem: it damages beneficial bacterial flora. Reduction or the absence of biological filtration may lead to the accumulation of organic waste, which, in the absence, or reduced presence, of beneficial bacteria will induce anaerobic processes that not only lead to toxic byproducts but also reduce oxygen levels in the exposure tank. While these processes are unlikely to have a measurable effect within a short time frame, that is, during acute alcohol exposure, for chronic exposure paradigms, they may represent an important confounding factor. Although attempts to minimize this confusion have been made by replenishing the exposure water with fresh alcohol solution (which has two important goals: (1) to remove organic waste and (2) to reinstate the constant chronic alcohol concentration), systematic analysis of how anaerobic processes, lack of biological filtration, and/or potentially reduced oxygen levels may contribute to functional changes in the brain during the chronic alcohol exposure period has not been performed.

Clearly, a lot of unanswered questions need to be addressed before the zebrafish may be considered as a true model for human alcoholism and alcohol abuse–related disorders. Nevertheless, despite the limitations, the numerous behavioral paradigms, the sophisticated molecular biology tools, the increasingly powerful pharmacological methods, and the strong neurobiology of the zebrafish combined with the amenability of this species for high-throughput behavioral screens are making the zebrafish a powerful future tool for alcohol addiction research.

References

American Psychiatric Association. (2013). *Diagnostic and statistical manual of mental disorders* (5th ed.) Washington, DC.

Bell, R. L., Rodd, Z. A., Lumeng, L., Murphy, J. M., & McBride, W. J. (2006). The alcohol-preferring P rat and animal models of excessive alcohol drinking. *Addiction Biology, 11*(3–4), 270–288.

Blaser, R. E., Koid, A., & Poliner, R. M. (2010). Context-dependent sensitization to ethanol in zebrafish (*Danio rerio*). *Pharmacology, Biochemistry and Behavior, 95*, 278–284.

Blaser, R. E., & Penalosa, Y. M. (2011). Stimuli affecting zebrafish (*Danio rerio*) behavior in the light/dark preference test. *Physiology & Behavior, 104*, 831–837.

Brabant, C., Guarnieri, D. J., & Quertemont, E. (2014). Stimulant and motivational effects of alcohol: Lesions from rodent and primate models. *Pharmacology, Biochemistry and Behavior, 122*, 37–52.

Cachat, J., Canavello, P., Elegante, M., Bartels, B., Hart, P., Bergner, C., ... Kalueff, A. V. (2010). Modeling withdrawal syndrome in zebrafish. *Behavioural Brain Research, 208*, 371–376.

Chatterjee, D., Shams, S., & Gerlai, R. (2014). Chronic and acute alcohol administration induced neurochemical changes in the brain: Comparison of distinct zebrafish populations. *Amino Acids, 46*, 921–930.

Collier, A. D., & Echevarria, D. J. (2013). The utility of the zebrafish model in conditioned place preference to assess the rewarding effects of drugs. *Behavioural Pharmcology, 24*, 375–383.

Collier, A. D., Khan, K. M., Caramillo, E. M., Mohn, R. S., & Echevarria, D. J. (2014). Zebrafish and conditioned place preference: A translational model of drug reward. *Progress in Neuro-psychopharmacology & Biological Psychiatry, 55*, 16–25.

Damodaran, S., Dlugos, C. A., Wood, T. D., & Rabin, R. A. (2006). Effects of chronic ethanol administration on brain protein levels: A proteomic investigation using 2-D DIGE system. *European Journal of Pharmacology, 547*, 75–82.

Dlugos, C. A., & Rabin, R. A. (2003). Ethanol effects on three strains of zebrafish: Model system for genetic investigations. *Pharmacology, Biochemistry and Behavior, 74*, 471–480.

Echevarria, D. J., Toms, C. N., & Jouandot, D. J. (2011). Alcohol-induced behavioral change in zebrafish models. *Reviews in Neurosciences, 22*, 85–93.

Egan, R. J., Bergner, C. L., Hart, P. C., Cachat, J. M., Canavello, P. R., Elegante, M. F., ... Kalueff, A. V. (2009). Understanding behavioral and physiological phenotypes of stress and anxiety in zebrafish. *Behavioural Brain Research, 205*, 38–44.

Franck, J., & Jayaram-Lindstrom, N. (2013). Pharmacotherapy for alcohol dependence: Status of current treatments. *Current Opinions in Neurobiology, 23*, 692–699.

Gerlai, R. (2012). Using zebrafish to unravel the genetics of complex brain disorders. *Current Topics in Behavioral Neurosciences, 12*, 3–24.

Gerlai, R., Chatterjee, D., Pereira, T., Sawashima, T., & Krishnannair, R. (2009). Acute and chronic alcohol dose: Population differences in behavior and neurochemistry of zebrafish. *Genes, Brain and Behavior, 8*, 586–599.

Gerlai, R., Lahav, M., Guo, S., & Rosenthal, A. (2000). Drinks like a fish: Zebrafish (*Danio rerio*) as a behavior genetic model to study alcohol effects. *Pharmacology, Biochemistry and Behavior, 67*, 773–782.

Haberstick, B. C., Young, S. E., Zeiger, J. S., Lessem, J. M., Kewitt, K. L., & Hopfer, C. J. (2014). Prevalence and correlates of alcohol and cannabis use disorders in the United States: Results from the national longitudinal study of adolescent health. *Drug and Alcohol Dependence, 136*, 158–161.

Holdstock, L., King, A. C., & de Wit, H. (2000). Subjective and objective responses to ethanol in moderate/heavy and light social drinkers. *Alcoholism: Clinical and Experimental Research, 24*, 789–794.

Jewett, A., Shults, R. A., Banerjee, T., & Bergen, G. (2015). Alcohol-impaired driving among adults – United States, 2012. *Morbidity and Mortality Weekly Report, 64*, 814–817.

Jones, J. D., Comer, S. D., & Kranzler, H. R. (2015). The pharmacogenetics of alcohol use disorder. *Alcoholism: Clinical and Experimental Research, 39*(3), 391–402.

Keane, B., & Leonard, B. E. (1989). Rodent models of alcoholism: A review. *Alcohol and Alcoholism, 24*(4), 299–309.

Kily, L. J., Cowe, Y. C., Hussain, O., Patel, S., McElwaine, S., Cotter, F. E., & Brennan, C. H. (2008). Gene expression changes in a zebrafish model of drug dependency suggest conservation of neuro-adaptation pathways. *Journal of Experimental Biology, 211*, 1623–1634.

Koob, G. F. (2009). Neurobiological substrates for the dark side of compulsivity in addiction. *Neuropharmacology, 56*, 18–31.

Martin-Fardon, R., & Weiss, F. (2013). Modeling relapse in animals. *Current Topics in Behavioral Neurosciences, 13*, 403–432.

Martin, C. S., Earleywine, M., Musty, R. E., Perrine, M. W., & Swift, R. M. (1993). Development and validation of the Biphasic Alcohol Effects Scale. *Alcoholism: Clinical and Experimental Research, 17*(1), 140–146.

Mathur, P., Berberoglu, M. A., & Guo, S. (2011). Preference for ethanol in zebrafish following a single exposure. *Behavioural Brain Research, 217*, 128–133.

Mathur, P., & Guo, S. (2011). Differences of acute versus chronic ethanol exposure on anxiety-like behavioral responses in zebrafish. *Behavioural Brain Research, 219*, 234–239.

Mathur, P., Lau, B., & Guo, S. (2011). Conditioned place preference behavior in zebrafish. *Nature Protocols, 6*, 338–345.

Maximino, C., da Silva, A. W., Gouveia, A., Jr., & Herculano, A. M. (2011). Pharmacological analysis of zebrafish (*Danio rerio*) scototaxis. *Progress in Neuro-psychopharmacology & Biological Psychiatry, 35*(2), 624–631.

Morean, M. E., & Corbin, W. R. (2010). Subjective response to alcohol: A critical review of the literature. *Alcoholism: Clinical & Experimental Research, 34*(3), 385–395.

Naimi, T. S., Brewer, R. D., Mokdad, A., Denny, C., Serdula, M. K., & Marks, J. S. (2003). Bing drinking among US adults. *JAMA, 289*, 70–75.

Naranjo, C. A., & Knoke, D. M. (2001). The role of selective serotonin reuptake inhibitors in reducing alcohol consumption. *Journal of Clinical Psychiatry, 62*, 18–25.

National Institute on Drug Abuse. (2008). *Drugs, brains, and behaviour: The science of addiction* (revised ed.). Washington, DC: National Institute on Drug Abuse.

Nowicki, M., Tran, S., Chatterjee, D., & Gerlai, R. (2015). Inhibition of phosphorylated tyrosine hydroxylase attenuates ethanol-induced hyperactivity in adult zebrafish (*Danio rerio*). *Pharmacology, Biochemistry and Behavior, 138*, 32–39.

Oliveira, T. A., Koalski, G., Kreutz, L. C., Ferreira, D., da Rosa, J. G., de Abreu, M. S., ... Barcellos, L. J. (2013). Alcohol impairs predation risk response and communication in zebrafish. *PLoS One, 8*, e75780.

Pan, Y., Kaiguo, M., Razak, Z., Westwood, J. T., & Gerlai, R. (2011). Chronic alcohol exposure induced gene expression changes in the zebrafish brain. *Behavioural Brain Research, 216*, 66–76.

Pannia, E., Tran, S., Rampersad, M., & Gerlai, R. (2014). Acute ethanol exposure induces behavioral differences in two zebrafish (*Danio rerio*) strains: A time course analysis. *Behavioural Brain Research, 259*, 174–185.

Phillips, T. J., Roberts, A. J., & Lessov, C. N. (1997). Behavioral sensitization to ethanol: Genetics and the effects of stress. *Pharmacology, Biochemistry and Behavior, 57*, 487–493.

Phillips, T. J., & Shen, E. H. (1996). Neurochemical bases of locomotion and ethanol stimulant effects. *International Review of Neurobiology, 243*, 243–282.

Pittman, J., & Hylton, A. (2015). Behavioral, endocrine, and neuronal alterations in zebrafish (*Danio rerio*) following sub-chronic coadministration of fluoxetine and ketamine. *Pharmacology, Biochemistry and Behavior, 139*, 158–162.

Pittman, J. T., & Ichikawa, K. M. (2013). iPhone applications as versatile video tracking tools to analyze behavior in zebrafish (*Danio rerio*). *Pharmacology, Biochemistry and Behavior, 106*, 137–142.

Puttonen, H. A., Sundvik, M., Rozov, S., Chen, Y. C., & Panula, P. (2013). Acute ethanol treatment upregulates Th1, Th2, and Hdc in larval zebrafish in stable networks. *Frontiers in Neural Circuits, 7*, 1–10.

Quoilin, C., Didone, V., Tirelli, E., & Quertemont, E. (2013). Chronic tolerance to ethanol-induced sedation: Implication for age-related differences in locomotor sensitization. *Alcohol, 47*(4), 317–322.

Radlow, R. (1994). A quantitative theory of acute tolerance to alcohol. *Psychopharmacology, 114*(1), 1–8.

Rico, E. P., Rosemberg, D. B., Seibt, K. J., Capiotti, K. M., Da Silva, R. S., & Bonan, C. D. (2011). Zebrafish neurotransmitter systems as potential pharmacological and toxicological targets. *Neurotoxicology and Teratology, 33*(6), 608–617.

Robinson, T. E., & Berridge, K. C. (2008). The incentive sensitization theory of addiction: Some current issues. *Philosophical Transactions of the Royal Society of London B: Biological Sciences, 363*, 3137–3146.

Rosemberg, D. B., Braga, M. M., Rico, E. P., Loss, C. M., Cordova, S. D., Mussulini, B. H., ... Souza, D. O. (2012). Behavioral effects of taurine pretreatment in zebrafish acute exposed to ethanol. *Neuropharmacology, 63*, 613–623.

Sacks, J. J., Gonzales, K. R., Bouchery, E. E., Tomedi, L. E., & Brewer, R. D. (2015). 2010 National and state costs of excessive alcohol consumption. *American Journal of Preventive Medicine, 49*, e73–e79.

Scholtz, H., & Mustard, J. A. (2013). Invertebrate models of alcoholism. *Current Topics in Behavioral Neurosciences, 13*, 433–457.

Smoothy, R., & Berry, M. S. (1985). Time course of the locomotor stimulant and depressant effects of a single low dose of ethanol in mice. *Psychopharmacology, 85*(1), 57–61.

Snelleman, M., Schoenmakers, T. M., & van de Mheen, D. (2014). The relationship between perceived stress and cue sensitivity for alcohol. *Addiction Behaviors, 39*(12), 1884–1889.

Sterling, M. E., Chang, G. Q., Karatyev, O., Chang, S. Y., & Leibowitz, S. F. (2016). Effects of embryonic ethanol exposure at low doses on neuronal development, voluntary ethanol consumption and related behaviors in larval and adult zebrafish: Role of hypothalamic orexigenic peptides. *Behavioral Brain Research, 304*, 125–138.

Sterling, M. E., Karatayev, O., Chang, G. Q., Algava, D. B., & Leibowitz, S. F. (2015). Model of voluntary ethanol intake in zebrafish: Effect on behavior and hypothalamic orexigenic peptides. *Behavioural Brain Research, 278*, 29–39.

Tran, S., Nowicki, M., Muraleetharan, A., Chatterjee, D., & Gerlai, R. (2015). Differential effects of acute administration of SCH-23390, a D_1 receptor antagonist, and of ethanol on swimming activity, anxiety-related responses, and neurochemistry of zebrafish. *Psychopharmacology, 232*, 3709–3718.

Tran, S., Chatterjee, D., & Gerlai, R. (2015). An integrative analysis of ethanol tolerance and withdrawal in zebrafish (*Danio rerio*). *Behavioural Brain Research, 276*, 161–170.

Tran, S., Facciol, A., & Gerlai, R. (2016a). The zebrafish, a novel model organism for screening compounds affecting acute and chronic ethanol-induced effects. *International Review in Neurobiology, 126*, 467–484.

Tran, S., Facciol, A., & Gerlai, R. (2016b). Alcohol-induced behavioral changes in zebrafish: The role of dopamine D_2-like receptors. *Psychopharmacology, 233*, 2119–2128.

Tran, S., & Gerlai, R. (2013). Time-course of behavioral changes induced by ethanol in zebrafish (*Danio rerio*). *Behavioural Brain Research, 252*, 204–213.

Tran, S., & Gerlai, R. (2014). Recent advances with a novel model organism: Alcohol tolerance and sensitization in zebrafish (*Danio rerio*). *Progress in Neuro-psychopharmacology & Biological Psychiatry, 55*, 87–93.

Tran, S., Nowicki, M., Chatterjee, D., & Gerlai, R. (2015). Acute and chronic ethanol exposure differentially alters alcohol dehydrogenase and aldehyde dehydrogenase activity in the zebrafish liver. *Progress in Neuro-psychopharmacology & Biological Psychiatry, 56*, 221–226.

Tropepe, V., & Sive, H. L. (2003). Can zebrafish be used as a model to study the neurodevelopmental causes of autism? *Genes Brain and Behavior, 2*(5), 268–281.

Vorspan, F., Mehtelli, W., Dupuy, G., Bloch, V., & Lepine, J. P. (2015). Anxiety and substance use disorders: Co-occurrence and clinical issues. *Current Psychiatry Reports, 17*(2), 4.

CHAPTER

8

Effect of Alcohol on the Regulation of α-Synuclein in the Human Brain

P. Janeczek, J.M. Lewohl

Griffith University, Gold Coast Campus, QLD, Australia

OVERVIEW

Alcoholism is a chronic, relapsing condition resulting in neuronal damage, which in turn reduces the behavioral control over alcohol intake and further cycles of alcohol misuse and neurodegeneration (Crews & Nixon, 2009). There are few treatments for alcoholism, and a clear understanding of the mechanisms underlying the disease is needed for the development of new therapies. The effects of excessive alcohol consumption can be seen throughout the body; one of the commonly known pathologies being cirrhosis of the liver. Long-term alcohol abuse also results in pathological damage to the brain, including but not limited to structural changes, neuronal loss, and loss of brain weight (de la Monte, 1988; Harding, Halliday, Ng, Harper, & Kril, 1996; Harper, 2009; Harper & Kril, 1985, 1989). These changes are not uniform and some areas of the brain, such as the prefrontal cortex, are more susceptible to the neurotoxic damage caused by alcohol (Harper & Kril, 1989; Kril, Halliday, Svoboda, & Cartwright, 1997). Neuronal loss in this region has been associated with the development and persistence of alcohol addiction (Abernathy, Chandler, & Woodward, 2010), and alcohol-induced neurotoxic damage in this region is likely to amplify the reinforcing effects of alcohol (Abernathy et al., 2010; Crews & Boettiger, 2009). The neuropathological abnormalities also include alterations to neuronal function and adaptions to alcohol, which are mediated, at least in part, by gene expression (Nestler & Aghajanian, 1997). It is well accepted that gene expression, and therefore gene regulation, is important in disease control and manifestation. There are several possible mechanisms for gene regulation; genetic variation within or near the promoter region of a candidate gene, as well as regulation via micro-RNA (miRNA) interaction with the 3′-untranslated region (3′-UTR) of a candidate gene.

Previous studies in which global gene and protein expression profiling techniques have been used to conduct an unbiased assessment of changes in gene expression that occur as part of the adaptive response of neurons in the prefrontal cortex, have identified a number of synaptic genes and proteins which are differentially regulated in the alcoholic brain (Etheridge, Lewohl, Mayfield, Harris, & Dodd, 2009; Lewohl et al., 2004, 2000; Liu et al., 2004; Liu, Lewohl, Harris, Dodd, & Mayfield, 2007; Liu et al., 2006; Mayfield et al., 2002). One of these genes is α-synuclein; a protein known to have a role in mediating dopaminergic neurotransmission (Baptista et al., 2003; Yavich, Tanila, Vepsalainen, & Jakala, 2004).

Dopaminergic neurotransmission mediates mechanisms involved in craving, withdrawal, and the reinforcing effects of alcohol and other drugs of abuse (Self and Nestler, 1998). The dopaminergic reward pathway projects from the nucleus accumbens to the ventral tegmentum and the prefrontal cortex, the region of the brain responsible for executive functions that are disrupted following chronic alcohol abuse in humans (Koob & Volkow, 2010). This provides further evidence for the involvement of α-synuclein in the pathophysiology of chronic alcohol abuse, with altered α-synuclein expression having the potential to change the function of the dopaminergic reward pathway.

α-Synuclein (*SNCA*), a protein abundantly expressed in neurons (Iwai et al., 1995; Jakes, Spillantini, & Goedert, 1994; Withers, George, Banker, & Clayton, 1997; Zhang et al., 2008), has a well-established role in neurodegenerative and neuropsychological disorders (Goedert, 2001), and is a candidate gene for alcohol

Addictive Substances and Neurological Disease
http://dx.doi.org/10.1016/B978-0-12-805373-7.00008-6

abuse (Levey et al., 2014). Changes to α-synuclein expression may therefore have severe consequences on these key pathways and may increase susceptibility to alcohol addiction. In this chapter we discuss the role of α-synuclein in mediating the neurotoxic effects of alcohol, the potential for sequence variations in the α-synuclein gene to influence the risk of developing an alcohol use phenotype as well as the role miRNAs play in the regulation of this important transcript.

α-SYNUCLEIN FUNCTION

α-Synuclein is predominately found in presynaptic terminals (Iwai et al., 1995; Jakes et al., 1994; Withers et al., 1997; Zhang et al., 2008) where it is a key mediator of dopaminergic neurotransmission; influencing the synthesis, storage, release, and reuptake of the neurotransmitter (Baptista et al., 2003; Yavich et al., 2004). In general, α-synuclein is a negative regulator of dopaminergic neurotransmission (Abeliovich et al., 2000; Yavich et al., 2004). The suppression of α-synuclein results in a reduction of the surface expression and maximal uptake velocity of the dopamine transporter (Fountaine & Wade-Martins, 2007; Lee, Liu, Pristupa, & Niznik, 2001; Wersinger & Sidhu, 2003). Alterations in α-synuclein expression have been shown to result in dopamine-induced apoptosis (Fountaine & Wade-Martins, 2007; Lee et al., 2001; Wersinger & Sidhu, 2003) and likely affect dopamine-mediated neuronal signaling. As dopaminergic neurotransmission is thought to be the key mediator of craving, withdrawal, and reinforcement pathways in alcohol addiction (Self and Nestler, 1998), changes to α-synuclein expression may affect the function of the dopaminergic reward pathway. Thus, an individual may be more susceptible to addiction or various psychiatric diseases depending on the levels of α-synuclein in their brain (Oksman, Tanila, & Yavich, 2006).

α-Synuclein is a negative regulator of dopamine release (Abeliovich et al., 2000; Perez et al., 2002) and utilizes transcriptional regulation of key genes involved in dopamine synthesis to regulate dopamine expression (Baptista et al., 2003). Dopamine reuptake by the dopamine transporter is also influenced by α-synuclein (Wersinger & Sidhu, 2003). Studies have shown that partial or complete knockdown of α-synuclein in laboratory animals up-regulates dopamine release (Abeliovich et al., 2000), reduces the uptake of dopamine (Fountaine & Wade-Martins, 2007), and exacerbates the brain reward system (Oksman et al., 2006). Both animal models and human studies have shown moderated α-synuclein expression induced by exposure to alcohol and other drugs of abuse. Increased craving for alcohol was correlated with elevated α-synuclein serum levels in actively drinking and recently withdrawn alcoholics (Bonsch, Greifenberg, et al., 2005; Bonsch et al., 2004). A threefold increase in α-synuclein expression is seen in primates that self-administer alcohol compared to the alcohol-naive controls (Walker & Grant, 2006). Expression of α-synuclein is also influenced by genetic variation in the 3′-UTR, where a single polymorphic variant resulted in altered expression in rats selectively bred for high alcohol preference (Liang et al., 2003). α-Synuclein is also differentially expressed in the prefrontal cortex of human alcoholics (Etheridge et al., 2009; Lewohl et al., 2004, 2000).

α-Synuclein is up-regulated in response to excitotoxicity and oxidative stress, this suggests that α-synuclein may have a role in protecting neuronal cells against damage (Sidhu, Wersinger, Moussa, & Vernier, 2004). Both low and high levels of α-synuclein expression result in cytotoxicity, suggesting a dual role in neurotoxicity and neuroprotection that is dependent on expression levels (Seo et al., 2002; Sidhu et al., 2004). The individual α-synuclein splice variants may have defined roles in this response.

Down-regulation of α-synuclein is seen in the prefrontal cortex of alcoholics (Janeczek, Brooker, Dodd, & Lewohl, 2015; Liu et al., 2006). Lower levels of α-synuclein may therefore increase susceptibility of neurons to the neurotoxic effects of alcohol due to a loss of its neuroprotective effects. As α-synuclein also plays a role in trafficking of neurotransmitter transporters, neurons with lowered α-synuclein levels may lose the ability to adapt to the continual presence of alcohol in the brain. Therefore, it is thought that α-synuclein may play a part in the alcohol addiction pathway. Normal physiological regulation of α-synuclein may be affected by genetic variation increasing the susceptibility of individuals to the disease.

The Effect of Genetic Variation on α-Synuclein Expression

The α-synuclein transcript is highly polymorphic, containing a number of sequence variants that have been associated with alcohol craving and dependence, as well as influencing gene expression. Previous studies have investigated the effect of genetic variability on α-synuclein expression and whether this contributes to the susceptibility of an alcohol abuse phenotype. The allele length of the microsatellite repeat marker, α-synuclein-repeat 1 (*SNCA*-Rep1), located about 10-kb upstream from the translational start site (Touchman et al., 2001) has previously been associated with expression levels of α-synuclein in cell lines, blood and brain (Fuchs et al., 2008; Linnertz et al., 2009; McCarthy et al., 2011). The longer *SNCA*-Rep1 allele lengths were correlated with increased expression levels in neuroblastoma cell lines (Chiba-Falek & Nussbaum, 2001).

SNCA-Rep1 allele length has also been associated with alcohol use phenotypes, however the findings are inconsistent. Alcohol-dependent individuals were found to have a greater frequency of the longer (271 and 273 bp) alleles, and this was correlated with higher α-synuclein expression in blood (Bonsch, Lederer, et al., 2005). A 2014 study correlated mRNA expression levels of the wildtype α-synuclein splice variant *SNCA*-140 in the prefrontal cortex of human alcoholics and controls to determine if the microsatellite marker *SNCA*-Rep1 altered the regulation of the expression of this gene and whether this contributed to the susceptibility of an alcohol abuse phenotype (Janeczek, MacKay, Lea, Dodd, & Lewohl, 2014). Alcoholics had a greater frequency of the shortest allele found in the Caucasian population (267 bp) which was associated with reduced expression of α-synuclein in the prefrontal cortex of alcoholics. In addition, individuals with at least one copy of the 267-bp allele were more likely to have shown an alcohol abuse phenotype. The results suggested that individuals with the 267-bp allele may have decreased expression of α-synuclein and may be more susceptible to developing an alcohol abuse phenotype. These conflicting expression patterns of α-synuclein in alcoholics could result from differences in expression levels of the individual α-synuclein splice variants.

In addition, previous studies have also explored the influence of genetic variation on α-synuclein expression by investigating the effect of several candidate single nucleotide polymorphisms (SNPs) throughout the gene on both expression and susceptibility to the disease. Some of these SNPs have been associated with alcohol abuse phenotypes. Eight SNPs located throughout the 5′- and 3′-ends of *SNCA* have been previously found to be associated with alcohol dependence in a subpopulation of individuals who also experienced craving (Foroud et al., 2007). In this subpopulation, a haplotype block in the 3′-UTR of *SNCA* consisting of these SNPs was over-transmitted to individuals who experienced craving, whereas the complementary haplotype was over-transmitted to individuals who did not experience craving. A 2015 study explored these eight SNPs in a Caucasian population and investigated whether there was an association with α-synuclein expression in autopsy samples of the prefrontal cortex of human alcoholics (Janeczek et al., 2015). Although three of the SNPs, rs356221, rs356219, and rs2736995 had significant differences in allele and genotype frequencies, these SNPs were not associated with increased risk. The results did, however, suggest that the rs356219/356221 G-A haplotype may reduce the risk of developing an alcohol abuse phenotype. These SNPs did not appear to influence the expression of α-synuclein. Due to the complexity of the alcohol addiction pathway it is unlikely that an SNP would have a vital effect on expression levels, these findings do, however, suggest that genetic variation near the *SNCA* promoter region can down-regulate gene expression and increase susceptibility to alcohol addiction.

α-Synuclein Splice Variants

According to the online *Ensembl* database (www.ensembl.org) there are a number of distinct α-synuclein splice variants including two protein coding variants, *SNCA*-140 and *SNCA*-112. These variants differ by the number of exons they contain with the sequence for the *SNCA*-140 variant (ENST00000394986) (Jakes et al., 1994; Ueda et al., 1993) encoding a 140 amino acid protein, containing all six exons. The *SNCA*-112 variant encodes for the shorter 112 amino acid protein that excludes exon 5 (ENST00000420646) (Ueda, Saitoh, & Mori, 1994). These two variants share the same sequence but differ in length, in particular the length of their 3′-UTR. In addition, *SNCA*-115 is a novel predicted variant in the *Ensembl* database that encodes a 115 amino acid protein and includes exons 1–4 and 393 nucleotides of intron 4 (ENST00000502987). The *SNCA*-115 3′-UTR is located in intron 4 and is composed of a completely unique sequence compared to the other two splice variants.

A 2015 study and the only study of its kind to date, measured the expression levels of three α-synuclein splice variants, *SNCA*-140, *SNCA*-112, and *SNCA*-115 in the prefrontal cortex of uncomplicated alcoholics, alcoholics with concomitant cirrhosis of the liver, and controls (Janeczek et al., 2015). Previous studies reported conflicting expression patterns of α-synuclein in alcoholics (Etheridge et al., 2009; Janeczek et al., 2014; Lewohl et al., 2004, 2000; Liu et al., 2004, 2006; Mayfield et al., 2002), with some studies reporting increased expression of α-synuclein and others reporting decreased expression of this same gene. The results of our study may explain the discrepancies seen in previous studies, as two of the splice variants *SNCA*-140 and *SNCA*-112 were found to have reduced expression levels in the prefrontal cortex of cirrhotic alcoholics, whereas the novel *SNCA*-115 was increased. These findings suggest that the α-synuclein splice variants are differentially expressed in human alcoholic brain and may therefore have different roles in the brain. The mechanisms of α-synuclein regulation in the brain, in particular the regulation of the individual α-synuclein splice variants and how this is mediated by alcohol are yet to be elucidated.

The Effect of miRNAs on α-Synuclein Expression

miRNAs are short noncoding RNAs, highly abundant in the brain, that regulate gene expression by binding to the 3′-UTR region of their target mRNAs. Each miRNA binds to a specific target sequence in the 3′-UTR referred to as the miRNA-binding site that contains a ~8-nucleotide seed sequence and can be complementary or partially complementary. Once the miRNA binds to the 3′-UTR of a target, translation of the mRNA is inhibited. Some miRNAs also target the mRNA for degradation (Bushati & Cohen, 2008). A single miRNA can regulate the expression of several genes and the 3′-UTR of a target gene can contain several binding sites for different miRNAs. miRNA-mediated regulation is a well-known mechanism for fine-tuning expression of genes in brain (Bartel & Chen, 2004; Georges, Coppieters, & Charlier, 2007; Hornstein & Shomron, 2006). Therefore, miRNAs have the capacity to regulate a large network of genes involved in neuronal plasticity, such as genes involved in the reward pathway in the prefrontal cortex (Schratt et al., 2006).

A number of alcohol-responsive miRNAs have been identified through studies in rodent and cell culture models, as well as in human alcoholic brain (Balaraman, Winzer-Serhan, & Miranda, 2012; Guo, Chen, Carreon, & Qiang, 2012; Lewohl et al., 2011; Sathyan, Golden, & Miranda, 2007; Tapocik et al., 2013; van Steenwyk, Janeczek, & Lewohl, 2013; Wang et al., 2009; Yadav et al., 2011). Rats with a history of alcohol dependence were found to have dysregulated expression levels of specific miRNAs and their mRNA target transcripts in the medial prefrontal cortex (Tapocik et al., 2013). Further, about 35 miRNAs were found to be significantly up-regulated in the prefrontal cortex of human alcoholics compared with controls (Lewohl et al., 2011). The predicted targets for these miRNAs showed a large degree of overlap with down-regulated genes identified in previous cDNA microarray studies (Liu et al., 2006) suggesting that up-regulation of miRNAs in the prefrontal cortex of human alcoholics may contribute to the deterioration and concomitant adaptation of neuronal functioning observed in cases of alcohol abuse.

A number of alcohol-responsive miRNAs (Lewohl et al., 2011) are predicted to target α-synuclein: miR-7, miR-153, miR-144, miR-203, miR-454, miR-152, and miR-374b according to microRNA.org. Two of these miRNAs, miR-7 and miR-153, have been experimentally validated. MiR-7, a miRNA predominately expressed in neurons, has been shown to regulate α-synuclein, by binding to the 3′-UTR and repressing its translation in a dose-dependent manner (Junn et al., 2009). MiR-153 down-regulates α-synuclein mRNA and protein levels posttranscriptionally, by binding to the specific sites on the 3′-UTR (Doxakis, 2010). Thus, miR-7 acts by inhibiting the translation of α-synuclein, whereas miR-153 acts by promoting the degradation of the mRNA, and they work synergistically to lower α-synuclein levels. Other miRNAs may also target α-synuclein, but these are yet to be experimentally validated.

The up-regulation of miR-7 and miR-153 in the prefrontal cortex of alcoholics is likely to result in the down-regulation of their predicted targets such as α-synuclein (Janeczek et al., 2015; Liu et al., 2006). To date, only the interaction between these miRNAs and the wildtype (*SNCA*-140) variant has been investigated. As such, it is not known whether each of the α-synuclein splice variants are targeted by the same miRNAs. However, it is likely that each variant is differentially targeted based on their unique 3′-UTR sequence and, as such, differences in miRNA-binding sites within each of the splice variants may affect the way the variants are expressed. This type of selective regulation of splice variants has been identified previously with alternatively spliced mRNAs encoding the large-conductance calcium- and voltage-activated potassium channel (BK), which is a known target of alcohol's actions in mediating molecular alcohol tolerance (Pietrzykowski et al., 2008). Alcohol caused a rapid up-regulation in miR-9 expression, resulting in selective degradation of BK mRNAs containing a miR-9 target site in their 3′-UTRs. The selective degradation of some splice variants but not others altered the profile of BK channels, consistent with the development of tolerance to alcohol (Pietrzykowski et al., 2008). This represents a new mechanism of gene regulation of splice variants that may underlie the neuroadaptive changes that occur at a cellular level to long-term alcohol exposure.

The Effect of Genetic Variation on miRNA-Mediated α-Synuclein Regulation

The 3′-UTR of *SNCA* contains many SNPs that have the potential to alter miRNA-mediated regulation of α-synuclein. A polymorphism may affect the binding of a miRNA to its binding site within the 3′-UTR of α-synuclein, by altering the secondary structure or unfolding of the 3′-UTR and inhibiting access to the site (Kertesz et al., 2007). Changes to a miRNA-binding site in the *SNCA* 3′-UTR may abolish or strengthen the binding affinity a miRNA has for its target, and also has the potential to create a new miRNA-binding site. Thus, genetic variation within a miRNA-binding site has the potential to alter gene expression and can influence the susceptibility of an individual developing an alcohol abuse phenotype (Chen et al., 2008; Georges et al., 2007).

As previously mentioned two SNPs, rs356221 and rs356219, were found to be associated with alcohol dependence in a subpopulation of individuals who also experienced craving (Foroud et al., 2007). Results in a 2015 study suggested that the rs356219/356221

G-A haplotype may reduce the risk of developing an alcohol abuse phenotype; however, this did not appear to influence the expression of α-synuclein in the prefrontal cortex of human alcoholics (Janeczek et al., 2015). Although the rs356219 and rs356221 are located further downstream of the *SNCA* 3′-UTR, they may still have the potential to influence α-synuclein expression. There are several SNPs within the *SNCA* 3′-UTR of each of the individual splice variants, future studies should investigate whether any SNPs within or near any known miRNA-binding sites have the potential to influence α-synuclein expression and are associated with alcohol abuse phenotypes. Although it is unlikely that an SNP would have a significant impact on α-synuclein gene expression, it may be one factor contributing to the altered expression changes seen in alcoholic brain. Genetic variation within the 3′-UTR and the effect of this on miRNA-mediated regulation of α-synuclein is a topic that requires further investigation as more miRNA: *SNCA* interactions are validated.

Since the α-synuclein splice variants are predicted to be targeted by different miRNAs, it may be possible to regulate the amounts of each of these variants in vivo using synthetic miRNAs. miRNAs may have therapeutic potential in protecting the brain against alcohol-induced damage by altering selective α-synuclein splice variants to promote neuroprotection. Based on previous studies, it has been suggested that α-synuclein may play a role in neuroprotection (Seo et al., 2002), and that one of the splice variants may be specifically involved in protecting the brain from damage. Thus, synthetic miRNAs could be used to selectively reduce expression levels of the α-synuclein splice variants or antagomirs could be used to knock down specific miRNAs allowing the up-regulation of selective splice variants. MiR-7 and miR-153 are potential candidates for such studies.

Down-regulation of α-synuclein is seen in the prefrontal cortex of alcoholics. Lower levels of α-synuclein may therefore increase susceptibility of neurons to the neurotoxic effects of alcohol due to a loss of its neuroprotective effects. Since α-synuclein also plays a role in trafficking of neurotransmitter transporters, neurons with lowered α-synuclein levels may lose the ability to adapt to the continual presence of alcohol in the brain. Therefore, it is thought that α-synuclein may play a part in the alcohol addiction pathway. Normal physiological regulation of α-synuclein may be affected by genetic variation increasing the susceptibility of individuals who consume alcohol long term at high levels to the disease.

CONCLUSION

The prefrontal cortex is particularly susceptible to the neurotoxic effects of alcohol and neuronal loss in this region has been associated with alcohol abuse (Harper & Kril, 1989; Kril et al., 1997). Gene expression studies using the prefrontal cortex have shown significant variations in gene expression of several key synaptic proteins thought to be involved in alcohol addiction (Liu et al., 2006). Several of the genes found to be down-regulated in the prefrontal cortex are predicted targets of about 35 alcohol-responsive miRNAs which are up-regulated in the prefrontal cortex of alcoholic brain (Lewohl et al., 2011). This suggests that miRNA-mediated regulation may be involved in the expression changes and cellular adaptions seen in the alcoholic brain. The prefrontal cortex forms part of the dopaminergic reward pathway and is responsible for executive functions. Alterations to gene expression, cellular adaptions, and essentially damage to this region of the brain, induced by long-term alcohol abuse are likely to contribute to the development and persistence of alcohol abuse and other drug addictions (Robison & Nestler, 2011). Changes to miRNA levels in this region following chronic alcohol abuse, may lead to gene dysregulation and induce the neuronal cellular adaptions that result in the functional and structural changes seen in alcoholic brain.

Gene expression and therefore gene regulation is central in disease control and manifestation. Studies are now highlighting the importance of α-synuclein in alcohol addiction and expanding our understanding of the regulation of this vital synaptic gene. This also highlights the potential impact of genetic variation and miRNAs in the development of alcohol-related gene expression changes in human brain. Studies have provided preliminary evidence of increased susceptibility of developing an alcohol abuse phenotype due to genetic variation near the promoter region of α-synuclein, and proposed selective miRNA-mediated regulation of the α-synuclein splice variants in brain which may be effected by alcohol exposure and further endorse α-synuclein as a candidate gene for alcoholism.

Chronic exposure to alcohol may result in changes to α-synuclein expression which in turn effects neuronal functions and may contribute to the mechanisms involved in craving, alcohol dependence, and neuronal loss. miRNAs may have roles in synapse formation and plasticity, which may be involved in the reward circuitry of the brain. As α-synuclein is a mediator of dopaminergic neurotransmission with a significant role in the reward system this provides further evidence of the potential involvement of α-synuclein in the changes observed in alcoholic brain following long-term alcohol exposure. miRNAs may therefore mediate these cellular adaptions that occur in response to alcohol exposure. This also provides a new area for investigation with the potential for therapeutic modulation in the treatment of addiction as well as other neurodegenerative diseases.

Determining the regulatory mechanisms of the individual α-synuclein splice variants will aid in discovering their biological functions in neurons, and their roles in not only alcohol addiction but other neurodegenerative diseases where α-synuclein dysfunction is present.

References

Abeliovich, A., Schmitz, Y., Farinas, I., Choi-Lundberg, D., Ho, W. H., Castillo, P. E., ... Rosenthal, A. (2000). Mice lacking alpha-synuclein display functional deficits in the nigrostriatal dopamine system. *Neuron, 25*, 239–252.

Abernathy, K., Chandler, L. J., & Woodward, J. J. (2010). Alcohol and the prefrontal cortex. *International Review of Neurobiology, 91*, 289–320.

Balaraman, S., Winzer-Serhan, U. H., & Miranda, R. C. (2012). Opposing actions of ethanol and nicotine on microRNAs are mediated by nicotinic acetylcholine receptors in fetal cerebral cortical-derived neural progenitor cells. *Alcoholism: Clinical and Experimental Research, 36*, 1669–1677.

Baptista, M. J., O'Farrell, C., Daya, S., Ahmad, R., Miller, D. W., Hardy, J., ... Cookson, M. R. (2003). Co-ordinate transcriptional regulation of dopamine synthesis genes by α-synuclein in human neuroblastoma cell lines. *Journal of Neurochemistry, 85*, 957–968.

Bartel, D. P., & Chen, C. Z. (2004). Micromanagers of gene expression: The potentially widespread influence of metazoan microRNAs. *Nature Reviews Genetics, 5*, 396–400.

Bonsch, D., Greifenberg, V., Bayerlein, K., Biermann, T., Reulbach, U., Hillemacher, T., ... Bleich, S. (2005). Alpha-synuclein protein levels are increased in alcoholic patients and are linked to craving. *Alcoholism: Clinical and Experimental Research, 29*, 763–765.

Bonsch, D., Lederer, T., Reulbach, U., Hothorn, T., Kornhuber, J., & Bleich, S. (2005). Joint analysis of the NACP-REP1 marker within the alpha synuclein gene concludes association with alcohol dependence. *Human Molecular Genetics, 14*, 967–971.

Bonsch, D., Reulbach, U., Bayerlein, K., Hillemacher, T., Kornhuber, J., & Bleich, S. (2004). Elevated alpha synuclein mRNA levels are associated with craving in patients with alcoholism. *Biological Psychiatry, 56*, 984–986.

Bushati, N., & Cohen, S. M. (2008). microRNAs in neurodegeneration. *Current Opinion in Neurobiology, 18*, 292–296.

Chen, K., Song, F., Calin, G. A., Wei, Q., Hao, X., & Zhang, W. (2008). Polymorphisms in microRNA targets: A gold mine for molecular epidemiology. *Carcinogenesis, 29*, 1306–1311.

Chiba-Falek, O., & Nussbaum, R. L. (2001). Effect of allelic variation at the NACP-Rep1 repeat upstream of the α-synuclein gene (*SNCA*) on transcription in a cell culture luciferase reporter system. *Human Molecular Genetics, 10*, 3101–3109.

Crews, F. T., & Boettiger, C. A. (2009). Impulsivity, frontal lobes and risk for addiction. *Pharmacology Biochemistry and Behavior, 93*, 237–247.

Crews, F. T., & Nixon, K. (2009). Mechanisms of neurodegeneration and regeneration in alcoholism. *Alcohol and Alcoholism, 44*, 115–127.

Doxakis, E. (2010). Post-transcriptional regulation of alpha-synuclein expression by mir-7 and mir-153. *The Journal of Biological Chemistry, 285*, 12726–12734.

Etheridge, N., Lewohl, J. M., Mayfield, R. D., Harris, R. A., & Dodd, P. R. (2009). Synaptic proteome changes in the superior frontal gyrus and occipital cortex of the alcoholic brain. *Proteomics Clinical Applications, 3*, 730–742.

Foroud, T., Wetherill, L. F., Liang, T., Dick, D. M., Hesselbrock, V., Kramer, J., ... Edenberg, H. J. (2007). Association of alcohol craving with α-synuclein (*SNCA*). *Alcoholism: Clinical and Experimental Research, 31*, 537–545.

Fountaine, T. M., & Wade-Martins, R. (2007). RNA interference-mediated knockdown of α-synuclein protects human dopaminergic neuroblastoma cells from MPP^+ toxicity and reduces dopamine transport. *Journal of Neuroscience Research, 85*, 351–363.

Fuchs, J., Tichopad, A., Golub, Y., Munz, M., Schweitzer, K. J., Wolf, B., ... Gasser, T. (2008). Genetic variability in the *SNCA* gene influences α-synuclein levels in the blood and brain. *FASEB Journal, 22*, 1327–1334.

Georges, M., Coppieters, W., & Charlier, C. (2007). Polymorphic miRNA-mediated gene regulation: Contribution to phenotypic variation and disease. *Current Opinion in Genetics & Development, 17*, 166–176.

Goedert, M. (2001). Alpha-synuclein and neurodegenerative diseases. *Nature Reviews Neuroscience, 2*, 492–501.

Guo, Y., Chen, Y., Carreon, S., & Qiang, M. (2012). Chronic intermittent ethanol exposure and its removal induce a different miRNA expression pattern in primary cortical neuronal cultures. *Alcoholism: Clinical and Experimental Research, 36*, 1058–1066.

Harding, A. J., Halliday, G. M., Ng, J. L., Harper, C. G., & Kril, J. J. (1996). Loss of vasopressin-immunoreactive neurons in alcoholics is dose-related and time-dependent. *Neuroscience, 72*, 699–708.

Harper, C. (2009). The neuropathology of alcohol-related brain damage. *Alcohol and Alcoholism, 44*, 136–140.

Harper, C., & Kril, J. (1985). Brain atrophy in chronic alcoholic patients: A quantitative pathological study. *Journal of Neurology, Neurosurgery & Psychiatry, 48*, 211–217.

Harper, C., & Kril, J. (1989). Patterns of neuronal loss in the cerebral cortex in chronic alcoholic patients. *Journal of the Neurological Sciences, 92*, 81–89.

Hornstein, E., & Shomron, N. (2006). Canalization of development by microRNAs. *Nature Genetics, 38*(Suppl.), S20–S24.

Iwai, A., Masliah, E., Yoshimoto, M., Ge, N., Flanagan, L., Rohan de Silva, H. A., ... Saitoh, T. (1995). The precursor protein of non-Aβ component of Alzheimer's disease amyloid is a presynaptic protein of the central nervous system. *Neuron, 14*, 467–475.

Jakes, R., Spillantini, M. G., & Goedert, M. (1994). Identification of two distinct synucleins from human brain. *FEBS Letters, 345*, 27–32.

Janeczek, P., Brooker, C., Dodd, P. R., & Lewohl, J. M. (2015). Differential expression of alpha-synuclein splice variants in the brain of alcohol misusers: Influence of genotype. *Drug and Alcohol Dependence, 155*, 284–292.

Janeczek, P., MacKay, R. K., Lea, R. A., Dodd, P. R., & Lewohl, J. M. (2014). Reduced expression of alpha-synuclein in alcoholic brain: Influence of *SNCA*-Rep1 genotype. *Addiction Biology, 19*, 509–515.

Junn, E., Lee, K. W., Jeong, B. S., Chan, T. W., Im, J. Y., & Mouradian, M. M. (2009). Repression of alpha-synuclein expression and toxicity by microRNA-7. *Proceedings of the National Academy of Sciences of the United States of America, 106*, 13052–13057.

Kertesz, M., Iovino, N., Unnerstall, U., Gaul, U., & Segal, E. (2007). The role of site accessibility in microRNA target recognition. *Nature Genetics, 39*, 1278–1284.

Koob, G. F., & Volkow, N. D. (2010). Neurocircuitry of addiction. *Neuropsychopharmacology, 35*, 217–238.

Kril, J. J., Halliday, G. M., Svoboda, M. D., & Cartwright, H. (1997). The cerebral cortex is damaged in chronic alcoholics. *Neuroscience, 79*, 983–998.

Lee, F. J., Liu, F., Pristupa, Z. B., & Niznik, H. B. (2001). Direct binding and functional coupling of α-synuclein to the dopamine transporters accelerate dopamine-induced apoptosis. *FASEB Journal, 15*, 916–926.

Levey, D. F., Le-Niculescu, H., Frank, J., Ayalew, M., Jain, N., Kirlin, B., ... Niculescu, A. B. (2014). Genetic risk prediction and neurobiological understanding of alcoholism. *Translational Psychiatry, 4*, e391.

Lewohl, J. M., Nunez, Y. O., Dodd, P. R., Tiwari, G. R., Harris, R. A., & Mayfield, R. D. (2011). Up-regulation of microRNAs in brain of human alcoholics. *Alcoholism: Clinical and Experimental Research, 35*, 1928–1937.

Lewohl, J. M., Van Dyk, D. D., Craft, G. E., Innes, D. J., Mayfield, R. D., Cobon, G., ... Dodd, P. R. (2004). The application of proteomics to the human alcoholic brain. *Annals of the New York Academy of Sciences, 1025*, 14–26.

Lewohl, J. M., Wang, L., Miles, M. F., Zhang, L., Dodd, P. R., & Harris, R. A. (2000). Gene expression in human alcoholism: Microarray analysis of frontal cortex. *Alcoholism: Clinical and Experimental Research, 24*, 1873–1882.

Liang, T., Spence, J., Liu, L., Strother, W. N., Chang, H. W., Ellison, J. A., ... Carr, L. G. (2003). Alpha-synuclein maps to a quantitative trait locus for alcohol preference and is differentially expressed in alcohol-preferring and -nonpreferring rats. *Proceedings of the National Academy of Sciences of the United States of America, 100*, 4690–4695.

Linnertz, C., Saucier, L., Ge, D., Cronin, K. D., Burke, J. R., Browndyke, J. N., ... Chiba-Falek, O. (2009). Genetic regulation of alpha-synuclein mRNA expression in various human brain tissues. *PLoS One, 4*, e7480.

Liu, J., Lewohl, J. M., Dodd, P. R., Randall, P. K., Harris, R. A., & Mayfield, R. D. (2004). Gene expression profiling of individual cases reveals consistent transcriptional changes in alcoholic human brain. *Journal of Neurochemistry, 90*, 1050–1058.

Liu, J., Lewohl, J. M., Harris, R. A., Dodd, P. R., & Mayfield, R. D. (2007). Altered gene expression profiles in the frontal cortex of cirrhotic alcoholics. *Alcoholism: Clinical and Experimental Research, 31*, 1460–1466.

Liu, J., Lewohl, J. M., Harris, R. A., Iyer, V. R., Dodd, P. R., Randall, P. K., & Mayfield, R. D. (2006). Patterns of gene expression in the frontal cortex discriminate alcoholic from nonalcoholic individuals. *Neuropsychopharmacology, 31*, 1574–1582.

Mayfield, R. D., Lewohl, J. M., Dodd, P. R., Herlihy, A., Liu, J., & Harris, R. A. (2002). Patterns of gene expression are altered in the frontal and motor cortices of human alcoholics. *Journal of Neurochemistry, 81*, 802–813.

McCarthy, J. J., Linnertz, C., Saucier, L., Burke, J. R., Hulette, C. M., Welsh-Bohmer, K. A., & Chiba-Falek, O. (2011). The effect of *SNCA* 3′ region on the levels of *SNCA*-112 splicing variant. *Neurogenetics, 12*, 59–64.

de la Monte, S. M. (1988). Disproportionate atrophy of cerebral white matter in chronic alcoholics. *Archives of Neurology, 45*, 990–992.

Nestler, E. J., & Aghajanian, G. K. (1997). Molecular and cellular basis of addiction. *Science, 278*, 58–63.

Oksman, M., Tanila, H., & Yavich, L. (2006). Brain reward in the absence of alpha-synuclein. *NeuroReport, 17*, 1191–1194.

Perez, R. G., Waymire, J. C., Lin, E., Liu, J. J., Guo, F., & Zigmond, M. J. (2002). A role for α-synuclein in the regulation of dopamine biosynthesis. *The Journal of Neuroscience, 22*, 3090–3099.

Pietrzykowski, A. Z., Friesen, R. M., Martin, G. E., Puig, S. I., Nowak, C. L., Wynne, P. M., ... Treistman, S. N. (2008). Posttranscriptional regulation of BK channel splice variant stability by miR-9 underlies neuroadaptation to alcohol. *Neuron, 59*, 274–287.

Robison, A. J., & Nestler, E. J. (2011). Transcriptional and epigenetic mechanisms of addiction. *Nature Reviews Neuroscience, 12*, 623–637.

Sathyan, P., Golden, H. B., & Miranda, R. C. (2007). Competing interactions between micro-RNAs determine neural progenitor survival and proliferation after ethanol exposure: Evidence from an ex vivo model of the fetal cerebral cortical neuroepithelium. *The Journal of Neuroscience, 27*, 8546–8557.

Schratt, G. M., Tuebing, F., Nigh, E. A., Kane, C. G., Sabatini, M. E., Kiebler, M., & Greenberg, M. E. (2006). A brain-specific microRNA regulates dendritic spine development. *Nature, 439*, 283–289.

Self, D. W., & Nestler, E. J. (1998). Relapse to drug-seeking: Neural and molecular mechanisms. *Drug and Alcohol Dependence, 51*, 49–60.

Seo, J. H., Rah, J. C., Choi, S. H., Shin, J. K., Min, K., Kim, H. S., ... Suh, Y. H. (2002). Alpha-synuclein regulates neuronal survival via Bcl-2 family expression and PI3/Akt kinase pathway. *FASEB Journal, 16*, 1826–1828.

Sidhu, A., Wersinger, C., Moussa, C. E., & Vernier, P. (2004). The role of α-synuclein in both neuroprotection and neurodegeneration. *Annals of the New York Academy of Sciences, 1035*, 250–270.

van Steenwyk, G., Janeczek, P., & Lewohl, J. M. (2013). Differential effects of chronic and chronic-intermittent ethanol treatment and its withdrawal on the expression of miRNAs. *Brain Sciences, 3*, 744–756.

Tapocik, J. D., Solomon, M., Flanigan, M., Meinhardt, M., Barbier, E., Schank, J. R., ... Heilig, M. (2013). Coordinated dysregulation of mRNAs and microRNAs in the rat medial prefrontal cortex following a history of alcohol dependence. *The Pharmacogenomics Journal, 13*, 286–296.

Touchman, J. W., Dehejia, A., Chiba-Falek, O., Cabin, D. E., Schwartz, J. R., Orrison, B. M., ... Nussbaum, R. L. (2001). Human and mouse α-synuclein genes: Comparative genomic sequence analysis and identification of a novel gene regulatory element. *Genome Research, 11*, 78–86.

Ueda, K., Fukushima, H., Masliah, E., Xia, Y., Iwai, A., Yoshimoto, M., ... Saitoh, T. (1993). Molecular cloning of cDNA encoding an unrecognized component of amyloid in Alzheimer disease. *Proceedings of the National Academy of Sciences of the United States of America, 90*, 11282–11286.

Ueda, K., Saitoh, T., & Mori, H. (1994). Tissue-dependent alternative splicing of mRNA for NACP, the precursor of non-A beta component of Alzheimer's disease amyloid. *Biochemical and Biophysical Research Communications, 205*, 1366–1372.

Walker, S. J., & Grant, K. A. (2006). Peripheral blood α-synuclein mRNA levels are elevated in cynomolgus monkeys that chronically self-administer ethanol. *Alcohol, 38*, 1–4.

Wang, L. L., Zhang, Z., Li, Q., Yang, R., Pei, X., Xu, Y., ... Li, Y. (2009). Ethanol exposure induces differential microRNA and target gene expression and teratogenic effects which can be suppressed by folic acid supplementation. *Human Reproduction, 24*, 562–579.

Wersinger, C., & Sidhu, A. (2003). Attenuation of dopamine transporter activity by α-synuclein. *Neuroscience Letters, 340*, 189–192.

Withers, G. S., George, J. M., Banker, G. A., & Clayton, D. F. (1997). Delayed localization of synelfin (synuclein, NACP) to presynaptic terminals in cultured rat hippocampal neurons. *Brain Research Developmental Brain Research, 99*, 87–94.

Yadav, S., Pandey, A., Shukla, A., Talwelkar, S. S., Kumar, A., Pant, A. B., & Parmar, D. (2011). miR-497 and miR-302b regulate ethanol-induced neuronal cell death through BCL2 protein and cyclin D2. *The Journal of Biological Chemistry, 286*, 37347–37357.

Yavich, L., Tanila, H., Vepsalainen, S., & Jakala, P. (2004). Role of alpha-synuclein in presynaptic dopamine recruitment. *The Journal of Neuroscience, 24*, 11165–11170.

Zhang, L., Zhang, C., Zhu, Y., Cai, Q., Chan, P., Ueda, K., ... Yang, H. (2008). Semi-quantitative analysis of alpha-synuclein in subcellular pools of rat brain neurons: An immunogold electron microscopic study using a C-terminal specific monoclonal antibody. *Brain Research, 1244*, 40–52.

CHAPTER

9

Consumption of Ethanol and Tissue Changes in the Central Nervous System

L.O. Bittencourt, F.B. Teixeira, K.L. Vieira, D.P. Santos, C.S.F. Maia, R.R. Lima

Federal University of Pará, Belém, Brazil

INTRODUCTION

The ethanol (EtOH) is a widely used psychoactive drug all over the world. This is due to its easy acquisition and wide acceptance by most of the societies. Acute EtOH use decreases the anxiety and stress, which provides an increased feeling of well-being after its consumption (Schuckit, 2009).

According to the World Health Organization (2011), EtOH abuse is responsible for over 2.5 million deaths every year and for almost 4% of all morbidities in the world. According to some projections, EtOH consumption will keep increasing in the coming decades, and the use of this drug by young people will increase consistently the rates of EtOH consumption in the world.

The EtOH can promote different modulations in the central nervous system (CNS) that are directly associated with the consumption pattern. This modulation is dependent on the amount, frequency, and duration of consumption, and also changes according to the subject variables, such as gender and age.

The depressant action of EtOH on CNS may change considerably its structure and its operating mode, promoting structural and functional injuries. In vitro studies showed that EtOH is a potent cellular toxic, capable of promoting cell death, as well as affecting the proliferative capacity, having effect on neural cells under development and differentiated cells.

This chapter aims to present a systematic approach toward the main cellular changes in CNS caused by EtOH consumption, as well as to describe the main tissue changes in motor and cognition control regions in brain.

NEURONAL AND GLIAL CHANGES

The chronic EtOH consumption affects various cell populations in the CNS, among which, one of the most affected are the neurons. These cells are submitted to a process of neurodegeneration mediated by a cascade directed by some proinflammatory factors that cause the liberation of superoxide radicals and, consequently, the cell death (Crews & Nixon, 2008; Zou & Crews, 2005).

This neurodegenerative process has been verified during the exposure to EtOH (in vivo and in vitro) and during a withdrawal period (in vitro studies) affecting the neuronal population in different regions of CNS, such as cerebral cortex, hypothalamus, cerebellum, and especially the limbic associative system, reducing significantly the number of neurons per area in these regions (Baker, Halliday, Kril, & Harper, 1996; Crews et al., 2004; Harding, Halliday, Caine, & Kril, 2000; Harding, Halliday, & Kril, 1996; Harper, Kril, & Daly, 1987; Kubota et al., 2001).

The light, moderate, and heavy exposure to EtOH of neuron culture showed that the increase of dose is directly related to morphological changes, having as one of the most observed changes, the atrophy of dendritic spines (Qiu, Yan, Tang, Zeng, & Liu, 2012).

This atrophy of dendritic spines is the first morphological modification that changes the normal functioning of neurons and causes functional implications of cognitive and motor deficits in alcoholics (Sullivan, Rosenbloom, & Pfefferbaum, 2000). Researchers suggest that the cause of reduction of this structure is due to changes in the cytoskeleton of neurons. However, the exact mechanism of how this drug works and which

Addictive Substances and Neurological Disease
http://dx.doi.org/10.1016/B978-0-12-805373-7.00009-8

proteins are involved are still unknown (Crews & Nixon, 2008).

The EtOH competes with the phospholipase D receptors, which promotes the formation of phosphatidylethanol. This last substance inhibits the activity of phosphatidic acid, which is responsible, among other functions, for survival and cell proliferation (Burkhardt, Wojcik, Zimmermann, & Klein, 2014; Guizzetti, Thompson, Kim, VanDeMark, & Costa, 2004). In astrocytes, it was observed in in vivo and in vitro models that EtOH exposure, at different stages of neural development, causes a decrease in astrocytic population (Miller & Potempa, 1990; Perez-Torrero et al., 1997). These data corroborate with some researches in astrocyte cultures, which showed inhibition of cell proliferation after exposure to EtOH (Guizzetti & Costa, 1996; Resnicoff, Rubini, Baserga, & Rubin, 1994).

The cytoskeleton of the astrocytes is an important structure in several cellular functions and has as the main forming protein of intermediate filament the glial fibrillary acidic protein (GFAP). Exposure models to EtOH, reveal changes in the content and distribution of GFAP in these glial cells (Davies & Cox, 1991; Gressens, Lammens, Picard, & Evrard, 1992; Sáez, Burgal, Renau-Piqueras, Marqués, & Guerri, 1991; Renau-Piqueras et al., 1989).

The oligodendrocytes are glial cells whose main function is the myelination of the CNS. They are also affected in several experimental models of exposure to EtOH. Some reviews that addressed in vivo studies revealed that 10 days old animals when exposed to EtOH had damages in these glial cells, once this period is equivalent to the process of myelination and maturation of oligodendrocytes (Lancaster, 1994; Phillips, 1994).

The myelin basic protein (MBP) is the main protein in the process of myelination promoted by oligodendrocytes and the exposure to EtOH promotes a decreased in expression of this protein as well as retards the process of cell maturation (Chiappelli, Taylor, Espinosa de los Monteros, & de Vellis, 1991). Moreover, the retardation of oligodendrocytes maturation is also a cause to induce apoptosis of these cells in the white matter region in neural development period during gestation (Creeley, Dikranian, Johnson, Farber, & Olney, 2013).

The microglial cells are phagocytes that act in the CNS immune protection. The morphology of these cells is given by the state of activity; when inactive, it has ramifications of the cytoplasm and while active, their structure is globular. Activation is given by biochemical mechanisms, which trigger a neuroinflammation condition.

The EtOH promotes the activation of proinflammatory mediators such as tumor necrosis factor (TNF), interleukin-1β (IL-1β), and monocyte chemotactic protein 1 (MCP-1) in mice exposed to EtOH (Lippai, Bala, Csak, Kurt-Jones, & Szabo, 2013). In mice exposed to EtOH for a period of 10 days at a dose of 5 g/kg, an increase of MCP-1 levels in the brain was observed (Qin et al., 2008). However, even at lower doses equivalent to 3 g/kg and for a shorter period of 4 days of exposure, it is possible to observe microglial damage in mice, as the reduction on microglial activation ability in the parietal association cortex, entorhinal cortex, and hippocampus, in addition to increase in the proinflammatory factors levels, characterizes a condition of neuroinflammation.

In studies conducted by our group, we obtained results that confirm the decrease in microglial cell density in the motor cortex and hippocampus of rats chronically exposed to a daily dose of 6.5 g/kg from adolescence to adulthood. We noticed that in the hippocampus of animals exposed, the decrease was statistically significant in the CA1, CA3, and hilar regions. We also observed a decrease in number between ramified and globular cell types, showing a larger number of active microglia in the evaluated areas. In addition, in motor cortex we found a reduction in the number of microglia in the primary motor area M1 (Oliveira et al., 2015).

MODULATIONS IN CELL PROLIFERATION AND NEUROGENESIS

The cell proliferation in CNS is due to the ability of the nervous tissue to form new cells, including neurons from neuroblasts, a process called neurogenesis. In mice, three known structures are responsible for this process: the hippocampus, dentate gyrus (DG), and olfactory bulb. The effects of EtOH in these regions are well described in the literature and results show that low dose in a single exposure is enough to trigger a slowdown on neurogenic process in the brain.

Nixon and Crews (2002) showed that a dose equivalent to the concentration of 230 mg/dL in human blood is capable to decrease the cell proliferation by 40%. Five hours after the exposure, is possible to observe a reduction in number of positive cells to BrdU (cell proliferation marker), which reiterates the idea that the EtOH promotes the reduction of number of cells that get into S phase on cell division.

Short periods and different doses of EtOH, when administered to experimental animals, show that this compound is capable of altering the physiology and morphology of the neuronal circuitry in regard to the development of new cells in CNS, as well as in differentiation of new glial cells and mature neurons.

Studies in experimental models have observed that chronic exposure to EtOH results in an inhibition of neurogenesis and growth in adult organisms (Nixon & Crews, 2002). The neurogenesis process includes four steps comprising the proliferation, differentiation, migration, and survival of new neurons, success of all these steps being necessary to establish the new cell (Kempermann, Wiskott, & Gage, 2004).

The first three steps usually occur in the prenatal period, constituting the brain structures, as the neocortex; however, there are two neurogenic regions in brain where the process continues until adulthood. The first is the subventricular zone of the lateral ventricle, producing neurons in the olfactory bulb. The second region is the subgranular zone of the DG of the hippocampal granule cells producing DG (Altman & Das, 1965; Alvarez-Buylla & Garcia-Verdugo, 2002; Curtis, Faull, & Eriksson, 2007; Eriksson et al., 1998).

Several studies (in vivo and in vitro) have found that EtOH causes changes in the ability of progenitors to perform the basic functions of the neurogenic process (Crews & Braun, 2003; Hao, Parker, Zhao, Barami, & Lyman, 2003a, 2003b), reporting that the decrease in proliferation of new neurons is about 40–60%, being evident in adolescent organisms (Jang, Shin, Jung, et al., 2002; Jang, Shin, Kim, & Kin, 2002; Nixon & Crews, 2002).

In addition, EtOH affects the growth of dendritic tree progenitor cells, changing the survival step of neurogenesis (He, Nixon, Shetty, & Crews, 2005). Thus, the decrease of progenitor cells is the primary mechanism by which EtOH acts in the process of neurogenesis of adult individuals and in growth (Crews et al., 2006; He & Crews, 2008; Herrera et al., 2003; Nixon & Crews, 2002).

The mechanism of how EtOH acts on progenitor cells is not yet established. Authors suggest that this effect only occurs during intoxication, which assists in decreasing brain volume and weight, increasing the changes that occur in the CNS (Crews & Nixon, 2008).

MOST SIGNIFICANT REGIONAL ALTERATIONS

Motor Cortex

The cerebral cortex is a structure of the CNS related to sensory, motor, and associative functions, which is divided into regions with distinct functions. On the other hand, the cerebral cortex mapped region involved in planning, control, and execution of voluntary movements, corresponds to the motor cortex, which integrates, processes, and manages the arrival of sensory stimuli and the output of the central nervous system motor stimuli (Barret, Barman, Boitano, & Brooks, 2014).

The motor cortex presents a cortical zone composed of gray matter that involves a central spinal cord white matter, in which predominant are the neuronal bodies with their dendrites, whereas the white matter, as well as glial cells, are present in the myelinated nerve fibers that conduct neural signals from subcortical regions in the cortex and vice versa. Among the cortical cells, the pyramidals are the most important ones to output informations from cortex due to originate most of the efferent fibers to the spinal circuit (Guyton & Hall, 2014).

EtOH promotes long-term motor coordination deficits, as well as decrease in spontaneous locomotion and balance. Ethanol-induced changes in brain structures have been studied over the years in rodents and humans. A variety of histological postmortem analysis, and image analysis, suggest that chronic EtOH intake alters the structure of the brain with loss of gray and white matter, resulting in dilation of the grooves and cerebral ventricles, thinner gyrus, and reduction of the dendritic arborization of pyramidal cells when compared with the controls (Crews, 1999).

Imaging studies in human brains associated with EtOH abuse in adolescence detected tissue loss or remodeling that may be related to the inhibition of generation and cell survival (Jacobus & Tarpet, 2013). These data reinforce the idea that inflammatory mediators induced by EtOH affect the integrity of white matter, causing demyelination and neurotoxicity (Bava, Jacobus, Thayer, & Tarpet, 2013; Pascual, Pla, Miñarro, & Guerri, 2014).

Neurodegeneration animal models induced by EtOH along with alcoholic humans, have been proposed for some mechanisms to justify the neural loss, as the excitotoxicity of glutamate, oxidative stress, and neuroinflammation (Marshall, 2013).

The neural excitatory degeneration occurs when the excitatory signalizing of glutamate and others excitatory amino acids are overcome, mainly, the gamma aminobutyric acid (GABA), followed by an influx of calcium to the neural cytoplasm, triggering a cascade of intracellular events which conduce to neural damages (Jaatinen & Rintala, 2008).

In general, the whole nervous tissue is vulnerable to EtOH effects, however, the astrocytes, oligodendrocytes, and synaptic terminals in response to the EtOH toxicity, cause white matter atrophy, neural inflammation, and synaptogenic deficiency (De la Monte & Krill, 2014).

The astrocytic cells, when damaged, affect the neuro-astroglial interactions (González & Salido, 2009), given that its hypertrophy may be related to the release of proinflammatory cytokines and, consequently, with neuroinflammatory process, neurodegeneration, and cellular apoptosis (Alfonso-Loeches, Pascual-Lucas, Blanco, Sanchez-Vera, & Guerri, 2010).

Another pathological indication of neuroinflammation is the presence of activated microglia, which expresses heterogeneity through its morphology, and liberation of neurotoxic and proinflammatory factors, such as cytokines and cell proteins, causing neural and glial lesion (Lima, Costa, & Souza, 2007).

The metabolism of EtOH is associated with the production of reactive oxygen species (ROS) that accentuate the oxidative state of the cells. However, the formation of oxidative stress affects the cell as a whole, as well as the proteins, lipids, and DNA, promoting neurotoxicity or neurodegeneration (Hernández, López-Sanchez, & Rendón-Ramírez, 2016). In a previous study, our group observed some parameters related to oxidative stress, such as a significant increase in nitrite and lipidic peroxidation levels in adolescent rats exposed to EtOH (Teixeira et al., 2014).

The mechanism of ethanol-induced neuropathology is not yet fully understood, because many factors such as neurotoxicity of EtOH molecule or of its metabolic products, thiamine deficiency, repeated abstinences, as well as the ethanol-induced neuroinflammation are among the factors that constitute a context to the neurodegeneration (Sullivan & Zahr, 2008).

Hippocampus

The hippocampus, DG, entorhinal cortex, and subiculum are structures that comprise the hippocampal formation and belong to the limbic system. Formed by cortical layers localized on medial portion of temporal lobe, has trilaminate cytoarchitecture. This internal organization comprises two types of main cells: granule cells of DG and cornu ammonis (CA) pyramidal cells (from Latin *cornu*—horn of Ammon lamb—due its similarity, in human brain, with the lamb's horn present on the head of the Egyptian god Ammon) divided into the sectors CA1, CA2, CA3, and CA4.

The Hippocampus presents important functions related to behavior and memory. It is particularly involved with memory phenomena, especially with the formation of the long-term memory (one that persists, sometimes, forever) and spatial exploration. When both hemispheres (right and left) are destroyed, nothing is stored in memory. Still, hippocampal damages of any kind cause serious mnemonic damage that is often permanent. The EtOH intoxication and its relationship with the hippocampus are discussed in the following.

Structural brain abnormalities mediated by EtOH consumption are evidenced (De Bellis et al., 2005; Medina et al., 2008; Risher, Fleming, et al., 2015; Risher, Sexton, et al., 2015), especially in the areas involved in processing memory and learning, such as the hippocampus and prefrontal cortex (PFC) of adolescents with alcohol use disorder (AUD). Adolescents who began the EtOH consumption earlier, when compared with individuals who started later, have presented smaller hippocampal volumes (De Bellis et al., 2000; Nagel, Schweinsburg, Phan, & Tapert, 2005). The EtOH effect in these groups seems unrelated to other comorbid conduct disorders (Nagel et al., 2005).

Hippocampal cellular loss and reduction of neurogenesis in DG, a neurogenic region in adult brain, are observed in animal models of EtOH exposure. These effects, cellular death plus reduction of neurogenesis, are associated with a substantial loss of granule neurons from DG after a single exposure (Maynard & Leasure, 2013).

When ingested in high doses and chronically, EtOH can induce cognitive loss. According to Givens and McMacahon (1995), the EtOH may impair the long-term potentiation (LTP) and also can interfere in the hippocampal place cells through blockade of NMDA receptors, which partially explains their deleterious effects on memory and other cognitive functions and irreversible neurological damage, as loss of cortical mass and increase of ventricular volume, which leads to irreversible dementia as in Wernicke–Korsakoff syndrome.

According to several experimental studies, the hippocampus seems to be particularly vulnerable to exposure to EtOH during adolescence. Long-term exposure to EtOH during this period induces severe morphological changes in hippocampal system and most of these changes persist into adulthood (Risher, Fleming, et al., 2015; Risher, Sexton, et al., 2015). Pyramidal hippocampal neurons are generated during the late gestation in the subventricular zone (Bayer, 1982) and can remain vulnerable when exposed to different neurotoxins during the early life (Hort, Brożek, Mares, Langmeier, & Komárek, 1999; Langmeier, Folbergrová, Haugvicová, Pokorný, & Mares, 2003; Miki, 2004; Milotová, 2006; Riljak, Milotová, Jandová, Pokorný, & Langmeier, 2007).

The long-term postnatal development of the hippocampus, resulting from prolonged proliferation of granule cells (Bayer, 1980), allows to study changes that arise during prenatal development and early postnatal period. Among the structural changes observed resulting from the EtOH exposure, the research shows a reduction of CA1 and CA3 areas of hippocampus and in dorsal and ventral layers of DG, having, consequently, intense cell death by apoptosis (Ramar & Saraswathi, 2013; Risher, Fleming, et al., 2015; Risher, Sexton, et al., 2015). Dead neurons were characterized by fragmented and condensed nucleus, which is associated many times with the cell death by apoptosis (Milotová, 2006; Ramar & Saraswathi, 2013).

Neurotoxicity studies on the mechanisms of EtOH have focused on its interaction with the neurons and glial cells. Considering glial parameters investigated, even under a moderate dose, EtOH caused astrocytic alterations in both morphology and function. A number of in vitro studies have demonstrated that the EtOH can inhibit the proliferation of several glial cells, especially astrocytes. These inhibitory effects of EtOH can contribute to its neurotoxicity to development observed following exposure in vivo. Animal models have shown that the EtOH causes microcephaly when administered during the brain growth spurt, a period of the development of the brain characterized by astroglial proliferation and maturation.

Cerebellum

The cerebellum represents about 10% of the brain volume, containing about 50% of all neurons of the brain. It receives excitatory information from somatosensory system and cerebral cortex by cerebellar peduncles, formed by beams of afferent and efferent fibers of cerebellum.

Although the cerebellum does not initiate the motor activity, it interacts with different brain regions in order to have planned and coordinated movements (Alekseeva et al., 2014). It is also critical to motor learning, acquisition of language and memory, and control of behavior and emotions.

It is distinguished in the cerebellum the vermis, ímpar and median portion located in corticonuclear area, linked to two cerebellar hemispheres. The cerebellar cortex is comprised by many cerebellar sheets, in which the white matter is in central and gray matter is in peripheral region. Inside this thin gray layer is the Purkinje neuron, excitatory and dominant in the efferent cerebellar system, modulating afferent information that arrives at the cerebellar cortex (Apfel, Ésberad, Rodrigues, Bahamad, & Sillero, 2002).

The EtOH promotes simultaneous changes in several neural pathways, which may cause large neurological impact. The cerebellum is vulnerable to harmful effects of EtOH, which after crossing the blood–brain barrier, interacts with a large number of neurochemical receptors causing behavioral and physiological changes such as hypotonia, dysmetria, ataxia, and others (Andersen, 2004).

Among some of the mechanisms related to the cerebellum neuropathology are excitotoxicity, apoptosis, oxidative stress, and changes in glial cells. In the chronic exposure to EtOH, the overactivity of NMDA and glutamate enhances the Ca^{2+} influx into the cytosol, triggering a cascade of events, and, consequently, neurodegeneration, which means that the action of EtOH in calcium regulation might serve as a potential mechanism to reduce survival of Purkinje cells (Thomas, Goodlett, & West, 1998).

Cellular apoptosis, although it is a physiological process of maintaining homeostasis, due to excessive EtOH exposure, may result in an increase of production of ROS and/or suppression of antioxidant defense mechanisms, thus, causing oxidative stress and apoptotic neurodegeneration. Histological analysis and oxidative stress in neonatal rats showed decrease of Purkinje cells in the cerebellar vermis, probably by the decrease in the activities of glutathione peroxidase (GPx) and increase in lipid peroxidation (Ramezani et al., 2012).

Among histopathological changes, a degeneration of the cerebellar cortex, especially in the anterior vermal region, has been described, probably caused by the death and/or atrophy of Purkinje cells (Oliveira et al., 2014). Studies in animals and humans treated with EtOH-signaled cell degeneration in Purkinje cells (Andersen, 2004; Apfel et al., 2002; Gonzalez-Maciel et al., 1994; Thomas et al., 1998). Moreover, alcoholics with Korsakoff syndrome have shown a significant decrease in the density of Purkinje cells in the vermis and molecular layer (Baker, Harding, Halliday, Krill, & Harper, 1999).

In the neuropathology, loss of spinal cord white matter volume is observed, probably due to the neurotoxicity of EtOH that affects the proliferation, differentiation, migration, and survival of glia. Studies in animals and humans, alive and postmortem, indicate the presence of neuroinflammation in white matter, which can impair the glial–neuronal interactions in order to induce neuronal degeneration (Marshall, 2013).

References

Alekseeva, N., McGee, J., Kelley, R. E., Maghzi, A. H., Gonzalez-Toledo, E., & Minagar, A. (November 2014). Toxic-metabolic, nutritional, and medicinal-induced disorders of cerebellum. *Neurologic Clinics, 32*(4), 901–911. http://dx.doi.org/10.1016/j.ncl.2014.07.001. Epub 2014 Sep. 11.

Alfonso-Loeches, S., Pascual-Lucas, M., Blanco, A. M., Sanchez-Vera, I., & Guerri, C. (2010). Pivotal role of TLR4 receptors in alcohol-induced neuroinflammation and brain damage. *The Journal of Neuroscience, 30*, 8285–8295.

Altman, J., & Das, G. D. (1965). Autoradiographic and histological evidence of postnatal hippocampal neurogenesis in rats. *Journal of Comparative Neurology, 124*, 319–335.

Alvarez-Buylla, A., & Garcia-Verdugo, J. M. (2002). Neurogenesis in adult subventricular zone. *The Journal of Neuroscience, 22*, 629–634.

Andersen, B. B. (2004). Reduction of Purkinje cell volume in cerebellum of alcoholics. *Brain Research, 1007*, 10–18.

Apfel, M. I. R., Ésberad, C. A., Rodrigues, F. K. P., Bahamad, F. M. J. R., & Sillero, R. O. (2002). Estudo estereológico das células de Purkinje cerebelares submetidas à intoxicação alcoólica em ratos Wistar. *Arquivos de Neuropsiquiatria, 60*(2A), 258–263.

Baker, K., Halliday, G. M., Kril, J. J., & Harper, C. G. (1996). Chronic alcoholism in the absence of Wernicke–Korsakoff syndrome and cirrhosis does not result in the loss of serotonergic neurons in the median raphe nucleus. *Metabolic Brain Disease, 11*, 217–227.

Baker, K., Harding, A., Halliday, G., Krill, J., & Harper, C. (1999). Neuronal em zonas funcionais do cerebelo de alcoólatras crônicos com e sem encefalopatia de Wernicke. *Neuroscience, 91*, 429–438.

Barret, K. E., Barman, S. M., Boitano, S., & Brooks, H. L. (2014). In *Fisiologia médica de Ganong* (24th ed.). Porto Alegre: AMGH.

Bava, S., Jacobus, J., Thayer, R. E., & Tarpet, S. F. (2013). Longitudinal changes in white matter integrity among adolescent substance users. *Alcoholism: Clinical and Experimental Research, 37*(Suppl. 1), E181–E189. http://dx.doi.org/10.1111/j.1530-0277.2012.01920.x.

Bayer, S. A. (1980). Development of the hippocampal region in the rat II. Morphogenesis during embryonic and early postnatal life. *Journal of Comparative Neurology, 190*, 115–134.

Bayer, S. A. (1982). Changes in the total number of dentate granule cells in juvenile and adult rats: A correlated volumetric and 3H·thymidine autoradiographic study. *Experimental Brain Research, 46*, 315–323.

Burkhardt, U., Wojcik, B., Zimmermann, M., & Klein, J. (2014). Phospholipase D is a target for inhibition of astroglial proliferation by ethanol. *Neuropharmacology, 79*, 1–9. http://dx.doi.org/10.1016/j.neuropharm.2013.11.002.

Chiappelli, F., Taylor, A. N., Espinosa de los Monteros, A., & de Vellis, J. (1991). Fetal alcohol delays the developmental expression of myelin basic protein and transferrin in rat primary oligodendrocyte cultures. *International Journal of Developmental Neuroscience, 9*, 67–75. http://dx.doi.org/10.1016/0736-5748(91)90074-V.

Creeley, C. E., Dikranian, K. T., Johnson, S. A., Farber, N. B., & Olney, J. W. (2013). Alcohol-induced apoptosis of oligodendrocytes in the fetal macaque brain. *Acta Neuropathologica Communications, 1*, 23. http://dx.doi.org/10.1186/2051-5960-1-23.

Crews, F. T. (1999). Alcohol and neurodegeneration. *CNS Drug Reviews, 5*, 379–394.

Crews, F. T., & Braun, C. J. (2003). Binge ethanol treatment causes greater brain damage in alcohol-preferring P rats than in alcohol nonpreferring NP rats. *Alcoholism: Clinical and Experimental Research, 27*, 1075–1082.

Crews, F. T., Collins, M. A., Dlugos, C., Littleton, J., Wilkins, L., & Neafsey, E. J. (2004). Alcohol-induced neurodegeneration: When, where and why? *Alcoholism: Clinical and Experimental Research, 28*, 350–364.

Crews, F. T., & Nixon, K. (2008). Mechanisms of neurodegeneration and regeneration in alcoholism. *Alcohol & Alcoholism, 44*(2), 115–127.

Crews, F., Nixon, K., Kim, D., Joseph, J., Shukitt-Hale, B., Qin, L., & Zou, J. (2006). BHT blocks NF-kappaB activation and ethanol-induced brain damage. *Alcoholism: Clinical and Experimental Research, 30*, 1938–1949.

Curtis, M. A., Faull, R. L., & Eriksson, P. S. (2007). The effect of neurodegenerative diseases on the subventricular zone. *Nature Review Neuroscience, 8*, 712–723.

Davies, D. L., & Cox, W. E. (1991). Delayed growth and maturation of astrocytic cultures following exposure to ethanol: Electron microscopic observations. *Brain Research, 547*(1), 53–61.

De Bellis, M. D., Clark, D. B., Beers, S. R., Soloff, P. H., Boring, A. M., Hall, J., & Keshavan, M. S. (2000). Hippocampal volume in adolescent-onset alcohol use disorders. *The American Journal of Psychiatry, 157*, 737–744.

De Bellis, M. D., Narasimhan, A., Thatcher, D. L., Keshavan, M. S., Soloff, P., & Clark, D. B. (2005). Prefrontal cortex, thalamus, and cerebellar volumes in adolescents and young adults with adolescent onset alcohol use disorders and comorbid mental disorders. *Alcoholism: Clinical and Experimental Research, 29*, 1590–1600.

De la Monte, S. M., & Krill, J. J. (2014). Human alcohol-related neuropathology. *Acta Neuropathologica, 127*(1), 71–90.

Eriksson, P. S., Perfilieva, E., Bjork-Eriksson, T., Alborn, A. M., Nordborg, C., & Peterson, D. A. (1998). Neurogenesis in the adult human hippocampus. *Nature Medicine, 4*, 1313–1317.

Givens, B., & McMacahon, K. (1995). Ethanol suppresses the induction of long-term potentiation in-vivo [S. l] *Brain Research, 688*, 27–33.

González, A., & Salido, G. M. (2009). Ethanol alters the physiology of neuron-glia communication. *International Review of Neurobiology, 88*, 167–198. http://dx.doi.org/10.1016/S0074-7742(09)88007-0.

Gonzalez-Maciel, A., Romero-Velazquez, R. M., Hernandez-Islas, J. L., Sicilia-Argumeddo, G., Fragoso-Soriano, R., & Cravioto, J. (1994). Purkinje cell density in cerebella of alcoholized and non-alcoholized male rat offspring. *Archives of Medical Research, 25*, 427–434.

Gressens, P., Lammens, M., Picard, J. J., & Evrard, P. (1992). Ethanol-induced disturbances of gliogenesis and neuronogenesis in the developing murine brain: An in vitro and in vivo immunohistochemical and ultrastructural study. *Alcohol and Alcoholism, 27*(3), 219–226.

Guizzetti, M., & Costa, L. G. (1996). Inhibition of muscarinic receptor-stimulated glial cell proliferation by ethanol. *Journal of Neurochemistry, 67*, 2236–2245. http://dx.doi.org/10.1046/j.1471-4159.1996.67062236.x.

Guizzetti, M., Thompson, B. D., Kim, Y., VanDeMark, K., & Costa, L. G. (2004). Role of phospholipase D signaling in ethanol-induced inhibition of carbachol-stimulated DNA synthesis of 1321N1 astrocytoma cells. *Journal of Neurochemistry, 90*, 646–653. http://dx.doi.org/10.1111/j.1471-4159.2004.02541.x.

Guyton, A. C., & Hall, J. E. (2014). *Tratado de Fisiologia Médica* (12th ed.).

Hao, H. N., Parker, G. C., Zhao, J., Barami, K., & Lyman, W. D. (2003a). Differential responses of human neural and hematopoietic stem cells to ethanol exposure. *Journal of Hematotherapy & Stem Cell Research, 12*, 389–399.

Hao, H. N., Parker, G. C., Zhao, J., Barami, K., & Lyman, W. D. (2003b). Human neural stem cells are more sensitive than astrocytes to ethanol exposure. *Alcoholism: Clinical and Experimental Research, 27*, 1310–1317.

Harding, A., Halliday, G., Caine, D., & Kril, J. (2000). Degeneration of anterior thalamic nuclei differentiates alcoholics with amnesia. *Brain, 123*, 141–154.

Harding, A., Halliday, G. M., & Kril, J. J. (1996). Loss of vasopressin immunoreactive neurons in alcoholics is dose-related and time dependent. *Neuroscience, 72*, 699–708. http://dx.doi.org/10.1016/0306-4522(95)00577-3.

Harper, C. G., Kril, J. J., & Daly, J. (1987). Are we drinking our neurones away? *British Medical Journal, 294*, 534–536.

He, J., & Crews, F. T. (2008). Increased MCP-1 and microglia in various regions of the human alcoholic brain. *Experimental Neurology, 210*, 349–358.

He, J., Nixon, K., Shetty, A. K., & Crews, F. T. (2005). Chronic ethanol reduces dendritic growth of newborn neurons. *The European Journal of Neuroscience, 21*, 2711–2720.

Hernández, J. A., López-Sanchez, R. C., & Rendón-Ramírez, A. (2016). Lipids and oxidative stress associated with ethanol-induced neurological damage. *Oxidative Medicine and Cellular Longevity, 2016*. http://dx.doi.org/10.1155/2016/1543809. Article ID. 1543809. 15 pages.

Herrera, D. G., Yague, A. G., Johnsen-Soriano, S., Bosch-Morell, F., Collad-Morente, L., Muriach, M., … Garcia-Verdugo, J. M. (2003). Selective impairment of hippocampal neurogenesis by chronic alcoholism: Protective effects of an antioxidant. *Proceedings of the National Academy of Sciences of the United States of America, 100*, 7919–7924.

Hort, J., Brożek, G., Mares, P., Langmeier, M., & Komárek, V. (1999). Cognitive functions after pilocarpine-induced status epilepticus: Changes during silent period precede appearance of spontaneous recurrent seizures. *Epilepsia, 40*, 1177–1183.

Jaatinen, P., & Rintala, J. (2008). Mechanisms of ethanol-induced degeneration in the developing, mature, and aging cerebellum. *The Cerebellum*, 332–347.

Jacobus, J., & Tarpet, S. (2013). Neurotoxic effects of alcohol in adolescence. *Annual Review of Clinical Psychology, 9*.

Jang, M. H., Shin, M. C., Jung, S. B., Lee, T. H., Bahn, G. H., Kwon, Y. K., ... Kim, C. J. (2002). Alcohol and nicotine reduce cell proliferation and enhance apoptosis in dentate gyrus. *Neuroreport, 13*, 1509–1513.

Jang, M. H., Shin, M. C., Kim, E. H., & Kin, C. J. (2002). Acute alcohol intoxication decreases cell proliferation and nitric oxide synthase expression in dentate gyrus of rats. *Toxicology Letters, 133*, 255–262.

Kempermann, G., Wiskott, L., & Gage, F. H. (2004). Functional significance of adult neurogenesis. *Current Opinion in Neurobiology, 14*, 186–191.

Kubota, M., Nakazaki, S., Hirai, S., Saeki, N., Yamaura, A., & Kusaka, T. (2001). Alcohol consumption and frontal lobe shrinkage: Study of 1432 non-alcoholic subjects. *Journal of Neurology, Neurosurgery & Psychiatry, 71*, 104–106.

Lancaster, F. E. (1994). Alcohol and white matter development – A review. *Alcoholism: Clinical and Experimental Research, 18*, 644–647. http://dx.doi.org/10.1111/j.1530-0277.1994.tb00924.x.

Langmeier, M., Folbergrová, J., Haugvicová, R., Pokorný, J., & Mares, P. (2003). Neuronal cell death in hippocampus induced by homocysteic acid in immature rats. *Epilepsia, 44*, 299–304.

Lima, R. R., Costa, A. M. R., & Souza, R. D. (2007). Inflammation in neurodegenerative disease. *Revista Paraense de Medicina, 21*, 29–34.

Lippai, D., Bala, S., Csak, T., Kurt-Jones, E. A., & Szabo, G. (2013). Chronic alcohol-induced microRNA-155 contributes to neuroinflammation in a TLR4-dependent manner in mice. *PLoS One, 8*, e70945.

Marshall, S. A. (2013). *Microglia activation in a rodent model of an alcohol use disorder: The importance of phenotype, initiation and duration of activation* (theses and dissertations – pharmacy). Paper 25.

Maynard, M., & Leasure, L. (2013). Exercise enhances hippocampal recovery following binge ethanol exposure. *PLoS One, 8*(9).

Medina, K. L., McQueeny, T., Nagel, B. J., Hanson, K. L., Scweinsburg, A. D., & Tapert, S. F. (2008). Prefrontal cortex volumes in adolescents with alcohol use disorders: Unique gender effects. *Alcoholism: Clinical and Experimental Research, 32*, 386–394.

Miki, T. (2004). Effect of age and alcohol exposure during early life on pyramidal cell numbers in the CA1-CA3 region of the rat hippocampus. *Hippocampus, 14*, 124–134.

Miller, M. W., & Potempa, G. (1990). Numbers of neurons and glia in mature rat somatosensory cortex: Effects of prenatal exposure to ethanol. *Journal of Comparative Neurology, 293*, 92–102. http://dx.doi.org/10.1002/cne.902930108 100.

Milotová, M. (2006). Alcohol abuse in mothers during gravidity and changes of hippocampal neurons in their offspring. *Physiological Research, 55*, 34P.

Nagel, B. J., Schweinsburg, A. D., Phan, V., & Tapert, S. F. (2005). Reduced hippocampal volume among adolescents with alcohol use disorders without psychiatric comorbidity. *Psychiatry Research, 139*, 181–190.

Nixon, K., & Crews, F. T. (2002). Binge ethanol exposure decreases neurogenesis in adult rat hippocampus. *Journal of Neurochemistry, 83*, 1087–1093.

Oliveira, S. A., Chuffa, I. G. A., Fioruci-Fontanelli, B. A., Neto, F. S. L., Novais, P. C., Tirapelli, L. F., ... Martinez, F. E. (2014). Apoptosis of Purkinje and granular cells of the cerebellum following chronic ethanol intake. *The Cerebellum, 13*(6), 728–738.

Oliveira, A. C., Pereira, M. C., Santana, L. N. D. S., Fernandes, R. M., Teixeira, F. B., Oliveira, G. B., ... Maia, C. D. S. F. (2015). Chronic ethanol exposure during adolescence through early adulthood in female rats induces emotional and memory deficits associated with morphological and molecular alterations in hippocampus. *Journal of Psychopharmacology (Oxford), 29*, 712–724.

Pascual, M., Pla, A., Miñarro, J., & Guerri, C. (2014). Neuroimmune activation and myelin changes in adolescent rats exposed to high-dose alcohol and associated cognitive dysfunction: A review with reference to human adolescent drinking. *Alcohol and Alcoholism, 49*(2), 187–192.

Perez-Torrero, E., Duran, P., Granados, L., Gutierez-Ospina, G., Cintra, L., & Diaz Cintra, S. (1997). Effects of acute prenatal ethanol exposure on Bergmann glia cells early postnatal development. *Brain Research, 746*, 305–308. http://dx.doi.org/10.1016/S0006-8993(96)01235-8.

Phillips, D. E. (1994). Research monograph no. 27. Effects of alcohol on glial development in vivo: Morphological studies. In F. E. Lancaster (Ed.), *Alcohol and glial cells* (pp. 195–214). Bethesda, MD: National Institute of Health, NIAAA.

Qin, L., He, J., Hanes, R. N., Pluzarev, O., Hong, J. S., & Crews, F. T. (2008). Increased systemic and brain cytokine production and neuroinflammation by endotoxin following ethanol treatment. *Journal of Neuroinflammation, 5*, 10.

Qiu, H., Yan, H., Tang, J., Zeng, Z., & Liu, P. (2012). A study on the influence of ethanol over the primary cultured rat cortical neurons by using the scanning microscopy. *Micron, 43*, 125–140.

Ramar, S., & Saraswathi, P. (2013). Histomorphometric study of fructus psoralea on ethanol induced neurodegeneration of hippocampus in rat. *Journal of Clinical and Diagnostic Research, 7*(8), 1561–1564.

Ramezani, A., Goudarzi, I., Lashkarboluki, T., Ghorbanian, M. T., Abrari, K., & Elahdadi Salmani, M. (July 2012). Role of oxidative stress in ethanol-induced neurotoxicity in the developing cerebellum. *Iranian Journal of Basic Medical Sciences, 15*(4), 965–974.

Renau-Piqueras, J., Zaragoza, R., De Paz, P., Baguena-Cervellera, R., Megias, L., & Guerri, C. (1989). Effects of prolonged ethanol exposure on the glial fibrillary acidic protein-containing intermediate filaments of astrocytes in primary culture: A quantitative immunofluorescence and immunogold electron microscopic study. *Journal of Histochemistry & Cytochemistry, 37*(2), 229–240.

Resnicoff, M., Rubini, M., Baserga, R., & Rubin, R. (1994). Ethanol inhibits insulin-like growth factor-1-mediated signalling and proliferation of C6 rat glioblastoma cells. *Laboratory Investigation, 71*, 657–662.

Riljak, V., Milotová, M., Jandová, K., Pokorný, J., & Langmeier, M. (2007). Morphological changes in the hippocampus following nicotine and kainic acid administration. *Physiological Research, 56*, 641–649.

Risher, M. L., Fleming, R., Risher, C., Miller, M., Klein, C., Wills, T., ... Swartzwelder, H. (2015). Adolescent intermittent alcohol exposure: Persistence of structural and functional hippocampal abnormalities into adulthood. *Alcoholism: Clinical and Experimental Research, 39*(6), 989–997.

Risher, M. L., Sexton, H. G., Risher, W. C., Wilson, W. A., Fleming, R. L., Madison, R. D., ... Swartzwelder, H. S. (2015). Adolescent intermittent alcohol exposure: Dysregulation of thrombospondins and synapse formation are associated with decreased neuronal density in the adult hippocampus. *Alcoholism: Clinical and Experimental Research, 39*(12), 2403–2413. http://dx.doi.org/10.1111/acer.12913.

Sáez, R., Burgal, M., Renau-Piqueras, J., Marqués, A., & Guerri, C. (1991). Evolution of several cytoskeletal proteins of astrocytes in primary culture: Effect of prenatal alcohol exposure. *Neurochemical Research, 16*(7), 737–747.

Schuckit, M. A. (2009). Alcohol-use disorders. *Lancet, 373*(9662), 492–501. http://dx.doi.org/10.1016/S0140-6736(09)60009-X.

Sullivan, E. V., Rosenbloom, M. J., & Pfefferbaum, A. (2000). Pattern of motor and cognitive deficits in detoxified alcoholic men. *Alcoholism: Clinical and Experimental Research, 24*, 611–621.

Sullivan, E. V., & Zahr, N. M. (2008). Increased MCP-1 and microglia in various regions of human alcoholic brain. *Experimental Neurology, 213*(1), 10–17. http://dx.doi.org/10.1016/j.expneurol.2008.05.016.

Teixeira, F. B., Santana, L. N. S., Bezerra, F. R., Carvalho, S., Fontes-Junior, E. A., Pregider, R. D., ... Lima, R. R. (2014). Chronic ethanol exposure during adolescence in rats induces motor impairments and cerebral cortex damage associated with oxidative stress. *PLoS One, 9*, e101074.

Thomas, J. D., Goodlett, C. R., & West, J. R. (1998). Alcohol-induced Purkinje cell loss depends on developmental timing of alcohol exposure and correlates with motor performance. *Brain Research. Developmental Brain Research, 105*(2), 159–166.

World Health Organization. (2011). *Global status report on alcohol and health*. Geneva, Switzerland: World Health Organization Press.

Zou, J. Y., & Crews, F. T. (2005). TNF alpha potentiates glutamate neurotoxicity by inhibiting glutamate uptake in organotypic brain slice cultures: Neuroprotection by NF kappa B inhibition. *Brain Research, 1034*, 11–24.

CHAPTER

10

Ethanol Consumption and Cerebellar Disorders

A.C.A. de Oliveira, B. Puty, L.K.R. Leão, R.M. Fernandes, C.S.F. Maia, R.R. Lima

Federal University of Pará, Belém, Brazil

INTRODUCTION

The World Health Organization (WHO) has identified alcoholism as the third leading cause of premature mortality and disability. The abuse of ethanol (EtOH) can lead to negative consequences for the well-being and health, increasing the risk for various diseases. The central nervous system (CNS) area is most vulnerable to the toxic effects of EtOH, and this effect may be associated with attention and learning deficits, memory loss, and changes in spatial processing (WHO, 2014).

In the CNS, the cerebellum has histological and cellular characteristics that make it particularly sensitive to this substance. Research has shown that simple drinking episodes are capable of eliciting physiological and behavioral changes that persist for long periods of time (McClain et al., 2011).

This chapter discusses the effects of EtOH on the cerebellar nerve tissue, beginning with a review of its anatomical structure and its signaling pathways. Apart from that, we describe how EtOH can affect those pathways and trigger clinically perceptible changes.

CEREBELLUM: AN ANATOMO-PHYSIOLOGICAL REVIEW

The cerebellum is a small organ located in the cerebellar fossa of the occipital bone, divided into hemispheres. It is an important component of the motor control, essential for planning and coordination of movement. It has a complex organization and may be represented according to its anatomy, physiology, and functionality (Albus, 1971; Chambers & Sprague, 1955; Ito, 1972; Pirnik & Kiss, 2002; Roostaei, Nazeri, Sahraian, & Minagar, 2014).

The cerebellum is divided into anatomical central region, known as vermis, and two side regions, called hemispheres. The posterior–inferior cerebellar hemisphere could be subdivided into three lobes: anterior lobe, posterior lobe, and flocculonodular lobe. Larsell and Whitlock (1952) have shown a comparison between the cerebellum of birds and mammals. According to them, the posterolateral fissure delimits the flocculonodular lobe of the cerebellum body. The prima fissure divides the corpus cerebelli into anterior and posterior lobes. In the posterior lobe are secondary fissure, intra-follicular fissure (corresponding to inter-floccular and intra-tonsilar groove), and the posterior superior fissure (described in human anatomy). The two main fissures of the anterior lobe of the cerebellar body correspond to the precentral fissure and preculminata of the human cerebellum. The mice cerebellum at vermis region are divided into leaves, I to X, and some of these leaves match with those in the human cerebellum.

The cerebellar cells are organized into three layers: molecular layer, Purkinje cell layer, and granular layer. In the granular layer are granular cells, with four to five short dendrites that receive excitatory synapses of mossy fibers (main input via the cerebellum, together with the climbing fibers) and a long axon that stimulates Purkinje cells and may reach molecular layer where it bifurcates in T-shape, known as parallel fibers (Mugnaini, 1983; Nunzi, Birnstiel, Bhattacharyya, Slater, & Mugnaini, 2001; Pirnik & Kiss, 2002; Roostaei et al., 2014).

The excitability of granule cells are controlled by GABAergic currents by Golgi cell, also present in the granular layer. Apart from that, in the molecular layer are present two neurons responsible for the inhibitory modulation of Purkinje cells, the stellate cells and the

Addictive Substances and Neurological Disease
http://dx.doi.org/10.1016/B978-0-12-805373-7.00010-4

basket cells. Under stimulation, Purkinje cells send inhibitory signals through their axons and form inhibitory synapses in the central cerebellar nuclei (Allen & Tsukahara, 1974; Wolpert, Miall, & Kawato, 1998).

The mossy and climbing fibers are responsible for bringing information from the spinal cord about the movements of all parts of the body and for planning movements before they are performed by different motor centers. According to the function, the cerebellar body is divided into medial, intermediate, and lateral cerebellar zones (Mugnaini, 1983; Nunzi et al., 2001; Wolpert et al., 1998).

The medial area is responsible for processing information coming from vestibular, somatosensory, visual, and auditory areas. After processing, the information flows out of fastigium, cores through the fastigial–vestibular and –reticular fibers, being responsible for controlling the axial and proximal muscles, which are fundamental to maintain the balance and posture. In the intermediate zone, arrives proprioceptive and somatosensory information from spinal cord and the cerebral cortex. The cerebellum exerts control on distal muscles, which are responsible for delicate movements, by inhibitory connections between the axons of Purkinje cells and interpositus core, from where fibers go toward red nucleus and thalamus (Mugnaini, 1983; Wetts, Kalaska, & Smith, 1985; Pirnik & Kiss, 2002; Roostaei et al., 2014).

The lateral area receives information from the cerebral cortex. The dentate nucleus processes that information and sends to efferent fiber of thalamus. Thereafter, the information reaches the opposite side of the motor cortex, mediating control of the overall planning of movements (Mugnaini, 1983; Wetts et al., 1985; Nunzi et al., 2001; Roostaei et al., 2014; Wolpert et al., 1998).

It is well known that the cerebellum functionality has a very complex organization that is responsible for normal planning and coordination of movements. Hence, many research groups have taken remarkable interest in the study of the cerebellum under different conditions, since those changes in the cerebellar functionality, anatomy, and physiology can induce irreversible deficits in cerebellar performance. Furthermore, it is important to know that cellular communication is essential for the physiological maintenance of the various areas of the CNS, particularly the cerebellum. Therefore, it is essential to understand what types of changes can be induced in neurophysiological cerebellar system after acute or chronic exposure to EtOH.

EtOH EFFECTS ON THE CEREBELLUM NEUROPHYSIOLOGY

Cerebellar changes caused by EtOH remain present even after long periods of abstinence. The main cellular targets of EtOH are the connections among mossy fiber–granule cells–Golgi cells, and between parallel fiber and Purkinje cells, that under EtOH effects, disrupt their circuitry information, causing cerebellar ataxia and motor dysfunction (Luo, 2015). In addition, the chronic exposure to EtOH has been related to cell degeneration in the dendritic region and to a significant decrease in the number of synapses formed by these cells (Dlugos, 2008; Nevo & Hamon, 1995).

The Purkinje cells, the only efferent cells from cerebellar cortex, are the most affected by the consumption of EtOH. Even considering the importance of EtOH effects in the mossy fibers and granule cells, since these are the main afferent pathways in the cerebellum, the Purkinje cells must have a special attention since the physiology of cellular communication in the different cerebellar layers seems to be organized in order to control the activation of these cells (Luo, 2012).

It is know that cerebellar efferent connections signaling, mediated by synapses between neurons and Purkinje cells, present in the central core occurs through the release of gamma-aminobutyric acid (GABA), the major inhibitory neurotransmitter in the CNS. Several studies have shown that consumption of EtOH is responsible for an increase in GABA release in Purkinje cells, granular cells, and interneurons in the molecular layer. This change in neurotransmitter levels is deeply related with a decrease in motor control mediated by the cerebellum (Nevo & Hamon, 1995).

In order to understand the effects of EtOH in GABAergic modulation in the cerebellum two hypotheses were formulated. The first hypothesis suggests that the EtOH can act directly on the potentiation of GABAergic extrasynaptic receptors, causing an increase in inhibitory synapses with cells that stimulate granule cells to release glutamate on the Golgi cells, thus enhancing the inhibition caused in cerebellar circuitry. The second hypothesis suggests that EtOH can act indirectly by increasing the action potential of Golgi cells by inhibiting the sodium–potassium pump (Luo, 2015). In both cases, the increase in the inhibitory modulation in Purkinje cells, mediated by granule cells, may be responsible for excessive excitatory inputs in the central core (Jaatinen & Rintala, 2008).

However, it is important to note that not only the increase in inhibitory inputs is harmful to the motor control performed by the cerebellum. Dar (2015) suggested that changes in glutamate levels, the main excitatory neurotransmitter in the CNS, and the adenosine levels may also be related with the cellular mechanisms of cerebellar dysfunction caused by EtOH. A decrease in the glutamate levels at synapses between granule cells and Purkinje cells may result in an intense activation of Purkinje cells, resulting in increased release of GABA in the central nuclei of the cerebellum and consequently in a decrease in the efferent excitatory inputs (Fig. 10.1). However, the exact molecular mechanism of the effects of EtOH has not been fully elucidated.

FUNCTIONAL AND CELLULAR CHANGES FROM INGESTING EtOH

The physiological basis of changes resulting from exposure to EtOH includes the neurochemical pathways previously described, as well as structural and functional changes in various cell populations of this tissue, as we deal in this section.

In the cerebellum, the effects of EtOH generate neuropathological changes more directly related to cerebellar atrophy due to loss and/or atrophy in Purkinje cells, and decrease the volume of these cells and dendritic network in the molecular layer. EtOH administration in 7-day old rats promote reduction in the cerebellum weight by 25%. After 12 days' exposure, EtOH promotes reduction of the number of neurons in the internal granular layer (Green, 2004).

Studies show that a prenatal EtOH exposure reduces brain volume, and in the cerebellum where the reduction is more intense, causes a framework change of cognitive and behavioral functions (De Smet et al., 2013). During

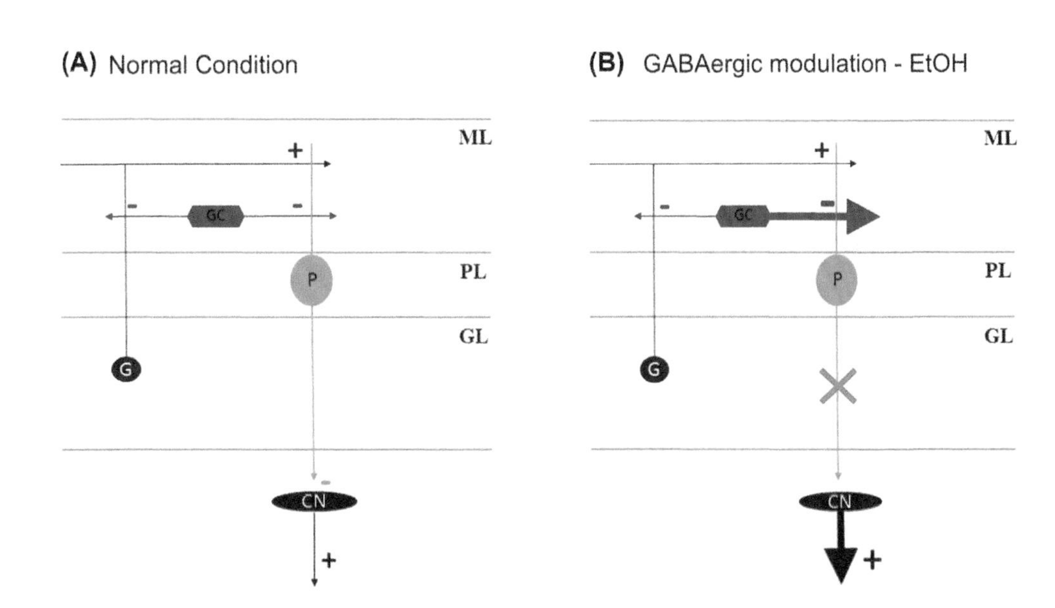

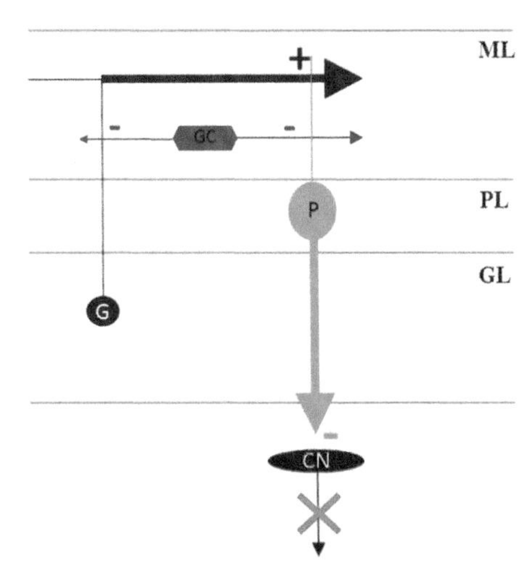

FIGURE 10.1 EtOH effect in the communication between granule cells (G), Golgi cells (GG), Purkinje cells (P), and central nuclei (CN) of the cerebellum, mediated by GABA action (−) and glutamate (+). (A) Communication cerebellar under normal conditions. (B) EtOH modulation in via GABAergic. (C) EtOH modulation via the glutamatergic. *Large arrows* mean increase in the release of neurotransmitters. *CM*, molecular layer; *CP*, Purkinje layer; *CG*, granular layer.

development, the cerebellum shows vulnerability to EtOH consumption in large quantities leading to damage to Purkinje cells, glia, and granule (Maier & West, 2001; Sakata-Haga, Sawada, Hisano, & Fukui, 2001).

Neurons in Purkinje and granule cells, in an EtOH exposure frame, can undergo apoptosis (Luo, 2012; Oliveira et al., 2014), leading to changes of frames in motor coordination (Sarna & Hawkes, 2003; Jaatinen & Rintala, 2008; Luo, 2012; Huang, Yen, Tsao, Tsai, Huang, 2014).

There is a correlation between exposure to EtOH and age of the animal with regard to possible degeneration. Exposure for 1 or 2 days during 4 or 5 days postnatal life of rats leads to a loss of 50% in Purkinje cells, whereas no or moderate loss of this cell population is observed after the same exposure time in later phase of life.

The effects of EtOH are particularly noticeable and disturbing during cerebellar development period, which can be observed in synaptogenesis, which is characterized by the development of neurites, synapse formation, and early neuronal signaling. This process begins in humans during the third trimester of pregnancy and lasts for all first years of life (Dobbing & Sands, 1973), while in mice, this period corresponds to 4–9 days, postnatal. Changes in the geometry of dendritic Purkinje cells can occur in up to 3 months old animal. From there on, the cerebellum can be considered fully mature.

Injuries resulting from prenatal exposure to EtOH are widely reported in the scientific literature. During development, this substance causes an increase in cell death by apoptosis, both in granule cells and in Purkinje cells in which it causes severe motor deficits. The first 10 days of life constitute the growing dendrite of Purkinje cells and synaptogenesis, justifying the higher sensitivity to EtOH. There is evidence that granule cells and interneurons of the molecular layer can be the most vulnerable structures of the mature cerebellum to chronic exposure to EtOH.

Bonthius, Karacay, Dai, and Pantazis (2003) also found a regional difference in vulnerability to the effects of EtOH. Granule cells and Purkinje have higher rate of reduction in the dorsal lobes. This difference may be related to maturation stage in each region: cerebellar lobes that mature early are more susceptible to cell loss than later maturation of lobules (Bauer-Moffett & Altman, 1977). Reports of single dose of EtOH in mice 5 days after birth found a significant loss of Purkinje cells in the lobules I of rat cerebellum.

From transmission electron microscopical observations insufficient intra-axonal space, accompanied by degradation of the myelin sheath and deformation, has been noted in rats. The damage to the white matter tracts can lead to motor and cognitive deficits related to alcoholism. Delays in transmission, the result of degeneration of the myelin sheath, has been identified as a basis for a cognitive slowness, since its speed is dependent on the integrity of white matter tracts.

There is a strong correlation between loss of granule cells and Purkinje cells, suggesting that granule cells are lost due to a lack of their postsynaptic targets, Purkinje cells. Recent studies, however, have identified significant loss of granular cell without loss of Purkinje cells after neonatal exposure to EtOH challenging the theory of correlation between reductions in both populations. Granule cells undergo apoptosis induced by reactive oxygen species resulting from oxidative stress (Kumar, Singh, Lavoie, Dipette, & Singh, 2011).

Apart from these two cell groups, the neurons of the cerebellar nuclei also show vulnerability to exposure to EtOH during the third-quarter equivalent in rats. In day 2–11 display window, at a dose of 3 g/kg/day caused a decrease in the number of neurons in interposito cores (Garro, McBeth, Lima, & Lieber, 1991). These centers play a key role in learning mediated by the cerebellum.

In neonatal model of fetal alcohol syndrome, there was a significant depletion of the microglial population. The surviving activated cells exhibit morphology.

CLINICAL CHANGES RESULTING FROM INGESTION OF EtOH

Intake of EtOH can cause serious damage to the cerebellum, through both acute and chronic exposure. The most frequently reported conditions are difficulties to remain orthostatic and in ambulation. Coordination deficit in the lower limbs is also reported.

Consumption of EtOH during pregnancy causes in the fetus, considerable damage to its development, which may manifest itself as a set of abnormalities known as fetal alcohol syndrome; however, the severity and range of results in the literature prefigures a broader term: fetal alcohol spectrum disorders (FASDs), especially affecting the CNS, leading to structural and behavioral deficits (Riley & Mcgee, 2005).

Clinically, motion injuries are the most important consequences of exposure to EtOH, an indication of intoxication of EtOH. Cerebellar ataxia is the first and the most clearly known effect observed in experimental animals after consumption of EtOH. The consequences of ethanol-induced cerebellar ataxia (EICA) include severe effects on motor function, stability, balance, and posture, which are covered in greater detail in the following chapters. This signal tends to become chronic as the EtOH intake endures (Dar & Al-Rejaie, 2013).

One of the common clinical changes mentioned is ataxia and gait deficits with signs of peripheral neuropathy and having a contrast with a change at a very small scale in the upper limbs. Along with the last nystagmus and dysarthria are cited low incidence of chronic consumption of EtOH (Zahr, Kaufman, & Harper, 2011).

Another manifestation of this combination is the deficit in vitamin B1 (thiamine) due to the use of EtOH, which generates a syndrome called Wernicke–Korsakoff syndrome, which in one-third of the cases with cerebellar degeneration, being an aggravating factor for this syndrome by increasing the rate and severity of neurodegeneration (Shirpoor et al., 2009).

Other studies showed the presence of cognitive impairment in patients who have cerebellar degeneration induced by EtOH, but did not suggest changes in higher brain functions (Johnson-Greene et al., 1997).

Classical research demonstrated, however, that some of these changes are almost entirely reversible with abstinence period (Zahr et al., 2011), and ataxia, a thiamine replacement feature associated with the removal of EtOH exposure. What is not observed in cases of nystagmus and ataxic gait where about 60–70% of patients exhibit an irreversible framework, the main explanation being irreversible damage to circuits and neural populations (Butterworth, 1993).

Thus, the EtOH injury to cerebellum of animals and man, in different ages displays different types. Research shows the impact of EtOH in the cerebellum; apoptosis is the main mechanism of cell death resulting from the use of this drug (Idrus, McGough, Riley, & Thomas, 2011; Luo, 2012; Rogers, Parks, Nickel, Katwal, & Martin, 2012; Van Skike et al., 2010; Zeeuw, Zwart, Schrama, van Engeland, & Durston, 2012). However, it will take still more studies to show the EtOH action mechanisms in cerebellar lesions and functional deficits resulting from the consumption of this substance in animals and in humans.

References

Albus, J. S. (1971). A theory of cerebellar function. *Mathematical Biosciences, 10*, 25–61.

Allen, G. I., & Tsukahara, N. (1974). Cerebrocerebellar communication systems. *Physiological Reviews, 54*, 957–1006.

Bauer-Moffett, C., & Altman, J. (1977). The effect of ethanol chronically administered to preweanling rats on cerebellar development: A morphological study. *Brain Research, 119*(2), 249–268.

Bonthius, D. J., Karacay, B., Dai, D., & Pantazis, N. J. (2003). FGF-2, NGF and IGF-1, but not BDNF, utilize a nitric oxide pathway to signal neurotrophic and neuroprotective effects against alcohol toxicity in cerebellar granule cell cultures. *Brain Research. Developmental Brain Research, 140*(1), 15–28.

Butterworth, R. F. (1993). Pathophysiology of cerebellar dysfunction in the Wernicke–Korsakoff syndrome. *The Canadian Journal of Neurological Sciences, 20*(Suppl. 3), S123–S126.

Chambers, W. W., & Sprague, J. M. (1955). Functional localization in the cerebellum. II. Somatotopic organization in cortex and nuclei. *A.M.A. Archives of Neurology and Psychiatry, 74*, 653–680.

Dar, M. S. (2015). Ethanol-induced cerebellar ataxia: Cellular and molecular mechanisms. *The Cerebellum, 14*(4), 447–465.

Dar, M. S., & Al-Rejaie, S. (2013). Tonic modulatory role of mouse cerebellar alpha- and beta-adrenergic receptors in the expression of ethanol-induced ataxia: Role of AC-cAMP. *Behavioural Brain Research, 241*, 154–160.

De Smet, H. J., Paquier, P., Verhoeven, J., & Marien, P. (2013). The cerebellum: Its role in language and related cognitive and affective functions. *Brain and Language, 127*(3), 334–342.

Dlugos, C. A. (2008). Ethanol-related increases in degenerating bodies in the Purkinje neuron dendrites of aging rats. *Brain Research, 1221*, 98–107.

Dobbing, J., & Sands, J. (1973). Quantitative growth and development of human brain. *Archives of Disease in Childhood, 48*(10), 757–767.

Garro, A. J., McBeth, D. L., Lima, V., & Lieber, C. S. (1991). Ethanol consumption inhibits fetal DNA methylation in mice: Implications for the fetal alcohol syndrome. *Alcoholism: Clinical and Experimental Research, 15*(3), 395–398.

Green, J. T. (2004). The effects of ethanol on the developing cerebellum and eyeblink classical conditioning. *The Cerebellum, 3*(3), 178–187.

Huang, J. J., Yen, C. T., Tsao, H. W., Tsai, M. L., & Huang, C. (2014). Neuronal oscillations in Golgi cells and Purkinje cells are accompanied by decreases in Shannon information entropy. *The Cerebellum, 13*(1), 97–108.

Idrus, N. M., McGough, N. N., Riley, E. P., & Thomas, J. D. (2011). Administration of memantine during ethanol withdrawal in neonatal rats: Effects on long-term ethanol-induced motor incoordination and cerebellar Purkinje cell loss. *Alcoholism: Clinical and Experimental Research, 35*(2), 355–364.

Ito, M. (1972). Neural design of the cerebellar control system. *Brain Research, 40*, 80–82.

Jaatinen, P., & Rintala, J. (2008). Mechanisms of ethanol-induced degeneration in the developing, mature, and aging cerebellum. *The Cerebellum, 7*(3), 332–347.

Johnson-Greene, D., Adams, K. M., Gilman, S., Kluin, K. J., Junck, L., Martorello, S., & Heumann, M. (1997). Impaired upper limb coordination in alcoholic cerebellar degeneration. *Archives of Neurology, 54*(4), 436–439.

Kumar, A., Singh, C. K., Lavoie, H. A., Dipette, D. J., & Singh, U. S. (2011). Resveratrol restores Nrf2 level and prevents ethanol-induced toxic effects in the cerebellum of a rodent model of fetal alcohol spectrum disorders. *Molecular Pharmacology, 80*(3), 446–457.

Larsell, O., & Whitlock, D. G. (1952). Further observations on the cerebellum of birds. *Journal of Comparative Neurology, 97*(3), 545–566.

Luo, J. (2012). Mechanisms of ethanol-induced death of cerebellar granule cells. *The Cerebellum, 11*(1), 145–154.

Luo, J. (2015). Effects of ethanol on the cerebellum: Advances and prospects. *The Cerebellum, 14*(4), 383–385.

Maier, S. E., & West, J. R. (2001). Regional differences in cell loss associated with binge-like alcohol exposure during the first two trimesters equivalent in the rat. *Alcohol, 23*(1), 49–57.

McClain, J. A., Morris, S. A., Deeny, M. A., Marshall, S. A., Hayes, D. M., Kiser, Z. M., & Nixon, K. (2011). Adolescent binge alcohol exposure induces long-lasting partial activation of microglia. *Brain, Behavior, and Immunity, 25*(Suppl. 1), S120–S128.

Mugnaini, E. (1983). The length of cerebellar parallel fibers in chicken and rhesus monkey. *Journal of Comparative Neurology, 220*, 7–15.

Nevo, I., & Hamon, M. (1995). Neurotransmitter and neuromodulatory mechanisms involved in alcohol abuse and alcoholism. *Neurochemistry International, 26*(4), 305–336.

Nunzi, M. G., Birnstiel, S., Bhattacharyya, B. J., Slater, N. T., & Mugnaini, E. (2001). Unipolar brush cells form a glutamatergic projection system within the mouse cerebellar cortex. *Journal of Comparative Neurology, 434*(3), 329–341.

Oliveira, S. A., Chuffa, L. G., Fioruci-Fontanelli, B. A., Lizarte Neto, F. S., Novais, P. C., Tirapelli, L. F., … Martinez, F. E. (2014). Apoptosis of Purkinje and granular cells of the cerebellum following chronic ethanol intake. *The Cerebellum, 13*(6), 728–738.

Pirnik, Z., & Kiss, A. (2002). The cerebellum: Anatomy, distribution of mediators and their receptors, communication with hypothalamic structures and comparison with the hypothalamic paraventricular nucleus under conditions of stress. *Ceskoslovenska fysiologie, 51*(2), 47–60.

Riley, E. P., & Mcgee, C. L. (2005). Fetal alcohol spectrum disorders: An overview with emphasis on changes in brain and behavior. *Experimental Biology and Medicine (Maywood), 230*(6), 357–365.

Rogers, B. P., Parks, M. H., Nickel, M. K., Katwal, S. B., & Martin, P. R. (2012). Reduced fronto-cerebellar functional connectivity in chronic alcoholic patients. *Alcoholism: Clinical and Experimental Research, 36*(2), 294–301.

Roostaei, T., Nazeri, A., Sahraian, M. A., & Minagar, A. (2014). The human cerebellum: A review of physiologic neuroanatomy. *Neurologic Clinics, 32*(4), 859–869.

Sakata-Haga, H., Sawada, K., Hisano, S., & Fukui, Y. (2001). Abnormalities of cerebellar foliation in rats prenatally exposed to ethanol. *Acta Neuropathologica, 102*(1), 36–40.

Sarna, J. R., & Hawkes, R. (2003). Patterned Purkinje cell death in the cerebellum. *Progress in Neurobiology, 70*(6), 473–507.

Shirpoor, A., Minassian, S., Salami, S., Khadem-Ansari, M. H., Ghaderi-Pakdel, F., & Yeghiazaryan, M. (2009). Vitamin E protects developing rat hippocampus and cerebellum against ethanol-induced oxidative stress and apoptosis. *Food Chemistry, 113*(1), 115–120.

Van Skike, C. E., Botta, P., Chin, V. S., Tokunaga, S., McDaniel, J. M., Venard, J., … Matthews, D. B. (2010). Behavioral effects of ethanol in cerebellum are age dependent: Potential system and molecular mechanisms. *Alcoholism: Clinical and Experimental Research, 34*(12), 2070–2208.

Wetts, R., Kalaska, J. F., & Smith, A. M. (1985). Cerebellar nuclear cell activity during antagonist cocontraction and reciprocal inhibition of forearm muscles. *Journal of Neurophysiology, 54*, 231–244.

WHO. (2014). *Global status report on alcohol and health 2014*. http://www.who.int/entity/substance_abuse/publications/global_alcohol_report/msb_gsr_2014_1.pdf?ua=1.

Wolpert, D. M., Miall, C., & Kawato, M. (1998). Internal models in the cerebellum. *Trends in Cognitive Sciences, 2*(9), 338–347.

Zahr, N. M., Kaufman, K. L., & Harper, C. G. (2011). Clinical and pathological features of alcohol-related brain damage. *Nature Reviews. Neurology, 7*(5), 284–294.

Zeeuw, P., Zwart, F., Schrama, R., van Engeland, H., & Durston, S. (2012). Prenatal exposure to cigarette smoke or alcohol and cerebellum volume in attention-deficit/hyperactivity disorder and typical development. *Translational Psychiatry, 2*, e84.

CHAPTER

11

Gene Expression in CNS Regions of Genetic Rat Models of Alcohol Abuse

W.J. McBride

Indiana University School of Medicine, Indianapolis, IN, United States

INTRODUCTION

Microarray technology offers a way to study the impact of alcohol drinking on the expression of thousands of genes in discrete brain regions and subregions in order to better understand the widespread and complex actions of ethanol on cellular function. However, alcohol drinking is a complex behavior and gene expression is sensitive to a variety of factors, all of which have to be carefully controlled.

In order to study alcohol drinking, it is important that the animal model will voluntarily consume sufficient ethanol to produce relevant blood alcohol concentrations (BACs). There are multiple drinking protocols that can produce different effects on gene expression. For example, the two general protocols most commonly used are 24-h free-choice access to an ethanol solution (5–20% ethanol), and scheduled access (1–2 h/day) during the dark phase of the light–dark cycle. The amount of daily intake of ethanol, blood alcohol levels attained, and duration of ethanol drinking could all influence gene expression. In addition, the timing of when samples are taken, in terms of the interval of last access to ethanol, can also influence gene expression. Finally, the effects of alcohol drinking on gene expression are also dependent upon the CNS region or subregion examined, as well as the genetic background of the subject and the age and sex of the subject.

This chapter focuses on the effects of alcohol drinking by alcohol-preferring (P) rats on gene expression in selected brain reward regions. The P rat is selectively bred for voluntary high alcohol consumption that produces pharmacologically relevant BACs. The P line satisfies the criteria for a suitable animal model of alcoholism and offers an experimental model for identifying changes in gene expression associated with alcohol abuse. In addition, innate differences between several rat line-pairs that have been selectively bred for high or low alcohol-drinking behavior will are reviewed. These subjects are classified as family history positive (FHP) or family history negative (FHN) for alcoholism. This approach offers the potential of identifying genes or gene networks contributing to the disparate alcohol-drinking behaviors of subjects with an innate genetic background for high or low alcohol consumption.

SELECTIVELY BRED HIGH AND LOW ETHANOL-CONSUMING LINES OF RATS

Studies from animals (Li & McBride, 1995) and humans (Cloninger, Bohman, Sigvardsson, & Von Knorring, 1985, Cloninger et al., 1989; Heath, 1995; Pickens et al., 1991; Sigvardsson, Bohman, & Cloninger, 1996) indicate that genetic factors have a significant impact on alcohol-drinking behavior. Gene expression studies can contribute to the identification of genes associated with high alcohol preference and help elucidate mechanisms contributing to a predisposition for high alcohol-drinking behavior. Gene expression studies with human tissue have been conducted using autopsy samples from individuals who have had a history of alcohol abuse (Flatscher-Bader et al., 2005; Flatscher-Bader, Zuvela, Landis, & Wilce, 2008; Lewohl et al., 2000; Mayfield et al., 2002) and comparing these results with samples from a control population. To potentially disentangle genetic differences related to susceptibility from those resulting from long-term excess alcohol consumption, it

Addictive Substances and Neurological Disease
http://dx.doi.org/10.1016/B978-0-12-805373-7.00011-6

is necessary to have subjects who are genetically susceptible to high alcohol drinking, but have not had any previous exposure to alcohol. Since this is not possible in humans, animal studies with selectively bred rat lines offer an alternative.

This section reviews differences in gene expression between five pairs of selectively bred high ethanol-consuming (HEC) and low ethanol-consuming (LEC) rat lines: (1) the alcohol-preferring (P) and non-preferring (NP) rats (Murphy et al., 2002); (2) two replicate lines of high alcohol-drinking (HAD) and low alcohol-drinking (LAD) rats (Murphy et al., 2002); (3) Alko alcohol (AA) and nonalcohol (ANA) rats (Eriksson, 1968); and (4) Sardinian alcohol-preferring (sP) and non-preferring (sNP) rats (Colombo, 1997). The selection criteria for the HEC lines were ethanol intakes equal to or greater than 5 g/kg per day and a 10% ethanol-to-water preference ratio equal to or greater than 2:1, using a 24-h two-bottle free-choice drinking procedure. The LEC lines had ethanol intakes less than 1 g/kg per day.

The CNS regions chosen to be examined were the ventral tegmental area (VTA), the nucleus accumbens (Acb) shell, and the central nucleus of the amygdala (CeA). The VTA area is a central region within the so-called "brain reward system," The VTA is a CNS site supporting the rewarding actions of alcohol (Rodd-Henricks, McKinzie, Crile, Murphy, & McBride, 2000), and regulating alcohol drinking (Czachowski, Delory, & Pope, 2012; Hodge, Haraguchi, Erickson, & Samson, 1993; Rodd et al., 2010). The Acb receives major dopamine (DA) innervation from the VTA (reviewed in McGinty, 1999). The Acb-shell is also a CNS subregion supporting the rewarding actions of alcohol (Engleman et al., 2009), and is also involved in regulating alcohol drinking (Besheer et al., 2010; Cozzoli et al., 2012). The CeA also receives DA innervation from the VTA; this region is involved in regulating alcohol drinking of alcohol-dependent animals (Gilpin, Herman, & Roberto, 2015; Roberts, Cole, & Koob, 1996).

INNATE DIFFERENCES IN GENE EXPRESSION IN THE VTA, Acb-SHELL, AND CeA BETWEEN HEC AND LEC RATS

Each CNS region has its own unique neurocircuitry. The main neurons within the VTA are DA projection neurons and GABA interneurons. The main neurons within the Acb-shell are GABA projection neurons and cholinergic interneurons. The CeA mainly has GABA interneurons and GABA projection neurons. Each region has different inputs and the neuronal activity within each region is regulated differently.

The VTA is a heterogeneous structure comprised of multiple nuclei (Oades & Halliday, 1987); it is a main CNS site where several drugs of abuse have rewarding actions, for example, ethanol (Rodd-Henricks et al., 2000), nicotine (Hauser et al., 2014; Ikemoto, Qin, & Liu, 2006), cocaine (Ding et al., 2012), and morphine (reviewed in McBride, Murphy, & Ikemoto, 1999). In addition, there is evidence that the VTA of the P rat is more sensitive to the rewarding actions of ethanol than is the VTA of either NP (Gatto, McBride, Murphy, Lumeng, & Li, 1994) or Wistar (Rodd et al., 2004) rats, suggesting there may be innate differences in the VTA of P rats that make them vulnerable to high alcohol-drinking behavior.

There were no genes differentially expressed in the VTA across all five line-pairs. There were 22 genes differentially expressed (in the same direction) in at least three of the five line-pairs; there were eight genes in common across four line-pairs (Table 11.1). *Ankrd12*, *Gsta4*, *Ncaph*, and *Zfp212* were differentially expressed in one subset of four line-pairs; *Psd3*, *Rt1-T24-4*, and *RT1-DMb* in another subset; and *Slc38a10* in a third subset (Table 11.1). Gsta4 is a glutathione S-transferase that functions primarily to detoxify lipid peroxidation products, but may have a role in ethanol preference (Bjork et al., 2006; Liang et al., 2004). Zfp212 is a zinc-finger protein with presumed transcriptional ability (Hossain, Barrow, Shen, Haq, & Bungert, 2015), although a specific role for this protein is not known. The Ankyrin repeat domain (Ankrd) proteins function as protein-interaction modules in a diverse range of cellular processes, including interacting with membrane transporters, ion-channels, cell-adhesion, and transcription (reviewed in Werbeck & Itzhaki, 2007). Overall, these results suggest that different combinations of genes within the VTA may contribute to a predisposition for high ethanol consumption across the five line-pairs.

No single molecular or cellular pathway within the VTA appeared to account for the disparate alcohol-drinking characteristics across all five line-pairs of HEC and LEC rats (McBride et al., 2012). Instead, the interactions of different combinations of multiple biological systems mediating transcription (e.g., glucocorticoid receptor signaling, NFkB signaling, and Wnt/B-catenin signaling), synaptic function (e.g., ephrin receptor signaling, integrin signaling, and clathrin-mediated endocytosis), oxidative stress production (e.g., retinoic acid receptor activation, glutathione systems, and Ox40 signaling), and intracellular signaling and transduction (e.g., cAMP-dependent protein kinase A and CREB) may contribute to the disparate alcohol-drinking behaviors within subsets of different line-pairs.

Comparison of the results for differences between the HEC and LEC line-pairs (McBride et al., 2012) with findings from meta-analyses and comprehensive genomic studies have identified a small number of gene candidates and associated pathways that may be involved in alcohol abuse, for example, *Nfkb*, *Il6*, and *Mapk*

TABLE 11.1 Genes That Were Significantly Different and Changed in the Same Direction in the VTA (of at Least Four Line-Pairs), or in the Nucleus Accumbens Shell or Central Nucleus of the Amygdala (of at Least Three Line-Pairs) of Alcohol-Naive Line-Pairs Selectively Bred for High Ethanol Consumption (HEC) or Low Ethanol Consumption (LEC)

Symbol	Description	Direction Versus LEC	Line-Pairs (Only HEC Line Indicated)
VTA			
Ankrd12	Ankyrin repeat domain 12	Higher in HEC	HAD1, HAD2, P, sP
Gsta4	Glutathione S-transferase alpha 4	Lower in HEC	HAD1, HAD2, P, sP
Ncaph	Non-SMC condensing 1 complex, H	Higher in HEC	HAD1, HAD2, P, sP
Psd3	Pleckstrin and Sec7 domain containing 3	Lower in HEC	HAD1, HAD2, P, AA
RT1-T24-4	RT1 class 1, locus T24, gene 4	Higher in HEC	HAD1, HAD2, P, AA
RT1-DMb	RT1 class II, locus DMb	Lower in HEC	HAD1, HAD2, P, AA
Slc38a10	Solute carrier family 38, member 10	Lower in HEC	HAD2, P, AA, sP
Zfp212	Zinc finger protein 212	Lower in HEC	HAD1, HAD2, P, sP
Acb-SH			
Avil	Advillin	Higher in HEC	HAD2, P, AA
Azi2	5-Azacytidine induced 2	Higher in HEC	HAD2, P, sP
Mef2c	Myocyte enhancer factor 2C	Lower in HEC	HAD2, P, sP
Nek1	NIMA-related kinase 1	Lower in HEC	HAD1, HAD2, sP
RT1-T24-4	RT1 class 1, locus T24, gene 4	Higher in HEC	HAD1, HAD2, P
Zcchc9	Zinc finger, CCHC domain containing 9	Lower in HEC	HAD2, P, sP
CeA			
Ankrd12	Ankyrin repeat domain 12	Higher in HEC	HAD1, HAD2, sP
Gsta4	Glutathione S-transferase alpha 4	Lower in HEC	HAD1, HAD2, P, sP
Plekhh1	Pleckstrin homology domain	Higher in HEC	HAD1, HAD2, sP
RT1-A2	RT1 class 1a, locus A2	Higher in HEC	HAD1, HAD2, P
Svep1	Sushi, von Willebrand factor type A, EGF and pentraxin domain containing 1	Lower in HEC	HAD1, HAD2, P

Data from McBride, W.J., Kimpel, M.W., McClintick, J.N., Ding, Z.-M., Hyytia, P., Colombo, C., ... Bell, R.L. (2012). Gene expression in the ventral tegmental area of 5 pairs of rat lines selectively bred for high or low ethanol consumption. Pharmacology, Biochemistry, and Behavior, *102, 275–285; McBride, W.J., Kimpel, M.W., McClintick, J.N., Ding, Z.-M., Hyytia, P., Colombo, C., ... Bell, R.L. (2013). Gene expression within the extended amygdala of 5 pairs of rat lines selectively bred for high or low ethanol consumption.* Alcohol, *47, 517–529.*

pathways (Aroor & Shukula, 2004; Crews et al., 2006; Mulligan et al., 2006; Zou & Crews, 2010). Furthermore, several studies have also associated members of the glutathione-S-transferase (*Gst*) family with excessive alcohol intake (Bjork et al., 2006; Hashimoto, Forquer, Tanchuck, Finn, & Wiren, 2011; Liang et al., 2004; Mulligan et al., 2006).

The Acb can be divided into two main subregions, the shell and the core. The core appears to be mainly involved in motor function, whereas the shell appears to be mainly involved in processing goal-directed behaviors (reviewed in Zahm, 1999). Therefore, because of the involvement of the shell (but not the core) in mediating the rewarding actions of ethanol (Engleman et al., 2009), gene expression differences were examined in the shell. In addition, the Acb-shell of P rats is more sensitive and responsive to ethanol than is the Acb-shell of Wistar rats (Engleman et al., 2009), suggesting innate differences within this subregion that may be contributing to the high alcohol intake of the P rats.

In the Acb-shell, there were no common genes that significantly differed across all five individual line-pairs (McBride, Kimpel, McClintick, Ding, Hyytia, et al., 2013). However, there were several genes (Table 11.1) that differed across at least three line-pairs, although not across the same subset of line-pairs. These included: (1) Advillin (*Avil*), a member of the gelsolin/villin family of actin regulatory proteins (Marks, Arai, Bandura, & Kwiatkowski, 1998) that are involved in neurite-like outgrowth (Shibata et al., 2004); (2) 5-azacytidine induced 2 (*Azi2*), which activates IKK-related kinases (Fujita et al., 2003) that are involved in many cellular and transcription processes; (3) myocyte enhancer factor 2C (*Mef2c*), a transcription factor involved in short-term synaptic plasticity (Akhtar et al., 2012) and neuronal response to cellular stress (She, Yang, & Mao, 2012); (4) NIMA-related expressed kinase 1 (*Nek1*), which plays an important role in preventing cell death induced by DNA damage (Chen, Gaczynska, Osmulski, Polci, & Riley, 2010; Pelegrini et al., 2010); (5) RT1 class 1 locus T24, gene 4 (*RT1-T24-4*), which produces a protein that processes immune responses in astrocytes (Muotri et al., 2005); and (6) zinc finger CCHC domain containing 9 (*Zcchc9*), a nuclear protein involved in regulating MAPK signaling pathways (Zhou et al., 2008).

These results suggest that differential expression of various combinations of genes within the Acb-shell may contribute to the disparate alcohol-drinking characteristics of each line-pair. The lack of common genes across all five line-pairs implies that there may be multiple mechanisms that can contribute to innate differences in vulnerability to high ethanol drinking behavior. The overall analysis indicated that there were differences in cell-to-cell signaling, cellular organization, cellular stress factors, and neurite outgrowth. Gene Ontology (GO) biological categories analysis revealed some overlap between the VTA and Acb-shell related to glucocorticoid receptor and retinoic acid receptor functions. In summary, the differences in gene expression indicated some general effects on cellular and biological pathways, but no overall differences within the VTA and Acb-shell in genes or gene networks that appeared to be related to vulnerability to high alcohol drinking.

There is evidence that the CeA is involved in mediating the actions of alcohol (Koob & Le Moal, 2008) and dependence-induced alcohol drinking (Roberts et al., 1996). The CeA appears to be a region mediating anxiety and alcohol-use disorders, involving the interactions of corticotropin-releasing factor (CRF) and neuropeptide Y (NPY) with modulation of GABA activity within the CeA (Gilpin et al., 2015).

Similar to the results found for the Acb-shell, there were no common genes in the CeA that significantly differed across all five line-pairs, although there were several genes that differed across at least three of the five line-pairs and there was one gene that differed across four line-pairs (Table 11.1). *Gsta4* differed across four of the five line-pairs in the CeA, whereas *Ankrd12*, *Plekhh1*, *Rt1-A2*, and *Svep1* differed in three of the five line-pairs. Ankyrin repeat domain 12 may facilitate the formation of NF-kB complex in the nucleus (Ferreiro & Komives, 2010; Zhang et al., 2004) and help regulate the cellular stress response (Miller et al., 2003). Plekhh1 may be involved in regulating guanine nucleotide-exchange activity (Baumeister, Rossman, Sondek, & Lemmon, 2006; Cheng, Mahon, Kostenko, & Whitehead, 2004).

None of the genes that differed in the CeA were also different in the Acb-shell (Table 11.1). However, there were a few genes that differed in at least three line-pairs that were in common between the VTA and CeA, that is, *Ankrd12* and *Gsta4*, and the VTA and the Acb-shell, that is, *RT1-T24-4*. The overall paucity of common genes among the five line-pairs in the three CNS regions may be due to the different neuronal populations and innervations among the three regions, and the contributions of multiple mechanisms contributing to vulnerability to high ethanol drinking behavior. However, in the CeA, there were differences between the HEC and LEC lines with regard to complement and interleukin signaling that may indicate alterations in the neuroimmune systems within the CeA being associated with selective breeding for disparate alcohol-drinking behaviors. Comparisons with results from other labs indicated that none of the candidate genes identified in the Acb-shell/CeA study (McBride, Kimpel, McClintick, Ding, Hyytia, et al., 2013) were observed in QTL (Bice et al., 2010; Carr et al., 2007; Tabakoff et al., 2009), transcriptome meta-analyses (Mulligan et al., 2006; Saba et al., 2011), or Genome-Wide Association (GWAS) (Edenberg et al., 2010) studies.

One other rat line-pair, the UChA (ethanol nondrinkers) and UChB (ethanol drinkers), derived from a Wistar colony at the University of Chile, was not included in the studies comparing gene expression in CNS regions across multiple line-pairs selectively bred for disparate alcohol drinking (McBride et al., 2012, McBride, Kimpel, McClintick, Ding, Hyytia, et al., 2013). However, differences in expression of different alleles for *Aldh2* (aldehyde dehydrogenase-2) have been reported for this line-pair (Quintanilla, Israel, Sapag, & Tampier, 2006).

In summary, examination of differences in gene expression in the VTA, Acb-shell, and CeA indicated very few genes in common across the five line-pairs in any of the regions. Although there was some overlap in biological processes in common within a given region across subsets of line-pairs, there were no biological processes that could reasonably be related to the innate disparate alcohol-drinking behaviors among all of the selected line-pairs. However, the loss of metabotropic glutamate receptor 2 (*Grm2*) gene expression has been proposed to be a factor in the high alcohol-drinking behavior of P rats (Zhou et al., 2013).

MULTIPLE INTERACTING FACTORS CAN IMPACT THE EFFECTS OF ALCOHOL DRINKING ON GENE EXPRESSION IN THE CNS

In order to understand better the gene expression changes associated with alcohol drinking, it is important to use an animal model that will voluntarily consume alcohol under free-choice conditions and will produce relevant blood alcohol levels, that is, those that satisfy the criteria for binge drinking (80 mg% or higher), or meet intoxicating/motor impairing (~100 mg %), or sedating (200 mg% or higher) levels. The interval between the last alcohol-drinking episode and when the brain samples were taken, as well as the duration and amount of alcohol drinking can all influence gene expression, as can genetic background, age, and time of day when the samples were taken. In addition, it is likely there will be region-specific changes in gene expression associated with alcohol drinking. Unfortunately, little is known about how these multiple interacting factors may influence gene expression.

THE P RAT AS AN ANIMAL MODEL FOR STUDYING ALCOHOL DRINKING–INDUCED CHANGES IN GENE EXPRESSION

The P rat satisfies the criteria of a suitable animal model of alcoholism proposed by Cicero (1979) and Lester and Freed (1973). The P rat will voluntarily consume ethanol in sufficient quantities to produce relevant BACs exceeding 80 mg% (criterion for binge levels of hazardous drinking), meeting intoxicating/motor impairing levels (~100 mg%), and reaching sedative levels of 200 mg% and higher (reviewed in Bell, Rodd, Engleman, Toalston, & McBride, 2014; McBride, Rodd, Bell, Lumeng, & Li, 2014). The P rat consumes alcohol for its CNS pharmacological effects and not solely due to its taste, smell, or caloric value, will work to obtain alcohol, and will develop tolerance and dependence through free-choice drinking (reviewed in McBride et al., 2014). The P rat also demonstrates robust relapse drinking and exhibits high craving for alcohol even after several weeks of abstinence (reviewed in McBride et al., 2014).

Under 24-h free-choice drinking conditions with either 10 or 15% ethanol versus water, P rats will consume 5–6 g/kg per day of ethanol, with most of the consumption occurring during the dark cycle, attaining blood alcohol levels of 50–100 mg% (reviewed in Bell et al., 2014). After 8 weeks of free-choice drinking, these rats likely qualify as being alcohol dependent. The P rat also meets the criteria of binge drinking, attaining blood alcohol levels of 80 mg% or higher after 1–2 h of scheduled access ethanol drinking with water always available (reviewed in Bell et al., 2014). Furthermore, with repeated scheduled access to ethanol (three or four daily 1-h sessions) during the dark phase, P rats can attain blood alcohol levels over 200 mg% (Bell et al., 2014; McBride, Kimpel, McClintick, Ding, Hauser, et al., 2013). Therefore, the P rat offers an animal model of alcoholism that can be used to study the impact of different alcohol-drinking protocols on gene expression in multiple CNS regions.

GENE EXPRESSION CHANGES IN THE Acb FOLLOWING 24-H FREE-CHOICE ALCOHOL DRINKING: EFFECTS OF WITHDRAWAL

In this study, the subjects were adult male P rats given continuous (7 days/week) 24-h free choice access to 15% and 30% ethanol versus water for 8 weeks. Alcohol intakes during this period averaged 9.6 ± 0.9 g/kg per day (Bell et al., 2009). Rats were killed approximately 15 h after removing the ethanol solutions to ensure that blood ethanol levels were not detectable, and for comparison purposes with P rats that were on scheduled access. Comparison of this alcohol-drinking group with the water control group indicated that there were significant ($p < .01$; false discovery rate, FDR ≤ 0.15) altered expression of 374 unique named genes (Bell et al., 2009). Most of the differences were less than 1.20-fold change, suggesting very small changes in gene expression following chronic alcohol drinking. An abbreviated list of genes that significantly differed between the water and alcohol-drinking groups is shown in Table 11.2.

Increased expression of *Fos* and *Jun* indicates increased excitation-coupled transcription (George et al., 2012; Schiavone, Avalle, Dewilde, & Poli, 2011), suggesting that increased neuronal activity is occurring in the Acb of the alcohol group. *Arc* is another immediate early gene found in soma and dendrites and is involved in, or associated with, synaptic plasticity (Guszowski et al., 2006). These changes in gene expression in the Acb between the alcohol-drinking and water-control groups may be due to withdrawal since this is the first time the rats had been deprived of alcohol after 8 weeks of continuous access; physical signs of withdrawal were experienced within this alcohol-free period following continuous alcohol drinking (Kampov-Polevoy, Matthews, Gause, Morrow, & Overstreet, 2000; Waller, McBride, Lumeng, & Li, 1982). The up-regulation of *Crh* (Table 11.2) is also consistent with signs of withdrawal-induced anxiety (Gilpin et al., 2015). In addition, the up-regulation of gene network pathways involved in transcription, calcium signaling, oxidative stress response, and glucocorticoid receptor signaling were also observed in the alcohol group (Bell et al., 2009), all of which could reflect the negative state of withdrawal-induced increase in neuronal activity. However, there were a number of genes that had reduced expression in the Acb of the alcohol-group compared to the controls (Table 11.2). Although the data with *Arc*, *Fos*, and *Jun* suggest neuronal activation associated with withdrawal from alcohol, it is possible that some of the changes may be due to continuous chronic alcohol drinking. To resolve this issue, it would be necessary to repeat the experiment and examine shorter intervals (e.g., 2 h) between removal of ethanol and euthanizing the rat for brain tissue.

In summary, the main findings from this study are consistent with withdrawal-induced increase in neuronal activity within the Acb following a prolonged period of chronic alcohol drinking and the sudden removal of alcohol. The increased neuronal activity is indicated by increased expression of several immediate early genes (*Arc*, *Fos*, and *Jun*), and the up-regulation of genetic networks involved in calcium signaling, oxidative stress response, and glucocorticoid receptor signaling (Bell et al., 2009). A study by Smith et al. (2016) examined time-course changes in gene expression in several brain regions of C57 mice following chronic intermittent ethanol drinking and withdrawal. This study did not indicate any similar changes in gene expression in the Acb as found in the study of Bell et al. (2009). This may be due to a combination of factors, most notably the use of repeated cycles of vapor chamber exposure and withdrawal in the mice compared to a single withdrawal following chronic free-choice drinking in rats (Bell et al., 2009).

GENE EXPRESSION CHANGES IN THE Acb DURING OPERANT SCHEDULED ACCESS ALCOHOL DRINKING

Operant techniques are used to demonstrate that the subject will work to obtain the drug and that the drug is producing reinforcing effects. The P rat readily responds on the lever to deliver ethanol without any prior training even when water or saccharin is concurrently available (reviewed in Bell et al., 2014; McBride et al., 2014).

TABLE 11.2 Withdrawal Associated Genes in the Nucleus Accumbens That Were Significantly Different in the Nucleus Accumbens of Alcohol-Preferring (P) Rats (Fold Change ≥ 1.20)

Symbol	Description	Fold-Change
Arc	Activity-regulated cytoskeletal-associated protein	1.53
Btg2	B-cell translocation gene 2, anti-proliferative	1.35
Crh	Corticotropin-releasing hormone	1.24
Dusp1	Dual specificity phosphatase 1	1.42
Fos	FBJ murine osteosarcoma viral oncogene homolog	1.72
Hcn1	Hyperpolarization-activated cyclic nucleotide-gated potassium channel 1	1.33
Junb	Jun-B oncogene	1.27
Jun	Jun oncogene	1.23
Nr4a1	Nuclear receptor subfamily 4, group A, member 1	1.62
Nr4a3	Nuclear receptor subfamily 4, group A, member 3	1.46
Pkib	Protein kinase inhibitor beta, cAMP dependent, catalytic	1.36
Garnl1	GTPase-activating RANGAP domain-like 1	−1.25
Hspa1b	Heat shock 70-kD protein 1B (mapped)	−1.28
Lims2	LIM and senescent cell antigen-like domain 2	−1.23
Lsr	Lipolysis-stimulated lipoprotein receptor	−1.21
Mobp	Myelin-associated oligodendrocytic basic protein	−1.22
Ntrk2	Neurotropic tyrosine kinase, receptor, type 2	−1.67
Nos3	Nitric oxide synthase 3, endothelial cell	−1.21
Pthr1	Parathyroid hormone receptor 1	−1.20
Kcnq3	Potassium voltage-gated channel, subfamily Q, member 3	−1.22
Sipa1l1	Signal-induced proliferation associated 1 like 1	−1.21
Wnk4	WNK lysine-deficient protein kinase 4	−1.21

Negative sign indicates a reduction in gene expression in the ethanol compared to the ethanol-naive water group.
Data from Bell, R.L., Kimpel, M.W., McClintick, J.N., Strother, W.N., Carr, L.G., Liang, T., ... McBride, W.J. (2009). Gene expression changes in the nucleus accumbens of alcohol-preferring rats following chronic ethanol consumption. Pharmacology, Biochemistry, and Behavior, *94, 131–147.*

However, this requires coordinated motor activity and learning to discriminate the ethanol lever from the water (or saccharin) lever. These factors alone may influence gene expression. Therefore, examining changes in gene expression, using operant techniques, requires controls to account for lever pressing activity and learning, in addition to the water control group, which has very low levels of responding for water (Rodd et al., 2008).

For this study, adult male inbred P rats were self-trained in standard 2-lever operant chambers to receive water versus water, 15% ethanol versus water, or 0.0125% saccharin versus water in daily 1-h sessions (Rodd et al., 2008). Responses per session on the water, ethanol, and saccharin levers for the three groups were approximately 20, 300, and 500, respectively. After 10 weeks of operant self-administration, rats were euthanized about 24 h after the last operant session. The rats in this study were killed at the time when they would have been placed into the operant chamber.

There were 55 differences ($p < .01$) in gene expression between the saccharin and water groups; however, with an FDR of 0.87, these differences could have occurred by chance alone. Similar results were obtained in the amygdala of the saccharin versus the water group (Rodd et al., 2008). Therefore, these results suggest that learning the operant procedure and the high number of lever presses had very little impact on gene expression within the Acb and amygdala. Therefore, any changes observed between the ethanol and water groups in these regions are mainly due to the CNS effects of ethanol self-administration.

There were 215 named genes that differed between the ethanol and water groups, with 131 genes having higher expression in the ethanol group (Rodd et al., 2008). About 70% of the changes were less than 1.2-fold, possibly suggesting tight regulation of mRNA levels. There were eight genes that had changes in expression levels equal to or greater than 1.3-fold (Table 11.3). These genes suggest alterations in intracellular signaling (*Camk4*, *Gnaq*, *Pdpk1*, and *Homer1*) and possibly GABA transmission (*Gabrb2*). Homer1 genes are part of a family of synaptic scaffolding proteins that regulate the insertion of metabotropic glutamate receptors into the synaptic plasma membrane (Kammermeier, 2006; Tappe & Kuner, 2006). The combination of reduced GABA transmission and increased metabotropic glutamate function may indicate a net excitatory tone developed within the Acb following operant ethanol self-administration.

Significant GO categories for the ethanol versus water group included anion transport, calcium ion transport, chemical homeostasis, and synaptic transmission (Rodd et al., 2008). Among the genes involved in synaptic transmission, several had higher (*Cacna2d3*, *Homer1*, *P2ry13*, and *Sv2a*), whereas others had lower (*Htr2a*, *Gabrb2*, *Gad1*, *Sstr1*, and *Syt6*) expression in the ethanol compared to the water group. Ingenuity pathway analysis (IPA) supported alterations in intracellular signaling pathways with genes mainly having reduced expression in the ethanol versus the water group (Rodd et al., 2008). In contrast, other genes involved in pro-inflammatory responses or histone regulation appear to have higher expression in the ethanol versus water group. Overall, it is difficult to define the net effects of these changes, some of which may alter synaptic function and promote the reinforcing effects of ethanol, whereas others may reflect the consequences of binge drinking, promoting pro-inflammatory responses and alterations in histone regulation.

TABLE 11.3 Genes That Were Significantly Different ($p < .005$; FDR = 0.2–0.3) in the Nucleus Accumbens of Inbred Alcohol-Preferring (iP) Rats Under Operant Alcohol Self-Administration Conditions Between the Ethanol and Ethanol-Naive Water Groups With a Fold-Change ≥ 1.3

Symbol	Description	Fold-Change
Ahi1	Abelson helper integration site 1	−1.31
Camk4	Calcium/calmodulin-dependent protein kinase IV	−1.38
Gabrb2	Gamma-aminobutyric acid A (GABA-A) receptor, subunit beta 2	−1.32
Gnaq	Guanine nucleotide–binding protein, alpha q polypeptide	−1.33
Map1b	Microtubule-associated protein 1b	−1.34
Pdpk1	3-Phosphoinositide-dependent protein kinase-1	−1.45
Cacna2d3	Calcium channel, voltage-dependent, alpha 2/delta 3 subunit	1.33
Homer1	Homer homolog (*Drosophila*)	1.75

Negative sign indicates a reduction in gene expression in the ethanol compared to the ethanol-naive water group.
Data from Rodd, Z.A., Kimpel, M.W., Edenberg, H.J., Bell, R.L., Strother, W.N., McClintick, J.N., ... McBride, W.J. (2008). Differential gene expression in the nucleus accumbens with ethanol self-administration in inbred alcohol-preferring rats. Pharmacology, Biochemistry and Behavior, *89, 481–498.*

CHANGES IN GENE EXPRESSION IN THE Acb-SHELL AND CeA FOLLOWING DAILY MULTIPLE ALCOHOL BINGE DRINKING EPISODES

One study examined the effects of multiple daily binge-drinking episodes (operant techniques were not used) on gene expression within the Acb-shell and CeA. Adult male P rats were given daily three 1-h access periods to 15% and 30% ethanol separated by 2-h intervals during the dark cycle for 5 consecutive days each week (no ethanol access on weekends). Ethanol intakes were 1.5–2.0 g/kg per session (McBride et al., 2010), which would produce BACs greater than 80 mg % (Bell et al., 2014). In the 8th week of access, rats were euthanized 1, 6, or 24 h after the first drinking episode of the day (McBride et al., 2010). A water control group was also killed at each time period.

With an FDR = 0.25, there were no significant differences at any of the time points between the water and ethanol groups; in addition, there were no significant effects of time within the water control or ethanol-drinking groups (McBride et al., 2010). Therefore, an overall ethanol effect was determined by combining the data for the three time points into a single water control and a single ethanol group. With the pooled time points, there were 276 named genes that were significantly (FDR = 0.25) different in the Acb-shell of the ethanol versus water group, with almost 80% of the genes showing higher expression with alcohol drinking. GO biological processes showed greater than 10:1 ratio of higher expression versus lower expression of genes in the categories of synaptic transmission (including ionotropic glutamate receptors), regulation of nerve impulse and synaptic plasticity (Table 11.4), and neurite development, cell projection morphogenesis, and axonogenesis in the Acb-shell of the ethanol group (McBride et al., 2010). IPA indicated 2.5- to 4-fold higher number of genes with increased expression in networks involved in long-term potentiation (LTP) and Ca^{2+}/calmodulin signaling, and glucocorticoid receptor and NF-kB signaling. Overall, these changes suggest increased synaptic excitatory function within the Acb-shell with binge-like alcohol drinking by P rats.

Pooling the CeA samples for time points resulted in 402 significant differences (FDR = 0.10) in named genes between the ethanol and water groups, with only 60% of the genes having higher expression in the ethanol group. Similar to the results for the Acb-shell, the GO biological categories showed higher expression in the ethanol versus the water group in synaptic transmission and intracellular signaling processes (Table 11.4), as well as neurite development (McBride et al., 2010). The gene expression data (Table 11.4) suggest increased GABA signaling and increased expression of ionotropic glutamate receptors within the CeA. Increased expression of ionotropic glutamate receptors would result in increased activity of GABA inter-neurons and GABA projection neurons. In agreement with the results of

TABLE 11.4 Significant Genes Listed Within Selected Gene Ontology (GO) Biological Processes in the Nucleus Accumbens (Acb) Shell and Central Nucleus of the Amygdala (CeA) of Daily Multiple Alcohol Binge Drinking and Water Control Alcohol-Preferring (P) Rats

GO Biological Processes Categories	Increased Expression—Binge Drinking	Reduced Expression—Binge Drinking
Acb-SHELL		
Synaptic transmission, regulation of nerve impulse and synaptic plasticity	**21 genes:** *Apoe, Axin2, Camk2g, Drd1ip, Egfr, Gria1, Grin1, Hras, Kcnd2, Lin7a, Myo5a, Ncdn, Nrgn, Nrxn1, Ntrk3, Rab14, Rab27b, Snap29, Syn2, Syt10, Syt5*	**2 genes:** *Htr5b, Stx5a*
CeA		
Synaptic transmission, GABA signaling	**27 genes:** *Abat, Accn1, Apba2, Axin2, Camk4, Cd24, Fyn, Gabbr1, Gabrg1, Gad2, Gnai2, Gria1, Gria2, Grik2, Grik5, Grin3a, Hap1, Ncam1, Nrxn3, Nsf, Ntrk2, Oprk1, P2ry1, Sst, Stx3, Syt10, Vapb*	**8 genes:** *Acp1, Camk2n1, Chrna7, Cyp2d22, Gabra1, Gabrg2, Rab2b, Snapap*
Intracellular signaling: second messenger-mediated, receptor-linked and protein kinase cascades	**25 genes:** *Acvr2a, Cdk7, Dgkg, Fgfr1, Foxo1a, Gabbr1, Gipr, Gkap1, Gnai2, Grik2, Grik5, Lrrn3, Mapk10, Mc4r, Ntrk2, Pprk1, P2ry1, Pclo, Ptpm, Prkca, Prprf, Prpm, Rgs2, Smad1, Src*	**10 genes:** *Alk, Avpi1, Cdc25a, Cdkn1c, Chrna7, F2r, Flt3, Insr, Rgn, Shc1*

"Increased expression—binge drinking" indicates higher expression of these genes in the ethanol compared to the water control group. "Reduced expression—binge drinking" indicates lower expression in the ethanol compared to the water control group.

Data from McBride, W.J., Kimpel, M.W., Schultz, J.A., McClintick, J.N., Edenberg, H.J., Bell, R.L. (2010). Changes in gene expression in regions of the extended amygdala of alcohol-preferring rats after binge-like alcohol drinking. Alcohol, *44, 171–183.*

the GO analysis, IPA showed 16 genes with higher and 12 genes with lower expression in the LTP network, and 12 genes with higher and 10 genes with lower expression in the glucocorticoid receptor and calcium/calmodulin signaling network of the ethanol group compared to the water group (McBride et al., 2010). However, there were only two genes (*Gria1* and *Syt10*) in common in the synaptic transmission category between the Acb-shell and CeA (Table 11.4). Furthermore, comparison of the significant biological processes categories between the Acb-shell and CeA revealed only 9 of the 34–38 categories were in common, with very few genes in common within these categories (McBride et al., 2010). These overall findings suggest that region-specific changes in gene expression are associated with binge drinking.

In general, none of the gene expression changes observed in the study of Bell et al. (2009) were observed in autopsy samples from alcoholics (Flatscher-Bader et al., 2005) with the possible exception of *Ntrk2*, which was reported to be decreased in the frontal and motor cortices of alcoholics (Mayfield et al., 2002). Single nucleotide polymorphism (SNP)-based analyses implicated the *Ntrk2* gene in alcohol dependence (Xu et al., 2007).

Overall, the results suggest that alcohol binge drinking had a significant impact on increasing expression of genes associated with excitatory synaptic function within the Acb-shell and CeA, and also increased expression of genes associated with intracellular signaling processes within the CeA. Evidence for demyelination has been reported for human alcoholics (Lewohl et al., 2000; Liu et al., 2006), but was not found in the study with multiple binge drinking by P rats (McBride et al., 2010). However, some general overlap did exist for genes in certain categories (e.g., certain GABA receptor subtypes, glutathione S-transferase subtypes, and cytokine or pro-inflammatory factors) reported by Collaborative Studies on Genetics of Alcoholism (COGA) (Strat, Ramoz, Schumann, & Gorwood, 2008) and the alcohol binge–drinking study (McBride et al., 2010).

CHANGES IN GENE EXPRESSION IN THE VTA FOLLOWING DAILY MULTIPLE EXCESSIVE ALCOHOL BINGE–DRINKING EPISODES

In this study, the effects of "loss-of-control" over alcohol intake on gene expression changes within the VTA were determined (McBride, Kimpel, McClintick, Ding, Hauser, et al., 2013). Subjects were adult female P rats; they were self-trained in three-lever operant chambers to concurrently administer 10%, 20%, and 30% ethanol in daily 4 × 1-h sessions (given 5 consecutive days each week; no ethanol on the weekends), with water freely available. These rats had 14 weeks of ethanol self-administration and attained blood alcohol levels over 200 mg% by the end of the fourth daily session. Rats were euthanized 3 h after the fourth session. Age-matched alcohol-naive water control female P rats were euthanized and brains removed in a similar manner.

There were a total of 211 unique genes that were significantly different (FDR = 0.10) between the ethanol and water groups (McBride, Kimpel, McClintick, Ding, Hauser, et al., 2013). There were approximately equal numbers of genes with increased or decreased expression in the ethanol versus the water group. Nearly 80% of the genes had fold-changes of 1.20 or greater, which is in contrast to the findings with multiple binge drinking on gene expression in the Acb-shell and CeA, where small fold-changes were found (see the foregoing). There were 26 unique named genes that differed with a fold-change of 1.4 or higher; at this level of fold-change, there were three times more genes with reduced expression than increased expression in the ethanol compared to the water group (Table 11.5). The large-fold reduction in *Fos* expression in the VTA of binge drinking animals suggests reduced excitation-coupled transcription (George et al., 2012; Schiavone et al., 2011) in the alcohol group compared to controls. IPA indicated that 11 of 16 genes associated with Fos had reduced expression in the binge-drinking group (McBride, Kimpel, McClintick, Ding, Hauser, et al., 2013), supporting the idea that excessive binge drinking reduced excitation-coupled transcription in the VTA. The large fold-reductions in gene expression (Table 11.5) would also support a negative effect on cellular function with this level of binge alcohol intake.

Within the gene network centered on *Nfkbia* (nuclear factor of kappa light polypeptide gene enhancer in B-cells inhibitor, alpha), 10 of 14 genes were up-regulated (McBride, Kimpel, McClintick, Ding, Hauser, et al., 2013), suggesting promotion of excitotoxic neuronal damage (Himadri, Kumari, Chitharanjan, & Dhananjay, 2010; Koltsova et al., 2012). IPA showed that there was reduced expression of 13 out of 15 genes in a network associated with *Srebf1* (sterol regulatory element binding transcription factor 1), suggesting reduced cholesterol and fatty acid synthesis (McBride, Kimpel, McClintick, Ding, Hauser, et al., 2013). There were several genes (*Emr1, Ctss, Npc2, Eif4ebp1, Psme1, Pycard, Cd53, C3, Serpinb9, Anax3a, B2m, C1qa, Fcer1g,* and *Itgb2*) up-regulated with excessive binge drinking (McBride, Kimpel, McClintick, Ding, Hauser, et al., 2013) that promote pro-inflammatory responses. Overall, this level of alcohol binge drinking, which produces BACs of 200 mg% and higher on a daily basis, significantly altered expression of genes that reduced normal excitation-coupled transcription and myelin formation, and promoted excitotoxic neuronal damage in the VTA.

TABLE 11.5 List of Unique Named Genes in the VTA of Alcohol-Preferring (P) Rats That Were Significantly Different (FDR = 0.1) Under Operant Excessive Daily Alcohol Binge–Drinking Episodes Between the Ethanol and Water Groups With a Fold-Change ≥ 1.4

Symbol	Description	Fold-Change
C3	Complement component 3	2.56
Emr1	EGF-like module containing, mucin-like, hormone receptor-like 1	1.57
Fcgr2a	Fc fragment of IgG, low affinity IIa, receptor (CD32)	1.69
Gpr34	G-protein-coupled receptor 34	1.81
Grxcr1	Glutaredoxin, cysteine rich 1	2.64
Trem2	Triggering receptor expressed on myeloid cells 2	1.43
Acss2	Acyl-CoA synthetase short-chain family member 2	−1.40
Apln	Apelin	−1.42
Azin1	Antizyme inhibitor 1	−1.42
Ccl6	Chemokine (C–C motif) ligand 6	−1.71
Col1a1	Collagen, type I, alpha 1	−1.40
Col3a1	Collagen, type III, alpha 1	−1.60
Cyp51	Cytochrome P450, family 51	−1.40
Ddit4	DNA-damage-inducible transcript 4	−1.56
Dusp1	Dual specificity phosphatase 1	−1.40
Egr1	Early growth response 1	−1.54
Fos	FBJ osteosarcoma oncogene	−1.54
Gpd1	Glycerol-3-phosphate dehydrogenase 1 (soluble)	−1.76
Nfil3	Nuclear factor, interleukin 3 regulated	−1.40
Pcdh20	Protocadherin 20	−1.47
Pex11a	Peroxisomal biogenesis factor 11 alpha	−1.44
Scd1	Stearoyl-coenzyme A desaturase 1	−1.48
Sgk1	Serum/glucocorticoid regulated kinase 1	−1.60
Tm7sf2	Transmembrane 7 superfamily member 2	−1.47
Ugt8	UDP glycosyltransferase 8	−1.64
Usp2	Ubiquitin specific peptidase 2	−1.41

A negative sign indicates a reduction in gene expression in the ethanol compared to the water control group.

Data from McBride, W.J., Kimpel, M.W., McClintick, J.N., Ding, Z.-M., Hauser, S.R., Edenberg, H.J., ... Rodd, Z.A. (2013). Changes in gene expression within the ventral tegmental area following repeated excessive binge-like alcohol drinking by alcohol-preferring (P) rats. Alcohol, *47, 367–380.*

There were 62 genes that significantly differed in the VTA between the alcohol-naive P and NP rats and also differed in the VTA between the excessive binge drinking and water control groups (McBride, Kimpel, McClintick, Ding, Hauser, et al., 2013). The GO biological processes gene enrichment analysis of these 62 genes revealed several significant categories, including immune effector process, inflammatory response, induction programmed cell death, and leukocyte activation. Overall, the changes in gene expression within these biological processes support increased VTA neuroimmune activation with excessive binge drinking. These findings are compatible with the results of Crews, Qin, Sheedy, Vetreno, and Zou (2013) for increased brain neuroimmune activation associated with alcohol dependence.

Comparison of changes in gene expression with excessive binge drinking in the present study and changes observed with binge drinking in the Acb of inbred P rats (Rodd et al., 2008), and in the Acb-shell and CeA of P rats (McBride et al., 2010) did not indicate any overlap in the top annotated genes, with the exception of *B2m* (beta-2 microglobulin). The paucity of common overlapping genes across the studies likely is due to a combination of factors, including but not limited to different regions analyzed, duration, and level of binge drinking, and timing of the samples after ethanol exposure. Flatscher-Bader, Harrison, Matsumoto, and Wilce (2010) reported on differences in gene expression in the VTA of alcoholics compared to control subjects. Only one overlap was found between the human and rat study and that was in a canonical pathway related to regulation of the actin skeleton.

The overall summary of the different bioinformatics analyses (McBride, Kimpel, McClintick, Ding, Hauser, et al., 2013) indicated (1) reduced excitation-coupled transcription and promotion of excitotoxic neuronal damage on multiple genes centered around *Nfkbia*, *Fos*, and *Srebf1*; (2) reduced cholesterol and fatty acid synthesis, and increased protein degradation; (3) altered cell-to-cell interactions, and dendritic development; (4) pro-inflammatory response; and (5) altered glucocorticoid signaling. These combinations of changes suggest that this level of alcohol binge drinking may be producing cellular damage within the VTA. Changes in loss of white matter integrity (Harris et al., 2008) and demyelination (Lewohl et al., 2000; Liu et al., 2006) have been reported in human autopsy brain samples of alcoholics versus control subjects.

CONCLUSIONS

Microarray technology offers the ability to examine differences between experimental groups in the expression of over 20,000 genes. This approach offers the

TABLE 11.6 Summary of Major Findings of Gene Expression Differences/Changes Associated With Alcohol Drinking by the Alcohol-Preferring (P) Rats

Drinking Procedure	CNS Region	Main Findings
Innate; five line-pairs	VTA	No similar gene differences across all five line-pairs; eight genes in common across different subsets of four line-pairs; different combinations of gene networks (transcription, synaptic function, oxidative stress processes, and intracellular signaling)
	Acb-shell	Lack of common genes across all five line-pairs; different combinations of gene networks; glucocorticoid and retinoic acid receptor responses
	CeA	Lack of common genes across all five line-pairs; alterations in complement and interleukin indicate neuroimmune differences between HEC versus LEC
Withdrawal	Acb	Increased expression of *Arc*, *Fos*, and *Jun* suggest increased neuronal activity; increased expression of *Crh* suggest increased withdrawal-induced anxiety; increased Ca^{2+} signaling, oxidative stress response and glucocorticoid receptor response
Daily single binge drinking with operant responding	Acb	Most changes small (fold-change < 1.2); only eight genes with fold-change ≥ 1.3; net excitatory tone with increased expression of *Homer1* and decreased expression of *Gabrb2*; lever pressing activity alone did not produce significant changes in gene expression
Daily multiple binge drinking	Acb-shell	Increased excitatory synaptic transmission; increased expression of genes for ionotropic glutamate receptors
	CeA	Increased excitatory synaptic transmission; increased expression of genes for ionotropic glutamate receptors; different changes in gene expression compared to Acb-shell; also increased expression of genes involved in intracellular signaling
Daily excessive multiple binge drinking with operant responding	VTA	~26 unique named genes had fold-change ≥ 1.4; reduced *Fos* expression suggests reduced neuronal activity; gene networks centered around *Nfkbia*, *Fos*, and *Srebf1* suggest excitotoxic neuronal damage and reduced cholesterol and fatty acid synthesis; also several genes up-regulated that promote pro-inflammatory response

potential of examining differences in almost every aspect of the molecular events underlying cellular function. Combining controlled behaviors with measurements on expression of thousands of genes offers the potential of gaining insight into the multiple molecular events that contribute to the behavior. This chapter reviewed differences or changes in multiple molecular events associated with innate differences in five line-pairs of rats selectively bred for high or low ethanol consumption, and the effects of different alcohol-drinking conditions on gene expression within selected CNS regions of the alcohol-preferring P rat (Table 11.6). CNS regions involved in mediating the rewarding actions of ethanol and alcohol drinking were studied. The lack of common genes across the five line-pairs in the VTA, Acb-shell, and CeA suggests that different combinations of gene networks (and molecular events) underlie the predisposition for high alcohol-drinking behavior.

Increased expression of immediate early genes indicates that enhanced neuronal excitability within the Acb may be associated with alcohol withdrawal. Although daily single episodes of binge drinking produced small fold-changes in gene expression, the increased fold-changes in certain genes suggested that a net increase in the excitatory tone within the Acb is associated with alcohol intake. A similar increase in excitatory synaptic transmission was observed in the Acb-shell and CeA of P rats under daily multiple alcohol binge–drinking episodes. In contrast, daily excessive binge drinking (BACs > 200 mg%) produced changes in gene expression that indicated reduced neuronal activity, and produced changes that supported excitotoxic neuronal damage and the promotion of pro-inflammatory responses. Overall, there were region-specific changes in gene expression that depended upon the alcohol-drinking conditions and the timing of the sample collection.

References

Akhtar, M. W., Kim, M. S., Adachi, M., Morris, M. J., Qi, X., Richardson, J. A., ... Monteggia, L. M. (2012). In vivo analysis of MEF2 transcription factors in synapse regulation and neuronal survival. *PLoS One, 7*, e34863.

Aroor, A. R., & Shukula, S. D. (2004). MAP kinase signaling in diverse effects of ethanol. *Life Sciences, 74*, 2339–2364.

Baumeister, M. A., Rossman, K. L., Sondek, J., & Lemmon, M. A. (2006). The Dbs PH domain contributes independently to membrane targeting and regulation of guanine nucleotide-exchange activity. *Biochemical Journal, 400*, 563–572.

Bell, R. L., Kimpel, M. W., McClintick, J. N., Strother, W. N., Carr, L. G., Liang, T., ... McBride, W. J. (2009). Gene expression changes in the nucleus accumbens of alcohol-preferring rats following chronic ethanol consumption. *Pharmacology, Biochemistry, and Behavior, 94*, 131–147.

Bell, R. L., Rodd, Z. A., Engleman, E. A., Toalston, J. E., & McBride, W. J. (2014). Scheduled access alcohol drinking by alcohol-preferring (P) and high-alcohol-drinking (HAD) rats: Modeling adolescent and adult binge-like drinking. *Alcohol, 48*, 225–234.

Besheer, J., Grondin, J. M., Cannady, R., Sharko, A. C., Faccidomo, S., & Hodge, C. W. (2010). Metabotropic glutamate receptor 5 activity in the nucleus accumbens is required for the maintenance of ethanol self-administration in a rat genetic model of high alcohol intake. *Biological Psychiatry, 67*, 812–822.

Bice, P. J., Liang, T., Zhang, L., Graves, T. J., Carr, L. G., Lai, D., ... Foroud, T. (2010). Fine mapping and expression of candidate genes within the chromosome 10 QTL region of the high and low alcohol-drinking rats. *Alcohol, 44*, 477–485.

Bjork, K., Saarikoski, S. T., Arlinde, C., Kovanen, L., Osei-Hylaman, D., Ubaldi, M., ... Sommer, W. J. (2006). Glutathione-S-transferase expression in the brain: Possible role in ethanol preference and longevity. *Federation of the American Society of Experimental Biologists Journal, 20*, 1826–1835.

Carr, L. G., Kimpel, M. W., Liang, T., McClintick, J. N., McCall, K., Morse, M., & Edenberg, H. J. (2007). Identification of candidate genes for alcohol preference by expression profiling of congenic rat strains. *Alcoholism: Clinical & Experimental Research, 31*, 1089–1098.

Chen, Y., Gaczynska, M., Osmulski, P., Polci, R., & Riley, D. J. (2010). Phosphorylation by Nek1 regulates opening and closing of voltage dependent anion channel 1. *Biochemical and Biophysical Research Communications, 394*, 798–803.

Cheng, L., Mahon, G. M., Kostenko, E. V., & Whitehead, I. P. (2004). Pleckstrin homology domain-mediated activation of the rho-specific guanine nucleotide exchange factor Dbs by Rac1. *Journal of Biological Chemistry, 279*, 12786–12793.

Cicero, T. J. (1979). A critique of animal analogues of alcoholism. In E. Majchrowicz, & E. P. Noble (Eds.), *Biochemistry and pharmacology of ethanol* (Vol. 2, pp. 533–560). New York: Plenum Press.

Cloninger, C. R., Bohman, M., Sigvardsson, S., & Von Knorring, A. L. (1985). Psychopathology in adopted-out children of alcoholics. *Recent Developments in Alcoholism: An Official Publication of the American Medical Society on Alcoholism, the Research Society on Alcoholism, and the National Council on Alcoholism, 3*, 37–51.

Cloninger, C. R., Sigvardsson, S., Gilligan, S. B., Von Knorring, A. L., Reich, T., & Bohman, M. (1989). Genetic heterogeneity and the classification of alcoholism. *Advances on Alcohol & Substance Abuse, 7*, 3–16.

Colombo, G. (1997). Ethanol drinking behavior in Sardinian alcohol-preferring rats. *Alcohol Alcoholism, 32*, 443–453.

Cozzoli, D. K., Courson, J., Caruana, A. L., Miller, B. W., Greentree, D. I., Thompson, A. B., ... Szumlinski, K. K. (2012). Nucleus accumbens mGluR5-associated signaling regulates binge alcohol drinking under drinking-in-the-dark procedures. *Alcoholism: Clinical & Experimental Research, 36*, 1623–1633.

Crews, F. T., Bechara, R., Brown, L. A., Guidot, D. M., Mandrekar, P., Oak, S., ... Zou, J. (2006). Cytokines and alcohol. *Alcoholism: Clinical & Experimental Research, 30*, 720–730.

Crews, F. T., Qin, L., Sheedy, D., Vetreno, R. P., & Zou, J. (2013). High mobility group box 1/toll-like receptor danger signaling increases brain neuroimmune activation in alcohol dependence. *Biological Psychiatry, 73*, 602–612.

Czachowski, C. L., Delory, M. J., & Pope, J. D. (2012). Behavioral and neurotransmitter specific roles for the ventral tegmental area in reinforce-seeking and intake. *Alcoholism: Clinical & Experimental Research, 36*, 1659–1668.

Ding, Z.-M., Oster, S. M., Hauser, S. R., Toalston, J. E., Bell, R. L., McBride, W. J., & Rodd, Z. A. (2012). Synergistic self-administration of ethanol and cocaine directly into the posterior ventral tegmental area: Involvement of serotonin-3 receptors. *Journal of Pharmacology & Experimental Therapeutics, 340*, 202–209.

Edenberg, H. J., Koller, D. L., Xuei, X., McClintick, J. N., Almasy, L., Bierut, L. J., ... Foroud, T. (2010). Genome-wide association study of alcohol dependence implicates a region on chromosome 11. *Alcoholism: Clinical & Experimental Research, 34*, 840–852.

Engleman, E. A., Ding, Z.-M., Oster, S. M., Toalston, J. E., Bell, R. L., Murphy, J. M., ... Rodd, Z. A. (2009). Ethanol is self-administered into the nucleus accumbens shell, but not the core: Evidence of genetic sensitivity. *Alcoholism: Clinical & Experimental Research, 33*, 2162–2171.

Eriksson, K. (1968). Genetic selection for voluntary alcohol consumption in the albino rat. *Science, 159*, 739–741.

Ferreiro, D. U., & Komives, E. A. (2010). Molecular mechanisms of system control of NF-kappaB signaling by IkappaBalpha. *Biochemistry, 49*, 1560–1567.

Flatscher-Bader, T., Harrison, E., Matsumoto, I., & Wilce, P. A. (2010). Genes associated with alcohol abuse and tobacco smoking in the human nucleus accumbens and ventral tegmental area. *Alcoholism: Clinical & Experimental Research, 34*, 1291–1302.

Flatscher-Bader, T., van der Brug, M., Hwang, J. W., Gochee, P. A., Matsumoto, J., Niwa, S., & Wilce, P. A. (2005). Alcohol-responsive genes in the frontal cortex and nucleus accumbens of human alcoholics. *Journal of Neurochemistry, 93*, 359–370.

Flatscher-Bader, T., Zuvela, N., Landis, N., & Wilce, P. A. (2008). Smoking and alcoholism target genes associated with plasticity and glutamate transmission in the human ventral tegmental area. *Human Molecular Genetics, 17*, 38–51.

Fujita, F., Taniguchi, Y., Kato, T., Narita, Y., Furuya, A., Ogawa, T., ... Nakanishi, M. (2003). Identification of NAP1, a regulatory subunit of IkappaB kinase-related kinases that potentiates NF-kappaB signaling. *Molecular and Cellular Biology, 23*, 7780–7793.

Gatto, G. J., McBride, W. J., Murphy, J. M., Lumeng, L., & Li, T.-K. (1994). Ethanol self-infusion into the ventral tegmental area by alcohol-preferring rats. *Alcohol, 11*, 557–564.

George, O., Sanders, C., Freiling, J., Grigoryan, E., Vu, S., Allen, C. D., ... Koob, G. F. (2012). Recruitment of medial prefrontal cortex neurons during alcohol withdrawal predicts cognitive impairment and excessive alcohol drinking. *Proceedings of the National Academy of Sciences of the United States of America, 109*, 18156–18161.

Gilpin, N. W., Herman, M. A., & Roberto, M. (2015). The central amygdala as an intergrative hub for anxiety and alcohol use disorders. *Biological Psychiatry, 77*, 859–869.

Guszowski, J. F., Miyashita, T., Chawla, M. K., Sanderson, J., Maes, L. I., Houston, F. P., ... Barnes, C. A. (2006). Recent behavioral history modifies coupling between cell activity and Arc gene transcription in hippocampal CA1 neurons. *Proceedings of the National Academy of Sciences of the United States of America, 103*, 1077–1082.

Harris, G. J., Jaffin, S. K., Hodge, S. M., Kennedy, D., Caviness, V. S., Marinkovic, K., ... Oscar-Berman, M. (2008). Frontal white matter and cingulum diffusion tensor imaging deficits in alcoholism. *Alcoholism: Clinical & Experimental Research, 32*, 1001–1013.

Hashimoto, J. G., Forquer, M. R., Tanchuck, M. A., Finn, D. A., & Wiren, K. M. (2011). Importance of genetic background for risk of relapse shown in altered prefrontal cortex gene expression during abstinence following chronic alcohol intoxication. *Neuroscience, 173*, 57–75.

Hauser, S. R., Bracken, A. L., Deehan, G. A., Jr., Toalston, J. E., Ding, Z.-M., Truitt, W. A., ... Rodd, Z. A. (2014). Selective breeding for high alcohol preference increases the sensitivity of the posterior VTA to the reinforcing effects of nicotine. *Addiction Biology, 19*, 800–811.

Heath, A. C. (1995). Genetic influences on alcoholism risk: A review of adoption and twin studies. *Alcohol & Health Research World, 19*, 166–171.

Himadri, P., Kumari, S. S., Chitharanjan, S., & Dhananjay, S. (2010). Role of oxidative stress and inflammation in hypoxia-induced cerebral edema: A molecular approach. *High Altitude Medical Biology, 11*, 231–244.

Hodge, C. W., Haraguchi, M., Erickson, H., & Samson, H. H. (1993). Ventral tegmental microinjections of quinpirole decrease ethanol and sucrose-reinforced responding. *Alcoholism: Clinical & Experimental Research, 17*, 370–375.

Hossain, M. A., Barrow, J. J., Shen, Y., Haq, M. I., & Bungert, J. (2015). Artificial zinc finger DNA binding domains: Versatile tools for genome engineering and modulation of gene expression. *Journal of Cellular Biochemistry, 116*, 2435–2444.

Ikemoto, S., Qin, M., & Liu, Z. H. (2006). Primary reinforcing effects of nicotine are triggered from multiple regions both inside and outside the ventral tegmental area. *Journal of Neuroscience, 26*, 723–730.

Kammermeier, P. J. (2006). Surface clustering of metabotropic glutamate receptor I induced by long Homer proteins. *BMC Neuroscience, 7*, 1–10.

Kampov-Polevoy, A. B., Matthews, D. B., Gause, L., Morrow, A. L., & Overstreet, D. H. (2000). P rats develop physical dependence on alcohol via voluntary drinking: Changes in seizure thresholds, anxiety and patterns of alcohol drinking. *Alcoholism: Clinical & Experimental Research, 24*, 278–284.

Koltsova, S. V., Trurchina, Y., Haloui, M., Akimova, O. A., Tremblay, J., Harnet, P., & Orlov, S. N. (2012). Ubiquitous $[Na^+]_i/[K^+]_i$-sensitive transcriptome in mammalian cells: Evidence for $[Ca^{2+}]_i$-independent excitation-transcription coupling. *PLoS One, 7*, e38032.

Koob, G. F., & Le Moal, M. (2008). Neurobiological mechanisms for opponent motivational processes in addiction. *Philosophical Transactions Royal Society of London B Biological Sciences, 363*, 3113–3123.

Lester, D., & Freed, E. X. (1973). Criteria for an animal model of alcoholism. *Pharmacology Biochemistry & Behavior, 1*, 103–107.

Lewohl, J. M., Wang, L., Miles, M. F., Zhang, L., Dodd, P. R., & Harris, R. A. (2000). Gene expression in human alcoholism: Microarray analysis of frontal cortex. *Alcoholism: Clinical & Experimental Research, 24*, 1873–1882.

Li, T.-K., & McBride, W. J. (1995). Pharmacogenetic models of alcoholism. *Clinical Neuroscience, 3*, 182–187.

Liang, T., Habegger, K., Spence, J. P., Foroud, T., Ellison, J. A., Lumeng, L., ... Carr, L. G. (2004). Glutathione S-transferase 8-8 expression is lower in alcohol-preferring than in alcohol-nonpreferring rats. *Alcoholism: Clinical & Experimental Research, 28*, 1622–1628.

Liu, J., Lewohl, J. M., Harris, R. A., Iyer, V. R., Dodd, P. R., Randall, P. K., & Mayfield, R. D. (2006). Patterns of gene expression in the frontal cortex discriminate alcoholic from nonalcoholic individuals. *Neuropsychopharmacology: Official Publication of the American College of Neuropsychopharmacology, 31*, 1574–1582.

Marks, P. W., Arai, M., Bandura, J. L., & Kwiatkowski, D. J. (1998). Advillin (p92): A new member of the gelsolin/villin family of actin regulatory proteins. *Journal of Cell Sciences, 111*, 2129–2136.

Mayfield, R. A., Lewohl, J. M., Dodd, P. R., Herlihy, A., Liu, J., & Harris, R. A. (2002). Patterns of gene expression are altered in the frontal and motor cortices of human alcoholics. *Journal of Neurochemistry, 81*, 802–813.

McBride, W. J., Kimpel, M. W., McClintick, J. N., Ding, Z.-M., Hauser, S. R., Edenberg, H. J., ... Rodd, Z. A. (2013). Changes in gene expression within the ventral tegmental area following repeated excessive binge-like alcohol drinking by alcohol-preferring (P) rats. *Alcohol, 47*, 367–380.

McBride, W. J., Kimpel, M. W., McClintick, J. N., Ding, Z.-M., Hyytia, P., Colombo, C., ... Bell, R. L. (2012). Gene expression in the ventral tegmental area of 5 pairs of rat lines selectively bred for high or low ethanol consumption. *Pharmacology, Biochemistry, and Behavior, 102*, 275–285.

McBride, W. J., Kimpel, M. W., McClintick, J. N., Ding, Z.-M., Hyytia, P., Colombo, C., ... Bell, R. L. (2013). Gene expression within the extended amygdala of 5 pairs of rat lines selectively bred for high or low ethanol consumption. *Alcohol, 47*, 517–529.

McBride, W. J., Kimpel, M. W., Schultz, J. A., McClintick, J. N., Edenberg, H. J., & Bell, R. L. (2010). Changes in gene expression in regions of the extended amygdala of alcohol-preferring rats after binge-like alcohol drinking. *Alcohol, 44*, 171–183.

McBride, W. J., Murphy, J. M., & Ikemoto, S. (1999). Localization of brain reinforcement mechanisms: Intracranial self-administration and intracranial place-conditioning studies. *Behavioral Brain Research, 101*, 129–152.

McBride, W. J., Rodd, Z. A., Bell, R. L., Lumeng, L., & Li, T.-K. (2014). The alcohol-preferring (P) and high-alcohol-drinking (HAD) rats – animal models of alcoholism. *Alcohol, 48*, 209–215.

McGinty, J. F. (1999). Regulation of neurotransmitter interactions in the ventral striatum. *Annals of the New York Academy of Sciences, 877*, 129–139.

Miller, M. K., Bang, M. L., Witt, C. C., Labeit, D., Trombitas, C., Watanabe, K., ... Labeit, S. (2003). The muscle Ankyrin repeat proteins: CARP, ankrd2/Arpp and DARP as a family of titin filament-based stress response molecules. *Journal of Molecular Biology, 333*, 951–964.

Mulligan, M. K., Ponomarev, I., Hitzemann, R. J., Belknap, J. K., Tabakoff, B., Harris, R. A., ... Bergeson, S. E. (2006). Toward understanding the genetics of alcohol drinking through transcriptome meta-analysis. *Proceedings of the National Academy of Sciences of the United States America, 103*, 6368–6373.

Muotri, A. R., Chu, V. T., Marchetto, M. C., Deng, W., Moran, J. V., & Gage, F. H. (2005). Somatic mosaicism in neuronal precursor cells mediated by L1 retro-transposition. *Nature, 435*, 903–910.

Murphy, J. M., Stewart, R. B., Bell, R. L., Badia-Elder, N. E., Carr, L. G., McBride, W. J., ... Li, T.-K. (2002). Phenotypic and genotypic characterization of the Indiana University rat lines selectively bred for high and low alcohol preference. *Behavioral Genetics, 32*, 363–388.

Oades, R. A., & Halliday, G. M. (1987). Ventral tegmental (A10) system: Neurobiology. I. Anatomy and connectivity. *Brain Research Reviews, 12*, 117–165.

Pelegrini, A. L., Moura, D. J., Brenner, B. L., Ledur, P. F., Maques, G. P., Henriques, J. A., ... Lenz, G. (2010). Nek1 silencing slows down DNA repair and blocks DNA damage-induced cell cycle arrest. *Mutagenesis, 25*, 447–454.

Pickens, R. W., Svidis, D. S., McGue, M., Lykken, D. T., Heaton, L. L., & Clayton, P. J. (1991). Heterogeneity in the inheritance of alcoholism. *Archives General Psychiatry, 48*, 19–28.

Quintanilla, M. E., Israel, Y., Sapag, A., & Tampier, L. (2006). The UChA and UChB rat lines: Metabolic and genetic differences influencing ethanol intake. *Addiction Biology, 11*, 310–323.

Roberts, A. J., Cole, M., & Koob, G. F. (1996). Intra-amygdala muscimol decreases operant ethanol self-administration in dependent rats. *Alcoholism: Clinical & Experimental Research, 20*, 1289–1298.

Rodd, Z. A., Bell, R. L., Melendez, R. I., Kuc, K. A., Lumeng, L., Li, T.-K., ... McBride, W. J. (2004). Comparison of intracranial self-administration of ethanol within the posterior ventral tegmental area between alcohol-preferring and Wistar rats. *Alcoholism: Clinical & Experimental Research, 28*, 1212–1219.

Rodd, Z. A., Bell, R. L., Oster, S. M., Toalston, J. E., Pommer, T. J., McBride, W. J., & Murphy, J. M. (2010). Serotonin-3 receptors in the posterior ventral tegmental area regulate ethanol self-administration of alcohol-preferring (P) rats. *Alcohol, 44*, 245–255.

Rodd, Z. A., Kimpel, M. W., Edenberg, H. J., Bell, R. L., Strother, W. N., McClintick, J. N., … McBride, W. J. (2008). Differential gene expression in the nucleus accumbens with ethanol self-administration in inbred alcohol-preferring rats. *Pharmacology, Biochemistry and Behavior, 89*, 481–498.

Rodd-Henricks, Z. A., McKinzie, D. L., Crile, R. S., Murphy, J. M., & McBride, W. J. (2000). Regional heterogeneity for the intracranial self-administration of ethanol within the ventral tegmental area of female Wistar rats. *Psychopharmacology, 149*, 217–224.

Saba, L. M., Bennett, B., Hoffman, P. L., Barcomb, K., Ishii, T., Kechris, K., & Tabakoff, B. (2011). A systems genetic analysis of alcohol drinking by mice, rats and men: Influence of brain GABAergic transmission. *Neuropharmacology, 60*, 1269–1280.

Schiavone, D., Avalle, L., Dewilde, S., & Poli, V. (2011). The immediate early genes Fos and Egr1 become STAT1 transcriptional targets in the absence of STAT3. *Federation of Experimental Biological Society Letters, 585*, 2455–2460.

She, H., Yang, Q., & Mao, Z. (2012). Neurotoxin-induced selective ubiquitination and regulation of MEF2A isoform in neuronal stress response. *Journal of Neurochemistry, 122*, 1203–1210.

Shibata, M., Ishii, J., Koizumi, H., Shibata, N., Dohmae, N., Takio, K., … Arai, H. (2004). Type F scavenger receptor SREC-1 interacts with advillin, a member of the gelsolin/villin family, and induces neurite-like outgrowth. *Journal of Biological Chemistry, 279*, 40084–40090.

Sigvardsson, S., Bohman, M., & Cloninger, C. R. (1996). Replication of the Stockholm adoption study of alcoholism. *Archives General Psychiatry, 53*, 681–687.

Smith, M. L., Lopez, M. E., Archer, K. J., Wolen, A. R., Becker, H. C., & Miles, M. F. (2016). Time-course analysis of brain regional expression network responses to chronic intermittent ethanol and withdrawal: Implications for mechanisms underlying excessive ethanol consumption. *PLoS One, 11*, e0146257.

Strat, Y. L., Ramoz, N., Schumann, G., & Gorwood, P. (2008). Molecular genetics of alcohol dependence and related endophenotypes. *Current Genomics, 9*, 444–451.

Tabakoff, B., Saba, L., Printz, M., Flodman, P., Hodgkinson, C., Goldman, D., … Hoffman, P. L. (2009). Genetical genomic determinants of alcohol consumption in rats and humans. *BMC Biology, 7*, 70. http://dx.doi.org/10.1186/1741-7007-7-70.

Tappe, A., & Kuner, R. (2006). Regulation of motor performance and striatal function by synaptic scaffolding proteins of the Homer 1 family. *Proceedings of the National Academy of Sciences of the United States of America, 103*, 774–779.

Waller, M. B., McBride, W. J., Lumeng, L., & Li, T.-K. (1982). Induction of dependence on ethanol by free-choice drinking in alcohol-preferring rats. *Pharmacology, Biochemistry, and Behavior, 16*, 501–507.

Werbeck, N. D., & Itzhaki, L. S. (2007). Probing a moving target with a plastic unfolding intermediate of an Ankyrin-repeat protein. *Proceedings of the National Academy of Sciences of the United States of America, 104*, 7863–7868.

Xu, K., Anderson, T. R., Neyer, K. M., Lamparella, N., Jenkins, G., Zhou, Z., … Lipsky, R. H. (2007). Nucleotide sequence variation within the human tyrosine kinase B neurotrophin receptor gene: Association with antisocial alcohol dependence. *Pharmacogenomics Journal, 7*, 368–379.

Zahm, D. S. (1999). Functional-anatomical implications of the nucleus accumbens core and shell subterritories. *Annals of the New York Academy of Sciences, 877*, 113–128.

Zhang, A., Yeung, P. L., Li, C.-W., Dinh, G. K., Wu, X., Li, H., & Chen, J. D. (2004). Identification of a novel family of Ankyrin repeats containing cofactors for p160 nuclear receptor coactivators. *Journal of Biological Chemistry, 279*, 33799–33805.

Zhou, Z., Karlsson, C., Liang, T., Xiong, W., Kimura, M., Tapocik, J. D., … Goldman, D. (2013). Loss of metabotropic glutamate receptor 2 escalates alcohol consumption. *Proceedings of the National Academy of Sciences of the United States of America, 110*, 16963–16968.

Zhou, A., Zhou, J., Yang, L., Liu, M., Li, H., Xu, S., … Zhang, J. (2008). A nuclear localized protein ZCCH9 is expressed in cerebral cortex and suppresses the MAPK signal pathway. *Journal of Genetics & Genomics, 35*, 467–472.

Zou, J., & Crews, F. (2010). Induction of innate immune gene expression cascades in brain slice cultures by ethanol: Key role of NF-kB and proinflammatory cytokines. *Alcoholism: Clinical & Experimental Research, 34*, 777–789.

CHAPTER

12

Role of TLR4 in the Ethanol-Induced Modulation of the Autophagy Pathway in the Brain

A. Pla, M. Pascual, C. Guerri

Príncipe Felipe Research Center, Valencia, Spain

Ethanol is a toxic compound, and its abuse induces damage to different body organs, such as liver, brain, heart, pancreas, or lungs. Ethanol is especially noxious for the central nervous system (CNS), where it is capable of inducing gliosis, production of cytokines and neuroinflammation, myelin derangements, and ultimately neurodegeneration. Abusive ethanol consumption results in significant alterations to the cerebral structure, physiology, and function (Alfonso-Loeches & Guerri, 2011; Crews et al., 2005). Frontal lobes are one of the most sensitive brain areas to ethanol effects, although structural damage to other brain structures, such as the hippocampus, thalamus, corpus callosum, or cerebellum, has also been reported (Rosenbloom, Sullivan, & Pfefferbaum, 2003). These morphological changes correlate with functional impairment and neuropsychological deficits, which are related to prefrontal cortex-related cognitive functions. Emerging data from recent years have supported the role of the innate immune system response in many actions of ethanol in the brain, such as cellular damage, cognitive and behavioral dysfunctions, and even addiction (Alfonso-Loeches, Pascual-Lucas, Blanco, Sanchez-Vera, & Guerri, 2010; Gorini, Roberts, & Mayfield, 2013; Pascual, Balino, Alfonso-Loeches, Aragon, & Guerri, 2011).

Dysfunction during proteolytic processes has also been related to the pathogenesis of several inflammatory and neurodegenerative disorders (Ciechanover & Brundin, 2003; Rami, 2009). Intracellular proteolysis plays a crucial role in maintaining cellular homeostasis by participating in essential cellular processes, such as cell cycle regulation, transcriptional control, signal transduction, antigen presentation, and protein turnover (Pickart & Cohen, 2004). Indeed, an abnormal accumulation of mid-folded proteins or aggregation of disease-specific proteins can result from deficient clearance systems, such as the autophagy–lysosome pathway (ALP), which can produce toxic neuronal effects (Morimoto, 2008). All these findings indicate that this proteolytic process could be a target of the actions of ethanol on adult and developing brains.

This chapter reviews the evidence that indicates the differential roles of the autophagy pathway in the brain during acute and chronic ethanol treatment. It also discusses the findings that indicate the participation of the innate immune response and TLR4 receptors in ethanol-induced neuroinflammation and brain damage, and the involvement of the ALP pathway in the neuropathological changes induced by ethanol. Potential therapeutic approaches for the treatment of ethanol pathology by targeting autophagic pathways are also discussed.

AUTOPHAGY AS A PHYSIOLOGICAL SELF-DEGRADATIVE PROCESS

To ensure the quality of intracellular proteins and organelles, cells rely on two major proteolytic machineries: the ubiquitin-proteasome system (UPS) and the ALP. Both systems play important roles in cellular catabolism, and participate in the management of metabolic stress and cellular homeostasis. While the proteasome degrades unnecessary or damaged proteins by proteolysis, a chemical reaction that breaks peptide bonds, the autophagy comprises every cellular degradation pathway that consists in the

Addictive Substances and Neurological Disease
http://dx.doi.org/10.1016/B978-0-12-805373-7.00012-8

delivery of a cargo to the lysosome. Although it shares some functions with the UPS, autophagy is the main responsible mechanism for the degradation of long-lived proteins and organelle turnover.

Autophagy levels increase rapidly when intracellular nutrients are needed, for example, during periods of nutrient starvation or under metabolic stress situations, which facilitates cell survival under high-energy demand conditions (Shen & Codogno, 2012). Therefore, autophagy is a key process in cell survival, differentiation, development, or immunity. Initial autophagy steps include the formation (membrane nucleation) and expansion (vesicle elongation) of an isolation membrane known as phagophore (see Fig. 12.1). The fusion of both phagophore ends forms the autophagosome, a double membrane vesicle that sequesters the cytoplasmic material to be degraded. This particle then fuses with a lysosome to generate an autolysosome, the final structure in which degradation takes place.

At a molecular level, the whole process is controlled by the Atg proteins ("autophagy-related proteins"), which are highly conserved between species. During autophagy, several protein complexes regulate various process phases: the Atg1/ULK1 complex participates in the induction of the phagophore; Atg6/Beclin-1 mediates vesicle nucleation; two ubiquitin-like conjugation complexes (Atg5–Atg12–Atg16 and Atg8/LC3-phosphatidylethanolamine) induce vesicle expansion; finally, a recycling complex regulates the disassembling of Atg proteins from mature autophagosomes (Atg2–Atg9–Atg18). Upstream of the Atg proteins, the mechanistic target of rapamycin (mTOR) constitutes the main autophagy inhibitor, and is activated in the presence of growth factors or under nutrient-rich conditions.

Another important key player in the autophagic process is p62/SQSTM1, a protein that accumulates in the cytoplasm under low autophagy conditions. When autophagy needs to be induced, p62 physically interacts with LC3-II and is rapidly degraded within the autolysosome (Pankiv et al., 2007). Hence the down-regulation of p62 levels is indicative of an increased autophagy flux. Moreover, p62 is capable of interacting with ubiquitinated proteins and protein aggregates, and guides them toward proteolytic mechanisms, thereby acting as an adaptor molecule that serves for the selective degradation of specific substrates.

ETHANOL EFFECTS ON THE AUTOPHAGY PATHWAY

Ethanol is oxidized to acetaldehyde by enzymes alcohol dehydrogenase (ADH) and cytochrome P450 2E1 (CYP2E1) mainly in the liver, a process that also

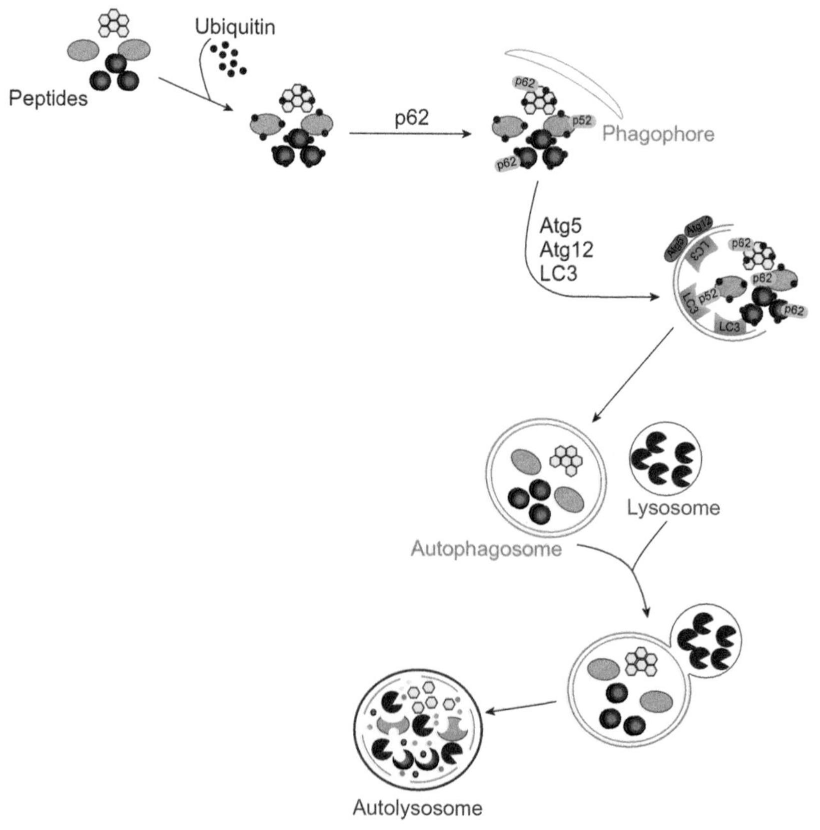

FIGURE 12.1 Schematic model of autophagy. Ubiquitinated proteins can bind to LC3 through protein p62 to induce the formation of the phagophore, which are the initial precursor structures of this transport pathway. These membrane cisterns are formed by Atg machinery. The closure of the expanding phagophore leads to the formation of a double-membrane vesicle called autophagosome, which contains the cargo targeted for degradation. Atg machinery is then released from the surface of complete autophagosomes. While the cargo material starts to be already turned over in autophagosomes, exposure to hydrolases by fusion with lysosomes to form autolysosomes allows its complete degradation.

generates reactive oxygen and nitrogen species. These compounds are highly toxic for cells, and apart from participating in the pathogenesis of several neurodegenerative diseases, they have also been related with proteolytic deficiencies. Oxidative modifications of proteins may result in an altered function and increased hydrophobicity, which favor protein aggregation. Therefore, autophagy has emerged as a vital pathway for the elimination of oxidative stress and toxicity generated by ethanol.

However, the link between ethanol and autophagy has remained a controversial subject for the last decade, mainly because of two different key subjects. On the one hand, the results as to whether ethanol induces or impairs autophagy are contradictory. While several studies have suggested that ethanol is able to impair autophagy in hepatic pathologies (Ding, Li, & Yin, 2011; Osna, Thomes, & Donohue, 2011; Thomes et al., 2013) or in several immortalized cell lines (von Haefen, Sifringer, Menk, & Spies, 2011), other studies have shown that autophagy is essential for the reduction of ethanol-caused ethanol (Ding et al., 2010).

On the other hand, the role of the autophagic response is a matter of heated debate. As ethanol usually triggers cell death, the autophagy response has been interpreted as either a detrimental cellular mechanism that favors cell death or as a way to prevent it. Indeed, autophagy has been proposed to remove damaged cells either directly through a process known as autophagic cell death (type II cell death) or indirectly through an interaction with apoptotic (type I) or necrotic (type III) cell death. On the contrary, different studies have described autophagy as a "self-defense" mechanism that attempts to avoid cell death. In line with these several neurodegenerative diseases, such as Alzheimer's disease (AD), Parkinson's disease (PD), Huntington's disease (HD), or multiple sclerosis (MS), present an accumulation of different defective proteins and disruption in the autophagic process (Yamamoto & Simonsen, 2011). It has also been demonstrated that induction of autophagy is beneficial for cell survival given its role in clearing accumulated aggregates (Giordano, Darley-Usmar, & Zhang, 2014; Wang & Mandelkow, 2012).

Because ethanol is metabolized mainly in the liver, this is where the relationship between ethanol and autophagy has been more extensively studied. However, ethanol is rapidly absorbed in the gastrointestinal tract and is distributed throughout the body, including the brain, because ethanol can easily cross the blood–brain barrier (BBB) due to its hydrophobicity. To date, there has been extensive literature on the role of proteolytic systems in the ethanol-induced liver pathology, but the molecular mechanisms implicated in the effect that ethanol has on the brain have only started being unveiled very recently.

Autophagy Dysfunctions in Neurodegenerative Disorders

One typical hallmark of neurodegeneration and neuroinflammation-related diseases is the accumulation of misfolded or defective proteins (tautopathies, synucleinopathies, or polyglutamine diseases), which can be either the cause or the consequence of the specific pathological condition. Along these lines, several neurodegenerative diseases (AD, PD, HD, or MS) present a disruption in the autophagic process and an accumulation of different defective proteins (Yamamoto & Simonsen, 2011), which generates a toxic environment which is commonly known as "proteotoxicity." Despite the controversy about the role of ethanol in autophagy and cell survival, it is generally assumed in neurodegenerative diseases that the induction of autophagy is beneficial given its role in clearing accumulated aggregates (Giordano et al., 2014; Wang & Mandelkow, 2012). Nevertheless, alcohol can cause induce or impair the autophagy process depending on alcohol treatment (see forthcoming section and Fig. 12.2).

Impairment of Autophagy in the Brain of Chronic Ethanol-Treated Mice

The effects of ethanol on brain autophagy have only recently begun to be studied. Using a mouse model fed with 10% ethanol in water for 4 months, it has been demonstrated that chronic ethanol consumption impairs the autophagy pathway, and this can lead to brain damage, and ultimately to neurodegeneration (Pla, Pascual, Renau-Piqueras, & Guerri, 2014).

Electron microscopy studies from cerebral cortices of chronic ethanol-treated mice have revealed that ethanol induces a size increase of the autophagic vacuoles (Fig. 12.3), which might be indicative of their inability to fuse with lysosomes and to, hence, trigger the effective degradation of their cargoes.

At the molecular level, chronic ethanol treatment changes the expression of autophagy-related proteins (Atg), which are ubiquitin-like proteins required for the formation of autophagosomal membranes. In particular, ethanol treatment inhibits the expression of different Atg proteins, such as Atg5, Atg12, and Atg8/LC3-II, as well as the protein levels of the major lysosomal hydrolase, cathepsin B, in the cerebral cortex of wild-type (WT) mice. These changes are associated with an increased amount of ubiquitinated proteins, which might reflect a general defect in proteostasis.

Moreover, Pla et al. (2014) proposed that the autophagy impairment induced by chronic ethanol treatment was mediated by changes in the master autophagy inhibitor mTOR. According to this hypothesis, in vivo administration of rapamycin, an mTOR inhibitor, to ethanol-treated mice was able to alleviate

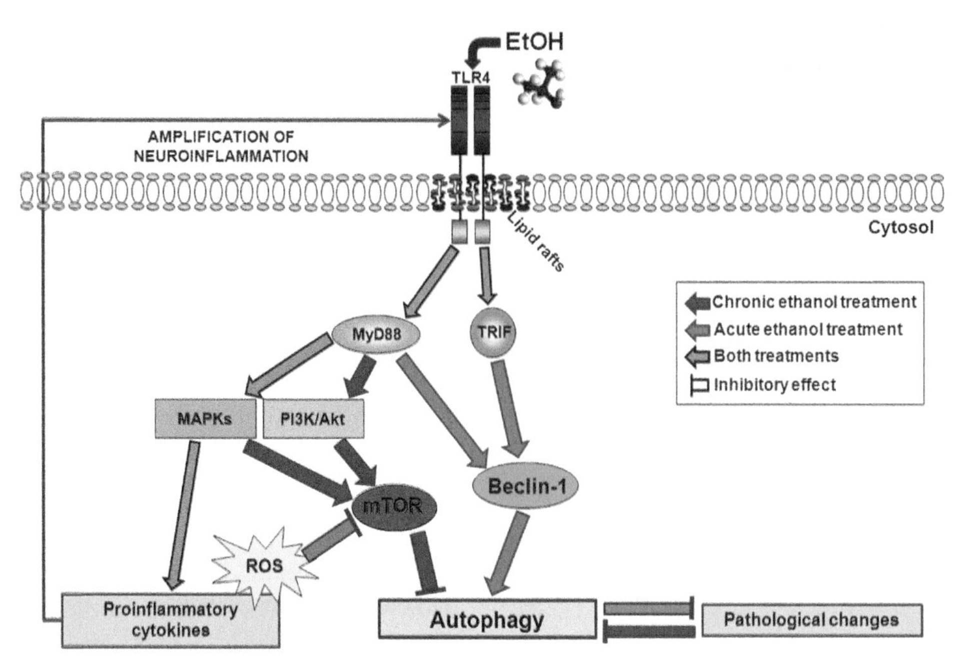

FIGURE 12.2 Diagram of the effects of ethanol (chronic vs. acute treatment) through TLR4 activation on the autophagy pathway. Chronic ethanol consumption (*red arrows*), by activating the TLR4 immune response, induces neuroinflammation and neuropathological changes, and impairs the autophagic pathway by altering its main molecular switch, mTOR, which might contribute to neurodegeneration. The neuropathological changes generated by ethanol toxicity in other cellular structures may also block the autophagy mechanism. Acute ethanol treatment (*blue arrows*) is able to activate an autophagy response in glial cells, intended to alleviate ethanol toxicity through the inhibition of mTOR by ROS, and beclin-1 activation by innate immunity receptor TLR4. *Gray arrows* point out the different proteins of the TLR4 pathway, which are induced by both treatments (chronic and acute). *MAPK*, mitogen-activated protein kinase; *MyD88*, myeloid differentiation primary response gene 88; *ROS*, reactive oxygen species; *TRIF*, TIR-domain-containing adaptor-inducing IFNβ.

the inflammation caused by ethanol in WT mice, as demonstrated by the decrease in cytokines, such as IL-1β or other inflammatory mediators like iNOS.

In summary, these results indicate that chronic alcohol intake impairs autophagy by impairing the fusion between autophagic vacuoles and lysosomes, and by preventing the effective degradation of damaged proteins and organelles, hence affecting cellular homeostasis. At the molecular level, the impairment of the autophagy pathway could be due to the activation of mTOR by phosphorylation, thereby affecting the expression levels of other Atg proteins.

Autophagy Protects Against Acute Ethanol Treatment

Contrarily to what is observed in chronic ethanol treatment, studies by Luo's group in 2012 have reported that acute ethanol treatment increases the autophagy pathway (Chen et al., 2012). They have shown that ethanol enhances the levels of LC3-II and the formation of LC3 punctae in SH-SY5Y neuroblastoma cells, and increases the protein levels of LC3-II and beclin-1 in the brain, while it lowers p62 expression. Notably, treatment with bafilomycin A1, which inhibits the fusion between autophagosomes and lysosomes, increases the levels of p62 after ethanol exposure, whereas co-treatment with bafilomycin A1 and rapamycin increases LC3-II formation, which suggests that ethanol promotes an autophagic flux. Treatment with rapamycin alone in the presence of ethanol alleviates ROS production and ameliorates neuronal death induced by ethanol both in vivo and in the brain. On the contrary, wortmannin, an autophagy inhibitor, increases oxidative stress and potentiates ethanol neurotoxicity.

Other studies that used the primary culture of cortical astrocytes and microglia cells exposed to 50 mM ethanol have demonstrated that ethanol is able to induce changes in the autophagy pathway as early as 30 min after exposure, and that these changes can persist even 24 h later (Pla, Pascual, & Guerri, 2016). At the molecular level, ethanol increases the levels of Atg12, LC3-II, and lysosomal hydrolase, cathepsin B, while down-regulating the expression levels of p62 in astrocytes and microglial cells.

Changes in cathepsin B, confirmed by confocal microscopy in mice WT astrocytes (Fig. 12.4), are accompanied by a basification of lysosomes, which could either be due to the engulfment of neutral pH material from the cytosol or be indicative of some sort of early damage caused by ethanol to lysosomes. Nevertheless, these lysosomes seem active and the autophagic flux is effective, as demonstrated by the increase in electrodense autolysosomes observed by electron microscopy (Fig. 12.3). The changes

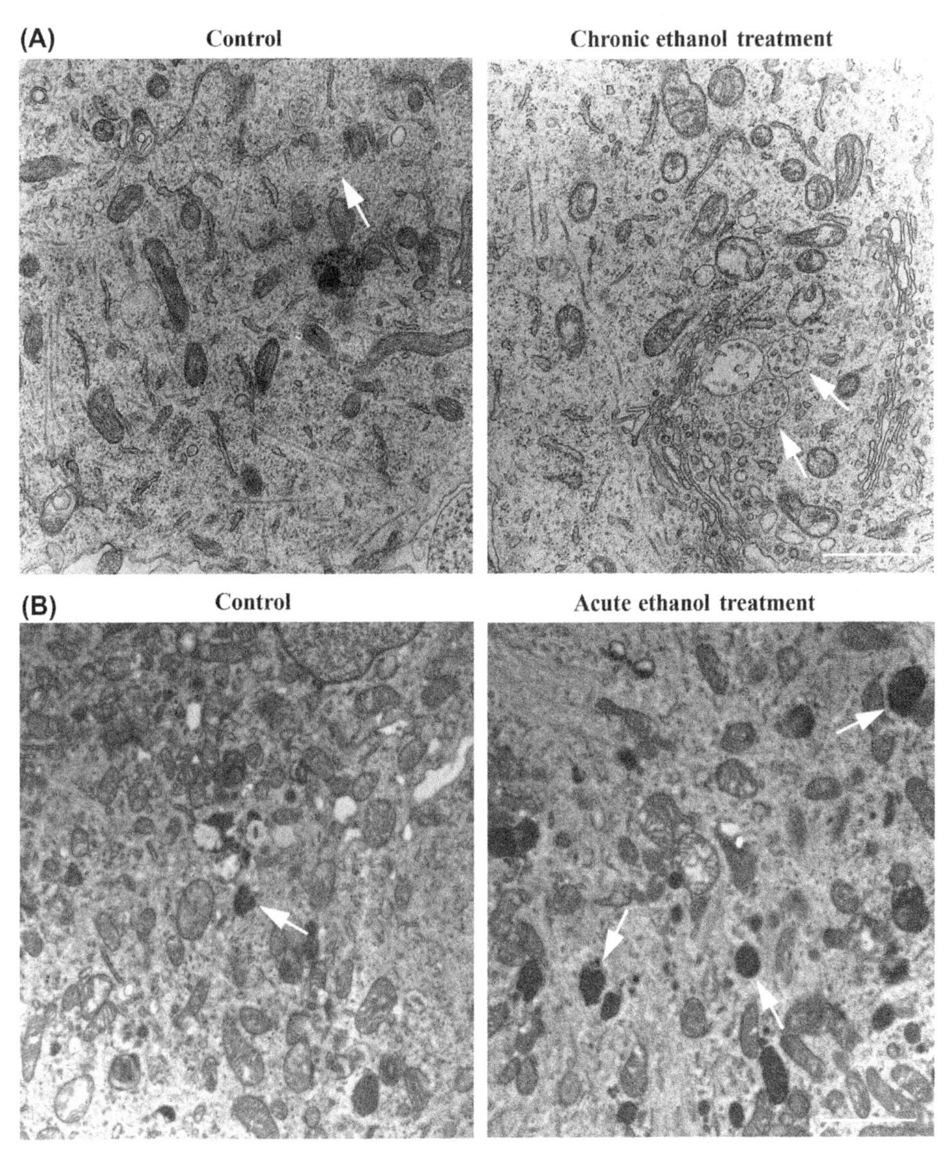

FIGURE 12.3 Electron microscopic examination of the autophagic components in ultrathin cortical sections of ethanol and untreated WT mice, and in cortical WT astrocytes in cultures treated with or without ethanol for 24 h. (A) Chronic ethanol treatment (10% of ethanol in drinking water for 4 months) increases the volume density of autophagic vacuoles in the cytoplasm of WT mice. (B) Acute ethanol treatment (50 mM for 24 h in astroglial cells in culture) increases the number of lysosomes in the WT astrocytes. Scale bar represents 2 μm.

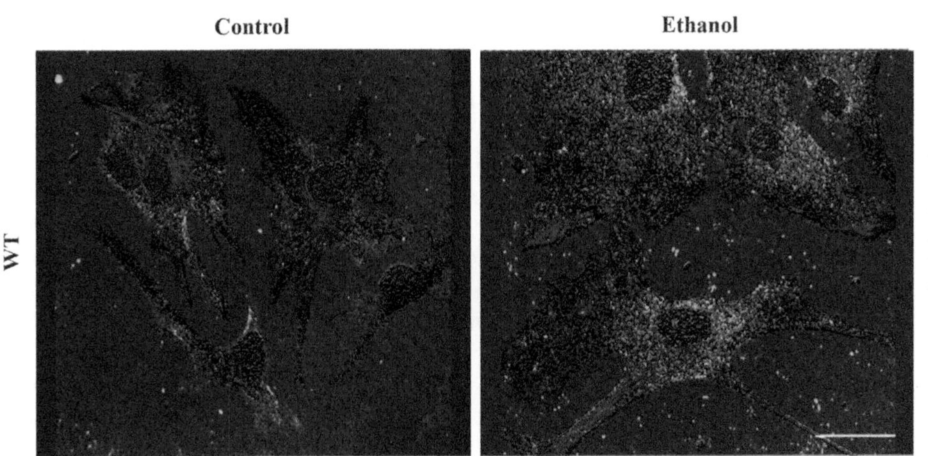

FIGURE 12.4 TLR4 contributes to the ethanol-induced changes in the expression of cathepsin B in cortical astrocytes. Representative photomicrograph of the double-labeling immunofluorescence of cathepsin B (in green, a marker of the lysosomal enzyme for the degradation of autophagolysosome content), GFAP (in red, an astrocyte marker) and nuclear marker (in blue) in the WT astrocytes treated with or without ethanol (50 mM) for 24 h. Scale bar represents 25 μm.

in the expression levels of Atg proteins were caused by a down-regulation of the phosphorylation levels of mTOR, as demonstrated by the consequent activation of the downstream ULK1 complex. Beclin-1 phosphorylation also increased, which contributed to the induction of autophagy.

In summary, depending on ethanol treatment, ethanol can increase or inhibit the autophagy pathway. An acute dose activates ALP in glial cells, while chronic ethanol treatment impairs this proteolytic pathway and contributes to brain damage. In fact, Ding et al. (2011) and Thomes, Trambly, Fox, Tuma, and Donohue (2015) have also observed the duality of ethanol's effects on hepatic autophagy. After acute ethanol consumption, autophagy is stimulated in the liver to remove two of the most contributing culprits of ethanol pathology, damaged mitochondria, and lipid droplets. However, after chronic ethanol consumption, accumulated cellular damage compromises the ability to launch an effective autophagic response, which results in increased liver injury (Ding et al., 2011). Regarding the role of autophagy after toxic insult in the brain, it seems to be sufficient consensus on the fact that autophagy is initially triggered as a protective mechanism, as confirmed by the many reports that have shown an ameliorated pathogenic process after autophagy induction under different neurodegenerative conditions. However, other authors have suggested the possibility of autophagy having dual functions, both of which are a beneficial process to cope with cellular stress and to induce pathogenic conditions after sustained overactivation, which ultimately results in cell death (Luo, 2014).

THE TLR4 IMMUNE RESPONSE

Emerging evidence has indicated that alcohol induces brain damage and neuroinflammation by activating the brain innate immune system, more specifically Toll-like receptor 4 (TLR4). TLRs are a family of pattern recognition receptors (PRRs) that were identified for the first time in the fruit fly *Drosophila*, and belong to the interleukin-1 (IL-1)/TLR superfamily. These receptors enable the recognition of structural motifs that are conserved across a wide range of pathogens (pathogen-associated molecular patterns, PAMPs) and other endogenous molecules produced during tissular damage (danger-associated molecular patterns, DAMPs). After recognition, TLR4 initiates two different signaling cascades (Fig. 12.5), myeloid differentiation primary-response protein 88 (MyD88)-dependent and an MyD88-independent TIR-domain-containing adapter-inducing interferon-β (TRIF)-dependent pathway, which lead to the activation of several nuclear factors to produce cytokines and other inflammatory mediators. In particular, TLR4 recognizes the lipopolysaccharide (LPS) present in Gram-negative bacteria. However, all TLRs must retain some grade of nonspecificity, as they constitute the first barrier of defense against infection, and must therefore, be able to respond against a wider range of pathogens and toxic compounds apart from canonical ones. By acting as an agonist of TLR4/IL-1RI receptors, it has been demonstrated that ethanol activates their signaling cascades similarly to canonical ligands LPS and IL-1β, which induces inflammation and can lead to neural damage,

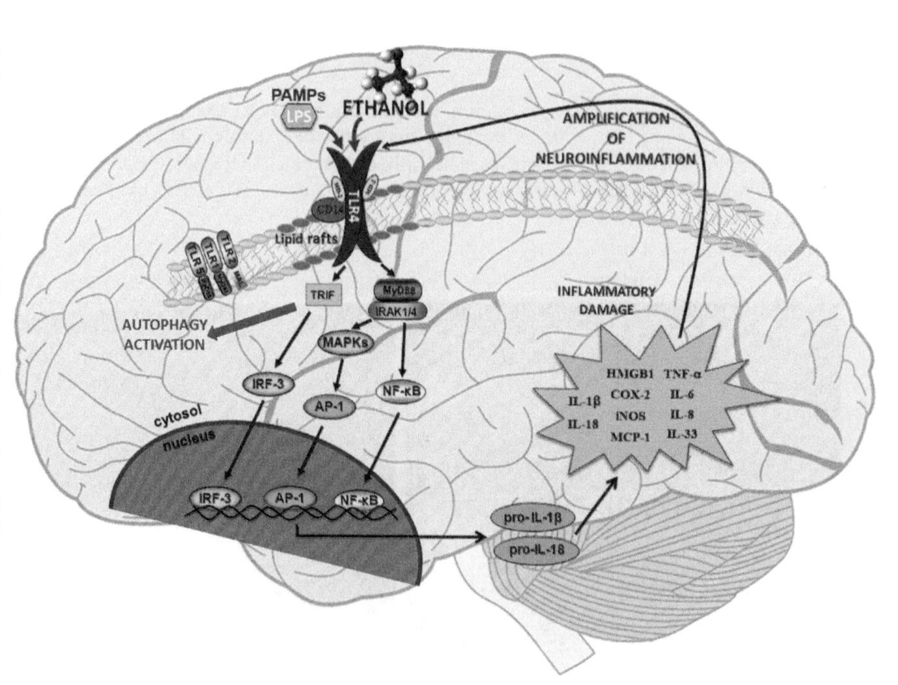

FIGURE 12.5 Diagram of TLR4 activation and alcohol-induced neuroinflammation. Activation of TLR4 by PAMPs (LPS) or ethanol is initiated by the recruitment of TLR4 with adaptor proteins CD14 and MD2 into *lipid rafts*. This activates and recruits several proteins and kinases (MyD88, IRAK1/4, TRIF), and triggers fast downstream signaling pathways (i.e., MAPKs) and transcription nuclear factors (NF-κB, AP-1, IRF-3), which culminate in the generation of cytokines (i.e., pro-IL-1β, IL-6, IL-8, and TNF-α), chemokines (i.e., MCP-1), and inflammatory mediators (iNOS and COX-2), which leads to neuroinflammation. *AP-1*, activator protein 1; *IRAK*, interleukin-1 receptor-associated kinase-1; *IRF-3*, interferon regulatory factor 3; *MAPK*, mitogen-activated protein kinase; *MyD88*, myeloid differentiation primary response gene 88; *TRIF*, TIR-domain-containing adaptor-inducing IFNβ.

and ultimately to neurodegeneration. Studies into the primary culture of astrocytes (Blanco, Perez-Arago, Fernandez-Lizarbe, & Guerri, 2008), microglial cells, and murine macrophages (Fernandez-Lizarbe, Pascual, Gascon, Blanco, & Guerri, 2008; Fernandez-Lizarbe, Pascual, & Guerri, 2009) have demonstrated that ethanol, at physiologically relevant conditions (10–50 mM), is capable of activating TLR4-mediated signaling pathways. In vivo studies have confirmed that chronic ethanol treatment causes neuroinflammation and brain damage, effects that have been eliminated using TLR4 knockout (KO) mice (Alfonso-Loeches et al., 2012, 2010).

Inflammation and proteostasis impairment are two closely linked processes that play a major role in the outcome of neuronal-related diseases. For instance, it has been shown that TLR4 is able to promote autophagy through the activation of beclin-1 by the two main adaptor molecules of the TLR4 signaling cascade, MyD88 and TRIF (Delgado, Elmaoued, Davis, Kyei, & Deretic, 2008; Shi & Kehrl, 2010). Moreover, the TRIF pathway has been proposed to induce autophagosome formation before their fusion to lysosomes (Xu et al., 2007), but the interaction between TLRs and autophagy can also take place in the opposite direction. Indeed, it has been suggested that autophagy machinery can deliver PAMPs to endosomal TLRs (Lee, Lund, Ramanathan, Mizushima, & Iwasaki, 2007), which reveals a role for autophagy in the recognition of toxic elements by TLRs.

Recent studies have associated the effects of ethanol on the autophagy pathway with the TLR4 response, since TLR4-KO mice treated with chronic or acute ethanol treatment, which present a reduced inflammatory response as compared to WT mice, displayed no changes in the autophagy machinery (Pla et al., 2016, 2014).

The induction of autophagy in glial cells after acute ethanol treatment was triggered in the presence of an inflammatory environment, characterized by the presence of cytokines and inflammatory mediators, such as iNOS and COX-2. This inflammation can be detrimental for cell survival, and is accompanied by a modest, but statistically significant, increase in the apoptotic cell death rate. Hence the question is: does the induction of autophagy favor cell death or, on the contrary, is it a response mechanism triggered to alleviate ethanol-induced toxicity and to avoid cell death? It has been demonstrated that a pharmacological inhibition of autophagy with wortmannin (a covalent inhibitor of phosphoinositide 3-kinases (PI3Ks)) and bafilomycin (which inhibits later autophagy steps by preventing the fusion between autophagosomes and lysosomes) does not only increase the amount of inflammatory mediators (iNOS and COX-2), but also raises apoptotic levels of mice astrocytes. Moreover, these inhibitors are also able to trigger necrotic cell death, as confirmed by cytomics studies and lactate dehydrogenase (LDH) release (Pla et al., 2016).

These data underpin the critical role of TLR4 in not only the release of cytokines and the consequent generation of an inflammatory environment, but also its importance in the production of protective responses, such as autophagy, intended to alleviate ethanol-induced toxicity. Induction of autophagy can be mediated directly by the action of TLR4 on autophagy proteins, such as beclin-1, and also indirectly by the inhibition of autophagy master regulator mTOR by the action of ROS. Conversely after chronic ethanol treatment, TLR4 might participate in long-term autophagy impairment, since the TLR4 downstream signaling cascade includes MAPKs that have been shown to activate mTOR, and could hence inhibit autophagy. Therefore, in the last few years, TLR4 has emerged as a promising target in the treatment of ethanol-related pathologies.

Ethanol Differentially Affects the Autophagy Pathway in Glial Cells and Neurons

Glial cells have been extensively characterized in the inflammatory response context in the brain caused by ethanol consumption since they express high levels of TLR4 receptors and are, therefore, capable of inducing the release of cytokines. These cells are also able to trigger autophagy to alleviate and counteract ethanol-induced toxicity and cell death. However, fewer studies have focused on the action of ethanol in neurons, the brain functional cell units, at the molecular level.

It has been well established that ethanol can damage and produce loss of gray matter in the brain, but it is less clear how, or if, these cells can cope with stress generated by ethanol, or on the contrary, of they rely solely on the action of glial cells. Hence, it is important to underpin that neurons also express TLR4 receptors (Okun, Griffioen, & Mattson, 2011), but to a much lesser extent than microglia or astrocytes.

A recent study has used neurons in primary culture to evaluate the effects of autophagy in ethanol-induced toxicity by measuring several autophagy proteins (Pla et al., 2016). The results have provided that neurons display an opposite phenotype to that of glial cells. In the presence of ethanol, neurons show a down-regulation of mTOR phosphorylation levels, lower expression levels of LC3-II, cathepsin B, and an increased accumulation of protein p62, which evidences that ethanol reduces the ALP pathway. This autophagy pathway impairment is associated with the induction of necrotic neuronal death after 24 h of ethanol treatment, which also confirms the hypothesis of ethanol constituting a protective mechanism against ethanol-induced neurotoxicity. Moreover, this necrotic process has been prevented by administering autophagy inducer rapamycin. Although other authors have obtained opposite results when they worked with neuroblastoma SH-SY5Y cells (Chen et al., 2012), this divergence could be explained by the traits derived from the self-renewal potential of immortalized cells, as opposed to the more physiologically relevant primary neurons.

The different autophagy phenotypes observed between glial cells and neurons are still unclear and need to be the focal point of further studies, although several possibilities could explain this opposite effect: (1) neurons exhibit a poorer autophagy response and hence present less capacity to cope with cellular stress; (2) neurons lack the strong inflammatory TLR4 response typical of microglia and astrocytes, which favors the autophagic process by activating TLRs in a feedback loop; and (3) neurons usually rely on glial cells not only for structural support, but also for physiological protection. In addition, neurons are postmitotic cells that prove more fragile against toxic stimuli. In line with this, it is to be noted that the sole presence of an inflammatory environment produced by glial cells is sufficient to trigger neuronal death, even in the protective autophagy response context as demonstrated by increased the necrotic neuronal death in the presence of ethanol-treated astrocyte-conditioned media in neuronal cultures.

CONCLUSIONS AND FUTURE DIRECTIONS

Despite some initial controversy about the effect of ethanol on cells regarding autophagy, there is sufficient evidence to assess that acute ethanol consumption is capable of triggering autophagy in different body organs, such as the liver, heart, and brain, and with different metabolic routes (mTOR, PI3K, TLR4, oxidative stress) implicated in the process, to alleviate and ameliorate the cells damaged by toxic products. However, chronic ethanol intake seems to induce pathological changes that are detrimental for correct autophagy execution, although the exact molecular mechanisms that lead to this situation are less clear. Autophagy should, therefore, be considered a promising potential target for treating ethanol-related pathologies. The primary focal point of this research should rely on finding an agent or a combination of agents that can work within the CNS and can modulate autophagy without affecting other metabolic pathways to thus avoid potential side effects as much as possible. Alternatively, more studies are needed to test if TLR4 could also be a good target to diminish ethanol-induced inflammation and, hence, avoid excessive autophagy induction.

Acknowledgments

This work has been supported by grants from the Spanish Ministry of Economics and Competitiveness (SAF2012-33747, SAF2015-69187R), the Spanish Ministry of Health: The Institute Carlos III and FEDER funds (RTA-Network RD12-0028-007), and PNSD (Ex. 20101037), GV-Conselleria d'Educació: PROMETEO II/2014-063.

References

Alfonso-Loeches, S., & Guerri, C. (2011). Molecular and behavioral aspects of the actions of alcohol on the adult and developing brain. *Critical Reviews in Clinical Laboratory Sciences, 48*, 19–47.

Alfonso-Loeches, S., Pascual, M., Gómez-Pinedo, U., Pascual-Lucas, M., Renau-Piqueras, J., & Guerri, C. (2012). Toll-like receptor 4 participates in the myelin disruptions associated with chronic alcohol abuse. *Glia, 60*, 948–964.

Alfonso-Loeches, S., Pascual-Lucas, M., Blanco, A. M., Sanchez-Vera, I., & Guerri, C. (2010). Pivotal role of TLR4 receptors in alcohol-induced neuroinflammation and brain damage. *Journal of Neuroscience, 30*, 8285–8295.

Blanco, A. M., Perez-Arago, A., Fernandez-Lizarbe, S., & Guerri, C. (2008). Ethanol mimics ligand-mediated activation and endocytosis of IL-1RI/TLR4 receptors via lipid rafts caveolae in astroglial cells. *Journal of Neurochemistry, 106*, 625–639.

Chen, G., Ke, Z., Xu, M., Liao, M., Wang, X., Qi, Y., … Luo, J. (2012). Autophagy is a protective response to ethanol neurotoxicity. *Autophagy, 8*, 1577–1589.

Ciechanover, A., & Brundin, P. (2003). The ubiquitin proteasome system in neurodegenerative diseases: Sometimes the chicken, sometimes the egg. *Neuron, 40*, 427–446.

Crews, F. T., Buckley, T., Dodd, P. R., Ende, G., Foley, N., Harper, C., … Sullivan, E. V. (2005). Alcoholic neurobiology: Changes in dependence and recovery. *Alcoholism: Clinical and Experimental Research, 29*, 1504–1513.

Delgado, M. A., Elmaoued, R. A., Davis, A. S., Kyei, G., & Deretic, V. (2008). Toll-like receptors control autophagy. *The EMBO Journal, 27*, 1110–1121.

Ding, W. X., Li, M., Chen, X., Ni, H. M., Lin, C. W., Gao, W., … Yin, X. M. (2010). Autophagy reduces acute ethanol-induced hepatotoxicity and steatosis in mice. *Gastroenterology, 139*, 1740–1752.

Ding, W. X., Li, M., & Yin, X. M. (2011). Selective taste of ethanol-induced autophagy for mitochondria and lipid droplets. *Autophagy, 7*, 248–249.

Fernandez-Lizarbe, S., Pascual, M., Gascon, M. S., Blanco, A., & Guerri, C. (2008). Lipid rafts regulate ethanol-induced activation of TLR4 signaling in murine macrophages. *Molecular Immunology, 45*, 2007–2016.

Fernandez-Lizarbe, S., Pascual, M., & Guerri, C. (2009). Critical role of TLR4 response in the activation of microglia induced by ethanol. *Journal of Immunology, 183*, 4733–4744.

Giordano, S., Darley-Usmar, V., & Zhang, J. (2014). Autophagy as an essential cellular antioxidant pathway in neurodegenerative disease. *Redox Biology, 2*, 82–90.

Gorini, G., Roberts, A. J., & Mayfield, R. D. (2013). Neurobiological signatures of alcohol dependence revealed by protein profiling. *PLoS One, 8*, e82656.

von Haefen, C., Sifringer, M., Menk, M., & Spies, C. D. (2011). Ethanol enhances susceptibility to apoptotic cell death via down-regulation of autophagy-related proteins. *Alcoholism: Clinical and Experimental Research, 35*, 1381–1391.

Lee, H. K., Lund, J. M., Ramanathan, B., Mizushima, N., & Iwasaki, A. (2007). Autophagy-dependent viral recognition by plasmacytoid dendritic cells. *Science, 315*, 1398–1401.

Luo, J. (2014). Autophagy and ethanol neurotoxicity. *Autophagy, 10*, 2099–2108.

Morimoto, R. I. (2008). Proteotoxic stress and inducible chaperone networks in neurodegenerative disease and aging. *Genes & Development, 22*, 1427–1438.

Okun, E., Griffioen, K. J., & Mattson, M. P. (2011). Toll-like receptor signaling in neural plasticity and disease. *Trends in Neurosciences, 34*, 269–281.

Osna, N. A., Thomes, P. G., & Donohue, T. M., Jr. (2011). Involvement of autophagy in alcoholic liver injury and hepatitis C pathogenesis. *World Journal of Gastroenterology, 17*, 2507–2514.

Pankiv, S., Clausen, T. H., Lamark, T., Brech, A., Bruun, J. A., Outzen, H., ... Johansen, T. (2007). p62/SQSTM1 binds directly to Atg8/LC3 to facilitate degradation of ubiquitinated protein aggregates by autophagy. *Journal of Biological Chemistry, 282*, 24131–24145.

Pascual, M., Balino, P., Alfonso-Loeches, S., Aragon, C. M., & Guerri, C. (2011). Impact of TLR4 on behavioral and cognitive dysfunctions associated with alcohol-induced neuroinflammatory damage. *Brain, Behavior, and Immunity, 25*(Suppl. 1), S80–S91.

Pickart, C. M., & Cohen, R. E. (2004). Proteasomes and their kin: Proteases in the machine age. *Nature Reviews Molecular Cell Biology, 5*, 177–187.

Pla, A., Pascual, M., & Guerri, C. (2016). Autophagy constitutes a protective mechanism against ethanol toxicity in mouse astrocytes and neurons. *PLoS One, 11*, e0153097.

Pla, A., Pascual, M., Renau-Piqueras, J., & Guerri, C. (2014). TLR4 mediates the impairment of ubiquitin-proteasome and autophagy-lysosome pathways induced by ethanol treatment in brain. *Cell Death & Disease, 5*, e1066.

Rami, A. (2009). Review: Autophagy in neurodegeneration: Firefighter and/or incendiarist? *Neuropathology and Applied Neurobiology, 35*, 449–461.

Rosenbloom, M., Sullivan, E. V., & Pfefferbaum, A. (2003). Using magnetic resonance imaging and diffusion tensor imaging to assess brain damage in alcoholics. *Alcohol Research & Health, 27*, 146–152.

Shen, H. M., & Codogno, P. (2012). Autophagy is a survival force via suppression of necrotic cell death. *Experimental Cell Research, 318*, 1304–1308.

Shi, C. S., & Kehrl, J. H. (2010). Traf6 and A20 differentially regulate TLR4-induced autophagy by affecting the ubiquitination of beclin 1. *Autophagy, 6*, 986–987.

Thomes, P. G., Ehlers, R. A., Trambly, C. S., Clemens, D. L., Fox, H. S., Tuma, D. J., & Donohue, T. M. (2013). Multilevel regulation of autophagosome content by ethanol oxidation in HepG2 cells. *Autophagy, 9*, 63–73.

Thomes, P. G., Trambly, C. S., Fox, H. S., Tuma, D. J., & Donohue, T. M., Jr. (2015). Acute and chronic ethanol administration differentially modulate hepatic autophagy and transcription factor EB. *Alcoholism: Clinical and Experimental Research, 39*, 2354–2363.

Wang, Y., & Mandelkow, E. (2012). Degradation of tau protein by autophagy and proteasomal pathways. *Biochemical Society Transactions, 40*, 644–652.

Xu, Y., Jagannath, C., Liu, X. D., Sharafkhaneh, A., Kolodziejska, K. E., & Eissa, N. T. (2007). Toll-like receptor 4 is a sensor for autophagy associated with innate immunity. *Immunity, 27*, 135–144.

Yamamoto, A., & Simonsen, A. (2011). The elimination of accumulated and aggregated proteins: A role for aggrephagy in neurodegeneration. *Neurobiology of Disease, 43*, 17–28.

CHAPTER

13

Ghrelinergic Signaling in Ethanol Reward

L.J. Zallar[1], H.M. Baumgartner[1], E.E. Garling[1], S. Abtahi[1], R. Pastor[2], P.J. Currie[1]

[1]Reed College, Portland, OR, United States; [2]Universitat Jaume I, Castellón, Spain

INTRODUCTION: GHRELIN PHYSIOLOGY

Ghrelin is a 28 amino acid peptide that is produced in the stomach and has been purified in rats and humans (Kojima et al., 1999; Kojima, Hosoda, & Kangawa, 2012). The peptide is encoded on the GHRL gene, which is the genetic origin of both acylated and unacylated ghrelin (Fig. 13.1) (Chow et al., 2012; Soares & Leite-Moreira, 2008). Ghrelin is converted into its active, acylated form through the incorporation of octanoic acid to the serine-3 amino acid residue by ghrelin O-acyltransferase (GOAT) (Gutierrez et al., 2008; Kouno et al., 2016; Yang, Brown, Liang, Grishin, & Goldstein, 2008). Acylated ghrelin acts as an active signaling peptide in the central nervous system (Broglio et al., 2003; Soares & Leite-Moreira, 2008). Within the brain, ghrelin receptor activation has been implicated in appetitive motivation, energy metabolism, and stress activation (Albarran-Zeckler, Sun, & Smith, 2011; Cone, Roitman, & Roitman, 2015; Currie et al., 2011; Dailey, Moran, Holland, & Johnson, 2016; Panagopoulos & Ralevski, 2014; Spencer et al., 2012; Wauson, Sarkodie, Schuette, & Currie, 2015), with investigations from mid-2000s onward targeting the peptide's role in drug seeking behavior and ethanol reward (Cepko et al., 2014; Davis, Wellman, & Clifford, 2007; Gomez et al., 2015; Jerlhag et al., 2009).

There are two classifications of ghrelin receptors, the growth hormone secretagogue 1a receptor (GHS-R1a) and the growth hormone secretagogue 1b receptor (GHS-R1b) (Chow et al., 2012; Mary et al., 2013). GHS-R1a is a G-protein–coupled receptor that is widely distributed throughout many neuroanatomical regions, including the hypothalamus, the amygdala, and areas associated with mesolimbic reward circuitry such as the nucleus accumbens (NAcc) and ventral tegmental area (VTA) (Figs. 13.2 and 13.3) (Chuang & Zigman, 2010; Howard et al., 1996; Mason, Wang, & Zigman, 2014; Sárvári et al., 2014). While previous reports suggest that the GHS-R1b is inactive, the intracellular concentration of GHS-R1b is notably distinct from GHS-R1a, with GHS-R1b primarily found in the endoplasmic reticulum, where it is able to heterodimerize with the GHS-R1a to reduce the constitutive activity of the GHS-R1a (Chow et al., 2012). Specifically, it is argued that the truncated ghrelin receptor polypeptide, GHS-R1b, forms heterodimers with GHS-R1a to attenuate cell surface 1a availability. This occurs in the endoplasmic reticulum where GHS-R1a/GHS-R1b heterodimers are concentrated. The dynamic between the GHS-R1a and GHS-R1b may function as a negative feedback mechanism in order to down-regulate neuronal GHS-R1a expression, as well as GHS-R1a availability at the synapse. Consequently, increased expression of GHS-R1b results in enhanced binding and heterodimerization with GHS-R1a. However, even in the presence of GHS-R1b, sufficient GHS-R1a homodimers are localized on the cell surface in order to mediate the physiological and behavioral actions of the ghrelin peptide (Chow et al., 2012; Mary et al., 2013).

The GHS-R1a is further associated with other signaling systems, including dopamine (Anderberg et al., 2016; Cone et al., 2015; Kern et al., 2015). Ghrelin's role in reward is documented (Cepko et al., 2014; Panagopoulos & Ralevski, 2014; St-Onge, Watts, & Abizaid, 2016; Wellman et al., 2011), and its impact on reward signaling appears to be due, at least in part, to GHS-R1a interactions with dopamine receptors, which are highly implicated in the neural control of reward-related behaviors (Abizaid et al., 2006; Jiang, Betancourt, & Smith, 2006; Kern et al., 2015; Sadeghzadeh, Babapour, & Haghparast, 2015). GHS-R1a is expressed on dopaminergic neurons in the VTA, and has been shown to heterodimerize with

Addictive Substances and Neurological Disease
http://dx.doi.org/10.1016/B978-0-12-805373-7.00013-X

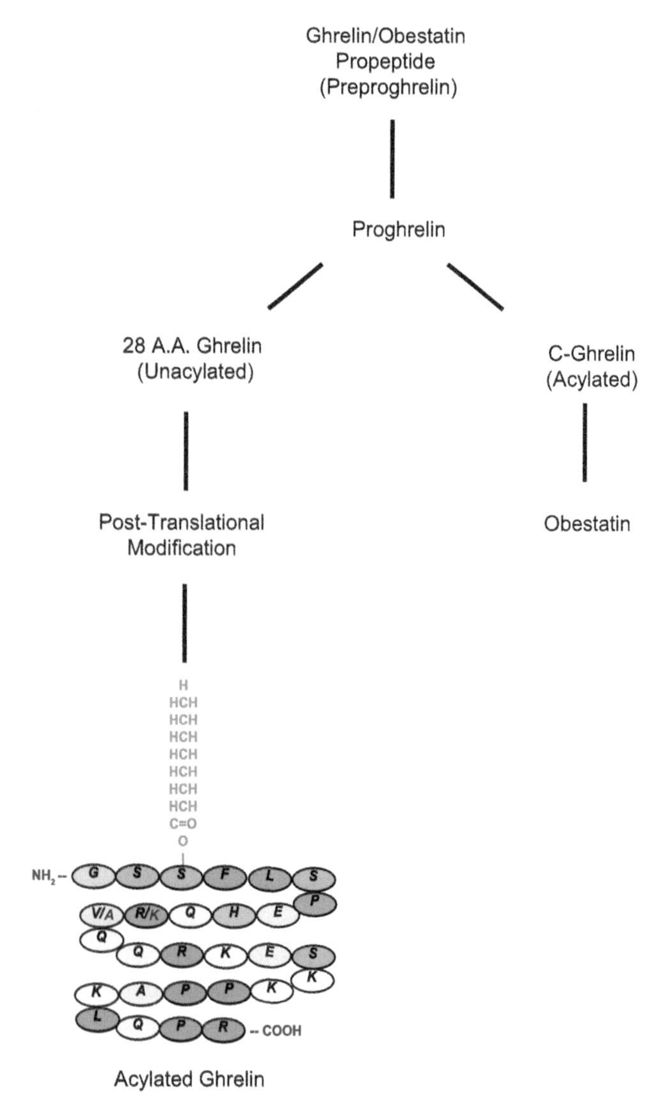

FIGURE 13.1 Schematic demonstrating the biochemical progression from preproghrelin to obestatin and the active signaling peptide acylated ghrelin. Rat and human acylated ghrelin are differentiated by two amino acids; the *pink* letters represent the amino acids that make up the rat ghrelin peptide as opposed to human ghrelin.

both the G-protein–coupled metabotropic D1 and D2 dopamine receptors (DRD1 and DRD2, respectively) (Jiang et al., 2006; Kern et al., 2015; Mary et al., 2013). The heterodimerization of GHS-R1a with DRD1, a receptor heavily implicated in signaling related to mesolimbic reward, is a likely mechanism for ghrelinergic modulation of dopamine (Abizaid et al., 2006; Cone et al., 2015; Jiang et al., 2006; Mary et al., 2013; Sadeghzadeh et al., 2015). In addition to interactions with dopaminergic signaling, the functional role of ghrelin in relation to drug seeking behavior may also be mediated by the peptide's interactions with neurochemicals such gamma-aminobutyric acid (GABA), a major inhibitory neurotransmitter in the mammalian brain (Cruz, Herman, Cote, Ryabinin, & Roberto, 2013).

As reviewed by Wellman and Abizaid (2015), there is evidence that the GHS-R1a interacts with other transmitter systems through heterodimerization, including endocannabinoid and serotonergic signaling. GHS-R1a appears to dimerize with the 5-hydroxytryptamine 2C (5-HT_{2C}) receptor in a manner that attenuates ghrelin activity, and in support of this, increased serotonergic signaling mitigates ghrelin's orexigenic and reward-related effects (Currie, John, Nicholson, Chapman, & Loera, 2010; Wellman & Abizaid, 2015). Furthermore, GHS-R1a may interact with the cannabinoid 1 (CB1) receptor (see Wellman & Abizaid, 2015). Although heterodimeric interactions between the two receptors have not been fully explored at this time, there is indication of interplay between the two transmitter systems. Farkas, Vastagh, Sárvári, and Liposits (2013) report that the activation of GHS-R1a leads to increased intracellular calcium availability, which in turn stimulates the synthesis of endocannabinoid ligands, and Ting, Chi, Li, and Chen (2015) have found that administration of the CB1 receptor inverse agonist, AM251, attenuates ghrelin's orexigenic action.

ENERGY METABOLISM AND HOMEOSTASIS

Indeed ghrelin has been extensively studied in relation to metabolic and homeostatic regulation, and has been demonstrated to be a potent orexigenic signal (Currie et al., 2011; Currie, Mirza, Fuld, Park, & Vasselli, 2005; Dailey et al., 2016; Mason et al., 2014; Sárvári et al., 2014; Sumithran et al., 2011; Wren et al., 2000). GHS-R1a is widely expressed throughout brain areas associated with eating and metabolism, including regions of the hypothalamus such as the paraventricular (PVN) and arcuate (ArcN) nuclei (see Fig. 13.3) (Currie et al., 2005; Howard et al., 1996; Kojima et al., 1999; Müller et al., 2015; Zigman, Jones, Lee, Saper, & Elmquist, 2006).

Clinical investigations have provided additional evidence of ghrelin's actions in appetitive mechanisms (Buss et al., 2014; Cummings, 2004). Gueorguiev et al. (2009) report that polymorphisms of the genes encoding the acylated ghrelin peptide or its receptor are correlated with disordered eating behavior and altered body weight. These polymorphisms are associated with the development of insulin resistance, which is in turn related to the development of obesity (Gueorguiev et al., 2009; Ye, 2013). Following diet-induced weight loss, physiological compensatory mechanisms arise to encourage increased appetite and weight gain, including increases in plasma ghrelin concentrations (Dixon, Lambert, & Lambert, 2015; Sumithran et al., 2011). Van Name et al. (2015) have found that adolescents with normal body weight demonstrate attenuated

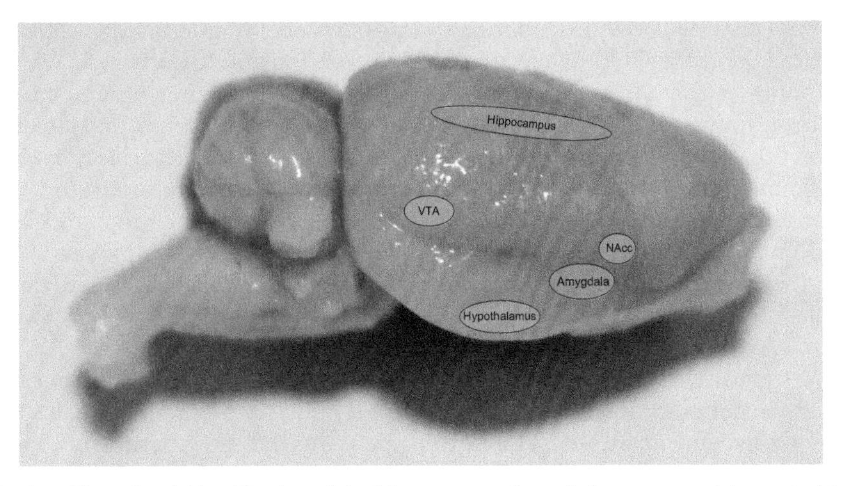

FIGURE 13.2 Whole brain with regional identification of the hippocampus, hypothalamus, amygdala, ventral tegmental area (VTA), and nucleus accumbens (NAcc).

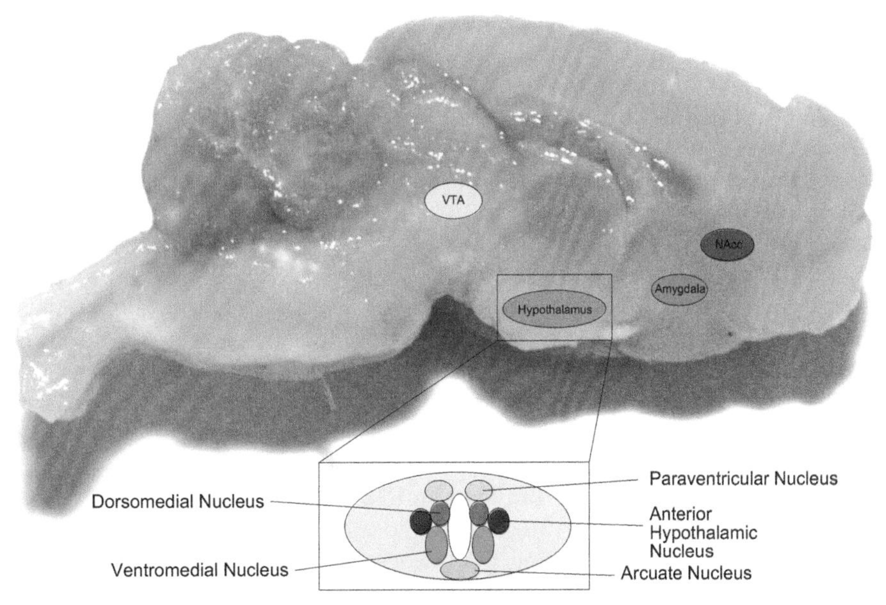

FIGURE 13.3 Sagittal section of a brain with identification of amygdala, hypothalamus, ventral tegmental area (VTA), and nucleus accumbens (NAcc). Bilateral placement of specific nuclei of the hypothalamus are represented in a zoomed-in schematic, including the dorsomedial, ventromedial, paraventricular, anterior hypothalamic, and arcuate nuclei.

plasma ghrelin levels in response to consumption of beverages sweetened with both sucrose and fructose, but adolescents with insulin resistance–related obesity do not show similar decreases in ghrelin. Together, these findings suggest that obesity and disordered eating behavior are associated with the dysregulation of ghrelin signaling.

Aside from human clinical work, research in rodent models has demonstrated ghrelin's mechanistic role in hypothalamically mediated food consumption and metabolic control. Direct injection of ghrelin into hypothalamic regions, specifically the ArcN and PVN, leads to a potentiation of food intake (Currie et al., 2011, 2005; Olszewski, Grace, Billington, & Levine, 2003; Wauson et al., 2015). Dailey et al. (2016) have found that systemic injections of a GHS-R1a antagonist attenuate food consumption, and a similar effect has been reported by Thomas, Ryu, and Bartness (2016), who also showed that GHS-R1a blockade reduced hypothalamic PVN neuronal activation. Further, food intake elicited as a result of prior food deprivation is blocked by treatment with GHS-R1a antagonism (Wei et al., 2015). In GHS-R1a knockout animals, intraventricular administration of ghrelin fails to stimulate appetite or elicit hypothalamic c-fos induction (Egecioglu et al., 2006; Zigman et al., 2005). GOAT knockout mice show both overall decreased food consumption and reduced susceptibility to obesity following exposure to a high caloric diet compared to wild type controls, providing more evidence in support of ghrelin and the GHS-R1a in mediating orexigenesis (Kouno et al., 2016; Wellman & Abizaid, 2015). In contrast, however, Sclafani, Touzani, and Ackroff (2015) report that GHS-R1a knockout mice do not demonstrate attenuated sugar or fat preference. These somewhat contradictory findings suggest that further research should investigate the nuances of ghrelinergic mechanisms of food consumption specifically in relation to food reward.

In connection to its role in the control of appetite and food consumption, ghrelin has been shown to interact with associated neural mechanisms in energy metabolism and homeostasis. One example is the hormone's role in relation to neuropeptide Y (NPY), a potent orexigenic peptide (Currie et al., 2011; Kim & Bi, 2016; Leibowitz, Sladek, Spencer, & Tempel, 1988). Within the ArcN, ghrelin acts on NPY/agouti-related peptide (NPY/AGRP) neurons to activate metabolic and orexigenic signaling (Currie et al., 2012; Fernandez et al., 2016; Kohno, Gao, Muroya, Kikuyama, & Yada, 2003). Furthermore, Kamegai et al. (2001) found that ventricularly administered ghrelin elicits increases in NPY and AGRP mRNA levels in the hypothalamus.

Ghrelin treatment has also been reported to shift substrate oxidation away from lipid utilization and to stimulate mRNA expression of lipid-storage promoting enzymes in white adipocytes (Albarran-Zeckler et al., 2011; Currie et al., 2010). As well, findings from our lab indicate that ghrelin increases the respiratory exchange ratio (RER) when injected into the PVN or ArcN, which signifies a transfer from lipid metabolism to carbohydrate energy utilization (Currie, Braver, Mirza, & Sricharoon, 2004; Currie et al., 2012). Our work demonstrates that this effect is robust during the active cycle, suggesting that ghrelin's actions are circadian dependent.

LIMBIC SYSTEM SIGNALING

In addition to its orexigenic and metabolic functions, ghrelin has been implicated in aspects of stress activation and limbic signaling (Asakawa et al., 2001; Currie et al., 2012; Spencer, Emmerzaal, Kozicz, & Andrews, 2015). Ghrelin plays a modulatory role in hypothalamic–pituitary–adrenal (HPA) axis regulation (Spencer et al., 2015, 2012). Human research supports this connection, as Raspopow, Abizaid, Matheson, and Anisman (2014) report that psychosocial stress leads to increased plasma ghrelin concentrations, which are not significantly attenuated after food consumption in individuals described as emotional eaters. This is consistent with other work demonstrating that stress, anxiety, and emotional eating are linked to ghrelinergic dysregulation in humans (Sarker, Franks, & Caffrey, 2013).

Research utilizing animal models further supports ghrelin's role in stress and anxiety related behavioral outcomes. Intraventricular ghrelin administration, along with direct injection into the dorsal raphe nucleus, amygdala, hypothalamus, and hippocampus increases anxiety-like behavior in rats (Asakawa et al., 2001; Carlini et al., 2002, 2004; Currie et al., 2012; Hansson et al., 2011). Our lab has previously demonstrated that injections of ghrelin into the PVN, ArcN, ventromedial hypothalamus (VMN), and perifornical hypothalamus (PFH) dose dependently reduce the percentage of time spent in the open arms of the elevated plus maze (EPM). The reduction in open arm exploration is indicative of anxiety-like behavior and is most robust following ArcN or PVN treatment (Currie et al., 2012). Research work in 2016 by Brockway, Krater, Selva, Wauson, and Currie found that food intake post ghrelin injection, but before EMP testing, reduces the anxiogenic effect of ghrelin in this behavioral paradigm. However, in mice, systemic ghrelin administration and over expression of the ghrelin receptor in the amygdala induces anxiolytic-like behavior (Jensen et al., 2015). In chronic stress paradigms in mice ghrelin has been reported to decrease stress and depressive symptoms (Lutter et al., 2008). While ghrelin's anxiogenic and anxiolytic effects might vary by species or route of administration, it remains possible that the peptide's effects on stress activation are a function of basal anxiety state and, in turn, may be influenced by circulating levels of corticosterone (Wauson et al., 2015). Ghrelin increases expression of corticotropin-releasing hormone (CRH) mRNA in the hypothalamus and increases activity in hypophysiotropic CRH neurons in the PVN (Cabral, Suescun, Zigman, & Perello, 2012; Wren et al., 2002). Additionally, ghrelin may interact with other stress-related neurotransmitters. The CRH-related peptide, urocortin 1, and serotoninergic receptor mechanisms have been shown to alter the peptide's effects on anxiety and stress activation (Currie, Schuette, Wauson, Voss, & Angeles, 2014; Wauson et al., 2015).

MESOLIMBIC REWARD PATHWAY

Since mid-2000s, human and animal investigations have focused on ghrelin's involvement in brain reward circuitry. GHS-R1a are found in regions of the brain mediating reward-related behavior, including mesolimbic or mesotelencephalic dopamine of the VTA and NAcc core and shell (Fig. 13.4) (Abizaid et al., 2006; Naleid, Grace, Cummings, & Levine, 2005; Sustkova-Fiserova, Jerabek, Havlickova, Kacer, & Krsiak, 2014). Abizaid et al. (2006) first demonstrated that GHS-R1a

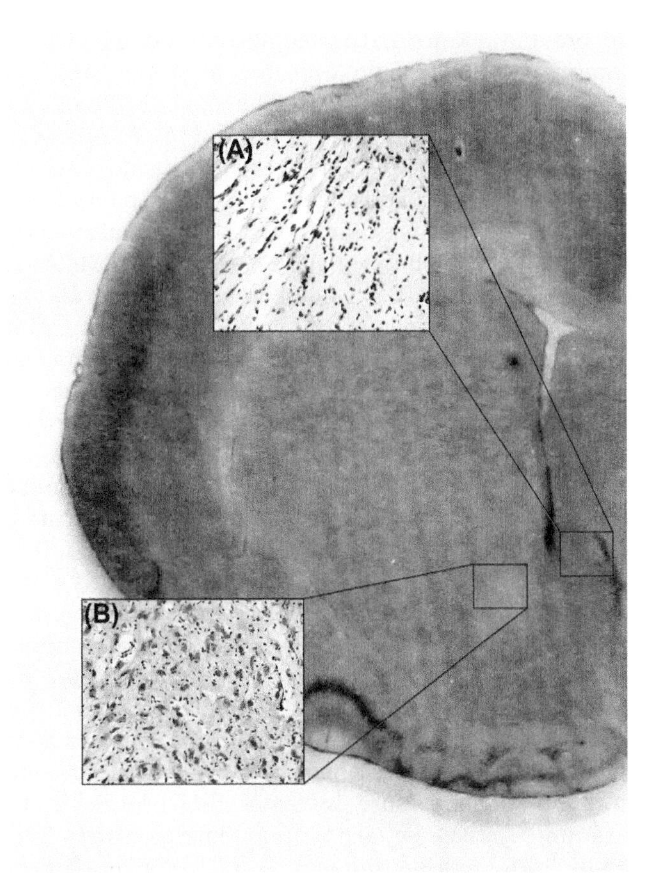

FIGURE 13.4 Cresyl violet staining in a coronal section of the striatum. The nucleus accumbens shell (A) and core (B) are differentiated in the respective upper and lower zoomed-in images.

binding regulates mesolimbic activity, as application of ghrelin increases the firing rate of dopaminergic neurons in the VTA and influences synaptic formation in a manner that increases dopamine transmission. Further, direct ghrelin administration to dopaminergic neurons expressing GHS-R1a in the VTA causes an increase in dopamine release in the NAcc (Abizaid et al., 2006; Cone et al., 2015; Jerlhag et al., 2007). Ghrelin is also associated with dopamine signaling in these mesolimbic cells through the demonstrated heterodimerization of GHS-R1a with DRD1 (Jiang et al., 2006; Kern et al., 2015; Mary et al., 2013; Schellekens, Dinan, & Cryan, 2013). This heterodimerization suggests that release of dopamine in the NAcc could be elicited by activation of either the GHS-R1a or a DRD1 in a heterodimer within the VTA. This would be consistent with the hypothesis that ghrelin modulates reward through dopamine transmission (Cone et al., 2015; Jerlhag et al., 2007; Sadeghzadeh et al., 2015).

Ghrelin neurons of the hypothalamus and mesolimbic system have been implicated in both homeostatic and reward-based modulation of food consumption (Menzies, Skibicka, Leng, & Dickson, 2013; Schellekens, Finger, Dinan, & Cryan, 2012; St-Onge et al., 2016; Wei et al., 2015). Clinical research has found that disordered ghrelin modulation and obesity are associated with altered dopamine activity (Savage et al., 2014). Indeed, emerging evidence implicates ghrelinergic and dopaminergic interactions in subjective palatability, and specifically the preference for salient foods (Kampov-Polevoy, Garbutt, & Janowsky, 1999; Nakazato et al., 2001). However, the increased preference for palatable food extends well beyond homeostatic mechanisms to include eating associated with hedonics and motivational characteristics (Dickson et al., 2011; Egecioglu et al., 2010; Perello et al., 2010; Skibicka, Hansson, Alvarez-Crespo, Friberg, & Dickson, 2011). Injections of ghrelin into the VTA have been shown to elicit an increase in appetitive and motivated behaviors, including an increase in the ingestion of palatable foods using a progressive ratio operant paradigm, suggesting that ghrelin modulates food intake by stimulating reward pathways (Egecioglu et al., 2010; King, Isaacs, O'Farrell, & Abizaid, 2011; Wei et al., 2015). Weinberg, Nicholson, and Currie (2011) demonstrated that lesioning of dopaminergic VTA neurons through treatment with 6-hydroxydopamine leads to an attenuation of ghrelin's food-reinforcing capabilities. Further, ghrelin antagonists delivered into the VTA decrease reward-based eating, and in addition, accumbal dopamine release associated with rewarding foods is not present in GHS-R1a knockout mice (Egecioglu et al., 2010; St-Onge et al., 2016; Wei et al., 2015). Supporting this, other research by Disse et al. (2010) has shown that ghrelin increases the consumption of sucrose regardless of its caloric content, where systemic injections of ghrelin increases the consumption of saccharin, a noncaloric artificial sweetener, independently of the availability of caloric food. The corroboration of these findings suggest that ghrelin-mediated sucrose consumption does not reflect an exclusive homeostatically driven process, but rather, incorporates reward mediating dopaminergic neurons of the mesolimbic system. These findings also suggest that ghrelinergic and dopaminergic actions within the VTA are required for the functional activation of appetitive behavior, specifically in regard to palatable foods. In contrast, however, other findings have suggested that ghrelin increases food consumption but may be of less importance in mediating sugar or fat flavor preferences in mice (Overduin, Figlewicz, Bennett-Jay, Kittleson, & Cummings, 2012; Sclafani et al., 2015). The contradictory findings of ghrelin's role in food palatability, and thereby the dopaminergic pathway, suggests that this circuit may be more complex than previously believed. Future work, therefore, should continue to elucidate the mechanisms underlying the interplay of homeostatic and non-homeostatic, reward-based motivation for food consumption.

ETHANOL AND DRUG REWARD

While it is clear that ghrelin plays a role in appetitive reward, emerging evidence now implicates the neuropeptide in drug seeking behavior (Clifford et al., 2012; Panagopoulos & Ralevski, 2014). One particular drug of abuse that has been extensively investigated in regard to ghrelin signaling is ethanol (Bahi et al., 2013; Cepko et al., 2014; Leggio, Addolorato, et al., 2011). A significant amount of work exploring the pharmacology of ethanol consumption is carried out using animal models (Gomez & Ryabinin, 2014; Phillips, Reed, & Pastor, 2015; Tabakoff & Hoffman, 2000). Due to ethanol's aversive taste, heterozygous rat and mouse models of ethanol consumption do not generally self-administer solutions of ethanol above a 6% concentration without prior exposure and experience of its hedonic effects (Brasser, Castro, & Feretic, 2015). One method to accommodate for this in preclinical ethanol research is to gradually expose the animals to ethanol over an extended habituation period (Cepko et al., 2014; Tabakoff & Hoffman, 2000). Once ethanol preference has been established, heterozygous animal models freely consume ethanol (Brockway et al., 2015; Cepko et al., 2014). Other studies utilize strains of mice exhibiting a robust ethanol preference, or alternatively, prairie voles, which freely consume high concentrations of ethanol (Font, Luján, & Pastor, 2013; Ledesma, Escrig, Pastor, & Aragon, 2014; Stevenson et al., 2016).

Ethanol is a central nervous system depressant, and its associated behavioral symptoms include relaxation and euphoria (Costardi et al., 2015; Tabakoff & Hoffman, 2013). As reviewed by Costardi et al. (2015), the ethanol molecule is small and simple, and has both hydrophilic and lipophilic chemical properties. These factors allow ethanol to be widely disseminated throughout the body, including the brain (Costardi et al., 2015; Pereira, Andrade, & Valentão, 2015). Ethanol's mechanism of action is largely mediated through inhibitory GABAergic projections and the facilitation of activation at the receptor subtype $GABA_A$ (Charlton et al., 1997; Davies, 2003). Prior evidence has found that ethanol exposure is linked to changes in expression of $GABA_A$ subunits, as well as increased GABA stimulation (Charlton et al., 1997; Costardi et al., 2015). The mechanism of GABA's inhibitory signaling is heavily mediated by $GABA_A$ receptor binding. These receptors are expressed on postsynaptic neurons and their activation increases chloride stores in the postsynaptic cell leading to neuronal hyperpolarization and signal inhibition (Davies, 2003). This underlies ethanol's subjective effects including sedation (Costardi et al., 2015). Indeed, GABA transmission may be a critical component of the mechanism through which ghrelin interacts with drugs of abuse, as Cruz et al. (2013) found that ghrelin increases GABA transmission in the central amygdala, a region of the brain implicated in the neural control of ethanol consumption (Dhaher, Finn, Snelling, & Hitzemann, 2008; Kissler & Walker, 2016; Vendruscolo et al., 2015). In addition to its mechanistic interaction with GABA, ethanol associates with other transmitter systems that, along with its inhibitory effects, mediate ethanol-induced physiological responses. Ethanol has been shown to affect glutamatergic signaling through the *N*-methyl-D-aspartate (NMDA) receptor (Davies, 2003; Santhakumar, Wallner, & Otis, 2007), and Möykkynen and Korpi (2012) have found that ethanol consumption leads to a decrease in glutamatergic NMDA receptor activation while ethanol dependence is associated with an upregulation of NMDA receptors.

Ethanol-induced reward is further mediated in part by ethanol's interference with the dopamine, opioid, and serotonergic (5-HT) transmitter systems (Costardi et al., 2015; Font et al., 2013; Zaleski, Morato, da Silva, & Lemos, 2004). Mesolimbic reward circuitry is extensively implicated in ethanol reinforcement, and neurons within this system are rich in opioid, dopamine, and 5-HT receptors (Kelley & Berridge, 2002; Kuo et al., 2016; Marcinkiewcz, 2015; Söderpalm, Löf, & Ericson, 2009). As reviewed by Marcinkiewcz (2015), 5-HT plays a significant role in the pharmacology of ethanol consumption. Ethanol exposure increases 5-HT levels in the striatum (Jamal et al., 2016; Marcinkiewcz, 2015; Yoshimoto, McBride, Lumeng, & Li, 1992), and a polymorphism of the 5-HT transporter, which leads to overall decreased concentrations of 5-HT in the central nervous system, has been previously associated with early onset ethanol use disorders (Ait-Daoud et al., 2009; Ishiguro et al., 1999; Schuckit et al., 1999). Ethanol-induced reward is also influenced or mediated via endogenous opioid activity. Ethanol facilitates *mu* opioid receptor binding (Zaleski et al., 2004) and animals administered with *mu* receptor antagonists demonstrate decreased ethanol consumption (Cruz, Bajo, Schweitzer, & Roberto, 2008; Zaleski et al., 2004). While interactions between ethanol and opioid and GABA systems have been shown to impact dopamine transmission (Costardi et al., 2015), past research has demonstrated ethanol's effects on dopamine signaling in mesolimbic reward-related structures. This includes the finding that ethanol exposure leads to increased dopamine transmission and extracellular dopamine stores in the NAcc (Boileau et al., 2003; Söderpalm et al., 2009).

The role of ghrelin in ethanol reward is similarly attributed to activation of neurons within the mesolimbic reward system. This, of course, includes an interaction with dopamine transmission but may involve more complex signaling pathways such as receptor heterodimerization as described in the preceding paragraphs (Jiang et al., 2006; Kern et al., 2015). Although the ethanol molecule induces dopaminergic signaling and hedonic reward, a significant aspect of ethanol consumption in humans is mediated by flavor preference (Bachmanov et al., 2003; Kampov-Polevoy et al., 1999). Research into the effects of flavored alcohol could provide a biologically relevant mechanism for examining the role of sucrose, alcohol, and ghrelin within mesolimbic neurons, as prior research has established an association between ghrelinergic signaling and ethanol consumption (Akkisi Kumsar & Dilbaz, 2015; Leggio, Addolorato, et al., 2011). Indeed, Landgren et al. (2008) have reported that single nucleotide polymorphisms on the gene encoding GHS-R1a are associated with high ethanol consuming behavior. Additionally, there is a positive correlation between plasma ghrelin levels and ethanol craving in individuals with ethanol use disorders, and subjects with ethanol use disorders have higher concentrations of plasma ghrelin during periods of abstinence compared to controls (Addolorato et al., 2006; Akkisi Kumsar & Dilbaz, 2015; Kim, 2004; Koopmann et al., 2012; Wurst et al., 2007). Further, Leggio, Ferrulli, et al. (2011) found that subjects who are ethanol dependent showed increased plasma ghrelin levels when abstaining from ethanol consumption, while nonabstaining ethanol-dependent individuals demonstrated a decrease in plasma ghrelin levels over the course of the 12-week study. Supporting these findings, intravenous ghrelin administered to ethanol-dependent subjects leads to increased ethanol craving (Haass-Koffler et al., 2015; Leggio et al., 2014). Also, a relationship between plasma ghrelin concentrations and ethanol has been demonstrated in nondependent social drinkers. Leggio, Schwandt, Oot, Dias, and Ramchandani (2013) observed that intravenously administered ethanol attenuated the increase in plasma ghrelin concentrations following fasting from food. Taken together, these findings suggest that ghrelin signaling is heavily implicated in ethanol consumption and craving, and may play an important role in mediating ethanol-induced euphoria.

Animal research has further explored the role of ghrelin in ethanol intake. Indeed, a cogent indicator of the significance of ghrelin signaling in ethanol-related reward is the positive correlation between high rates of ethanol preference and the concentration of GHS-R1a in the VTA and NAcc (Landgren et al., 2011). Studies examining the exogenous administration of ghrelin further support the peptide's role in the neural control of ethanol consumption. In rats, Cepko et al. (2014) reported that both systemic injection and direct administration of ghrelin into the VTA significantly potentiates ethanol intake. Injections into the NAcc or intraventricular administration elicit similar increases in ethanol consumption (Brockway et al., 2015; Cepko et al., 2014; Jerlhag et al., 2009). Furthermore, studies of ghrelin antagonism indicate the peptide's importance in mediating ethanol preference. Kaur and Ryabinin (2010) found that systemic administration of the ghrelin antagonist D-Lys3-GHRP-6 leads to a decrease in ethanol consumption in mice. In support of this, Stevenson et al. (2016, 2015) demonstrated in prairie voles that GHS-R1a antagonism with JMV2959 attenuates ethanol intake. This effect was similarly found in rats and mice, with JMV2959 reportedly reducing ethanol consumption in both species (Gomez et al., 2015; Gomez & Ryabinin, 2014). Consistent with the evidence from pharmacological manipulations, GHS-R1a knockout mice exhibit both decreased dopamine transmission in the mesolimbic pathway and attenuated ethanol intake (Bahi et al., 2013; Jerlhag, Landgren, Egecioglu, Dickson, & Engel, 2011). Szulc et al. (2013) demonstrated that ethanol-dependent rats have lower levels of plasma ghrelin compared to nondependent controls. This finding is intriguing, as Jerlhag, Ivanoff, Vater, and Engel (2014) have reported that peripheral circulating ghrelin does not mediate ethanol-induced reward nor ethanol intake in rodents. This stands in direct opposition to a number of other reports including those focusing on peripheral ghrelin activity and ethanol craving in human clinical research (Haass-Koffler et al., 2015; Leggio et al., 2014). Accordingly future research should continue to investigate the association between plasma ghrelin concentrations and ethanol-seeking behavior in both human and nonhuman animal models.

In addition to the neural control of ethanol consumption, it is apparent that ghrelin plays a role in mediating the effects of other drugs of abuse including psychostimulants and euphoriants (Cepko et al., 2014; Davis et al., 2007; Dickson et al., 2011; de Lartigue, Dimaline, Varro, & Dockray, 2007; Wellman, Davis, & Nation, 2005). Suchankova, Steensland, Fredriksson, Engel, and Jerlhag (2013) have reported that single nucleotide polymorphisms for loci that encode both the ghrelin gene and the GHS-R1a receptor are related to a propensity for methamphetamine dependence, suggesting that methamphetamine dependence is in fact associated with dysregulated ghrelinergic signaling. Animal research has further explored the relationship between ghrelin activity and other drugs of abuse. Davis et al. (2007) have shown that animals pretreated with systemic ghrelin demonstrate augmented cocaine conditioned place preference (CPP). Other work has shown that animals treated with ghrelin and cocaine exhibit altered locomotor function, with reduced locomotor responses to cocaine demonstrated in GHS-R1a knockouts (Abizaid et al., 2011; Jerlhag, Egecioglu, Dickson, & Engel, 2010), and enhanced locomotor activity induced by cocaine following ghrelin injection into the NAcc core (Jang, Kim, Cho, Lee, & Kim, 2013). Microinjection of ghrelin into the VTA potentiates cocaine-induced CPP (Schuette, Gray, & Currie, 2013), and while ghrelin administration may increase dopamine release in the forebrain, including the striatum and the amygdala (Palotai et al., 2013a, 2013b), the peptide also appears to modulate the rewarding effects of nicotine. In fact 2013 work suggests that ghrelin and nicotine may equally stimulate dopamine release in these brain regions (Palotai et al., 2013a, 2013b). Moreover, GHS-R1a antagonism has been shown to attenuate nicotine-induced dopamine release, CPP, and locomotor stimulation (Jerlhag & Engel, 2011), and this work is consistent with other findings implicating ghrelin and ghrelin receptors in the modulation of nicotine and psychostimulant action (Wellman, Clifford, & Rodriguez, 2013). Cepko et al. (2014) have found that ghrelin effectively potentiates cocaine's stimulatory effects on ethanol intake and this is in agreement with prior work showing that the reinforcing properties of ethanol and cocaine interact synergistically within the VTA (Ding et al., 2012).

CONCLUSION

As described in the foregoing, ghrelin's action on metabolic circuits within the hypothalamus is well documented, including its impact on food intake, energy expenditure, and substrate oxidation. Additionally, ghrelinergic receptor mechanisms mediating metabolic function appear to at least partially overlap with limbic synapses underlying stress activation and anxiety. However, more recent attention has focused on the peptide's role in the modulation of drug reward. Increasing evidence, from both human and animal investigations, indicates that ghrelin plays a critical role in mediating ethanol reinforcement and euphoria via actions on mesolimbic neurons. There is strong speculation that such effects may be mediated via heterodimeric interactions of the ghrelin receptor with receptors for dopamine in the ventral tegmental area and striatum. Finally, while the literature provides compelling evidence implicating mesolimbic ghrelin in ethanol reward, the fact that ghrelin signaling might contribute to the reinforcing properties of other psychoactive drugs suggests that targeting central ghrelin receptors may prove highly effective in the treatment of addiction and drug abuse.

Acknowledgment

This work was generously supported by a grant from the M.J. Murdock Charitable Trust to PJC.

References

Abizaid, A., Liu, Z.-W., Andrews, Z. B., Shanabrough, M., Borok, E., Elsworth, J. D., … Horvath, T. L. (2006). Ghrelin modulates the activity and synaptic input organization of midbrain dopamine neurons while promoting appetite. *Journal of Clinical Investigation, 116*(12), 3229–3239. http://doi.org/10.1172/JCI29867.

Abizaid, A., Mineur, Y. S., Roth, R. H., Elsworth, J. D., Sleeman, M. W., Picciotto, M. R., & Horvath, T. L. (2011). Reduced locomotor responses to cocaine in ghrelin-deficient mice. *Neuroscience, 192*, 500–506. http://doi.org/10.1016/j.neuroscience.2011.06.001.

Addolorato, G., Capristo, E., Leggio, L., Ferrulli, A., Abenavoli, L., Malandrino, N., … Gasbarrini, G. (2006). Relationship between ghrelin levels, alcohol craving, and nutritional status in current alcoholic patients. *Alcoholism: Clinical and Experimental Research, 30*(11), 1933–1937. http://doi.org/10.1111/j.1530-0277.2006.00238.x.

Ait-Daoud, N., Roache, J. D., Dawes, M. A., Liu, L., Wang, X. Q., Javors, M. A., & Johnson, B. A. (2009). Can serotonin transporter genotype predict craving in alcoholism? *Alcoholism: Clinical and Experimental Research, 33*(8), 1329–1335.

Akkisi Kumsar, N., & Dilbaz, N. (2015). Relationship between craving and ghrelin, adiponectin, and resistin levels in patients with alcoholism. *Alcoholism: Clinical and Experimental Research, 39*(4), 702–709. http://doi.org/10.1111/acer.12689.

Albarran-Zeckler, R. G., Sun, Y., & Smith, R. G. (2011). Physiological roles revealed by ghrelin and ghrelin receptor deficient mice. *Peptides, 32*(11), 2229–2235. http://doi.org/10.1016/j.peptides.2011.07.003.

Anderberg, R. H., Hansson, C., Fenander, M., Richard, J. E., Dickson, S. L., Nissbrandt, H., … Skibicka, K. P. (2016). The stomach-derived hormone ghrelin increases impulsive behavior. *Neuropsychopharmacology: Official Publication of the American College of Neuropsychopharmacology, 41*(5), 1199–1209.

Asakawa, A., Inui, A., Kaga, O., Yuzuriha, H., Nagata, T., Ueno, N., … Kasuga, M. (2001). Ghrelin is an appetite-stimulatory signal from stomach with structural resemblance to motilin. *Gastroenterology, 120*(2), 337–345. http://doi.org/10.1053/gast.2001.22158.

Bachmanov, A. A., Kiefer, S. W., Molina, J. C., Tordoff, M. G., Duffy, V. B., Bartoshuk, L. M., & Mennella, J. A. (2003). Chemosensory factors influencing alcohol perception, preferences, and consumption. *Alcoholism: Clinical and Experimental Research, 27*(2), 220–231. http://doi.org/10.1097/01.ALC.0000051021.99641.19.

Bahi, A., Tolle, V., Fehrentz, J.-A., Brunel, L., Martinez, J., Tomasetto, C.-L., & Karam, S. M. (2013). Ghrelin knockout mice show decreased voluntary alcohol consumption and reduced ethanol-induced conditioned place preference. *Peptides, 43*, 48–55. http://doi.org/10.1016/j.peptides.2013.02.008.

Boileau, I., Assaad, J.-M., Pihl, R. O., Benkelfat, C., Leyton, M., Diksic, M., … Dagher, A. (2003). Alcohol promotes dopamine release in the human nucleus accumbens. *Synapse, 49*(4), 226–231. http://doi.org/10.1002/syn.10226.

Brasser, S. M., Castro, N., & Feretic, B. (2015). Alcohol sensory processing and its relevance for ingestion. *Physiology & Behavior, 148*, 65–70. http://doi.org/10.1016/j.physbeh.2014.09.004.

Brockway, E. T., Krater, K. R., Selva, J. A., Wauson, S. E. R., & Currie, P. J. (2016). Impact of [D-Lys^3]-GHRP-6 and feeding status on hypothalamic ghrelin-induced stress activation. *Peptides, 79*, 95–102. http://doi.org/10.1016/j.peptides.2016.03.013.

Brockway, E. T., Selva, J. A., Zallar, L. J., Garling, E. E., Baumgartner, H. M., Sheskier, M. B., … Currie, P. J. (2015). Differential effects of ghrelin on alcohol intake in C57BL/6J mice and Sprague Dawley rats. *Program No.258.16. 2015 Neuroscience Meeting Planner.*

Broglio, F., Benso, A., Gottero, C., Prodam, F., Gauna, C., Filtri, L., … Ghigo, E. (2003). Non-acylated ghrelin does not possess the pituitaric and pancreatic endocrine activity of acylated ghrelin in humans. *Journal of Endocrinological Investigation, 26*(3), 192–196. http://doi.org/10.1007/BF03345156.

Buss, J., Havel, P. J., Epel, E., Lin, J., Blackburn, E., & Daubenmier, J. (2014). Associations of ghrelin with eating behaviors, stress, metabolic factors, and telomere length among overweight and obese women: Preliminary evidence of attenuated ghrelin effects in obesity? *Appetite, 76*, 84–94. http://doi.org/10.1016/j.appet.2014.01.011.

Cabral, A., Suescun, O., Zigman, J. M., & Perello, M. (2012). Ghrelin indirectly activates hypophysiotropic CRF neurons in rodents. *PLoS One, 7*(2), e31462. http://doi.org/10.1371/journal.pone.0031462.

Carlini, V. P., Monzón, M. E., Varas, M. M., Cragnolini, A. B., Schiöth, H. B., Scimonelli, T. N., & de Barioglio, S. R. (2002). Ghrelin increases anxiety-like behavior and memory retention in rats. *Biochemical and Biophysical Research Communications, 299*(5), 739–743. http://dx.doi.org/10.1016/S0006-291X(02)02740-7.

Carlini, V. P., Varas, M. M., Cragnolini, A. B., Schiöth, H. B., Scimonelli, T. N., & de Barioglio, S. R. (2004). Differential role of the hippocampus, amygdala, and dorsal raphe nucleus in regulating feeding, memory, and anxiety-like behavioral responses to ghrelin. *Biochemical and Biophysical Research Communications, 313*(3), 635–641. http://doi.org/10.1016/j.bbrc.2003.11.150.

Cepko, L. C. S., Selva, J. A., Merfeld, E. B., Fimmel, A. I., Goldberg, S. A., & Currie, P. J. (2014). Ghrelin alters the stimulatory effect of cocaine on ethanol intake following mesolimbic or systemic administration. *Neuropharmacology, 85*, 224–231. http://doi.org/10.1016/j.neuropharm.2014.05.030.

Charlton, M. E., Sweetnam, P. M., Fitzgerald, L. W., Terwilliger, R. Z., Nestler, E. J., & Duman, R. S. (1997). Chronic ethanol administration regulates the expression of $GABA_A$ receptor α_1 and α_5 subunits in the ventral tegmental area and hippocampus. *Journal of Neurochemistry, 68*(1), 121–127.

Chow, K. B. S., Sun, J., Chu, K. M., Tai Cheung, W., Cheng, C. H. K., & Wise, H. (2012). The truncated ghrelin receptor polypeptide (GHS-R1b) is localized in the endoplasmic reticulum where it forms heterodimers with ghrelin receptors (GHS-R1a) to attenuate their cell surface expression. *Molecular and Cellular Endocrinology, 348*(1), 247–254. http://doi.org/10.1016/j.mce.2011.08.034.

Chuang, J.-C., & Zigman, J. M. (2010). Ghrelin's roles in stress, mood, and anxiety regulation. *International Journal of Peptides, 2010*, 1–5. http://doi.org/10.1155/2010/460549.

Clifford, P. S., Rodriguez, J., Schul, D., Hughes, S., Kniffin, T., Hart, N., ... Wellman, P. J. (2012). Attenuation of cocaine-induced locomotor sensitization in rats sustaining genetic or pharmacologic antagonism of ghrelin receptors. *Addiction Biology, 17*(6), 956–963. http://dx.doi.org/10.1111/j.1369-1600.2011.00339.x.

Cone, J. J., Roitman, J. D., & Roitman, M. F. (2015). Ghrelin regulates phasic dopamine and nucleus accumbens signaling evoked by food-predictive stimuli. *Journal of Neurochemistry, 133*(6), 844–856. http://doi.org/10.1111/jnc.13080.

Costardi, J. V. V., Nampo, R. A. T., Silva, G. L., Ribeiro, M. A. F., Stella, H. J., Stella, M. B., & Malheiros, S. V. P. (2015). A review on alcohol: From the central action mechanism to chemical dependency. *Revista Da Associação Médica Brasileira, 61*(4), 381–387. http://doi.org/10.1590/1806-9282.61.04.381.

Cruz, M. T., Bajo, M., Schweitzer, P., & Roberto, M. (2008). Shared mechanisms of alcohol and other drugs. *Alcohol Research & Health, 31*(2), 137–147.

Cruz, M. T., Herman, M. A., Cote, D. M., Ryabinin, A. E., & Roberto, M. (2013). Ghrelin increases GABAergic transmission and interacts with ethanol actions in the rat central nucleus of the amygdala. *Neuropsychopharmacology: Official Publication of the American College of Neuropsychopharmacology, 38*(2), 364–375. http://doi.org/10.1038/npp.2012.190.

Cummings, D. E. (2004). Plasma ghrelin levels and hunger scores in humans initiating meals voluntarily without time- and food-related cues. *American Journal of Physiology. Endocrinology and Metabolism, 287*(2), E297–E304. http://doi.org/10.1152/ajpendo.00582.2003.

Currie, P. J., Braver, M., Mirza, A., & Sricharoon, K. (2004). Sex differences in the reversal of fluoxetine-induced anorexia following raphe injections of 8-OH-DPAT. *Psychopharmacology, 172*(4), 359–364. http://doi.org/10.1007/s00213-003-1681-x.

Currie, P. J., Coiro, C. D., Duenas, R., Guss, J. L., Mirza, A., & Tal, N. (2011). Urocortin I inhibits the effects of ghrelin and neuropeptide Y on feeding and energy substrate utilization. *Brain Research, 1385*, 127–134. http://doi.org/10.1016/j.brainres.2011.01.114.

Currie, P. J., John, C. S., Nicholson, M. L., Chapman, C. D., & Loera, K. E. (2010). Hypothalamic paraventricular 5-hydroxytryptamine inhibits the effects of ghrelin on eating and energy substrate utilization. *Pharmacology, Biochemistry, and Behavior, 97*(1), 152–155. http://doi.org/10.1016/j.pbb.2010.05.027.

Currie, P. J., Khelemsky, R., Rigsbee, E. M., Dono, L. M., Coiro, C. D., Chapman, C. D., & Hinchcliff, K. (2012). Ghrelin is an orexigenic peptide and elicits anxiety-like behaviors following administration into discrete regions of the hypothalamus. *Behavioural Brain Research, 226*(1), 96–105. http://doi.org/10.1016/j.bbr.2011.08.037.

Currie, P. J., Mirza, A., Fuld, R., Park, D., & Vasselli, J. R. (2005). Ghrelin is an orexigenic and metabolic signaling peptide in the arcuate and paraventricular nuclei. *American Journal of Physiology. Regulatory, Integrative and Comparative Physiology, 289*(2), R353–R358. http://doi.org/10.1152/ajpregu.00756.2004.

Currie, P. J., Schuette, L. M., Wauson, S. E. R., Voss, W. N., & Angeles, M. J. (2014). Activation of urocortin 1 and ghrelin signaling in the basolateral amygdala induces anxiogenesis. *Neuroreport*, (1). http://doi.org/10.1097/WNR.0000000000000047.

Dailey, M. J., Moran, T. H., Holland, P. C., & Johnson, A. W. (2016). The antagonism of ghrelin alters the appetitive response to learned cues associated with food. *Behavioural Brain Research, 303*, 191–200. http://doi.org/10.1016/j.bbr.2016.01.040.

Davies, M. (2003). The role of GABA (A) receptors in mediating the effects of alcohol in the central nervous system. *Journal of Psychiatry & Neuroscience: JPN, 28*(4), 263.

Davis, K. W., Wellman, P. J., & Clifford, P. S. (2007). Augmented cocaine conditioned place preference in rats pretreated with systemic ghrelin. *Regulatory Peptides, 140*(3), 148–152. http://doi.org/10.1016/j.regpep.2006.12.003.

Dhaher, R., Finn, D., Snelling, C., & Hitzemann, R. (2008). Lesions of the extended amygdala in C57BL/6J mice do not block the intermittent ethanol vapor-induced increase in ethanol consumption. *Alcoholism: Clinical and Experimental Research, 32*(2), 197–208. http://doi.org/10.1111/j.1530-0277.2007.00566.x.

Dickson, S. L., Egecioglu, E., Landgren, S., Skibicka, K. P., Engel, J. A., & Jerlhag, E. (2011). The role of the central ghrelin system in reward from food and chemical drugs. *Molecular and Cellular Endocrinology, 340*(1), 80–87. http://doi.org/10.1016/j.mce.2011.02.017.

Ding, Z.-M., Oster, S. M., Hauser, S. R., Toalston, J. E., Bell, R. L., McBride, W. J., & Rodd, Z. A. (2012). Synergistic self-administration of ethanol and cocaine directly into the posterior ventral tegmental area: Involvement of serotonin-3 receptors. *The Journal of Pharmacology and Experimental Therapeutics, 340*(1), 202–209. http://doi.org/10.1124/jpet.111.187245.

Disse, E., Bussier, A.-L., Veyrat-Durebex, C., Deblon, N., Pfluger, P. T., Tschöp, M. H., ... Rohner-Jeanrenaud, F. (2010). Peripheral ghrelin enhances sweet taste food consumption and preference, regardless of its caloric content. *Physiology & Behavior, 101*(2), 277–281. http://doi.org/10.1016/j.physbeh.2010.05.017.

Dixon, J. B., Lambert, E. A., & Lambert, G. W. (2015). Neuroendocrine adaptations to bariatric surgery. *Molecular and Cellular Endocrinology, 418*, 143–152. http://doi.org/10.1016/j.mce.2015.05.033.

Egecioglu, E., Bjursell, M., Ljungberg, A., Dickson, S. L., Kopchick, J. J., Bergström, G., ... Bohlooly-Y, M. (2006). Growth hormone receptor deficiency results in blunted ghrelin feeding response, obesity, and hypolipidemia in mice. *American Journal of Physiology.*

Endocrinology and Metabolism, 290(2), E317–E325. http://doi.org/10.1152/ajpendo.00181.2005.

Egecioglu, E., Jerlhag, E., Salomé, N., Skibicka, K. P., Haage, D., Bohlooly-Y, M., … Dickson, S. L. (2010). Ghrelin increases intake of rewarding food in rodents: Ghrelin and food reward. *Addiction Biology, 15*(3), 304–311. http://doi.org/10.1111/j.1369-1600.2010.00216.x.

Farkas, I., Vastagh, C., Sárvári, M., & Liposits, Z. (2013). Ghrelin decreases firing activity of gonadotropin-releasing hormone (GnRH) neurons in an estrous cycle and endocannabinoid signaling dependent manner. *PLoS One, 8*(10), e78178.

Fernandez, G., Cabral, A., Cornejo, M. P., De Francesco, P. N., Garcia-Romero, G., Reynaldo, M., & Perello, M. (2016). Des-acyl Ghrelin directly targets the arcuate nucleus in a ghrelin-receptor independent manner and impairs the orexigenic effect of ghrelin. *Journal of Neuroendocrinology*, 1–12.

Font, L., Luján, M.Á., & Pastor, R. (2013). Involvement of the endogenous opioid system in the psychopharmacological actions of ethanol: The role of acetaldehyde. *Frontiers in Behavioral Neuroscience, 7*. http://doi.org/10.3389/fnbeh.2013.00093.

Gomez, J. L., Cunningham, C. L., Finn, D. A., Young, E. A., Helpenstell, L. K., Schuette, L. M., … Ryabinin, A. E. (2015). Differential effects of ghrelin antagonists on alcohol drinking and reinforcement in mouse and rat models of alcohol dependence. *Neuropharmacology, 97*, 182–193. http://doi.org/10.1016/j.neuropharm.2015.05.026.

Gomez, J. L., & Ryabinin, A. E. (2014). The effects of ghrelin antagonists [D-Lys^3]-GHRP-6 or JMV2959 on ethanol, water, and food intake in C57BL/6J mice. *Alcoholism: Clinical and Experimental Research, 38*(9), 2436–2444. http://doi.org/10.1111/acer.12499.

Gueorguiev, M., Lecoeur, C., Meyre, D., Benzinou, M., Mein, C. A., Hinney, A., … Froguel, P. (2009). Association studies on ghrelin and ghrelin receptor gene polymorphisms with obesity. *Obesity, 17*(4), 745–754. http://doi.org/10.1038/oby.2008.589.

Gutierrez, J. A., Solenberg, P. J., Perkins, D. R., Willency, J. A., Knierman, M. D., Jin, Z., … Hale, J. E. (2008). Ghrelin octanoylation mediated by an orphan lipid transferase. *Proceedings of the National Academy of Sciences of the United States of America, 105*(17), 6320–6325.

Haass-Koffler, C. L., Aoun, E. G., Swift, R. M., de la Monte, S. M., Kenna, G. A., & Leggio, L. (2015). Leptin levels are reduced by intravenous ghrelin administration and correlated with cue-induced alcohol craving. *Translational Psychiatry, 5*(9), e646. http://doi.org/10.1038/tp.2015.140.

Hansson, C., Haage, D., Taube, M., Egecioglu, E., Salomé, N., & Dickson, S. L. (2011). Central administration of ghrelin alters emotional responses in rats: Behavioural, electrophysiological and molecular evidence. *Neuroscience, 180*, 201–211. http://doi.org/10.1016/j.neuroscience.2011.02.002.

Howard, A. D., Feighner, S. D., Cully, D. F., Arena, J. P., Liberator, P. A., Rosenblum, C. I., … Van der Ploeg, L. H. (1996). A receptor in pituitary and hypothalamus that functions in growth hormone release. *Science (New York, N.Y.), 273*(5277), 974–977.

Ishiguro, H., Saito, T., Akazawa, S., Mitushio, H., Tada, K., Enomoto, M., … Arinami, T. (1999). Association between drinking-related antisocial behavior and a polymorphism in the serotonin transporter gene in a Japanese population. *Alcoholism: Clinical and Experimental Research, 23*(7), 1281–1284. http://doi.org/10.1111/j.1530-0277.1999.tb04289.x.

Jamal, M., Ameno, K., Miki, T., Tanaka, N., Ito, A., Ono, J., & Kinoshita, H. (2016). Ethanol and acetaldehyde differentially alter extracellular dopamine and serotonin in Aldh2-knockout mouse dorsal striatum: A reverse microdialysis study. *Neurotoxicology, 52*, 204–209.

Jang, J. K., Kim, W. Y., Cho, B. R., Lee, J. W., & Kim, J.-H. (2013). Microinjection of ghrelin in the nucleus accumbens core enhances locomotor activity induced by cocaine. *Behavioural Brain Research, 248*, 7–11. http://doi.org/10.1016/j.bbr.2013.03.049.

Jensen, M., Ratner, C., Rudenko, O., Christiansen, S. H., Skov, L. J., Hundahl, C., … Holst, B. (2015). Anxiolytic-like effects of increased ghrelin receptor signaling in the amygdala. *International Journal of Neuropsychopharmacology.* pyv123 http://doi.org/10.1093/ijnp/pyv123.

Jerlhag, E., Egecioglu, E., Dickson, S. L., Douhan, A., Svensson, L., & Engel, J. A. (2007). Ghrelin administration into tegmental areas stimulates locomotor activity and increases extracellular concentration of dopamine in the nucleus accumbens. *Addiction Biology, 12*(1), 6–16. http://doi.org/10.1111/j.1369-1600.2006.00041.x.

Jerlhag, E., Egecioglu, E., Dickson, S. L., & Engel, J. A. (2010). Ghrelin receptor antagonism attenuates cocaine- and amphetamine-induced locomotor stimulation, accumbal dopamine release, and conditioned place preference. *Psychopharmacology, 211*(4), 415–422. http://doi.org/10.1007/s00213-010-1907-7.

Jerlhag, E., Egecioglu, E., Landgren, S., Salomé, N., Heilig, M., Moechars, D., … Engel, J. A. (2009). Requirement of central ghrelin signaling for alcohol reward. *Proceedings of the National Academy of Sciences of the United States of America, 106*(27), 11318–11323. http://doi.org/10.1073/pnas.0812809106.

Jerlhag, E., & Engel, J. A. (2011). Ghrelin receptor antagonism attenuates nicotine-induced locomotor stimulation, accumbal dopamine release and conditioned place preference in mice. *Drug and Alcohol Dependence, 117*(2–3), 126–131. http://doi.org/10.1016/j.drugalcdep.2011.01.010.

Jerlhag, E., Ivanoff, L., Vater, A., & Engel, J. A. (2014). Peripherally circulating ghrelin does not mediate alcohol-induced reward and alcohol intake in rodents. *Alcoholism: Clinical and Experimental Research, 38*(4), 959–968. http://doi.org/10.1111/acer.12337.

Jerlhag, E., Landgren, S., Egecioglu, E., Dickson, S. L., & Engel, J. A. (2011). The alcohol-induced locomotor stimulation and accumbal dopamine release is suppressed in ghrelin knockout mice. *Alcohol, 45*(4), 341–347. http://doi.org/10.1016/j.alcohol.2010.10.002.

Jiang, H., Betancourt, L., & Smith, R. G. (2006). Ghrelin amplifies dopamine signaling by cross talk involving formation of growth hormone secretagogue receptor/dopamine receptor subtype 1 heterodimers. *Molecular Endocrinology, 20*(8), 1772–1785. http://doi.org/10.1210/me.2005-0084.

Kamegai, J., Tamura, H., Shimizu, T., Ishii, S., Sugihara, H., & Wakabayashi, I. (2001). Chronic central infusion of ghrelin increases hypothalamic neuropeptide Y and Agouti-related protein mRNA levels and body weight in rats. *Diabetes, 50*(11), 2438–2443.

Kampov-Polevoy, A. B., Garbutt, J. C., & Janowsky, D. S. (1999). Association between preference for sweets and excessive alcohol intake: A review of animal and human studies. *Alcohol and Alcoholism, 34*(3), 386–395. http://doi.org/10.1093/alcalc/34.3.386.

Kaur, S., & Ryabinin, A. E. (2010). Ghrelin receptor antagonism decreases alcohol consumption and activation of perioculomotor urocortin-containing neurons. *Alcoholism: Clinical and Experimental Research, 34*(9), 1525–1534. http://doi.org/10.1111/j.1530-0277.2010.01237.x.

Kelley, A. E., & Berridge, K. C. (2002). The neuroscience of natural rewards: Relevance to addictive drugs. *The Journal of Neuroscience, 22*(9), 3306–3311.

Kern, A., Mavrikaki, M., Ullrich, C., Albarran-Zeckler, R., Brantley, A. F., & Smith, R. G. (2015). Hippocampal dopamine/DRD1 signaling dependent on the ghrelin receptor. *Cell, 163*(5), 1176–1190. http://doi.org/10.1016/j.cell.2015.10.062.

Kim, D.-J. (2004). Increased fasting plasma ghrelin levels during alcohol abstinence. *Alcohol and Alcoholism, 40*(1), 76–79. http://doi.org/10.1093/alcalc/agh108.

Kim, Y. J., & Bi, S. (2016). Knockdown of neuropeptide Y in the dorsomedial hypothalamus reverses high-fat diet-induced obesity and impaired glucose tolerance in rats. *American Journal of Physiology.*

Regulatory, Integrative and Comparative Physiology, 310(2), R134–R142.

King, S. J., Isaacs, A. M., O'Farrell, E., & Abizaid, A. (2011). Motivation to obtain preferred foods is enhanced by ghrelin in the ventral tegmental area. *Hormones and Behavior, 60*(5), 572–580. http://doi.org/10.1016/j.yhbeh.2011.08.006.

Kissler, J. L., & Walker, B. M. (2016). Dissociating motivational from physiological withdrawal in alcohol dependence: Role of central amygdala κ-opioid receptors. *Neuropsychopharmacology: Official Publication of the American College of Neuropsychopharmacology, 41*(2), 560–567. http://doi.org/10.1038/npp.2015.183.

Kohno, D., Gao, H.-Z., Muroya, S., Kikuyama, S., & Yada, T. (2003). Ghrelin directly interacts with neuropeptide-y-containing neurons in the rat arcuate nucleus Ca^{2+} signaling via protein kinase a and n-type channel-dependent mechanisms and cross-talk with leptin and orexin. *Diabetes, 52*(4), 948–956.

Kojima, M., Hosoda, H., Date, Y., Nakazato, M., Matsuo, H., & Kangawa, K. (1999). Ghrelin is a growth-hormone-releasing acylated peptide from stomach. *Nature, 402*(6762), 656–660. http://doi.org/10.1038/45230.

Kojima, M., Hosoda, H., & Kangawa, K. (2012). Purification of rat and human ghrelins. *Methods in Enzymology, 514*, 45–61.

Koopmann, A., von der Goltz, C., Grosshans, M., Dinter, C., Vitale, M., Wiedemann, K., & Kiefer, F. (2012). The association of the appetitive peptide acetylated ghrelin with alcohol craving in early abstinent alcohol dependent individuals. *Psychoneuroendocrinology, 37*(7), 980–986. http://doi.org/10.1016/j.psyneuen.2011.11.005.

Kouno, T., Akiyama, N., Ito, T., Okuda, T., Nanchi, I., Notoya, M., ... Yukioka, H. (2016). Ghrelin O-acyltransferase knockout mice show resistance to obesity when fed high-sucrose diet. *Journal of Endocrinology, 228*(2), 115–125. http://doi.org/10.1530/JOE-15-0330.

Kuo, C.-C., Shen, H., Harvey, B. K., Yu, S.-J., Kopajtic, T., Hinkle, J. J., ... Wang, Y. (2016). Differential modulation of methamphetamine-mediated behavioral sensitization by overexpression of Mu opioid receptors in nucleus accumbens and ventral tegmental area. *Psychopharmacology, 233*(4), 661–672. http://doi.org/10.1007/s00213-015-4134-4.

Landgren, S., Engel, J. A., Hyytiä, P., Zetterberg, H., Blennow, K., & Jerlhag, E. (2011). Expression of the gene encoding the ghrelin receptor in rats selected for differential alcohol preference. *Behavioural Brain Research, 221*(1), 182–188. http://doi.org/10.1016/j.bbr.2011.03.003.

Landgren, S., Jerlhag, E., Zetterberg, H., Gonzalez-Quintela, A., Campos, J., Olofsson, U., ... Engel, J. A. (2008). Association of pro-ghrelin and GHS-R1A gene polymorphisms and haplotypes with heavy alcohol use and body mass. *Alcoholism: Clinical and Experimental Research, 32*(12), 2054–2061. http://doi.org/10.1111/j.1530-0277.2008.00793.x.

de Lartigue, G., Dimaline, R., Varro, A., & Dockray, G. J. (2007). Cocaine- and amphetamine-regulated transcript: Stimulation of expression in rat vagal afferent neurons by cholecystokinin and suppression by ghrelin. *The Journal of Neuroscience, 27*(11), 2876–2882. http://doi.org/10.1523/JNEUROSCI.5508-06.2007.

Ledesma, J. C., Escrig, M. A., Pastor, R., & Aragon, C. M. (2014). The MAO-A inhibitor clorgyline reduces ethanol-induced locomotion and its volitional intake in mice. *Pharmacology Biochemistry and Behavior, 116*, 30–38.

Leggio, L., Addolorato, G., Cippitelli, A., Jerlhag, E., Kampov-Polevoy, A. B., & Swift, R. M. (2011). Role of feeding-related pathways in alcohol dependence: A focus on sweet preference, NPY, and ghrelin: Role of feeding-related pathways in alcohol dependence. *Alcoholism: Clinical and Experimental Research, 35*(2), 194–202. http://doi.org/10.1111/j.1530-0277.2010.01334.x.

Leggio, L., Ferrulli, A., Cardone, S., Nesci, A., Miceli, A., Malandrino, N., ... Addolorato, G. (2011). Ghrelin system in alcohol-dependent subjects: Role of plasma ghrelin levels in alcohol drinking and craving: Ghrelin and alcoholism. *Addiction Biology, 17*(2), 452–464. http://doi.org/10.1111/j.1369-1600.2010.00308.x.

Leggio, L., Schwandt, M. L., Oot, E. N., Dias, A. A., & Ramchandani, V. A. (2013). Fasting-induced increase in plasma ghrelin is blunted by intravenous alcohol administration: A within-subject placebo-controlled study. *Psychoneuroendocrinology, 38*(12), 3085–3091. http://doi.org/10.1016/j.psyneuen.2013.09.005.

Leggio, L., Zywiak, W. H., Fricchione, S. R., Edwards, S. M., de la Monte, S. M., Swift, R. M., & Kenna, G. A. (2014). Intravenous ghrelin administration increases alcohol craving in alcohol-dependent heavy drinkers: A preliminary investigation. *Biological Psychiatry, 76*(9), 734–741. http://doi.org/10.1016/j.biopsych.2014.03.019.

Leibowitz, S. F., Sladek, C., Spencer, L., & Tempel, D. (1988). Neuropeptide Y, epinephrine and norepinephrine in the paraventricular nucleus: Stimulation of feeding and the release of corticosterone, vasopressin and glucose. *Brain Research Bulletin, 21*(6), 905–912.

Lutter, M., Sakata, I., Osborne-Lawrence, S., Rovinsky, S. A., Anderson, J. G., Jung, S., ... Zigman, J. M. (2008). The orexigenic hormone ghrelin defends against depressive symptoms of chronic stress. *Nature Neuroscience, 11*(7), 752–753. http://doi.org/10.1038/nn.2139.

Marcinkiewcz, C. A. (2015). Serotonergic systems in the pathophysiology of ethanol dependence: Relevance to clinical alcoholism. *ACS Chemical Neuroscience, 6*(7), 1026–1039. http://doi.org/10.1021/cn5003573.

Mary, S., Fehrentz, J.-A., Damian, M., Gaibelet, G., Orcel, H., Verdie, P., ... Baneres, J.-L. (2013). Heterodimerization with its splice variant blocks the ghrelin receptor 1a in a non-signaling conformation: A study with a purified heterodimer assembled into lipid discs. *Journal of Biological Chemistry, 288*(34), 24656–24665. http://doi.org/10.1074/jbc.M113.453423.

Mason, B. L., Wang, Q., & Zigman, J. M. (2014). The central nervous system sites mediating the orexigenic actions of ghrelin. *Annual Review of Physiology, 76*(1), 519–533. http://doi.org/10.1146/annurev-physiol-021113-170310.

Menzies, J. R., Skibicka, K. P., Leng, G., & Dickson, S. L. (2013). Ghrelin, reward and motivation. *Endocrine Development, 25*, 101–111.

Möykkynen, T., & Korpi, E. R. (2012). Acute effects of ethanol on glutamate receptors. *Basic & Clinical Pharmacology & Toxicology, 111*(1), 4–13. http://doi.org/10.1111/j.1742-7843.2012.00879.x.

Müller, T. D., Nogueiras, R., Andermann, M. L., Andrews, Z. B., Anker, S. D., Argente, J., ... Tschöp, M. H. (2015). Ghrelin. *Molecular Metabolism, 4*(6), 437–460. http://doi.org/10.1016/j.molmet.2015.03.005.

Nakazato, M., Murakami, N., Date, Y., Kojima, M., Matsuo, H., Kangawa, K., & Matsukura, S. (2001). A role for ghrelin in the central regulation of feeding. *Nature, 409*(6817), 194–198. http://doi.org/10.1038/35051587.

Naleid, A. M., Grace, M. K., Cummings, D. E., & Levine, A. S. (2005). Ghrelin induces feeding in the mesolimbic reward pathway between the ventral tegmental area and the nucleus accumbens. *Peptides, 26*(11), 2274–2279. http://doi.org/10.1016/j.peptides.2005.04.025.

Olszewski, P. K., Grace, M. K., Billington, C. J., & Levine, A. S. (2003). Hypothalamic paraventricular injections of ghrelin: Effect on feeding and c-fos immunoreactivity. *Peptides, 24*(6), 919–923. http://dx.doi.org/10.1016/S0196-9781(03)00159-1.

Overduin, J., Figlewicz, D. P., Bennett-Jay, J., Kittleson, S., & Cummings, D. E. (2012). Ghrelin increases the motivation to eat, but does not alter food palatability. *American Journal of Physiology. Regulatory, Integrative and Comparative Physiology, 303*(3), R259–R269. http://doi.org/10.1152/ajpregu.00488.2011.

Palotai, M., Bagosi, Z., Jászberényi, M., Csabafi, K., Dochnal, R., Manczinger, M., ... Szabó, G. (2013a). Ghrelin amplifies the nicotine-induced dopamine release in the rat striatum.

Neurochemistry International, 63(4), 239–243. http://doi.org/10.1016/j.neuint.2013.06.014.

Palotai, M., Bagosi, Z., Jászberényi, M., Csabafi, K., Dochnal, R., Manczinger, M., ... Szabó, G. (2013b). Ghrelin and nicotine stimulate equally the dopamine release in the rat amygdala. *Neurochemical Research, 38*(10), 1989–1995. http://doi.org/10.1007/s11064-013-1105-1.

Panagopoulos, V. N., & Ralevski, E. (2014). The role of ghrelin in addiction: A review. *Psychopharmacology, 231*(14), 2725–2740. http://doi.org/10.1007/s00213-014-3640-0.

Pereira, R. B., Andrade, P. B., & Valentão, P. (2015). A comprehensive view of the neurotoxicity mechanisms of cocaine and ethanol. *Neurotoxicity Research, 28*(3), 253–267. http://doi.org/10.1007/s12640-015-9536-x.

Perello, M., Sakata, I., Birnbaum, S., Chuang, J.-C., Osborne-Lawrence, S., Rovinsky, S. A., ... Zigman, J. M. (2010). Ghrelin increases the rewarding value of high fat diet in an orexin-dependent manner. *Biological Psychiatry, 67*(9), 880–886. http://doi.org/10.1016/j.biopsych.2009.10.030.

Phillips, T. J., Reed, C., & Pastor, R. (2015). Preclinical evidence implicating corticotropin-releasing factor signaling in ethanol consumption and neuroadaptation. *Genes, Brain and Behavior, 14*(1), 98–135. http://dx.doi.org/10.1111/gbb.12189.

Raspopow, K., Abizaid, A., Matheson, K., & Anisman, H. (2014). Anticipation of a psychosocial stressor differentially influences ghrelin, cortisol and food intake among emotional and non-emotional eaters. *Appetite, 74*, 35–43. http://doi.org/10.1016/j.appet.2013.11.018.

Sadeghzadeh, F., Babapour, V., & Haghparast, A. (2015). Role of dopamine D1-like receptor within the nucleus accumbens in acute food deprivation- and drug priming-induced reinstatement of morphine seeking in rats. *Behavioural Brain Research, 287*, 172–181. http://doi.org/10.1016/j.bbr.2015.03.055.

Santhakumar, V., Wallner, M., & Otis, T. S. (2007). Ethanol acts directly on extrasynaptic subtypes of $GABA_A$ receptors to increase tonic inhibition. *Alcohol, 41*(3), 211–221. http://doi.org/10.1016/j.alcohol.2007.04.011.

Sarker, M. R., Franks, S., & Caffrey, J. (2013). Direction of post-prandial ghrelin response associated with cortisol response, perceived stress and anxiety, and self-reported coping and hunger in obese women. *Behavioural Brain Research, 257*, 197–200. http://doi.org/10.1016/j.bbr.2013.09.046.

Sárvári, M., Kocsis, P., Deli, L., Gajári, D., Dávid, S., Pozsgay, Z., ... Liposits, Z. (2014). Ghrelin modulates the fMRI BOLD response of homeostatic and hedonic brain centers regulating energy balance in the rat. *PLoS One, 9*(5), e97651. http://doi.org/10.1371/journal.pone.0097651.

Savage, S. W., Zald, D. H., Cowan, R. L., Volkow, N. D., Marks-Shulman, P. A., Kessler, R. M., & Dunn, J. P. (2014). Regulation of novelty seeking by midbrain dopamine D2/D3 signaling and ghrelin is altered in obesity. *Obesity, 22*(6), 1452–1457. http://doi.org/10.1002/oby.20690.

Schellekens, H., Dinan, T. G., & Cryan, J. F. (2013). Dimerization of G-protein coupled receptors (GPCRs) in appetite regulation and food reward. *The FASEB Journal, 27*(1), 881–883. MeetingAbstracts.

Schellekens, H., Finger, B. C., Dinan, T. G., & Cryan, J. F. (2012). Ghrelin signalling and obesity: At the interface of stress, mood and food reward. *Pharmacology & Therapeutics, 135*(3), 316–326. http://doi.org/10.1016/j.pharmthera.2012.06.004.

Schuckit, M. A., Mazzanti, C., Smith, T. L., Ahmed, U., Radel, M., Iwata, N., & Goldman, D. (1999). Selective genotyping for the role of 5-HT2A, 5-HT2C, and GABAα6 receptors and the serotonin transporter in the level of response to alcohol: A pilot study. *Biological Psychiatry, 45*(5), 647–651. http://dx.doi.org/10.1016/S0006-3223(98)00248-0.

Schuette, L. M., Gray, C. C., & Currie, P. J. (2013). Microinjection of ghrelin into the ventral tegmental area potentiates cocaine-induced conditioned place preference. *Journal of Behavioral and Brain Science, 3*(8), 576–580.

Sclafani, A., Touzani, K., & Ackroff, K. (2015). Ghrelin signaling is not essential for sugar or fat conditioned flavor preferences in mice. *Physiology & Behavior, 149*, 14–22. http://doi.org/10.1016/j.physbeh.2015.05.016.

Skibicka, K. P., Hansson, C., Alvarez-Crespo, M., Friberg, P. A., & Dickson, S. L. (2011). Ghrelin directly targets the ventral tegmental area to increase food motivation. *Neuroscience, 180*, 129–137. http://doi.org/10.1016/j.neuroscience.2011.02.016.

Soares, J.-B., & Leite-Moreira, A. F. (2008). Ghrelin, des-acyl ghrelin and obestatin: Three pieces of the same puzzle. *Peptides, 29*(7), 1255–1270. http://doi.org/10.1016/j.peptides.2008.02.018.

Söderpalm, B., Löf, E., & Ericson, M. (2009). Mechanistic studies of Ethanol's interaction with the mesolimbic dopamine reward system. *Pharmacopsychiatry, 42*(Suppl. 1), S87–S94. http://doi.org/10.1055/s-0029-1220690.

Spencer, S. J., Emmerzaal, T. L., Kozicz, T., & Andrews, Z. B. (2015). Ghrelin's role in the hypothalamic-pituitary-adrenal axis stress response: Implications for mood disorders. *Biological Psychiatry, 78*(1), 19–27.

Spencer, S. J., Xu, L., Clarke, M. A., Lemus, M., Reichenbach, A., Geenen, B., ... Andrews, Z. B. (2012). Ghrelin regulates the hypothalamic-pituitary-adrenal axis and restricts anxiety after acute stress. *Biological Psychiatry, 72*(6), 457–465. http://doi.org/10.1016/j.biopsych.2012.03.010.

St-Onge, V., Watts, A., & Abizaid, A. (2016). Ghrelin enhances cue-induced bar pressing for high fat food. *Hormones and Behavior, 78*, 141–149. http://doi.org/10.1016/j.yhbeh.2015.11.005.

Stevenson, J. R., Buirkle, J. M., Buckley, L. E., Young, K. A., Albertini, K. M., & Bohidar, A. E. (2015). GHS-R1A antagonism reduces alcohol but not sucrose preference in prairie voles. *Physiology & Behavior, 147*, 23–29. http://doi.org/10.1016/j.physbeh.2015.04.001.

Stevenson, J. R., Francomacaro, L. M., Bohidar, A. E., Young, K. A., Pesarchick, B. F., Buirkle, J. M., ... O'Bryan, C. M. (2016). Ghrelin receptor (GHS-R1A) antagonism alters preference for ethanol and sucrose in a concentration-dependent manner in prairie voles. *Physiology & Behavior, 155*, 231–236. http://doi.org/10.1016/j.physbeh.2015.12.017.

Suchankova, P., Steensland, P., Fredriksson, I., Engel, J. A., & Jerlhag, E. (2013). Ghrelin receptor (GHS-R1A) antagonism suppresses both alcohol consumption and the alcohol deprivation effect in rats following long-term voluntary alcohol consumption. *PLoS One, 8*(8), e71284. http://doi.org/10.1371/journal.pone.0071284.

Sumithran, P., Prendergast, L. A., Delbridge, E., Purcell, K., Shulkes, A., Kriketos, A., & Proietto, J. (2011). Long-term persistence of hormonal adaptations to weight loss. *New England Journal of Medicine, 365*(17), 1597–1604.

Sustkova-Fiserova, M., Jerabek, P., Havlickova, T., Kacer, P., & Krsiak, M. (2014). Ghrelin receptor antagonism of morphine-induced accumbens dopamine release and behavioral stimulation in rats. *Psychopharmacology, 231*(14), 2899–2908. http://doi.org/10.1007/s00213-014-3466-9.

Szulc, M., Mikolajczak, P. L., Geppert, B., Wachowiak, R., Dyr, W., & Bobkiewicz-Kozlowska, T. (2013). Ethanol affects acylated and total ghrelin levels in peripheral blood of alcohol-dependent rats. *Addiction Biology, 18*(4), 689–701. http://doi.org/10.1111/adb.12025.

Tabakoff, B., & Hoffman, P. L. (2000). Animal models in alcohol research. *Alcohol Research & Health, 24*(2).

Tabakoff, B., & Hoffman, P. L. (2013). The neurobiology of alcohol consumption and alcoholism: An integrative history. *Pharmacology Biochemistry and Behavior, 113*(77–84), 20–37. http://doi.org/10.1016/j.pbb.2013.10.009.

Thomas, M. A., Ryu, V., & Bartness, T. J. (2016). Central ghrelin increases food foraging/hoarding that is blocked by GHSR antagonism and attenuates hypothalamic paraventricular nucleus neuronal activation. *American Journal of Physiology. Regulatory, Integrative and Comparative Physiology, 310*(3), R275–R285. http://doi.org/10.1152/ajpregu.00216.2015.

Ting, C. H., Chi, C. W., Li, C. P., & Chen, C. Y. (2015). Differential modulation of endogenous cannabinoid CB1 and CB2 receptors in spontaneous and splice variants of ghrelin-induced food intake in conscious rats. *Nutrition, 31*(1), 230–235.

Van Name, M., Giannini, C., Santoro, N., Jastreboff, A. M., Kubat, J., Li, F., … Caprio, S. (2015). Blunted suppression of acyl-ghrelin in response to fructose ingestion in obese adolescents: The role of insulin resistance. *Obesity, 23*(3), 653–661. http://doi.org/10.1002/oby.21019.

Vendruscolo, L. F., Estey, D., Goodell, V., Macshane, L. G., Logrip, M. L., Schlosburg, J. E., … Mason, B. J. (2015). Glucocorticoid receptor antagonism decreases alcohol seeking in alcohol-dependent individuals. *The Journal of Clinical Investigation, 125*(8), 3193–3197. http://doi.org/10.1172/JCI79828.

Wauson, S. E., Sarkodie, K., Schuette, L. M., & Currie, P. J. (2015). Midbrain raphe 5-HT1A receptor activation alters the effects of ghrelin on appetite and performance in the elevated plus maze. *Journal of Psychopharmacology, 29*(7), 836–844.

Weinberg, Z. Y., Nicholson, M. L., & Currie, P. J. (2011). 6-Hydroxydopamine lesions of the ventral tegmental area suppress ghrelin's ability to elicit food-reinforced behavior. *Neuroscience Letters, 499*(2), 70–73. http://doi.org/10.1016/j.neulet.2011.05.034.

Wei, X. J., Sun, B., Chen, K., Lv, B., Luo, X., & Yan, J. Q. (2015). Ghrelin signaling in the ventral tegmental area mediates both reward-based feeding and fasting-induced hyperphagia on high-fat diet. *Neuroscience, 300*, 53–62. http://doi.org/10.1016/j.neuroscience.2015.05.001.

Wellman, M., & Abizaid, A. (2015). Knockdown of central ghrelin O-acyltransferase by vivo-morpholino reduces body mass of rats fed a high-fat diet. *Peptides, 70*, 17–22. http://doi.org/10.1016/j.peptides.2015.05.007.

Wellman, P. J., Clifford, P. S., & Rodriguez, J. A. (2013). Ghrelin and ghrelin receptor modulation of psychostimulant action. *Frontiers in Neuroscience, 7*. http://doi.org/10.3389/fnins.2013.00171.

Wellman, P. J., Clifford, P. S., Rodriguez, J., Hughes, S., Eitan, S., Brunel, L., … Martinez, J. (2011). Pharmacologic antagonism of ghrelin receptors attenuates development of nicotine induced locomotor sensitization in rats. *Regulatory Peptides, 172*(1–3), 77–80. http://doi.org/10.1016/j.regpep.2011.08.014.

Wellman, P. J., Davis, K. W., & Nation, J. R. (2005). Augmentation of cocaine hyperactivity in rats by systemic ghrelin. *Regulatory Peptides, 125*(1–3), 151–154. http://doi.org/10.1016/j.regpep.2004.08.013.

Wren, A. M., Small, C. J., Fribbens, C. V., Neary, N. M., Ward, H. L., Seal, L. J., … Bloom, S. R. (2002). The hypothalamic mechanisms of the hypophysiotropic action of ghrelin. *Neuroendocrinology, 76*(5), 316–324.

Wren, A. M., Small, C. J., Ward, H. L., Murphy, K. G., Dakin, C. L., Taheri, S., … Bloom, S. R. (2000). The novel hypothalamic peptide ghrelin stimulates food intake and growth hormone secretion. *Endocrinology, 141*(11), 4325–4328. http://doi.org/10.1210/endo.141.11.7873.

Wurst, F. M., Graf, I., Ehrenthal, H. D., Klein, S., Backhaus, J., Blank, S., … Junghanns, K. (2007). Gender differences for ghrelin levels in alcohol-dependent patients and differences between alcoholics and healthy controls. *Alcoholism: Clinical and Experimental Research, 31*(12), 2006–2011. http://doi.org/10.1111/j.1530-0277.2007.00527.x.

Yang, J., Brown, M. S., Liang, G., Grishin, N. V., & Goldstein, J. L. (2008). Identification of the acyltransferase that octanoylates ghrelin, an appetite-stimulating peptide hormone. *Cell, 132*(3), 387–396. http://doi.org/10.1016/j.cell.2008.01.017.

Ye, J. (2013). Mechanisms of insulin resistance in obesity. *Frontiers of Medicine, 7*(1), 14–24. http://doi.org/10.1007/s11684-013-0262-6.

Yoshimoto, K., McBride, W. J., Lumeng, L., & Li, T.-K. (1992). Alcohol stimulates the release of dopamine and serotonin in the nucleus accumbens. *Alcohol, 9*(1), 17–22. http://dx.doi.org/10.1016/0741-8329(92)90004-t.

Zaleski, M., Morato, G. S., da Silva, V. A., & Lemos, T. (2004). Neuropharmacological aspects of chronic alcohol use and withdrawal syndrome. *Revista Brasileira de Psiquiatria, 26*, 40–42. http://doi.org/10.1590/S1516-44462004000500010.

Zigman, J. M., Jones, J. E., Lee, C. E., Saper, C. B., & Elmquist, J. K. (2006). Expression of ghrelin receptor mRNA in the rat and the mouse brain. *The Journal of Comparative Neurology, 494*(3), 528–548. http://doi.org/10.1002/cne.20823.

Zigman, J. M., Nakano, Y., Coppari, R., Balthasar, N., Marcus, J. N., Lee, C. E., … Elmquist, J. K. (2005). Mice lacking ghrelin receptors resist the development of diet-induced obesity. *The Journal of Clinical Investigation, 115*(12), 3564–3572. http://doi.org/10.1172/JCI26002.

CHAPTER

14

Alcoholic Neurological Syndromes

B.G. Pinheiro[1], A.S. Melo[1], L.M.P. Fernandes[1], E. Fontes de Andrade, Jr.[1], R.D. Prediger[2], C.S.F. Maia[1]

[1]Federal University of Pará, Belém, Brazil; [2]Federal University of Santa Catarina, Florianópolis, Brazil

INTRODUCTION

It has been established that prolonged heavy alcohol consumption elicits neurological disorders. Epidemiological studies have pointed that the mortality induced by alcohol is about 3.3 million of deaths per year and about 5.1% of the global morbidity (World Health Organisation, 2014, pp. 1−392).

Alcohol-use disorders (AUDs) can display a broad range of disorders, which some of them have been assembled into a category designed alcohol-related brain damage (ARBD) and fetal alcohol spectrum disorders (FASDs) (Coriale et al., 2014; Qin & Crews, 2014). Among the ARBD, the central pontine (CPM) and extrapontine myelinolysis (EPM), Marchiafava−Bignami disease (MBD), and Wernicke−Korsakoff syndrome (WKS) have been diagnosed. The CPM/EPM consists of an osmotic demyelination syndrome that can be developed by excessive and prolonged alcohol intake (Alleman, 2014). In fact, CPM/EPM affects primarily alcoholics, and the lesions reach the pons or additional brain structures, as cerebellar peduncle, basal ganglia, diencephalic and limbic structures, and cortex (Martin, 2004).

MBD is reported as an unusual alcohol-induced neurological syndrome related to symmetrical demyelination, atrophy, and necrosis of the corpus callosum (Folescu et al., 2014). Such disorder is commonly allied to vitamin B1 deficiency that may be aggravated by additional ethanol-induced damage mechanisms (Folescu et al., 2014). Besides MBD, WKS also shares the thiamine deficiency and a background history of alcoholism as the probable pathophysiological mechanism (Sullivan & Pfefferbaum, 2009). Initial stages of WKS, the Wernicke's encephalopathy, occur among 35−80% chronic alcoholism population, in whom the principal regions disrupted are thalamus and mammillary bodies (for review, see Qin & Crews, 2014).

In this regard, this chapter summarizes the signs/symptoms and diagnosis of the three more incident alcoholic syndromes (CPM/EPM, MBD, and WKS), as well as the principal disorder related to prenatal alcohol exposure (i.e., fetal alcohol syndrome, FAS). Moreover, the pathophysiology and mechanisms of the damage are discussed.

FETAL ALCOHOL SYNDROME

Since mid-1970s, several researchers have demonstrated that alcohol is a powerful drug that is able to promote behavioral changes, as well as generating teratogenic effects (Murawski, Moore, Thomas, & Riley, 2015). Besides, a series of gestational consequences can last throughout the life of exposure of offspring. These effects may cause several complications and affect directly the individual, as well as the family and society (Riley, Infante, & Warren, 2011).

FAS was first described by Jones, Smith, Ulleland, and Streissguth (1973) as reference to a small group of children who were exposed to ethanol in utero during pregnancy. Abnormal cognitive functioning, body growth reduction, craniofacial alterations (i.e., short palpebral fissures and maxillary hypoplasia), joint and palmar fold abnormalities, and cardiac defects have been observed in the FAS (Jones & Smith, 1973; Jones, Smith, & Hanson, 1976; Jones et al., 1973).

Recently, other alterations related to prenatal alcohol exposure have been reported, including several complications. These new features were grouped in different clinical categories of the syndrome. Initially, scientists have named the general features

Addictive Substances and Neurological Disease
http://dx.doi.org/10.1016/B978-0-12-805373-7.00014-1

TABLE 14.1 Examples of Facial Characteristics Diagnosed in the FAS

Fundamental Facial Characteristics	Any Facial Characteristics
Short palpebral fissures	Epicanthal folds
Flat midface and short nose	Minor ear abnormalities
Low nasal bridge	Micrognathia
Indistinct philtrum	
Thin upper lip	

related to alcohol prenatal exposure as FASDs. The newborns affected by FASD can present a broad range of physical and behavioral alterations as microcephaly, facial abnormalities, short stature, low birthweight, dysmorphic characteristics, hyperactivity, and learning and memory impairments. According to the National Academy of Medicine (formerly Institute of Medicine, IOM), FASD includes FAS, partial FAS (pFAS), alcohol-related neurodevelopmental disorder (ARND), and alcohol-related birth defects.

In fact, FAS has been pointed as the most severe degree of FASD, with almost all features of prenatal chronic alcohol exposure damage. Among the FAS spectrum characteristics, the abnormalities on the face as small palpebral fissures, thin vermilion border, and abnormal philtrum are presented (Table 14.1). Regarding the ocular system, strabismus, ptosis, and myopia have been reported. Nevertheless, the numerous studies involving FAS describe the central nervous system (CNS) impairments as the most disturbing symptoms related to this syndrome. Microcephaly, mood changes, and cognitive and behavioral impairments such as irritability, attention deficits, poor learning and memory, hyperactivity, anxiety, and depression in different degrees have been documented (Coriale et al., 2014; Jacobson & Jacobson, 2003).

In addition to FAS, Stratton, Howe, and Bataglia (1996) reported that there are three forms of fetal alcohol effects (FAEs), which include the pFAS, alcohol-related birth defects, and ARND. The pFAS exhibit some, but not all, of FAS physical features. The principal disabilities exhibited by pFAS children are CNS morphological, anatomical, or neurobehavioral impairments. In the ARND features, children are diagnosed with ocular and auditory damage, as well as cardiorenal system alterations. Finally, the alcohol-related birth defect children display less severe CNS alterations than FAS, however CNS functions that are marked affected in the full FAS are also impaired in the ARBD (i.e., cognitive and socioemotional behavior) (for review see Davis, Desrocher, & Moore, 2011; Jacobson & Jacobson, 2003).

PHARMACOKINETIC MECHANISMS OF ALCOHOL TERATOGENY

The alcohol and its metabolites are able to induce damage in the fetus. The ethanol administered by the oral route is rapidly absorbed in the stomach mucosa (about 20%) and in the intestinal portion (about 80%). The amount and kinetic of alcohol absorption depends on many variables including food intake, alcohol concentration, types of ethanol consumption, blood flow, and mucosal integrity (Kalant, Grant, & Mitchell, 2007; Swift, 2003).

In fact, alcohol is a small uncharged molecule, soluble in water and rapidly distributed through the body water in about 50–60% in men and 45–55% in women. The volume of distribution depends on the gender, age, fitness, and blood flow (Kalant et al., 2007; Norberg, Jones, Hahn, & Gabrielsson, 2003). In the case of pregnant women, alcohol distribution is increased since the water body content is increased, allowing ethanol to cross the blood–placental barrier, resulting in fetal alcohol concentrations similar to that of maternal blood (Dawes & Chowienczyk, 2001).

Ethanol and acetaldehyde (the metabolic of the oxidative metabolism of ethanol) are able to reach the maternal and fetal bloodstream and to concentrate mainly in the amniotic fluid. Some authors suggest that the amniotic liquid may function as a reservoir (as a compartment) that contributes to reduce alcohol elimination, which extends the alcohol contact period with the fetus. Actually, the delay in the fetal alcohol elimination relies on the "recurrent cycle," that includes: (1) the limited fetal metabolism of alcohol; (2) the presence of alcohol in the fetal excretes released in the amniotic fluid, as urine and respiratory exudates; and (3) and closing the cycle, the fetal swallowing of the amniotic fluid that contains alcohol (Heller & Burd, 2014; Zelner & Koren, 2013) (Fig. 14.1).

TERATOGENIC EFFECTS OF ALCOHOL ON THE CNS

The main damage in the CNS of children affected by alcohol prenatal exposure involves the executive function, formed by circuits of the frontal lobe, the basal ganglia, and thalamus nuclei. The executive function involves a variety of cognitive and also integrative processes, such as memory and motor attention perception (Pennington & Ozonoff, 1996; Spadoni et al., 2009).

Hippocampus is one of the vulnerable CNS areas related to alcohol impairment. Such susceptibility displays learning and memory functions damage, including verbal and nonverbal skills (Livy, Miller, Maier, & West, 2003). In verbal ability, children with FAS have demonstrated impairment of the memory retention, as well as of the

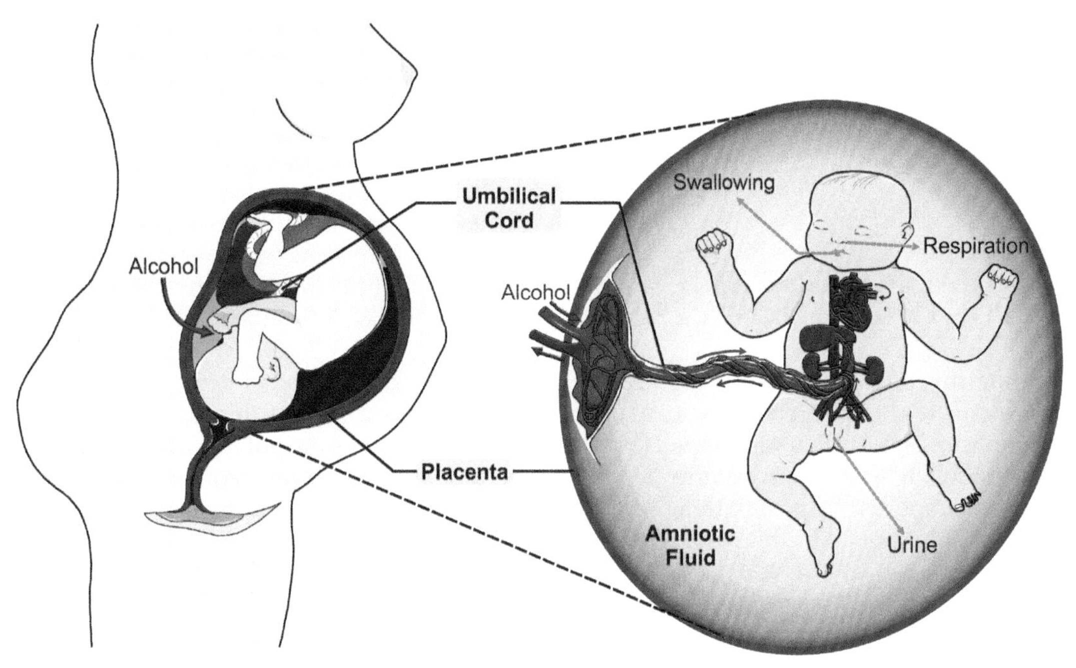

FIGURE 14.1 The amniotic fluid recycling system of alcohol. Alcohol is distributed to the fetus through umbilical cord bidirectional transport. In turn, the fetus excretes alcohol into the amniotic fluid through the urine and respiratory exudates. Lastly, the fetus swallows the amniotic fluid containing molecules of alcohol. *This figure is produced, in part, by using Servier Medical Art. The Creative Commons license (http://creativecommons.org/licenses/by/4.0/) and the symbol* .

retrieval and recognition (Crocker, Vaurio, Riley, & Mattson, 2011; Schonfeld, Mattson, Lang, Delis, & Riley, 2001). In the nonverbal skills involving visual–spatial memory, children and adolescents diagnosed with FAS exposed to a virtual Morris water maze (a hippocampal-dependent task) displayed poor performance in the task indicatives of spatial learning and memory deficits (Hamilton, Kodituwakku, Sutherland, & Savage, 2003).

In addition to the hippocampus, the cerebellum is other brain area affected in FAS. Considering that cerebellar area are related to the control of motor coordination, and children with FAS present deficits in these functions that persist during childhood (about 12 years old), including poor coordination in the eyes and hands (weak grips), muscle tremors, gait and balance deficits, gross and fine motor function impairment, postural instability, delayed motor reaction time, and poor performance of sensory–motor processing (Jirikowic, Kartin, & Olson, 2008; Kalberg et al., 2006; Roebuck, Simmons, Mattson, & Riley, 1998).

Responsible for connecting the cerebral hemispheres, the corpus callosum is also affected by alcohol prenatal exposure. Rich in white substance, the contralateral axon projections command the inter-hemispheric communication. Children with FAS present agenesis of the corpus callosum, which is an anomaly that promotes the absence of this CNS structure that exhibits diverse severity degrees (Montandon, Ribeiro, Lobo, Montandon Júnior, & Teixeira, 2003).

CLINICAL DIAGNOSIS OF FETAL ALCOHOL SPECTRUM DISORDERS

Several guidelines were published, establishing criteria of FAS and other neurological syndromes associated to prenatal alcohol exposure. The first publication was the National Academy of Medicine (formerly Institute of Medicine, IOM) in 1996 after a congress in the United States, which became known as IOM criteria, that was revised to better clinical practice in pediatrics by Hoyme et al. (2005).

In 2000, Astley and Clarren developed a more refined diagnostic criterion known as 4-digit code that takes into account effects in four areas (growth, face, CNS, and exposure to alcohol) resulting in 256 different codes and 22 diagnostic categories, where there is a specific pattern or levels of exposure to alcohol, requiring only confirmation of exposure or not to the teratogenic agent. In 2003, a National Task Force on Fetal Alcohol Syndrome and Fetal Alcohol Effect (NTF FAS/FAE), a federal advisory committee with members from various federal public health agencies also developed diagnostic criteria for FAS, taking into account evidence of alcohol exposure through clinical features (Bertrand, Floyd, & Weber, 2005). Additionally, a Canadian public health agency also published a set of FASD diagnostic standardization guidelines resulting from binge drinking (Chudley et al., 2005, Fig. 14.2).

CENTRAL PONTINE AND EXTRAPONTINE MYELINOLYSIS

Central pontine myelinolysis (CPM) was described by Victor and Adams (1949) that reported a case of patient with alcoholism history that developed quadriplegia and pseudobulbar palsy, and inability to chew, talk, or swallow was reported. A postmortem examination confirmed possible lesions in the pons basis with demyelinative characteristics. The lesions were large and symmetric, with neurons and axons relatively preserved, but was preserved pupillary reflexes, eye movements, corneal reflexes and facial sensation. In the course of 5 years, another four similar cases were notified among alcoholics and poor nutrition patient (Adams, 1959; Adams & Victor, 1993, pp. 891–892).

Afterward, several cases of CPM were described by Wright, Laureno, and Victor (1979), of which 10% of the cases showed myelinolysis outside the pons, characterizing the extrapontine myelinolysis (EPM). Subsequently, the relationship between the CPM/EPM with the hyponatremia was demonstrated in 1976. Thus, the terminology osmotic demyelination syndromes (ODS) was used when there are the association of CPM/EPM with hyponatremia (Burcar, Norenberg, & Yarnell, 1977; Tomlinson, Pierides, & Bradley, 1976).

Besides alcohol, other drugs are able to display hyponatremic state. In fact, thiazide diuretics, desmopressin, glucocorticoids, and inhibitors of serotonin can induce serum sodium level reduction and hence provoke CPM or EPM (Wilkinson, Begg, Winter, & Sainsbury, 1999). Furthermore, possible involvement of the CPM and EPM with other factors such as septic, malnourished including anorexic and bulimic disorders, as well as other comorbidities, for example, electrolyte serum abnormalities (Kumar, Fowler, Gonzalez-Toledo, & Jaffe, 2006) are observed.

Although CPM/EPM was initially described among alcoholics, nowadays it is related to states of

FAS	IOM Revised	4-Digits code	Canadian	NTF/FAS
Facial Alterations	Confirmed	Confirmed	Confirmed	Confirmed
Growth Retardation	Confirmed	Confirmed	Confirmed	Confirmed
CNS Damage	Confirmed	Confirmed	Confirmed	Confirmed
Alcohol exposure	Confirmed or not confirmed	Confirmed or not confirmed	Confirmed or not confirmed	Confirmed or not confirmed
pFAS				
Facial alterations	Confirmed	Confirmed	Confirmed	Confirmed
Growth Retardation	Confirmed	Not applicable or not requerited	Not applicable or not requerited	Not applicable or not requerited
CNS Damage	Confirmed	Confirmed	Confirmed	Not applicable or not requerited
Alcohol exposure	Confirmed or not confirmed	Confirmed	Confirmed	Not applicable or not requerited
ARND				
CNS Damage	Confirmed	Confirmed	Confirmed	Not applicable or not requerited
Alcohol exposure	Confirmed	Confirmed	Confirmed	Not applicable or not requerited
ARBD				
Congenital Deficit	Confirmed	Not applicable or not requerited	Confirmed	Not applicable or not requerited
Alcohol exposure	Confirmed	Not applicable or not requerited	Confirmed	Not applicable or not requerited

Confirmed Confirmed or not confirmed Not applicable or not requerited

FIGURE 14.2 Summary of clinical diagnostics for fetal alcohol spectrum disorders (FASDs) by the National Academy of Medicine (formerly Institute of Medicine, IOM), 4-digit code, Canadian public health agency, and National Task Force on Fetal Alcohol Syndrome and Fetal Alcohol Effect (NTF FAS/FAE). *FAS*, fetal alcohol syndrome; *pFAS*, partial FAS; *ARND*, alcohol-related neurodevelopmental disorder; *ARBD*, alcohol-related birth defects.

hyponatremia. In fact, osmotic demyelination diseases are described in both children and adults of all ages, but the incidence is difficult to define. Wright et al. (1979) reported the incidence of 1:300 after analysis of a series of 3500 adult necropsies, while Sterns, Silver, Kleinschmidt-DeMasters, and Rojiani (2007) reported the presence of neurological complications among 25% of patient after severe hyponatremic state correction. Actually, the CPM/EPM related to alcohol disorders are frequent (up to 40% of cases) and may be associated with poor nutrition as well as suppression of antidiuretic hormone (ADH) that affects level of sodium in blood serum (Martin, 2004).

CPM Signs and Symptoms

The myelinolysis process may occur in three categories: (1) CPM isolated, (2) EPM isolated, and (3) CPM associated to EPM. The CPM is defined as the acute demyelinating disease, caused by fast serum osmolality fluctuation, resulting in symmetric demyelination of the central part of the basal pons. On the other hand, EPM can affect symmetrically cerebellar peduncle, caudate, putamen, frontal and temporal white matter, fornix, external and extreme capsule, cloister, thalamus, subthalamic nucleus, internal capsule, amygdaloid nucleus, lateral geniculate body, deep layers of the cerebral cortex, hippocampus, and corpus callosum (Martin, 2004) (Fig. 14.3).

The CPM is classically presented as a biphasic clinical course. First, a generalized encephalopathy occurs, which can be associated to seizures (caused by the hyponatremia). During the first week, other symptoms related to the onset of myelinolysis emerge, as progressive obtundation that can lead to coma, diplegia with pseudobulbar palsy, dysarthria, ophthalmoplegia, dysphagia, nystagmus, and ataxia. These latter symptoms evidence that the second phase of CPM is in progress, and a "locked-in" syndrome (pseudocoma) or death can occur (Brown, 2000; Laureno & Karp, 1997).

The principal EPM symptoms consist of motor disorders, which is a consequence of the widespread nature of the lesions. The main disturbances are mutism, parkinsonism, dystonia, and catatonia. Actually, a variety of clinical symptoms can be observed, as spastic paraparesis, choreoathetosis, and permanent parkinsonism that are signs of pyramidal dysfunction. According to the study of Martin (2004), after the treatment, EPM manifestation can be substituted by oromandibular and arm dystonia associated to spasmodic dysphonia (Martin, 2004; Musana & Yale, 2005).

CPM Pathology

In the CPM pathology, lesions onset in the central part of the pons near the median raphe and rapidly converge to the basis pontis. The midbrain can be affected, but not the medulla. According to Martin (2004), the shape and location of the lesions are not well understood. The author affirms that the central area of lesions seems to reside around the brain stem, which consists of a region

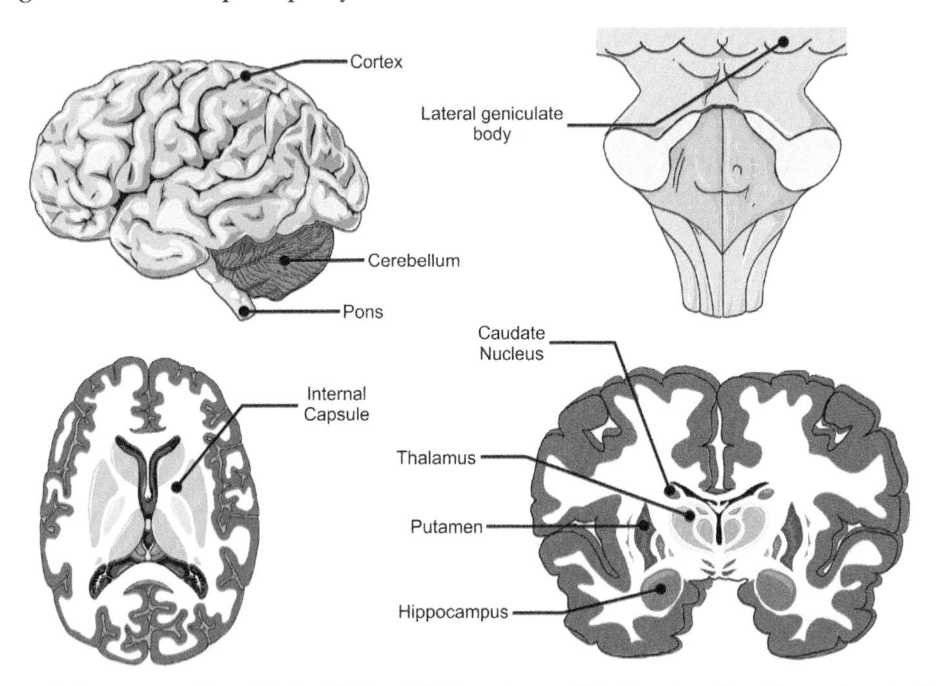

FIGURE 14.3 Anatomical structures affected in the CPM and EPM syndromes. *This figure is produced, in part, by using Servier Medical Art. The Creative Commons license (http://creativecommons.org/licenses/by/4.0/) and the symbol* .

of high gray and white matter constituents mixture (Martin, 2004; Pearce, 2009).

Histologically, the oligodendroglial cells involve the axons of large neurons in the central pons leading to CPM lesions. In the pons, the transverse pontocerebellar fibers are most frequently embedded, followed by the rostrocaudal tracts (i.e., corticobulbar and corticospinal fibers). The differential diagnosis from a central pontine infarcts lies in the fact that in the CPM basis pontis neurons and axons are preserved (Kumar et al., 2006).

Although is not mandatory that EPM be preceded of CPM disease, EPM lesions are similar and share the same gray–white apposition affinity. The extrapontine lesions have a distribution related to brain areas such as the thalamus and striatum. These regions share the same morphological characteristics, which axons and supporting oligodendroglia are in tight linear formation commanded by surrounding neurons. This region features limit the ability of the oligodendroglia to swell in the extracellular hypoosmolar environment, displaying the necessity of osmolytes-active extrusion for isotonicity homeostasis, and consequently increasing the shrinkage during the response to subsequent hyperosmolality (Kumar et al., 2006).

CPM Diagnosis

The neuroimaging techniques such as magnetic resonance imaging (MRI) are essential for CPM/EPM diagnostic confirmation. The MRI presents high sensitivity, and has been considered critical for the identification of these lesions. The CPM/EPM neuroimaging findings indicate areas of hypointense signal on T1-weighted sequences. In T2-weighted sequences, the hyperintense signal takes different shapes (i.e., triangular or round). The lesions can be large, which affects the basis pontis, but not pons ventral region (Ichikawa, Murakami, Katoh, Hieda, & Kawamura, 2008). Even after a month from acute CPM recovery, the lesions still can be noted on neuroimaging examinations. Therefore, the normal data in the onset of the disorder do not exclude the diagnosis of myelinolysis process (Alleman, 2014; Uchino et al., 2003). Apart from MRI, computed tomography (CT) can be performed, which reveals symmetrical hypodense regions on the pons or extrapontine areas (Ichikawa et al., 2008).

Another important point in the differential diagnosis is the clinical management of hyponatremic patients. Some authors claim that a safety level of CPM/EPM does not exist. However, symptomatic hyponatremia may be corrected by parenteral sodium that can be limited to 10–12 mEq/day (Germiniani, Roriz, Nabhan, Teive, & Werneck, 2002). Other investigations can be considered in order to differentiate the encephalopathy from hyponatremia and CPM/EPM. Thus, others techniques of diagnostics emerged for identification of CPM/EPM as the auditory evocation potential that presents a decreased driving, slowed pattern in electroencephalogram finding, and an increase in protein levels and myelin basic protein in the cerebrospinal fluid analysis (Laureno & Karp, 1997).

MARCHIAFAVA–BIGNAMI DISEASE

In 1898, Carducci reported the presence of sudden convulsions, loss of consciousness, coma, and eventual death of an alcoholic farmer related to necrosis of the corpus callosum on the autopsy. In 1903, Ettore Marchiafava and Amico Bignami reported six cases of corpus callosum changes in alcoholic individuals. Since then, several studies on this disease have been published (Lechevalier, Andersson, & Morin, 1977). The Marchiafava–Bignami disease (MBD) has been characterized by a rare toxic complication, mainly due to chronic alcoholism, which results in necrosis accompanied by symmetrical demyelination of the corpus callosum (Rawat, Pinto, Kulkarni, Muthusamy, & Dave, 2014). The injury can be found in the hemispheric white matter, rarely having cortical involvement (Tuntiyatorn & Laothamatas, 2008).

After observations of various degrees of clinical progression in neuroimage studies, Brion (1977) categorized the MBD as acute, subacute, and chronic forms, still widely used today. On the other hand, another MBD classification was proposed by Heinrich, Runge, and Khaw (2004), in which the patients were grouped into type A or type B, which the former evolve the worse prognosis with a loss of consciousness and extracallosal damage that were not present in the type B pattern. In fact, the prognosis and recovery depends on the MBD form, which according to Brion categories, the acute and subacute forms, elicit the worst outcome (for review see Wenz, Eisele, Artemis, Forster, & Brockmann, 2014).

MBD Pathophysiology

The main etiopathogenic hypothesis is that the disease occurs primarily due to deficiency of vitamins B complex followed by prolonged chronic alcohol intake, while many patients may obtain benefit from the administration of these compounds, other patients do not present the same kind of recovery (Jorge, 2013).

The MBD principal pathological feature is the occurrence of corpus callosum necrosis and demyelination, frequently symmetrical (Shiota, Nakano, Kawamura, & Hirayama, 1996). Histological changes can range from demyelination to other histological disruption, as severe bleeding, chronic hemosiderin deposits, cavitation, and cysts formation (Shiota et al., 1996). The necrosis commonly includes the entire length of the corpus callosum, or focused on some parts of this structure (Shiota et al., 1996). Besides, the MBD demyelination can also occur in extracallosal regions, as reported by Yang, Qin, Xu, & Hu (2015).

MBD Signs and Symptoms

Clinical features of MBD include neuropsychiatric disorders, dysarthria, tetraparesis, astasia–abasia, impaired consciousness, and symptoms of interhemispheric disconnection. Among psychiatric diseases, mania, depression, paranoia, and dementia have been described. Regarding the neurological symptoms, seizures, paresis, and ataxia have been described, which can progress to coma and death within few months (Kohler et al., 2000).

As mentioned earlier, the signs and symptoms depend on the MBD form. Acute MBD is characterized by rapid onset of unconsciousness, seizures, and eventual progression to coma. Neurological dysfunction that precedes coma may include mutism and diffuse muscle hypertonia with dysphagia. The prognosis is poor, and most patients with the acute MBD form die within few days after onset (Berek, Wagner, Chemelli, Aichner, & Benke, 1994).

In the subacute MBD form occurs a sudden onset of dementia that may develop into a vegetative state. In general, subacute MBD patients present dysarthria, appendicular hypertonia, and neck extension, which can constantly progress to death within few months after the onset (Valk & Van der Knaap, 1989).

The chronic MBD form became more often recognized after recent radiological technology improvement. Although in the past this type of MBD has been described only in 10% of cases, more recent cases have been diagnosed with chronic MBD. The controversial symptoms related to the chronic MBD form are signs of interhemispheric disconnection as well as behavioral/cognitive impairments (Castaigne, Buge, Cambier, Escourolle, & Rancurel, 1971). In addition to cognitive disruption, hallucinations and/or depression can last several months. The chronic MBD can also coexist with Wernicke encephalopathy, Korsakoff's syndrome, demyelinating syndrome, and osmotic and laminar necrosis (Geibprasert, Gallucci, & Krings, 2010).

According to Heinrich et al. (2004) study, the hospitalized MBD patients were predominantly male, with 46 years old, and the main symptoms observed were related to cognitive impairments (at 100% of the cases). Other symptoms that are less frequent are related to callosal disconnection syndromes, limb hypertonia, and motor speech disorder.

MBD Diagnosis

The definitive diagnosis of MBD is difficult to be established. The history of alcoholism associated to neuroimaging findings are the main approaches used to confirm the diagnosis. Actually, the advances of neuroimaging technology such as CT and MRI have improved the accuracy of MBD diagnosis. For instance, corpus callosum degeneration (hypodensity) is the main determinant in the establishment of MBD diagnosis. However, other surrounding brain structures and white matter tracts may be affected (Heinrich et al., 2004; Rawat et al., 2014; Yang et al., 2015).

WERNICKE–KORSAKOFF SYNDROME

In 1881, Carl Wernicke described an acute thiamine-deficiency encephalopathy characterized by eye movement paralysis, nystagmus, ataxia, and confusion, with higher incidence among alcoholics (Breathnach, 1991; Kopelman, Thomson, Guerrini, & Marshall, 2009). The onset of parenteral thiamine administration during the early stages can recover from Wernicke's encephalopathy, however about 85% of the Wernicke's encephalopathy patients may develop Korsakoff's syndrome (Day, Bentham, Callaghan, Kuruvilla, & George, 2013).

Korsakoff's syndrome is a long-lasting amnestic syndrome, primarily described by Sergei Sergueïevitch Korsakoff, who analyzed several cases of alcoholic polyneuropathy associated with mental disorders related to acute amnesia. Thus, the severe cognitive impairment is the most important feature of this syndrome (Breathnach, 1991; Pitel et al., 2008). In fact, both long- and short-term memories are affected, but not working memory. It seems that different stages of memory process including acquisition, consolidation, and retrieval are disrupted in the Korsakoff's syndrome (Kopelman, 1991; Kopelman et al., 2009). In this sense, due to the sharing of the same etiology, the Wernicke's encephalopathy and Korsakoff's syndrome are commonly described as Wernicke–Korsakoff syndrome (WKS) (Svanberg & Evans, 2013).

WKS Pathophysiology

Similar to other alcoholic syndromes, thiamine deficiency plays a pivotal role in the WKS pathophysiology. Chronic alcoholism reduces dietary thiamine absorption and activation, that promotes a thiamine deficiency profile. In the WKS, the mnemonic disruption is related primarily to focal lesions in the thalamus and the mammillary bodies (Kopelman et al., 2009).

Both the thalamus and the mammillary bodies are involved in the Papez circuit. The Papez circuit begins in the hippocampus, continues into the fornix to reach the mammillary body, that through the mammillothalamic tract projects to the anterior nucleus of the thalamus, which in turn connects to the cingulum, and after the entorhinal cortex, and finally returning to the hippocampus (Sullivan & Pfefferbaum, 2009). Thalamic and mammillary body lesions induced by thiamine deficiency disrupt the circuitry described, which induces amnesic state (Sullivan & Pfefferbaum, 2009).

In the thiamine deficiency-induced brain damage, the chronic ethanol exposure accelerates microglial activation, induction of CNS proinflammatory cytokine, neuronal death, and astrocyte dysfunction in the thalamus (He & Crews, 2008). Briefly, in the glucose metabolism, thiamine acts as a cofactor for pyruvate dehydrogenase, transketolase, and alpha-ketoglutarate dehydrogenase. The reduced level of the cofactor diminishes the enzyme activity, displaying the protein metabolism. Qin and Crews (2014) hypothesized that ethanol exposure associated to thiamine deficiency induces global metabolic insult independent of glucose. For instance, the acetate (a product of ethanol metabolism) becomes the principal source of brain energy among alcoholics, and the thalamus is one of the CNS structure affected by glucose metabolism impairment induced by ethanol (Volkow et al., 2013). The use of acetate as substrate to energy in the body systems protects the tissues from cell death. However, some brain regions (e.g., thalamus and mammillary bodies) display limited acetate metabolism, which could reduce the tissue protection induced by acetate metabolism, that leads to tissue insults (Qin & Crews, 2014). Therefore, additional studies are necessary to comprise the pathological mechanisms that underlies the focal lesions in the thalamus during WKS.

In addition to thiamine deficiency–induced cognitive impairment, authors claim that glutamatergic system alterations might represent synergic mechanisms involving thiamine deficiency and ethanol per se. It is well known that mnemonic process depends on the glutamatergic pathway. On the other hand, ethanol blocks glutamatergic receptors, including N-methyl-D-aspartate (NMDA) receptors. Thus, lesions in the Papez circuit allied to glutamatergic pathway blocks elicit cognitive impairment among active alcoholics. Besides, both the long-lasting thiamine deficiency that leads to excessive glutamate release and chronic ethanol exposure that elicits excitotoxicity may develop neuronal disorders in the vital structures related to cognition (Brust, 2010).

WKS Signs and Symptoms

As described earlier, Wernicke encephalopathy leads to Korsakoff's syndrome in almost 85% of survivors (Day et al., 2013). During the acute Wernicke encephalopathy, the main described symptoms are the involuntary eye movement, paralysis or weakness of the eye muscles (ophthalmoplegia), confusion, agitation, and ataxia (Breathnach, 1991). Due to the progress in Korsakoff's syndrome and the establishment of WKS, the symptoms associated to cognitive disturbance are revealed (Kopelman et al., 2009).

In fact, the WKS patients are conscious and able to communicate. However, the short-term memory may be disrupted, whereas the working memory disruption is controversial. Some WKS patients have difficulty to complete cognitive tasks (e.g., the text of a lecture), as well as recognize known people, but this harmful effect is refused by other authors (Kopelman et al., 2009; Svanberg & Evans, 2013). Actually, it seems that recent memories are more affected than the long-term memory (Kopelman et al., 2009). Another symptom of a presenile dementia status is the confabulation. In this sense, the WKS individuals create and believe in stories about experiences or situations to compensate memory lapses (Svanberg & Evans, 2013). In addition to CNS effects, any level of mild peripheral neuropathy and cardiovascular dysfunctions can occur (Zubaran, Fernandes, & Rodnight, 1997).

WKD Diagnosis

Unfortunately, only 20% of hospitalized patients with Wernicke's encephalopathy are identified before death (Kopelman et al., 2009). Initially, the clinical diagnosis is based on the classical triad of oculomotor dysfunction, gait ataxia, and global confusion behavior related to a background history of alcoholism. In addition, imaging approaches can be a useful tool to detect the Wernicke's encephalopathy. Among the imaging techniques, MRI, but not CT, presents an elevated sensitivity for the detection of Wernicke's encephalopathy. The MRI findings include symmetric lesions (bilateral hyperintensity) on anterior and medial nuclei of the thalamus, mammillary bodies and periventricular gray matter, tectal plate, inferior and superior colliculi, and periaqueductal gray (Sullivan & Pfefferbaum, 2009; Zuccoli & Pipitone, 2009).

In the WKS diagnosis, the clinical diagnosis related to Wernicke's encephalopathy triad and severe cognitive processes disruption (i.e., amnesia, disorientation, and confabulation) may be observed allied to alcoholism. In this stage of the illness, MRI may be of vital importance to confirm diagnosis, which reveals the atrophy of the mammillary bodies, cortical thinning, enlargement of the sulci, and dilatation of the ventricles, as well as reduction in the structural volume of the thalamus and mammillary bodies. The delay of WKS diagnosis progress to a state of stupor, coma, and death (Sullivan & Pfefferbaum, 2009).

FINAL COMMENTS

Since the 19th century, alcoholic syndromes have been reported. Unfortunately, even nowadays undiagnosed cases still consist of the majority (Martin, 2004; Sullivan & Pfefferbaum, 2009; Yang et al., 2015). Moreover, children who were exposed to ethanol by maternal consumption can elicit long-lasting physical and behavioral/cognitive impairments (Coriale et al., 2014). Both cases have become a public health issue, since alcohol-induced neurological disturbance may contribute to mortality or morbidity worldwide. In fact, a previous history of alcoholism allied to characteristics of harmful effects of alcohol and neuroimage technology are useful tools for clinicians for early diagnosis and recovery of the patients (Zuccoli & Pipitone, 2009).

References

Adams, R. D. (1959). Central pontine myelinolysis. *A.M.A. Archives of Neurology & Psychiatry, 81*(2), 154.

Adams, R. D., & Victor, M. (1993). *Principles of Neurology* (5th ed.). New York: McGraw-Hill.

Alleman, A. M. (2014). Osmotic demyelination syndrome: Central pontine myelinolysis and extrapontine myelinolysis. *Seminars in Ultrasound, CT, and MRI, 35*(2), 153–159.

Astley, S., & Clarren, S. (2000). Diagnosing the full spectrum of fetal alcohol-exposed individuals: Introducing the 4-digit diagnostic code. *Alcohol and Alcoholism, 35*(4), 400–410.

Berek, K., Wagner, M., Chemelli, A. P., Aichner, F., & Benke, T. (1994). Hemispheric disconnection in Marchiafava–Bignami disease: Clinical, neurophysiological and MRI findings. *Journal of the Neurological Sciences, 123*, 2–5.

Bertrand, J., Floyd, L., & Weber, M. (2005). Fetal alcohol syndrome. Guidelines for referral and diagnosis. *MMWR Recommendations and Report*, 1–14.

Breathnach, C. S. (1991). Wernicke. *Irish Journal of Psychological Medicine, 8*(2), 172.

Brion, S. (1977). Marchiafava–Bignami disease. In P. J. Vinken, & G. W. Bruyn (Eds.), *Handbook of clinical neurology* (pp. 317–329). Amsterdam: North Holland.

Brown, W. D. (2000). Osmotic demyelination disorders: Central pontine and extrapontine myelinolysis. *Current Opinion in Neurology, 13*(6), 691–697.

Brust, J. C. M. (2010). Ethanol and cognition: Indirect effects, neurotoxicity and neuroprotection: A review. *International Journal of Environmental Research and Public Health, 7*(4), 1540–1557.

Burcar, P. J., Norenberg, M. D., & Yarnell, P. R. (1977). Hyponatremia and central pontine myelinolysis. *Neurology, 27*, 223–226.

Carducci, A. (1898). Contributo allo studio delle encefaliti non suppurate. *Riv Psicol Psichiat Neuropat, 8*, 125–135.

Castaigne, P., Buge, A., Cambier, J., Escourolle, R., & Rancurel, G. (1971). La maladie de Marchiafava-Bignami. Etude anatomo-clinique de 10 observations. *Revue Neur- ologique (Paris), 125*, 179–186.

Chudley, A., Conry, J., Cook, J., Loock, C., Rosales, T., & LeBlanc, N. (2005). Fetal alcohol spectrum disorder: Canadian guidelines for diagnosis. *Canadian Medical Association Journal, 172*(5), 1–21.

Coriale, G., Fiorentino, D., Kodituwakku, P. W., Tarani, L., Parlapiano, G., Scalese, B., & Ceccanti, M. (2014). Identification of children with prenatal alcohol exposure. *Current Developmental Disorders Reports, 1*(3), 141–148.

Crocker, N., Vaurio, L., Riley, E. P., & Mattson, S. N. (2011). Comparison of verbal learning and memory in children with heavy prenatal alcohol exposure or attention-deficit/hyperactivity disorder. *Alcoholism: Clinical and Experimental Research, 35*(6), 1114–1121.

Davis, K., Desrocher, M., & Moore, T. (2011). Fetal alcohol spectrum disorder: A review of neurodevelopmental findings and interventions. *Journal of Developmental and Physical Disabilities, 23*(2), 143–167.

Dawes, M., & Chowienczyk, P. J. (2001). Drugs in pregnancy. Pharmacokinetics in pregnancy. *Best Practice & Research. Clinical Obstetrics & Gynaecology, 15*(6), 819–826.

Day, E., Bentham, P. W., Callaghan, R., Kuruvilla, T., & George, S. (2013). Thiamine for prevention and treatment of Wernicke–Korsakoff Syndrome in people who abuse alcohol. *Cochrane Database of Systematic Reviews*, (7), CD004033.

Folescu, R., Zamfir, C. L., Sisu, A. M., Motoc, A. G. M., Ilie, A. C., & Moise, M. (2014). Histopathological and imaging modifications in chronic ethanolic encephalopathy. *Romanian Journal of Morphology and Embryology, 55*(3), 797–801.

Geibprasert, S., Gallucci, M., & Krings, T. (2010). Alcohol-induced changes in the brain as assessed by MRI and CT. *European Radiology, 20*, 1492–1501.

Germiniani, F. M. B., Roriz, M., Nabhan, S. K., Teive, H. A. G., & Werneck, L. C. (2002). Mielinólise pontina central e extra-pontina em paciente alcoolista sem distúrbios hidro-eletrolíticos: Relato de caso. *Arquivos de Neuro-Psiquiatria, 60*(4), 1030–1033.

Hamilton, D. A., Kodituwakku, P., Sutherland, R. J., & Savage, D. D. (2003). Children with fetal alcohol syndrome are impaired at place learning but not cued-navigation in a virtual Morris water task. *Behavioural Brain Research, 143*(1), 85–94.

He, J., & Crews, F. T. (2008). Increased MCP-1 and microglia in various regions of the human alcoholic brain. *Experimental Neurology, 210*, 349–358.

Heinrich, A., Runge, U., & Khaw, A. V. (2004). Clinicoradiologic subtypes, of Marchiafava–Bignami disease. *Journal of Neurology, 251*(9), 1050–1059.

Heller, M., & Burd, L. (2014). Review of ethanol dispersion, distribution, and elimination from the fetal compartment. *Birth Defects Research Part A – Clinical and Molecular Teratology, 100*(4), 277–283.

Hoyme, H. E., May, P. A., Kalberg, W. O., Kodituwakku, P., Gossage, J. P., Trujillo, P. M., & Robinson, L. K. (2005). A practical clinical approach to diagnosis of fetal alcohol spectrum disorders: Clarification of the 1996 institute of medicine criteria. *Pediatrics, 115*(1), 39–47.

Ichikawa, H., Murakami, H., Katoh, H., Hieda, S., & Kawamura, M. (2008). Central pontine lesions observed with MRI in four diabetic patients. *Internal Medicine (Tokyo, Japan), 47*(15), 1425–1430.

Jacobson, S., & Jacobson, J. (2003). FAS/FAE and its impact on psychosocial child development. In R. E. Tremblay, R. G. Barr, & R. D. V. Peters (Eds.), *Encyclopedia on early childhood development* (pp. 1–7). Montreal, Quebec: Centre of Excellence for Early Childhood Development.

Jirikowic, T., Kartin, D., & Olson, H. C. (2008). Children with fetal alcohol spectrum disorders: A descriptive profile of adaptive function. *Canadian Journal of Occupational Therapy – Revue Canadienne D Ergotherapie, 75*(4), 238–248.

Jones, K. L., & Smith, D. W. (1973). Recognition of the fetal alcohol syndrome in early infancy. *Lancet, 2*(7836), 999–1001.

Jones, K. L., Smith, D. W., & Hanson, J. W. (1976). The fetal alcohol syndrome: Clinical delineation. *Annals of the New York Academy of Sciences, 273*, 130–139.

Jones, K. L., Smith, D. W., Ulleland, C. N., & Streissguth, A. P. (1973). Pattern of malformation in offspring of chronic alcoholic mothers. *Lancet, 1*(7815), 1267–1271.

Jorge, A. (2013). Doença de Marchiafava-Bignami: uma rara entidade com prognóstico sombrio. *Revista brasileira de terapia intensiva, 25*(4), 68–72.

Kalant, H., Grant, D., & Mitchell, J. (2007). *Principles of medical pharmacology.* Elsevier.

Kalberg, W. O., Provost, B., Tollison, S. J., Tabachnick, B. G., Robinson, L. K., Eugene Hoyme, H., & May, P. A. (2006). Comparison of motor delays in young children with fetal alcohol syndrome to those with prenatal alcohol exposure and with no prenatal alcohol exposure. *Alcoholism: Clinical & Experimental Research, 30*(12), 2037–2045.

Kohler, C. G. M. D., Ances, B. M., Coleman, R. A., Ragland, D. J., Lazarev, M. B. S., & Gur, R. C. (2000). Marchiafava–Bignami disease: Literature review and case report. *Neuropsychiatry, Neuropsychology, & Behavioral Neurology, 13*(1), 67–76.

Kopelman, M. D. (1991). Frontal dysfunction and memory deficits in the alcoholic Korsakoff syndrome and Alzheimer-type dementia. *Brain : A Journal of Neurology, 114*(1), 117–137.

Kopelman, M. D., Thomson, A. D., Guerrini, I., & Marshall, E. J. (2009). The Korsakoff syndrome: Clinical aspects, psychology and treatment. *Alcohol and Alcoholism, 44*(2), 148–154. http://dx.doi.org/10.1093/alcalc/agn118.

Kumar, S., Fowler, M., Gonzalez-Toledo, E., & Jaffe, S. L. (2006). Central pontine myelinolysis, an update. *Neurological Research, 28*(3), 360–366.

Laureno, R., & Karp, B. I. (1997). Myelinolysis after correction of hyponatremia. *Annals of Internal Medicine, 126*(1), 57–62.

Lechevalier, B., Andersson, J. C., & Morin, P. (1977). Hemispheric disconnection syndrome with a "crossed avoiding" reaction in a case of Marchiafava–Bignami disease. *Journal of Neurology, Neurosurgery, and Psychiatry, 40*(5), 483–497.

Livy, D. J., Miller, E. K., Maier, S. E., & West, J. R. (2003). Fetal alcohol exposure and temporal vulnerability: Effects of binge-like alcohol exposure on the developing rat hippocampus. *Neurotoxicology and Teratology, 25*(4), 447–458.

Martin, R. J. (2004). Central pontine and extrapontine myelinolysis: The osmotic demyelination syndromes. *Journal of Neurology, Neurosurgery & Psychiatry, 75*(3), 22–28.

Montandon, C., Ribeiro, F. A. D. S., Lobo, L. V. B., Montandon Júnior, M. E., & Teixeira, K.-I.-S. S. (2003). Disgenesia do corpo caloso e más-formações associadas: achados de tomografia computadorizada e ressonância magnética. *Radiologia Brasileira, 36*(5), 311–316.

Murawski, N. J., Moore, E. M., Thomas, J. D., & Riley, E. P. (2015). Advances in diagnosis and treatment of fetal alcohol spectrum disorders: From animal models to human studies. *Alcohol Research : Current Reviews, 37*(1), 97–108.

Musana, K., & Yale, S. H. (2005). Central pontine myelinolysis: Case series and review. *WMJ: Official Publication of the State Medical Society of Wisconsin, 104*(6), 56–60.

Norberg, Å., Jones, W. A., Hahn, R. G., & Gabrielsson, J. L. (2003). Role of variability in explaining ethanol pharmacokinetics: Research and forensic applications. *Clinical Pharmacokinetics, 42*(1), 1–31.

Pearce, J. M. S. (2009). Central pontine myelinolysis. *European Neurology, 61*(1), 59–62.

Pennington, B. F., & Ozonoff, S. (1996). Executive functions and developmental psychopathology. *Journal of Child Psychology and Psychiatry, and Allied Disciplines, 37*(1), 51–87.

Pitel, A. L., Beaunieux, H., Witkowski, T., Vabret, F., De La Sayette, V., Viader, F., & Eustache, F. (2008). Episodic and working memory deficits in alcoholic Korsakoff patients: The continuity theory revisited. *Alcoholism: Clinical and Experimental Research, 32*(7), 1229–1241.

Qin, L., & Crews, F. T. (2014). Focal thalamic degeneration from ethanol and thiamine deficiency is associated with neuroimmune gene induction, microglial activation, and lack of monocarboxylic acid transporters. *Alcoholism: Clinical and Experimental Research, 38*(3), 657–671.

Rawat, J. P., Pinto, C., Kulkarni, K. S., Muthusamy, M. A. K., & Dave, M. D. (2014). Marchiafava–Bignami disease possibly related to consumption of a locally brewed alcoholic beverage: Report of two cases. *Indian Journal of Psychiatry, 56*(1), 76–78.

Riley, E. P., Infante, M. A., & Warren, K. R. (2011). Fetal alcohol spectrum disorders: An overview. *Neuropsychology Review, 21*(2), 73–80.

Roebuck, T. M., Simmons, R. W., Mattson, S. N., & Riley, E. P. (1998). Prenatal exposure to alcohol affects the ability to maintain postural balance. *Alcoholism: Clinical and Experimental Research, 22*(1), 252–258.

Schonfeld, M., Mattson, S. N., Lang, R., Delis, D. C., & Riley, E. P. (2001). Verbal and nonverbal fluency in children with heavy prenatal alcohol exposure. *Journal of Studies on Alcohol, 62*(2), 239–246.

Shiota, J. I., Nakano, I., Kawamura, M., & Hirayama, K. (1996). An autopsy case of Marchiafava–Bignami disease with peculiar chronological CT changes in the corpus callosum: Neuroradiopathological correlations. *Journal of the Neurological Sciences, 136*(1–2), 90–93.

Spadoni, A. D., Bazinet, A. D., Fryer, S. L., Tapert, S. F., Mattson, S. N., & Riley, E. P. (2009). BOLD response during spatial working memory in youth with heavy prenatal alcohol exposure. *Alcoholism: Clinical and Experimental Research, 33*(12), 2067–2076.

Sterns, R. H., Silver, S., Kleinschmidt-DeMasters, B. K., & Rojiani, A. M. (2007). Current perspectives in the management of hyponatremia: Prevention of CPM. *Expert Review of Neurotherapeutics, 7*(12), 1791–1797.

Stratton, K., Howe, C., & Bataglia, F. (1996). *Fetal alcohol syndrome: Diagnosis, epidemiology, prevention, and treatment.* Washington, DC: National Academy Press.

Sullivan, E. V., & Pfefferbaum, A. (2009). Neuroimaging of the Wernicke–Korsakoff syndrome. *Alcohol and Alcoholism: International Journal of the Medical Council on, 44*, 155–165.

Svanberg, J., & Evans, J. J. (2013). Neuropsychological rehabilitation in alcohol-related brain damage: A systematic review. *Alcohol and Alcoholism, 48*(6), 704–711.

Swift, R. (2003). Direct measurement of alcohol and its metabolites. *Addiction, 98*(2), 73–80.

Tomlinson, B. E., Pierides, A. M., & Bradley, W. G. (1976). Central pontine myelinolysis. Two cases with associated electrolyte disturbance. *The Quarterly journal of medicine, 45*(179), 373–386.

Tuntiyatorn, L., & Laothamatas, J. (2008). Acute Marchiafava–Bignami disease with callosal, cortical, and white matter involvement. *Emergency Radiology, 15*(2), 137–140.

Uchino, A., Yuzuriha, T., Murakami, M., Endoh, K., Hiejima, S., Koga, H., & Kudo, S. (2003). Magnetic resonance imaging of sequelae of central pontine myelinolysis in chronic alcohol abusers. *Neuroradiology, 45*(12), 877–880.

Valk, J., & Van der Knaap, M. S. (1989). Marchiafava–Bignami syndrome. In J. Valk, & M. S. van der Knaap (Eds.), *Magnetic resonance of myelin, myelination and myelin disorders* (pp. 265–267). Berlin: Springer-Verlag.

Volkow, N. D., Kim, S. W., Wang, G. J., Alexoff, D., Logan, J., Muench, L., … Tomasi, D. (2013). Acute alcohol intoxication decreases glucose metabolism but increases acetate uptake in the human brain. *NeuroImage, 64*, 277–283.

Wenz, H., Eisele, P., Artemis, D., Forster, A., & Brockmann, M. A. (2014). Acute Marchiafava–Bignami disease with extensive diffusion restriction and early recovery: Case report and review of the literature. *Journal of Neuroimaging, 24*(4), 421–424.

Wilkinson, T. J., Begg, E. J., Winter, A. C., & Sainsbury, R. (1999). Incidence and risk factors for hyponatraemia following treatment

with fluoxetine or paroxetine in elderly people. *British Journal of Clinical Pharmacology, 47*(2), 211–217.
World Health Organisation. (2014). *Global status report on alcohol and health 2014*. http://doi.org//entity/substance_abuse/publications/global_alcohol_report/en/index.html.
Wright, D. G., Laureno, R., & Victor, M. (1979). Pontine and extrapontine myelinolisis. *Brain, 102*, 361–385.
Yang, L., Qin, W., Xu, J.-H., & Hu, W.-L. (2015). Letter to the editor Marchiafava–Bignami disease with asymmetric extracallosal lesions. *Archives of Medical Science, 4*, 895–898.
Zelner, I., & Koren, G. (2013). Pharmacokinetics of ethanol in the maternal-fetal unit. *Journal of Population Therapeutics and Clinical Pharmacology, 20*(3), 259–265.
Zubaran, C., Fernandes, J. G., & Rodnight, R. (1997). Wernicke–Korsakoff syndrome. *Postgraduate Medical Journal, 73*, 27–31.
Zuccoli, G., & Pipitone, N. (2009). Neuroimaging findings in acute Wernicke's encephalopathy: Review of the literature. *AJR, 192*, 501–508.

CHAPTER

15

Frontal Lobe Dysfunction After Developmental Alcohol Exposure: Implications From Animal Models

Z.H. Gursky, A.Y. Klintsova

University of Delaware, Newark, DE, United States

INTRODUCTION

Among the leading causes of preventable anatomical and neurological birth defects, fetal alcohol spectrum disorders (FASDs) are estimated to affect 2–5% of live births in the United States (Cannon et al., 2015; May et al., 2014) whereas international estimates are much higher, reaching over 10% in Australia and South Africa (Roozen et al., 2016). Critically, the fiscal and labor cost estimates for care of FASD-afflicted individuals have been steadily rising, with a median annual cost estimate of over 3 billion dollars in the United States alone—including an estimated cumulative cost of 2 million dollars per individual by age 65 (Lupton, Burd, & Harwood, 2004). To comprehensively and appropriately characterize the existence and severity of atypical development following prenatal alcohol exposure (PAE), multiple diagnostic domains should be addressed (i.e., psychological, neuropsychological, pediatric, and educational) (Hoyme et al., 2005); the many manifestations of FASD have been outlined in the literature up to this point (Astley & Clarren, 2000). In light of this epidemic's severity and prevalence, this chapter summarizes the current views of the implications of FASD on human cognition (both behaviorally and neuroanatomically), and review the cutting-edge research in animal models to both characterize the mechanisms of these deficits and design interventions to ameliorate these deficits throughout the lifespan.

FASDs IN HUMANS

Behavioral Consequences

Individuals afflicted with FASD have a comprehensive array of behavioral deficits, dependent on a variety of factors including the severity and frequency of PAE (Astley & Clarren, 2000). Although a four-digit code allows for the diagnostically relevant characterization of FASD, broader categories such as FAS, partial FAS (pFAS), alcohol-related birth defects (ARBD), and alcohol-related neurodevelopmental disorder (ARND), remain active terms within the literature (Chasnoff, Wells, Telford, Schmidt, & Messer, 2010; Riley, Infante, & Warren, 2011; Warren & Foudin, 2001).

While more severe FASDs (i.e., FAS, pFAS, and ARBD) often result in a range of physical or anatomical deficits, individuals with any diagnosable FASD are at odds for severe cognitive deficits at some point in the lifespan (reviewed in Gibbard, Wass, & Clarke, 2003; Harper, 2009; May et al., 2014). These deficits can manifest in numerous aspects of behavior including executive functioning (Bertrand & Interventions for Children with Fetal Alcohol Spectrum Disorders Research, 2009; Rasmussen & Bisanz, 2009), behavioral inhibition (Bertrand & Interventions for Children with Fetal Alcohol Spectrum Disorders Research, 2009; Coles, Kable, Taddeo, & Strickland, 2015; Paolozza, Treit, Beaulieu, & Reynolds, 2014; Ware et al., 2015), verbal skills (Adnams et al., 2007; Rasmussen

Addictive Substances and Neurological Disease
http://dx.doi.org/10.1016/B978-0-12-805373-7.00015-3

& Bisanz, 2009), and mathematical skills (Bertrand & Interventions for Children with Fetal Alcohol Spectrum Disorders Research, 2009; Lebel, Rasmussen, Wyper, Andrew, & Beaulieu, 2010) and may persist into adulthood if unaddressed (Moore & Riley, 2015).

As mentioned earlier, the most severe manifestation of PAE is FAS; this manifests a host of gross anatomical and neuropsychiatric abnormalities not typically present in other instances of FASD (Burd, Klug, Martsolf, & Kerbeshian, 2003; Chasnoff et al., 2010). The detrimental effects of FAS on executive function exist irrespective of decreases in IQ, with PAE retaining higher predictive power than IQ on a battery of tests of executive function (Connor, Sampson, Bookstein, Barr, & Streissguth, 2000). While pFAS and ARND are characterized as less severe than FAS, they still display increased risk for neuropsychiatric impairment (Burd et al., 2003).

Current attempts to minimize the impact of FASD-induced deficits in humans include both pharmacological and behavioral interventions (Peadon, Rhys-Jones, Bower, & Elliott, 2009), as well as preventative nutritional supplements for mothers with at-risk pregnancies (Coles, Kable, Keen, et al., 2015). Comprehensive characterization of an FASD-afflicted individual's learning profile should be assessed before prescribing the appropriate intervention (Kalberg & Buckley, 2007). At this point in time, behavioral interventions often target a specific deficit (Paley & O'Connor, 2011) however, this approach requires precise diagnosis of the individual's deficits, and not all impairments may be apparent at the time of diagnosis. Recent animal research has led to the development of promising interventions to ameliorate a broader range of cognitive and behavioral deficits, but due to the lack of clinical data, they are addressed with the animal literature to which they belong, later in this chapter.

Neurological Consequences

Exposure to alcohol during gestation can lead to significant decreases in total brain weight, a widely recognized diagnostic feature of severe FASDs (i.e., FAS, pFAS) (Clarren, Alvord, Sumi, Streissguth, & Smith, 1978; Ferrer & Galofre, 1987). This decrease in brain weight is accompanied by a decreased size (Wass, Persutte, & Hobbins, 2001) and cortical thickness (Sowell, Mattson, et al., 2008) of the frontal lobe, the latter of which correlates with performance on cognitive tests of verbal and visuospatial functioning (Sowell, Mattson, et al., 2008). Animal models (to be discussed later) have revealed that this reduction in brain size is due to both increases in cell death and impaired growth and maturation of new neurons. In addition to reductions in brain size, abnormal development of the corpus callosum is robustly observable even with imaging techniques that are relatively low resolution in the field of neuroimaging (i.e., ultrasound) (Bookstein et al., 2005, 2007).

These highly pronounced gross neuroanatomical abnormalities are accompanied (and likely mediated) by fine structural alterations to dendritic spines in neurons, with robust alterations to cortical regions (Ferrer & Galofre, 1987). During 2000s, there have been dramatic advances in neuroimaging, allowing clinicians and researchers to noninvasively characterize brain connectivity through the use of diffusion tensor imaging (DTI) and function using techniques such as functional magnetic resonance imaging (fMRI), functional near-infrared spectroscopy (fNIRS), and electroencephalography (EEG), resulting in an unprecedented depiction of the neurological underpinnings of FASD (Ehlis, Schneider, Dresler, & Fallgatter, 2014; Norman, Crocker, Mattson, & Riley, 2009).

The neuroanatomical bases of FASD-induced impairment in humans have been characterized extensively with recent imaging techniques (a substantial amount of this literature is reviewed in Norman et al., 2009). These studies corroborate previous findings that individuals afflicted with FASD have abnormal brain volume and densities; however, the reports on directionality of these changes are inconsistent. Alterations in connectivity are analyzed in vivo through the use of DTI, and reveal decreased efficiency of signal transfer (functional anisotropy, or FA) in white matter tracts that connect the frontal lobe with other cortical regions responsible for both visual perception and cognition (Lebel et al., 2010; Paolozza et al., 2014; Sowell, Johnson, et al., 2008), and increased frontoparietal activity during some working memory-dependent tasks (O'Hare et al., 2009), indicating aberrant connectivity of cortical regions following PAE. Developmental abnormalities in this circuit can lead to impaired processing speed (Turken et al., 2008), another hallmark of individuals with FASDs. Alterations to frontal tracts of the corpus callosum are implicated (in conjunction with cerebellar damage) to cause delayed reaction time on visual attention tasks (Green, Lebel, Rasmussen, Beaulieu, & Reynolds, 2013). It is apparent that structural changes measured with DTI must be cross-examined with functional measures, since FASD-induced changes in connectivity may alter brain microcircuits in a way that is not easily interpretable by structural analysis alone (Wozniak et al., 2006).

Research into novel functional recording techniques has developed to a tool that can be integrated into clinical evaluation of disorders: fNIRS (Ehlis et al., 2014). This technique has not been used extensively with FASD-afflicted individuals, but the combination of decreased price and increased resolution of fNIRS in recent years have shown promise for its use (Ferrari &

Quaresima, 2012). As mentioned earlier, the ability to correlate structural, behavioral, and functional evidence will help to a better understanding of microcircuit connectivity in the human brain following PAE. The mechanisms of cellular and subcellular changes that precipitate the structural and functional impairments in these imaging studies is derived from rodent models of FASDs.

ANIMAL MODELS OF FASDs

Methods of Alcohol Delivery

In contrast to the correlative nature of human studies on FASD where alcohol consumption is estimated only approximately (due to logical limitations), animal studies allow stringent regulation of the dose, time, and frequency of alcohol exposure in models of FASDs. While the "brain growth spurt," a period of rapid brain growth across multiple measures, occurs during the second and especially third trimesters of human development, it takes place across late prenatal development and the subsequent two postnatal weeks of the rat (Dobbing & Sands, 1979). Because of this, different methods of alcohol delivery address the effects of alcohol exposure on different stages of development. Models of FASD that aim to target early gestational development (analogous to exposure that results in FAS in humans) require prenatal exposure. Two common avenues of exposure are intragastric gavage of the pregnant dam (e.g., Helfer, White, & Christie, 2012) and maternal self-administration (e.g., Brady, Allan, & Caldwell, 2012). A wide variety of postnatal exposure paradigms are currently used including intragastric intubation (e.g., Helfer et al., 2009), vapor chamber (e.g., De Giorgio & Granato, 2015), and injection (intraperitoneal: Medina & Ramoa, 2005; subcutaneous: Young & Olney, 2006). The mechanics, benefits, and limitations of these different techniques have been thoroughly reviewed in the literature over the past several years (Kelly & Lawrence, 2008; Patten, Fontaine, & Christie, 2014). As mentioned previously, critical considerations when addressing the translatability of animal studies on FASDs are the severity and frequency of exposure (Bonthius & West, 1988, 1990; Smith, Guevremont, Williams, & Napper, 2015; Thomas, Wasserman, West, & Goodlett, 1996), as well as the developmental age at exposure (Goodlett & Lundahl, 1996; Ikonomidou et al., 2000; Thomas et al., 1996) to alcohol. The literature tries to be as consistent as possible with its interpretation of blood alcohol content and patterns of exposure relative to the human conditions that the models aim to replicate. As a result, the same model is often used across multiple laboratories.

Comparative Prefrontal Cortex Neuroanatomy

It is not surprising that researchers question the accuracy of rodent models as human analogues of cognitive behaviors and frontal cortex activity. Debates in favor of the relevance of the rodent prefrontal cortex (PFC) have been driven by similar neuroanatomical connectivity and behavioral function of the subdivisions of the PFC (Bissonette, Powell, & Roesch, 2013; Heidbreder & Groenewegen, 2003; Kesner & Churchwell, 2011; Preston & Eichenbaum, 2013). Therefore, it is important to understand the similarities and differences between rodent and human PFC. Any comparisons across species must carefully consider both cytoarchitectonic (neuroanatomical) similarities as well as functional roles across species rather than the anatomical locations or names of the regions.

Structural Deficits in Rodent Studies

Some of the most fundamental questions that rodent studies examine are region-specific changes in cell death, cell survival, and cell phenotyping. It is well accepted that neonatal alcohol exposure leads to widespread cell death (particularly apoptosis, or programmed cell death) in the central nervous system (reviewed in Creeley & Olney, 2013) and that this apoptosis occurs in a time- and region-specific manner (Bonthius & West, 1990; Ikonomidou et al., 2000). The PFC is particularly sensitive to alcohol insult during the rodent analogue of the human third trimester (particularly in the second postnatal week) (Ikonomidou et al., 2000). In addition to short-term increases in apoptosis, rodent models of FASD indicate significant cell number deficits in layers II and V of medial PFC (mPFC) extending into adulthood (Mihalick, Crandall, Langlois, Krienke, & Dube, 2001). A complete profile of cell death in the PFC does not exist at this point. Although widespread apoptosis is observed shortly after neonatal alcohol exposure and long-term cell deficits exist into adulthood, it is still unknown what types of cells are predominantly lost following PAE.

Although there is still uncertainty whether different populations of cells are more sensitive to alcohol-induced apoptosis in PFC, it has been well characterized that developmental alcohol exposure can lead to aberrant cortico-cortical connectivity (De Giorgio & Granato, 2015; Granato, Di Rocco, Zumbo, Toesca, & Giannetti, 2003; Hamilton, Whitcher, & Klintsova, 2010). Function is likely affected on the synaptic level as well. Lasting changes in long-term potentiation (LTP) of hippocampal pyramidal cells are observed following neonatal exposure to alcohol (Puglia & Valenzuela, 2010), while LTP is transiently increased in the mPFC following chronic alcohol exposure in adulthood (Kroener et al., 2012). Two critical mechanisms of effects of alcohol on neuronal neurophysiology are its agonism of gamma amino butyric acid ($GABA_A$) receptors and antagonism of N-methyl-D-aspartate (NMDA) glutamate receptors, comprehensively discussed by Ikonomidou and Olney (Creeley & Olney, 2013; Ikonomidou et al., 2000; Olney, 2014; Olney, Ishimaru, Bittigau, & Ikonomidou, 2000). Alcohol agonism of $GABA_A$ also affects neural development by premature development of neuronal networks in the neonatal brain (Galindo, Zamudio, & Valenzuela, 2005).

One mechanism by which alcohol severely impacts the developing brain is through neuroimmune activation and modulation (Ahlers, Karacay, Fuller, Bonthius, & Dailey, 2015; Boschen, Ruggiero, & Klintsova, 2016; Drew, Johnson, Douglas, Phelan, & Kane, 2015). The neuroimmune system consists of a population of immune cells (microglia) that are generated in the periphery during gestation, then colonize the brain and assess it for homeostatic disruption (Nimmerjahn, Kirchhoff, & Helmchen, 2005). Microglia phagocytose inactive presynaptic elements (Schafer et al., 2012) and apoptotic neurons (Fricker, Oliva-Martin, & Brown, 2012) to allow for further neurogenic events in some regions of the brain (Sierra et al., 2010). The body of literature implicating alcohol-induced neuroimmune alterations in the PFC is less extensive than the literature regarding this phenomenon in the hippocampus and cerebellum. Nonetheless, alterations to both pro- and anti-inflammatory cytokines are observed in the PFC following prenatal alcohol exposure (Bodnar, Hill, & Weinberg, 2016). Developmental alcohol exposure can lead to atypical activation of microglia that results in aberrant maintenance of the developing parenchyma and aberrant patterns of synaptic pruning, leading to abnormal patterns of cell survival and cortical thinning throughout the lifespan (Abbott, Kozanian, Kanaan, Wendel, & Huffman, 2016; El Shawa, Abbott, & Huffman, 2013; Mihalick et al., 2001). Developmentally atypical neuroimmune responses increase likelihood of neuropathologies that develop later in life following a subsequent neuroimmune or stress-induced "second hit" (Comeau, Lee, Anderson, & Weinberg, 2015; Williamson, Sholar, Mistry, Smith, & Bilbo, 2011).

Another critical mechanism of alcohol-induced developmental impairment is altered expression of neurotropic factors throughout the lifespan. Neonatal alcohol exposure leads to short-term increases in proapoptotic factors and decreases in antiapoptotic factors (Smith et al., 2015; Young et al., 2003). After an initial increase in ratios of proapoptotic (vs. antiapoptotic) factors, there is a region-specific change in the expression of neuroprotective factors, both short term (Boschen, Criss, Palamarchouk, Roth, & Klintsova, 2015) and later in life (Heaton, Paiva, Madorsky, & Shaw, 2003).

Recent studies have demonstrated significant effects of PAE on the mesocorticolimbic pathway (originating in the ventral tegmental area and projecting to nucleus accumbens and PFC) (Fabio et al., 2013; Fabio, Vivas, & Pautassi, 2015). PAE leads to significant changes in gene expression in PFC, including genes involved in apoptosis and neural development (Lussier et al., 2015).

Rodent models of FASD have illustrated the multifaceted mechanisms of alcohol insult that contribute to long-term brain impairment on a structural and subcellular level. As stated earlier, apoptosis is a robust mechanism by which alcohol can directly damage the central nervous system and PFC. Alcohol also impairs the function of synapses leading to inappropriate connectivity and neuroimmune dysregulation in the PFC. With these principles in mind, we discuss the work that has been done in classifying behavioral impairments in rodent models of FASD and their relation to these structural deficits and the human literature.

Behavioral Impairments in Rodents

As the field develops a more comprehensive understanding of functional and neuroanatomical alterations due to PAE, behavioral assays must become more sophisticated to appropriately model deficits seen in FASD-afflicted humans. In spite of the large volume of human literature indicating impairment on a variety of frontal lobe–dependent cognitive functions (reviewed earlier), currently the rodent literature on PFC-dependent behavior after developmental exposure to alcohol is lacking both depth and width. One major reason for this is the difficulty of modeling rodent behavioral tasks that accurately represent and map onto human "cognitive" assessments and neuroanatomically similar regions (reviewed in Kesner & Churchwell, 2011). The attentional set shifting task (ASST) has been developed to address the components of human "cognition" in animal models (reviewed in Bissonette et al., 2013). The specific processes that the task assesses are discrimination learning, reversal learning, and cognitive flexibility. There are multiple variants of the task, each contributing uniquely to our understanding of the role of the frontal lobe due to subtle differences in the task itself. The following section discusses behavioral impairments to PFC-dependent behaviors in the context of the components of the ASST. It is critical to note, however, that the field's understanding of behavioral impairments due to PFC damage following developmental alcohol exposure is not comprehensive at this point in time. This is due in part to the extreme heterogeneous connectivity between the subregions that comprise the PFC. Additionally, the PFC is capable of fulfilling the role of other brain structures following damage (e.g., Zelikowsky et al., 2013), illustrating potentially redundant functions of the PFC in addition to unique roles of its subregions (e.g., Rajasethupathy et al., 2015). Further understanding of the specific neural substrates of the PFC that are damages following PAE and their direct behavioral implications is paramount to our understanding of FASDs in humans.

Due to the increasing popularity of the ASST, a review in 2013 highlighted the multitude of task variants and their implementation (e.g., Bissonette et al., 2013). In short, the task is a forced-choice operant condition paradigm. After the animal is habituated to the arena and exposed to the rewarding stimulus (typically a highly rewarding food), the animal is taught to discriminate between two different contingencies to receive the food reward. Different contingencies can include turning right versus left in a plus maze, or digging in a cup filled with aquarium gravel versus plastic grass in a different variant. To test reversal learning, the previously rewarded contingency is no longer rewarded while the previously unrewarded contingency becomes rewarded (aquarium gravel being rewarding, whereas plastic grass was previously rewarded). Additionally, the ASST allows the experimenter to assess "cognitive flexibility," or the animal's ability to switch between rewarded and previously unrewarded rules rather than individual contingencies (e.g., the scent of the cup is now rewarding and the digging medium is no longer an accurate predictor of reward). This task is ideal for animal models of FASD due to its combined assessment of components of executive function defunct in humans with FASD (discussed in Behavioral Consequences section in FASDs in Humans section).

As mentioned in the previous paragraph, few publications address the connection between PFC damage and these components of executive function. The first study directly linking neuroanatomical deficits in the mPFC of rats, and reversal learning indicate a strong link between neuronal number and performance on reversal learning tasks in adulthood following prenatal ethanol exposure (Mihalick et al., 2001). Although reviews use timing and severity of exposure to group studies (e.g., Marquardt & Brigman, 2016), more work must be done to address the direct neuroanatomical and functional alterations by which developmental alcohol exposure affects these complex, rule-directed learning processes later in life. It is possible that first- versus third-trimester rodent models may be damaging the critical structure versus a downstream structure for these behaviors, and thus extensive neuroanatomical causality has yet to be demonstrated using this paradigm.

CONCLUSION

It is widely accepted that exposure to alcohol during development (in utero in humans, in utero and early postnatal periods in rodents) causes widespread damage to the central nervous system and frontal lobe. The magnitude of damage is dependent on the frequency, severity, and timing of exposure to alcohol as well as the region being examined. Traditionally, human studies have focused on a wide variety of diagnostic criteria. Recent advances in imaging techniques have led to an unprecedented ability to monitor neuroanatomical correlates of FASDs. Supplementing the correlational human research, animal models have been critical in establishing cellular and subcellular mechanisms underlying short- and long-term damage to the developing brain. The behavioral assays utilized in animal models of fetal alcohol exposure during pregnancy have been limited in their ability to model complex cognitive deficits observed in humans until the recent development of more comprehensive behavioral assays. Between advances in technology that have expanded our understanding of both humans and animal development following developmental alcohol exposure, we should now direct attention toward amelioration of these debilitating conditions.

The human literature has indicated the potential for interventions capable of ameliorating select deficits seen in individuals with FASDs (Adnams et al., 2007; Bertrand & Interventions for Children with Fetal Alcohol Spectrum Disorders Research, 2009; Coles, Kable, Taddeo, et al., 2015; Hallman, 2012; Kalberg & Buckley, 2007; Mandryk et al., 2013; Peadon et al., 2009). The rodent literature can assist in the development of novel therapeutic interventions, with interventions capable of rescuing certain neuroanatomical and behavioral deficits (reviewed in Klintsova, Hamilton, & Boschen, 2012). Choline supplementation is one effective pharmacological treatment that provided partial attenuation of behavioral deficits (including improvement of working memory) resulting from PAE (Schneider & Thomas, 2016; Thomas, Idrus, Monk, & Dominguez, 2010). Cardiovascular activity and enriched environments have been shown to provide neuroprotective (Brockett, LaMarca, & Gould, 2015; Klintsova, Dickson, Yoshida, & Greenough, 2004) and behavioral benefits (Brockett et al., 2015; Sampedro-Piquero, Zancada-Menendez, & Begega, 2015); and have shown promise in ameliorating both structural and behavioral deficits seen in rodent models of FASDs (Hamilton, Boschen, Goodlett, Greenough, & Klintsova, 2012; Hamilton, Criss, & Klintsova, 2015; Hamilton et al., 2014; Klintsova et al., 2002; Thomas, Sather, & Whinery, 2008). It is important to note that although neuroanatomical abnormalities may recover (to a certain extent), not all functional or behavioral correlates may be equally rescued (Abbott et al., 2016). The only absolute intervention to minimize the impact of FASDs in society is prevention; abstinence from alcohol consumption during pregnancy or while attempting to become pregnant is the only guaranteed preventative measure for FASD. Although the prevalence of FASD in the USA and abroad is high, education and prevention are the first steps to treating this epidemic. Additional collaboration is necessary between rodent and clinical researchers on the topic to develop interventions and management plans to maximize the quality of life for the high number of individuals afflicted by this preventable phenomenon.

References

Abbott, C. W., Kozanian, O. O., Kanaan, J., Wendel, K. M., & Huffman, K. J. (2016). The impact of prenatal ethanol exposure on neuroanatomical and behavioral development in mice. *Alcoholism: Clinical and Experimental Research, 40*(1), 122–133. http://dx.doi.org/10.1111/acer.12936.

Adnams, C. M., Sorour, P., Kalberg, W. O., Kodituwakku, P., Perold, M. D., Kotze, A., ... May, P. A. (2007). Language and literacy outcomes from a pilot intervention study for children with fetal alcohol spectrum disorders in South Africa. *Alcohol, 41*(6), 403–414. http://dx.doi.org/10.1016/j.alcohol.2007.07.005.

Ahlers, K. E., Karacay, B., Fuller, L., Bonthius, D. J., & Dailey, M. E. (2015). Transient activation of microglia following acute alcohol exposure in developing mouse neocortex is primarily driven by BAX-dependent neurodegeneration. *Glia, 63*(10), 1694–1713. http://dx.doi.org/10.1002/glia.22835.

Astley, S. J., & Clarren, S. K. (2000). Diagnosing the full spectrum of fetal alcohol-exposed individuals: Introducing the 4-digit diagnostic code. *Alcohol and Alcoholism, 35*(4), 400–410.

Bertrand, J., & Interventions for Children with Fetal Alcohol Spectrum Disorders Research, C. (2009). Interventions for children with fetal alcohol spectrum disorders (FASDs): Overview of findings for five innovative research projects. *Research in Developmental Disabilities, 30*(5), 986–1006. http://dx.doi.org/10.1016/j.ridd.2009.02.003.

Bissonette, G. B., Powell, E. M., & Roesch, M. R. (2013). Neural structures underlying set-shifting: Roles of medial prefrontal cortex and anterior cingulate cortex. *Behavioural Brain Research, 250*, 91–101. http://dx.doi.org/10.1016/j.bbr.2013.04.037.

Bodnar, T. S., Hill, L. A., & Weinberg, J. (2016). Evidence for an immune signature of prenatal alcohol exposure in female rats. *Brain, Behavior, and Immunity.* http://dx.doi.org/10.1016/j.bbi.2016.05.022.

Bonthius, D. J., & West, J. R. (1988). Blood alcohol concentration and microencephaly: A dose-response study in the neonatal rat. *Teratology, 37*(3), 223–231. http://dx.doi.org/10.1002/tera.1420370307.

Bonthius, D. J., & West, J. R. (1990). Alcohol-induced neuronal loss in developing rats: Increased brain damage with binge exposure. *Alcoholism: Clinical and Experimental Research, 14*(1), 107–118.

Bookstein, F. L., Connor, P. D., Covell, K. D., Barr, H. M., Gleason, C. A., Sze, R. W., ... Streissguth, A. P. (2005). Preliminary evidence that prenatal alcohol damage may be visible in averaged ultrasound images of the neonatal human corpus callosum. *Alcohol, 36*(3), 151–160. http://dx.doi.org/10.1016/j.alcohol.2005.07.007.

Bookstein, F. L., Connor, P. D., Huggins, J. E., Barr, H. M., Pimentel, K. D., & Streissguth, A. P. (2007). Many infants prenatally exposed to high levels of alcohol show one particular anomaly of the corpus callosum. *Alcoholism: Clinical and Experimental Research, 31*(5), 868–879. http://dx.doi.org/10.1111/j.1530-0277.2007.00367.x.

Boschen, K. E., Criss, K. J., Palamarchouk, V., Roth, T. L., & Klintsova, A. Y. (2015). Effects of developmental alcohol exposure vs. intubation stress on BDNF and TrkB expression in the hippocampus and frontal cortex of neonatal rats. *International Journal of Developmental Neuroscience, 43*, 16–24. http://dx.doi.org/10.1016/j.ijdevneu.2015.03.008.

Boschen, K. E., Ruggiero, M. J., & Klintsova, A. Y. (2016). Neonatal binge alcohol exposure increases microglial activation in the developing rat hippocampus. *Neuroscience, 324*, 355–366. http://dx.doi.org/10.1016/j.neuroscience.2016.03.033.

Brady, M. L., Allan, A. M., & Caldwell, K. K. (2012). A limited access mouse model of prenatal alcohol exposure that produces long-lasting deficits in hippocampal-dependent learning and memory. *Alcoholism: Clinical and Experimental Research, 36*(3), 457–466. http://dx.doi.org/10.1111/j.1530-0277.2011.01644.x.

Brockett, A. T., LaMarca, E. A., & Gould, E. (2015). Physical exercise enhances cognitive flexibility as well as astrocytic and synaptic markers in the medial prefrontal cortex. *PLoS One, 10*(5), e0124859. http://dx.doi.org/10.1371/journal.pone.0124859.

Burd, L., Klug, M. G., Martsolf, J. T., & Kerbeshian, J. (2003). Fetal alcohol syndrome: Neuropsychiatric phenomics. *Neurotoxicology and Teratology, 25*(6), 697–705.

Cannon, M. J., Guo, J., Denny, C. H., Green, P. P., Miracle, H., Sniezek, J. E., & Floyd, R. L. (2015). Prevalence and characteristics of women at risk for an alcohol-exposed pregnancy (AEP) in the United States: Estimates from the National Survey of Family Growth. *Maternal and Child Health Journal, 19*(4), 776–782. http://dx.doi.org/10.1007/s10995-014-1563-3.

Chasnoff, I. J., Wells, A. M., Telford, E., Schmidt, C., & Messer, G. (2010). Neurodevelopmental functioning in children with FAS, pFAS, and ARND. *Journal of Developmental and Behavioral Pediatrics, 31*(3), 192–201. http://dx.doi.org/10.1097/DBP.0b013e3181d5a4e2.

Clarren, S. K., Alvord, E. C., Jr., Sumi, S. M., Streissguth, A. P., & Smith, D. W. (1978). Brain malformations related to prenatal exposure to ethanol. *The Journal of Pediatrics, 92*(1), 64–67.

Coles, C. D., Kable, J. A., Keen, C. L., Jones, K. L., Wertelecki, W., Granovska, I. V., ... CIFASD. (2015). Dose and timing of prenatal alcohol exposure and maternal nutritional supplements: Developmental effects on 6-Month-Old infants. *Maternal and Child Health Journal, 19*(12), 2605–2614. http://dx.doi.org/10.1007/s10995-015-1779-x.

Coles, C. D., Kable, J. A., Taddeo, E., & Strickland, D. C. (2015). A metacognitive strategy for reducing disruptive behavior in children with fetal alcohol spectrum disorders: GoFAR pilot. *Alcoholism: Clinical and Experimental Research, 39*(11), 2224–2233. http://dx.doi.org/10.1111/acer.12885.

Comeau, W. L., Lee, K., Anderson, K., & Weinberg, J. (2015). Prenatal alcohol exposure and adolescent stress increase sensitivity to stress and gonadal hormone influences on cognition in adult female rats. *Physiology & Behavior, 148*, 157–165. http://dx.doi.org/10.1016/j.physbeh.2015.02.033.

Connor, P. D., Sampson, P. D., Bookstein, F. L., Barr, H. M., & Streissguth, A. P. (2000). Direct and indirect effects of prenatal alcohol damage on executive function. *Developmental Neuropsychology, 18*(3), 331–354. http://dx.doi.org/10.1207/S1532694204Connor.

Creeley, C. E., & Olney, J. W. (2013). Drug-induced apoptosis: Mechanism by which alcohol and many other drugs can disrupt brain development. *Brain Sciences, 3*(3), 1153–1181. http://dx.doi.org/10.3390/brainsci3031153.

De Giorgio, A., & Granato, A. (2015). Reduced density of dendritic spines in pyramidal neurons of rats exposed to alcohol during early postnatal life. *International Journal of Developmental Neuroscience, 41*, 74–79. http://dx.doi.org/10.1016/j.ijdevneu.2015.01.005.

Dobbing, J., & Sands, J. (1979). Comparative aspects of the brain growth spurt. *Early Human Development, 3*(1), 79–83.

Drew, P. D., Johnson, J. W., Douglas, J. C., Phelan, K. D., & Kane, C. J. (2015). Pioglitazone blocks ethanol induction of microglial activation and immune responses in the hippocampus, cerebellum, and cerebral cortex in a mouse model of fetal alcohol spectrum disorders. *Alcoholism: Clinical and Experimental Research, 39*(3), 445–454. http://dx.doi.org/10.1111/acer.12639.

Ehlis, A. C., Schneider, S., Dresler, T., & Fallgatter, A. J. (2014). Application of functional near-infrared spectroscopy in psychiatry. *NeuroImage, 85*(Pt 1), 478–488. http://dx.doi.org/10.1016/j.neuroimage.2013.03.067.

El Shawa, H., Abbott, C. W., 3rd, & Huffman, K. J. (2013). Prenatal ethanol exposure disrupts intraneocortical circuitry, cortical gene expression, and behavior in a mouse model of FASD. *Journal of Neuroscience, 33*(48), 18893–18905. http://dx.doi.org/10.1523/JNEUROSCI.3721-13.2013.

Fabio, M. C., March, S. M., Molina, J. C., Nizhnikov, M. E., Spear, N. E., & Pautassi, R. M. (2013). Prenatal ethanol exposure increases ethanol intake and reduces c-Fos expression in infralimbic cortex of adolescent rats. *Pharmacology, Biochemistry, and Behavior, 103*(4), 842–852. http://dx.doi.org/10.1016/j.pbb.2012.12.009.

Fabio, M. C., Vivas, L. M., & Pautassi, R. M. (2015). Prenatal ethanol exposure alters ethanol-induced Fos immunoreactivity and dopaminergic activity in the mesocorticolimbic pathway of the adolescent brain. *Neuroscience, 301*, 221–234. http://dx.doi.org/10.1016/j.neuroscience.2015.06.003.

Ferrari, M., & Quaresima, V. (2012). A brief review on the history of human functional near-infrared spectroscopy (fNIRS) development and fields of application. *NeuroImage, 63*(2), 921–935. http://dx.doi.org/10.1016/j.neuroimage.2012.03.049.

Ferrer, I., & Galofre, E. (1987). Dendritic spine anomalies in fetal alcohol syndrome. *Neuropediatrics, 18*(3), 161–163. http://dx.doi.org/10.1055/s-2008-1052472.

Fricker, M., Oliva-Martin, M. J., & Brown, G. C. (2012). Primary phagocytosis of viable neurons by microglia activated with LPS or Abeta is dependent on calreticulin/LRP phagocytic signalling. *Journal of Neuroinflammation, 9*, 196. http://dx.doi.org/10.1186/1742-2094-9-196.

Galindo, R., Zamudio, P. A., & Valenzuela, C. F. (2005). Alcohol is a potent stimulant of immature neuronal networks: Implications for fetal alcohol spectrum disorder. *Journal of Neurochemistry, 94*(6), 1500–1511.

Gibbard, W. B., Wass, P., & Clarke, M. E. (2003). The neuropsychological implications of prenatal alcohol exposure. *The Canadian Child and Adolescent Psychiatry Review, 12*(3), 72–76.

Goodlett, C. R., & Lundahl, K. R. (1996). Temporal determinants of neonatal alcohol-induced cerebellar damage and motor performance deficits. *Pharmacology, Biochemistry, and Behavior, 55*(4), 531–540.

Granato, A., Di Rocco, F., Zumbo, A., Toesca, A., & Giannetti, S. (2003). Organization of cortico-cortical associative projections in rats exposed to ethanol during early postnatal life. *Brain Research Bulletin, 60*(4), 339–344.

Green, C. R., Lebel, C., Rasmussen, C., Beaulieu, C., & Reynolds, J. N. (2013). Diffusion tensor imaging correlates of saccadic reaction time in children with fetal alcohol spectrum disorder. *Alcoholism: Clinical and Experimental Research, 37*(9), 1499–1507. http://dx.doi.org/10.1111/acer.12132.

Hallman, D. W. (2012). 19-Channel neurofeedback in an adolescent with FASD. *Journal of Neurotherapy, 16*(2), 150–154.

Hamilton, G. F., Boschen, K. E., Goodlett, C. R., Greenough, W. T., & Klintsova, A. Y. (2012). Housing in environmental complexity following wheel running augments survival of newly generated hippocampal neurons in a rat model of binge alcohol exposure during the third trimester equivalent. *Alcoholism: Clinical and Experimental Research, 36*(7), 1196–1204. http://dx.doi.org/10.1111/j.1530-0277.2011.01726.x.

Hamilton, G. F., Criss, K. J., & Klintsova, A. Y. (2015). Voluntary exercise partially reverses neonatal alcohol-induced deficits in mPFC layer II/III dendritic morphology of male adolescent rats. *Synapse, 69*(8), 405–415. http://dx.doi.org/10.1002/syn.21827.

Hamilton, G. F., Jablonski, S. A., Schiffino, F. L., St Cyr, S. A., Stanton, M. E., & Klintsova, A. Y. (2014). Exercise and environment as an intervention for neonatal alcohol effects on hippocampal adult neurogenesis and learning. *Neuroscience, 265*, 274–290. http://dx.doi.org/10.1016/j.neuroscience.2014.01.061.

Hamilton, G. F., Whitcher, L. T., & Klintsova, A. Y. (2010). Postnatal binge-like alcohol exposure decreases dendritic complexity while increasing the density of mature spines in mPFC Layer II/III pyramidal neurons. *Synapse, 64*(2), 127–135. http://dx.doi.org/10.1002/syn.20711.

Harper, C. (2009). The neuropathology of alcohol-related brain damage. *Alcohol and Alcoholism, 44*(2), 136–140. http://dx.doi.org/10.1093/alcalc/agn102.

Heaton, M. B., Paiva, M., Madorsky, I., & Shaw, G. (2003). Ethanol effects on neonatal rat cortex: Comparative analyses of neurotrophic factors, apoptosis-related proteins, and oxidative processes during vulnerable and resistant periods. *Brain Research. Developmental Brain Research, 145*(2), 249–262.

Heidbreder, C. A., & Groenewegen, H. J. (2003). The medial prefrontal cortex in the rat: Evidence for a dorso-ventral distinction based upon functional and anatomical characteristics. *Neuroscience and Biobehavioral Reviews, 27*(6), 555–579.

Helfer, J. L., Calizo, L. H., Dong, W. K., Goodlett, C. R., Greenough, W. T., & Klintsova, A. Y. (2009). Binge-like postnatal alcohol exposure triggers cortical gliogenesis in adolescent rats. *Journal of Comparative Neurology, 514*(3), 259–271. http://dx.doi.org/10.1002/cne.22018.

Helfer, J. L., White, E. R., & Christie, B. R. (2012). Enhanced deficits in long-term potentiation in the adult dentate gyrus with 2nd trimester ethanol consumption. *PLoS One, 7*(12), e51344. http://dx.doi.org/10.1371/journal.pone.0051344.

Hoyme, H. E., May, P. A., Kalberg, W. O., Kodituwakku, P., Gossage, J. P., Trujillo, P. M., ... Robinson, L. K. (2005). A practical clinical approach to diagnosis of fetal alcohol spectrum disorders: Clarification of the 1996 institute of medicine criteria. *Pediatrics, 115*(1), 39–47. http://dx.doi.org/10.1542/peds.2004-0259.

Ikonomidou, C., Bittigau, P., Ishimaru, M. J., Wozniak, D. F., Koch, C., Genz, K., ... Olney, J. W. (2000). Ethanol-induced apoptotic neurodegeneration and fetal alcohol syndrome. *Science, 287*(5455), 1056–1060.

Kalberg, W. O., & Buckley, D. (2007). FASD: What types of intervention and rehabilitation are useful? *Neuroscience and Biobehavioral Reviews, 31*(2), 278–285.

Kelly, S. J., & Lawrence, C. R. (2008). Intragastric intubation of alcohol during the perinatal period. In L. E. Nagy (Ed.), *Alcohol: Methods and protocols*.

Kesner, R. P., & Churchwell, J. C. (2011). An analysis of rat prefrontal cortex in mediating executive function. *Neurobiology of Learning and Memory, 96*(3), 417–431. http://dx.doi.org/10.1016/j.nlm.2011.07.002.

Klintsova, A. Y., Dickson, E., Yoshida, R., & Greenough, W. T. (2004). Altered expression of BDNF and its high-affinity receptor TrkB in response to complex motor learning and moderate exercise. *Brain Research, 1028*(1), 92–104. http://dx.doi.org/10.1016/j.brainres.2004.09.003.

Klintsova, A. Y., Hamilton, G. F., & Boschen, K. E. (2012). Long-term consequences of developmental alcohol exposure on brain structure and function: Therapeutic benefits of physical activity. *Brain Sciences, 3*(1), 1–38. http://dx.doi.org/10.3390/brainsci3010001.

Klintsova, A. Y., Scamra, C., Hoffman, M., Napper, R. M., Goodlett, C. R., & Greenough, W. T. (2002). Therapeutic effects of complex motor training on motor performance deficits induced by neonatal binge-like alcohol exposure in rats: II. A quantitative stereological study of synaptic plasticity in female rat cerebellum. *Brain Research, 937*(1–2), 83–93.

Kroener, S., Mulholland, P. J., New, N. N., Gass, J. T., Becker, H. C., & Chandler, L. J. (2012). Chronic alcohol exposure alters behavioral and synaptic plasticity of the rodent prefrontal cortex. *PLoS One, 7*(5), e37541. http://dx.doi.org/10.1371/journal.pone.0037541.

Lebel, C., Rasmussen, C., Wyper, K., Andrew, G., & Beaulieu, C. (2010). Brain microstructure is related to math ability in children with fetal alcohol spectrum disorder. *Alcoholism: Clinical and Experimental Research, 34*(2), 354–363. http://dx.doi.org/10.1111/j.1530-0277.2009.01097.x.

Lupton, C., Burd, L., & Harwood, R. (2004). Cost of fetal alcohol spectrum disorders. *American Journal of Medical Genetics. Part C, Seminars in Medical Genetics, 127C*(1), 42–50. http://dx.doi.org/10.1002/ajmg.c.30015.

Lussier, A. A., Stepien, K. A., Neumann, S. M., Pavlidis, P., Kobor, M. S., & Weinberg, J. (2015). Prenatal alcohol exposure alters steady-state and activated gene expression in the adult rat brain. *Alcoholism: Clinical and Experimental Research, 39*(2), 251–261. http://dx.doi.org/10.1111/acer.12622.

Mandryk, R. L., Dielschneider, S., Kalyn, M. R., Bertram, C. P., Gaetz, M., Doucette, A., … Keiver, K. (2013). *Games as neurofeedback training for children with FASD*. http://dx.doi.org/10.1145/2485760.2485762.

Marquardt, K., & Brigman, J. L. (2016). The impact of prenatal alcohol exposure on social, cognitive and affective behavioral domains: Insights from rodent models. *Alcohol, 51*, 1–15. http://dx.doi.org/10.1016/j.alcohol.2015.12.002.

May, P. A., Baete, A., Russo, J., Elliott, A. J., Blankenship, J., Kalberg, W. O., … Hoyme, H. E. (2014). Prevalence and characteristics of fetal alcohol spectrum disorders. *Pediatrics, 134*(5), 855–866. http://dx.doi.org/10.1542/peds.2013-3319.

Medina, A. E., & Ramoa, A. S. (2005). Early alcohol exposure impairs ocular dominance plasticity throughout the critical period. *Brain Research. Developmental Brain Research, 157*(1), 107–111. http://dx.doi.org/10.1016/j.devbrainres.2005.03.012.

Mihalick, S. M., Crandall, J. E., Langlois, J. C., Krienke, J. D., & Dube, W. V. (2001). Prenatal ethanol exposure, generalized learning impairment, and medial prefrontal cortical deficits in rats. *Neurotoxicology and Teratology, 23*(5), 453–462.

Moore, E. M., & Riley, E. P. (2015). What happens when children with fetal alcohol spectrum disorders become adults? *Current Developmental Disorders Reports, 2*(3), 219–227.

Nimmerjahn, A., Kirchhoff, F., & Helmchen, F. (2005). Resting microglial cells are highly dynamic surveillants of brain parenchyma in vivo. *Science, 308*(5726), 1314–1318. http://dx.doi.org/10.1126/science.1110647.

Norman, A. L., Crocker, N., Mattson, S. N., & Riley, E. P. (2009). Neuroimaging and fetal alcohol spectrum disorders. *Developmental Disabilities Research Reviews, 15*(3), 209–217. http://dx.doi.org/10.1002/ddrr.72.

O'Hare, E. D., Lu, L. H., Houston, S. M., Bookheimer, S. Y., Mattson, S. N., O'Connor, M. J., & Sowell, E. R. (2009). Altered frontal-parietal functioning during verbal working memory in children and adolescents with heavy prenatal alcohol exposure. *Human Brain Mapping, 30*(10), 3200–3208. http://dx.doi.org/10.1002/hbm.20741.

Olney, J. W. (2014). Focus on apoptosis to decipher how alcohol and many other drugs disrupt brain development. *Frontiers in Pediatrics, 2*, 81. http://dx.doi.org/10.3389/fped.2014.00081.

Olney, J. W., Ishimaru, M. J., Bittigau, P., & Ikonomidou, C. (2000). Ethanol-induced apoptotic neurodegeneration in the developing brain. *Apoptosis, 5*(6), 515–521.

Paley, B., & O'Connor, M. J. (2011). Behavioral interventions for children and adolescents with fetal alcohol spectrum disorders. *Alcohol Research & Health, 34*(1), 64–75.

Paolozza, A., Treit, S., Beaulieu, C., & Reynolds, J. N. (2014). Response inhibition deficits in children with Fetal Alcohol Spectrum Disorder: Relationship between diffusion tensor imaging of the corpus callosum and eye movement control. *NeuroImage Clinal, 5*, 53–61. http://dx.doi.org/10.1016/j.nicl.2014.05.019.

Patten, A. R., Fontaine, C. J., & Christie, B. R. (2014). A comparison of the different animal models of fetal alcohol spectrum disorders and their use in studying complex behaviors. *Frontiers in Pediatrics, 2*, 93. http://dx.doi.org/10.3389/fped.2014.00093.

Peadon, E., Rhys-Jones, B., Bower, C., & Elliott, E. J. (2009). Systematic review of interventions for children with fetal alcohol spectrum disorders. *BMC Pediatrics, 9*(1), 35. http://dx.doi.org/10.1186/1471-2431-9-35.

Preston, A. R., & Eichenbaum, H. (2013). Interplay of hippocampus and prefrontal cortex in memory. *Current Biology, 23*(17), R764–R773. http://dx.doi.org/10.1016/j.cub.2013.05.041.

Puglia, M. P., & Valenzuela, C. F. (2010). Repeated third trimester-equivalent ethanol exposure inhibits long-term potentiation in the hippocampal CA1 region of neonatal rats. *Alcohol, 44*(3), 283–290. http://dx.doi.org/10.1016/j.alcohol.2010.03.001.

Rajasethupathy, P., Sankaran, S., Marshel, J. H., Kim, C. K., Ferenczi, E., Lee, S. Y., … Deisseroth, K. (2015). Projections from neocortex mediate top-down control of memory retrieval. *Nature, 526*(7575), 653–659. http://dx.doi.org/10.1038/nature15389.

Rasmussen, C., & Bisanz, J. (2009). Executive functioning in children with fetal alcohol spectrum disorders: Profiles and age-related differences. *Child Neuropsychology, 15*(3), 201–215. http://dx.doi.org/10.1080/09297040802385400.

Riley, E. P., Infante, M. A., & Warren, K. R. (2011). Fetal alcohol spectrum disorders: An overview. *Neuropsychology Review, 21*(2), 73–80. http://dx.doi.org/10.1007/s11065-011-9166-x.

Roozen, S., Peters, G. J., Kok, G., Townend, D., Nijhuis, J., & Curfs, L. (2016). Worldwide prevalence of fetal alcohol spectrum disorders: A systematic literature review including meta-analysis. *Alcoholism: Clinical and Experimental Research, 40*(1), 18–32. http://dx.doi.org/10.1111/acer.12939.

Sampedro-Piquero, P., Zancada-Menendez, C., & Begega, A. (2015). Housing condition-related changes involved in reversal learning and its c-Fos associated activity in the prefrontal cortex. *Neuroscience, 307*, 14–25. http://dx.doi.org/10.1016/j.neuroscience.2015.08.038.

Schafer, D. P., Lehrman, E. K., Kautzman, A. G., Koyama, R., Mardinly, A. R., Yamasaki, R., … Stevens, B. (2012). Microglia sculpt postnatal neural circuits in an activity and complement-dependent manner. *Neuron, 74*(4), 691–705. http://dx.doi.org/10.1016/j.neuron.2012.03.026.

Schneider, R. D., & Thomas, J. D. (2016). Adolescent choline supplementation attenuates working memory deficits in rats exposed to alcohol during the third trimester equivalent. *Alcoholism: Clinical and Experimental Research, 40*(4), 897–905. http://dx.doi.org/10.1111/acer.13021.

Sierra, A., Encinas, J. M., Deudero, J. J., Chancey, J. H., Enikolopov, G., Overstreet-Wadiche, L. S., … Maletic-Savatic, M. (2010). Microglia shape adult hippocampal neurogenesis through apoptosis-coupled phagocytosis. *Cell Stem Cell, 7*(4), 483–495. http://dx.doi.org/10.1016/j.stem.2010.08.014.

Smith, C. C., Guevremont, D., Williams, J. M., & Napper, R. M. (2015). Apoptotic cell death and temporal expression of apoptotic proteins Bcl-2 and Bax in the hippocampus, following binge ethanol in the neonatal rat model. *Alcoholism: Clinical and Experimental Research, 39*(1), 36–44. http://dx.doi.org/10.1111/acer.12606.

Sowell, E. R., Johnson, A., Kan, E., Lu, L. H., Van Horn, J. D., Toga, A. W., … Bookheimer, S. Y. (2008). Mapping white matter integrity and neurobehavioral correlates in children with fetal alcohol spectrum disorders. *Journal of Neuroscience, 28*(6), 1313–1319. http://dx.doi.org/10.1523/JNEUROSCI.5067-07.2008.

Sowell, E. R., Mattson, S. N., Kan, E., Thompson, P. M., Riley, E. P., & Toga, A. W. (2008). Abnormal cortical thickness and brain-behavior correlation patterns in individuals with heavy prenatal alcohol exposure. *Cerebral Cortex, 18*(1), 136–144. http://dx.doi.org/10.1093/cercor/bhm039.

Thomas, J. D., Idrus, N. M., Monk, B. R., & Dominguez, H. D. (2010). Prenatal choline supplementation mitigates behavioral alterations associated with prenatal alcohol exposure in rats. *Birth Defects Research. Part A, Clinical and Molecular Teratology, 88*(10), 827–837. http://dx.doi.org/10.1002/bdra.20713.

Thomas, J. D., Sather, T. M., & Whinery, L. A. (2008). Voluntary exercise influences behavioral development in rats exposed to alcohol during the neonatal brain growth spurt. *Behavioral Neuroscience, 122*(6), 1264–1273. http://dx.doi.org/10.1037/a0013271.

Thomas, J. D., Wasserman, E. A., West, J. R., & Goodlett, C. R. (1996). Behavioral deficits induced by bingelike exposure to alcohol in neonatal rats: Importance of developmental timing and number of episodes. *Developmental Psychobiology, 29*(5), 433–452. http://dx.doi.org/10.1002/(SICI)1098-2302(199607)29:5<433::AID-DEV3>3.0.CO;2-P.

Turken, A., Whitfield-Gabrieli, S., Bammer, R., Baldo, J. V., Dronkers, N. F., & Gabrieli, J. D. (2008). Cognitive processing speed and the structure of white matter pathways: Convergent evidence from normal variation and lesion studies. *NeuroImage, 42*(2), 1032–1044. http://dx.doi.org/10.1016/j.neuroimage.2008.03.057.

Ware, A. L., Infante, M. A., O'Brien, J. W., Tapert, S. F., Jones, K. L., Riley, E. P., & Mattson, S. N. (2015). An fMRI study of behavioral response inhibition in adolescents with and without histories of heavy prenatal alcohol exposure. *Behavioural Brain Research, 278*, 137–146. http://dx.doi.org/10.1016/j.bbr.2014.09.037.

Warren, K. R., & Foudin, L. L. (2001). Alcohol-related birth defects—the past, present, and future. *Alcohol Research & Health, 25*(3), 153–158.

Wass, T. S., Persutte, W. H., & Hobbins, J. C. (2001). The impact of prenatal alcohol exposure on frontal cortex development in utero. *American Journal of Obstetrics and Gynecology, 185*(3), 737–742. http://dx.doi.org/10.1067/mob.2001.117656.

Williamson, L. L., Sholar, P. W., Mistry, R. S., Smith, S. H., & Bilbo, S. D. (2011). Microglia and memory: Modulation by early-life infection. *Journal of Neuroscience, 31*(43), 15511–15521. http://dx.doi.org/10.1523/JNEUROSCI.3688-11.2011.

Wozniak, J. R., Mueller, B. A., Chang, P. N., Muetzel, R. L., Caros, L., & Lim, K. O. (2006). Diffusion tensor imaging in children with fetal alcohol spectrum disorders. *Alcoholism: Clinical and Experimental Research, 30*(10), 1799–1806. http://dx.doi.org/10.1111/j.1530-0277.2006.00213.x.

Young, C., Klocke, B. J., Tenkova, T., Choi, J., Labruyere, J., Qin, Y. Q., … Olney, J. W. (2003). Ethanol-induced neuronal apoptosis in vivo requires BAX in the developing mouse brain. *Cell Death and Differentiation, 10*(10), 1148–1155. http://dx.doi.org/10.1038/sj.cdd.4401277.

Young, C., & Olney, J. W. (2006). Neuroapoptosis in the infant mouse brain triggered by a transient small increase in blood alcohol concentration. *Neurobiology of Disease, 22*(3), 548–554. http://dx.doi.org/10.1016/j.nbd.2005.12.015.

Zelikowsky, M., Bissiere, S., Hast, T. A., Bennett, R. Z., Abdipranoto, A., Vissel, B., & Fanselow, M. S. (2013). Prefrontal microcircuit underlies contextual learning after hippocampal loss. *Proceedings of the National Academy of Sciences of the United States of America, 110*(24), 9938–9943. http://dx.doi.org/10.1073/pnas.1301691110.

C H A P T E R

16

Ethanol's Action Mechanisms in the Brain: From Lipid General Alterations to Specific Protein Receptor Binding

M.T. Marin, G. Morais-Silva
São Paulo State University (UNESP), Araraquara, Brazil

INTRODUCTION

Ethanol, popularly known as alcohol, is the psychoactive substance found in alcoholic beverages. This is a low molecular weight alcohol whose chemical formula is C_2H_6O. Due to its small and relatively simple structure, it passes through the cellular membrane by simple diffusion and access many body tissues to bind targets both in the cell surface and intracellular. Ethanol effects in the central nervous system (CNS) causing positive reinforcement induces drug seeking and consumption (Begleiter & Kissin, 1996). Its easiness to obtain and the use dissemination around the globe make it the most used addictive psychoactive substance in the word (Winstock, 2014).

Alcohol mainly has a depressant effect on the CNS; however, it has a biphasic effect on behavior (Pohorecky, 1977), which occurs as its blood levels increase (Fig. 16.1). Low-to-moderate doses of ethanol have an anxiolytic effect, increasing sociability and reducing tension. In rodents, it also increases locomotor activity. Although it may be interpreted as a stimulant effect, these behavioral alterations occur due to ethanol depressant actions in neurons. High doses of ethanol impair motor, sensorial, and cognitive functions. The highest doses could lead to coma and be lethal due to respiratory depression (Pohorecky & Brick, 1988).

In addition to ethanol's action on different neuronal cells and brain areas, the variety of effects of ethanol in the CNS is due to its relatively unspecific action. This chapter describes a number of studies depicting ethanol's action mechanisms in the lipid neuronal membranes and also on specific binding sites in protein receptors.

ETHANOL ACTIONS ON LIPID MEMBRANE

This is one of the oldest hypotheses about ethanol's action mechanism in the CNS. It postulates that the psychoactive actions of ethanol are an outcome of the interaction of this molecule with the neuronal membrane. It is easy to understand this supposition since ethanol is a simple molecule with a relatively high liposolubility. Another point relates to its diverse and biphasic effects, which could be the consequence of a nonspecific action.

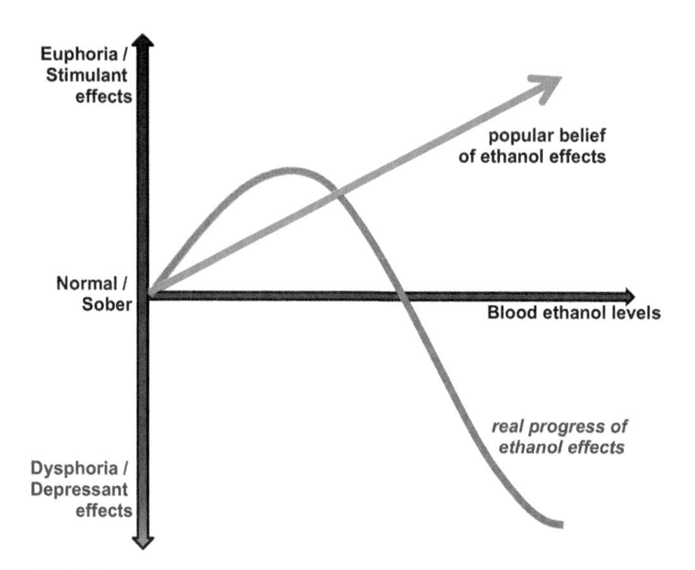

FIGURE 16.1 Ethanol biphasic effect. Graph showing the progress of ethanol effects according to blood ethanol level. The real progression of effects, reaching negative effects, differs a lot from popular belief of this drug effects based on the recreational ethanol use.

Addictive Substances and Neurological Disease
http://dx.doi.org/10.1016/B978-0-12-805373-7.00016-5

The primary effect of ethanol on membranes results in a disordering effect on this structure. Numerous reports show that ethanol has disordering effect on neuronal (Bae et al., 2005; Beaugé, Fleuret-Balter, Barin, Nordmann, & Nordmann, 1982; Fleuret-Balter, Beaugé, Barin, Nordmann, & Nordmann, 1983; Goldstein & Chin, 1981; Logan & Laverty, 1987), sarcoplasmic (Almeida, Vaz, Stuempel, & Madeira, 1986), mitochondrial (Sanchez-Amate, Carrasco, Zurera, Segovia, & Marco, 1995), erythrocyte (Chin & Goldstein, 1977; Hrelia et al., 1986), and intestinal (Bikle, Gee, & Munson, 1986; Gastaldi, Casirola, Ferrari, Casasco, & Calligaro, 1991; Hunter et al., 1983), as well as yeast (Jones & Greenfield, 1987), protozoal (Nandini-Kishore, Mattox, Martin, & Thompson, 1979), and bacterial (Dombek & Ingram, 1984) membranes, increasing the motion of the lipids in the bilayer. Ethanol also alters the phase-transition temperature in which lipids enter in a gel state (more rigid state) (Kamaya, Ma, & Lin, 1986; Tamura, Kaminoh, Kamaya, & Ueda, 1991).

According to this hypothesis, the disordering effect of ethanol and other anesthetic drugs alters the synaptic transmission through conformational changes in proteins anchored in the membrane due to the conformational changes in the lipid bilayer (Gruner & Shyamsunder, 1991; Seeman, 1972). This means that, when the lipid bilayer is disordered, it alters the interaction between the membrane and its proteins, altering their activity. There are some issues concerning the relevance of these alterations to behavioral effects of ethanol. Although some behavioral changes, especially those related to age (Egawa, Pearn, Lemkuil, Patel, & Head, 2016) and schizophrenia (Eckert, Schaeffer, Schmitt, Maras, & Gattaz, 2011), are related to membrane lipids and fluidity, these changes do not explain all the behavioral effects of ethanol. Membrane fluidizing agents do not replicate ethanol's behavioral effects (Buck, Allan, & Adron Harris, 1989). Furthermore, the effects of ethanol on fluidity and membrane composition occur only in high concentrations, far from the range that is physiologically relevant (Peoples, Chaoying, & Weight, 1996). Also, ethanol's effects on neuronal proteins can be observed without significant alterations on membrane fluidity (Tapia, Aguilar, Sotomayor, & Aguayo, 1998) and *n*-alcohols with more than five carbon atoms have high lipid solubility and membrane lipid disordering potency while having low intoxicating effects (Peoples & Weight, 1995).

In spite of the difficulty to explain the effects of ethanol based on its interaction with the lipid membrane, this property probably accounts for some chronic alterations. Cellular membranes adapt to the fluidizing effects after prolonged exposure and thus become less susceptible to this action (Chin, Parsons, & Goldstein, 1978; Gutiérrez-Ruiz, Gómez, Souza, & Bucio, 1995; Maturu, Vaddi, Pannuru, & Nallanchakravarthula, 2010). Lipid membranes from ethanol-tolerant mice (Chin et al., 1978) and ethanol-dependent mice (Lyon & Goldstein, 1983) show reduced fluidity.

PROTEIN TARGETS RELATED TO ETHANOL EFFECTS

The focus on lipid theories of ethanol and anesthetics' action changed after the purification and characterization of the firefly (*Photinus pyralis*) enzyme luciferase, which facilitates the bioluminescence process from the luminescent substrate luciferin (Bitler & McElroy, 1957; Green & McElroy, 1956a; McElroy & Green, 1956). This enzyme is sensitive to anesthetics and ethanol action in lipid-free luciferin–luciferase isolates from fireflies (Chiou & Ueda, 1994; Ueda, 1956; Ueda & Kamaya, 1973) at subclinical concentrations. Moreover, their direct interaction with the protein structure and the modification of its function fits better with the mathematical models than their interaction with membrane lipids (Franks & Lieb, 1982, 1994; Green & McElroy, 1956b). These finds stimulated studies involving the direct interaction of ethanol with protein receptors, after which numerous reports appeared showing that this drug could directly modulate several protein receptors in the CNS.

Ethanol Action on GABA Receptors

GABA (gamma-aminobutyric acid) is the main inhibitory neurotransmitter in mammalian brain. Its main receptor, the type A GABA receptor ($GABA_A$-R), is a pentameric ligand-gated ion channel (whose family, the "Cys-loop" superfamily, includes glycine receptors, nicotinic–acetylcholine receptors and the type 3 serotonin receptors), which selectively conducts Cl^- currents. They are widely expressed throughout the brain, in the synaptic cleft–mediating rapid phasic inhibition and in the extrasynaptic and perisynaptic space, mediating tonic inhibition (Olsen & Sieghart, 2009). There are also metabotropic receptors for GABA, called $GABA_B$ receptors ($GABA_B$-Rs), which are a G-protein-coupled receptor that increase K^+ conductance and decrease Ca^{2+} conductance (Bowery, 1989; Emson, 2007).

$GABA_A$ Receptors

There are 19 known genes (α1–6, β1–3, γ1–3, δ, ε, θ, π, and ρ1–3) that encode subunits for $GABA_A$-R, with different expression patterns throughout the CNS, which can be organized as a heteromer or as a monomer to form the ion channel. Most of the $GABA_A$-Rs are composed of two α, two β, and one of the other subunits, being γ subunit the most common. Monomers are formed by ρ subunits (formerly known as $GABA_C$ receptors) (Bormann, 2000; Knoflach, Hernandez, & Bertrand, 2016). The wide distribution of its receptors makes GABA important in the control of many brain functions, from general mechanisms important to homeostasis to specific behaviors.

Since early studies show that many behavioral ethanol effects are altered by GABAergic drugs (Allan & Harris, 1986; Martz, Dietrich, & Harris, 1983) and that ethanol potentiates GABA neurotransmission (Nestoros, 1980), many results have shown that ethanol increases GABA-mediated Cl^- currents (Aguayo, 1990; Celentano, Gibbs, & Farb, 1988; Nishio & Narahashi, 1990; Reynolds & Prasad, 1991; Suzdak, Schwartz, Skolnick, & Paul, 1986; Ticku, Lowrimore, & Lehoullier, 1986).

This increased $GABA_A$-R function only occurs in presence of the GABA molecule, ethanol being a positive allosteric modulator of these receptors. It is important to note that the response pattern is different for different cellular types. The mechanism of the action of ethanol on $GABA_A$-Rs is the mediation of the channel-opening events, raising the mean open time, the frequency of opening bursts, and the mean burst percentage (Tatebayashi, Motomura, & Narahashi, 1998). Together, these effects increase Cl^- currents in the ion channel in the presence of the neurotransmitter. Although important, these results do not explain an intriguing question: where does the ethanol molecule act on the GABA receptor to exert its influence?

A significant step toward answering this question was achieved in the classic study of Mihic et al., published in 1997. Using chimeric receptor constructs expressed in *Xenopus* oocytes, they identified a region of 45 amino acid residues from $GABA_A$-R transmembrane domains TM2 and TM3 that is critical for the ethanol and volatile anesthetics' action on $GABA_A$-Rs and glycine receptors (GlyRs) (Mihic et al., 1997). Subsequent studies have shown that this binding site is relatively small; estimated to be between 250 and 370 Å (Jenkins et al., 2001; Wick et al., 1998).

Another important factor that complicates the interaction between ethanol and GABA neurotransmission is the heterogeneity of $GABA_A$-Rs composition. Depending on the subunit composition of the receptor, its sensitivity to allosteric ligands could be drastically changed. For example, $GABA_A$-Rs containing α4 or α6 subunits are insensitive to classical benzodiazepines (Derry, Dunn, & Davies, 2004; Wieland, Lüddens, & Seeburg, 1992). Besides positive results mentioned earlier, many reports from the end of 1980s and beginning of 1990s did not replicate the potentiating effect of ethanol on GABA-mediated Cl^- currents (Morrisett & Swartzwelder, 1993; Proctor, Allan, & Dunwiddie, 1992; Siggins, Pittman, & French, 1987; Soldo, Proctor, & Dunwiddie, 1994), being this effect in general dependent of the cell type used.

Most of the $GABA_A$-Rs are, at some concentration, sensitive to ethanol's enhancement effect (Olsen, Hanchar, Meera, & Wallner, 2007). However, at low physiologically relevant concentrations, $GABA_A$-Rs containing δ subunit, especially when combined to α4 or α6, seems to be the most sensitive (Glykys et al., 2007; Hanchar et al., 2006; Mihalek et al., 2001; Wallner, Hanchar, & Olsen, 2003, 2006a, 2006b). A subset of $GABA_A$-Rs containing a long splice variant of the γ2 subunit (γ2L), when phosphorylated, is also responsive to ethanol at low concentrations (Wafford et al., 1991; Wafford & Whiting, 1992).

The diversity of some behavioral effects of ethanol could be explained by its high affinity to $GABA_A$-Rs containing δ subunit and differential affinity to others (Proctor et al., 1992; Soldo et al., 1994). These receptors are located extrasynaptically, have a low expression at CNS and control mainly tonic inhibition of the neurons (Kaur, Baur, & Sigel, 2009; Nusser & Mody, 2002; Zheleznova, Sedelnikova, & Weiss, 2009). In accordance with this property, ethanol acts preferentially at the extrasynaptic space increasing tonic inhibition, instead of phasic inhibition. Many experimental results corroborate this idea (Carlen, Gurevich, & Durand, 1982; Durand, Corrigall, Kujtan, & Carlen, 1981; Santhakumar, Wallner, & Otis, 2007). For example, primary targets of ethanol on CNS, as evaluated by cFos expression, are regions where δ subunit is highly expressed (Olsen & Sieghart, 2009; Vilpoux, Warnault, Pierrefiche, Daoust, & Naassila, 2009; Zheleznova et al., 2009).

Receptors formed by ρ subunits possess distinct functioning from other $GABA_A$-Rs. In these receptors, ethanol has an inhibitory action (Blednov et al., 2014; Borghese et al., 2016; Mihic & Harris, 1996). They seem to have counteracting effects, since ρ1 mutant mice show increased sedative effects of ethanol (Blednov et al., 2014).

$GABA_B$ Receptors

$GABA_B$-Rs does not appear to be affected by ethanol (Frye & Fincher, 1996; Frye, Taylor, Trzeciakowski, & Griffith, 1991), although these receptors may account for the behavioral effects of ethanol through a nondirect interaction (Allan, Burnett, & Harris, 1991; Ariwodola & Weiner, 2004).

Ethanol Action on Glutamatergic Receptors

Glutamate is a nonessential amino acid that also works as a neurotransmitter, which mediates most of the excitatory neurotransmission in vertebrate's CNS. There are two classes of glutamatergic receptors, the metabotropic receptors (mGluRs—coupled to G-protein signal transduction) and ionotropic receptors (iGluRs—ion channels) (Ozawa, 1998).

The iGluRs are tetrameric complexes that form a pore permeable to cationic (Ca^{2+}, Na^{+}) ions. There are three types of receptors in this class: N-methyl-D-aspartate (NMDA), α-amino-3-hydroxy-5-methyl-4-isoxazole proprionic acid (AMPA) and kainate (KA) receptors, named based on the agonist selectivity (Niciu, Kelmendi, & Sanacora, 2012). Being that these receptors are responsible for most of the excitatory neurotransmission in mammals, they were seen as good candidates to explain ethanol's effects on the CNS.

The mGluRs are G-protein-coupled receptors widely distributed in the mammalian brain. There are eight subtypes, namely mGluR1–8, classified in three groups according to which G-protein they are coupled: group I mGluR (mGluR1 and mGluR5 that are coupled to G-protein that activates phospholipase C, G_q/G_{11}), group II mGluR (mGluR2 and mGluR3 that are coupled to inhibitory G-protein G_i/G_o), and group III mGluR (mGluR4, mGluR6, mGluR7, and mGluR8 that are coupled to inhibitory G-protein G_i/G_o) (Nicoletti et al., 2011; Witkin, Marek, Johnson, & Schoepp, 2007).

NMDA Receptors

The first evidence of ethanol inhibition of glutamate neurotransmission appeared in the beginning of 1990s (Lovinger, White, & Weight, 1989, 1990; White, Lovinger, & Weight, 1990; Wirkner et al., 1999), especially for NMDA receptor, which are inhibited by low physiologically relevant ethanol concentrations. Ethanol seems to be a noncompetitive allosteric modulator, decreasing the channel-opening frequency and mean open time, instead of blocking the channel directly (Lima-Landman & Albuquerque, 1989; Wirkner, Eberts, Poelchen, Allgaier, & Illes, 2000; Wright, Peoples, & Weight, 1996).

Ethanol interacts with NMDA receptors through a modulatory site exposed to the extracellular space (Peoples & Stewart, 2000) that differs from the classical modulatory sites (Cebers, Cebere, Zharkovsky, & Liljequist, 1996; Chu, Anantharam, & Treistman, 1995; Gothert & Fink, 1989; Peoples & Weight, 1992). This evidence is not absolute, since glycine alters ethanol effects in some cellular types (Woodward & Gonzales, 1990) and in vivo models (Kiefer et al., 2003). However, ethanol interacts with glycine receptor, which can explain some results (Mihic, 1999).

The NMDA receptor is composed of two GluN1 subunits and two GluN2 subunits (which can be one of the four homologs called N2A–D) (Traynelis et al., 2010). Receptors composed of N2A or N2B subunits are the most sensitive to ethanol inhibition, and the site of interaction is located at M3 and M4 domains of GluN1 and GluN2A and B subunits (Honse, Ren, Lipsky, & Peoples, 2004; Ren, Zhao, Wu, & Peoples, 2007, 2012, 2013; Ronald, 2001; Smothers, Jin, & Woodward, 2013; Smothers & Woodward, 2006; Xu, Smothers, & Woodward, 2015; Zhao, Ren, Dwyer, & Peoples, 2015).

AMPA and Kainate Receptors

Ethanol appears to inhibit non-NMDA currents, as shown by various studies (Möykkynen & Korpi, 2012; Roberto et al., 2004). However, in some reports it is difficult to differentiate between AMPA or kainate receptors.

Critical changes occur in glutamatergic receptors during brain development. It is not surprising that ethanol interactions could be different in neonatal ages. Generally, AMPA receptors are not responsive to ethanol. However, AMPA receptors in developing brains are acutely inhibited by ethanol, effect which disappears in receptors obtained from adult brains (Frye & Fincher, 2000; Mameli, Zamudio, Carta, & Valenzuela, 2005; Puglia & Valenzuela, 2010; Wirkner et al., 2000). In some experimental conditions, where desensitization processes of AMPA receptors are increased, ethanol at low concentrations inhibits AMPA currents, suggesting that this drug acts to increase this process (Möykkynen, Coleman, Keinänen, Lovinger, & Korpi, 2009; Möykkynen & Korpi, 2012; Moykkynen, Korpi, & Lovinger, 2003).

Kainate receptors seems to be very sensitive to ethanol (Carta, Ariwodola, Weiner, & Valenzuela, 2003; Costa, Soto, Cardoso, Olivera, & Valenzuela, 2000), especially those from the CA3 region of the hippocampus (Weiner, Dunwiddie, & Valenzuela, 1999) and basolateral amygdala (Läck, Ariwodola, Chappell, Weiner, & McCool, 2008). Subunit composition of this receptor does not increase the inhibitory effects of ethanol (Valenzuela & Cardoso, 1999). Furthermore, alcohol's effects on protein kinase C (PKC) activity does not increase its inhibition of kainate receptors (Dildy-Mayfield & Harris, 1995; Valenzuela, Cardoso, Lickteig, Browning, & Nixon, 1998).

Metabotropic Glutamatergic Receptors

There are little data available regarding the acute ethanol effects on mGluR. Some works suggest an inhibitory effect on mGluR5, possibly through the activation of PKC (Minami, Gereau, Minami, Heinemann, & Harris, 1998; Narahashi et al., 2001). However, evidences of a direct interaction between ethanol and mGluR have not been demonstrated yet.

Ethanol Action on Glycine Receptors

Together with GABA, glycine is an inhibitory neurotransmitter in the CNS of adult mammalians. Glycinergic neurotransmission occurs mainly in the brain stem and spinal cord, controlling both motor and sensory functions (Dutertre, Becker, & Betz, 2012). In addition to $GABA_A$-Rs, glycine receptors (GlyRs) are a pentameric ligand-gated ion channel that selectively conducts Cl^- currents. They are located mostly in the postsynaptic neuron, in which they inhibit neuronal firing (Lynch, 2009). Five subunits have been identified (α1–4, β). α subunits can form homomeric receptors that are functional, while β subunits always form heteromeric receptors with α subunits (2α/3β or 3α/2β) (Webb & Lynch, 2007). There are developmental differences regarding the expression of GlyRs subunits. α2 subunits are preferentially found in neonates, while α1 and β are highly expressed in adults (Aguayo, van Zundert, Tapia, Carrasco, & Alvarez, 2004; Becker, Hoch, & Betz, 1988; Singer, Talley, Bayliss, & Berger, 1998).

Evidences of ethanol's interaction with GlyRs date from the early studies from Celentano and coworkers, in which Cl^- currents induced by glycine in chick spinal cord neurons were potentiated by ethanol at low physiological concentrations (Celentano et al., 1988). Following this, others have replicated ethanol related potentiation of glycine-mediated Cl^- currents in many neuronal lineages (Aguayo & Pancetti, 1994; Aguayo, Tapia, & Pancetti, 1996; Eggers, O'Brien, & Berger, 2000; Engblom & Akerman, 1991; Mascia, Machu, & Harris, 1996, Mascia, Mihic, Valenzuela, Schofield, & Harris, 1996, Mascia, Wick, Martinez, & Harris, 1998).

As for $GABA_A$-Rs, ethanol potentiates glycine's affinity for GlyRs, increasing the open channel probability (Eggers & Berger, 2004; Yevenes, Moraga-Cid, Peoples, Schmalzing, & Aguayo, 2008). These effects are stronger in α1 subunit containing receptors, which are more sensitive to ethanol in low concentrations (Eggers et al., 2000; Mascia, Machu et al., 1996, Mascia, Mihic et al., 1996; Sebe, Eggers, & Berger, 2003). Regarding a binding site for ethanol interaction, there are some evidences that suggest a hydrophobic region formed by the transmembrane domains 2–3, and/or a modulation through the intracellular space mediated by G-protein βγ dimmers (Burgos, Muñoz, Guzman, & Aguayo, 2015).

Ethanol Action on Nicotinic Acetylcholine Receptors

Another member of the "Cys-loop" superfamily ligand-gated ion channels are the nicotinic acetylcholine receptors (nAchRs). They mediate fast synaptic transmission conducting cationic ions through the cell membranes in the ganglionic neurons of the autonomic nervous system. Additionally, they are the receptors of the neuromuscular junction and preferentially control neurotransmitter release in CNS, due to being located mainly in presynaptic sites (Fasoli & Gotti, 2015). There are 17 subunits identified that can compose nAchRs: α1–10, β1–4, γ, δ, and ε, being α1, β1, γ, δ, and ε exclusively found in muscular nAchRs. Vertebrate neuronal nAchRs is found as homomeric combinations of α7, 8, 9, or 10 subunits or as heteromeric combinations of α and β subunits (Dani, 2015; Fasoli & Gotti, 2015; Wu, Cheng, Jiang, Melcher, & Xu, 2015).

In sensitive nAchRs, ethanol acts as a co-agonist (Marszalec, Aistrup, & Narahashi, 1999), increasing the affinity of the receptor to nicotine and acetylcholine (Forman, Righi, & Miller, 1989; Wood, Forman, & Miller, 1991), stabilizing the open channel state (Wu, Tonner, & Miller, 1994), and increasing burst frequency and rate of opening (Bradley, Peper, & Sterz, 1980; Gage, McBurney, & Schneider, 1975; Linder, Pennefather, & Quastel, 1984). In addition to the sensitive nAchRs, there are some receptors that are not altered or inhibited by ethanol, depending of the subunit composition. The homomeric α7 nAchRs are inhibited by ethanol, even in low physiologically relevant doses (Yu et al., 1996). The highly centrally expressed α4β2 nAchRs are the most sensitive nicotinic receptors naturally found that is excited by ethanol (Marszalec et al., 1999), although α4β4 and α2β2 combinations were slightly less sensitive to ethanol in oocytes expressing these receptors (Cardoso et al., 1999). Muscular nAchRs are also affected by ethanol, but in a lesser degree than α4β2 receptors (Narahashi, Aistrup, Marszalec, & Nagata, 1999).

Ethanol Action on $5HT_3$ Serotonergic Receptors

Serotonin (5-HT) is an important neurotransmitter related to behavioral control. There are many types of 5-HT receptors, but only the $5\text{-}HT_3$ receptors ($5\text{-}HT_3Rs$) are ligand-gated ion channels (Giulietti et al., 2014).

The $5\text{-}HT_3Rs$ are pentameric ligand-gated ion channels that also belong to "Cys-loop" superfamily, permeable to cationic ions (mainly to Na^+ and K^+). There are five subunits identified in humans ($5\text{-}HT_3A$, a homomeric type, $5\text{-}HT_3B$, a heteromeric type, $5\text{-}HT_3D$, $5\text{-}HT_3E$, and $5\text{-}HT_3F$ which structure has not been elucidated yet), but only two have been found in rodents ($5\text{-}HT_3A$ and $5\text{-}HT_3B$) (Gupta, Prabhakar, & Radhakrishnan, 2016).

The first evidence of ethanol's action on $5\text{-}HT_3Rs$ dates to the beginning of the 1990s, when ethanol potentiated $5\text{-}HT_3$-mediated currents in neuroblastoma cells and isolated mammalian neurons (Lovinger, 1991; Lovinger & White, 1991). In the following years, others have found potentiation of $5\text{-}HT_3$ currents by ethanol in other systems expressing these receptors (Barann, Ruppert, Göthert, & Bönisch, 1995; Lovinger & Zhou, 1993, 1994; Machu & Harris, 1994).

Ethanol acts to increase influx through the ion channel (Barann et al., 1995), stabilizing the open channel state, increasing open channel probability and agonist affinity to the receptor (Lovinger, Sung, & Zhou, 2000; Zhou, Verdoorn, & Lovinger, 1998), these effects being more pronounced at 5-HT_3A homomeric receptor than 5-HT_3AB heteromeric receptor (Hayrapetyan, Jenschke, Dillon, & Machu, 2005; Rusch, Musset, Wulf, Schuster, & Raines, 2007).

Ethanol Action on Voltage-Gated Ion Channels

A superfamily of ion channels imperative for electrochemical signaling in excitable cells is the voltage-gated ion channels. They are rapidly activated by changes in the membrane electrical potential, conducting ion currents inside and outside the cell. There are three ion channels in this superfamily of ion channels: the voltage-gated calcium channels (Ca^{2+}Vs), the voltage-gated potassium channels (K^+Vs) and the voltage-gated sodium channels (Na^+Vs) (Moran, Barzilai, Liebeskind, & Zakon, 2015).

Voltage-Gated Calcium Channels

There are seven types of Ca^{2+}Vs, as follows: two L-type Ca^{2+}Vs (activated by high-voltage, slow inactivation), one T-type Ca^{2+}Vs (activated by low-voltage, rapid inactivation), N-P-Q-types Ca^{2+}Vs (activated by high-voltage, intermediate to slow inactivation) and an R-type Ca^{2+}Vs (activated by high-voltage, resistant to organic inactivation) (Walter & Messing, 1999). Much evidence exists regarding ethanol's influence on L-type, whereas much less is known about non-L-type channels. Ethanol acutely inhibits L-type Ca^{2+}Vs (Mullikin-Kilpatrick & Treistman, 1995; Twombly, Herman, Kye, & Narahashi, 1990; Wang, Wang, Lemos, & Treistman, 1994) promoting channel inactivation and decreasing open channel probability (Huang & McArdle, 1994; Wang et al., 1994). In non-L-type Ca^{2+}Vs ethanol inhibition appear to be due to its effects on protein kinase A (PKA). These ion channels are also less sensitive to ethanol than L-type channels (Solem, McMahon, & Messing, 1997; Twombly et al., 1990).

Voltage-Gated Potassium Channels

Potassium channels form a huge superfamily of ion channels with diverse properties. Among them, Ca^{2+}-activated potassium channels (K^+Ca^{2+}) are very interesting due to their dual characteristic of being activated by cytosolic increases of Ca^{2+} and by alterations in the membrane electric potential (Guéguinou et al., 2014).

There are three types of K^+Ca^{2+} channels: big-conductance K^+Ca^{2+} channels (BK—sensitive to alterations in membrane potential), small-conductance K^+Ca^{2+} channels (SK—insensitive to alterations in membrane potential), and intermediate-conductance K^+Ca^{2+} channels (IK—show conductance proprieties between SK and BK channels). Among them, there is evidence of ethanol's interaction with BK channels, which are potentiated by relevant physiological concentrations easily achieved in few drinks (Dopico, Lemos, & Treistman, 1996; Jakab, Weiger, & Hermann, 1997). As for other ligand-gated ion channels, ethanol acts as a co-agonist, increasing the effects of calcium and membrane potential alterations on BK channels (Liu, Vaithianathan, Manivannan, Parrill, & Dopico, 2008).

Voltage-Gated Sodium Channels

The Na^+Vs are responsible for the depolarization phase of the action potential in excitable cells. They are composed of α subunits (Na^+1.1−1.9) associated to β subunits (Catterall & Swanson, 2015; Kruger & Isom, 2016). Due to its substantial role in action potentials, it is a target for local anesthetic effects (Fozzard, Lee, & Lipkind, 2005).

First reports concerning ethanol effects on Na^+Vs found inhibitory effects, which were associated with ethanol's effects on membrane lipids (Harris & Bruno, 1985; Mullin & Hunt, 1985, 1987). Later studies replicated this inhibitory effect, showing that ethanol likely disrupts protein hydrogen bonds that form the ion pore of the channel (Krylov, Vilin, Katina, & Podzorova, 2000), increase in channel inactivation and open channel block (Horishita & Harris, 2008). The influence of ethanol on Na^+Vs occur in high ethanol concentrations, in which ethanol is already anesthetic.

Evidence of Other Possible Targets for Direct Ethanol Action

The small and relatively simple structure of ethanol molecules increases the number of possible targets for ethanol interaction. Thus, it is not surprising that evidence of new targets continue to appear. Molecular and electrophysiological studies have been drastically increasing our understanding of ethanol's targets in the CNS, opening many new opportunities to investigate its effects. For example, excitatory ion channels activated by extracellular adenosine 5′-triphosphate (P2X receptors—P2XRs) are inhibited by ethanol (Weight, Li, & Peoples, 1999). This inhibition is dependent on the subunit composition of these receptors, $P2X_4$ being the most sensitive subgroup (Davies et al., 2005, 2002; Ostrovskaya et al., 2011). On these receptors ethanol seems to act like an allosteric modulator, decreasing the agonist's affinity to the receptor (Li, Peoples, & Weight, 1998).

Protein kinases are also possible targets for ethanol action. PKC is acutely inhibited by ethanol and other anesthetics (Slater et al., 1993), especially the α isoform

(Slater et al., 1997). The activity of PKA is enhanced by ethanol, but through an indirect pathway, since this drug could increase adenylyl cyclase activity (Maas et al., 2005). Once both PKC and PKA could regulate the function of many receptors through phosphorylation (Smart, 1997), they could be an important regulator of ethanol's activity.

Ethanol's targets in the CNS are not completely understood and many other evidences of new sites of action probably will appear in the next years.

CONCLUSIONS

Since the first reports suggesting ethanol's effects on lipid membranes, much evidence has emerged showing a direct interaction of this drug with protein receptors, of which we can highlight $GABA_A$-Rs, NMDA receptors, GlyRs, nAchRs, $5\text{-}HT_3$Rs, L-type Ca^{2+}Vs, and BK channels. Most of them belong to the "Cys-loop" ligand-gated ion channels superfamily, what is expected due to the homology among them. These appear to be the most sensitive targets, which are responsible for the behavioral and physiological ethanol-induced alterations (Vengeliene, Bilbao, Molander, & Spanagel, 2009). Fig. 16.2 depicts most of ethanol's sites of action in the CNS and respective necessary concentrations to initiate these mechanisms and possible effects. Behavioral and physiological effects were organized according to blood ethanol level (CDC, 2016). The sites of action were then organized according to data about the effective concentration of ethanol listed in the references cited earlier. It is only a proposal for organizing ethanol action mechanism and resulting effects.

Although much of the evidence cited here used heterologous systems to show direct ethanol actions on receptors, they provide valuable tools to the study of these interactions. The physiological relevance of these interactions, on the other hand, has to be elucidated in more complex systems, such as laboratory rodents, which, unfortunately, complicate the interpretation of the results. We can conclude that, although each one is important, the elucidation of a specific behavioral effect of ethanol is far more complicate than a simple interaction with one or two isolated receptor types.

An interesting feature of ethanol's action on its many targets is the characteristic region specificity, which is related to differential receptor's subunit expression and sensitivity. For example, only regions that express ethanol-sensitive NMDA and $GABA_A$-Rs are altered during an ethanol challenge (Criswell et al., 1993; Knapp et al., 2001; Soldo et al., 1994; Yang, Criswell, Simson, Moy, & Breese, 1996).

Some behavioral and neurochemical alterations also seem to be region and receptor specific. Ventral tegmental area (VTA) activation and consequent dopamine release in nucleus accumbens (Acb) are related to $5\text{-}HT_3$Rs, instead of nAch or NMDA receptors (Ericson, Molander, Löf, Engel, & Söderpalm, 2003; Larsson, Svensson, Söderpalm, & Engel, 2002;

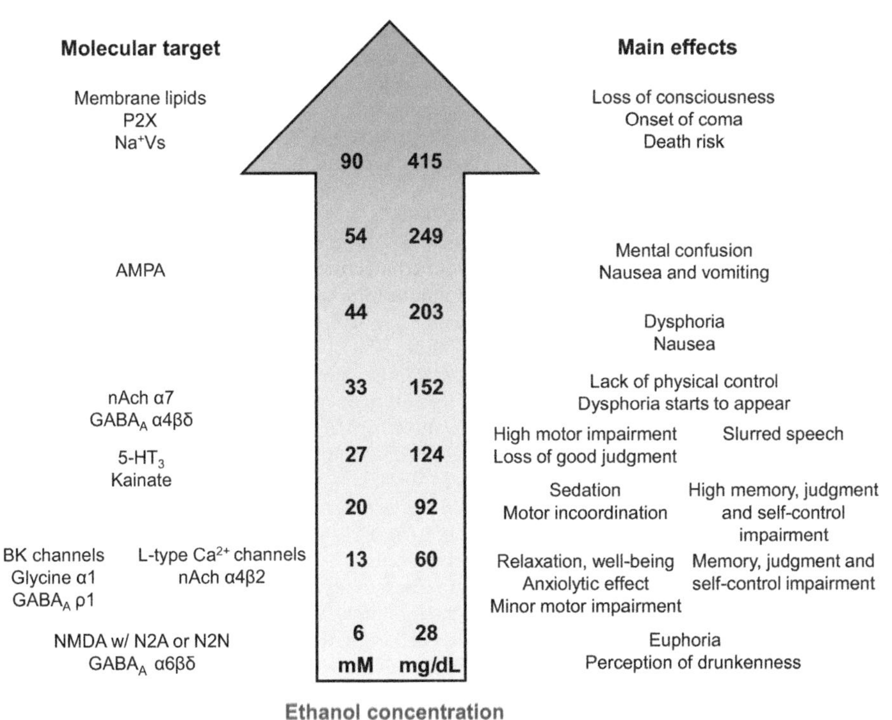

FIGURE 16.2 Overview of main ethanol targets on the central nervous system and its relation to local (mM) or blood (mg/dL) ethanol concentration (IC50 for inhibitory effects or EC50 for excitatory effects) and behavioral alterations. *Information used to produce this figure was gathered from the original papers cited through the text and the CDC website from CDC. (2016).* Impaired driving: Get the facts. *[WWW document] URL http://www.cdc.gov/motorvehiclesafety/impaired_driving/impaired-drv_factsheet.html.*

Yoshimoto, McBride, Lumeng, & Li, 1992). Ethanol's effects on neurohormone release and the analgesic effect of ethanol are more related to its action on BK channels and L-type Ca^{2+}Vs (Brodie, Scholz, Weiger, & Dopico, 2007; Dopico, Bukiya, & Martin, 2014; Gruss et al., 2001; Walter & Messing, 1999).

As blood and tissue ethanol levels increase, it starts to interact with less sensitive sites of action, increasing behavioral alterations and risk of dangerous effects, such as adverse cardiac and respiratory effects. In this sense, only high ethanol doses affect cardiac currents (Bébarová et al., 2010). As previously stated, Na^{+}Vs inhibition by ethanol appears at high drug concentrations, which is correlated to its anesthetic action.

In conclusion, ethanol has many targets in the CNS, which account for its acute effects. They differ in sensitivity and modulation in presence of ethanol. These characteristics could explain biphasic effects of ethanol; once different targets are modulated depending on the concentration of the drug in its tissues.

References

Aguayo, L. G. (1990). Ethanol potentiates the $GABA_A$-activated Cl^- current in mouse hippocampal and cortical neurons. *European Journal of Pharmacology, 187*, 127–130.

Aguayo, L. G., & Pancetti, F. C. (1994). Ethanol modulation of the gamma-aminobutyric acidA- and glycine-activated Cl^- current in cultured mouse neurons. *The Journal of Pharmacology and Experimental Therapeutics, 270*, 61–69.

Aguayo, L. G., Tapia, J. C., & Pancetti, F. C. (1996). Potentiation of the glycine-activated Cl^- current by ethanol in cultured mouse spinal neurons. *The Journal of Pharmacology and Experimental Therapeutics, 279*, 1116–1122.

Aguayo, L. G., van Zundert, B., Tapia, J. C., Carrasco, M. A., & Alvarez, F. J. (2004). Changes on the properties of glycine receptors during neuronal development. *Brain Research Reviews, 47*, 33–45.

Allan, A. M., Burnett, D., & Harris, R. A. (1991). Ethanol-induced changes in chloride flux are mediated by both $GABA_A$ and $GABA_B$ receptors. *Alcoholism, Clinical and Experimental Research, 15*, 233–237.

Allan, A. M., & Harris, R. A. (1986). Gamma-aminobutyric acid and alcohol actions: Neurochemical studies of long sleep and short sleep mice. *Life Sciences, 39*, 2005–2015.

Almeida, L. M., Vaz, W. L. C., Stuempel, J., & Madeira, V. M. C. (1986). Effect of short-chain primary alcohols on fluidity and activity of sarcoplasmic reticulum membranes. *Biochemistry, 25*, 4832–4839.

Ariwodola, O. J., & Weiner, J. L. (2004). Ethanol potentiation of GABAergic synaptic transmission may be self-limiting: Role of presynaptic $GABA_B$ receptors. *The Journal of Neuroscience: The Official Journal of the Society for Neuroscience, 24*, 10679–10686.

Bae, M., Jeong, D., Park, N., Lee, C., Cho, B., Jang, H., & Yun, I. (2005). The effect of ethanol on the physical properties of neuronal membranes. *Molecules and Cells, 19*, 356–364.

Barann, M., Ruppert, K., Göthert, M., & Bönisch, H. (1995). Increasing effect of ethanol on 5-HT3 receptor-mediated 14C-guanidinium influx in N1E-115 neuroblastoma cells. *Naunyn-Schmiedeberg's Archives of Pharmacology, 352*, 149–156.

Beaugé, F., Fleuret-Balter, C., Barin, F., Nordmann, J., & Nordmann, R. (1982). Brain membrane disordering related to acute ethanol administration in naive and short-term ethanol-intoxicated rats. *Drug and Alcohol Dependence, 10*, 143–151.

Bébarová, M., Matejovič, P., Pásek, M., Ohlídalová, D., Jansová, D., Šimurdová, M., & Šimurda, J. (2010). Effect of ethanol on action potential and ionic membrane currents in rat ventricular myocytes. *Acta Physiologica, 200*, 301–314.

Becker, C. M., Hoch, W., & Betz, H. (1988). Glycine receptor heterogeneity in rat spinal cord during postnatal development. *The EMBO Journal, 7*, 3717–3726.

Begleiter, H., & Kissin, B. (1996). *The pharmacology of alcohol and alcohol dependence* (1st ed.). Oxford: Oxford University Press.

Bikle, D. D., Gee, E. A., & Munson, S. J. (1986). Effect of ethanol on intestinal calcium transport in chicks. *Gastroenterology, 91*, 870–876.

Bitler, B., & McElroy, W. D. (1957). The preparation and properties of crystalline firefly luciferin. *Archives of Biochemistry and Biophysics, 72*, 358–368.

Blednov, Y. A., Benavidez, J. M., Black, M., Leiter, C. R., Osterndorff-Kahanek, E., Johnson, D., ... Harris, R. A. (2014). $GABA_A$ receptors containing ρ1 subunits contribute to in vivo effects of ethanol in mice. *PLoS One, 9*, e85525.

Borghese, C. M., Ruiz, C. I., Lee, U. S., Cullins, M. A., Bertaccini, E. J., Trudell, J. R., & Harris, R. A. (2016). Identification of an inhibitory alcohol binding site in $GABA_A$ ρ1 receptors. *ACS Chemical Neuroscience, 7*, 100–108.

Bormann, J. (2000). The "ABC" of GABA receptors. *Trends in Pharmacological Sciences, 21*, 16–19.

Bowery, N. (1989). $GABA_B$ receptors and their significance in mammalian pharmacology. *Trends in Pharmacological Sciences, 10*, 401–407.

Bradley, R. J., Peper, K., & Sterz, R. (1980). Postsynaptic effects of ethanol at the frog neuromuscular junction. *Nature, 284*, 60–62.

Brodie, M. S., Scholz, A., Weiger, T. M., & Dopico, A. M. (2007). Ethanol interactions with calcium-dependent potassium channels. *Alcoholism, Clinical and Experimental Research, 31*, 1625–1632.

Buck, K. J., Allan, A. M., & Adron Harris, R. (1989). Fluidization of brain membranes by A_2C does not produce anesthesia and does not augment muscimol-stimulated $^{36}Cl^-$ influx. *European Journal of Pharmacology, 160*, 359–367.

Burgos, C. F., Muñoz, B., Guzman, L., & Aguayo, L. G. (2015). Ethanol effects on glycinergic transmission: From molecular pharmacology to behavior responses. *Pharmacological Research: The Official Journal of the Italian Pharmacological Society, 101*, 18–29.

Cardoso, R. A., Brozowski, S. J., Chavez-Noriega, L. E., Harpold, M., Valenzuela, C. F., & Harris, R. A. (1999). Effects of ethanol on recombinant human neuronal nicotinic acetylcholine receptors expressed in *Xenopus* oocytes. *The Journal of Pharmacology and Experimental Therapeutics, 289*, 774–780.

Carlen, P., Gurevich, N., & Durand, D. (1982). Ethanol in low doses augments calcium-mediated mechanisms measured intracellularly in hippocampal neurons. *Science, 215*, 306–309 (80).

Carta, M., Ariwodola, O. J., Weiner, J. L., & Valenzuela, C. F. (2003). Alcohol potently inhibits the kainate receptor-dependent excitatory drive of hippocampal interneurons. *Proceedings of the National Academy of Sciences of the United States of America, 100*, 6813–6818.

Catterall, W. A., & Swanson, T. M. (2015). Structural basis for pharmacology of voltage-gated sodium and calcium channels. *Molecular Pharmacology, 88*, 141–150.

CDC. (2016). *Impaired driving: Get the facts* [WWW document] URL http://www.cdc.gov/motorvehiclesafety/impaired_driving/impaired-drv_factsheet.html.

Cebers, G., Cebere, A., Zharkovsky, A., & Liljequist, S. (1996). Glycine does not reverse the inhibitory actions of ethanol on NMDA receptor functions in cerebellar granule cells. *Naunyn-Schmiedeberg's Archives of Pharmacology, 354*, 736–745.

Celentano, J. J., Gibbs, T. T., & Farb, D. H. (1988). Ethanol potentiates GABA- and glycine-induced chloride currents in chick spinal cord neurons. *Brain Research, 455*, 377–380.

Chin, J. H., & Goldstein, D. B. (1977). Effects of low concentrations of ethanol on the fluidity of spin-labeled erythrocyte and brain membranes. *Molecular Pharmacology, 13*, 435–441.

Chin, J. H., Parsons, L. M., & Goldstein, D. B. (1978). Increased cholesterol content of erythrocyte and brain membranes in ethanol-tolerant mice. *Biochimica et Biophysica Acta (BBA) – Biomembranes, 513*, 358–363.

Chiou, J.-S., & Ueda, I. (1994). Ethanol unfolds firefly luciferase while competitive inhibitors antagonize unfolding: DSC and FTIR analyses. *Journal of Pharmaceutical and Biomedical Analysis, 12*, 969–975.

Chu, B., Anantharam, V., & Treistman, S. N. (1995). Ethanol inhibition of recombinant heteromeric NMDA channels in the presence and absence of modulators. *Journal of Neurochemistry, 65*, 140–148.

Costa, E. T., Soto, E. E., Cardoso, R. A., Olivera, D. S., & Valenzuela, C. F. (2000). Acute effects of ethanol on kainate receptors in cultured hippocampal neurons. *Alcoholism, Clinical and Experimental Research, 24*, 220–225.

Criswell, H. E., Simson, P. E., Duncan, G. E., McCown, T. J., Herbert, J. S., Morrow, A. L., & Breese, G. R. (1993). Molecular basis for regionally specific action of ethanol on gamma-aminobutyric acida receptors: Generalization to other ligand-gated ion channels. *The Journal of Pharmacology and Experimental Therapeutics, 267*, 522–537.

Dani, J. A. (2015). Neuronal nicotinic acetylcholine receptor structure and function and response to nicotine. In *International review of neurobiology* (pp. 3–19). Elsevier Inc.

Davies, D. L., Kochegarov, A. A., Kuo, S. T., Kulkarni, A. A., Woodward, J. J., King, B. F., & Alkana, R. L. (2005). Ethanol differentially affects ATP-gated $P2X_3$ and $P2X_4$ receptor subtypes expressed in *Xenopus* oocytes. *Neuropharmacology, 49*, 243–253.

Davies, D. L., Machu, T. K., Guo, Y., & Alkana, R. L. (2002). Ethanol sensitivity in ATP-gated P2X receptors is subunit dependent. *Alcoholism, Clinical and Experimental Research, 26*, 773–778.

Derry, J. M. C., Dunn, S. M. J., & Davies, M. (2004). Identification of a residue in the γ-aminobutyric acid type A receptor α subunit that differentially affects diazepam-sensitive and -insensitive benzodiazepine site binding. *Journal of Neurochemistry, 88*, 1431–1438.

Dildy-Mayfield, J. E., & Harris, R. A. (1995). Ethanol inhibits kainate responses of glutamate receptors expressed in *Xenopus* oocytes: Role of calcium and protein kinase C. *The Journal of Neuroscience: The Official Journal of the Society for Neuroscience, 15*, 3162–3171.

Dombek, K. M., & Ingram, L. O. (1984). Effects of ethanol on the *Escherichia coli* plasma membrane. *Journal of Bacteriology, 157*, 233–239.

Dopico, A. M., Bukiya, A. N., & Martin, G. E. (2014). Ethanol modulation of mammalian BK channels in excitable tissues: Molecular targets and their possible contribution to alcohol-induced altered behavior. *Frontiers in Physiology, 5*, 1–13.

Dopico, A. M., Lemos, J. R., & Treistman, S. N. (1996). Ethanol increases the activity of large conductance, Ca^{2+}-activated K^+ channels in isolated neurohypophysial terminals. *Molecular Pharmacology, 49*, 40–48.

Durand, D., Corrigall, W. A., Kujtan, P., & Carlen, P. L. (1981). Effects of low concentrations of ethanol on CA1 hippocampal neurons in vitro. *Canadian Journal of Physiology and Pharmacology, 59*, 979–984.

Dutertre, S., Becker, C.-M., & Betz, H. (2012). Inhibitory glycine receptors: An update. *The Journal of Biological Chemistry, 287*, 40216–40223.

Eckert, G. P., Schaeffer, E. L., Schmitt, A., Maras, A., & Gattaz, W. F. (2011). Increased brain membrane fluidity in schizophrenia. *Pharmacopsychiatry, 44*, 161–162.

Egawa, J., Pearn, M. L., Lemkuil, B. P., Patel, P. M., & Head, B. P. (2016). Membrane/lipid rafts and neurobiology: Age-related changes in membrane lipids and loss of neuronal function. *The Journal of Physiology, 594*, 4565–4579.

Eggers, E. D., & Berger, A. J. (2004). Mechanisms for the modulation of native glycine receptor channels by ethanol. *Journal of Neurophysiology, 91*, 2685–2695.

Eggers, E. D., O'Brien, J. A., & Berger, A. J. (2000). Developmental changes in the modulation of synaptic glycine receptors by ethanol. *Journal of Neurophysiology, 84*, 2409–2416.

Emson, P. C. (2007). $GABA_B$ receptors: Structure and function. *Progress in Brain Research, 160*, 43–57.

Engblom, A. C., & Akerman, K. E. (1991). Effect of ethanol on gamma-aminobutyric acid and glycine receptor-coupled Cl^- fluxes in rat brain synaptoneurosomes. *Journal of Neurochemistry, 57*, 384–390.

Ericson, M., Molander, A., Löf, E., Engel, J. A., & Söderpalm, B. (2003). Ethanol elevates accumbal dopamine levels via indirect activation of ventral tegmental nicotinic acetylcholine receptors. *European Journal of Pharmacology, 467*, 85–93.

Fasoli, F., & Gotti, C. (2015). Structure of neuronal nicotinic receptors. In D. J. K. Balfour, & M. R. Munafò (Eds.), *Current topics in behavioral neurosciences* (pp. 1–17). Cham: Springer International Publishing.

Fleuret-Balter, C., Beaugé, F., Barin, F., Nordmann, J., & Nordmann, R. (1983). Brain membrane disordering by administration of a single ethanol dose. *Pharmacology, Biochemistry, and Behavior, 18*, 25–29.

Forman, S. A., Righi, D. L., & Miller, K. W. (1989). Ethanol increases agonist affinity for nicotinic receptors from *Torpedo*. *Biochimica et Biophysica Acta (BBA) – Biomembranes, 987*, 95–103.

Fozzard, H., Lee, P., & Lipkind, G. (2005). Mechanism of local anesthetic drug action on voltage-gated sodium channels. *Current Pharmaceutical Design, 11*, 2671–2686.

Franks, N. P., & Lieb, W. R. (1982). Molecular mechanisms of general anaesthesia. *Nature, 300*, 487–493.

Franks, N. P., & Lieb, W. R. (1994). Molecular and cellular mechanisms of general anaesthesia. *Nature, 367*, 607–614.

Frye, G. D., & Fincher, A. (1996). Sensitivity of postsynaptic $GABA_B$ receptors on hippocampal CA1 and CA3 pyramidal neurons to ethanol. *Brain Research, 735*, 239–248.

Frye, G. D., & Fincher, A. (2000). Sustained ethanol inhibition of native AMPA receptors on medial septum/diagonal band (MS/DB) neurons. *British Journal of Pharmacology, 129*, 87–94.

Frye, G. D., Taylor, L., Trzeciakowski, J. P., & Griffith, W. H. (1991). Effects of acute and chronic ethanol treatment on pre- and postsynaptic responses to baclofen in rat hippocampus. *Brain Research, 560*, 84–91.

Gage, P. W., McBurney, R. N., & Schneider, G. T. (1975). Effects of some aliphatic alcohols on the conductance change caused by a quantum of acetylcholine at the toad end-plate. *The Journal of Physiology, 244*, 409–429.

Gastaldi, G., Casirola, D., Ferrari, G., Casasco, A., & Calligaro, A. (1991). The effect of ethanol and other alcohols on morphometric parameters of rat small intestinal microvillous vesicles. *European Journal of Basic and Applied Histochemistry, 35*, 185–193.

Giulietti, M., Vivenzio, V., Piva, F., Principato, G., Bellantuono, C., & Nardi, B. (2014). How much do we know about the coupling of G-proteins to serotonin receptors? *Molecular Brain, 7*, 49.

Glykys, J., Peng, Z., Chandra, D., Homanics, G. E., Houser, C. R., & Mody, I. (2007). A new naturally occurring $GABA_A$ receptor subunit partnership with high sensitivity to ethanol. *Nature Neuroscience, 10*, 40–48.

Goldstein, D. B., & Chin, J. H. (1981). Disordering effect of ethanol at different depths in the bilayer of mouse brain membranes. *Alcoholism, Clinical and Experimental Research, 5*, 256–258.

Gothert, M., & Fink, K. (1989). Inhibition of *N*-methyl-D-aspartate (NMDA)- and L-glutamate-induced noradrenaline and acetylcholine release in the rat brain by ethanol. *Naunyn-Schmiedeberg's Archives of Pharmacology, 340*, 516–521.

Green, A. A., & McElroy, W. D. (1956a). Crystalline firefly luciferase. *Biochimica et Biophysica Acta, 20*, 170–176.

Green, A. A., & McElroy, W. D. (1956b). A molecular mechanism of general anesthesia. *Biochimica et Biophysica Acta, 20*, 170–176.

Gruner, S. M., & Shyamsunder, E. (1991). Is the mechanism of general anesthesia related to lipid membrane spontaneous curvature? *Annals of the New York Academy of Sciences, 625*, 685–697.

Gruss, M., Henrich, M., Konig, P., Hempelmann, G., Vogel, W., & Scholz, A. (2001). Ethanol reduces excitability in a subgroup of primary sensory neurons by activation of BKCa channels. *The European Journal of Neuroscience, 14*, 1246–1256.

Guéguinou, M., Chantôme, A., Fromont, G., Bougnoux, P., Vandier, C., & Potier-Cartereau, M. (2014). KCa and Ca^{2+} channels: The complex thought. *Biochimica et Biophysica Acta (BBA) – Molecular Cell Research, 1843*, 2322–2333.

Gupta, D., Prabhakar, V., & Radhakrishnan, M. (2016). 5HT3 receptors: Target for new antidepressant drugs. *Neuroscience and Biobehavioral Reviews, 64*, 311–325.

Gutiérrez-Ruiz, M. C., Gómez, J. L., Souza, V., & Bucio, L. (1995). Chronic and acute ethanol treatment modifies fluidity and composition in plasma membranes of a human hepaic cell line (WRL-68). *Cell Biology and Toxicology, 11*, 69–78.

Hanchar, H. J., Chutsrinopkun, P., Meera, P., Supavilai, P., Sieghart, W., Wallner, M., & Olsen, R. W. (2006). Ethanol potently and competitively inhibits binding of the alcohol antagonist Ro15-4513 to $\alpha_4/_6\beta_3\delta$ $GABA_A$ receptors. *Proceedings of the National Academy of Sciences of the United States of America, 103*, 8546–8551.

Harris, R. A., & Bruno, P. (1985). Effects of ethanol and other intoxicant-anesthetics on voltage-dependent sodium channels of brain synaptosomes. *The Journal of Pharmacology and Experimental Therapeutics, 232*, 401–406.

Hayrapetyan, V., Jenschke, M., Dillon, G. H., & Machu, T. K. (2005). Co-expression of the $5\text{-}HT_{3B}$ subunit with the $5\text{-}HT_{3A}$ receptor reduces alcohol sensitivity. *Molecular Brain Research, 142*, 146–150.

Honse, Y., Ren, H., Lipsky, R. H., & Peoples, R. W. (2004). Sites in the fourth membrane-associated domain regulate alcohol sensitivity of the NMDA receptor. *Neuropharmacology, 46*, 647–654.

Horishita, T., & Harris, R. A. (2008). *n*-Alcohols inhibit voltage-gated Na^+ channels expressed in *Xenopus* oocytes. *The Journal of Pharmacology and Experimental Therapeutics, 326*, 270–277.

Hrelia, S., Lercker, G., Biagi, P. L., Bordoni, A., Stefanini, F., Zunarelli, P., & Rossi, C. A. (1986). Effect of ethanol intake on human erythrocyte membrane fluidity and lipid composition. *Biochemistry International, 12*, 741–750.

Huang, G.-J., & McArdle, J. J. (1994). Role of the GTP-binding protein G_o in the suppressant effect of ethanol on voltage-activated calcium channels of murine sensory neurons. *Alcoholism, Clinical and Experimental Research, 18*, 608–615.

Hunter, C. K., Treanor, L. L., Gray, J. P., Halter, S. A., Hoyumpa, A., & Wilson, F. A. (1983). Effects of ethanol in vitro on rat intestinal brush-border membranes. *Biochimica et Biophysica Acta (BBA) – Biomembranes, 732*, 256–265.

Jakab, M., Weiger, T. M., & Hermann, A. (1997). Ethanol activates maxi Ca^{2+}-activated K^+ channels of clonal pituitary (GH3) cells. *The Journal of Membrane Biology, 157*, 237–245.

Jenkins, A., Greenblatt, E. P., Faulkner, H. J., Bertaccini, E., Light, A., Lin, A., … Harrison, N. L. (2001). Evidence for a common binding cavity for three general anesthetics within the $GABA_A$ receptor. *The Journal of Neuroscience: The Official Journal of the Society for Neuroscience, 21*, RC136.

Jones, R. P., & Greenfield, P. F. (1987). Ethanol and the fluidity of the yeast plasma membrane. *Yeast, 3*, 223–232.

Kamaya, H., Ma, S., & Lin, S. H. (1986). Dose-dependent nonlinear response of the main phase-transition temperature of phospholipid membranes to alcohols. *The Journal of Membrane Biology, 90*, 157–161.

Kaur, K. H., Baur, R., & Sigel, E. (2009). Unanticipated structural and functional properties of δ-Subunit-containing $GABA_A$ receptors. *Journal of Biological Chemistry, 284*, 7889–7896.

Kiefer, F., Jahn, H., Koester, A., Montkowski, A., Reinscheid, R. K., & Wiedemann, K. (2003). Involvement of NMDA receptors in alcohol-mediated behavior: Mice with reduced affinity of the NMDA R1 glycine binding site display an attenuated sensitivity to ethanol. *Biological Psychiatry, 53*, 345–351.

Knapp, D. J., Braun, C. J., Duncan, G. E., Qian, Y., Fernandes, A., Crews, F. T., & Breese, G. R. (2001). Regional specificity of ethanol and NMDA action in brain revealed with FOS-like immunohistochemistry and differential routes of drug administration. *Alcoholism, Clinical and Experimental Research, 25*, 1662–1672.

Knoflach, F., Hernandez, M.-C., & Bertrand, D. (2016). (2016). $GABA_A$ receptor-mediated neurotransmission: Not so simple after all. *Biochemical Pharmacology, 115*, 10–17.

Kruger, L. C., & Isom, L. L. (2016). Voltage-gated Na^+ channels: Not Just for conduction. *Cold Spring Harbor Perspectives in Biology, 8*, a029264.

Krylov, B. V., Vilin, Y. Y., Katina, I. E., & Podzorova, S. A. (2000). Ethanol modulates the ionic permeability of sodium channels in rat sensory neurons. *Neuroscience and Behavioral Physiology, 30*, 331–337.

Läck, A. K., Ariwodola, O. J., Chappell, A. M., Weiner, J. L., & McCool, B. A. (2008). Ethanol inhibition of kainate receptor-mediated excitatory neurotransmission in the rat basolateral nucleus of the amygdala. *Neuropharmacology, 55*, 661–668.

Larsson, A., Svensson, L., Söderpalm, B., & Engel, J. A. (2002). Role of different nicotinic acetylcholine receptors in mediating behavioral and neurochemical effects of ethanol in mice. *Alcohol, 28*, 157–167.

Lima-Landman, M. T. R., & Albuquerque, E. X. (1989). Ethanol potentiates and blocks NMDA-activated single-channel currents in rat hippocampal pyramidal cells. *FEBS Letters, 247*, 61–67.

Linder, T. M., Pennefather, P., & Quastel, D. M. (1984). The time course of miniature endplate currents and its modification by receptor blockade and ethanol. *The Journal of General Physiology, 83*, 435–468.

Li, C., Peoples, R. W., & Weight, F. F. (1998). Ethanol-induced inhibition of a neuronal P2X purinoceptor by an allosteric mechanism. *British Journal of Pharmacology, 123*, 1–3.

Liu, J., Vaithianathan, T., Manivannan, K., Parrill, A., & Dopico, A. M. (2008). Ethanol modulates BKCa channels by acting as an adjuvant of calcium. *Molecular Pharmacology, 74*, 628–640.

Logan, B. J., & Laverty, R. (1987). Comparative effects of ethanol and other depressant drugs on membrane order in rat synaptosomes using ESR spectroscopy. *Alcohol and Drug Research, 7*, 11–24.

Lovinger, D. M. (1991). Ethanol potentiation of $5\text{-}HT_3$ receptor-mediated ion current in NCB-20 neuroblastoma cells. *Neuroscience Letters, 122*, 57–60.

Lovinger, D. M., Sung, K.-W., & Zhou, Q. (2000). Ethanol and trichloroethanol alter gating of $5\text{-}HT_3$ receptor-channels in NCB-20 neuroblastoma cells. *Neuropharmacology, 39*, 561–570.

Lovinger, D. M., & White, G. (1991). Ethanol potentiation of 5-hydroxytryptamine3 receptor-mediated ion current in neuroblastoma cells and isolated adult mammalian neurons. *Molecular Pharmacology, 40*, 263–270.

Lovinger, D., White, G., & Weight, F. (1989). Ethanol inhibits NMDA-activated ion current in hippocampal neurons. *Science, 243*, 1721–1724 (80).

Lovinger, D. M., White, G., & Weight, F. F. (1990). NMDA receptor-mediated synaptic excitation selectively inhibited by ethanol in hippocampal slice from adult rat. *The Journal of Neuroscience: The Official Journal of the Society for Neuroscience, 10*, 1372–1379.

Lovinger, D. M., & Zhou, Q. (1993). Trichloroethanol potentiation of 5-hydroxytryptamine3 receptor-mediated ion current in nodose ganglion neurons from the adult rat. *The Journal of Pharmacology and Experimental Therapeutics, 265*, 771–776.

Lovinger, D. M., & Zhou, Q. (1994). Alcohols potentiate ion current mediated by recombinant $5\text{-}HT_3RA$ receptors expressed in a mammalian cell line. *Neuropharmacology, 33*, 1567–1572.

Lynch, J. W. (2009). Native glycine receptor subtypes and their physiological roles. *Neuropharmacology, 56*, 303–309.

Lyon, R. C., & Goldstein, D. B. (1983). Changes in synaptic membrane order associated with chronic ethanol treatment in mice. *Molecular Pharmacology, 23*, 86–91.

Maas, J. W., Vogt, S. K., Chan, G. C. K., Pineda, V. V., Storm, D. R., & Muglia, L. J. (2005). Calcium-stimulated adenylyl cyclases are critical modulators of neuronal ethanol sensitivity. *The Journal of Neuroscience: The Official Journal of the Society for Neuroscience, 25*, 4118–4126.

Machu, T. K., & Harris, R. A. (1994). Alcohols and anesthetics enhance the function of 5-hydroxytryptamine3 receptors expressed in *Xenopus laevis* oocytes. *The Journal of Pharmacology and Experimental Therapeutics, 271*, 898–905.

Mameli, M., Zamudio, P. A., Carta, M., & Valenzuela, C. F. (2005). Developmentally regulated actions of alcohol on hippocampal glutamatergic transmission. *The Journal of Neuroscience: The Official Journal of the Society for Neuroscience, 25*, 8027–8036.

Marszalec, W., Aistrup, G. L., & Narahashi, T. (1999). Ethanol-nicotine interactions at α-bungarotoxin-insensitive nicotinic acetylcholine receptors in rat cortical neurons. *Alcoholism, Clinical and Experimental Research, 23*, 439–445.

Martz, A., Dietrich, R. A., & Harris, R. A. (1983). Behavioral evidence for the involvement of γ-aminobutyric acid in the actions of ethanol. *European Journal of Pharmacology, 89*, 53–62.

Mascia, M. P., Machu, T. K., & Harris, R. A. (1996). Enhancement of homomeric glycine receptor function by longchain alcohols and anaesthetics. *British Journal of Pharmacology, 119*, 1331–1336.

Mascia, M. P., Mihic, S. J., Valenzuela, C. F., Schofield, P. R., & Harris, R. A. (1996). A single amino acid determines differences in ethanol actions on strychnine-sensitive glycine receptors. *Molecular Pharmacology, 50*, 402–406.

Mascia, M. P., Wick, M. J., Martinez, L. D., & Harris, R. A. (1998). Enhancement of glycine receptor function by ethanol: Role of phosphorylation. *British Journal of Pharmacology, 125*, 263–270.

Maturu, P., Vaddi, D. R., Pannuru, P., & Nallanchakravarthula, V. (2010). Alterations in erythrocyte membrane fluidity and Na^+/K^+-ATPase activity in chronic alcoholics. *Molecular and Cellular Biochemistry, 339*, 35–42.

McElroy, W. D., & Green, A. (1956). Function of adenosine triphosphate in the activation of luciferin. *Archives of Biochemistry and Biophysics, 64*, 257–271.

Mihalek, R. M., Bowers, B. J., Wehner, J. M., Kralic, J. E., VanDoren, M. J., Morrow, A. L., & Homanics, G. E. (2001). $GABA_A$-receptor delta subunit knockout mice have multiple defects in behavioral responses to ethanol. *Alcoholism, Clinical and Experimental Research, 25*, 1708–1718.

Mihic, S. J. (1999). Acute effects of ethanol on $GABA_A$ and glycine receptor function. *Neurochemistry International, 35*, 115–123.

Mihic, S. J., & Harris, R. A. (1996). Inhibition of rho1 receptor GABAergic currents by alcohols and volatile anesthetics. *The Journal of Pharmacology and Experimental Therapeutics, 277*, 411–416.

Mihic, S. J., Ye, Q., Wick, M. J., Koltchine, V. V., Krasowski, M. D., Finn, S. E., … Harrison, N. L. (1997). Sites of alcohol and volatile anaesthetic action on $GABA_A$ and glycine receptors. *Nature, 389*, 385–389.

Minami, K., Gereau, R. W., Minami, M., Heinemann, S. F., & Harris, R. A. (1998). Effects of ethanol and anesthetics on type 1 and 5 metabotropic glutamate receptors expressed in *Xenopus laevis* oocytes. *Molecular Pharmacology, 53*, 148–156.

Moran, Y., Barzilai, M. G., Liebeskind, B. J., & Zakon, H. H. (2015). Evolution of voltage-gated ion channels at the emergence of Metazoa. *The Journal of Experimental Biology, 218*, 515–525.

Morrisett, R. A., & Swartzwelder, H. S. (1993). Attenuation of hippocampal long-term potentiation by ethanol: A patch-clamp analysis of glutamatergic and GABAergic mechanisms. *The Journal of Neuroscience: The Official Journal of the Society for Neuroscience, 13*, 2264–2272.

Möykkynen, T. P., Coleman, S. K., Keinänen, K., Lovinger, D. M., & Korpi, E. R. (2009). Ethanol increases desensitization of recombinant GluR-D AMPA receptor and TARP combinations. *Alcohol, 43*, 277–284.

Möykkynen, T., & Korpi, E. R. (2012). Acute effects of ethanol on glutamate receptors. *Basic & Clinical Pharmacology & Toxicology, 111*, 4–13.

Moykkynen, T., Korpi, E. R., & Lovinger, D. M. (2003). Ethanol inhibits α-Amino-3-hydyroxy-5-methyl-4-isoxazolepropionic acid (AMPA) receptor function in central nervous system neurons by stabilizing desensitization. *The Journal of Pharmacology and Experimental Therapeutics, 306*, 546–555.

Mullikin-Kilpatrick, D., & Treistman, S. N. (1995). Inhibition of dihydropyridine-sensitive Ca^{++} channels by ethanol in undifferentiated and nerve growth factor-treated PC12 cells: Interaction with the inactivated state. *The Journal of Pharmacology and Experimental Therapeutics, 272*, 489–497.

Mullin, M. J., & Hunt, W. A. (1985). Actions of ethanol on voltage-sensitive sodium channels: Effects on neurotoxin-stimulated sodium uptake in synaptosomes. *The Journal of Pharmacology and Experimental Therapeutics, 232*, 413–419.

Mullin, M. J., & Hunt, W. A. (1987). Actions of ethanol on voltage-sensitive sodium channels: Effects on neurotoxin binding. *The Journal of Pharmacology and Experimental Therapeutics, 242*, 536–540.

Nandini-Kishore, S. G., Mattox, S. M., Martin, C. E., & Thompson, G. A. (1979). Membrane changes during growth of Tetrahymena in the presence of ethanol. *Biochimica et Biophysica Acta, 551*, 315–327.

Narahashi, T., Aistrup, G. L., Marszalec, W., & Nagata, K. (1999). Neuronal nicotinic acetylcholine receptors: A new target site of ethanol. *Neurochemistry International, 35*, 131–141.

Narahashi, T., Kuriyama, K., Illes, P., Wirkner, K., Fischer, W., Muhlberg, K., … Sato, N. (2001). Neuroreceptors and ion channels as targets of alcohol. *Alcoholism, Clinical and Experimental Research, 25*, 182S–188S.

Nestoros, J. (1980). Ethanol specifically potentiates GABA-mediated neurotransmission in feline cerebral cortex. *Science, 209*, 708–710 (80).

Niciu, M. J., Kelmendi, B., & Sanacora, G. (2012). Overview of glutamatergic neurotransmission in the nervous system. *Pharmacology, Biochemistry, and Behavior, 100*, 656–664.

Nicoletti, F., Bockaert, J., Collingridge, G. L., Conn, P. J., Ferraguti, F., Schoepp, D. D., … Pin, J. P. (2011). Metabotropic glutamate receptors: From the workbench to the bedside. *Neuropharmacology, 60*, 1017–1041.

Nishio, M., & Narahashi, T. (1990). Ethanol enhancement of GABA-activated chloride current in rat dorsal root ganglion neurons. *Brain Research, 518*, 283–286.

Nusser, Z., & Mody, I. (2002). Selective modulation of tonic and phasic inhibitions in dentate gyrus granule cells. *Journal of Neurophysiology, 87*, 2624–2628.

Olsen, R. W., Hanchar, H. J., Meera, P., & Wallner, M. (2007). $GABA_A$ receptor subtypes: The "one glass of wine" receptors. *Alcohol, 41*, 201–209.

Olsen, R. W., & Sieghart, W. (2009). $GABA_A$ receptors: Subtypes provide diversity of function and pharmacology. *Neuropharmacology, 56*, 141–148.

Ostrovskaya, O., Asatryan, L., Wyatt, L., Popova, M., Li, K., Peoples, R. W., … Davies, D. L. (2011). Ethanol is a fast channel inhibitor of $P2X_4$ receptors. *The Journal of Pharmacology and Experimental Therapeutics, 337*, 171–179.

Ozawa, S. (1998). Glutamate receptors in the mammalian central nervous system. *Progress in Neurobiology, 54*, 581–618.

Peoples, R. W., Chaoying, L., & Weight, F. F. (1996). Lipid vs protein theories of alcohol action in the nervous system. *Annual Review of Pharmacology and Toxicology, 36*, 185–201.

Peoples, R. W., & Stewart, R. R. (2000). Alcohols inhibit *N*-methyl-D-aspartate receptors via a site exposed to the extracellular environment. *Neuropharmacology, 39*, 1681–1691.

Peoples, R. W., & Weight, F. F. (1992). Ethanol inhibition of *N*-methyl-D-aspartate-activated ion current in rat hippocampal neurons is not competitive with glycine. *Brain Research, 571*, 342–344.

Peoples, R. W., & Weight, F. F. (1995). Cutoff in potency implicates alcohol inhibition of *N*-methyl-D-aspartate receptors in alcohol intoxication. *Proceedings of the National Academy of Sciences of the United States of America, 92*, 2825–2829.

Pohorecky, L. A. (1977). Biphasic action of ethanol. *Biobehavioral Reviews, 1*, 231–240.

Pohorecky, L. A., & Brick, J. (1988). Pharmacology of ethanol. *Pharmacology & Therapeutics, 36*, 335–427.

Proctor, W. R., Allan, A. M., & Dunwiddie, T. V. (1992). Brain region-dependent sensitivity of $GABA_A$ receptor-mediated responses to modulation by ethanol. *Alcoholism, Clinical and Experimental Research, 16*, 480–489.

Puglia, M. P., & Valenzuela, C. F. (2010). Ethanol acutely inhibits ionotropic glutamate receptor-mediated responses and long-term potentiation in the developing CA1 Hippocampus. *Alcoholism, Clinical and Experimental Research, 34*, 594–606.

Ren, H., Salous, A. K., Paul, J. M., Lipsky, R. H., & Peoples, R. W. (2007). Mutations at F637 in the NMDA receptor NR2A subunit M3 domain influence agonist potency, ion channel gating and alcohol action. *British Journal of Pharmacology, 151*, 749–757.

Ren, H., Zhao, Y., Dwyer, D. S., & Peoples, R. W. (2012). Interactions among positions in the third and fourth membrane-associated domains at the intersubunit interface of the *N*-methyl-D-aspartate receptor forming sites of alcohol action. *Journal of Biological Chemistry, 287*, 27302–27312.

Ren, H., Zhao, Y., Wu, M., & Peoples, R. W. (2013). A novel alcohol-sensitive position in the *N*-methyl-D-Aspartate receptor GluN2A subunit M3 domain regulates agonist affinity and ion channel gating. *Molecular Pharmacology, 84*, 501–510.

Reynolds, J. N., & Prasad, A. (1991). Ethanol enhances $GABA_A$ receptor-activated chloride currents in chick cerebral cortical neurons. *Brain Research, 564*, 138–142.

Roberto, M., Schweitzer, P., Madamba, S. G., Stouffer, D. G., Parsons, L. H., & Siggins, G. R. (2004). Acute and chronic ethanol alter glutamatergic transmission in rat central amygdala: An in vitro and in vivo analysis. *The Journal of Neuroscience: The Official Journal of the Society for Neuroscience, 24*, 1594–1603.

Ronald, K. M. (2001). Ethanol inhibition of *N*-methyl-D-aspartate receptors is reduced by site-directed mutagenesis of a transmembrane domain phenylalanine residue. *Journal of Biological Chemistry, 276*, 44729–44735.

Rusch, D., Musset, B., Wulf, H., Schuster, A., & Raines, D. E. (2007). Subunit-dependent modulation of the 5-Hydroxytryptamine type 3 receptor open-close equilibrium by *n*-alcohols. *The Journal of Pharmacology and Experimental Therapeutics, 321*, 1069–1074.

Sanchez-Amate, M. C., Carrasco, M. P., Zurera, J. M., Segovia, J. L., & Marco, C. (1995). Persistence of the effects of ethanol in vitro on the lipid order and enzyme activities of chick-liver membranes. *European Journal of Pharmacology: Environmental Toxicology and Pharmacology, 292*, 215–221.

Santhakumar, V., Wallner, M., & Otis, T. S. (2007). Ethanol acts directly on extrasynaptic subtypes of $GABA_A$ receptors to increase tonic inhibition. *Alcohol, 41*, 211–221.

Sebe, J. Y., Eggers, E. D., & Berger, A. J. (2003). Differential effects of ethanol on $GABA_A$ and glycine receptor-mediated synaptic currents in brain stem motoneurons. *Journal of Neurophysiology, 90*, 870–875.

Seeman, P. (1972). The membrane actions of anesthetics and tranquilizers. *Pharmacological Reviews, 24*, 583–655.

Siggins, G. R., Pittman, Q. J., & French, E. D. (1987). Effects of ethanol on CA1 and CA3 pyramidal cells in the hippocampal slice preparation: An intracellular study. *Brain Research, 414*, 22–34.

Singer, J. H., Talley, E. M., Bayliss, D. A., & Berger, A. J. (1998). Development of glycinergic synaptic transmission to rat brain stem motoneurons. *Journal of Neurophysiology, 80*, 2608–2620.

Slater, S. J., Cox, K. J. A., Lombardi, J. V., Ho, C., Keily, M. B., Rubin, E., & Stubbs, C. D. (1993). Inhibition of protein kinase C by alcohols and anaesthetics. *Nature, 364*, 82–84.

Slater, S. J., Kelly, M. B., Larkin, J. D., Ho, C., Mazurek, A., Taddeo, F. J., ... Stubbs, C. D. (1997). Interaction of alcohols and anesthetics with protein kinase C. *Journal of Biological Chemistry, 272*, 6167–6173.

Smart, T. G. (1997). Regulation of excitatory and inhibitory neurotransmitter-gated ion channels by protein phosphorylation. *Current Opinion in Neurobiology, 7*, 358–367.

Smothers, C. T., Jin, C., & Woodward, J. J. (2013). Deletion of the N-Terminal domain alters the ethanol inhibition of *N*-methyl-D-aspartate receptors in a subunit-dependent manner. *Alcoholism, Clinical and Experimental Research, 37*, 1882–1890.

Smothers, C. T., & Woodward, J. J. (2006). Effects of amino acid substitutions in transmembrane domains of the NR1 subunit on the ethanol inhibition of recombinant *N*-methyl-D-aspartate receptors. *Alcoholism, Clinical and Experimental Research, 30*, 523–530.

Soldo, B. L., Proctor, W. R., & Dunwiddie, T. V. (1994). Ethanol differentially modulates $GABA_A$ receptor-mediated chloride currents in hippocampal, cortical, and septal neurons in rat brain slices. *Synapse, 18*, 94–103.

Solem, M., McMahon, T., & Messing, R. O. (1997). Protein kinase A regulates regulates inhibition of N- and P/Q-type calcium channels by ethanol in PC12 cells. *The Journal of Pharmacology and Experimental Therapeutics, 282*, 1487–1495.

Suzdak, P. D., Schwartz, R. D., Skolnick, P., & Paul, S. M. (1986). Ethanol stimulates gamma-aminobutyric acid receptor-mediated chloride transport in rat brain synaptoneurosomes. *Proceedings of the National Academy of Sciences of the United States of America, 83*, 4071–4075.

Tamura, K., Kaminoh, Y., Kamaya, H., & Ueda, I. (1991). High pressure antagonism of alcohol effects on the main phase-transition temperature of phospholipid membranes: Biphasic response. *Biochimica et Biophysica Acta (BBA) — Biomembranes, 1066*, 219–224.

Tapia, J. C., Aguilar, L. F., Sotomayor, C. P., & Aguayo, L. G. (1998). Ethanol affects the function of a neurotransmitter receptor protein without altering the membrane lipid phase. *European Journal of Pharmacology, 354*, 239–244.

Tatebayashi, H., Motomura, H., & Narahashi, T. (1998). Alcohol modulation of single $GABA_A$ receptor-channel kinetics. *Neuroreport, 9*, 1769–1775.

Ticku, M. K., Lowrimore, P., & Lehoullier, P. (1986). Ethanol enhances GABA-induced ^{36}Cl-influx in primary spinal cord cultured neurons. *Brain Research Bulletin, 17*, 123–126.

Traynelis, S. F., Wollmuth, L. P., McBain, C. J., Menniti, F. S., Vance, K. M., Ogden, K. K., ... Dingledine, R. (2010). Glutamate receptor ion channels: Structure, regulation, and function. *Pharmacological Reviews, 62*, 405–496.

Twombly, D. A., Herman, M. D., Kye, C. H., & Narahashi, T. (1990). Ethanol effects on two types of voltage-activated calcium channels. *The Journal of Pharmacology and Experimental Therapeutics, 254*, 1029–1037.

Ueda, I. (1956). Effects of diethyl ether and halothane on firefly luciferin bioluminescence. *Biochimica et Biophysica Acta, 20*, 170–176.

Ueda, I., & Kamaya, H. (1973). Kinetic and thermodynamic aspects of the mechanism of general anesthesia in a model system of firefly luminescence in vitro. *Anesthesiology, 38*, 425–436.

Valenzuela, C. F., & Cardoso, R. A. (1999). Acute effects of ethanol on kainate receptors with different subunit compositions. *The Journal of Pharmacology and Experimental Therapeutics, 288*, 1199–1206.

Valenzuela, C. F., Cardoso, R. A., Lickteig, R., Browning, M. D., & Nixon, K. M. (1998). Acute effects of ethanol on recombinant kainate receptors: Lack of role of protein phosphorylation. *Alcoholism, Clinical and Experimental Research, 22*, 1292–1299.

Vengeliene, V., Bilbao, A., Molander, A., & Spanagel, R. (2009). Neuropharmacology of alcohol addiction. *British Journal of Pharmacology, 154*, 299–315.

Vilpoux, C., Warnault, V., Pierrefiche, O., Daoust, M., & Naassila, M. (2009). Ethanol-sensitive brain regions in rat and mouse: A cartographic review, using immediate early gene expression. *Alcoholism, Clinical and Experimental Research, 33*, 945–969.

Wafford, K. A., Burnett, D. M., Leidenheimer, N. J., Burt, D. R., Wang, J. B., Kofuji, P., ... Sikela, J. M. (1991). Ethanol sensitivity of the $GABA_A$ receptor expressed in *Xenopus* oocytes requires 8 amino acids contained in the γ2L subunit. *Neuron, 7*, 27–33.

Wafford, K. A., & Whiting, P. J. (1992). Ethanol potentiation of $GABA_A$ receptors requires phosphorylation of the alternatively spliced variant of the γ2 subunit. *FEBS Letters, 313*, 113–117.

Wallner, M., Hanchar, H. J., & Olsen, R. W. (2003). Ethanol enhances $\alpha_4\beta_3\delta$ and $\alpha_6\beta_3\delta$ γ-aminobutyric acid type A receptors at low concentrations known to affect humans. *Proceedings of the National Academy of Sciences of the United States of America, 100*, 15218–15223.

Wallner, M., Hanchar, H. J., & Olsen, R. W. (2006a). Low-dose alcohol actions on $\alpha_4\beta_3\delta$ $GABA_A$ receptors are reversed by the behavioral alcohol antagonist Ro15-4513. *Proceedings of the National Academy of Sciences of the United States of America, 103*, 8540–8545.

Wallner, M., Hanchar, H. J., & Olsen, R. W. (2006b). Low dose acute alcohol effects on $GABA_A$ receptor subtypes. *Pharmacology & Therapeutics, 112*, 513–528.

Walter, H. J., & Messing, R. O. (1999). Regulation of neuronal voltage-gated calcium channels by ethanol. *Neurochemistry International, 35*, 95–101.

Wang, X., Wang, G., Lemos, J. R., & Treistman, S. N. (1994). Ethanol directly modulates gating of a dihydropyridine-sensitive Ca^{2+} channel in neurohypophysial terminals. *The Journal of Neuroscience: The Official Journal of the Society for Neuroscience, 14*, 5453–5460.

Webb, T. I., & Lynch, J. W. (2007). Molecular pharmacology of the glycine receptor chloride channel. *Current Pharmaceutical Design, 13*, 2350–2367.

Weight, F. F., Li, C., & Peoples, R. W. (1999). Alcohol action on membrane ion channels gated by extracellular ATP (P2X receptors). *Neurochemistry International, 35*, 143–152.

Weiner, J. L., Dunwiddie, T. V., & Valenzuela, C. F. (1999). Ethanol inhibition of synaptically evoked kainate responses in rat hippocampal CA3 pyramidal neurons. *Molecular Pharmacology, 56*, 85–90.

White, G., Lovinger, D. M., & Weight, F. F. (1990). Ethanol inhibits NMDA-activated current but does not alter GABA-activated current in an isolated adult mammalian neuron. *Brain Research, 507*, 332–336.

Wick, M. J., Mihic, S. J., Ueno, S., Mascia, M. P., Trudell, J. R., Brozowski, S. J., ... Harris, R. A. (1998). Mutations of γ-aminobutyric acid and glycine receptors change alcohol cutoff: Evidence for an alcohol receptor? *Proceedings of the National Academy of Sciences of the United States of America, 95*, 6504–6509.

Wieland, H. A., Lüddens, H., & Seeburg, P. H. (1992). A single histidine in $GABA_A$ receptors is essential for benzodiazepine agonist binding. *Journal of Biological Chemistry, 267*, 1426–1429.

Winstock, A. R. (2014). *The global drug survey 2014 findings* [WWW document] URL http://www.globaldrugsurvey.com/the-global-drug-survey-2014-findings/.

Wirkner, K., Eberts, C., Poelchen, W., Allgaier, C., & Illes, P. (2000). Mechanism of inhibition by ethanol of NMDA and AMPA receptor channel functions in cultured rat cortical neurons. *Naunyn-Schmiedeberg's Archives of Pharmacology, 362*, 568–576.

Wirkner, K., Poelchen, W., Köles, L., Mühlberg, K., Scheibler, P., Allgaier, C., ... Illes, P. (1999). Ethanol-induced inhibition of NMDA receptor channels. *Neurochemistry International, 35*, 153–162.

Witkin, J., Marek, G., Johnson, B., & Schoepp, D. (2007). Metabotropic glutamate receptors in the control of mood disorders. *CNS & Neurological Disorders Drug Targets, 6*, 87–100.

Wood, S. C., Forman, S. A., & Miller, K. W. (1991). Short chain and long chain alkanols have different sites of action on nicotinic acetylcholine receptor channels from *Torpedo*. *Molecular Pharmacology, 39*, 332–338.

Woodward, J. J., & Gonzales, R. A. (1990). Ethanol inhibition of *N*-methyl-D-aspartate-stimulated endogenous dopamine release from rat striatal slices: Reversal by Glycine. *Journal of Neurochemistry, 54*, 712–715.

Wright, J. M., Peoples, R. W., & Weight, F. F. (1996). Single-channel and whole-cell analysis of ethanol inhibition of NMDA-activated currents in cultured mouse cortical and hippocampal neurons. *Brain Research, 738*, 249–256.

Wu, Z., Cheng, H., Jiang, Y., Melcher, K., & Xu, H. E. (2015). Ion channels gated by acetylcholine and serotonin: Structures, biology, and drug discovery. *Acta Pharmacologica Sinica, 36*, 895–907.

Wu, G., Tonner, P. H., & Miller, K. W. (1994). Ethanol stabilizes the open channel state of the *Torpedo* nicotinic acetylcholine receptor. *Molecular Pharmacology, 45*, 102–108.

Xu, M., Smothers, C. T., & Woodward, J. J. (2015). Cysteine substitution of transmembrane domain amino acids alters the ethanol inhibition of GluN1/GluN2A *N*-methyl-D-aspartate receptors. *The Journal of Pharmacology and Experimental Therapeutics, 353*, 91–101.

Yang, X., Criswell, H. E., Simson, P., Moy, S., & Breese, G. R. (1996). Evidence for a selective effect of ethanol on *N*-methyl-D-aspartate responses: Ethanol affects a subtype of the ifenprodil-sensitive *N*-methyl-D-aspartate receptors. *The Journal of Pharmacology and Experimental Therapeutics, 278*, 114–124.

Yevenes, G. E., Moraga-Cid, G., Peoples, R. W., Schmalzing, G., & Aguayo, L. G. (2008). A selective Gβγ-linked intracellular mechanism for modulation of a ligand-gated ion channel by ethanol. *Proceedings of the National Academy of Sciences of the United States of America, 105*, 20523–20528.

Yoshimoto, K., McBride, W., Lumeng, L., & Li, T. (1992). Alcohol stimulates the release of dopamine and serotonin in the nucleus accumbens. *Alcohol, 9*, 17–22.

Yu, D., Zhang, L., Eiselé, J. L., Bertrand, D., Changeux, J. P., & Weight, F. F. (1996). Ethanol inhibition of nicotinic acetylcholine type alpha 7 receptors involves the amino-terminal domain of the receptor. *Molecular Pharmacology, 50*, 1010–1016.

Zhao, Y., Ren, H., Dwyer, D. S., & Peoples, R. W. (2015). Different sites of alcohol action in the NMDA receptor GluN2A and GluN2B subunits. *Neuropharmacology, 97*, 240–250.

Zheleznova, N. N., Sedelnikova, A., & Weiss, D. S. (2009). Function and modulation of δ-containing $GABA_A$ receptors. *Psychoneuroendocrinology, 34*, S67–S73.

Zhou, Q., Verdoorn, T. A., & Lovinger, D. M. (1998). Alcohols potentiate the function of $5\text{-}HT_3$ receptor-channels on NCB-20 neuroblastoma cells by favouring and stabilizing the open channel state. *The Journal of Physiology, 507*, 335–352.

C H A P T E R

17

Antioxidant Vitamins and Brain Dysfunction in Alcoholics

E. González-Reimers, G. Quintero-Platt, M.C. Martín-González, L. Romero-Acevedo, F. Santolaria-Fernández

Universidad de La Laguna, Tenerife, Canary Islands, Spain

INTRODUCTION

Many alcoholics display a wide spectrum of organic brain diseases (Charness, 1993) and functional alterations (Erdozain & Callado, 2014), including hippocampal atrophy (Wilhelm et al., 2008), frontal lobe cortical atrophy, and cerebellar atrophy. These alterations lead to cognitive dysfunction, which affects 50–80% of alcoholics (Bates, Bowden, & Barry, 2002; Fitzpatrick & Crowe, 2013), and can also cause movement disorders due to cerebellar alterations (observed in about 42% of nonsenile alcoholics) (Torvik & Torp, 1986). More uncommon manifestations include centropontine myelinolysis (Gille et al., 1993), Marchiafava–Bignami disease (Rawat, Pinto, Kulkarni, Muthusamy, & Dave, 2014), thiamine-deficiency derived Wernicke encephalopathy, pellagra, and other vitamin-deficient states (Ridley, Draper, & Withall, 2013). Comorbidity of several of these alterations is not uncommon (Butterworth, 1995; Lee, Jung, Na, Park, & Kim, 2005). Alcoholics also show cerebral blood flow alterations and ischemic stroke, and intraparenchymal hemorrhage is common in this group of patients (Nicolás et al., 1993).

As mentioned previously, hippocampal atrophy, frontal lobe cortical atrophy, and cerebellar atrophy are the most common manifestations of alcoholic brain damage. Both decreased neurogenesis and increased neurodegeneration contribute to brain injury.

Since mid-2000s, a bulk of evidence has incriminated oxidative damage as a major mechanism leading to brain alterations in the alcoholic. Cognitive impairment, together with atrophy of the brain that affects both gray matter and white matter (more or less marked in different areas of the brain), and possibly in association with the concurrent effect of specific micronutrient deficiencies, is the final result of a process in which reduced neurogenesis is coupled with increased neurodegeneration. Briefly, the metabolism of ethanol to acetaldehyde involves chemical reactions that are accompanied by increased production of reactive oxygen species (ROS). This is the case of CYP 2E1, a pathway associated with increased activity of nicotinamide adenine dinucleotide phosphate (NADPH) oxidase, that constitutes an important source of ROS (Cederbaum, 2006), since it is able to generate a superoxide anion when an electron is transferred from NADPH to oxygen. NADPH oxidase activity is enhanced in microglia, neurons, and astrocytes by a direct, up-regulating effect of ethanol (Qin & Crews, 2012).

Increased ROS may induce local synthesis of proinflammatory cytokines, which, in turn, promote generation of more ROS. These cytokines not only derive from local accumulation of ROS in the glia, but also from the effect of ethanol on intestinal permeability that allows Gram-negative bacilli to reach the brain and activate glial production of proinflammatory cytokines (Qin, Liu, Hong, & Crews, 2013). Additional mechanisms are involved in oxidative damage, including those related to micro-RNA induction and the subsequent activation of tumor necrosis factor alpha (TNF-α) and monocyte chemotactic protein-1 (MCP-1) by cerebellar microglia (Lippai, Bala, Csak, Kurt-Jones, & Szabo, 2013); mechanisms related to ethanol-mediated iron overload and altered permeability at the blood–brain barrier, that allows free iron to come into contact with brain parenchyma (Hua, Keep, Hoff, & Xi, 2007); and mechanisms related to ethanol-derived toxic lipid synthesis (de la Monte, Longato, Tong, DeNucci, & Wands, 2009). As reviewed elsewhere (González-Reimers et al., 2015), in addition to other concomitant causes that may

Addictive Substances and Neurological Disease
http://dx.doi.org/10.1016/B978-0-12-805373-7.00017-7

accompany heavy alcoholism, such as protein-calorie malnutrition (Santolaria et al., 2000), the effect of liver disease (Sutherland, Sheedy, Sheahan, Kaplan, & Kril, 2014), hepatic encephalopathy (Butterworth, 1995), or concurrent tobacco consumption (Durazzo et al., 2014).

As mentioned previously, in all of these mechanisms, oxidative damage is significantly involved (Crews & Nixon, 2009). Therefore, preservation of ROS scavenger machinery is of profound importance. These mechanisms include cellular enzymatic systems such as superoxide dismutases (SODs; Mailloux, 2015); catalase and glutathione peroxidase; ferritin, heme-oxygenase, and ceruloplasmin (Finazzi & Arosio, 2014; Müllebner, Moldzio, Redl, Kozlov, & Duvigneau, 2015; Samygina et al., 2013); and several other systems and compounds that act as cofactors for antioxidant enzymes (Lu & Cederbaum, 2008, p. 12 bis 5). Among the latter, antioxidant vitamins are of undisputed importance. The aim of the present work is to review some relevant aspects of the role of the main antioxidant vitamins, specifically vitamin A, vitamin E, folate, vitamin B12, vitamin B6, vitamin D, and vitamin C on brain alterations observed in alcoholic patients.

VITAMIN E DEFICIENCY

Vitamin E includes several compounds that have a chromanol ring system and a polyprenyl side chain in common. These compounds may be saturated (tocopherols) or threefold unsaturated (tocotrienols). The relative position of methyl or hydrogen radicals at the chromanol ring determines the existence of several homologues, termed alpha, beta, gamma, or delta (Falk & Munne–Bosch, 2010). Alpha tocopherol and gamma tocopherol are the most abundant in the western diet. Tocopherols are actively absorbed in the duodenum and incorporated into chylomicrons in order to be transported to the liver. In the liver, a cytochrome catabolic pathway (CYP450) shows a low affinity for alpha tocopherol, but a high affinity for gamma tocopherol (Sontag & Parker, 2002). Therefore, alpha tocopherol is not degraded, rather it is exported to peripheral organs, bound to carrier proteins, mostly to tocopherol transfer protein (TTP; Ulatowski & Manor, 2015). In peripheral organs, alpha tocopherol is typically localized in cell membranes and is able to quench peroxy radicals that are potentially harmful to polyunsaturated fatty acids. The action of vitamin E is likely coupled with ascorbate, a molecule that is able to recycle the oxidized molecule of tocopherol generated during this reaction (Packer, Slater, & Willson, 1979). Significantly, TTP is also expressed in the brain, especially in the Bergmann glial cells localized beneath Purkinje cells (Hosomi et al., 1998). The expression of TTP is dependent upon oxidative stress and on vitamin E availability (Ulatowski et al., 2012), and is highly variable in different areas of the brain, perhaps reflecting different sensitivity to the oxidative insult, so that TTP expression is a major regulator of the antioxidant action of vitamin E.

As explained earlier, vitamin E protects membrane phospholipids and polyunsaturated fatty acids from oxidation. TTP expression is highest in cerebellar astrocytes. In accordance with this fact, lesions associated with vitamin E deficiency are more marked in the cerebellum, especially in Purkinje cells (Ulatowski et al., 2014), providing a histological basis for spinocerebellar ataxia, a major manifestation of vitamin E deficiency. In addition, the protective effects of vitamin E also extends to other parts of the brain, including prevention of apoptosis of hippocampal cells induced by a high cholesterol diet (Reisi, Dashti, Shabrang, & Rashidi, 2014), and possibly on amygdala, Indeed, vitamin E deficiency is related to impaired learning and memory, and altered cognition (Ulatowski & Manor, 2015). The cognitive impairment observed in children with cystic fibrosis and vitamin E deficiency (Koscik et al., 2005) also supports this assertion. Indeed, in a study of 50 severely alcoholic patients admitted to our unit for alcohol-withdrawal syndrome, we found a significant relationship between vitamin E levels and mini-mental state examination scores ($\rho = 0.28$; $p = .046$), especially among men ($\rho = 0.32$; $p = .031$), a relationship that was independent on age.

Vitamin E deficiency is also involved in the prevention of arteriosclerosis (Azzi, Ricciarelli, & Zingg, 2002) by a non-antioxidant mechanism, and it also exerts a protective effect against Alzheimer's disease (Giraldo, Lloret, Fuchsberger, & Viña, 2014). In this sense, apoptosis and deposition of β amyloid proteins take place in vitamin E–deficient animals, especially affecting CA-1 pyramidal hippocampal cells (Fukui et al., 2005). Although it may be possible that these alterations are related to increased oxidative damage (Fukui et al., 2012), other altered metabolic pathways may also be involved (Saldeen, Li, & Mehta, 1999).

Several mechanisms may account for the low levels of vitamin E described in alcoholics (Tanner et al., 1986), including reduced intake, malabsorption clearly occurs in patients with primary biliary cirrhosis (Arria, Tarter, Warty, & Van Thiel, 1990), and possibly an increased demand of vitamin E by the liver (Kawase, Kato, & Lieber, 1989; Nordmann, 1994). Tiwari, Kuhad, and Chopra (2009) showed that treatment with tocopherol or tocotrienol reversed the ethanol-mediated increase in proinflammatory cytokines involved in ROS generation. This was accompanied by an improvement in cerebral performance. In the cerebellum, an increase in cytosolic iron has been documented in ethanol-treated animals. The increase of intracellular iron, in the face of low tocopherol and ascorbate, as well as decreased activity of antioxidant enzymes such as SOD and glutathione levels may be responsible for cerebellar damage.

In conclusion, vitamin E deficiency may exert several effects on the brain of alcoholics, especially on the cerebellum and hippocampus. Most of these effects are probably related to deficiency in antioxidant capacity, although anti-inflammatory, anti-arteriosclerotic, and antithrombotic effects cannot be fully excluded.

VITAMIN A DEFICIENCY

Alterations provoked by ethanol consumption on vitamin A metabolism are complex and has undergone plenty of research during the past decades. Despite this, some uncertainty still persists, due in part to some apparently contradictory results obtained by different research groups.

Vitamin A is ingested either as beta carotene, a compound widely available in vegetables, or as preformed vitamin A (including retinol, retinal, retinoic acid, and retinyl esters), especially abundant in liver, butter, kidney, and egg yolk.

As for other dietary lipids, proper absorption requires emulsification with biliary salts and fatty acids in the intestinal lumen, a process that is previous to the active transport mechanism from the lumen to the enterocyte.

Transport mechanisms differ among the type of vitamin A ingested. Beta carotene requires scavenger receptor class B type I for its absorption (Van Bennekum et al., 2005). Within the enterocyte, beta carotene is either cleaved into retinal, which is transformed into retinol by retinal reductase, or it is transferred to chylomicrons. Retinol is directly taken up by the enterocytes, but retinyl esters require the presence of pancreatic lipase–related peptide 1 and/or pancreatic triglyceride lipase, or retinyl ester hydrolase of the brush border (Van Bennekum, Fisher, Blaner, & Harrison, 2000).

Retinol is re-esterified within the enterocyte to form retinyl esters and then is incorporated into chylomicrons that are secreted into lymph. Lipoprotein lipase removes triglycerides from chylomicrons, leading to the formation of chylomicron remnants. The liver takes up between 66% and 75% of dietary retinoids contained in these lipoproteins (Goodman, Huang, & Shiratori, 1965). In this organ, retinoids are stored almost exclusively in hepatic stellate cells, in the form of retinyl esters, packaged into cytoplasmic lipid droplets (Blaner et al., 2009). If dietary deficit ensues, this form of vitamin A is transferred to hepatocytes as retinol after hydrolysis by a retinyl ester hydrolase (Clugston & Blaner, 2012).

Retinol is transported in serum bound to retinol-binding protein (RBP), and it is taken up by the peripheral cells after binding to a plasma membrane retinoid-binding protein receptor. Once within the cells, retinol is bound to cytoplasmic RBP type 1, and may be esterified by lecithin:retinol acyltransferase, or it may be transformed into retinal, and this last compound is transformed into retinoic acid. Retinoic acid (all-trans-retinoic acid) acts as a transcription factor after binding to nuclear receptors [retinoic acid receptors (RARs) alpha, beta, and gamma, to a different type of retinoid receptors named RXR alpha, beta, and gamma, and also to peroxisome proliferator–activated receptors (PPAR) beta and delta]. RAR forms heterodimers with RXR. Binding of a ligand to the receptor activates the complex, acting as gene transcriptors by binding to specific retinoic acid response elements, termed RAREs (Marill, Idres, Capron, Nguyen, & Chabot, 2003), a mechanism generally, but not universally, observed regarding the mode of action of vitamin A (Xia, Hu, Ketterer, & Taylor, 1996). More than 500 genes become activated in such a way, many of them involved in cell development, cellular differentiation, and cancer repression.

In addition to these effects, and to the well-known function of phototransduction, vitamin A also exerts antioxidant effects. For instance, in rat aorta, it was shown that vitamin A deficiency was associated with an increased oxidative stress (Gatica et al., 2005). Increased lipid peroxidation has also been shown in retinol deficient rats (Kaul & Krishnakantha, 1997). It is possible that these effects may explain in part some of the consequences derived from reduced vitamin A levels and/or impaired signaling in the brain. Indeed, age-related retinoid hyposignaling impairs spatial memory (Etchamendy et al., 2001), and in rats, vitamin A deficiency leads to decreased hippocampal plasticity (Misner et al., 2001) and neurogenesis (Bonnet et al., 2008). Although these effects may be unrelated to vitamin A antioxidant properties, they are indeed related to decreased vitamin A. Other studies also report an inhibitory effect of all-trans-retinoic acid on oxidative stress in the brains of alcoholized rats (Nair, Prathibha, Syam Das, Kavitha, & Indira, 2015), as well as on brain alterations mediated by other kinds of pro-oxidants (Banala & Karnati, 2015). A general agreement does exist regarding the finding of impaired memory and spatial learning in rats deprived from vitamin A (Cocco et al., 2002).

In contrast, prolonged treatment with either 13-cis- or all-trans-retinoic acid suppresses adult neurogenesis both in the hippocampal subgranular zone and in the subventricular zone (Crandall et al., 2004). In a similar sense, high concentrations of retinoic acid seem to have a negative effect on dendrite morphology and arborization, in contrast with low doses (Liu et al., 2008). Studies also showed that administration of retinyl palmitate is associated with deleterious changes in the brain that are related to altered redox status and are independent on liver dysfunction (Schnorr et al., 2015). These alterations are especially evident in the hippocampus (Schnorr et al., 2011). Therefore, it seems that the effects of vitamin A on some areas of the brain adapt to a U-shaped curve, although considerable debate exists regarding these contradictory results, not sustained by all studies (Touyarot et al., 2013).

As shown in Fig. 17.1, ethanol depletes liver retinol and retinyl ester contents (Leo & Lieber, 1982) in all the main cell types, but especially in hepatic stellate cells (Adachi et al., 1991), although it is unclear whether vitamin A supplementation exerts a positive effect or not on ethanol-mediated liver injury (Clugston & Blaner, 2012). A great deal of controversy exists about the effect of ethanol on serum and brain concentrations of the different metabolites of vitamin A. A detailed study by Kane, Folias, Wang, and Napoli (2010) has shown that ethanol, by promoting efflux of vitamin A from the liver, increases several fold all-trans-retinoic acid levels in the hippocampus, and less consistently, in other areas of the brain, as well as in other organs, also including serum levels. It was hypothesized that very high all-trans-retinoic acid concentrations in the brain may contribute to cognitive dysfunction; perhaps increases in all-trans-retinoic acid in the hippocampus obeys to an ethanol-induced increase in RAR alpha receptor. Ethanol may also alter cytoplasmic retinoid–binding protein expression and activity (Grummer & Zachman, 1995), and it is possible that ethanol inhibits the formation of retinoic acid from retinol (Sauvant et al., 2002), although there is also controversy regarding this item (Wolf, 2010). In any case, changes in vitamin A liver content are already observable during the early stages of ethanol-mediated liver injury, although these changes are more marked in advanced stages of the disease (Sato & Lieber, 1981).

In addition to changes in brain availability of vitamin A secondary to liver alterations, retinoic acid can be synthesized by the cerebellum (Yamamoto, Ullman, Dräger, & McCaffery, 1999), and ethanol may exert a direct, increasing effect on local cerebellar synthesis of

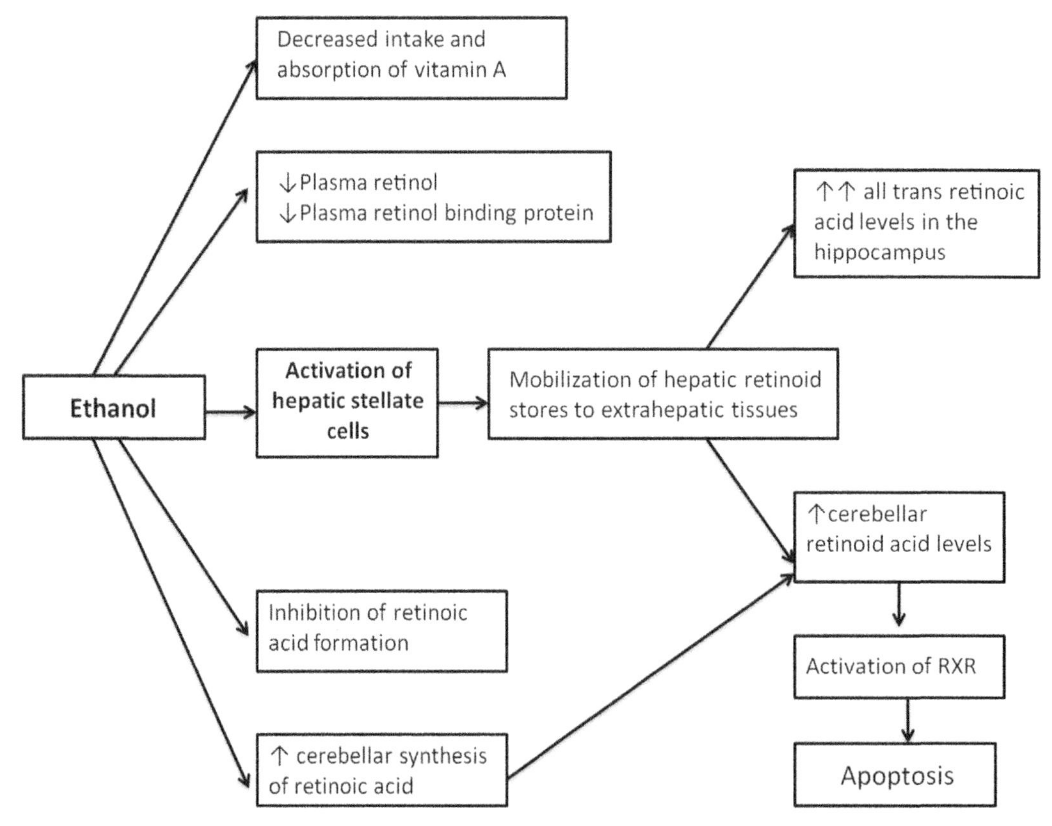

FIGURE 17.1 Putative mechanisms of the effects of ethanol on vitamin A metabolism and consequences in brain. In alcoholics, there may be a defective intake and absorption of vitamin A. Serum vitamin A levels are decreased and retinol-binding protein (RBP) levels are also decreased. Retinoic acid, the active metabolite of dietary vitamin A, may also be low in liver and plasma, partly due to an inhibitory effect of ethanol on retinoic synthesis from retinol. Additionally, ethanol depletes liver retinol and retinyl ether contents in all the main cell types, but especially in hepatic stellate cells. A great deal of controversy exists about the effect of ethanol on serum and brain concentrations of the different metabolites of vitamin A. Ethanol increases several fold all-trans-retinoic acid levels in the hippocampus, and this may contribute to cognitive dysfunction. Ethanol induces retinoic acid synthesis by the cerebellum. The cerebellum also expresses a high number of RXRs, the activation of which leads to increased apoptosis

this compound (McCaffery et al., 2004). The cerebellum also expresses a high number of RXRs, the activation of which leads to increased apoptosis. Ethanol differentially modulates retinoid receptors, reducing the expression of RAR alpha/gamma and increasing RXR alpha/gamma (Kumar, Singh, DiPette, & Singh, 2010).

In addition to the altered tissue distribution of retinoids purely caused by chronic ethanol ingestion, in alcoholics there may be a defective intake and absorption of vitamin A, which complicates the scenario. Serum vitamin A levels are decreased, and RBP levels are also decreased, and these two alterations also take place in patients with nonalcoholic forms of liver injury (Smith & Goodman, 1971). Decreased zinc may contribute to these changes, since this element is necessary for RBP mobilization (Smith, McDaniel, Fan, & Halsted, 1973). Retinoic acid, the active metabolite of dietary vitamin A, may also be low in liver and plasma, partly due to an inhibitory effect of ethanol on retinoic synthesis from retinol (Sauvant et al., 2002) and partly due to accelerated catabolism in chronic alcoholics with strong microsomal induction (Lieber, 1993). Although this view has been challenged, increased liver, extrahepatic tissues, and plasma retinoic acid levels have all been reported (Wolff, 2010). A similar debate exists regarding beta carotene levels: ethanol may actually increase beta carotene levels (Leo, Kim, Lowe, & Lieber, 1992), although this contradicts the reported finding of low beta carotene in alcoholics (Leo & Lieber, 1999). The net effect of altered carotene levels is even more obscure, as some studies show that beta carotene (Bowen & Omaye, 1998) may exert some pro-oxidant effects (Behr et al., 2012), while opposite results have been obtained regarding carotenoids have also been described, protecting unsaturated fatty acids from oxidative damage (Burton, 1989).

As mentioned earlier, both vitamin A deficiency and excess have negative effects on brain function in alcoholics. Disturbances in vitamin A metabolism can also alter redox status and contribute to brain damage. The role of this vitamin in brain dysfunction in alcoholics is a field of active research but contradictory results have hindered the development of a theory of brain damage induced by alterations in vitamin A. Therefore, in a similar fashion to that observed in tocopherol deficiency there is reduced hippocampal neurogenesis accompanied by functional impairment in vitamin A deficiency, especially affecting memory and spatial learning. Hippocampal and functional recovery was achieved, at least partially, with vitamin A supplementation (Cocco et al., 2002). Additionally, a 2015 report also showed that vitamin A supplementation is able to inhibit Krüppel-like factor 11, an inductor of neuronal cell death (Nair et al., 2015). However, there are contradictory results, such as those reported by Crandall et al. (2004). In accordance with the former, by 2014 we found low vitamin A levels among alcoholics that were related to brain atrophy (González- Reimers et al., 2014). Further research has led to the finding of a strong, inverse correlation between vitamin A levels and malondialdehyde ($\rho = -0.63$, $p < .001$), and also a relationship between vitamin A and mini-mental state examination scores ($\rho = 0.29$, $p = .039$), therefore supporting the fact that, in alcoholics, low serum vitamin A levels are related to cognitive impairment, probably through a pro-oxidant effect.

VITAMIN D DEFICIENCY

Vitamin D Metabolism

Although traditionally associated with bone metabolism, in the past decades it has been shown that vitamin D has pleiomorphic roles, and its' deficiency has been described in cancer and degenerative diseases (Christakos, Dhawan, Verstuyf, Verlinden, & Carmeliet, 2016). Vitamin D is a liposoluble prohormone that is obtained from dietary sources and from dermal synthesis. Exposure to ultraviolet rays catalyzes the synthesis of previtamin D3 from 7-dehydrocholesterol. 25-Hydroxyvitamin D [25(OH) D, calcidiol] is then synthesized in the liver by adding a hydroxyl group to the vitamin D molecule. This compound is inactive and is transformed to the active form of vitamin D—1,25-dihydroxyvitamin D—by renal 1-αhydroxylase. As discussed elsewhere (Wilkens-Knudsen et al., 2014), vitamin D deficiency is common among alcoholics and can be attributed to poor diet, lack of sun exposure, reduced hepatic 25-hydroxylase activity, increased renal 24-hydroxylase activity, malabsorption associated with portal hypertension, and pancreatic insufficiency.

Once the active form of vitamin D is synthesized, it traverses the cell membrane to reach the nuclear vitamin D receptor (VDR). This receptor has a DNA-binding domain and acts through vitamin D–responsive elements (VDREs), regulating gene expression. VDR expression has been described in several types of cells, which explains the broad range of effects of vitamin D. Eyles, Smith, Kinobe, Hewison, and McGrath (2005) described the distribution of VDR in the human brain. They found that VDR was expressed in both neurons and glial cells and that it was restricted to the cell nucleus. They found strong immuno-histochemical staining for VDR in the hypothalamus and in large neurons in the substantia nigra. The distribution of VDR in animal models has also been extensively described.

Thus, it is no surprise that vitamin D has been shown to be involved in several processes within the central nervous system, including brain development. Eyles, Brown, Mackay-Sim, McGrath, and Feron (2003) found that the brains of vitamin D–deficient newborn rats were larger than controls, had larger lateral ventricles, and reduced cortical thickness. They also found reduced expression of nerve growth factor and glial cell line–derived neurotrophic factor.

Neuroprotective Effects of Vitamin D

Vitamin D may act as an antioxidant in the brain, and it has been shown to exert a protective effect against endotoxin-induced neuroinflammation and against excitotoxicity (Huang, Ho, Lai, Chiu, & Wang, 2015). Excitotoxicity involves several aspects (Lewerenz & Maher, 2015): (1) glutamate-mediated activation of NMDA receptors; (2) subsequent increase in intracellular calcium concentration that leads to enzyme activation and derangement of cell structure and function; and (3) production of ROS such as nitric oxide (NO).

Taniura et al. (2006) found that vitamin D protected against neuronal cell death induced by glutamate exposure and also upregulated the expression of VDR in cultured neurons. In their study, chronic treatment of rat cortical neurons with vitamin D3 had neurotrophic effects: it increased the expression of microtubule-associated protein 2, growth-associated protein 43, and synapsin 1. Other neurotrophic effects of vitamin D have been described in other studies: it modulates the production of neurotrophin 3, glial cell–derived neurotrophic factor, nerve growth factor; it inhibits inducible nitric oxide synthase (iNOS) and choline acyltransferase (Balion et al., 2012); and it enhances amyloid beta peptide clearance (Lu'o'ng & Nguyen, 2013). In addition, it was observed that vitamin D supplementation—at low doses—increases neuronal glutathione levels, an effect that strongly supports an antioxidant role of vitamin D (Shinpo, Kikuchi, Sasaki, Moriwaka, & Tashiro, 2000).

Vitamin D is essential in maintaining the adequate levels of calcium within the cells. An excess of calcium within the nerve cell contributes to excitotoxicity and increased generation of ROS (Harms, Burne, Eyles, & McGrath, 2011). Brewer et al. (2001) hypothesized that vitamin D not only regulates calcium homeostasis in peripheral tissues but also in the central nervous system and that a possible target for vitamin D may be the L-type voltage-sensitive calcium channel (L-VSCC) in hippocampal neurons. They found that in rat hippocampal cultures, treatment with vitamin D at low concentrations led to neuroprotection against excitotoxic insults and also to lower L-VSCC currents. On the other hand, high concentrations of vitamin D (nonphysiological: 500–1000 nM) were not neuroprotective and increased L-VSCC currents.

Garcion et al. (1998) describe a model of brain inflammation in rats in which expression of iNOS is induced by intracellular injection of lipopolysaccharide (LPS). They found that injection of 1,25-dihydroxyvitamin D3 along with LPS inhibits expression of iNOS and decreases the proportion of apoptotic cells with an increase in macrophages that can, in turn, synthesize more vitamin D. Huang et al. (2015) studied cortical and glial cell cultures and found that LPS injection induced production of NO, ROS, interleukin-6 (IL-6), and macrophage inflammatory protein (MIP-2), and this was in turn inhibited by treatment with 1,25-dihydroxyvitamin D3. They found that the vitamin D–mediated reduction in inflammatory molecules was a result of the inhibition of MAPK pathways.

Alcohol-Induced Brain Damage and Clinical Observations

In alcoholics, renal metabolism may be diverted to the synthesis of the less active 24–25 dihydroxyvitamin D (Shankar et al., 2008). This fact, together with nutritional disturbances, malabsorption, and decreased sun exposure may explain the frequently observed low vitamin D levels in alcoholics (Naude, Carey, Laubscher, Fein, & Senekal, 2012) which might play a role in brain oxidative damage. As discussed earlier, vitamin D has a protective role in neuroinflammation, since it inhibits microglial production of TNF-α and IL-6 (Wrzosek et al., 2013). All of this has implications for alcohol-induced brain damage.

The neuroprotective effects of vitamin D have been studied in several types of neurological diseases. Low levels of vitamin D have been related to cognitive impairment (Chei et al., 2014; Dickens, Lang, Langa, Kos, & Llewellyn, 2011; Etgen, Sander, Bickel, Sander, & Förstl, 2012). Littlejohns et al. (2014) found an increased risk for all-cause dementia and Alzheimer's disease in patients with vitamin D deficiency compared to patients with normal vitamin D levels. Stein, Scherer, Ladd, and Harrison (2011) carried out a randomized controlled trial in patients with Alzheimer's disease but found that high-dose vitamin D provided no benefit for cognitive impairment. In a post hoc analysis of a randomized double-blind placebo-controlled trial Rossom et al. (2012) found no significant differences in incident dementia between patients who received vitamin D supplementation and those who received placebo. However, more studies are needed to define the effect of vitamin D supplementation on cognitive function.

Jorde, Sneve, Figenschau, Svartberg, and Waterloo (2008) have studied the link between vitamin D deficiency and depression. They found that in overweight and obese patients, vitamin D supplementation led to an improvement in depressive traits. The authors propose a possible causal relationship between vitamin D deficiency and depressive traits. However, more studies are needed that analyze the relationship between monoamines and vitamin D. Cass, Smith, and Peters (2006) found that repeated methamphetamine administration resulted in a decrease in brain dopamine and serotonin but that calcitriol protects against this effect in male rats.

Few studies have analyzed the relationship between altered vitamin D and brain atrophy in alcoholics. In the preliminary results of an ongoing study we have failed to find any relationship between vitamin D levels, altered cognition, or cerebellar or frontal atrophy among 47 alcoholic patients, although, indeed, calcidiol levels were inversely related to MDA ($\rho = -0.41$; $p = .045$) and TNF-α ($\rho = -0.46$; $p = .014$), underscoring the role of vitamin D deficiency on inflammation and oxidative damage.

VITAMIN B12, B6, AND FOLATE ALTERATIONS: HYPERHOMOCYSTEINEMIA

Cobalamin or vitamin B12 is heavily involved in DNA synthesis, so that its deficiency leads to megaloblastic changes in several cells, especially notorious in red blood cell precursors, enterocytes, and epidermal cells, among others (Kannan & Ng, 2008). In addition to megaloblastic changes, cyanocobalamin deficiency is associated with demyelination (Smith & Refsum, 2009), especially involving large myelinated nerve fibers, brain atrophy, and cognitive impairment (Briani et al., 2013). In demyelination associated with vitamin B12 deficiency, increased TNF-α and IL-6 values have been reported, suggesting the existence of a link between vitamin B12 deficiency and neuroinflammation (Scalabrino, Buccellato, Veber, & Mutti, 2003).

Together with folate and vitamin B6 (pyridoxine), vitamin B12 is also involved in homocysteine metabolism, a compound that plays a crucial role in increased risk of thrombosis, accelerated atherogenesis, and brain atrophy. It is therefore appropriate to review in conjunction the involvement of these three vitamins in oxidative damage and brain alterations in the alcoholic patient.

Cobalamin is abundant in animal products, especially meat. Body stores of this vitamin (2000–5000 μg) are enough to meet the daily metabolic requirements (6–9 μg) during some years. Cobalamin is absorbed at the ileum coupled to intrinsic factor, a protein secreted by the gastric parietal cells to which cobalamin binds when arriving at the duodenum. Gastric acid secretion is necessary to liberate cobalamin from binding to ingested proteins. After liberation, dietary cobalamin is bound to R proteins; this complex is broken by pancreatic proteases so that cobalamin can bind to intrinsic factor.

In plasma, it circulates to bound to transcobalamin II, which allows binding to specific cell receptors (Tefferi & Pruthi, 1994). However, it may also be absorbed by passive diffusion throughout the entire small intestine. High doses of cyanocobalamin in the absence of intrinsic factor may suffice to meet the requirements of a normal adult (Stabler, 2013).

Folate is present in many vegetables although a great proportion of it may be destroyed through cooking. Ingested folate is in the form of pteroylpolyglutamates, that are transformed in the enterocyte brush border to pteroylmonoglutamate by glutamate carboxypeptidase II. Pteroylmonoglutamate enters the portal circulation and is transported across the basolateral membrane of the hepatocyte to become re-polyglutamylated intracellularly, and stored as polyglutamates in the liver. Within the hepatocyte, when the stores are mobilized, methylated pteroylmonoglutamate is released and enters either enterohepatic circulation or reaches the systemic circulation. The kidney also regulates folate metabolism by resorbing about 90% of the filtered methylpteroylglutmate in the proximal tubule (Medici & Halsted, 2013).

Chronic alcoholics show decreased absorption of folate (Wani & Kaur, 2011), in contrast with the lack of reduction following an acute ethanol exposure (Halsted, Robles, & Mezey, 1971). In addition, alcoholic liver disease severely impairs folate metabolism and storage, also altering enterohepatic circulation. Moreover, acute ethanol exposure increases urinary folate excretion (Medici & Halsted, 2013). Obviously, poor intake also contributes but the importance of this is decreasing since the recently introduced recommendation of fortification of meals with folate.

As shown in Fig. 17.2, the metabolism of folate, vitamin B12, and vitamin B6 are intimately coupled. Vitamin B6 is markedly deficient in alcoholics (Lumeng & Li, 1974). The main reason may reside in the ability of acetaldehyde to displace vitamin B6 from its protein carrier (Lumeng, 1978), although low dietary intake may also contribute. Indeed, pyridoxal 5 phosphate, the active form of vitamin B6, if unbound to its carrier, is rapidly hydrolyzed by alkaline phosphatase (Li, Lumeng, & Veitch, 1974). One of the main consequences of B6 deficiency is that homocysteine cannot be degraded properly to cystathionine and cysteine, so that hyperhomocysteinemia ensues. From Fig. 17.1, it is also evident that vitamin B12 deficiency or folate deficiency are also main causes of hyperhomocysteinemia. However, hyperhomocysteinemia in alcoholics is more closely related to folate and/or vitamin B6 deficiency than to vitamin B12 deficiency (Heese et al., 2012). In fact, it is not uncommon to find high that vitamin B12 levels among advanced alcoholic cirrhotics

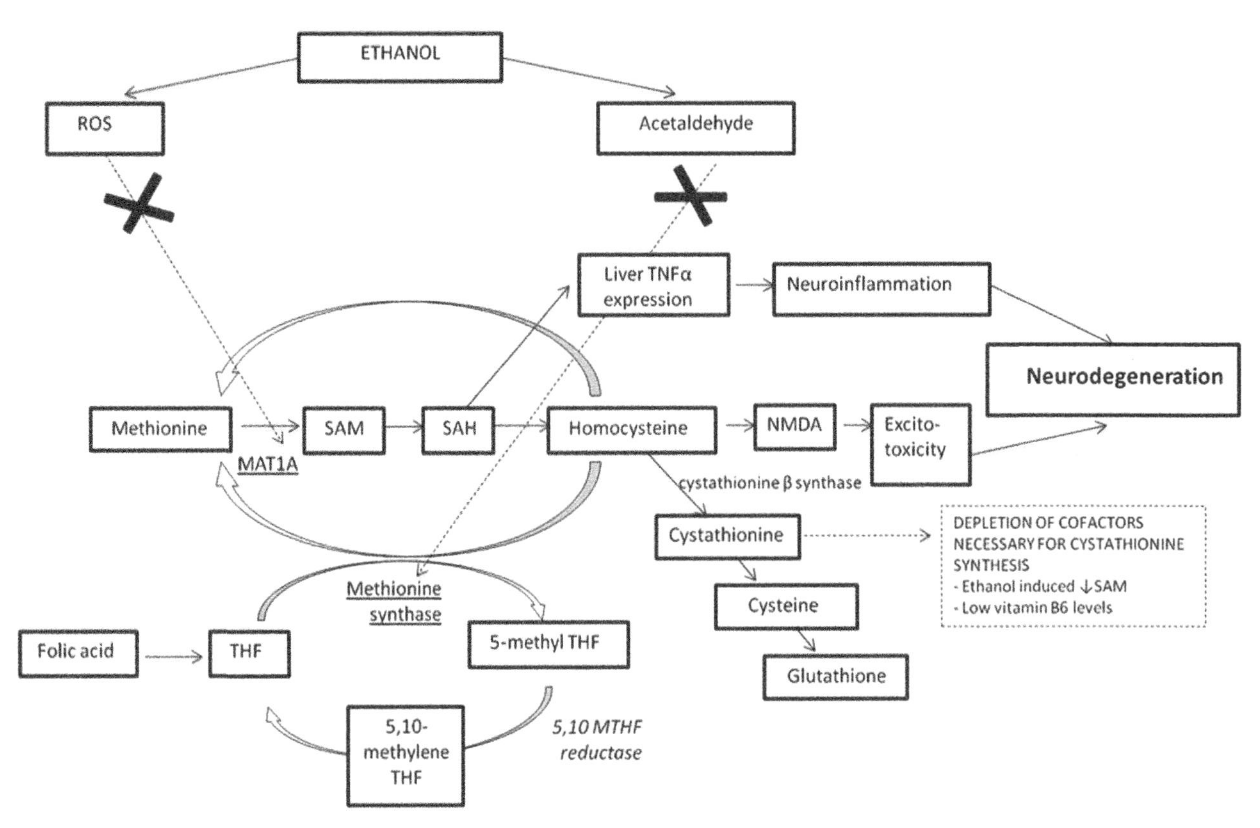

FIGURE 17.2 The effects of alcohol on homocysteine metabolism. Increased generation of ROS by ethanol metabolism impairs the activity of methionine adenosyl transferase (MAT1A), blocking the formation of *S*-adenosyl methionine (SAM). Since SAM is a cofactor for cystathionine beta synthase, and eventually B6 deficiency may further decrease the activity of this enzyme, catabolism of homocysteine is impaired. On the other hand, acetaldehyde inhibits methionine synthase activity. All of these mechanisms lead to increased homocysteine and renders an increased formation of *S*-adenosyl homocysteine, a compound that is able to enhance liver TNF-α expression. Increased TNF-α production may promote neuroinflammation. *MAT1A*, methionine adenosyl transferase; *MTHF*, methytetrahydrofolate; *SAH*, *S*-adenosyl homocysteine; *SAM*, *S*-adenosyl methionine; *THF*, tetrahydrofolate.

(Lambert et al., 1997). In a study on 113 alcoholics we found that serum vitamin B12 levels were significantly higher among patients categorized as Child class C (Fig. 17.3A), showing a direct, significant relationship with MDA levels (ρ = 0.534; $p < .001$, Fig. 17.3B). It is likely that vitamin B12–associated cognitive impairment adapts to a U-shaped curve (Castillo-Lancellotti et al., 2015). In the aforementioned study (unpublished data), patients with cerebellar atrophy showed higher vitamin B12 values than those without cerebellar atrophy (Fig. 17.3C).

Several reports have pointed out that hyperhomocysteinemia is a risk factor for thromboembolic disease (Cattaneo, 1999). In the meta-analysis performed by Ray (1998), the odds ratio for venous thromboembolism was 2.95 (2.08–4.17) when hyperhomocysteinemia was present, and this figure was higher (4.37) when only younger individuals (<60 years) were included. The high incidence of either arterial or venous thrombosis may be related to the well-described pro-oxidant effects of this compound which is able to induce sustained injury of arterial smooth muscle and endothelial cells, acting both on the vascular wall structure and the blood coagulation system (Guilland, Favier, Potier de Courcy, Galan, & Hercberg, 2003). In accordance with these observations, hyperhomocysteinemia has been associated with brain atrophy and cognitive impairment in elderly patients (Sachdev, 2005). Later, Yang et al. (2007), in 89 patients who suffered a stroke, found that brain atrophy was directly related to homocysteine, and, inversely, to serum folate levels, with an odds ratio of brain atrophy nearing 10 in the low folate group and hyperhomocysteinemia [9.6, CI (1.1–81.3) and 9.8 CI (1.7–56.4), respectively].

Folate is intimately related to cyanocobalamin and homocysteine metabolism: folate depletion causes demyelinating lesions in white matter in the spinal cord and brain (Ramaekers & Blau, 2004). In a community-based study Vogiatzoglou et al. (2008) report a relationship between low vitamin B12 and an increased rate of brain volume loss along a 5-year follow-up period in 107 volunteers aged 61–87 years, and to cognitive deficits, even in well-nourished people, as shown in a large cohort of 1935, community-dwelling

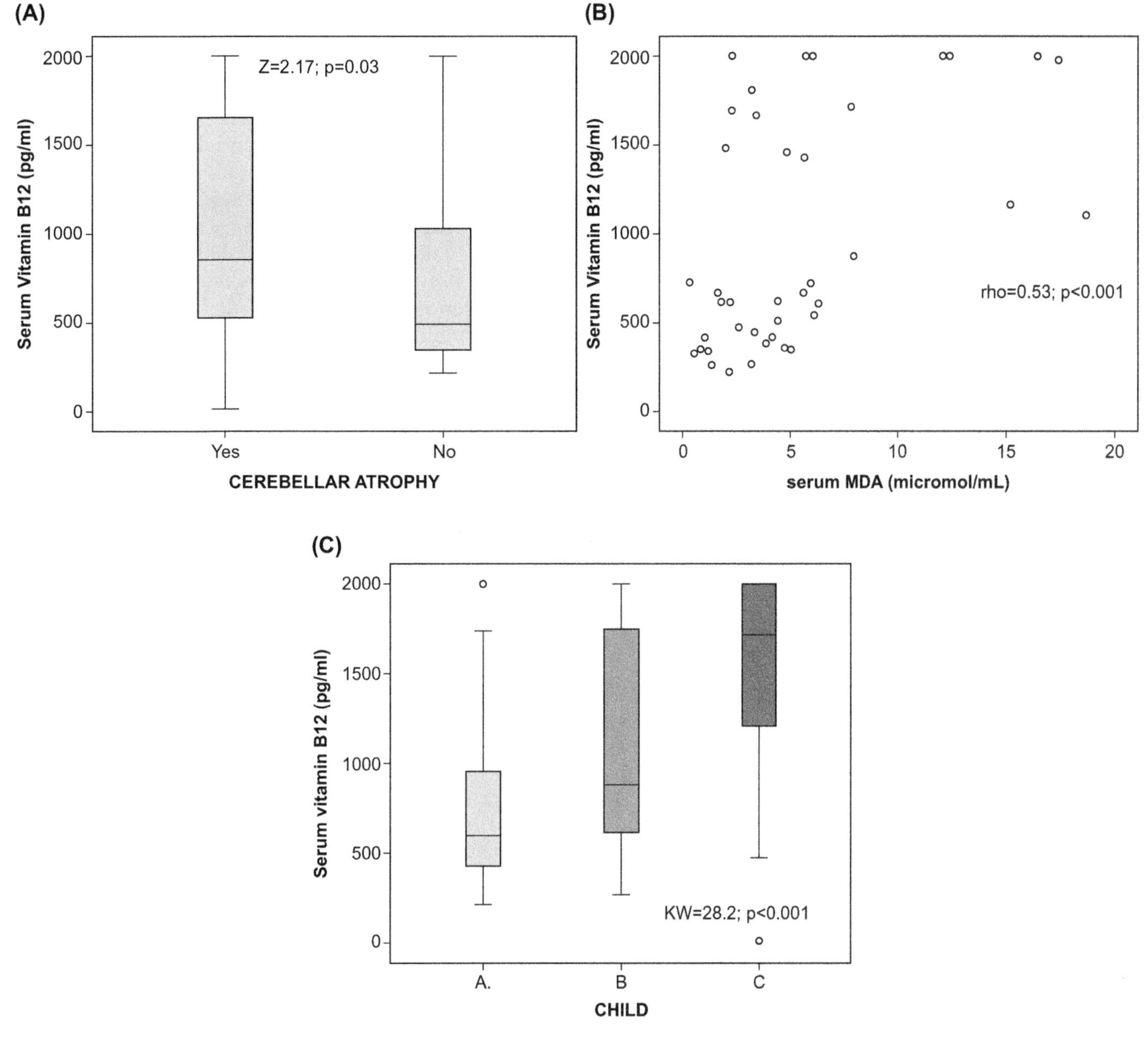

FIGURE 17.3 In a study on 113 alcoholics we found that serum vitamin B12 levels were significantly higher among patients categorized as Child class C (panel A), showing a direct, significant relationship with MDA levels ($\rho = 0.534$; $p < .001$, panel B). Patients with cerebellar atrophy showed higher vitamin B12 values than those without cerebellar atrophy (panel C).

Norwegian individuals aged 71–74 years (Vogiatzoglou et al., 2013).

In alcoholics, hyperhomocysteinemia is also involved in the development of brain atrophy (Bleich et al., 2003). As shown in Fig. 17.2 several factors account for increased homocysteine levels and brain damage in these patients. Increased generation of ROS by ethanol metabolism impairs the activity of methionine adenosyl transferase, blocking the formation of *S*-adenosyl methionine (SAM) (Kharbanda, 2009). Since SAM is a cofactor for cystathionine beta synthase, and eventually B6 deficiency may further decrease the activity of this enzyme, catabolism of homocysteine is impaired. On the other hand, acetaldehyde inhibits methionine synthase activity. Given the central role of the liver in methionine metabolism, ethanol-mediated liver function impairment could also alter homocysteine levels (Ferré et al., 2002). All of these mechanisms lead to increased homocysteine, and renders an increased formation of *S*-adenosyl homocysteine, a compound that is able to enhance TNF-α expression, at least in the liver (Song et al., 2004). Increased TNF-α production may promote neuroinflammation (Qin et al., 2006). On the other hand, homocysteine down-regulates the expression of glutathione peroxidase, impairing the defense against enhanced ROS generation (Handy, Zhang, & Loscalzo, 2005). Increased homocysteine levels may also be linked to excitotoxicity. Homocysteine is an agonist of the NMDA receptor which becomes up-regulated during chronic ethanol consumption but is blocked by ethanol. During withdrawal, the blocking effect of ethanol disappears, so that in the presence of raised homocysteine, excitation and seizures are more frequent, and cognitive impairment (Wilhelm et al., 2006) and more intense brain atrophy (Bleich et al., 2003) will probably develop.

THIAMINE DEFICIENCY AND WERNICKE ENCEPHALOPATHY

Thiamine is another water-soluble vitamin, present in large quantities in many food products, both of vegetal or animal origin. It is absorbed along the whole small intestine, especially at the jejunum and ileum, both by active transport and passive diffusion, and it becomes widely distributed in many organs, such as skeletal muscles, brain, liver, kidneys, and heart. This wide distribution is probably related to the main effects of thiamine as a cofactor of several enzymes involved in carbohydrate and amino acid metabolism. Body stores of thiamine only comprises around 30–50 mg, so that reduced intake and/or absorption lead to clinically relevant deficiency, since daily needs reach 1–2 mg, depending on the amount of carbohydrates consumed. Alcoholics are especially prone to develop thiamine deficiency. Poor intake, malabsorption, and decreased liver storage probably all contribute (Thomson et al., 2012). It is also possible that subtle changes in the transport system hampers the arrival of thiamine to brain cells (Guerrini, Thomson, & Gurling, 2009). Three additional mechanisms may be involved. For thiamine to become active, phosphorylation of thiamine is a required step that may be altered among alcoholics Thiamine-dependent enzymes are usually dependent on magnesium availability, which is frequently deficient in alcoholics (Thomson & Marshall, 2006). Moreover, ethanol alters cerebellar metabolism of thiamine (Laforenza, Patrini, Gastaldi, & Rindi, 1990).

Oxidative damage is also one of the effects of thiamine deficiency in the brain (Langlais, Anderson, Guo, & Bondy, 1997). Increased ROS generation is in association with the alpha ketoglutarate dehydrogenase complex which is defective in thiamine deficiency (Kruse, Navarro, Desjardins, & Butterworth, 2004). In addition to increased ROS generation, defective alpha ketoglutarate dehydrogenase complex activity may lead to increased glutamate levels (by blocking the transformation of oxo-glutarate into succinyl CoA). In normal conditions glia takes up extracellular glutamate and aspartate, thus maintaining an adequate balance of excitatory amino acids. As shown in diverse experiences, thiamine deficiency provokes early changes in glial morphology and function (Watanabe, Tomita, Hung, & Iwasaki, 1981), possibly altering its capacity to reuptake excitatory amino acids (Langlais & Mair, 1990). Moreover, ethanol consumption blocks NMDA receptors, but enhances its expression and avidity to bind glutamate (Thomson & Marshall, 2006). Therefore, withdrawal syndrome may exacerbate thiamine deficiency. After cessation of alcohol intake, NMDA receptor is no longer blocked, and glutamate-mediated excitotoxicity may contribute to neuronal damage (Langlais & Mair, 1990). Thiamine deficiency also leads to increased permeability in the blood–brain barrier (Harata & Iwasaki, 1995), although this seems to be a more delayed and less consistent effect. In any case, in advanced cases of Wernicke–Korsakoff syndrome there are focal hemosiderin deposits and hemosiderin-laden macrophages in mammillary bodies and thalamic nuclei, strongly suggesting an increased blood–brain barrier permeability, and possibly, iron-mediated oxidative damage.

Thiamine deficiency is observed among 29.7% to more than 50% of alcoholic patients, depending on the diagnostic criteria utilized (Baines, 1978). The classic, full-blown manifestation of thiamine deficiency is the so-called Wernicke–Korsakoff encephalopathy, an acute situation characterized by ophthalmoplegia, ataxia, and impaired consciousness, suffered by alcoholics with variable degrees of previous brain alterations. Indeed, it is possible that thiamine deficiency exerts a synergistic effect with ethanol, at least regarding some pathophysiological aspects such as increased blood–brain barrier permeability or a direct effect on cerebellar atrophy. Impairment of pyruvate dehydrogenase, transketolase, α-keto acid decarboxylase, and α-ketoglutarate dehydrogenase—thiamine-dependent enzymes—severely impair energy generation and lead to increased ROS generation and further damage to mitochondria. Increased blood–brain barrier permeability, in addition to provoking brain edema, allows iron to escape to the interstitium, and the generation of a more intense oxidative damage and enhanced ROS formation. The impossibility to convert pyruvate to acetyl coenzyme A leads to lactic acidosis, which also causes cytotoxic cerebral edema and induces neuronal death (Sechi & Serra, 2007). Impaired synthesis of acetyl coenzyme A may also lead to defective acetylcholine synthesis. In any case, cholinergic neurons are especially sensitive to glucose deprivation, and cholinergic neurons death is a characteristic feature of thiamine deficiency (Nardone et al., 2013). In addition, thiamine deficiency causes neuroinflammation, since it is involved in increased transcription of genes coding for proinflammatory cytokines and chemokines

(Hazell & Butterworth, 2009). Therefore, several pathways lead to altered brain structure and function, not all of them directly dependent on oxidative damage.

VITAMIN C DEFICIENCY

Ascorbic acid is abundant in fresh, uncooked, vegetables, especially citrus fruits, and is absorbed in the distal small intestine by active transport. Human beings are unable to synthesize vitamin C, but it is indeed synthesized from glucose by most animals species.

Ascorbic acid acts as a scavenger of ROS: it is able to react with a free radical, donating one electron. By this way ascorbic acid is transformed into an oxidized semidehydroascorbic acid, a stable, nonreactive compound, that can again be reduced or transformed into dehydroascorbic acid (Rinnerthaler, Bischof, Streubel, Trost, & Richter, 2015), that is later deoxidized by the glutathione reductase activity, linking its function to selenium stores.

Vitamin C probably acts in conjunction with vitamin E in cell membranes (Fig. 17.4). As mentioned previously, vitamin E may react with a peroxy radical, forming tocopheroxyl radicals, that are recycled again to tocopherol by the action of ascorbate (Packer et al., 1979). In addition, ascorbate is necessary for norepinephrine synthesis (May, Qu, Nazarewicz, & Dikalov, 2013), Although decreased vitamin C levels have been reported in patients with dementia (Charlton, Rabinowitz, Geffen, & Dhansay, 2004), its role in brain atrophy in alcoholics is unclear. However, experimental data do support a beneficial effect on ethanol-induced hippocampal neurodegeneration (Ambadath, Venu, & Madambath, 2010; Badshah et al., 2015; Naseer et al., 2011), and on ethanol-induced microglia and astrocyte activation, apoptosis, and neurodegeneration in the developing brain. Probably, the most relevant effect in alcoholics is the coupled reaction tocopherol/ascorbate against lipid peroxidation in cerebellar cell membranes (Rouach, Park, Orfanelli, Janvier, & Nordmann, 1987).

CONCLUSIONS AND FUTURE PROSPECTS

The role of the deficiency of the main antioxidant vitamins on alcohol-related brain damage is of paramount importance, but there is a sharp contrast between the pathogenetic theories and the reported beneficial effects in most studies dealing with laboratory animals, and the relatively small benefit obtained with vitamin supplementation in humans.

Tiwari et al. (2009) have shown that the administration of two isoforms of vitamin E (α-tocopherol and tocotrienol) to rats who were chronically exposed to ethanol-prevented deficits in learning and behavior, tocotrienol being more potent in preventing cognitive dysfunction. A vitamin E derivative called U-83836E reduced lipid peroxidation in an animal model of myocardial ischemia/reperfusion injury (Campo et al., 1997), but Grisel and Chen (2005) studied the effect of U-83836E on cerebellar Purkinje cell injury in developing rat pups exposed to alcohol and found no benefit of the vitamin E derivative on these lesions.

In humans, our knowledge about the effect of antioxidants on cognitive health has been mostly derived from studies performed on patients with Alzheimer disease. Devore et al. (2010) published data after a mean follow-up of 10 years obtained from the Rotterdam study, a prospective cohort study based in the Netherlands. They followed up 5395 patients without dementia aged 55 years and older. They found that higher dietary intake of vitamin E, but not vitamin C, beta carotene, or flavonoids, was associated with a decrease in the long-term risk of dementia. More specifically, they found that those in the highest tertile of vitamin E levels were 25% less likely to develop dementia, even after adjusting for confounders such as alcohol intake. Earlier results published

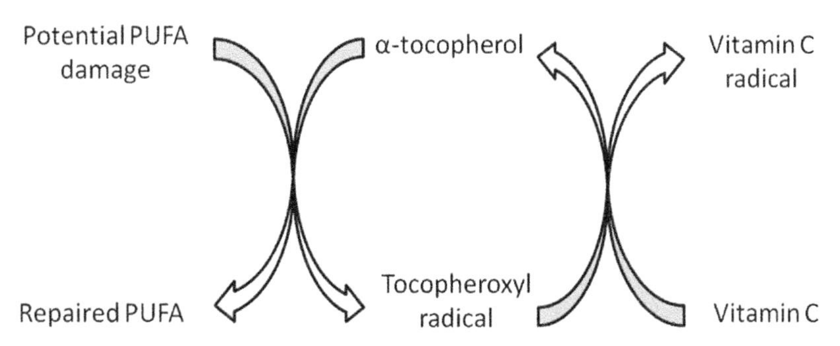

FIGURE 17.4 Vitamin C acts in conjunction with vitamin E in cell membranes in order to repair oxidative damage in polyunsaturated fatty acids (PUFA) in cell membranes. Vitamin E may react with a peroxy radical, forming tocopheroxyl radicals, that are recycled again to tocopherol by the action of ascorbate (vitamin C).

from this study in 2002 by Engelhart et al. (2002) described that after a mean follow-up of 6 years, both high intake of vitamin C and vitamin D decrease the incidence of Alzheimer disease.

As mentioned previously, the studies that are currently available do not show a clear benefit of vitamin D supplementation for the prevention of dementia. However, there is an ongoing clinical trial in the United States that is investigating the effects of vitamin D3 and omega-3 fatty acid supplementation on the risk of cancer, heart disease, and stroke (VITamin D and OmegA-3 TriaL or VITAL). The VITAL-Cog leg of this study will study the development of cognitive decline in these subjects. The estimated completion date of this study is October 2017. In contrast with the lack of benefit observed with vitamin supplementation on brain atrophy in alcoholics, the majority of authors agree in the observation that brain atrophy and ethanol-related brain dysfunction improve with ethanol abstinence. Alcohol abstinence, together with adequate nutrition, constitutes the most effective therapeutic approach in these patients (Bartels et al., 2007).

References

Adachi, S., Moriwaki, H., Muto, Y., Yamada, Y., Fukutomi, Y., Shimazaki, M., ... Ninomiya, M. (1991). Reduced retinoid content in hepatocellular carcinoma with special reference to alcohol consumption. *Hepatology, 14*, 776–780.

Ambadath, V., Venu, R. G., & Madambath, I. (2010). Comparative study of the efficacy of ascorbic acid, quercetin, and thiamine for reversing ethanol-induced toxicity. *Journal of Medicinal Food, 13*, 1485–1489. http://dx.doi.org/10.1089/jmf.2009.1387.

Arria, A. M., Tarter, R. E., Warty, V., & Van Thiel, D. H. (1990). Vitamin E deficiency and psychomotor dysfunction in adults with primary biliary cirrhosis. *American Journal of Clinical Nutrition, 52*, 383–390.

Azzi, A., Ricciarelli, R., & Zingg, J. M. (2002). Non-antioxidant molecular functions of alpha-tocopherol (vitamin E). *FEBS Letters, 519*, 8–10.

Badshah, H., Ali, T., Ahmad, A., Kim, M. J., Abid, N. B., Shah, S. A., ... Kim, M. O. (2015). Co-treatment with anthocyanins and vitamin C ameliorates ethanol- induced neurodegeneration via modulation of $GABA_B$ receptor signaling in the adult rat brain. *CNS & Neurological Disorders- Drug Targets, 14*, 791–803.

Baines, M. (1978). Detection and incidence of B and C vitamin deficiency in alcohol-related illness. *Annals of Clinical Biochemistry, 15*, 307–312.

Balion, C., Griffith, L. E., Strifler, L., Henderson, M., Patterson, C., Heckman, G., ... Raina, P. (2012). Vitamin D, cognition, and dementia: A systematic review and meta-analysis. *Neurology, 79*, 1397–1405.

Banala, R. R., & Karnati, P. R. (2015). Vitamin A deficiency: An oxidative stress marker in sodium fluoride (NaF) induced oxidative damage in developing rat brain. *International Journal of Developmental Neuroscience, 47*(Pt B), 298–303. http://dx.doi.org/10.1016/j.ijdevneu.2015.08.010.

Bartels, C., Kunert, H. J., Stawicki, S., Kröner-Herwig, B., Ehrenreich, H., & Krampe, H. (2007). Recovery of hippocampus-related functions in chronic alcoholics during monitored long-term abstinence. *Alcohol and Alcoholism, 42*, 92–102.

Bates, M. E., Bowden, S. C., & Barry, D. (2002). Neurocognitive impairment associated with alcohol use disorders: Implications for treatment. *Experimental and Clinical Psychopharmacology, 10*, 193–212.

Behr, G. A., Schnorr, C. E., Simões-Pires, A., da Motta, L. L., Frey, B. N., & Moreira, J. C. (2012). Increased cerebral oxidative damage and decreased antioxidant defenses in ovariectomized and sham-operated rats supplemented with vitamin A. *Cell Biology and Toxicology, 28*, 317–330. http://dx.doi.org/10.1007/s10565-012-9226-x.

Blaner, W. S., O'Byrne, S. M., Wongsiriroj, N., Kluwe, J., D'Ambrosio, D. M., Jiang, H., ... Libien, J. (2009). Hepatic stellate cell lipid droplets: A specialized lipid droplet for retinoid storage. *Biochimica et Biophysica Acta, 1791*, 467–473. http://dx.doi.org/10.1016/j.bbalip.2008.11.001.

Bleich, S., Bandelow, B., Javaheripour, K., Müller, A., Degner, D., Wilhelm, J., ... Kornhuber, J. (2003). Hyperhomocysteinemia as a new risk factor for brain shrinkage in patients with alcoholism. *Neuroscience Letter, 335*, 179–182.

Bonnet, E., Touyarot, K., Alfos, S., Pallet, V., Higueret, P., & Abrous, D. N. (2008). Retinoic acid restores adult hippocampal neurogenesis and reverses spatial memory deficit in vitamin A deprived rats. *PLoS One, 3*, e3487. http://dx.doi.org/10.1371/journal.pone.0003487.

Bowen, H. T., & Omaye, S. T. (1998). Oxidative changes associated with beta-carotene and alpha-tocopherol enrichment of human low-density lipoproteins. *The Journal of the American College of Nutrition, 17*, 171–179.

Brewer, L. D., Thibault, V., Chen, K. C., Langub, M. C., Landfield, P. W., & Porter, N. M. (2001). Vitamin D hormone confers neuroprotection in parallel with downregulation of L-type calcium channel expression in hippocampal neurons. *Journal of Neuroscience, 21*, 98–108.

Briani, C., Dalla Torre, C., Citton, V., Manara, R., Pompanin, S., Binotto, G., & Adami, F. (2013). Cobalamin deficiency: Clinical picture and radiological findings. *Nutrients, 5*, 4521–4539. http://dx.doi.org/10.3390/nu5114521.

Burton, G. W. (1989). Antioxidant action of carotenoids. *Journal of Nutrition, 119*, 109–111.

Butterworth, R. F. (1995). Pathophysiology of alcoholic brain damage: Synergistic effects of ethanol, thiamine deficiency and alcoholic liver disease. *Metabolic Brain Disease, 10*, 1–8.

Campo, G. M., Squadrito, F., Campo, S., Altavilla, D., Avenoso, A., Ferlito, M., ... Caputi, A. P. (1997). Antioxidant activity of U-83836E, a second generation lazaroid, during myocardial ischemia/reperfusion injury. *Free Radical Research, 27*, 577–590. PMID: 9455693.

Cass, W. A., Smith, M. P., & Peters, L. E. (2006). Calcitriol protects against the dopamine- and serotonin-depleting effects of neurotoxic doses of methamphetamine. *Annals of the New York Academy of Sciences, 1074*, 261–271.

Castillo-Lancellotti, C., Margozzini, P., Valdivia, G., Padilla, O., Uauy, R., Rozowski, J., & Tur, J. A. (2015). Serum folate, vitamin B12 and cognitive impairment in Chilean older adults. *Public Health Nutrition, 20*, 1–9.

Cattaneo, M. (1999). Hyperhomocysteinemia, atherosclerosis and thrombosis. *Thrombosis and Haemostasis, 81*, 165–176.

Cederbaum, A. I. (2006). Cytochrome P450 2E1-dependent oxidant stress and upregulation of anti-oxidant defense in liver cells. *Journal of Gastroenterology and Hepatology, 21*(Suppl. 3), S22–S25.

Charlton, K. E., Rabinowitz, T. L., Geffen, L. N., & Dhansay, M. A. (2004). Lowered plasma vitamin C, but not vitamin E, concentrations in dementia patients. *The Journal of Nutrition, Health & Aging, 8*, 99–107.

Charness, M. E. (1993). Brain lesions in alcoholics. *Alcoholism: Clinical and Experimental Research, 17*, 2–11.

Chei, C. L., Raman, P., Yin, Z. X., Shi, X. M., Zeng, Y., & Matchar, D. B. (2014). Vitamin D levels and cognition in elderly adults in China. *Journal of the American Geriatrics Society, 62*, 2125–2129. http://dx.doi.org/10.1111/jgs.13082.

Christakos, S., Dhawan, P., Verstuyf, A., Verlinden, L., & Carmeliet, G. (2016). Vitamin D: Metabolism, molecular mechanism of action, and pleiotropic effects. *Physiological Reviews, 96*, 365–408. http://dx.doi.org/10.1152/physrev.00014.2015.

Clugston, R. D., & Blaner, W. S. (2012). The adverse effects of alcohol on vitamin A metabolism. *Nutrients, 4*, 356–371. http://dx.doi.org/10.3390/nu4050356.

Cocco, S., Diaz, G., Stancampiano, R., Diana, A., Carta, M., Curreli, R., ... Fadda, F. (2002). Vitamin A deficiency produces spatial learning and memory impairment in rats. *Neuroscience, 115*, 475–482.

Crandall, J., Sakai, Y., Zhang, J., Koul, O., Mineur, Y., Crusio, W. E., & McCaffery, P. (2004). 13-cis-retinoic acid suppresses hippocampal cell division and hippocampal-dependent learning in mice. *Proceedings of the National Academy of Sciences of the United States of America, 101*, 5111–5116.

Crews, C. T., & Nixon, K. (2009). Mechanisms of neurodegeneration and regeneration in alcoholism. *Alcohol and Alcoholism, 44*, 115–127.

Devore, E. E., Grodstein, F., van Rooij, F. J. A., Hofman, A., Stampfer, M. J., Witteman, J. C. M., & Breteler, M. M. B. (2010). Dietary antioxidants and long-term risk of dementia. *Archives of Neurology, 67*, 819–825. http://dx.doi.org/10.1001/archneurol.2010.144.

Dickens, A. P., Lang, I. A., Langa, K. M., Kos, K., & Llewellyn, D. J. (2011). Vitamin D, cognitive dysfunction and dementia in older adults. *CNS Drugs, 25*, 629–639. http://dx.doi.org/10.2165/11593080-000000000-00000.

Durazzo, T. C., Mon, A., Pennington, D., Abé, C., Gazdzinski, S., & Meyerhoff, D. J. (2014). Interactive effects of chronic cigarette smoking and age on brain volumes in controls and alcohol-dependent individuals in early abstinence. *Addiction Biology, 19*, 132–143. http://dx.doi.org/10.1111/j.1369-1600.2012.00492.x.

Engelhart, M. J., Geerlings, M. I., Ruitenberg, A., van Swieten, J. C., Hofman, A., Witteman, J. C., & Breteler, M. M. (2002). Dietary intake of antioxidants and risk of Alzheimer disease. *JAMA, 287*, 3223–3229.

Erdozain, A. M., & Callado, L. F. (2014). Neurobiological alterations in alcohol addiction: A review. *Adicciones, 26*, 360–370.

Etchamendy, N., Enderlin, V., Marighetto, A., Vouimba, R. M., Pallet, V., Jaffard, R., & Higueret, P. (2001). Alleviation of a selective age-related relational memory deficit in mice by pharmacologically induced normalization of brain retinoid signaling. *The Journal of Neuroscience, 21*, 6423–6429.

Etgen, T., Sander, D., Bickel, H., Sander, K., & Förstl, H. (2012). Vitamin D deficiency, cognitive impairment and dementia: A systematic review and meta-analysis. *Dementia and Geriatrics Cognitive Disorders, 33*, 297–305. http://dx.doi.org/10.1159/000339702.

Eyles, D., Brown, J., Mackay-Sim, A., McGrath, J., & Feron, F. (2003). Vitamin D3 and brain development. *Neuroscience, 118*, 641–653.

Eyles, D. W., Smith, S., Kinobe, R., Hewison, M., & McGrath, J. J. (2005). Distribution of the vitamin D receptor and 1 alpha-hydroxylase in human brain. *Journal of Chemical Neuroanatomy, 29*, 21–30.

Falk, J., & Munne–Bosch, S. (2010). Tocochromanol functions in plants: Antioxidation and beyond. *Journal of Experimental Botany, 61*, 1549–1566.

Ferré, N., Gómez, F., Camps, J., Simó, J. M., Murphy, M. M., Fernández-Ballart, J., & Joven, J. (2002). Plasma homocysteine concentrations in patients with liver cirrhosis. *Clinical Chemistry, 48*, 183–185.

Finazzi, D., & Arosio, P. (2014). Biology of ferritin in mammals: An update on iron storage, oxidative damage and neurodegeneration. *Archives of Toxicology, 88*, 1787–1802. http://dx.doi.org/10.1007/s00204-014-1329-0.

Fitzpatrick, L. E., & Crowe, S. F. (2013). Cognitive and emotional deficits in chronic alcoholics: A role for the cerebellum? *Cerebellum, 12*, 520–533. http://dx.doi.org/10.1007/s12311-013-0461-3.

Fukui, K., Kawakami, H., Honjo, T., Ogasawara, R., Takatsu, H., Shinkai, T., ... Urano, S. (2012). Vitamin E deficiency induces axonal degeneration in mouse hippocampal neurons. *Journal of Nutritional Science and Vitaminology (Tokyo), 58*, 377–383.

Fukui, K., Takatsu, H., Shinkai, T., Suzuki, S., Abe, K., & Urano, S. (2005). Appearance of amyloid beta-like substances and delayed-type apoptosis in rat hippocampus CA1 region through aging and oxidative stress. *Journal of Alzheimer's Disease, 8*, 299–309.

Garcion, E., Sindji, L., Montero-Menei, C., Andre, C., Brachet, P., & Darcy, F. (1998). Expression of inducible nitric oxide synthase during rat brain inflammation: Regulation by 1,25-dihydroxyvitamin D_3. *Glia, 22*, 282–294.

Gatica, L., Alvarez, S., Gomez, N., Zago, M. P., Oteiza, P., Oliveros, L., & Gimenez, M. S. (2005). Vitamin A deficiency induces prooxidant environment and inflammation in rat aorta. *Free Radical Research, 39*, 621–628.

Gille, M., Jacquemin, C., Kiame, G., Delbecq, J., Guilmot, D., & Depré, A. (1993). Central pontine myelinolysis with cerebellar ataxia and dystonia. *Revue Neurologique (Paris), 149*, 344–346.

Giraldo, E., Lloret, A., Fuchsberger, T., & Viña, J. (2014). Aβ and tau toxicities in Alzheimer's are linked via oxidative stress-induced p38 activation: Protective role of vitamin E. *Redox Biology, 2*, 873–877. http://dx.doi.org/10.1016/j.redox.2014.03.002.

González-Reimers, E., Fernández-Rodríguez, C. M., Candelaria Martín-González, M., Hernández-Betancor, I., Abreu-González, P., de la Vega-Prieto, M. J., ... Santolaria-Fernández, F. (2014). Antioxidant vitamins and brain dysfunction in alcoholics. *Alcohol and Alcoholism, 49*, 45–50. http://dx.doi.org/10.1093/alcalc/agt150.

González-Reimers, E., Quintero-Platt, G., Fernández-Rodríguez, C., González-Arnay, E., Martín-González, M. C., & Pérez-Hernández, O. (2015). Oxidative damage and brain atrophy in alcoholics. *Oxidants and Antioxidants in Medical Science, 4*, 1–12. http://dx.doi.org/10.5455/oams.020615.rv.017.

Goodman, D. W., Huang, H. S., & Shiratori, T. (1965). Tissue distribution and metabolism of newly absorbed vitamin A in the rat. *Journal of Lipid Research, 6*, 390–396.

Grisel, J. J., & Chen, W. J. (2005). Antioxidant pretreatment does not ameliorate alcohol-induced Purkinje cell loss in the developing rat cerebellum. *Alcoholism Clinical and Experimental Research, 29*, 1223–1229. PMID:16046878.

Grummer, M. A., & Zachman, R. D. (1995). Prenatal ethanol consumption alters the expression of cellular retinol binding protein and retinoic acid receptor mRNA in fetal rat embryo and brain. *Alcoholism: Clinical and Experimental Research, 19*, 1376–1381.

Guerrini, I., Thomson, A. D., & Gurling, H. M. (2009). Molecular genetics of alcohol-related brain damage. *Alcohol and Alcoholism, 44*, 166–170. http://dx.doi.org/10.1093/alcalc/agn101.

Guilland, J. C., Favier, A., Potier de Courcy, G., Galan, P., & Hercberg, S. (2003). Hyperhomocysteinemia: An independent risk factor or a simple marker of vascular disease?. 1. Basic data. *Pathologie Biologie (Paris), 51*, 101–110.

Halsted, C. H., Robles, E. A., & Mezey, E. (1971). Decreased jejunal uptake of labeled folic acid (3 H-PGA) in alcoholic patients: Roles of alcohol and nutrition. *The New England Journal of Medicine, 285*, 701–706.

Handy, D. E., Zhang, Y., & Loscalzo, J. (2005). Homocysteine down-regulates cellular glutathione peroxidase (GPx1) by decreasing translation. *Journal of Biological Chemistry, 280*, 15518–15525.

Harata, N., & Iwasaki, Y. (1995). Evidence for early blood–brain barrier breakdown in experimental thiamine deficiency in the mouse. *Metabolic Brain Disease, 10*, 159–174.

Harms, L. R., Burne, T. H., Eyles, D. W., & McGrath, J. J. (2011). Vitamin D and the brain. *Best Practice & Research: Clinical Endocrinology & Metabolism, 25*, 657–669. http://dx.doi.org/10.1016/j.beem.2011.05.009.

Hazell, A. S., & Butterworth, R. F. (2009). Update of cell damage mechanisms in thiamine deficiency: Focus on oxidative stress, excitotoxicity and inflammation. *Alcohol and Alcoholism, 44*, 141–147. http://dx.doi.org/10.1093/alcalc/agn120.

Heese, P., Linnebank, M., Semmler, A., Muschler, M. A., Heberlein, A., Frieling, H., … Hillemacher, T. (2012). Alterations of homocysteine serum levels during alcohol withdrawal are influenced by folate and riboflavin: Results from the German investigation on neurobiology in alcoholism (GINA). *Alcohol and Alcoholism, 47*, 497–500. http://dx.doi.org/10.1093/alcalc/ags058.

Hosomi, A., Goto, K., Kondo, H., Iwatsubo, T., Yokota, T., Ogawa, M., … Inoue, K. (1998). Localization of alpha-tocopherol transfer protein in rat brain. *Neuroscience Letters, 256*, 159–162.

Hua, Y., Keep, R. F., Hoff, J. T., & Xi, G. (2007). Brain injury after intracerebral hemorrhage: The role of thrombin and iron. *Stroke; A Journal of Cerebral Circulation, 38*, 759–762.

Huang, Y. N., Ho, Y. J., Lai, C. C., Chiu, C. T., & Wang, J. Y. (2015). 1,25-Dihydroxyvitamin D_3 attenuates endotoxin-induced production of inflammatory mediators by inhibiting MAPK activation in primary cortical neuron-glia cultures. *Journal of Neuroinflammation, 12*, 147. http://dx.doi.org/10.1186/s12974-015-0370-0.

Jorde, R., Sneve, M., Figenschau, Y., Svartberg, J., & Waterloo, K. (2008). Effects of vitamin D supplementation on symptoms of depression in overweight and obese subjects: Randomized double blind trial. *Journal of Internal Medicine, 264*, 599–609. http://dx.doi.org/10.1111/j.1365-2796.2008.02008.x.

Kane, M. A., Folias, A. E., Wang, C., & Napoli, J. L. (2010). Ethanol elevates physiological all-trans-retinoic acid levels in select loci through altering retinoid metabolism in multiple loci: A potential mechanism of ethanol toxicity. *FASEB Journal: Official Publication of the Federation of American Societies for Experimental Biology, 24*, 823–832. http://dx.doi.org/10.1096/fj.09-141572.

Kannan, R., & Ng, M. J. (2008). Cutaneous lesions and vitamin B12 deficiency: An often-forgotten link. *Canadian Family Physician, 54*, 529–532.

Kaul, S., & Krishnakantha, T. P. (1997). Influence of retinol deficiency and curcumin/turmeric feeding on tissue microsomal membrane lipid peroxidation and fatty acids in rats. *Molecular and Cellular Biochemistry, 175*, 43–48.

Kawase, T., Kato, S., & Lieber, C. S. (1989). Lipid peroxidation and antioxidant defense systems in rat liver after chronic ethanol feeding. *Hepatology, 10*, 815–821.

Kharbanda, K. K. (2009). Alcoholic liver disease and methionine metabolism. *Seminars in Liver Diseases, 29*, 155–165. http://dx.doi.org/10.1055/s-0029-1214371.

Koscik, R. L., Lai, H. J., Laxova, A., Zaremba, K. M., Kosorok, M. R., Douglas, J. A., … Farrell, P. M. (2005). Preventing early, prolonged vitamin E deficiency: An opportunity for better cognitive outcomes via early diagnosis through neonatal screening. *The Journal of Pediatrics, 147*(3 Suppl), S51–S56.

Kruse, M., Navarro, D., Desjardins, P., & Butterworth, R. F. (2004). Increased brain endothelial nitric oxide synthase expression in thiamine deficiency: Relationship to selective vulnerability. *Neurochemistry International, 45*, 49–56.

Kumar, A., Singh, C. K., DiPette, D. D., & Singh, U. S. (2010). Ethanol impairs activation of retinoic acid receptors in cerebellar granule cells in a rodent model of fetal alcohol spectrum disorders. *Alcoholism Clinical and Experimental Research, 34*, 928–937. http://dx.doi.org/10.1111/j.1530-0277.2010.01166.x.

Laforenza, U., Patrini, C., Gastaldi, G., & Rindi, G. (1990). Effects of acute and chronic ethanol administration on thiamine metabolizing enzymes in some brain areas and in other organs of the rat. *Alcohol and Alcoholism, 25*, 591–603.

Lambert, D., Benhayoun, S., Adjalla, C., Gélot, M. M., Renkes, P., Gérard, P., … Nicolas, J. P. (1997). Alcoholic cirrhosis and cobalamin metabolism. *Digestion, 58*, 64–71.

Langlais, P. J., Anderson, G., Guo, S. X., & Bondy, S. C. (1997). Increased cerebral free radical production during thiamine deficiency. *Metabolic Brain Disease, 12*, 137–143.

Langlais, P. J., & Mair, R. G. (1990). Protective effects of the glutamate antagonist MK-801 on pyrithiamine-induced lesions and amino acid changes in rat brain. *Journal of Neuroscience, 10*, 1664–1674.

Lee, S. T., Jung, Y. M., Na, D. L., Park, S. H., & Kim, M. (2005). Corpus callosum atrophy in Wernicke's encephalopathy. *Journal of Neuroimaging, 15*, 367–372.

Leo, M. A., Kim, C., Lowe, N., & Lieber, C. S. (1992). Interaction of ethanol with beta-carotene: Delayed blood clearance and enhanced hepatotoxicity. *Hepatology, 15*, 883–891.

Leo, M. A., & Lieber, C. S. (1982). Hepatic vitamin A depletion in alcoholic liver injury. *The New England Journal of Medicine, 307*, 597–601.

Leo, M. A., & Lieber, C. S. (1999). Alcohol, vitamin A, and beta-carotene: Adverse interactions, including hepatotoxicity and carcinogenicity. *American Journal of Clinical Nutrition, 69*, 1071–1085.

Lewerenz, J., & Maher, P. (2015). Chronic glutamate toxicity in neurodegenerative diseases-what is the evidence? *Frontiers in Neuroscience, 9*, 469. http://dx.doi.org/10.3389/fnins.2015.00469. eCollection 2015.

Lieber, C. S. (1993). Biochemical factors in alcoholic liver disease. *Seminars in Liver Diseases, 13*, 136–153.

Li, T. K., Lumeng, L., & Veitch, R. L. (1974). Regulation of pyridoxal 5′-phosphate metabolism in liver. *Biochemical and Biophysical Research Communications, 61*, 677–684.

Lippai, D., Bala, S., Csak, T., Kurt-Jones, E. A., & Szabo, G. (2013). Chronic alcohol-induced microRNA-155 contributes to neuroinflammation in a TLR4-dependent manner in mice. *PLoS One, 8*, e70945. http://dx.doi.org/10.1371/journal.pone.0070945.

Littlejohns, T. J., Henley, W. E., Lang, I. A., Annweiler, C., Beauchet, O., Chaves, P. H., … Llewellyn, D. J. (2014). Vitamin D and the risk of dementia and Alzheimer disease. *Neurology, 83*, 920–928. http://dx.doi.org/10.1212/WNL.0000000000000755.

Liu, Y., Kagechika, H., Ishikawa, J., Hirano, H., Matsukuma, S., Tanaka, K., & Nakamura, S. (2008). Effects of retinoic acids on the dendritic morphology of cultured hippocampal neurons. *Journal of Neurochemistry, 106*, 1104–1116. http://dx.doi.org/10.1111/j.1471-4159.2008.05445.x.

Lu'o'ng, K. V., & Nguyen, L. T. (2013). The role of vitamin D in Alzheimer's disease: Possible genetic and cell signaling mechanisms. *American Journal of Alzheimer's Disease and Other Dementias, 28*, 126–136. http://dx.doi.org/10.1177/1533317512473196.

Lu, Y., & Cederbaum, A. I. (2008). CYP2E1 and oxidative liver injury by alcohol. *Free Radical Biology and Medicine, 44*, 723–738.

Lumeng, L. (1978). The role of acetaldehyde in mediating the deleterious effect of ethanol on pyridoxal 5′-phosphate metabolism. *Journal of Clinical Investigation, 62*, 286–293.

Lumeng, L., & Li, T. K. (1974). Vitamin B6 metabolism in chronic alcohol abuse. Pyridoxal phosphate levels in plasma and the effects of acetaldehyde on pyridoxal phosphate synthesis and degradation in human erythrocytes. *Journal of Clinical Investigation, 53*, 693–704.

Mailloux, R. J. (2015). Teaching the fundamentals of electron transfer reactions in mitochondria and the production and detection of reactive oxygen species. *Redox Biology, 4*, 381–398. http://dx.doi.org/10.1016/j.redox.2015.02.001.

Marill, J., Idres, N., Capron, C. C., Nguyen, E., & Chabot, G. G. (2003). Retinoic acid metabolism and mechanism of action: A review. *Current Drug Metabolism, 4*, 1–10.

May, J. M., Qu, Z. C., Nazarewicz, R., & Dikalov, S. (2013). Ascorbic acid efficiently enhances neuronal synthesis of norepinephrine from dopamine. *Brain Research Bulletin, 90*, 35–42. http://dx.doi.org/10.1016/j.brainresbull.2012.09.009.

McCaffery, P., Koul, O., Smith, D., Napoli, J. L., Chen, N., & Ullman, M. D. (2004). Ethanol increases retinoic acid production in cerebellar astrocytes and in cerebellum. *Brain Research. Developmental Brain Research, 153*, 233–241.

Medici, V., & Halsted, C. H. (2013). Folate, alcohol, and liver disease. *Molecular Nutrition & Food Research, 57*, 596–606. http://dx.doi.org/10.1002/mnfr.201200077.

Misner, D. L., Jacobs, S., Shimizu, Y., de Urquiza, A. M., Solomin, L., Perlmann, T., ... Evans, R. M. (2001). Vitamin A deprivation results in reversible loss of hippocampal long-term synaptic plasticity. *Proceedings of the National Academy of Sciences of the United States of America, 98*, 11714–11719.

de la Monte, S. M., Longato, L., Tong, M., DeNucci, S., & Wands, J. R. (2009). The liver-brain axis of alcohol-mediated neurodegeneration: Role of toxic lipids. *International Journal of Environmental Research and Public Health, 6*, 2055–2075. http://dx.doi.org/10.3390/ijerph6072055.

Müllebner, A., Moldzio, R., Redl, H., Kozlov, A. V., & Duvigneau, J. C. (2015). Heme degradation by heme oxygenase protects mitochondria but induces ER stress via formed bilirubin. *Biomolecules, 5*, 679–701. http://dx.doi.org/10.3390/biom5020679.

Nair, S. S., Prathibha, P., Syam Das, S., Kavitha, S., & Indira, M. (2015). All trans retinoic acid (ATRA) mediated modulation of N-methyl D-aspartate receptor (NMDAR) and Kruppel like factor 11 (KLF11) expressions in the mitigation of ethanol induced alterations in the brain. *Neurochemistry International, 83–84*, 41–47. http://dx.doi.org/10.1016/j.neuint.2015.02.007.

Nardone, R., Höller, Y., Storti, M., Christova, M., Tezzon, F., Golaszewski, S., ... Brigo, F. (2013). Thiamine deficiency induced neurochemical, neuroanatomical, and neuropsychological alterations: A reappraisal. *The Scientific World Journal, 2013*, 309143. http://dx.doi.org/10.1155/2013/309143.

Naseer, M. I., Ullah, N., Ullah, I., Koh, P. O., Lee, H. Y., Park, M. S., & Kim, M. O. (2011). Vitamin C protects against ethanol and PTZ-induced apoptotic neurodegeneration in prenatal rat hippocampal neurons. *Synapse, 65*, 562–571. http://dx.doi.org/10.1002/syn.20875.

Naude, C. E., Carey, P. D., Laubscher, R., Fein, G., & Senekal, M. (2012). Vitamin D and calcium status in South African adolescents with alcohol use disorders. *Nutrients, 4*, 1076–1094.

Nicolás, J. M., Catafau, A. M., Estruch, R., Lomeña, F. J., Salamero, M., Herranz, R., ... Urbano-Marquez, A. (1993). Regional cerebral blood flow-SPECT in chronic alcoholism: Relation to neuropsychological testing. *Journal of Nuclear Medicine, 34*, 1452–1459.

Nordmann, R. (1994). Alcohol and antioxidant systems. *Alcohol and Alcoholism, 29*, 513–522.

Packer, J. E., Slater, T. F., & Willson, R. L. (1979). Direct observation of a free radical interaction between vitamin E and vitamin C. *Nature, 278*, 737–738.

Qin, L., & Crews, F. T. (2012). NADPH oxidase and reactive oxygen species contribute to alcohol-induced microglial activation and neurodegeneration. *Journal of Neuroinflammation, 9*, 5. http://dx.doi.org/10.1186/1742-2094-9-5.

Qin, L., Liu, Y., Hong, J. S., & Crews, F. T. (2013). NADPH oxidase and aging drive microglial activation, oxidative stress, and dopaminergic neurodegeneration following systemic LPS administration. *Glia, 61*, 855–868. http://dx.doi.org/10.1002/glia.22479.

Qin, L., Wu, X., Block, M. L., Liu, Y., Breese, G. R., Hong, J. S., ... Crews, F. T. (2006). Systemic LPS causes chronic neuroinflammation and progressive neurodegeneration. *Glia, 55*, 453–462. http://dx.doi.org/10.1002/glia.20467.

Ramaekers, V. T., & Blau, N. (2004). Cerebral folate deficiency. *Developmental Medicine and Child Neurology, 46*, 843–851.

Rawat, J. P., Pinto, C., Kulkarni, K. S., Muthusamy, M. A., & Dave, M. D. (2014). Marchiafava-Bignami disease possibly related to consumption of a locally brewed alcoholic beverage: Report of two cases. *Indian Journal of Psychiatry, 56*, 76–78. http://dx.doi.org/10.4103/0019-5545.124720.

Ray, J. G. (1998). Meta-analysis of hyperhomocysteinemia as a risk factor for venous thromboembolic disease. *Archives of Internal Medicine, 158*, 2101–2106.

Reisi, P., Dashti, G. R., Shabrang, M., & Rashidi, B. (2014). The effect of vitamin E on neuronal apoptosis in hippocampal dentate gyrus in rabbits fed with high-cholesterol diets. *Advanced Biomedical Research, 3*, 42. http://dx.doi.org/10.4103/2277-9175.125731.

Ridley, N. J., Draper, B., & Withall, A. (2013). Alcohol-related dementia: An update of the evidence. *Alzheimer's Research and Therapy, 5*, 3. http://dx.doi.org/10.1186/alzrt157.

Rinnerthaler, M., Bischof, J., Streubel, M. K., Trost, A., & Richter, K. (2015). Oxidative stress in aging human skin. *Biomolecules, 5*, 545–589. http://dx.doi.org/10.3390/biom5020545.

Rossom, R. C., Espeland, M. A., Manson, J. E., Dysken, M. W., Johnson, K. C., Lane, D. S., ... Margolis, K. L. (2012). Calcium and vitamin D supplementation and cognitive impairment in the women's health initiative. *Journal of the American Geriatrics Society, 60*, 2197–2205. http://dx.doi.org/10.1111/jgs.12032.

Rouach, H., Park, M. K., Orfanelli, M. T., Janvier, B., & Nordmann, R. (1987). Ethanol-induced oxidative stress in the rat cerebellum. *Alcohol and Alcoholism*, (Suppl. 1), 207–211.

Sachdev, P. (2005). Homocysteine and brain atrophy. *Progress in Neuro-psychopharmacology & Biological Psychiatry, 29*, 1152–1161.

Saldeen, T., Li, D., & Mehta, J. L. (1999). Differential effects of alpha- and gamma-tocopherol on low-density lipoprotein oxidation, superoxide activity, platelet aggregation and arterial thrombogenesis. *Journal of the American College of Cardiology, 34*, 1208–1215.

Samygina, V. R., Sokolov, A. V., Bourenkov, G., Petoukhov, M. V., Pulina, M. O., Zakharova, E. T., ... Svergun, D. I. (2013). Ceruloplasmin: Macromolecular assemblies with iron-containing acute phase proteins. *PLoS One, 8*, e67145. http://dx.doi.org/10.1371/journal.pone.0067145.

Santolaria, F., Pérez-Manzano, J. L., Milena, A., González-Reimers, E., Gómez-Rodríguez, M. A., Martínez-Riera, A., ... de la Vega-Prieto, M. J. (2000). Nutritional assessment in alcoholic patients. Its relationship with alcoholic intake, feeding habits, organic complications and social problems. *Drug and Alcohol Dependence, 59*, 295–304.

Sato, M., & Lieber, C. S. (1981). Hepatic vitamin A depletion after chronic ethanol consumption in baboons and rats. *Journal of Nutrition, 111*, 2015–2023.

Sauvant, P., Sapin, V., Abergel, A., Schmidt, C. K., Blanchon, L., Alexandre-Gouabau, M. C., ... Azaïs-Braesco, V. (2002). PAV-1, a new rat hepatic stellate cell line converts retinol into retinoic acid, a process altered by ethanol. *The International Journal of Biochemistry and Cell Biology, 34*, 1017–1029.

Scalabrino, G., Buccellato, F. R., Veber, D., & Mutti, E. (2003). New basis of the neurotrophic action of vitamin B12. *Clinical Chemistry and Laboratory Medicine, 41*, 1435–1437.

Schnorr, C. E., Bittencourt, L. S., Petiz, L. L., Gelain, D. P., Zeidán-Chuliá, F., & Moreira, J. C. (2015). Chronic retinyl palmitate supplementation to middle-aged Wistar rats disrupts the brain redox homeostasis and induces changes in emotional behavior. *Molecular Nutrition and Food Research, 59*, 979–990. http://dx.doi.org/10.1002/mnfr.201400637.

Schnorr, C. E., da Silva Morrone, M., Simões-Pires, A., da Rocha, R. F., Behr, G. A., & Moreira, J. C. (2011). Vitamin A supplementation in rats under pregnancy and nursing induces behavioral changes and oxidative stress upon striatum and hippocampus of dams and their offspring. *Brain Research, 1369*, 60–73. http://dx.doi.org/10.1016/j.brainres.2010.11.042.

Sechi, G., & Serra, A. (2007). Wernicke's encephalopathy: New clinical settings and recent advances in diagnosis and management. *Lancet Neurology, 6*, 442–455.

Shankar, K., Liu, X., Singhal, R., Chen, J. R., Nagarajan, S., Badger, T. M., & Ronis, M. J. (2008). Chronic ethanol consumption leads to disruption of vitamin D3 homeostasis associated with induction of renal 1,25 dihydroxyvitamin D3-24-hydroxylase (CYP24A1). *Endocrinology, 149*, 1748–1756.

Shinpo, K., Kikuchi, S., Sasaki, H., Moriwaka, F., & Tashiro, K. (2000). Effect of 1,25-dihydroxyvitamin D(3) on cultured mesencephalic dopaminergic neurons to the combined toxicity caused by L-buthionine sulfoximine and 1-methyl-4-phenylpyridine. *Journal of Neuroscience Research, 62*, 374–382.

Smith, F. R., & Goodman, D. S. (1971). The effects of diseases of the liver, thyroid, and kidneys on the transport of vitamin A in human plasma. *Journal of Clinical Investigation, 50*, 2426–2436.

Smith, J. C., Jr., McDaniel, E. G., Fan, F. F., & Halsted, J. A. (1973). Zinc: A trace element essential in vitamin A metabolism. *Science, 181*, 954–955.

Smith, A. D., & Refsum, H. (2009). Vitamin B-12 and cognition in the elderly. *American Journal of Clinical Nutrition, 89*, 707S–711S. http://dx.doi.org/10.3945/ajcn.2008.26947D.

Song, Z., Zhou, Z., Uriarte, S., Wang, L., Kang, Y. J., Chen, T., ... McClain, C. J. (2004). S-adenosylhomocysteine sensitizes to TNF-alpha hepatotoxicity in mice and liver cells: A possible etiological factor in alcoholic liver disease. *Hepatology, 40*, 989–997.

Sontag, T. J., & Parker, R. S. (2002). Cytochrome P450 omega-hydroxylase pathway of tocopherol catabolism. Novel mechanism of regulation of vitamin E status. *The Journal of Biological Chemistry, 277*, 25290–25296.

Stabler, S. P. (2013). Clinical practice. Vitamin B12 deficiency. *The New England Journal of Medicine, 368*, 149–160. http://dx.doi.org/10.1056/NEJMcp1113996.

Stein, M. S., Scherer, S. C., Ladd, K. S., & Harrison, L. C. (2011). A randomized controlled trial of high-dose vitamin D2 followed by intranasal insulin in Alzheimer's disease. *Journal of Alzheimers Disease, 26*, 477–484. http://dx.doi.org/10.3233/JAD-2011-110149.

Sutherland, G. T., Sheedy, D., Sheahan, P. J., Kaplan, W., & Kril, J. J. (2014). Comorbidities, confounders, and the white matter transcriptome in chronic alcoholism. *Alcoholism: Clinical and Experimental Research, 38*, 994–1001. http://dx.doi.org/10.1111/acer.12341.

Taniura, H., Ito, M., Sanada, N., Kuramoto, N., Ohno, Y., Nakamichi, N., & Yoneda, Y. (2006). Chronic vitamin D3 treatment protects against neurotoxicity by glutamate in association with upregulation of vitamin D receptor mRNA expression in cultured rat cortical neurons. *Journal of Neuroscience Research, 83*, 1179–1189.

Tanner, A. R., Bantock, I., Hinks, L., Lloyd, B., Turner, N. R., & Wright, R. (1986). Depressed selenium and vitamin E levels in an alcoholic population. Possible relationship to hepatic injury through increased lipid peroxidation. *Digestive Diseases and Sciences, 31*, 1307–1312.

Tefferi, A., & Pruthi, R. K. (1994). The biochemical basis of cobalamin deficiency. *Mayo Clinic Proceedings, 69*, 181–186.

Thomson, A. D., Guerrini, I., Bell, D., Drummond, C., Duka, T., Field, M., ... Marshall, E. J. (2012). Alcohol-related brain damage: Report from a medical council on alcohol symposium, June 2010. *Alcohol and Alcoholism, 47*, 84–91. http://dx.doi.org/10.1093/alcalc/ags009.

Thomson, A. D., & Marshall, E. J. (2006). The natural history and pathophysiology of Wernicke's Encephalopathy and Korsakoff's Psychosis. *Alcohol and Alcoholism, 41*, 151–158.

Tiwari, V., Kuhad, A., & Chopra, K. (2009). Suppression of neuro-inflammatory signaling cascade by tocotrienol can prevent chronic alcohol-induced cognitive dysfunction in rats. *Behavioural Brain Research, 203*, 296–303.

Torvik, A., & Torp, S. (1986). The prevalence of alcoholic cerebellar atrophy. A morphometric and histological study of an autopsy material. *Journal of the Neurological Sciences, 75*, 43–51.

Touyarot, K., Bonhomme, D., Roux, P., Alfos, S., Lafenêtre, P., Richard, E., ... Pallet, V. (2013). A mid-life vitamin A supplementation prevents age-related spatial memory deficits and hippocampal neurogenesis alterations through CRABP-I. *PLoS One, 8*, e72101. http://dx.doi.org/10.1371/journal.pone.0072101. eCollection 2013.

Ulatowski, L., Dreussi, C., Noy, N., Barnholtz-Sloan, J., Klein, E., & Manor, D. (2012). Expression of the α-tocopherol transfer protein gene is regulated by oxidative stress and common single-nucleotide polymorphisms. *Free Radical Biology and Medicine, 53*, 2318–2326. http://dx.doi.org/10.1016/j.freeradbiomed.2012.10.528.

Ulatowski, L. M., & Manor, D. (2015). Vitamin E and neurodegeneration. *Neurobiology of Disease, 84*, 78–83. http://dx.doi.org/10.1016/j.nbd.2015.04.002.

Ulatowski, L., Parker, R., Warrier, G., Sultana, R., Butterfield, D. A., & Manor, D. (2014). Vitamin E is essential for Purkinje neuron integrity. *Neuroscience, 260*, 120–129. http://dx.doi.org/10.1016/j.neuroscience.2013.12.001.

Van Bennekum, A. M., Fisher, E. A., Blaner, W. S., & Harrison, E. H. (2000). Hydrolysis of retinyl esters by pancreatic triglyceride lipase. *Biochemistry, 39*, 4900–4906.

Van Bennekum, A., Werder, M., Thuahnai, S. T., Han, C. H., Duong, P., Williams, D. L., ... Hauser, H. (2005). Class B scavenger receptor-mediated intestinal absorption of dietary beta-carotene and cholesterol. *Biochemistry, 44*, 4517–4525.

Vogiatzoglou, A., Refsum, H., Johnston, C., Smith, S. M., Bradley, K. M., de Jager, C., ... Smith, A. D. (2008). Vitamin B12 status and rate of brain volume loss in community-dwelling elderly. *Neurology, 71*, 826–832. http://dx.doi.org/10.1212/01.wnl.0000325581.26991.f2.

Vogiatzoglou, A., Smith, A. D., Nurk, E., Drevon, C. A., Ueland, P. M., Vollset, S. E., ... Refsum, H. (2013). Cognitive function in an elderly population: Interaction between vitamin B12 status, depression, and apolipoprotein E ε4: The Hordaland Homocysteine Study. *Psychosomatic Medicine, 75*, 20–29. http://dx.doi.org/10.1097/PSY.0b013e3182761b6c.

Wani, N. A., & Kaur, J. (2011). Reduced levels of folate transporters (PCFT and RFC) in membrane lipid rafts result in colonic folate malabsorption in chronic alcoholism. *Journal of Cellular Physiology, 226*, 579–587. http://dx.doi.org/10.1002/jcp.22525.

Watanabe, I., Tomita, T., Hung, K. S., & Iwasaki, Y. (1981). Edematous necrosis in thiamine-deficient encephalopathy of the mouse. *Journal of Neuropathology and Experimental Neurology, 40*, 454–471.

Wilhelm, J., Bayerlein, K., Hillemacher, T., Reulbach, U., Frieling, H., Kromolan, B., ... Bleich, S. (2006). Short-term cognition deficits during early alcohol withdrawal are associated with elevated plasma homocysteine levels in patients with alcoholism. *Journal of Neural Transmission (Vienna), 113*, 357–363.

Wilhelm, J., Frieling, H., Hillemacher, T., Degner, D., Kornhuber, J., & Bleich, S. (2008). Hippocampal volume loss in patients with alcoholism is influenced by the consumed type of alcoholic beverage. *Alcohol and Alcoholism, 43*, 296–299. http://dx.doi.org/10.1093/alcalc/agn002.

Wilkens-Knudsen, A., Jensen, J. E., Nordgaard-Lassen, I., Almdal, T., Kondrup, J., & Becker, U. (2014). Nutritional intake and status in persons with alcohol dependency: Data from an outpatient treatment programme. *European Journal of Nutrition, 53*, 1483–1492. http://dx.doi.org/10.1007/s00394-014-0651-x. Epub January 20, 2014.

Wolf, G. (2010). Tissue-specific increases in endogenous all-trans retinoic acid: Possible contributing factor in ethanol toxicity. *Nutrition Reviews, 68*, 689–692. http://dx.doi.org/10.1111/j.1753-4887.2010.00323.x.

Wrzosek, M., Łukaszkiewicz, J., Wrzosek, M., Jakubczyk, A., Matsumoto, H., Piątkiewicz, P., ... Nowicka, G. (2013). Vitamin D and the central nervous system. *Pharmacology Reports, 65*, 271–278.

Xia, C., Hu, J., Ketterer, B., & Taylor, J. B. (1996). The organization of the human GSTP1-1 gene promoter and its response to retinoic acid and cellular redox status. *The Biochemical Journal, 313*(Pt 1), 155–161.
Yamamoto, M., Ullman, D., Dräger, U. C., & McCaffery, P. (1999). Postnatal effects of retinoic acid on cerebellar development. *Neurotoxicology and Teratology, 21*, 141–146.
Yang, L. K., Wong, K. C., Wu, M. Y., Liao, S. L., Kuo, C. S., & Huang, R. F. (2007). Correlations between folate, B12, homocysteine levels, and radiological markers of neuropathology in elderly post-stroke patients. *Journal of the American College of Nutrition, 26*, 272–278.

C H A P T E R

18

Serotonin Deficiency and Alcohol Use Disorders

B.D. Sachs, K. Dodson
Villanova University, Villanova, PA, United States

INTRODUCTION

Over 16 million Americans meet the criteria for an alcohol use disorder (AUD) (SAMHSA, 2015), and the excessive consumption of alcohol costs the US economy an estimated $250 billion annually (Sacks, Gonzales, Bouchery, Tomedi, & Brewer, 2015). Research has shown that both genetic and environmental factors contribute to the development of AUDs, but the precise neurobiological underpinnings of alcoholism have proven difficult to uncover. This fact derives in part from the current conceptualization of alcoholism as a single entity. Substantial evidence suggests that what we commonly call "alcoholism" or "AUD" is actually a collection of multiple disorders with partially overlapping characteristics and varying etiology (Babor et al., 1992). Despite this, the Diagnostic and Statistical Manual of Mental Disorders, Fifth Edition, *DSM-5*, lists a single entry for AUD, which is further classified only by severity, not by symptom profile (American Psychiatric Association, 2013). Subtyping AUD patients on the basis of shared personality traits, comorbidities, chronicity, drinking patterns, and/or functional neuroimaging data could identify populations of AUD patients who exhibit similar neurobiological disturbances. Because underlying pathology may be a key determinant of treatment response, such a categorization scheme could ultimately improve treatment selection and outcomes. Here, we review the evidence that dysfunction of the brain serotonin (5-hydroxytryptamine, 5-HT) system, particularly 5-HT deficiency, occurs in a subset of AUD patients and contributes to their pathological behavior. Reviewing clinical and preclinical data, we describe the behavioral alterations that result from brain 5-HT deficiency and discuss how reduced 5-HT could impact the development and treatment of AUDs and comorbid disorders.

ALCOHOL TYPOLOGIES

Given the wide variety of drinking patterns, personality traits, and neuropsychiatric comorbidities of individuals with AUDs, there have been multiple attempts to categorize patients into subtypes (Babor et al., 1992). One prominent example of alcoholism typology was proposed by Cloninger, Bohman, and Sigvardsson, (1981). In this system, Type 2 alcoholics exhibit an earlier age of onset and are more likely to engage in violent and aggressive antisocial behaviors compared to Type 1 alcoholics (Cloninger, Sigvardsson, & Bohman, 1996). Although Type 2 Alcoholism and antisociality are not synonymous, the two overlap frequently, with roughly 60–80% of individuals with antisocial personality disorder (APD) suffering from comorbid AUD (Virkkunen & Linnoila, 2002). Due to their unique characteristics, Type 1 and Type 2 alcoholisms have been hypothesized to derive from distinct causes. Specifically, Type 1 alcoholism has been primarily associated with dopaminergic dysfunction and Type 2 alcoholism has been theorized to stem from serotonergic dysfunction (Cloninger et al., 1996; Tupala & Tiihonen, 2004). Importantly, neurobiological abnormalities are not considered when making AUD diagnoses or for Cloninger's subtyping. Thus, while several studies have documented serotonergic dysfunction in AUD patients on a group level, particularly in Type 2 alcoholics (Fils-Aime et al., 1996), the actual percentage of Type 1 and Type 2 alcoholics who exhibit serotonergic dysregulation remains unknown.

5-HT DEPLETION STUDIES

Some of the first studies to examine the relationship between 5-HT deficiency and drinking behavior involved acute or sub-chronic depletion of brain 5-HT

Addictive Substances and Neurological Disease
http://dx.doi.org/10.1016/B978-0-12-805373-7.00018-9

using pharmacological, dietary, or ablative techniques. One of these methods, acute tryptophan depletion (ATD), has been applied in humans by providing subjects a diet lacking the 5-HT precursor, tryptophan (Moreno et al., 1999). Studies of ATD have provided mixed results regarding the impact of low 5-HT on alcohol consumption. For example, it has been reported that ATD, but not placebo treatment, significantly increases the desire to drink in alcoholic participants with comorbid major depressive disorder (MDD) (Pierucci-Lagha et al., 2004). However, other studies report no significant effects of ATD on the number of drinks consumed, the subjective level of intoxication, or self-reported alcohol craving in AUD patients (Petrakis et al., 2002, 2001). It is possible that the conflicting results relate to the presence or absence of MDD in the study samples. The Petrakis studies, which reported negative findings, excluded patients that had been diagnosed with Axis I disorders. In contrast, the Pierucci–Lagha study found significant effects specifically in AUD patients with comorbid MDD. It is possible that neurobiological alterations present in MDD patients lead them to be more sensitive to ATD, but future work would be required to evaluate this.

Animal studies employing pharmacological or ablative methods to reduce brain 5-HT have also reported inconsistent effects on drinking behavior. For example, using *p*-chlorophenylalanine (PCPA) or *p*-chloroamphetamine (PCA) to deplete 5-HT levels has been reported to reduce alcohol consumption in rodents (Myers & Cicero, 1969; Myers & Tytell, 1972; Myers & Veale, 1968), but other groups found no effect of 5-HT depletion (Holman, Hoyland, & Shillito, 1975; Kiianmaa, 1976), and still others have reported increased alcohol consumption (Geller, 1973). It is possible that differences in the extent of 5-HT deficiency, or inconsistencies in behavioral testing procedures (including the timeline of testing) could contribute to these discrepant findings. In addition, strain differences could have led to some of the conflicting data. There is evidence that PCPA specifically reduces alcohol consumption in alcohol-preferring strains, but not in strains that display low baseline alcohol consumption (Contreras, Alvarado, & Mardones, 1990).

Another potential explanation for the mixed findings of 5-HT depletion studies is that these studies induce sub-chronic or acute 5-HT depletion, whereas chronic 5-HT depletion may be required to significantly influence drinking behavior. Studies examining antidepressant effects have shown that only chronic, not acute, elevation of 5-HT levels is sufficient to improve mood (Wong & Licinio, 2001), thus suggesting that some of the behavioral consequences that result from manipulating 5-HT levels emerge only after several weeks. Even if 5-HT depletion were maintained for several weeks, such studies in adult subjects would still not reproduce any behavioral changes resulting from developmental disturbances in serotonergic neurotransmission. This could significantly limit our ability to interpret negative findings in light of the profound long-term behavioral effects of developmental manipulations of 5-HT signaling. For example, administering selective serotonin reuptake inhibitors (SSRIs) to mice during the first 3 weeks of life increases anxiety-like behavior in adult animals (Ansorge, Zhou, Lira, Hen, & Gingrich, 2004), whereas similar treatments in adult animals induces the opposite effect (Bodnoff, Suranyi-Cadotte, Quirion, & Meaney, 1989; Sachs, Jacobsen, et al., 2013; Santarelli et al., 2003). Genetic 5-HT-depletion studies in which low levels of brain 5-HT can be achieved throughout the lifespan or during particular developmental windows are required to understand the consequences of developmental serotonergic dysfunction.

5-HT DYSFUNCTION AND ALCOHOL CONSUMPTION

Human Studies

Multiple lines of evidence implicate serotonergic dysfunction in AUDs (LeMarquand, Pihl, & Benkelfat, 1994b). For example, reduced levels of 5-HT metabolites (i.e., 5-hydroxyindoleacetic acid, 5-HIAA) have been identified in AUD patients (Ballenger, Goodwin, Major, & Brown, 1979; Banki, 1981). Molecular neuroimaging and post-mortem studies have also revealed serotonergic dysfunction in AUD patients (Maron, Nutt, & Shlik, 2012; Storvik, Tiihonen, Haukijarvi, & Tupala, 2007). Although these types of serotonergic dysfunction could potentially be the result (rather than cause) of chronic alcohol consumption, the fact that differences in the extent and/or type of serotonergic abnormalities have been reported in Type 1 versus Type 2 alcoholics suggests that some of these alterations may exist prior to chronic alcohol consumption (Fils-Aime et al., 1996; Roy, Virkkunen, & Linnoila, 1987; Sander et al., 1998; Storvik, Tiihonen, Haukijarvi, & Tupala, 2006; Storvik et al., 2007). In keeping with this, dysregulation of serotonergic neurotransmission has been hypothesized to underlie several psychiatric disorders that are frequently comorbid with AUDs, including MDD (Delgado, 2000) and obsessive–compulsive disorder (Murphy et al., 1989). Because AUDs are thought to occasionally reflect attempts at self-medication for various psychiatric disorders, it is possible that serotonergic dysfunction leads to AUDs indirectly by increasing the incidence of mental illness (Crutchfield & Gove, 1984; Weiss, Griffin, & Mirin, 1992).

Genetic studies have further supported the idea that at least some of the serotonergic dysfunction observed in AUD patients occurs independently of alcohol consumption. One widely studied serotonergic polymorphism in AUDs and mental illness is a bi-allelic variation in the serotonin transporter (SERT) gene promoter (Lesch et al., 1996). The two alleles differ in whether they contain a 44 base pair insertion. The short SERT isoform (the *s* allele) has been reported to have reduced expression and activity compared to the long allele (the *l* allele), suggesting that the long isoform might decrease extracellular 5-HT levels and thus impair serotonergic neurotransmission (Lesch et al., 1996). Schuckit et al. (1999) reported that *l*/*l* individuals are significantly more likely to become alcoholics compared to *s*/*s* or *s*/*l* individuals, suggesting that low 5-HT function could contribute to AUDs. However, Ishiguro et al. (1999) reported no significant overall differences in the *l* and *s* allele frequencies in AUD patients compared to controls. Interestingly, AUD patients with a history of antisocial behavior and drinking-related arrests are significantly less likely to express the *s* allele than AUD patients without antisocial features, suggesting that AUD patients with antisocial tendencies may be more likely to exhibit low brain 5-HT (Ishiguro et al., 1999). Importantly, not all studies support the notion that the long isoform confers increased susceptibility to AUDs. Sander et al. (1998) actually reported a trend toward increased frequency of the short *s* allele in dissocial AUD patients compared to controls, but this report did not achieve significance when corrected for multiple comparisons. While the reasons for these discrepancies are unknown, it is possible that improved subtyping will lead to more consistent results.

Evidence also suggests that mutations in the brain 5-HT-synthesis enzyme, tryptophan hydroxylase-2 (Tph2), could play a role in subpopulations of AUD patients. For example, several SNPs in Tph2 were nominally significantly associated with AUDs in a sample of young women, but not in a college drinking sample (Agrawal et al., 2011). Another study also reported significant associations between Tph2 polymorphisms and AUDs (Plemenitas et al., 2015). However, because the functional significance of most identified SNPs in Tph2 remains unknown, our ability to make definitive conclusions regarding the relationship between 5-HT synthesis and AUD development based on genetic association studies remains limited.

Animal Studies

Preclinical work has shown that chronic exposure to alcohol leads to significant alterations in the expression of key serotonergic regulators in the brain, such as reductions in hippocampal SERT binding (Burnett, Davenport, Grant, & Friedman, 2012) and increased Tph2 expression (Shibasaki, Inoue, Kurokawa, Ogou, & Ohkuma, 2010). Thus, it is likely that at least some of the dysregulation of the serotonergic system observed in AUD patients is the result of chronic alcohol consumption. However, the relationships between alcohol consumption and serotonergic dysfunction are bidirectional, because preclinical studies have shown that manipulating 5-HT signaling significantly impacts alcohol consumption.

One genetic mouse model of brain 5-HT deficiency is the Tph2(R439H)knock-in (Tph2KI) (Beaulieu et al., 2008), which harbors a single point mutation in tryptophan hydroxylase 2, the enzyme responsible for brain 5-HT synthesis (Walther & Bader, 2003). Tph2KI mice display 60–80% reductions in brain 5-HT and have been reported to exhibit increased alcohol preference and consumption (Sachs, Salahi, & Caron, 2013). These mice also display decreased sensitivity to the sedative and locomotor-inhibitory effects of alcohol (Sachs, Salahi, et al., 2013). Another study examining the Tph2KI mouse line has shown that these animals are more likely than wild-type littermates to maintain high levels of alcohol consumption following the addition of quinine to the alcohol solution to decrease its palatability (Lemay, Dore, & Beaulieu, 2015). This second study did not report a significant increase in baseline alcohol preference in Tph2KI animals, an effect that the authors theorized might result from differences in experimental procedure (i.e., using sweetened vs. unsweetened alcohol solutions) between the studies (Lemay et al., 2015).

The C1473G polymorphism in Tph2, which results in about 50% reductions in 5-HT, has also been reported to influence alcohol responses. Specifically, alcohol administration significantly reduces locomotor activity in 1473C mice (which have "normal" levels of brain 5-HT), but not in hyposerotonergic 1473G mice (Bazovkina, Lichman, & Kulikov, 2015). While this result is consistent with the results obtained in Tph2KI mice (Sachs, Salahi, et al., 2013), the C1473G polymorphism did not affect alcohol preference or alcohol-induced recovery of righting reflex (Bazovkina et al., 2015), two behaviors in which phenotypes were observed in Tph2(R439H)KI mice (Sachs, Salahi, et al., 2013). Importantly, the extent of 5-HT deficiency differs between the two lines. Bazovkina et al. (2015) studied mice on a pure c57BL6 background, whereas the Tph2KI studies used animals on a mixed background (Sachs, Salahi, et al., 2013), and the alcohol preference tests for mice with the C1473G polymorphism did not use sweeteners (Bazovkina et al., 2015). Although not all results have been consistent, multiple lines of evidence point to decreased 5-HT neurotransmission leading to increased alcohol consumption in animal models.

5-HT DEFICIENCY, IMPULSIVITY, AND AGGRESSION

Research has consistently demonstrated strong correlations between alcohol consumption and violence, and alcohol abuse is referenced in 50–80% of reports of domestic violence, intense verbal abuse, and child abuse (Yudko, Blanchard, Henrie, & Blanchard, 2002). Longitudinal regression studies have even shown that childhood levels of aggression are better predictors of adult AUD than frequency or intensity of underage drinking (Virkkunen & Linnoila, 2002; Yudko et al., 2002). This naturally raises questions regarding whether impulsivity indirectly predisposes to alcoholism, or whether alcohol consumption exacerbates preexisting impulsivity and aggression in certain individuals. Preclinical research and research on human populations indicates that the relationship between AUD and impulsive/aggressive phenotype (IAP) may be bidirectional and, in some cases, perpetuating. Both 5-HT deficiency (Angoa-Perez et al., 2012; Beaulieu et al., 2008; Mosienko et al., 2012) and alcohol (Miczek & Barry, 1977) have been shown to increase aggression and impulsivity in mice, but whether 5-HT deficiency exacerbates alcohol-induced impulsivity or aggression has not been determined.

Most of the evidence supporting serotonergic disturbances in AUD patients has been found in Type 2 alcoholics or other patient populations characterized by impulsive or aggressive behavior (Virkkunen & Linnoila, 2002). If 5-HT plays a causal role in these disorders, one might predict that 5-HT dysfunction would increase impulsivity or aggression. Numerous studies of genetically modified mice, including Tph2 knock-out (KO) (Angoa-Perez et al., 2012; Mosienko et al., 2012) and Tph2(R439H)KI mice (Beaulieu et al., 2008), have revealed that impaired brain 5-HT synthesis promotes IAP. Given the known relationship between impulsivity and AUDs (Bjork, Hommer, Grant, & Danube, 2004), it is possible that 5-HT deficiency influences AUD risk by promoting impulsive behavior. Human studies support such a link between impaired 5-HT synthesis and impulsivity. For example, an intronic polymorphism in the Tph2 gene has been associated with poor response inhibition on a stop signal task (Stoltenberg et al., 2006), and a polymorphism in the promoter region of Tph2 (TPH2 703G/T) has been associated with poor performance on the executive control component of the attention network test (Reuter, Ott, Vaitl, & Hennig, 2007). In line with these observations, homozygosity for the T allele at the Tph2 703G/T SNP and other Tph2 "risk" haplotypes are observed commonly in individuals with cluster B personality disorders (e.g., borderline personality disorder or APD), which are traditionally characterized by impulsive and erratic behavior (Gutknecht et al., 2007; Perez-Rodriguez et al., 2010).

Additional support for the link between serotonergic dysfunction and impulsivity can be found in studies examining suicide. For example, genetic studies have identified several SNPs of Tph2 that are associated with suicide completion in the context of MDD (Zhou et al., 2005; Zill et al., 2004; Zill, Preuss, Koller, Bondy, & Soyka, 2007). Another SNP in Tph2 (rs1386483) has been associated with completed suicide, especially in individuals with a history of multiple attempts (Fudalej et al., 2010), and biomarkers of low 5-HT neurotransmission have been associated with impulsive or violent suicide. For example, low 5-HIAA levels have been observed in suicide victims, especially individuals who make repeated attempts at impulsive, rather than planned, suicide (Krakowski, 2003; Stanley et al., 2000; Zhou et al., 2005). Similarly, impulsive suicidality has been associated with low prolactin response to fenfluramine (FEN) (Krakowski, 2003), and 5-HT deficiency has been shown to reduce prolactin release following FEN administration (Jacobsen et al., 2012).

Impairments in impulse control are thought to characterize multiple psychiatric and substance abuse disorders, and potential biomarkers of 5-HT deficiency have been identified in patients with many of these disorders, including AUDs (LeMarquand, Pihl, & Benkelfat, 1994a), impulsive aggression (Linnoila et al., 1983), and APD (Moss, Yao, & Panzak, 1990). Although clinical association studies do not provide causal information, one experimental study reported that nonalcoholic young men with a family history of alcoholism (but not controls) exhibit a significant increase in impulsive behavior following ATD, suggesting that 5-HT deficiency may lead to behavioral disinhibition in individuals at risk for AUDs (LeMarquand, Benkelfat, Pihl, Palmour, & Young, 1999). FEN-induced prolactin release is particularly low in cluster B patients with IAP, particularly males (Soloff, Kelly, Strotmeyer, Malone, & Mann, 2003; Virkkunen & Linnoila, 2002). Positron emission tomography (PET) studies have also shown a diminished response to FEN in areas critical to impulse control, such as the prefrontal cortex, in patients with IAP or cluster B personality disorders (Soloff, Meltzer, Greer, Constantine, & Kelly, 2000). In line with the association between low 5-HT and IAP, ATD has been associated with increases in aggressive and impulsive behavior. This is true of healthy subjects (Walderhaug et al., 2002), but the effects are more pronounced in those with a history of aggression (Dougherty, Bjork, Marsh, & Moeller, 1999; Young & Leyton, 2002). Low levels of 5-HIAA are also associated with IAP or antisocial behaviors and crime, such as arson, murder, and battery (Krakowski, 2003; Stanley et al., 2000; Virkkunen & Linnoila, 2002). Furthermore, low 5-HIAA is more strongly related to unplanned violent crimes and recidivism than to premeditated violent crime or single offenses (Krakowski, 2003; Virkkunen & Linnoila, 2002). Overall, clinical studies reveal a very strong association between low 5-HT and IAP, but as with most work in humans, causation is difficult to establish.

5-HT DEFICIENCY AND AUD TREATMENT RESPONSES

Several selective serotonin reuptake inhibitor (SSRI) antidepressants, including sertraline (Naranjo, Poulos, Bremner, & Lanctot, 1992) and fluoxetine (Naranjo, Poulos, Bremner, & Lanctot, 1994) have been reported to reduce drinking in AUD patients. However, negative findings have also been reported (Gorelick & Paredes, 1992), and some studies have even found that SSRI-treated groups exhibit poorer drinking responses compared to placebo (Charney, Heath, Zikos, Palacios-Boix, & Gill, 2015). Interestingly, some studies have demonstrated that the subtypes of alcoholics most commonly associated with biomarkers of 5-HT deficiency (i.e., Babor's Type B or Cloninger's Type 2) are significantly less likely to benefit from SSRIs than other subtypes (Dundon, Lynch, Pettinati, & Lipkin, 2004). This failure of SSRIs to induce therapeutic effects in putative 5-HT-deficient populations is consistent with other lines of evidence. For example, mutations in Tph2 are associated with poor responses to antidepressants (Peters, Slager, McGrath, Knowles, & Hamilton, 2004; Tsai et al., 2009; Tzvetkov, Brockmoller, Roots, & Kirchheiner, 2008), suggesting that impaired 5-HT synthesis may prevent responses to SSRIs.

SSRIs have been shown to significantly reduce drinking behavior in rodents (Maurel, De Vry, & Schreiber, 1999; Simon O'Brien et al., 2011) and nonhuman primates (Higley, Hasert, Suomi, & Linnoila, 1998), suggesting that 5-HT levels can directly influence drinking behavior in the absence of comorbid psychiatric conditions. Similarly, a number of specific 5-HT_r agonists and antagonists have been shown to reduce alcohol consumption in rodent models (de Bruin et al., 2013; Rezvani, Cauley, & Levin, 2014; Wilson, Neill, & Costall, 1998). To date, no work has investigated the effects of SSRIs on drinking behavior in 5-HT deficient mice. However, the aforementioned Tph2KI mouse line, which exhibits aggression and increased alcohol consumption, exhibits reduced responses to several antidepressant-like effects of SSRIs (Sachs, Jacobsen, et al., 2013; Sachs, Ni, & Caron, 2015) and does not exhibit the typical increases in hippocampal neurogenesis or BDNF levels following chronic SSRI administration (Sachs, Jacobsen, et al., 2013). Thus, one might predict that by limiting SSRI-induced increases in extracellular 5-HT (Jacobsen et al., 2012; Sachs, Jacobsen, et al., 2013), brain 5-HT deficiency may also impair the ability of SSRIs to reduce drinking, but future research would be required to evaluate this.

CONCLUDING REMARKS

The current absence of validated noninvasive methods to test for brain 5-HT deficiency precludes our ability to conclusively identify the subset of alcoholics who exhibit low brain 5-HT. However, the many similarities exhibited by 5-HT-deficient mouse models and Type 2 alcoholics (putative low 5-HT biomarker abnormalities, aggression, impulsivity, and impaired SSRI responses) suggest that 5-HT deficiency may be a common feature of Type 2 alcoholism. Developing methods to identify individuals with low brain 5-HT could improve our understanding of AUD etiology and has the potential to facilitate the selection of appropriate treatment options for 5-HT-deficient individuals. Given the known importance of environmental factors and stress in the development of AUDs, it would be an overstatement to claim that low 5-HT is sufficient to cause AUDs. However, it may increase the likelihood of developing an AUD for individuals who consume alcohol regularly. In addition, preclinical work suggests that brain 5-HT deficiency increases susceptibility to stress, which could reflect yet another mechanism whereby low 5-HT could predispose to AUDs (Sachs et al., 2015). Overall, through its ability to increase alcohol consumption, impulsivity, aggression, and vulnerability to stress, brain 5-HT deficiency appears likely to play a major role in the development of AUDs, particularly in individuals with high levels of aggression and impulsivity.

Glossary

2C-B 2,5-Dimethoxy-4-bromophenethylamine, a psychedelic drug of phenethylamines class.

2C-E 2,5-Dimethoxy-4-ethylphenethylamine, a psychedelic drug of phenethylamines class.

2C-I 2,5-Dimethoxy-4-iodophenethylamine, a psychedelic drug of phenethylamines class.

4-OH-DALT 4-Hydroxy-5-methoxydimethyltryptamine, a psychedelic drug of tryptamines class with hallucinogenic effects.

5-MeO-AMT 5-Methoxy-α-methyltryptamine, a psychedelic drug of tryptamines class.

5-MeO-DALT 5-Methoxy-*N*,*N*-diallyltryptamine, a psychedelic drug of tryptamines class.

5-MeO-DIPT 5-Methoxy-diisopropyltryptamine, a psychedelic drug of tryptamines class.

5-MeO-5-MeO-DMT 5-Methoxy-*N,N*-dimethyltryptamine, a naturally occurring psychedelic drug of tryptamines class.

5-MeO-DPT337 5-Methoxy-*N,N*-dimethyltryptamine, a tryptamine derivate.

11-Sterone Brand name for substances marketed as food-supplements and/or for bodybuilders.

AM-2201 1-(5-Fluoropentyl)-3-(1-naphthoyl)indole, a synthetic cannabinoid.

AM-7921 An opioid analgesic drug acting as μ-opioid receptor agonist.

AMT Alpha-methyltryptamine, a psychedelic and stimulant drug of tryptamines class.

Aroma/Artic Spice Brand names for synthetic cannabinoids.

Ayahuasca A beverage used originally for religious purpose containing dimethyl-tryptamine (DMT) derived from *Banisteriopsis caapi* and inducing spiritual experiences and deeper insight.

Baclofen A $GABA_B$ receptor agonist.

Benzylpiperazine Or BZP, a piperazine that induces euphoria, alertness, and general sensation of well-being.

Blue Silk Brand name for synthetic cathinones.

Bromo-DragonFLY 3C-Bromo-Dragonfly, a psychedelic drug of phenethylamines class.

Bufotenine 5-OH-DMT, an alkaloid of the tryptamines class that induces hallucinogenic effects, euphoria, distorted sense of time, and mind-altering effects.

Butylone β-keto-*N*-methylbenzodioxolylbutanamine, a synthetic cathinone with stimulant properties.

(C8)-CP47,497 A bicyclic compound of the synthetic cannabinoid class.

Charge+ Brand name for synthetic cathinones.

CPP 1-(3-Chlorophenyl) piperazines, a stimulant drug of piperazines class.

DALT *N,N*-Diallyltryptamine, is a tryptamines class derivate that induces euphoria and hallucinogenic effects.

Dimethocaine Synthetic cocaine with local anesthetic propriety.

DiPT Diisopropyltryptamine, a psychedelic hallucinogenic drug of tryptamines class.

DMT Dimethyltryptamine, a psychedelic compound of tryptamines class.

DND Diphenidine, a dissociative drug of ketamine and phencyclidine like class.

DNP 2,4-Dinitrophenol, an organic compound used as anti-obesity drug that inhibits the production of ATP.

DOET 2,5-Dimethoxy-4-ethylamphetamin, a psychedelic drug of phenethylamines class.

DOI 2,5-Dimethoxy-4-iodoamphetamine, a psychedelic drug of phenethylamines class.

DOC 2,5-Dimethoxy-4-chloroamphetamine, a psychedelic drug of phenethylamines class.

DOM 2,5-Dimethoxy-4-methylamphetamine, a psychedelic drug of phenethylamines class.

DPT Dipropyltryptamine, a psychedelic drug of tryptamines class.

Dream Brand name for synthetic cannabinoids.

DXM Dextromethorphan, a dissociative anesthetic and a nonopioid derivative of morphine.

Ethylone 3,4-Methylenedioxy-*N*-ethylcathinone, also known as MDEC or bk-MDEA, belongs to the class of synthetic cathinones.

pFBT 3-(*p*-Fluorobenzoyloxy)tropane, synthetic cocaine with stimulant effects.

Foxy/Foxy Methoxy Street names for 5-MeO-DIPT.

Genie Brand name for synthetic cannabinoids.

GHB Gamma-hydroxybutyric acid, a central nervous system depressant that induces euphoria and increases sexual arousal.

HU-210 (6aR)-trans-3-(1,1-dimethylheptyl)-6a,7,10,10a-tetrahydro-1-hydroxy-6,6-dimethyl-6H-dibenzo[b,d]pyran-9-methanol, a tricyclic compound of the synthetic cannabinoid class.

Ibogaina The main alkaloid present in *Tabernanthe iboga*, a naturally occurring psychoactive substance with dissociative properties.

IC-26 Methiodone, a synthetic opioid with analgesic property similar to methadone.

Ivory Snow/Ivory Wave Brand names for synthetic cathinones.

Jipper Street name for methoxetamine (MXE).

JWH-018 1-Pentyl-3-(1-naphthoyl)indole, an aminoalkylindole analogue of the class of synthetic cannabinoids.

JWH-073 Naphthalen-1-yl-(1-butylindol-3-yl)methanone, a naphthoylindole analogue of the class of synthetic cannabinoid.

JWH-081 4-Methoxynaphthalen-1-yl-(1-pentylindol-3-yl)methanone, a naphthoylindole analogue of the class of synthetic cannabinoids.

JWH-122 (4-Methyl-1-naphthyl)-(1-pentylindol-3-yl)methanone, a synthetic cannabinoid derived by methylation of JWH-018.

JWH-200 (1-(2-Morpholin-4-ylethyl)indol-3-yl)-naphthalen-1-ylmethanone, an aminoalkylindole analogue of the class of synthetic cannabinoids.

K2/K3 Brand names for synthetic cannabinoids.

Kanna *Sceletium tortuosum*, a plant used as mood-altering substances.

Kava *Piper methysticum*, a plant with recreational effects.

Kmax Street name for methoxetamine (MXE).

Kratom *Mitragyna speciosa*, a plant used for recreational purpose containing mitragynine and 7-hydroxymitragynine (very active alkaloids).

Legal Ketamine Street name for methoxetamine (MXE).

Lilly Street name for olanzapine.

LSD Lysergic acid diethylamide, a psychedelic drug closely related to LSA (Lysergic acid amide).

MA Street name for methoxetamine (MXE).

MBDB *N*-Methyl-1,3-benzodioxolylbutanamine, a closely related analog of MDMA.

MBDP *N*-Methyl-1,3-benzodioxolylpentanamine, a psychoactive drug that increases body temperature that may lead to death.

MBZP 1-Methyl-4-benzylpiperazine, a benzylpiperazines (piperazines class).

MDEA 3,4-Methylenedioxy-*N*-ethyl-amphetamine, a psychoactive drug that increases body temperature that may lead to death.

MDMA 3,4-Methylenedioxymethamphetamine, better known as ecstasy.

MDPBP 3′,4′-Methylenedioxy-α-pyrrolidinobutyrophenone, a synthetic cathinone.

MDPPP 3′,4′-Methylenedioxy-α-pyrrolidinopropiophenone, a synthetic cathinone.

MDPV 3,4-Methylenedioxyprovalerone, a synthetic cathinone.

Melonotan I/Melonotan II Synthetic peptides acting as melanocortin receptor agonists.

MeOPP *para*-Methoxyphenylpiperazine, a piperazine.

Mephedrone 4-Methylmethcathinone, a synthetic cathinone.

Methcathinone α-Methylamino-propiophenone, better known as ephedrone, a synthetic cathinone similar to amphetamine.

Methylone 3,4-Methylenedioxy-*N*-methylcathinone, a synthetic cathinone.

Minx/M-Ket/Mexxy Street names for methoxetamine (MXE).

MT-45 Analogue of lefetamine with analgesic propriety.

MXE Methoxetamine, a ketamine-like hallucinogenic, dissociative drug.

MXP Methoxphenidine, a ketamine-like hallucinogenic, dissociative drug.

Nortilidine A synthetic opioid with analgesic property.

Ocean Burst Brand name for synthetic cathinones.

Olanzapine Atypical antipsychotic used for the treatment of schizophrenia and bipolar disorder, more recently self-prescribed and used as "trip terminator" substance for NPS-induced psychosis.

Pentylone β-Keto-methylbenzodioxolylpentanamine, a synthetic cathinone.

Phenazepam 1-4-Benzodiazepine, a benzodiazepine.

Phenibut 4-Amino-3-phenyl-butyric acid, a central nervous system depressant drug.
Phenylpiperazine A piperazine that induces euphoria, alertness, and a general sense of well-being.
Piracetam A cyclic analogue of GABA used to enhance cognitive functions and work performance.
P-MAG Brand name for substances marketed as food supplements and/or for bodybuilders.
α-PPP α-Pyrrolidinopropiophenone, a synthetic cathinone.
Pregabalin An analog of GABA that induces visual hallucinations, increases energy and sociability, and reduces anxiety.
Protodrol Brand name for substances marketed as food-supplements and/or for bodybuilders.
Psilocin 4-OH-DMT, an alkaloid of the tryptamines class.
Psilocybin A naturally occurring tryptamine rapidly converted into psilocin.
Pure Ivory/Purple Wave Brand names for synthetic cathinones.
Quetiapine Atypical antipsychotic used for the treatment of schizophrenia and bipolar disorder, more recently self-prescribed and used as "come off psychedelic trip" substance for NPS- induced psychosis.
Roflcopter Street name for methoxetamine (MXE).
Salvia divinorum A psychoactive plant with anxiolytic and sedative properties.
Scene/Silver/Smoke Brand names for synthetic cannabinoids.
Snow Leopard Brand name for synthetic cathinones.
Special K/Special M Street names for methoxetamine (MXE).
Spice/Spice Diamond/Spice Gold Brand names for synthetic cannabinoids.
Stardust Brand name for synthetic cathinones.
Straight Epi/Super Halo/Super Tren/MG Brand names for substances marketed as food-supplements and/or for bodybuilders.
TFMPP 3-Trifluoromethylphenylpiperazine, a piperazine that induces mild hallucinogenic effects.
THH Tetrahydroharmine, a tryptamine that induces hallucinogenic effects, euphoria, and feeling of well-being.
TMA Trimethoxyamphetamines, a mescaline derived psychedelic drug of the phenethylamines class.
Tren Bomb/Ultra Mass Brand name for substances marketed as food-supplements and/or for bodybuilders.
Vanilla Sky Brand name for synthetic cathinones.
W15/W18 Synthetic opioids with analgesic property.
White Knight/White Lightening Brand names for synthetic cathinones.
Yucatan Fire Brand name for synthetic cannabinoids.

References

Agrawal, A., Lynskey, M. T., Todorov, A. A., Schrage, A. J., Littlefield, A. K., Grant, J. D., & Heath, A. C. (2011). A candidate gene association study of alcohol consumption in young women. *Alcoholism, Clinical and Experimental Research, 35*(3), 550–558.

American Psychiatric Association. (2013). Substance-related and addictive disorders. In *Diagnostic and statistical manual of mental disorders: DSM-5* (5th ed.). Washington, DC: American Psychiatric Association.

Angoa-Perez, M., Kane, M. J., Briggs, D. I., Sykes, C. E., Shah, M. R., Francescutti, D. M., & Kuhn, D. M. (June 2012). Genetic depletion of brain 5HT reveals a common molecular pathway mediating compulsivity and impulsivity. *Journal of Neurochemistry, 121*(6), 974–984.

Ansorge, M. S., Zhou, M., Lira, A., Hen, R., & Gingrich, J. A. (2004). Early-life blockade of the 5-HT transporter alters emotional behavior in adult mice. *Science, 306*(5697), 879–881.

Babor, T. F., Hofmann, M., DelBoca, F. K., Hesselbrock, V., Meyer, R. E., Dolinsky, Z. S., & Rounsaville, B. (1992). Types of alcoholics, I. Evidence for an empirically derived typology based on indicators of vulnerability and severity. *Archives of General Psychiatry, 49*(8), 599–608.

Ballenger, J. C., Goodwin, F. K., Major, L. F., & Brown, G. L. (1979). Alcohol and central serotonin metabolism in man. *Archives of General Psychiatry, 36*(2), 224–227.

Banki, C. M. (1981). Factors influencing monoamine metabolites and tryptophan in patients with alcohol dependence. *Journal of Neural Transmission, 50*(2–4), 89–101.

Bazovkina, D. V., Lichman, D. V., & Kulikov, A. V. (2015). The C1473G polymorphism in the tryptophan hydroxylase-2 gene: Involvement in ethanol-related behavior in mice. *Neuroscience Letters, 589*, 79–82.

Beaulieu, J. M., Zhang, X., Rodriguiz, R. M., Sotnikova, T. D., Cools, M. J., Wetsel, W. C., … Caron, M. G. (2008). Role of GSK3 beta in behavioral abnormalities induced by serotonin deficiency. *Proceedings of the National Academy of Sciences of the United States of America, 105*(4), 1333–1338.

Bjork, J. M., Hommer, D. W., Grant, S. J., & Danube, C. (2004). Impulsivity in abstinent alcohol-dependent patients: Relation to control subjects and type 1-/type 2-like traits. *Alcohol, 34*(2–3), 133–150.

Bodnoff, S. R., Suranyi-Cadotte, B., Quirion, R., & Meaney, M. J. (1989). A comparison of the effects of diazepam versus several typical and atypical anti-depressant drugs in an animal model of anxiety. *Psychopharmacology, 97*(2), 277–279.

de Bruin, N. M., McCreary, A. C., van Loevezijn, A., de Vries, T. J., Venhorst, J., van Drimmelen, M., & Kruse, C. G. (2013). A novel highly selective 5-HT6 receptor antagonist attenuates ethanol and nicotine seeking but does not affect inhibitory response control in Wistar rats. *Behavioral Brain Research, 236*(1), 157–165.

Burnett, E. J., Davenport, A. T., Grant, K. A., & Friedman, D. P. (2012). The effects of chronic ethanol self-administration on hippocampal serotonin transporter density in monkeys. *Frontiers in Psychiatry, 3*, 38.

Charney, D. A., Heath, L. M., Zikos, E., Palacios-Boix, J., & Gill, K. J. (2015). Poorer drinking outcomes with citalopram treatment for alcohol dependence: A randomized, double-blind, placebo-controlled trial. *Alcoholism, Clinical and Experimental Research, 39*(9), 1756–1765.

Cloninger, C. R., Bohman, M., & Sigvardsson, S. (1981). Inheritance of alcohol abuse. Cross-fostering analysis of adopted men. *Archives of General Psychiatry, 38*(8), 861–868.

Cloninger, C. R., Sigvardsson, S., & Bohman, M. (1996). Type I and type II alcoholism – An update. *Alcohol Health and Research World, 20*(1), 18–23.

Contreras, S., Alvarado, R., & Mardones, J. (1990). Effects of *p*-chlorophenylalanine on the voluntary consumption of ethanol, water and solid food by UChA and UChB rats. *Alcohol, 7*(5), 403–407.

Crutchfield, R. D., & Gove, W. R. (1984). Determinants of drug use: A test of the coping hypothesis. *Social Science & Medicine, 18*(6), 503–509.

Delgado, P. L. (2000). Depression: The case for a monoamine deficiency. *The Journal of Clinical Psychiatry, 61*(Suppl. 6), 7–11.

Dougherty, D. M., Bjork, J. M., Marsh, D. M., & Moeller, F. G. (1999). Influence of trait hostility on tryptophan depletion-induced laboratory aggression. *Psychiatry Research, 88*(3), 227–232.

Dundon, W., Lynch, K. G., Pettinati, H. M., & Lipkin, C. (2004). Treatment outcomes in type A and B alcohol dependence 6 months after serotonergic pharmacotherapy. *Alcoholism, Clinical and Experimental Research, 28*(7), 1065–1073.

Fils-Aime, M. L., Eckardt, M. J., George, D. T., Brown, G. L., Mefford, I., & Linnoila, M. (1996). Early-onset alcoholics have lower cerebrospinal fluid 5-hydroxyindoleacetic acid levels than late-onset alcoholics. *Archives of General Psychiatry, 53*(3), 211–216.

Fudalej, S., Ilgen, M., Fudalej, M., Kostrzewa, G., Barry, K., Wojnar, M., ... Ploski, R. (2010). Association between tryptophan hydroxylase 2 gene polymorphism and completed suicide. *Suicide and Life Threatening Behavior, 40*(6), 553–560.

Geller, I. (1973). Effects of para-chlorophenylalanine and 5-hydroxytryptophan on alcohol intake in the rat. *Pharmacology Biochemistry and Behavior, 1*(3), 361–365.

Gorelick, D. A., & Paredes, A. (1992). Effect of fluoxetine on alcohol consumption in male alcoholics. *Alcoholism, Clinical and Experimental Research, 16*(2), 261–265.

Gutknecht, L., Jacob, C., Strobel, A., Kriegebaum, C., Muller, J., Zeng, Y., & Lesch, K. P. (2007). Tryptophan hydroxylase-2 gene variation influences personality traits and disorders related to emotional dysregulation. *International Journal of Neuropsychopharmacology, 10*(3), 309–320.

Higley, J., Hasert, M., Suomi, S., & Linnoila, M. (1998). The serotonin reuptake inhibitor sertraline reduces excessive alcohol consumption in nonhuman primates: Effect of stress. *Neuropsychopharmacology, 18*(6), 431–443.

Holman, R. B., Hoyland, V., & Shillito, E. E. (1975). The failure of *p*-chlorophenylalanine to affect voluntary alcohol consumption in rats. *British Journal of Pharmacology, 53*(2), 299–304.

Ishiguro, H., Saito, T., Akazawa, S., Mitushio, H., Tada, K., Enomoto, M., & Arinami, T. (1999). Association between drinking-related antisocial behavior and a polymorphism in the serotonin transporter gene in a Japanese population. *Alcoholism, Clinical and Experimental Research, 23*(7), 1281–1284.

Jacobsen, J. P., Siesser, W. B., Sachs, B. D., Peterson, S., Cools, M. J., Setola, V., ... Caron, M. G. (2012). Deficient serotonin neurotransmission and depression-like serotonin biomarker alterations in tryptophan hydroxylase 2 (Tph2) loss-of-function mice. *Molecular Psychiatry, 17*(7), 694–704.

Kiianmaa, K. (1976). Alcohol intake in the rat after lowering brain 5-hydroxtryptamine content by electrolytic midbrain raphe lesions, 5,6-dihydroxytryptamine or *p*-chlorophenylalanine. *Medical Biology, 54*(3), 203–209.

Krakowski, M. (2003). Violence and serotonin: Influence of impulse control, affect regulation, and social functioning. *The Journal of Neuropsychiatry and Clinical Neurosciences, 15*(3), 294–305.

LeMarquand, D. G., Benkelfat, C., Pihl, R. O., Palmour, R. M., & Young, S. N. (1999). Behavioral disinhibition induced by tryptophan depletion in nonalcoholic young men with multigenerational family histories of paternal alcoholism. *American Journal of Psychiatry, 156*(11), 1771–1779.

LeMarquand, D., Pihl, R. O., & Benkelfat, C. (1994a). Serotonin and alcohol intake, abuse, and dependence: Clinical evidence. *Biological Psychiatry, 36*(5), 326–337.

LeMarquand, D., Pihl, R. O., & Benkelfat, C. (1994b). Serotonin and alcohol intake, abuse, and dependence: Findings of animal studies. *Biological Psychiatry, 36*(6), 395–421.

Lemay, F., Dore, F. Y., & Beaulieu, J. M. (2015). Increased ethanol consumption despite taste aversion in mice with a human tryptophan hydroxylase 2 loss of function mutation. *Neuroscience Letters, 609*, 194–197.

Lesch, K. P., Bengel, D., Heils, A., Sabol, S. Z., Greenberg, B. D., Petri, S., & Murphy, D. L. (1996). Association of anxiety-related traits with a polymorphism in the serotonin transporter gene regulatory region. *Science, 274*(5292), 1527–1531.

Linnoila, M., Virkkunen, M., Scheinin, M., Nuutila, A., Rimon, R., & Goodwin, F. K. (1983). Low cerebrospinal fluid 5-hydroxyindoleacetic acid concentration differentiates impulsive from nonimpulsive violent behavior. *Life Sciences, 33*(26), 2609–2614.

Maron, E., Nutt, D., & Shlik, J. (2012). Neuroimaging of serotonin system in anxiety disorders. *Current Pharmaceutical Design, 18*(35), 5699–5708.

Maurel, S., De Vry, J., & Schreiber, R. (1999). Comparison of the effects of the selective serotonin-reuptake inhibitors fluoxetine, paroxetine, citalopram and fluvoxamine in alcohol-preferring cAA rats. *Alcohol, 17*(3), 195–201.

Miczek, K. A., & Barry, H., 3rd (1977). Comparison of the effects of alcohol, chlordiazepoxide, and delta9-tetrahydrocannabinol on intraspecies aggression in rats. *Advances in Experimental Medicine and Biology, 85B*, 251–264.

Moreno, F. A., Gelenberg, A. J., Heninger, G. R., Potter, R. L., McKnight, K. M., Allen, J., & Delgado, P. L. (1999). Tryptophan depletion and depressive vulnerability. *Biological Psychiatry, 46*(4), 498–505.

Mosienko, V., Bert, B., Beis, D., Matthes, S., Fink, H., Bader, M., & Alenina, N. (2012). Exaggerated aggression and decreased anxiety in mice deficient in brain serotonin. *Translational Psychiatry, 2*.

Moss, H. B., Yao, J. K., & Panzak, G. L. (1990). Serotonergic responsivity and behavioral dimensions in antisocial personality disorder with substance abuse. *Biological Psychiatry, 28*(4), 325–338.

Murphy, D. L., Zohar, J., Benkelfat, C., Pato, M. T., Pigott, T. A., & Insel, T. R. (1989). Obsessive–compulsive disorder as a 5-HT subsystem-related behavioural disorder. *The British Journal of Psychiatry. Supplement, 8*, 15–24.

Myers, R. D., & Cicero, T. J. (1969). Effects of serotonin depletion on the volitional alcohol intake of rats during a condition of psychological stress. *Psychopharmacologia, 15*(5), 373–381.

Myers, R. D., & Tytell, M. (1972). Volitional consumption of flavored ethanol solution by rats: The effects of pCPA, and the absence of tolerance. *Physiology & Behavior, 8*(3), 403–408.

Myers, R. D., & Veale, W. L. (1968). Alcohol preference in the rat: Reduction following depletion of brain serotonin. *Science, 160*(3835), 1469–1471.

Naranjo, C. A., Poulos, C. X., Bremner, K. E., & Lanctot, K. L. (1992). Citalopram decreases desirability, liking, and consumption of alcohol in alcohol-dependent drinkers. *Clinical Pharmacology and Therapeutics, 51*(6), 729–739.

Naranjo, C. A., Poulos, C. X., Bremner, K. E., & Lanctot, K. L. (1994). Fluoxetine attenuates alcohol intake and desire to drink. *International Clinical Psychopharmacology, 9*(3), 163–172.

Perez-Rodriguez, M. M., Weinstein, S., New, A. S., Bevilacqua, L., Yuan, Q., Zhou, Z., & Siever, L. J. (2010). Tryptophan-hydroxylase 2 haplotype association with borderline personality disorder and aggression in a sample of patients with personality disorders and healthy controls. *Journal of Psychiatric Research, 44*(15), 1075–1081. http://dx.doi.org/10.1016/j.jpsychires.2010.03.014.

Peters, E. J., Slager, S. L., McGrath, P. J., Knowles, J. A., & Hamilton, S. P. (2004). Investigation of serotonin-related genes in antidepressant response. *Molecular Psychiatry, 9*(9), 879–889.

Petrakis, I. L., Buonopane, A., O'Malley, S., Cermik, O., Trevisan, L., Boutros, N. N., & Krystal, J. H. (2002). The effect of tryptophan depletion on alcohol self-administration in non-treatment-seeking alcoholic individuals. *Alcoholism, Clinical and Experimental Research, 26*(7), 969–975.

Petrakis, I. L., Trevisan, L., Boutros, N. N., Limoncelli, D., Cooney, N. L., & Krystal, J. H. (2001). Effect of tryptophan depletion on alcohol cue-induced craving in abstinent alcoholic patients. *Alcoholism, Clinical and Experimental Research, 25*(8), 1151–1155.

Pierucci-Lagha, A., Feinn, R., Modesto-Lowe, V., Swift, R., Nellissery, M., Covault, J., & Kranzler, H. R. (2004). Effects of rapid tryptophan depletion on mood and urge to drink in patients with co-morbid major depression and alcohol dependence. *Psychopharmacology (Berlin), 171*(3), 340–348. http://dx.doi.org/10.1007/s00213-003-1588-6.

Plemenitas, A., Kores Plesnicar, B., Kastelic, M., Porcelli, S., Serretti, A., & Dolzan, V. (2015). Genetic variability in tryptophan hydroxylase 2 gene in alcohol dependence and alcohol-related psychopathological symptoms. *Neuroscience Letters, 604*, 86–90.

Reuter, M., Ott, U., Vaitl, D., & Hennig, J. (2007). Impaired executive control is associated with a variation in the promoter region of the tryptophan hydroxylase 2 gene. *Journal of Cognitive Neuroscience, 19*(3), 401–408.

Rezvani, A. H., Cauley, M. C., & Levin, E. D. (2014). Lorcaserin, a selective 5-HT(2C) receptor agonist, decreases alcohol intake in female alcohol preferring rats. *Pharmacology Biochemistry and Behavior, 125*, 8–14.

Roy, A., Virkkunen, M., & Linnoila, M. (1987). Reduced central serotonin turnover in a subgroup of alcoholics? *Progress in Neuropsychopharmacology and Biological Psychiatry, 11*(2–3), 173–177.

Sachs, B. D., Jacobsen, J. P., Thomas, T. L., Siesser, W. B., Roberts, W. L., & Caron, M. G. (2013). The effects of congenital brain serotonin deficiency on responses to chronic fluoxetine. *Translational Psychiatry, 3*.

Sachs, B. D., Ni, J. R., & Caron, M. G. (2015). Brain 5-HT deficiency increases stress vulnerability and impairs antidepressant responses following psychosocial stress. *Proceedings of the National Academy of Sciences of the United States of America, 112*(8), 2557–2562.

Sachs, B. D., Salahi, A. A., & Caron, M. G. (2013). Congenital brain serotonin deficiency leads to reduced ethanol sensitivity and increased ethanol consumption in mice. *Neuropharmacology, 77C*, 177–184.

Sacks, J. J., Gonzales, K. R., Bouchery, E. E., Tomedi, L. E., & Brewer, R. D. (2015). 2010 National and state costs of excessive alcohol consumption. *American Journal of Preventive Medicine, 49*(5), e73–e79.

Sander, T., Harms, H., Dufeu, P., Kuhn, S., Hoehe, M., Lesch, K. P., & Schmidt, L. G. (1998). Serotonin transporter gene variants in alcohol-dependent subjects with dissocial personality disorder. *Biological Psychiatry, 43*(12), 908–912.

Santarelli, L., Saxe, M., Gross, C., Surget, A., Battaglia, F., Dulawa, S., & Hen, R. (2003). Requirement of hippocampal neurogenesis for the behavioral effects of antidepressants. *Science, 301*(5634), 805–809.

Schuckit, M. A., Mazzanti, C., Smith, T. L., Ahmed, U., Radel, M., Iwata, N., & Goldman, D. (1999). Selective genotyping for the role of 5-HT2A, 5-HT2C, and GABA alpha 6 receptors and the serotonin transporter in the level of response to alcohol: A pilot study. *Biological Psychiatry, 45*(5), 647–651.

Shibasaki, M., Inoue, M., Kurokawa, K., Ogou, S., & Ohkuma, S. (2010). Expression of serotonin transporter in mice with ethanol physical dependency. *Journal of Pharmacological Science, 114*(2), 234–237.

Simon O'Brien, E., Legastelois, R., Houchi, H., Vilpoux, C., Alaux-Cantin, S., Pierrefiche, O., ... Naassila, M. (2011). Fluoxetine, desipramine, and the dual antidepressant milnacipran reduce alcohol self-administration and/or relapse in dependent rats. *Neuropsychopharmacology, 36*(7), 1518–1530.

Soloff, P. H., Kelly, T. M., Strotmeyer, S. J., Malone, K. M., & Mann, J. J. (2003). Impulsivity, gender, and response to fenfluramine challenge in borderline personality disorder. *Psychiatry Research, 119*(1–2), 11–24.

Soloff, P. H., Meltzer, C. C., Greer, P. J., Constantine, D., & Kelly, T. M. (2000). A fenfluramine-activated FDG-PET study of borderline personality disorder. *Biological Psychiatry, 47*(6), 540–547.

Stanley, B., Molcho, A., Stanley, M., Winchel, R., Gameroff, M. J., Parsons, B., & Mann, J. J. (2000). Association of aggressive behavior with altered serotonergic function in patients who are not suicidal. *Americal Journal of Psychiatry, 157*(4), 609–614.

Stoltenberg, S. F., Glass, J. M., Chermack, S. T., Flynn, H. A., Li, S., Weston, M. E., & Burmeister, M. (2006). Possible association between response inhibition and a variant in the brain-expressed tryptophan hydroxylase-2 gene. *Psychiatric Genetics, 16*(1), 35–38.

Storvik, M., Tiihonen, J., Haukijarvi, T., & Tupala, E. (2006). Nucleus accumbens serotonin transporters in alcoholics measured by whole-hemisphere autoradiography. *Alcohol, 40*(3), 177–184.

Storvik, M., Tiihonen, J., Haukijarvi, T., & Tupala, E. (2007). Amygdala serotonin transporters in alcoholics measured by whole hemisphere autoradiography. *Synapse, 61*(8), 629–636.

Substance Abuse, and Mental Health Services Administration, (SAMHSA). (2015). *Behavioral health trends in the United States: Results from the 2014 national survey on drug use and health.* Rockville, MD: Substance Abuse and Mental Health Services Administration. NSDUH Series H-50, HHS Publication No. (SMA) 15-4927.

Tsai, S. J., Hong, C. J., Liou, Y. J., Yu, Y. W., Chen, T. J., Hou, S. J., & Yen, F. C. (2009). Tryptophan hydroxylase 2 gene is associated with major depression and antidepressant treatment response. *Progress in Neuropsychopharmacology and Biological Psychiatry, 33*(4), 637–641.

Tupala, E., & Tiihonen, J. (2004). Dopamine and alcoholism: Neurobiological basis of ethanol abuse. *Progress in Neuropsychopharmacology and Biological Psychiatry, 28*(8), 1221–1247.

Tzvetkov, M. V., Brockmoller, J., Roots, I., & Kirchheiner, J. (2008). Common genetic variations in human brain-specific tryptophan hydroxylase-2 and response to antidepressant treatment. *Pharmacogenetics and Genomics, 18*(6), 495–506.

Virkkunen, M., & Linnoila, M. (2002). Serotonin in early-onset alcoholism. *Recent Developments in Alcoholism, 13*, 173–189.

Walderhaug, E., Lunde, H., Nordvik, J. E., Landro, N. I., Refsum, H., & Magnusson, A. (2002). Lowering of serotonin by rapid tryptophan depletion increases impulsiveness in normal individuals. *Psychopharmacology (Berlin), 164*(4), 385–391. http://dx.doi.org/10.1007/s00213-002-1238-4.

Walther, D. J., & Bader, M. (2003). A unique central tryptophan hydroxylase isoform. *Biochemical Pharmacology, 66*(9), 1673–1680.

Weiss, R. D., Griffin, M. L., & Mirin, S. M. (1992). Drug abuse as self-medication for depression: An empirical study. *American Journal of Drug and Alcohol Abuse, 18*(2), 121–129.

Wilson, A. W., Neill, J. C., & Costall, B. (1998). An investigation into the effects of 5-HT agonists and receptor antagonists on ethanol self-administration in the rat. *Alcohol, 16*(3), 249–270.

Wong, M. L., & Licinio, J. (2001). Research and treatment approaches to depression. *Nature Reviews Neuroscience, 2*(5), 343–351.

Young, S. N., & Leyton, M. (2002). The role of serotonin in human mood and social interaction. Insight from altered tryptophan levels. *Pharmacology Biochemistry and Behavior, 71*(4), 857–865.

Yudko, E., Blanchard, D. C., Henrie, J. A., & Blanchard, R. J. (2002). Emerging themes in preclinical research on alcohol and aggression. *Recent Developments in Alcoholism, 13*, 123–138.

Zhou, Z., Roy, A., Lipsky, R., Kuchipudi, K., Zhu, G., Taubman, J., ... Goldman, D. (2005). Haplotype-based linkage of tryptophan hydroxylase 2 to suicide attempt, major depression, and cerebrospinal fluid 5-hydroxyindoleacetic acid in 4 populations. *Archives of General Psychiatry, 62*(10), 1109–1118.

Zill, P., Buttner, A., Eisenmenger, W., Moller, H. J., Bondy, B., & Ackenheil, M. (2004). Single nucleotide polymorphism and haplotype analysis of a novel tryptophan hydroxylase isoform (TPH2) gene in suicide victims. *Biological Psychiatry, 56*(8), 581–586.

Zill, P., Preuss, U. W., Koller, G., Bondy, B., & Soyka, M. (2007). SNP- and haplotype analysis of the tryptophan hydroxylase 2 gene in alcohol-dependent patients and alcohol-related suicide. *Neuropsychopharmacology, 32*(8), 1687–1694.

PART II

ADDICTIVE SUBSTANCES AND BEHAVIORAL HEALTH

CHAPTER

19

Functional Reorganization of Reward- and Habit-Related Brain Networks in Addiction

Y. Yalachkov[1,2], *J. Kaiser*[2], *M.J. Naumer*[2]

[1]University Hospital Frankfurt, Frankfurt am Main, Germany; [2]Goethe University, Frankfurt am Main, Germany

INTRODUCTION

Addiction has been recognized as a significant health burden with a major impact on society. Health consequences for the individual range from increased risk for cardiovascular (e.g., myocardial infarction and stroke) to various other diseases (e.g., tumor-related and liver or infectious diseases), as well as involve highly relevant socioeconomic aspects. Developing successful therapies for addiction depends on elucidating the nature of the processes responsible for the development and persistence of addictive behavior. Fortunately, during the recent decade there has been significant progress in understanding the structures targeted by addictive substances and their reorganization during the course of the disease. Nevertheless, many of the mechanisms involved in addiction remain unknown and are still being studied intensively.

This chapter gives a brief overview of the reorganizational processes in the reward and habit systems contributing essentially to the emergence and persistence of addiction. While we recognize that further brain systems and processes (e.g., executive and stress/homeostasis) also play a major role in the development of the disease, we refer the reader to excellent reviews covering the topics not included in this chapter due to space limitations (Goldstein & Volkow, 2011; Koob et al., 2014).

Paradigms such as cue reactivity have been particularly helpful in investigating the reorganizational processes taking place on a systems level in addiction. The assessment usually includes behavioral measures and the application of modern neuroimaging methods such as functional magnetic resonance imaging (fMRI) or positron emission tomography (PET) that reflect the neural correlates of cue reactivity. In these paradigms drug users are exposed to stimuli associated with their respective drug of abuse. The stimuli may be presented in different sensory modalities, for example, the visual (words/pictures/videos) (Janes et al., 2010; Luijten et al., 2011), auditory (drug-associated sounds, scripts) (Kilts et al., 2001; Seo et al., 2011), audiovisual (Childress et al., 1999; Garavan et al., 2000; Maas et al., 1998), tactile or haptic (handling the corresponding paraphernalia) (Filbey, Schacht, Myers, Chavez, & Hutchison, 2009; Wilson, Creswell, Sayette, & Fiez, 2013; Wilson, Sayette, Delgado, & Fiez, 2005; Yalachkov, Kaiser, Gorres, Seehaus, & Naumer, 2012), olfactory or gustatory (smelling or tasting) modalities (Claus, Ewing, Filbey, Sabbineni, & Hutchison, 2011; Schneider et al., 2001). Multisensory drug cues can also be employed (e.g., holding a cigarette while watching audio-videos of smoking) (Brody et al., 2007; Franklin et al., 2007; Grant et al., 1996) and have been shown to be even more successful in inducing cue reactivity in drug users (Yalachkov, Kaiser, & Naumer, 2012). Subjects may either passively view the cues or may be required to actively respond to the stimuli. Drug cues themselves may be presented subliminally or above perceptual threshold (Childress et al., 2008) and may be either task-related targets (Wilcox, Teshiba, Merideth, Ling, & Mayer, 2011; Zhang et al., 2011) or distractors (Artiges et al., 2009; Due, Huettel, Hall, & Rubin, 2002; Fryer et al., 2012; McClernon, Hiott, Huettel, & Rose, 2005). Accordingly, neutral or significant nondrug-related stimuli (e.g., positive, neutral, or negative cues) are presented as control stimuli using similar timing, sensory modality, and response paradigm. Behavioral and neural cue reactivity measures are collected during the task. The final and crucial comparison is then between the conditions "drug-related cues" and "control cues"

Addictive Substances and Neurological Disease
http://dx.doi.org/10.1016/B978-0-12-805373-7.00019-0

within the patients. Often an additional comparison is computed between patients and healthy controls, ensuring a higher validity of the results. Sometimes the employed paradigms are more complex, requiring from subjects, for example, to control their cue-induced craving, thus allowing conclusions on the brain regions involved in cognitive control mechanisms (Brody et al., 2007; Hartwell et al., 2011; Kober et al., 2010). Importantly, behavioral and neural measures can be correlated, which allows linking important aspects of drug-related behavior (e.g., subjective craving or urge, reaction time, or degree of dependence) to particular brain structures and neural processes.

The behavioral and neural differences provide some insight into how the addicted brain processes drug cues and what functional changes might have occurred in the course of addiction. If a higher fMRI response in a reward-related brain region and shorter reaction times for drug cues are detected in patients compared to control cues and healthy participants, one possible interpretation is that during the course of disease the reward-related system has undergone a functional reorganization leading to an altered brain response to a drug cue, which is then associated with a more pronounced neural processing and an altered behavioral response. If there were also a correlation between the cue-elicited brain response and, for example, severity of dependence or duration of the disease, this would be a further indication of consumption-dependent adaptational processes occurring in this particular brain region. Of course, all these measurements are purely correlative. To establish a causal relationship between drug use and neuroplastic changes in a particular brain system, experimental approaches are necessary. In humans, there are methodological and ethical limitations to the classical experimental approach of systematically changing the variables (e.g., drug use) and observing their effects on the subjects. Furthermore, long-term neuroimaging studies comparing brain responses in patients at the beginning and during their disease are also hard to conduct. One possible approach are modern minimally invasive methods such as transcranial magnetic stimulation (TMS) or transcranial direct-current stimulation (tDCS) which can stimulate or inhibit the activity of particular brain regions. However, their application has a number of limitations—for example, subcortical regions are barely affected, time and space resolution are limited, and duration and strength of stimulation are also restricted. Animal studies could play a major role in delivering methodologically sound evidence for the neuroadaptational processes proceeding and causing addiction. Often conclusions from brain activation differences found in neuroimaging studies are drawn based on what is known about the major function of this brain region from animal experiments. Therefore, while this chapter focuses on human behavioral and neuroimaging studies, we corroborate results from human studies with knowledge from animal experiments wherever possible.

FUNCTIONAL REORGANIZATION PROCESSES IN BRAIN CIRCUITS OF REWARD AND MOTIVATION

One important common characteristic of drugs of abuse is the drug-elicited increase of the extracellular dopamine (DA) concentration in the mesocorticolimbic system. This change in dopaminergic signaling has been demonstrated for the ventral striatum (VS), extended amygdala, hippocampus, anterior cingulate cortex (ACC), prefrontal cortex (PFC), and insula, which are innervated by dopaminergic projections predominantly from the ventral tegmental area (VTA) (Hyman, Malenka, & Nestler, 2006; Nestler, 2005). The exact mechanisms of this process have been understood for some but not all of the drugs. For example, nicotine administration increases the firing rate of VTA DA neurons (Calabresi, Lacey, & North, 1989) and causes more prominent DA release in the nucleus accumbens core (NAc) (Imperato, Mulus, & DiChiara, 1986). Its binding to nicotinic acetylcholine receptors located on the DA neurons projecting from the VTA to NAc leads to increased DA release (Clarke & Pert, 1985; Deutch, Holliday, Roth, Chun, & Hawrot, 1987). Similar effects have been observed for nicotine binding to glutamatergic and GABAergic neurons that modulate these DA neurons (Mansvelder, Keath, & McGehee, 2002; Wooltorton, Pidoplichko, Broide, & Dani, 2003). Cocaine inhibits the DA transporter and thus increases the extracellular DA concentration (Torres, Gainetdinov, & Caron, 2003). Drug-induced increases in extracellular DA have also been found for amphetamine, heroin, marijuana, and alcohol (Hyman et al., 2006).

The mesocorticolimbic system is usually involved in the processing of natural rewards such as food, water, and sex, responding to them with increased firing of DA neurons. However, drugs of abuse induce similar, and most importantly, far more prominent dopaminergic surges in the reward system compared to the physiological response to natural rewards (Jay, 2003; Kelley, 2004; Nestler, 2005). By inducing larger and longer dopaminergic responses, drugs of abuse "hijack" the natural neurobiological mechanisms by which the brain responds to reward, induce strongly consolidated drug-associated memories and powerfully reinforce actions associated with the drug and its consumption (Everitt & Robbins, 2005; Kalivas & O'Brien, 2008).

Repeated drug use establishes an association between the drug itself, its effects and the drug-related stimuli (i.e., environment, drug paraphernalia, and other stimuli predictive of drug use). Thus, drug cues become conditioned stimuli and are able to elicit DA release and craving (Goldstein & Volkow, 2011; Tang, Fellows, Small, & Dagher, 2012; Volkow et al., 2006). It has also been shown that incentive salience of drug cues increases over time (Robinson & Berridge, 1993), which might be attributed to the increased cue-elicited dopaminergic signaling. Thus, drug contexts, environments, or paraphernalia evolve to powerful cues capable of producing physiological arousal and robust attentional biases, and acting as a potent trigger of drug-seeking and drug-taking behaviors.

Evidence for increased incentive salience of drug cues has been demonstrated both by behavioral and by neuroimaging studies. Drug cues produce an attentional bias (Field & Cox, 2008), they are associated with shorter reaction times and approach bias (Cousijn et al., 2012a, 2012b) and are capable of steering behavior even if they are presented below the perceptual threshold (Childress et al., 2008). These powerful characteristics of drug cues are closely related with cue-induced activation of mesocorticolimbic circuitry. This has been demonstrated repeatedly in human neuroimaging studies (for recent meta-analyses, see Chase, Eickhoff, Laird, & Hogarth, 2011; Engelmann et al., 2012; Kuhn & Gallinat, 2011; Schacht, Anton, & Myrick, 2012). Compared to neutral control cues, drug-related cues elicit higher brain activations within the mesocorticolimbic circuits, including VTA, VS, amygdala, ACC, PFC, insula, and hippocampus in drug users (Brody et al., 2007; Childress et al., 2008; Childress et al., 1999; Claus et al., 2011; Due et al., 2002; Franklin et al., 2007; Grüsser et al., 2004; Kilts et al., 2001; Luijten et al., 2011; Smolka et al., 2006; Volkow et al., 2006; Vollstadt-Klein et al., 2010; Yalachkov, Kaiser, & Naumer, 2009). Indirect evidence for a causal link between drug-induced neuroplasticity processes in these brain regions and drug seeking as well as drug taking elicited by associated cues in humans can be derived from preclinical research in rodent and nonhuman primates. It has been shown that phasic firing of DA neurons projecting from VTA to VS is crucial for behavioral conditioning (Tsai et al., 2009), and activity in these brain regions reflects the reward value predicted by discriminative cues (Schultz, 2007a, 2007b; Schultz, Dayan, & Montague, 1997). Furthermore, amygdala and hippocampus are also crucial for conditioning processes: the amygdala is important for establishing drug–cue associations and the hippocampus for linking drug effects to specific contexts (Robbins, Ersche, & Everitt, 2008). Thus, their activation in neuroimaging experiments probably reflects learned reward values of conditioned cues and contexts.

The orbitofrontal cortex (OFC) and the ventromedial PFC (VMPFC) partially overlap and have been recognized as essential not only for merging sensory inputs but also for integrating these inputs with learned reward values as well as with current needs of the organism, thus steering motivated behavior (Lucantonio, Stalnaker, Shaham, Niv, & Schoenbaum, 2012; Schoenbaum, Roesch, & Stalnaker, 2006; Schoenbaum, Roesch, Stalnaker, & Takahashi, 2009). Projections from the amygdala and OFC to the VS contribute to drug seeking over long delays bridged by conditioned reinforcers (Everitt & Robbins, 2005). For the VS, this is only one possible source of information about the respective motivational values and incentive drives of stimuli apart from a broad network of cortical and subcortical regions with which it is directly or indirectly connected, governing the basal ganglia's final action output (Haber & Knutson, 2010).

Neuroimaging experiments have also implied the ACC and the insula in drug cue reactivity. While the ACC is usually involved in cognitive control, conflict, or error monitoring (e.g., Dosenbach et al., 2006; Garavan, Ross, Murphy, Roche, & Stein, 2002; Nee, Wager, & Jonides, 2007), it is also activated by salient stimuli (e.g., Liu, Hairston, Schrier, & Fan, 2011), including reward-related stimuli, and also stimuli that elicit pain or negative effects (for a review on the integrative role of this region, see Shackman et al., 2011). The insula has been associated primarily with interoception, or the awareness of bodily states and internal homeostasis (for a review, see Craig, 2003). However, in a close parallel to the ACC, the insula and the adjacent inferior frontal gyrus are also often engaged during tasks requiring cognitive control (e.g., Wager et al., 2005) and in response to salient external stimuli (e.g., Liu et al., 2011). Indeed, the ACC and the insula are commonly regarded as parts of a common large-scale brain network, variously referred to as the cingulo-opercular, fronto-insular, or salience network (Dosenbach et al., 2006; Seeley et al., 2007), whose function may be to integrate internal and external signals of salience and to initiate interactions between large-scale brain networks to best meet the current demands for control (Menon & Uddin, 2010; Sridharan, Levitin, & Menon, 2008; Sutherland, McHugh, Pariyadath, & Stein, 2012).

The impact of drug-related modulation of the mesocorticolimbic circuitry also extends to sensory representations of drug cues. Rewards enhance the sensory representations of cues associated with these rewards in the occipital, temporal, and parietal regions (Serences, 2008; Yalachkov, Kaiser, & Naumer, 2010). In particular, due to their acute reinforcing effects mediated by increases in DA and other neurotransmitter signaling, drugs of abuse are thought to facilitate the sensory processing of drug cues and to promote a broad range of learning and plasticity processes (Devonshire et al., 2004; Devonshire, Mayhew, & Overton, 2007). Arguably, such drug-induced enhancement of sensory processing of drug cues is an early manifestation of increased incentive salience of these cues. Because of this enhanced early processing, the sensory representations of drug cues are easily activated and trigger robust attentional biases in drug users. These processing biases may then be used by the decision-making and motor control systems to increase the chances of drug-seeking behavior. These mechanisms may explain the strong responses in sensory cortices often observed in human neuroimaging studies of drug cue reactivity (Due et al., 2002; Luijten et al., 2011; Yalachkov et al., 2010).

FUNCTIONAL REORGANIZATION PROCESSES IN BRAIN CIRCUITS OF HABITS AND AUTOMATICITY

Apart from the mesocorticolimbic system comprising VTA, VS, amygdala, and PFC, a further dopaminergic ascending pathway connects the substantia nigra (SN) with the caudate, putamen, (both referred to as the dorsal striatum, DS) and globus pallidus. While the VTA–VS projections are seen as essential for reward- and goal-oriented learning and behavior, the latter structures have been often reported in the context of habit learning and automaticity.

The majority of the studies revealing the involvement of DS in habit learning have been conducted with rodents and have demonstrated that the DS can be divided anatomically and functionally into dorsomedial striatum (DMS, corresponding to the dorsal caudate nucleus in humans) and dorsolateral striatum (DLS, corresponding to the dorsal putamen in humans). Meanwhile, the DMS has been shown to play a prominent role in action-outcome learning and the acquisition of instrumental responding (Belin, Jonkman, Dickinson, Robbins, & Everitt, 2009). However, the DLS is thought to be more involved in the development and expression of habits. Habits can be defined as a robust stimulus-response link emerging as a product of intensive learning processes where reinforcers strengthen the behavioral and neural associations between cue and response. While the behavior is initially goal oriented and reward dependent, after extensive training control over behavior is transferred to the stimulus. At this stage of learning reinforcers no longer steer the behavior, and their devaluation (i.e., by satiation or inducing nausea in the case of food stimuli) would remain without effect: the behavioral responses would be conducted automatically upon the presentation of the stimulus and their future performance can be maintained solely by cue presentation (Belin et al., 2009; Everitt & Robbins, 2005). This change from goal-oriented actions to robust habits is reflected by a shift of neural control over the behavior from the ventral to DLS (Belin et al., 2009; Everitt & Robbins, 2005).

Research since 2000s has broadened our knowledge about the exact mechanisms of drug-related habits. Although habit development has primarily been seen as a function of DLS, recent animal studies have shown that there is a complex brain network involved in the development and expression of habits. Several experiments have demonstrated that there are spiraling striato-nigro-striatal interconnections between VTA, VS, and DS which are targeted by drugs of abuse. Bilateral DA blockade in the DLS (Murray, Belin, & Everitt, 2012) or bilateral glutamate receptor blockade/lesions in the NAc core (i.e., VS) (Di Ciano & Everitt, 2001; El-Amamy & Holland, 2007) result in the same effects as the disconnection of the ventral from the DLS (Belin & Everitt, 2008; Belin et al., 2009). A 2015 study provided more insight into the mechanisms linking associative processes in the amygdala and robust drug-related habits: Murray et al. (2015) showed that disconnecting the basolateral amygdala (BLA) from the DLS in the early stages of a cocaine self-administration training does not influence cue-controlled cocaine-seeking behavior. This is in line with previous findings demonstrating that goal-directed cocaine seeking in the early stages of training is attributable to the NAc core and the posterior DMS (Murray et al., 2012). However, when rats had at least an intermediate history of training, disconnecting the BLA from the DLS reduced cocaine seeking similarly to bilateral lesions of the DLS. This finding highlights the interplay between DLS and BLA necessary to transfer the control over behavior from initially goal directed to intermediate, more habitually governed mechanisms. Murray et al. provided even more differentiated insight into those mechanisms by showing that disconnecting the central nucleus of the amygdala (CeN) from DLS also diminished cocaine seeking. However, this effect was observed only at later but not early or intermediate stages of training, which implies that CeN and its connection to DLS are recruited and influence drug-seeking habits only later during the course of addiction. Further results from this experiment demonstrated a sequential involvement of BLA and CeN in drug-related habits: bilateral inactivation of BLA was effective in reducing early but not well-established cocaine seeking, while the opposite effect was observed for bilateral inactivation of CeN. The nature of the interconnections between those two parts of the amygdala and DLS seems to be rather complex: while CeN is linked to DLS by direct projections to SN dopaminergic neurons (El-Amamy & Holland, 2007), BLA does not directly project to DLS but recruits polysynaptic routes and depends on antecedent glutamatergic mechanisms in the NAc (Murray et al., 2015). To sum up, this study showed that conditioned incentive values processed by BLA can be transmitted to DLS to recruit the DA-dependent habit system. However, control over habits transfers from BLA to CeN once cocaine seeking is robust and habitual, leaving manipulations of BLA without effect on behavior at this stage but stressing the role of CeN in maintaining the habits during the later course of the disease.

Results from brain imaging studies have confirmed the involvement of striatal structures in drug seeking in humans. Volkow et al. (2006) found cocaine cue-induced increases in DA release in the dorsal but not VS. Furthermore, a number of studies have shown increases in DS activity in response to drug cues relative to neutral cues in drug users (Claus et al., 2011; Schacht et al., 2011; Vollstadt-Klein et al., 2010; Wilson et al., 2013). Claus et al. demonstrated a particularly robust cue-induced activation in the DS and in the VS in response to gustatory alcohol cues in 326 heavy drinkers. Vollstadt-Klein et al. (2010) found that heavy drinkers (5.0 ± 1.5 drinks/day) showed higher cue-induced activations in the DS compared to light social drinkers (0.4 ± 0.4 drinks/day), while light drinkers showed higher cue-induced activation in the VS and PFC compared to heavy drinkers. Furthermore, the DS activation to drug cues was positively correlated with drug craving in all participants, whereas the VS activation was negatively correlated with such craving in heavy drinkers (Vollstadt-Klein et al., 2010). In addition, a greater cue-induced activity in the DS (putamen) but not in the VS was predictive for relapse in smokers who tried to remain abstinent (Janes et al., 2010). Evidently, during the course of addiction there seems to be a transition from the initial hedonic, controlled drug use (mediated by the VS/DMS and PFC) to a more habit-driven and eventually compulsive drug abuse and dependence (mediated by the DS, more specifically by DLS and its interconnections with VTA, VS, BLA, and CeN).

Recently, further cortical and subcortical structures have shifted in the focus of addiction science, pointing at further mechanisms of drug-related habits. These brain regions might also be associated with the striatal habit system, as DS is known to project to thalamic as well as cortical circuits involved in planning and execution of motor responses. Premotor cortex (PMC) and motor cortex (MC), as well as supplementary motor area (SMA), superior and inferior parietal lobules (SPL and IPL, respectively), posterior middle temporal gyrus (pMTG), and inferior temporal cortex (ITC) have been shown to store and process action knowledge and tool use skills (Buxbaum, Kyle, Grossman, & Coslett, 2007; Calvo-Merino, Glaser, Grezes, Passingham, & Haggard, 2005; Calvo-Merino, Grezes, Glaser, Passingham, & Haggard, 2006; Chao & Martin, 2000; Creem-Regehr & Lee, 2005; Johnson-Frey, 2004; Johnson-Frey, Newman-Norlund, & Grafton, 2005; Lewis, 2006). Consequently, subjects with lesions in one or several of these brain regions show apraxia or general action planning and execution difficulties (Lewis, 2006). Behavioral tasks involving tool use skills and object manipulation knowledge typically activate the abovementioned circuitry (Grezes & Decety, 2002; Grezes, Tucker, Armony, Ellis, & Passingham, 2003; Yalachkov et al., 2009). Essentially, higher activation in this brain network has been observed also for drug cues compared to neutral cues, as well as for drug addicts in comparison to healthy controls (Kosten et al., 2006; Smolka et al., 2006; Wagner, Dal Cin, Sargent, Kelley, & Heatherton, 2011; Yalachkov et al., 2009, 2010). This implies that this circuitry may represent not only everyday tool use skills, such as using a hammer, a pen, or scissors but also drug-taking skills such as smoking a cigarette, using a lighter, and rolling a joint. While drug-use skills are the core of drug acquisition and consumption behavior, they have been hypothesized to become highly automatized after repeated practice (Tiffany, 1990). These increasingly automatized (and thus eventually easily triggered by drug cues) motor skills are reflected by higher fMRI activation of PMC, MC, SMA, SPL, IPL, pMTG, ITC, and cerebellum (Kosten et al., 2006; Smolka et al., 2006; Wagner et al., 2011; Yalachkov et al., 2009, 2010), as well as positive correlations between the responses in these regions with the severity of dependence and the degree of automaticity of the behavioral responses toward drug cues (Smolka et al., 2006; Yalachkov et al., 2009). Thus, in addition to reward, motivational, and goal-directed mechanisms, drug cues may trigger drug taking by activating the corresponding drug-taking skills in drug users (Yalachkov et al., 2009).

CONCLUSION

Recent findings from behavioral and neuroimaging studies have provided an elaborated view of addiction as a brain disorder developing on the basis of maladaptive functional, reorganizational processes in the reward-related, habit, executive control, and stress-related systems. In this chapter we gave a concise overview of the most important results from animal and human neuroimaging experiments that reveal the reward and habit mechanisms hijacked by drugs of abuse to establish a robust and often treatment-resistant addictive behavior.

This work was supported by the Deutsche Forschungsgemeinschaft (DFG YA 335/2-1).

References

Artiges, E., Ricalens, E., Berthoz, S., Krebs, M. O., Penttila, J., Trichard, C., & Martinot, J. L. (2009). Exposure to smoking cues during an emotion recognition task can modulate limbic fMRI activation in cigarette smokers. *Addiction Biology, 14*, 469–477.

Belin, D., & Everitt, B. J. (2008). Cocaine seeking habits depend upon dopamine-dependent serial connectivity linking the ventral with the dorsal striatum. *Neuron, 57*, 432–441.

Belin, D., Jonkman, S., Dickinson, A., Robbins, T. W., & Everitt, B. J. (2009). Parallel and interactive learning processes within the basal ganglia: Relevance for the understanding of addiction. *Behavioural Brain Research, 199*, 89–102.

Brody, A. L., Mandelkern, M. A., Olmstead, R. E., Jou, J., Tiongson, E., Allen, V., … Cohen, M. S. (2007). Neural substrates of resisting craving during cigarette cue exposure. *Biological Psychiatry, 62*, 642–651.

Buxbaum, L. J., Kyle, K., Grossman, M., & Coslett, H. B. (2007). Left inferior parietal representations for skilled hand-object interactions: Evidence from stroke and corticobasal degeneration. *Cortex, 43*, 411–423.

Calabresi, P., Lacey, M. G., & North, R. (1989). Nicotinic excitation of rate ventral tegmental neurons in vitro studies by intracellular recording. *British Journal of Pharmacology, 98*, 135–149.

Calvo-Merino, B., Glaser, D. E., Grezes, J., Passingham, R. E., & Haggard, P. (2005). Action observation and acquired motor skills: An fMRI study with expert dancers. *Cerebral Cortex, 15*, 1243–1249.

Calvo-Merino, B., Grezes, J., Glaser, D. E., Passingham, R. E., & Haggard, P. (2006). Seeing or doing? Influence of visual and motor familiarity in action observation. *Current Biology, 16*, 1905–1910.

Chao, L. L., & Martin, A. (2000). Representation of manipulable man-made objects in the dorsal stream. *NeuroImage, 12*, 478–484.

Chase, H. W., Eickhoff, S. B., Laird, A. R., & Hogarth, L. (2011). The neural basis of drug stimulus processing and craving: An activation likelihood estimation meta-analysis. *Biological Psychiatry, 70*, 785–793.

Childress, A. R., Ehrman, R. N., Wang, Z., Li, Y., Sciortino, N., Hakun, J., … O'Brien, C. P. (2008). Prelude to passion: Limbic activation by "unseen" drug and sexual cues. *PLoS One, 3*, e1506.

Childress, A. R., Mozley, P. D., McElgin, W., Fitzgerald, J., Reivich, M., & O'Brien, C. P. (1999). Limbic activation during cue-induced cocaine craving. *The American Journal of Psychiatry, 156*, 11–18.

Clarke, P. B., & Pert, A. (1985). Autoradiographic evidence for nicotine receptors on nigrostriatal and mesolimbic dopaminergic neurons. *Brain Research, 348*, 355–358.

Claus, E. D., Ewing, S. W., Filbey, F. M., Sabbineni, A., & Hutchison, K. E. (2011). Identifying neurobiological phenotypes associated with alcohol use disorder severity. *Neuropsychopharmacology, 36*, 2086–2096.

Cousijn, J., Goudriaan, A. E., Ridderinkhof, K. R., van den Brink, W., Veltman, D. J., & Wiers, R. W. (2012a). Approach-bias predicts development of cannabis problem severity in heavy cannabis users: Results from a prospective fMRI study. *PLoS One, 7*, e42394.

Cousijn, J., Goudriaan, A. E., Ridderinkhof, K. R., van den Brink, W., Veltman, D. J., & Wiers, R. W. (2012b). Neural responses associated with cue-reactivity in frequent cannabis users. *Addiction Biology, 18*.

Craig, A. D. (2003). Interoception: The sense of the physiological condition of the body. *Current Opinion in Neurobiology, 13*, 500–505.

Creem-Regehr, S. H., & Lee, J. N. (2005). Neural representations of graspable objects: Are tools special? *Brain Research. Cognitive Brain Research, 22*, 457–469.

Deutch, A. Y., Holliday, J., Roth, R. H., Chun, L. L., & Hawrot, E. (1987). Immunohistochemical localization of a neuronal nicotinic acetylcholine receptor in mammalian brain. *Proceedings of the National Academy of Sciences of the United States of America, 84*, 8697–8701.

Devonshire, I. M., Berwick, J., Jones, M., Martindale, J., Johnston, D., Overton, P. G., & Mayhew, J. E. (2004). Haemodynamic responses to sensory stimulation are enhanced following acute cocaine administration. *NeuroImage, 22*, 1744–1753.

Devonshire, I. M., Mayhew, J. E., & Overton, P. G. (2007). Cocaine preferentially enhances sensory processing in the upper layers of the primary sensory cortex. *Neuroscience, 146*, 841–851.

Di Ciano, P., & Everitt, B. J. (2001). Dissociable effects of antagonism of NMDA and AMPA/KA receptors in the nucleus accumbens core and shell on cocaine-seeking behavior. *Neuropsychopharmacology, 25*, 341–360.

Dosenbach, N. U., Visscher, K. M., Palmer, E. D., Miezin, F. M., Wenger, K. K., Kang, H. C., … Petersen, S. E. (2006). A core system for the implementation of task sets. *Neuron, 50*, 799–812.

Due, D. L., Huettel, S. A., Hall, W. G., & Rubin, D. C. (2002). Activation in mesolimbic and visuospatial neural circuits elicited by smoking cues: Evidence from functional magnetic resonance imaging. *The American Journal of Psychiatry, 159*, 954–960.

El-Amamy, H., & Holland, P. C. (2007). Dissociable effects of disconnecting amygdala central nucleus from the ventral tegmental area or substantia nigra on learned orienting and incentive motivation. *The European Journal of Neuroscience, 25*, 1557–1567.

Engelmann, J. M., Versace, F., Robinson, J. D., Minnix, J. A., Lam, C. Y., Cui, Y., … Cinciripini, P. M. (2012). Neural substrates of smoking cue reactivity: A meta-analysis of fMRI studies. *NeuroImage, 60*, 252–262.

Everitt, B. J., & Robbins, T. W. (2005). Neural systems of reinforcement for drug addiction: From actions to habits to compulsion. *Nature Neuroscience, 8*, 1481–1489.

Field, M., & Cox, W. M. (2008). Attentional bias in addictive behaviors: A review of its development, causes, and consequences. *Drug and Alcohol Dependence, 97*, 1–20.

Filbey, F. M., Schacht, J. P., Myers, U. S., Chavez, R. S., & Hutchison, K. E. (2009). Marijuana craving in the brain. *Proceedings of the National Academy of Sciences of the United States of America, 106*, 13016–13021.

Franklin, T. R., Wang, Z., Wang, J., Sciortino, N., Harper, D., Li, Y., … Childress, A. R. (2007). Limbic activation to cigarette smoking cues independent of nicotine withdrawal: A perfusion fMRI study. *Neuropsychopharmacology, 32*, 2301–2309.

Fryer, S. L., Jorgensen, K. W., Yetter, E. J., Daurignac, E. C., Watson, T. D., Shanbhag, H., … Mathalon, D. H. (2012). Differential brain response to alcohol cue distractors across stages of alcohol dependence. *Biological Psychology, 92*.

Garavan, H., Pankiewicz, J., Bloom, A., Cho, J. K., Sperry, L., Ross, T. J., … Stein, E. A. (2000). Cue-induced cocaine craving: Neuroanatomical specificity for drug users and drug stimuli. *The American Journal of Psychiatry, 157*, 1789–1798.

Garavan, H., Ross, T. J., Murphy, K., Roche, R. A., & Stein, E. A. (2002). Dissociable executive functions in the dynamic control of behavior: Inhibition, error detection, and correction. *NeuroImage, 17*, 1820–1829.

Goldstein, R. Z., & Volkow, N. D. (2011). Dysfunction of the prefrontal cortex in addiction: Neuroimaging findings and clinical implications. *Nature Reviews. Neuroscience, 12*, 652–669. http://dx.doi.org/10.1038/nrn3119.

Grant, S., London, E. D., Newlin, D. B., Villemagne, V. L., Liu, X., Contoreggi, C., … Margolin, A. (1996). Activation of memory circuits during cue-elicited cocaine craving. *Proceedings of the National Academy of Sciences of the United States of America, 93*, 12040–12045.

Grezes, J., & Decety, J. (2002). Does visual perception of object afford action? Evidence from a neuroimaging study. *Neuropsychologia, 40*, 212–222.

Grezes, J., Tucker, M., Armony, J., Ellis, R., & Passingham, R. E. (2003). Objects automatically potentiate action: An fMRI study of implicit processing. *The European Journal of Neuroscience, 17*, 2735–2740.

Grüsser, S. M., Wrase, J., Klein, S., Hermann, D., Smolka, M. N., Ruf, M., … Heinz, A. (2004). Cue-induced activation of the striatum and medial prefrontal cortex is associated with subsequent relapse in abstinent alcoholics. *Psychopharmacology (Berl), 175*, 296–302.

Haber, S. N., & Knutson, B. (2010). The reward circuit: Linking primate anatomy and human imaging. *Neuropsychopharmacology, 35*, 4–26.

Hartwell, K. J., Johnson, K. A., Li, X., Myrick, H., LeMatty, T., George, M. S., & Brady, K. T. (2011). Neural correlates of craving and resisting craving for tobacco in nicotine dependent smokers. *Addiction Biology, 16*, 654–666.

Hyman, S. E., Malenka, R. C., & Nestler, E. J. (2006). Neural mechanisms of addiction: The role of reward-related learning and memory. *Annual Review of Neuroscience, 29*, 565–598.

Imperato, A., Mulus, A., & DiChiara, G. (1986). Nicotine preferentially stimulates dopamine released in the limbic system of freely moving rats. *European Journal of Pharmacology, 132*, 337–338.

Janes, A. C., Pizzagalli, D. A., Richardt, S., deB Frederick, B., Chuzi, S., Pachas, G., ... Kaufman, M. J. (2010). Brain reactivity to smoking cues prior to smoking cessation predicts ability to maintain tobacco abstinence. *Biological Psychiatry, 67*, 722–729.

Jay, T. M. (2003). Dopamine: A potential substrate for synaptic plasticity and memory mechanisms. *Progress in Neurobiology, 69*, 375–390.

Johnson-Frey, S. H. (2004). The neural bases of complex tool use in humans. *Trends in Cognitive Sciences, 8*, 71–78.

Johnson-Frey, S. H., Newman-Norlund, R., & Grafton, S. T. (2005). A distributed left hemisphere network active during planning of everyday tool use skills. *Cerebral Cortex, 15*, 681–695.

Kalivas, P. W., & O'Brien, C. (2008). Drug addiction as a pathology of staged neuroplasticity. *Neuropsychopharmacology, 33*, 166–180.

Kelley, A. E. (2004). Memory and addiction: Shared neural circuitry and molecular mechanisms. *Neuron, 44*, 161–179.

Kilts, C. D., Schweitzer, J. B., Quinn, C. K., Gross, R. E., Faber, T. L., Muhammad, F., ... Drexler, K. P. (2001). Neural activity related to drug craving in cocaine addiction. *Archives of General Psychiatry, 58*, 334–341.

Kober, H., Mende-Siedlecki, P., Kross, E. F., Weber, J., Mischel, W., Hart, C. L., & Ochsner, K. N. (2010). Prefrontal-striatal pathway underlies cognitive regulation of craving. *Proceedings of the National Academy of Sciences of the United States of America, 107*, 14811–14816.

Koob, G. F., Buck, C. L., Cohen, A., Edwards, S., Park, P. E., Schlosburg, J. E., ... George, O. (2014). Addiction as a stress surfeit disorder. *Neuropharmacology, 76*, 370–382. http://dx.doi.org/10.1016/j.neuropharm.2013.05.024. Epub 2013 June 06.

Kosten, T. R., Scanley, B. E., Tucker, K. A., Oliveto, A., Prince, C., Sinha, R., ... Wexler, B. E. (2006). Cue-induced brain activity changes and relapse in cocaine-dependent patients. *Neuropsychopharmacology, 31*, 644–650.

Kuhn, S., & Gallinat, J. (2011). Common biology of craving across legal and illegal drugs – A quantitative meta-analysis of cue-reactivity brain response. *The European Journal of Neuroscience, 33*, 1318–1326.

Lewis, J. W. (2006). Cortical networks related to human use of tools. *Neuroscientist, 12*, 211–231.

Liu, X., Hairston, J., Schrier, M., & Fan, J. (2011). Common and distinct networks underlying reward valence and processing stages: A meta-analysis of functional neuroimaging studies. *Neuroscience and Biobehavioral Reviews, 35*, 1219–1236.

Lucantonio, F., Stalnaker, T. A., Shaham, Y., Niv, Y., & Schoenbaum, G. (2012). The impact of orbitofrontal dysfunction on cocaine addiction. *Nature Neuroscience, 15*, 358–366.

Luijten, M., Veltman, D. J., van den Brink, W., Hester, R., Field, M., Smits, M., & Franken, I. H. (2011). Neurobiological substrate of smoking-related attentional bias. *NeuroImage, 54*, 2374–2381.

Maas, L. C., Lukas, S. E., Kaufman, M. J., Weiss, R. D., Daniels, S. L., Rogers, V. W., ... Renshaw, P. F. (1998). Functional magnetic resonance imaging of human brain activation during cue-induced cocaine craving. *The American Journal of Psychiatry, 155*, 124–126.

Mansvelder, H. D., Keath, J. R., & McGehee, D. S. (2002). Synaptic mechanisms underlie nicotine-induced excitability of brain reward areas. *Neuron, 33*, 905–919.

McClernon, F. J., Hiott, F. B., Huettel, S. A., & Rose, J. E. (2005). Abstinence-induced changes in self-report craving correlate with event-related fMRI responses to smoking cues. *Neuropsychopharmacology, 30*, 1940–1947.

Menon, V., & Uddin, L. Q. (2010). Saliency, switching, attention and control: A network model of insula function. *Brain Structure & Function, 214*, 655–667.

Murray, J. E., Belin-Rauscent, A., Simon, M., Giuliano, C., Benoit-Marand, M., Everitt, B. J., & Belin, D. (2015). Basolateral and central amygdala differentially recruit and maintain dorsolateral striatum-dependent cocaine-seeking habits. *Nature Communications, 6*, 10088. http://dx.doi.org/10.1038/ncomms10088.

Murray, J. E., Belin, D., & Everitt, B. J. (2012). Double dissociation of the dorsomedial and dorsolateral striatal control over the acquisition and performance of cocaine seeking. *Neuropsychopharmacology, 37*, 2456–2466. http://dx.doi.org/10.1038/npp.2012.104. Epub 2012 June 27.

Nee, D. E., Wager, T. D., & Jonides, J. (2007). Interference resolution: Insights from a meta-analysis of neuroimaging tasks. *Cognitive, Affective & Behavioral Neuroscience, 7*, 1–17.

Nestler, E. J. (2005). Is there a common molecular pathway for addiction? *Nature Neuroscience, 8*, 1445–1449.

Robbins, T. W., Ersche, K. D., & Everitt, B. J. (2008). Drug addiction and the memory systems of the brain. *Annals of the New York Academy of Sciences, 1141*, 1–21.

Robinson, T. E., & Berridge, K. C. (1993). The neural basis of drug craving: An incentive-sensitization theory of addiction. *Brain Research. Brain Research Reviews, 18*, 247–291.

Schacht, J. P., Anton, R. F., & Myrick, H. (2012). Functional neuroimaging studies of alcohol cue reactivity: A quantitative meta-analysis and systematic review. *Addiction Biology, 18*.

Schacht, J. P., Anton, R. F., Randall, P. K., Li, X., Henderson, S., & Myrick, H. (2011). Stability of fMRI striatal response to alcohol cues: A hierarchical linear modeling approach. *NeuroImage, 56*, 61–68.

Schneider, F., Habel, U., Wagner, M., Franke, P., Salloum, J. B., Shah, N. J., ... Zilles, K. (2001). Subcortical correlates of craving in recently abstinent alcoholic patients. *The American Journal of Psychiatry, 158*, 1075–1083.

Schoenbaum, G., Roesch, M. R., & Stalnaker, T. A. (2006). Orbitofrontal cortex, decision-making and drug addiction. *Trends in Neurosciences, 29*, 116–124.

Schoenbaum, G., Roesch, M. R., Stalnaker, T. A., & Takahashi, Y. K. (2009). A new perspective on the role of the orbitofrontal cortex in adaptive behaviour. *Nature Reviews. Neuroscience, 10*, 885–892.

Schultz, W. (2007a). Behavioral dopamine signals. *Trends in Neurosciences, 30*, 203–210.

Schultz, W. (2007b). Multiple dopamine functions at different time courses. *Annual Review of Neuroscience, 30*, 259–288.

Schultz, W., Dayan, P., & Montague, P. R. (1997). A neural substrate of prediction and reward. *Science, 275*, 1593–1599.

Seeley, W. W., Menon, V., Schatzberg, A. F., Keller, J., Glover, G. H., Kenna, H., ... Greicius, M. D. (2007). Dissociable intrinsic connectivity networks for salience processing and executive control. *The Journal of Neuroscience, 27*, 2349–2356.

Seo, D., Jia, Z., Lacadie, C. M., Tsou, K. A., Bergquist, K., & Sinha, R. (2011). Sex differences in neural responses to stress and alcohol context cues. *Human Brain Mapping, 32*, 1998–2013.

Serences, J. T. (2008). Value-based modulations in human visual cortex. *Neuron, 60*, 1169–1181.

Shackman, A. J., Salomons, T. V., Slagter, H. A., Fox, A. S., Winter, J. J., & Davidson, R. J. (2011). The integration of negative affect, pain and cognitive control in the cingulate cortex. *Nature Reviews. Neuroscience, 12*, 154–167.

Smolka, M. N., Buhler, M., Klein, S., Zimmermann, U., Mann, K., Heinz, A., & Braus, D. F. (2006). Severity of nicotine dependence modulates cue-induced brain activity in regions involved in motor preparation and imagery. *Psychopharmacology (Berl), 184*, 577–588.

Sridharan, D., Levitin, D. J., & Menon, V. (2008). A critical role for the right fronto-insular cortex in switching between central-executive and default-mode networks. *Proceedings of the National Academy of Sciences of the United States of America, 105*, 12569–12574.

Sutherland, M. T., McHugh, M. J., Pariyadath, V., & Stein, E. A. (2012). Resting state functional connectivity in addiction: Lessons learned and a road ahead. *NeuroImage, 62*, 2281–2295.

Tang, D. W., Fellows, L. K., Small, D. M., & Dagher, A. (2012). Food and drug cues activate similar brain regions: A meta-analysis of functional MRI studies. *Physiology & Behavior, 106*, 317–324.

Tiffany, S. T. (1990). A cognitive model of drug urges and drug-use behavior: Role of automatic and nonautomatic processes. *Psychological Review, 97*, 147–168.

Torres, G. E., Gainetdinov, R. R., & Caron, M. G. (2003). Plasma membrane monoamine transporters: Structure, regulation and function. *Nature Reviews. Neuroscience, 4*, 13–25.

Tsai, H. C., Zhang, F., Adamantidis, A., Stuber, G. D., Bonci, A., de Lecea, L., & Deisseroth, K. (2009). Phasic firing in dopaminergic neurons is sufficient for behavioral conditioning. *Science, 324*, 1080–1084.

Volkow, N. D., Wang, G. J., Telang, F., Fowler, J. S., Logan, J., Childress, A. R., … Wong, C. (2006). Cocaine cues and dopamine in dorsal striatum: Mechanism of craving in cocaine addiction. *Journal of Neuroscience, 26*, 6583–6588.

Vollstadt-Klein, S., Wichert, S., Rabinstein, J., Buhler, M., Klein, O., Ende, G., … Mann, K. (2010). Initial, habitual and compulsive alcohol use is characterized by a shift of cue processing from ventral to dorsal striatum. *Addiction, 105*, 1741–1749.

Wager, T. D., Sylvester, C. Y., Lacey, S. C., Nee, D. E., Franklin, M., & Jonides, J. (2005). Common and unique components of response inhibition revealed by fMRI. *NeuroImage, 27*, 323–340.

Wagner, D. D., Dal Cin, S., Sargent, J. D., Kelley, W. M., & Heatherton, T. F. (2011). Spontaneous action representation in smokers when watching movie characters smoke. *The Journal of Neuroscience, 31*, 894–898.

Wilcox, C. E., Teshiba, T. M., Merideth, F., Ling, J., & Mayer, A. R. (2011). Enhanced cue reactivity and fronto-striatal functional connectivity in cocaine use disorders. *Drug and Alcohol Dependence, 115*, 137–144.

Wilson, S. J., Creswell, K. G., Sayette, M. A., & Fiez, J. A. (2013). Ambivalence about smoking and cue-elicited neural activity in quitting-motivated smokers faced with an opportunity to smoke. *Addictive Behaviors, 38*, 1541–1549.

Wilson, S. J., Sayette, M. A., Delgado, M. R., & Fiez, J. A. (2005). Instructed smoking expectancy modulates cue-elicited neural activity: A preliminary study. *Nicotine & Tobacco Research, 7*, 637–645.

Wooltorton, J. R., Pidoplichko, V. I., Broide, R. S., & Dani, J. A. (2003). Differential desensitization and distribution of nicotinic acetylcholine receptor subtypes in midbrain dopamine areas. *The Journal of Neuroscience, 23*, 3176–3185.

Yalachkov, Y., Kaiser, J., Gorres, A., Seehaus, A., & Naumer, M. J. (2012). Sensory modality of smoking cues modulates neural cue reactivity. *Psychopharmacology (Berl), 225*.

Yalachkov, Y., Kaiser, J., & Naumer, M. J. (2009). Brain regions related to tool use and action knowledge reflect nicotine dependence. *Journal of Neuroscience, 29*, 4922–4929.

Yalachkov, Y., Kaiser, J., & Naumer, M. J. (2010). Sensory and motor aspects of addiction. *Behavioural Brain Research, 207*, 215–222.

Yalachkov, Y., Kaiser, J., & Naumer, M. J. (2012). Functional neuroimaging studies in addiction: Multisensory drug stimuli and neural cue reactivity. *Neuroscience and Biobehavioral Reviews, 36*, 825–835.

Zhang, X., Salmeron, B. J., Ross, T. J., Gu, H., Geng, X., Yang, Y., & Stein, E. A. (2011). Anatomical differences and network characteristics underlying smoking cue reactivity. *NeuroImage, 54*, 131–141.

CHAPTER

20

Ethanol: Neurotoxicity and Brain Disorders

L.M.P. Fernandes[1], E. Fontes de Andrade, Jr. [1], M.C. Monteiro[1], S.C. Cartágenes[1], R.R. Lima[1], R.D. Prediger[2], C.S.F. Maia[1]

[1]Federal University of Pará, Belém, Brazil; [2]Federal University of Santa Catarina, Florianópolis, Brazil

INTRODUCTION

Ethanol (EtOH or ethyl alcohol—C_2H_5OH) is one of the oldest psychoactive substances used worldwide, and it is the most widely consumed drug in the Western societies with well-known dependence-producing properties (Boutros, Semenova, & Markou, 2014; Crego et al., 2010; Sanchis & Aragón, 2007; WHO, 2014). Besides the cultural background, EtOH is accepted in almost all organized societies and its consumption is supported by its legality, low cost, and wide availability (Johnston, O'Malley, Bachman, & Schulenberg, 2009; Johnston, O'Malley, Miech, Bachman, & Schulenberg, 2014; WHO, 2014).

Additionally, alcohol consumption is encouraged by celebrations, social and business situations, religious ceremonies, and cultural events (Meloni & Laranjeira, 2004). However, the abuse of this substance can trigger a cascade of acute health problems such as automobile accidents, social interaction problems, including domestic violence, child abuse, emergency and hospital care, reduced work production (Manzo-Avalos & Saavedra-Molina, 2010), crime, and public disorder (Rehm et al., 2009).

Besides the acute health problems, the loss of control over alcohol intake results in a progressive and chronic psychiatric illness, the chronic alcoholism or alcohol dependence (Petit, Kornreich, Verbanck, & Campanella, 2013). This more severe disorder can be an indication to reliably identify people for whom drinking causes major physiological consequences and persistent impairment in the quality of life (Schuckit, 2009). Moreover, the harmful use of alcohol causes health and social consequences for the drinker, the people around the drinker and society in general. Furthermore, the pattern of drinking determines the risk of adverse health outcomes (WHO, 2010).

In this way, the alcohol-use disorders (AUDs) comprise the alcohol abuse or harmful use and alcohol dependence (also known as chronic alcoholism or alcohol-dependence syndrome). The first is defined as a pattern of alcohol use that displays physical or mental damage. The alcohol dependence is defined as behavioral, cognitive, and physiological alterations that was developed after repeated alcohol intake, linked to a strong desire to consume alcohol (craving), difficulties in controlling its use, increased tolerance, and sometimes a physiological withdrawal state (Schuckit, 2009; WHO, 2014).

It is noteworthy that the AUDs cause substantial morbidity and mortality (Schuckit, 2009; WHO, 2014). According to the World Health Organization (WHO, 2014), the harmful use of alcohol is about 3.3 million deaths each year (5.9% of all deaths worldwide) and about 5.1% of the global burden of morbidity.

In general, the AUDs are associated with neuropsychiatric conditions (i.e., depressive episodes and severe anxiety); cognitive deficits; motor impairment; cardiovascular (i.e., ischemic heart disease and ischemic stroke), and gastrointestinal (i.e., liver cirrhosis and pancreatitis) diseases; cancers; salivary gland atrophy; decreased bone density; and among offspring the fetal alcohol syndrome (FAS) (Bannach et al., 2015; Fernandes et al., 2015; Oliveira et al., 2014, 2015; Schuckit, 2009; WHO, 2014).

Alcohol uses have continued their long declines and are now at the lowest levels recorded in the history of the survey (Johnston et al., 2014). However, in Brazil, the National Survey on Alcohol and Other Drugs reveals that heaviest drinking pattern had increased among females (36%), especially the younger ones, at higher levels than total rates of gender (31.1%), indicating that females have become the population at

Addictive Substances and Neurological Disease
http://dx.doi.org/10.1016/B978-0-12-805373-7.00020-7

risk in the modern society by increasing the EtOH consumption in its most harmful form (INPAD, 2013).

In this context, the women achieving higher blood alcohol concentrations than men for the same amount of EtOH intake, it may be explained by female have lower body weight, smaller liver capacity to metabolize alcohol, and a higher proportion of body fat (WHO, 2014).

Furthermore, the alcohol abuse effects are more severe during pregnancy, since offspring is susceptible to develop the fetal alcohol spectrum disorder (FASD), which includes the FAS (WHO, 2014). In this sense, alcohol induces disorders at different degrees, which are associated with physical, cognitive, and behavioral deficits that persist until adulthood (Fernandes, Lima, Monteiro, Gomes-Leal, & Maia, 2014).

In this sense, this chapter discusses the pharmacokinetics, as well as the main structural and behavioral changes resulted from chronic ethanol exposure. In addition to the general issues, we reveal the alcohol kinetics and damage differences according to the gender and age.

ALCOHOL CHEMISTRY AND KINETICS

EtOH is a primary monoalcohol, a small polar molecule (with a molar mass of 46). The characteristic of the alcohol function is the presence of a hydroxyl group bound directly to a saturated carbon, which directly influences their pharmacokinetic properties (Goullé & Guerbet, 2015; Pohorecky & Brick, 1988).

The hydroxyl group is related to its hydrophilic properties, since it determines the formation of hydrogen bonds between EtOH and water. The strength of this connection ensures complete miscibility (Ferreira & Willoughby, 2008). The EtOH aliphatic carbon chain, display a lipid solubility, which despite being low—EtOH does not dissolve well in fat or oil—facilitates their passage through biological membranes by passive diffusion (Cederbaum, 2012; Kent, 2012). This set of chemical characteristics summarizes the EtOH amphiphilicity and explain alcohol distribution in the body.

Routes of Administration and Absorption

Despite the evidence that EtOH absorption can occur by various routes (lung, colorectal, and so on), oral administration is the most common. In standard conditions, after oral intake, EtOH is largely and rapidly absorbed by simple diffusion in the stomach (~20%). However, the largest extent of absorption occurs in the duodenum and proximal jejunum (~80%), reaching the peak plasma levels between 30 and 90 min (Goullé & Guerbet, 2015; Marek & Kraft, 2014; Pohorecky & Brick, 1988; Ramchandani, Bosron, & Li, 2001).

This pattern of absorption can be modified by several factors such as the presence of food in the stomach (lipids, proteins, and glucides, particularly), the speed of the gastrointestinal transit, enteric blood flow, and individual genetic factors. The alcohol content also interferes significantly with its kinetic features. Beverages with an alcohol content below 20% or above 30% produces a lower absorption rate. The lower concentrations, according to Fick's laws of diffusion, reduce the diffusion rate, while very concentrated solution promoting gastric mucosa irritation and increased mucus secretion and cause spasm of the stomach and the pylorus (Kent, 2012; Paton, 2005).

Distribution

In the blood circulation, EtOH does not bind to any plasma proteins. At this stage, the main limiting factor for its bioavailability is the first-pass effect in the liver. After liver drug metabolism, the alcohol concentration is greatly reduced, and it reaches the systemic circulation and tissues (Ramchandani et al., 2001).

Due to their amphipathic characteristics and its strong interaction with the water, EtOH distribution pattern follows the water content of each tissue and blood flow, easily reaching the lung, kidney, liver, brain, and skeletal muscle, for example. Thus, the alcohol volume of distribution is similar to the total content of body water (0.7 L/kg for men and 0.6 L/kg for women) (Maudens et al., 2014; Pohorecky & Brick, 1988). As adipose and bone tissues have low aqueous constituent, the penetration of EtOH in these compartments is insignificant (Ferreira & Willoughby, 2008). Following a standard two-compartment distribution, hormones, factors of vessels constriction or dilatation, skeletal muscle activity, body temperature, and other factors that regulate peripheral circulation can affect EtOH distribution (Norberg, Jones, Hahn, & Gabrielsson, 2003).

These characteristics explain the variation in the BAC in accordance with the volume constitution of the individual. In women, for example, BAC tends to be higher than in men, because women generally have smaller muscle mass and fatter, with proportionally lower aqueous composition. Similarly, variations are observed according to age and weight (Graham, Wilsnack, Dawson, & Vogeltanz, 1998; Marshall, Kingstone, Boss, & Morgan, 1983; Mirand & Welte, 1994).

In pregnant woman, EtOH readily crosses the placenta, distributing the fetal circulation at concentrations proportional to the BAC maternal (~60%). The BAC in the umbilical circulation and mammary glands is equivalent to the maternal circulation, while in the amniotic fluid is in about half and may vary according to the dose (Brien, Loomis, Tranmer, & McGrath, 1983; Haastrup, Pottegård, & Damkier, 2014; Idanpaan-Heikkila et al., 1972; Nava-Ocampo, Velazquez-Armenta, Brien, & Koren, 2004).

Metabolism and Excretion

The process of EtOH elimination relies primarily in the biotransformation mechanisms. About 10% of ingested EtOH can be metabolized in the stomach by a gastric dehydrogenase, often being called first-pass metabolism (Vonghia et al., 2008). Absorbed fraction of EtOH is predominantly metabolized by liver enzymes. A small portion (~1%) undergoes conjugation mechanisms with endogenous substrates (e.g., glucuronic acid); however, the oxidative metabolism is principally responsible for its elimination (Schmitt, Aderjan, Keller, & Wu, 1995; Zakhari, 2006).

In fact, three alternative routes for EtOH oxidation are recognized: (1) alcohol dehydrogenase (ADH); (2) microsomal ethanol oxidation system (MEOS); and (3) catalase. In all cases, the main reaction product is the acetaldehyde (Lieber, 1991; Zakhari, 2006).

In fact, the most important EtOH biotransformation mechanism is the dehydrogenation by ADH route, accounted for about 90% of EtOH metabolism. ADH is a dimeric zinc-dependent metalloenzyme, not inducible, located in the cytosol that promotes the oxidation of EtOH to acetaldehyde by reaction with nicotinamide dinucleotide (NAD^+). This metalloenzyme is found in the cytosol of cells in various organs, but the hepatocyte is the most important location. The ADH pathway involves the presence of aldehyde dehydrogenase (ALDH) that converts acetaldehyde to acetate that can be used in several metabolic reactions or converted to CO_2 and water in the peripheral tissue (Kent, 2012; Lieber, 1991) (Fig. 20.1).

The cytochrome P450 (CYP) is an EtOH oxidation secondary route. Previously called "microsomal EtOH oxidation system," the CYP2E1 enzyme activity accounts for up to 10% of hepatic EtOH clearance in moderate doses. In the heavy consumption, CYP2E1 suffers alcohol induction, which plays an important role in the adaptation to this pattern of consumption. Moreover, the increase in the CYP2E1 activity may result in increased metabolism of other drugs (e.g., acetaminophen) and some vitamins (e.g., retinol) that are

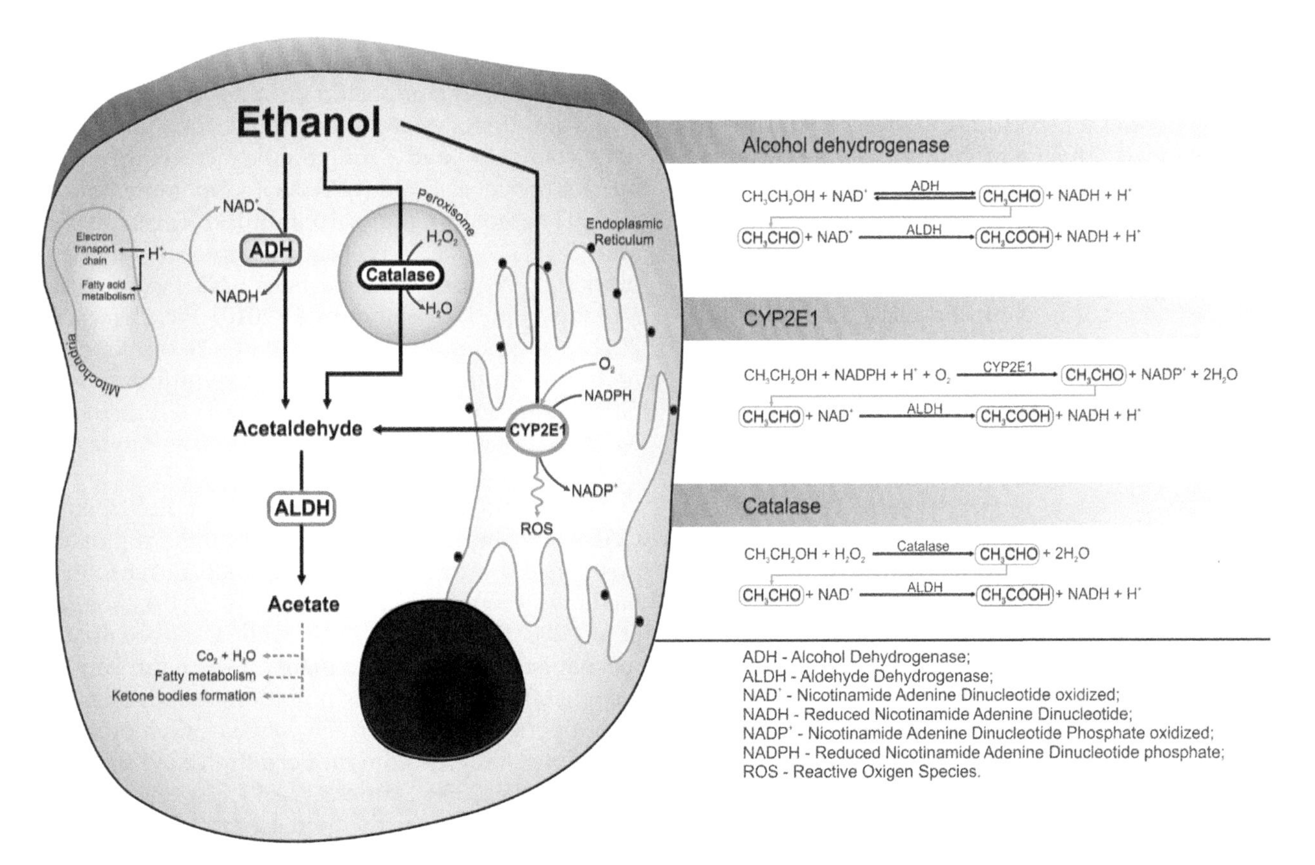

FIGURE 20.1 Ethanol metabolism pathways. The first ethanol biotransformation stage results in the formation of acetaldehyde through three main pathways. The first and most important is the alcohol dehydrogenase (ADH) route. Other secondary mechanisms consist of the cytochrome P450 (CYP)-2E1 and catalase pathways. Regarding to the last step, the acetaldehyde is converted to acetate by aldehyde dehydrogenase (ALDH).

CYP2E1 substrates. Other consequence of EtOH metabolism is the increased formation of reactive oxygen species (ROS) associated with cell damage (Cederbaum, 2012; Fraser, 1997).

The catalase is a heme enzyme present in the cell peroxisomes, the main activity of which is the conversion of H_2O_2 to O_2 and water. However, it consists of an alternative route that catalyzes the reaction that requires hydrogen peroxide (H_2O_2) to oxidize alcohol to produce acetaldehyde. This route consists of no more than 2% of the hepatic metabolism of EtOH; however, it has been attributed to catalase pathway, which plays an important role in the EtOH oxidation in the brain (Handler & Thurman, 1990; Vonghia et al., 2008; Zimatkin & Deitrich, 1997).

Although there exist three EtOH metabolism routes, a small fraction of the drug absorbed can be excreted in unaltered state through breathing (~0.7%), sweat (~0.1%), and urine (~0.3%). In fact, the pulmonary excretion presents strong correlation with plasma concentration, which can be used for alcohol intake detection purposes (Goullé & Guerbet, 2015; Holford, 1987).

MECHANISMS OF NEUROTOXICITY

The EtOH alters various organs and systems, and this chapter focuses on the effects on central nervous system (CNS) (Oliveira et al., 2014, 2015; Teixeira et al., 2014). Some of the main EtOH effects in this system are mnemonic process damage, motor impairments, and neurogenesis inhibition that are associated, at least in part, to neuroinflammation, oxidative stress, and excitotoxicity events (McClain, Morris, Marshall, & Nixon, 2014; Oliveira et al., 2014, 2015; Pascual, Blanco, Cauli, Miñarro, & Guerri, 2007; Teixeira et al., 2014).

In fact, each of these alcohol-related harm is determined by the amount, period, and frequency of EtOH exposure (Brown & Tapert, 2004; Yang et al., 2014; Zeigler et al., 2005). The pattern of drinking consists conceptually of heavy, moderate, and light drinking, which can be continuous or intermittent, as well as acute or chronic forms (NIAAA, 2004; WHO, 2014).

The consumption of large amounts of EtOH in a short time, followed by a period of abstinence, represents the dominant type of alcohol misuse in adolescents and young adults (Crego et al., 2010; Jacobus & Tapert, 2013; NIAAA, 2004; Petit et al., 2013). This drinking pattern is known as binge drinking that reaches a BAC of 0.08 g% or above (Briones & Woods, 2013; Jacobus & Tapert, 2013; NIAAA, 2004; Petit et al., 2013).

In addition, the heaviest drinking pattern peaks occur during adolescence (Kuperman et al., 2005). Among adolescents, high patterns of emotionality/anxiety as well as risk-taking behavior may explain the initiation pattern of EtOH and other drugs consumption (Guerri & Pascual, 2010). Indeed, in this phase of development occurs the maturation of CNS that involves alterations in the neurotransmission and plasticity that are associated with structural changes in some regions of the brain. This risk factor contributes to the fact that heavy drinking seems to be more devastating when the alcohol intake begins at puberty (Beenstock, Adams, & White, 2011; Skala & Walter, 2013).

The CNS disorders provoked by EtOH exposure seems to reflect differences in the brain neurotransmitter systems affected that may influence the alcohol pharmacodynamics and is related to alcohol-seeking behavior that generate alcohol dependence (Vengeliene, Bilbao, Molander, & Spanagel, 2008). Furthermore, the binge drinking during adolescence produces changes in alcohol intake and preference, which could be mediated by alterations in dopaminergic and glutamatergic neurotransmission (Witt, 2010).

Neurotransmission and Intracellular Pathways

The EtOH moves freely across the lipid bilayer cells and acts by disrupting distinct receptor or effector proteins, through direct or indirect interactions (Witt, 2010). At very high concentrations, alcohol might even change the lipid composition in the surrounding membrane and directly interfere with the function of several ion channels and receptors, increasing gamma-aminobutyric acid A ($GABA_A$), dopamine (both D1 and D2 receptors), and serotonin [5-HT; mainly 5-HT3 and 5-HT(1A) autoreceptor] receptors activity, and blocking *N*-methyl-D-aspartate (NMDA) receptors (Adams, Short, & Lawrence, 2010; Pickens & Calu, 2011; Vengeliene et al., 2008; Witt, 2010). Among the neurotransmitter systems, the most affected are the GABAergic and glutamatergic (Guerri & Pascual, 2010; Kumar, LaVoie, DiPette, & Singh, 2013; Kumar et al., 2009; Schuckit, 2009).

The use of high or acute doses of EtOH enhance GABAergic inhibitory activity, since this drug produces positive allosteric modulation on $GABA_A$ receptors and facilitates the presynaptic release of GABA (Vengeliene et al., 2008). Nowadays, it is well described that such mechanism contributes to many of the neurobehavioral alcohol effects, such as disinhibition, sedative–hypnotic activity, cognitive impairment, ataxia, and motor incoordination (Heilig, Goldman, Berrettini, & O'Brien, 2011; Kumar et al., 2009; Schuckit, 2009). However, the probable consequences of the increased GABAergic system activity in the developing brain seems to be the generation of widespread apoptotic neurodegeneration in different regions of the CNS (Young, Straiko, Johnson, Creeley, & Olney, 2008).

On the other hand, the glutamate is the major excitatory neurotransmitter in the brain. It has been shown that EtOH may affect both ionotropic (iGluR) and metabotropic (mGluR) glutamatergic receptors. It is well documented that the glutamatergic system generates postsynaptic excitatory potential that mediates a wide variety of physiological processes in the CNS, including neurotrophic activity and synaptic plasticity (Casillas-Espinosa, Powell, & O'Brien, 2012; Möykkynen & Korpi, 2012; Zhuo, 2009). Besides, the NMDA receptors play a pivotal role in many processes of the CNS such as cognitive functions, emotionality, motor control, learning and memory, and excitotoxicity during neurodegenerative diseases (Möykkynen & Korpi, 2012). These iGluRs mediate some of the effects of acute and chronic EtOH intoxication in the CNS, including cognitive defects, seizures, and neuronal degeneration (Dodd, Beckmann, Davidson, & Wilce, 2000; Kumari & Ticku, 2000). The chronic EtOH exposure induces upregulation of NMDA receptors both in vivo and in vitro, which may result in increased neuronal vulnerability to glutamate-induced excitotoxicity (Guerri & Pascual, 2010; Kumar et al., 2013; Nagy, 2004).

In summary, the acute EtOH consumption displays the blockade of NMDA glutamatergic receptors and excessive activation of $GABA_A$ receptors (Ikonomidou, Stefovska, & Turski, 2000; Miller, 2006). Therefore, adaptations in these systems seems to play a major role in the development of alcohol tolerance, reduced density of GABA receptors in the brain, and consequent GABAergic neurotransmission downregulation, associated with excitotoxic events by enhanced glutamatergic activity (Jung & Metzger, 2010; Vengeliene et al., 2008). This context contributes to anxiety and insomnia during acute and protracted alcohol withdrawal (Schuckit, 2009).

Furthermore, dopamine is a neurotransmitter primarily involved in mesolimbic system, which projects from the brain's ventral tegmental area (VTA) to the nucleus accumbens (Guerri & Pascual, 2010; Pickens & Calu, 2011; Schuckit, 2009). This reward circuitry is activated during initial alcohol use and early stages of the dependence, because the EtOH consumption releases dopamine and increases activity at related synapses contributing to the rewarding effects related to motivation, craving, and disinhibition (Gilpin & Koob, 2008; Guerri & Pascual, 2010; Schuckit, 2009). Finally, EtOH withdrawal produces decreases in the dopamine pathway that may contribute to withdrawal symptoms and alcohol relapse (Melis, Spiga, & Diana, 2005; Volkow et al., 2007).

In the serotonergic system, EtOH potentiates 5-HT3 receptor function by increasing channel activation action (Schuckit, 2009). In addition, low concentrations of 5-HT in the synapse, also elicited by EtOH-induced 5-HT1A autoreceptor supersensitivity, are associated with a reduced effect of alcohol, and perhaps a propensity to EtOH consumption and to the addictive effects (Gilpin & Koob, 2008; Schuckit, 2009; Vengeliene et al., 2008). However, both dopamine and 5-HT neurotransmitters play only a minor role in mediating the EtOH high doses sensitivity. On the other hand, such neurotransmitter systems are critically involved in the initiation of EtOH reinforcement processes (Gilpin & Koob, 2008; Vengeliene et al., 2008).

Similarly, opioid systems can activate dopaminergic neurons in VTA, releasing dopamine in the nucleus accumbens (Camarini & Pautassi, 2016; Vengeliene et al., 2008). In turn, EtOH can release dopamine in the nucleus accumbens by action on endogenous opioids systems, activating two subtypes of opioid receptors, mu (μ) and kappa (κ) (Camarini & Pautassi, 2016). In this context, it has been postulated that endogenous opioids system plays a crucial role in the alcohol-induced reinforcement process and consumption maintenance, since EtOH increases dopaminergic activity in the reward circuitry, in the mesolimbic dopaminergic system (Camarini & Pautassi, 2016; Gilpin & Koob, 2008; Vengeliene et al., 2008).

To summarize, EtOH acts on GABA and glutamate receptors as an allosteric agonist and antagonist, respectively. These main effects also affect many other molecular targets as dopaminergic, serotonergic, and opioidergic systems, as shown in Fig. 20.2.

Neuroinflammation

During an immune response, the brain and the immune system interrelate, a process that is essential for maintaining homoeostasis. In the natural immune response, phagocytic cells, such as macrophages and microglia, are recruited to release several mediators as pro-inflammatory cytokines and chemokines, in order to attract other immune system cells to the affected area (Alikunju, Abdul Muneer, Zhang, Szlachetka, & Haorah, 2011; Ward et al., 2009).

It has been known that EtOH abuse, both chronic and acute pattern, is a modulator of immune function (Ward et al., 2009). Beyond of deleterious effects caused by EtOH in the neurotransmissions pathway mentioned earlier, the neuroinflammation has been proposed as one of the trigger events involved in the EtOH misuse-induced neuropathological mechanisms (Oliveira et al., 2014, 2015; Pascual et al., 2007; Teixeira et al., 2014). For instance, increased levels of microglial markers are observed in the brains of both postmortem human alcoholics and various alcohol-treated rodents (Yang et al., 2014). Additionally, intermittent binge-type drinking generates microglial activation in cerebral cortex and hippocampus of adolescent and adult rats (Zhao et al., 2013).

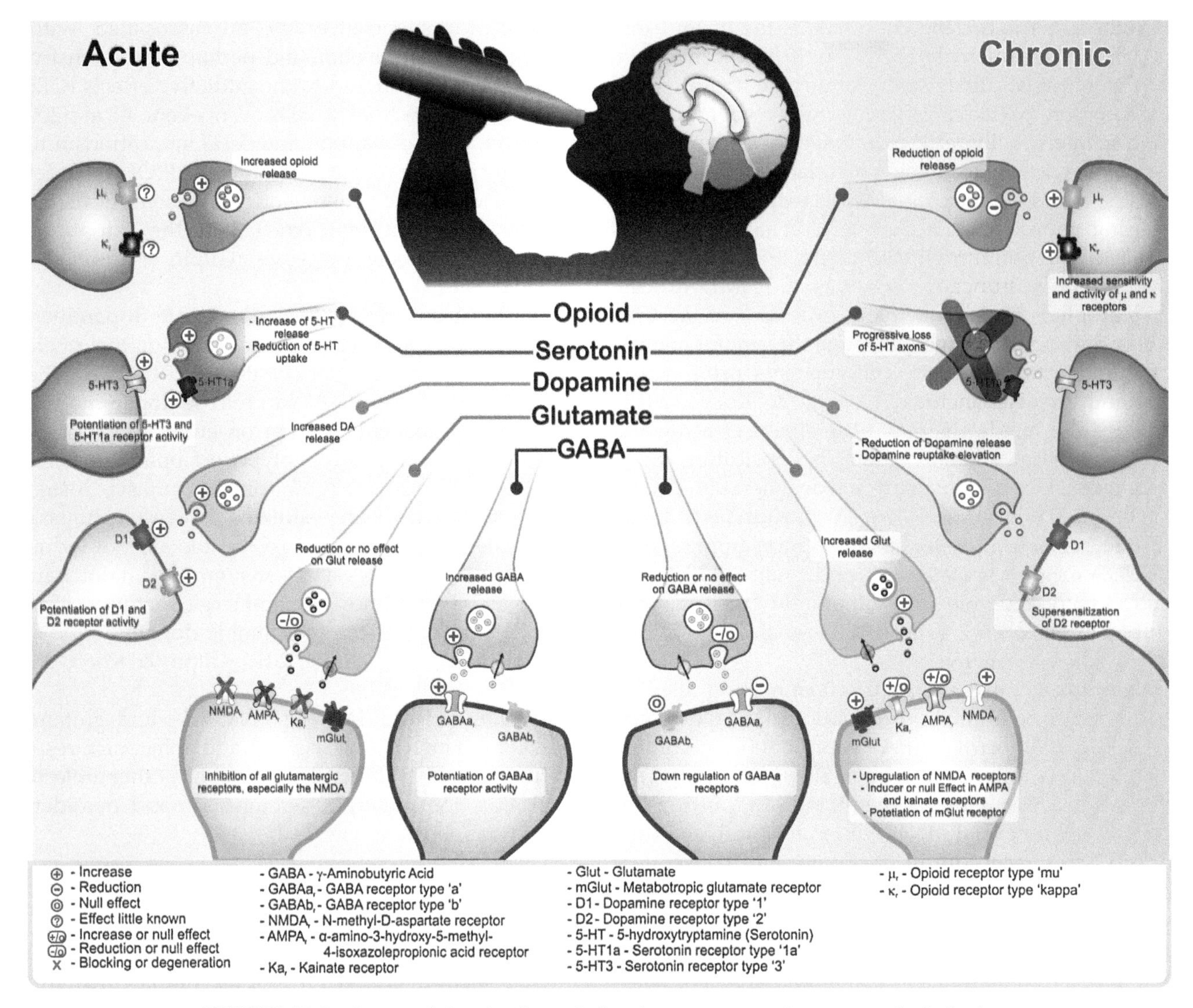

FIGURE 20.2 Acute and chronic effects of ethanol on neurotransmitter systems in the brain.

However, it seems that the relationship between EtOH and microglia activity are dependent on some factors as period of exposure and drinking pattern. In fact, chronic heavy-drinking consumption in female rats from adolescence to adulthood provoke reduced numbers of microglia with fine processes in hippocampal subregions CA1, CA3, and hilus (Oliveira et al., 2015). Besides, this same chronic heavy drinking paradigm (6.5 g/kg/day) shows a reduction in the number of microglial cells in both motor cortex of adult rats by immunostaining of Iba-1$^+$ cells labeled (Teixeira et al., 2014).

The etiology that justifies the EtOH-induced activated microglia remains uncertain, however seems to be provoked by its major metabolite, the acetaldehyde (Ward et al., 2009). Remarkably, the increase of hypertrophied microglia number, densely stained with enlarged somas and thickened processes feature the activated microglia, which occur concomitantly with production of pro-inflammatory cytokines, neuronal neurodegeneration, cognitive deficits, anxiety, and motor impairment (Oliveira et al., 2015; Pascual et al., 2007; Teixeira et al., 2014; Ward et al., 2009; Zhao et al., 2013). This process occurs because the microglia respond rapidly to pathological changes in the brain by expressing pathogen recognition receptors such the toll-like receptor-4 (TLR-4) that activates signaling cascades in microglia, including nuclear factor kappa B (NF-κB) and mitogen-activated protein kinase (MAPK) pathways, which eventually induce the transcription of pro-inflammatory mediators (Pascual, Baliño, Alfonso-Loeches, Aragón, & Guerri, 2011; Yang et al., 2014).

The microglial activation after intermittent EtOH exposure induces inflammatory mediators and stimulates intracellular signaling pathways that trigger the induction pro-inflammatory cytokines (i.e., tumor necrosis factor-α, TNF-α, and interleukin 1β, IL-1β), cyclooxygenase-2 (COX-2), and inducible nitric oxide synthase (iNOS) in the hippocampus, cerebellum, parietal association cortex, and entorhinal cortex of rats, which have a detrimental effect on the neighboring neurons (Pascual et al., 2011, 2007; Zhao et al., 2013).

Furthermore, Zhao et al. (2013) reported that the microglia with a hypertrophied shape disappeared, accompanied by a declination of inflammatory cytokines levels, and the numbers of microglia with fine processes gradually increased during the EtOH withdrawal associated with an improvement in spinal degeneration and cognitive dysfunction on nonspatial tasks. In this way, microglia have been involved during EtOH intoxication and withdrawal, which plays an important role in the pro-inflammatory state of neuropathological processes and may be included in the recovery during abstinence.

Oxidative Stress

The EtOH neurotoxicity may be related to several mechanisms, but it is generally accepted that oxidative stress plays a key role in this process (Chen et al., 2008; Haorah, Ramirez, et al., 2008; Harper, 2009). The oxidative stress refers to the oxidative imbalance in consequences of an excess levels of free radicals and reduced levels of antioxidants in the cells, leading to lipid and protein peroxidation and DNA damage. The terms "free radicals" and "reactive oxygen species (ROS)" are used as equivalents. However, free radicals are molecules with an unpaired electron, which are produced by most tissues in the normal course of their activity (Mantle & Preedy, 1999). Meanwhile, ROS refers to small, highly reactive, oxygen-containing molecules such as superoxide radical (O_2•), hydroxyl radical (HO•), H_2O_2, and hypochlorous acid (HOCl), and peroxides, like lipid peroxides, and peroxides of proteins, and nucleic acids (Galicia-Moreno & Gutiérrez-Reyes, 2014).

On the other hand, the reactive nitrogen species (RNS) refer to nitric oxide and the molecules derived, such as peroxynitrite and nitrogen dioxide. Both ROS and RNS have dual biological actions, which oxidize important structures and macromolecules in the cells, and also act as a part of defense/signaling mechanisms (Das & Vasudevan, 2007). In addition, the RNS has a longer half-life than the ROS, thus more devastate. In this regard, the ROS and RNS products can react and damage complex cellular molecules, such as fats, proteins, or DNA, and finally lead to cell injury. On the other hand, all tissues, including CNS, have the capacity to neutralize oxygen radicals by antioxidants mechanisms, which include superoxide dismutase (SOD) and catalase enzymes that convert the superoxide anion to hydrogen peroxide and hydrogen peroxide to water, respectively.

Other important antioxidant mechanism is the glutathione (GSH) system, in which enzymatic components dependent upon selenium (glutathione reductase, GR, and glutathione peroxidases, GPx) provide a mechanism to reduce oxidized molecules, wherein the reduction in cellular levels of GSH and increased levels of glutathione disulfide (GSSG) are used as an indication of oxidative stress in tissue (Das & Vasudevan, 2007). Therefore, this oxidative imbalance has been implicated in several physiological and pathological processes, such as DNA mutations, carcinogenesis, aging, atherosclerosis, radiation damage, inflammation, ischemia reperfusion injury, diabetes mellitus, neurodegenerative diseases, and toxic injuries, including acute and chronic alcohol toxicity.

In this sense, the excessive alcohol consumption causes profound damage to human organ systems including the liver, brain, heart, pancreas, lungs, endocrine, and immune systems, as well as bone and skeletal muscles. These damages are associated with alcohol-induced oxidative stress during metabolism, which can alter the levels of certain metals and the production of ROS, and at the same time to reduced activity of the protective antioxidant mechanisms, resulting in oxidative stress (Wu & Cederbaum, 2003). In addition, others factors, such as hypoxia, endotoxemia, and cytokine release may lead to the formation of an environment favorable to oxidative stress in the body (Sergent, Griffon, Cillard, & Cillard, 2001).

The CNS is particularly sensitive to alcohol-induced oxidative stress; due to its high oxygen consumption rate, elevated levels of polyunsaturated fatty acids, and relatively low content of antioxidant enzymes (Cohen-Kerem & Koren, 2003). In addition, the oxidative stress and mitochondrial dysfunction mechanisms are implicated in EtOH-induced neurotoxicity that lead to the tissue injury. As described earlier, EtOH readily crosses the blood–brain barrier (BBB), and it is metabolized in the brain by enzymes, such as catalase, ADH, or CYP2E1. These processes produce acetaldehyde and increase ROS levels, including O_2•, HO•, and H_2O_2, disturbing cellular normal redox state and may cause BBB damage, neuroinflammation, and neurological diseases (Hampton & Orrenius, 1998; Haorah, Ramirez, et al., 2008, Haorah, Schall, Ramirez, & Persidsky, 2008; Zakhari, 2006).

Furthermore, neurons in the cerebral cortex, as well as in the cerebellum (i.e., Purkinje and granule cell layers) constitutively express CYP2E1. The chronic EtOH consumption upregulates the CYP2E1 activity in the rat or human brain, as well as elevate the ROS production and BBB dysfunction (Haorah, Knipe, Gorantla, Zheng, & Persidsky, 2007; Haorah, Knipe, Leibhart, Ghorpade, & Persidsky, 2005; Haorah, Ramirez, et al., 2007). Moreover, Haorah, Ramirez, et al. (2008) showed that the ADH and CYP2E1 enzymes upregulate the production of ROS and NO in neurons during the EtOH metabolism, via the activation of NADPH/xanthine oxidase (NOX/XOX) and iNOS activities by acetaldehyde. Thus, these authors suggested that brain cells including neurons can metabolize alcohol, which contribute to elevated oxidative stress commonly observed in alcoholics, thereby supporting the idea that the EtOH metabolism results in end-organ injury. This process leads to the cellular oxidative process, with high levels of lipid peroxidation product, accompanied by diminished expression of neuronal markers (neurofilaments) and enhanced neuronal death (Haorah, Ramirez, et al., 2008; Haorah, Schall, et al., 2008).

BBB, constituted by brain microvascular endothelial cells (BMVECs), pericytes, and astrocytes (Alikunju et al., 2011; Rubin & Staddon, 1999), regulates the trafficking of ions, molecules, and leukocytes into and out of the brain (Hawkins & Davis, 2005). Thereby, the loss of BBB integrity is a critical event in the development and progression of neurological disorders (Fiala et al., 2002). Neuropathological findings in chronic alcoholics reveal degeneration of neurons, neuronal cell death, and white matter abnormalities (Haorah, Ramirez, et al., 2007; Harper, 1998; Mann et al., 2001). The BBB dysfunction can be caused by alcohol-induced oxidative stress, which led to neurological disorders, due to an activation of myosin light-chain kinase (MLCK) with subsequent phosphorylation of myosin light chain (MLC) and tight junction proteins, tight junction phosphorylation, activation of inositol 1,4,5-triphosphate receptor-gated intracellular Ca^{2+} release (IP3R), and the activation of matrix metalloproteinases by protein tyrosine kinases (Haorah, Heilman, et al., 2005; Haorah, Knipe, et al., 2007, 2005; Haorah, Ramirez, et al., 2008, 2007; Haorah, Schall, et al., 2008). In addition, some studies reported that Ca^{2+} channels are affected by ROS as a part of normal signaling pathways or under pathological conditions (Haorah, Ramirez, et al., 2007; Waring, 2005). In this regard, Na^{+}/Ca^{2+} exchange blocker can modulate the BBB disruption, implicating the regulation of BBB integrity mediated by intracellular Ca^{2+} channel (Bhattacharjee, Nagashima, Kondoh, & Tamaki, 2001).

Moreover, EtOH-treated rats have increased IP3R mRNA and protein levels, indicating that the enhanced expression of IP3R protein induced by EtOH/ROS can be caused by activation of transcription factors such as NF-κB, activated proteins 1 and 2 (AP-1 and -2) or the activated protein kinases (Haorah, Ramirez, et al., 2007). Thus, these authors showed that changes in IP3R mRNA and protein levels were correlated with the enhanced activity of IP3R-gated intracellular Ca^{2+} release in response to EtOH, acetaldehyde, or ROS in a dose-dependent manner. These mechanisms can result in an increased permeability and enhancement of leukocyte migration across the BBB caused by EtOH (Haorah, Heilman, et al., 2005; Haorah, Knipe, et al., 2005).

Role of Acetaldehyde in the EtOH-Mediated Neurotoxicity

The importance of acetaldehyde for the alcohol-induced CNS damage was controversial for a long time, mainly because acetaldehyde produced by hepatic metabolism presents limited passage across the BBB (Zimatkin, 1991). This issue was clarified after the characterization of acetaldehyde formation pathway in the CNS. As described earlier, in the brain tissue, the ADH activity is considered despicable. Catalase, on the other hand, is responsible for about 50% of the acetaldehyde formed in the brain. Besides, the CYP2E1 plays an important role in the brain tissue (Aragon, Rogan, & Amit, 1992; Vonghia et al., 2008; Zimatkin & Deitrich, 1997; Fig. 20.3).

First, the contribution of acetaldehyde for the neurotoxicity is related to the CYP2E1 enzyme expression induction, which consists of the characteristic event of heavy drinking, resulting in increased production of ROS (Zimatkin, Pronko, Vasiliou, Gonzalez, & Deitrich, 2006). The elevation of CYP2E1 activity is linked to the induction of the enzymatic activity of other groups, such as NOX, XOX, and NOS, contributing to the oxidative mechanisms for brain injury (Haorah, Ramirez, et al., 2008).

Moreover, acetaldehyde reacts with nucleophilic groups of proteins, forming unstable protein adducts, that in turn interferes with cellular functions (Nakamura et al., 2003, 2000). These acetaldehyde adducts have been associated with EtOH addiction, induction of immune response (production of cytokines and T-cell induction),

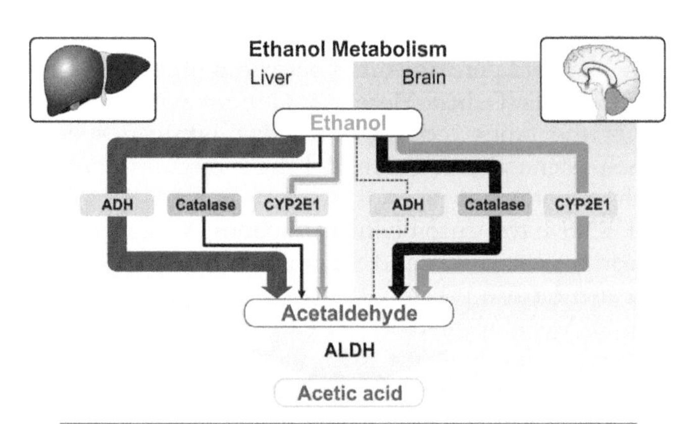

FIGURE 20.3 Differences in ethanol metabolism in the liver and brain. In the liver metabolism, alcohol dehydrogenase (ADH) is responsible for over 90% of ethanol biotransformation. Cytochrome P450 (CYP2E1) has a small contribution, which rises with the enzyme induction process. Catalase, on the other hand, has despicable participation in alcoholic metabolism in this tissue. In the brain, the presence of ADH is negligible, and the ethanol metabolism is mainly performed by way of catalase and CYP2E1.

enzymes conformational changes related to increased protein degradation and pathological features, and reticular stress (Duryee et al., 2004; Ji, 2012; Willis, Thiele, Tuma, & Klassen, 2003).

Finally, acetaldehyde also seems to influence the neurochemical activity of the CNS. There is evidence that acetaldehyde increases dopaminergic activity in the VTA and mesolimbic system, which are involved in learning, motivated behaviors and reinforcement effects associated with the consumption of alcoholic beverages (Rodd-Henricks et al., 2002). In fact, such neurochemical alterations do not produce directly neurotoxicity or cell death, however it contributes to the development of EtOH addiction.

CNS STRUCTURAL CHANGES AND BEHAVIORAL EFFECTS

Excessive EtOH consumption may cause structural abnormalities in the brain (Oliveira et al., 2015; Teixeira et al., 2014). However, the structural changes depend on the time of exposure and alcohol consumption pattern, affecting brain regions differently, as well as promoting changes of different intensities between the white and gray matter (Brown & Tapert, 2004; Kroenke et al., 2013).

During mid-2010s our research group demonstrated that chronic EtOH exposure in young rats promoted a decrease on neuronal density, as well as the reduction in the number of astrocytes and resident microglia, both in the motor cortex and in the hippocampus (Oliveira et al., 2015; Teixeira et al., 2014). Using the same chronic EtOH exposure protocol, we observed that the density of motor neurons decreased in the spinal cord. Besides, the integrity of myelin sheath was affected, both the medullar tract and the brain white matter were loosely compacted (unpublished data).

In fact, the chronic continuous EtOH exposure is intrinsically associated with the widespread neuronal and glial death process. In this sense, several brain regions can be affected by EtOH intake, that in turn can display different brain disorders (i.e., neurodegenerative and psychiatric diseases) (Bertotto, Bustos, Molina, & Martijena, 2006; De Witte, Pinto, Ansseau, & Verbanck, 2003; Gilpin & Koob, 2008).

It has been reported that regions such as cortex, hippocampus, and cerebellum are more susceptible to alcohol-induced damage and the withdrawal effects (Jung & Metzger, 2010). In this sense, excessive EtOH intake produces many significant behavioral impairments in humans and rodents including anxiety, motor coordination, and cognitive damage, as well as depressive disorders (García-Moreno & Cimadevilla, 2012; Oliveira et al., 2015; Pascual et al., 2011, 2007; Sullivan & Pfefferbaum, 2005; Teixeira et al., 2014; Yang et al., 2014; Zhao et al., 2013).

Anxiety

Low and moderate doses of acute EtOH exposure produces anxiolytic effects (Silberman et al., 2009). However, repeated alcohol exposure and ethanol withdrawal are related to neuroadaptive changes that may lead to persistent increases in a range of anxiety symptoms (Silberman et al., 2009). The intensity of these effects depends on the dose, period and time of exposure, behavioral test protocol, and species and strains analyzed (Acevedo, Nizhnikov, Molina, & Pautassi, 2014).

According to Acevedo et al. (2014), EtOH anxiolytic effects are associated with low and moderate doses and are often assessed by tests as elevated plus-maze or light–dark box, which take advantage of the rodent natural avoidance of open and brightly lit spaces.

The hippocampus plays a pivotal role in many brain functions such as memory and anxiety, and many studies have demonstrated its vulnerability to ethanol effects (Oliveira et al., 2015). In this way, chronic EtOH consumption can modify emotional behavior and cognition in humans and rats (Yang et al., 2014). The abrupt cessation of chronic EtOH administration induces an alcohol-withdrawal syndrome, which is associated with a negative affective state including negative emotions such as anxiety and depression (Bertotto et al., 2006; De Witte et al., 2003; Gilpin & Koob, 2008).

It has already become known that heavy chronic EtOH exposure during adolescence through early adulthood increases anxiety-like behavior in female rats that was accompanied by morphological alterations in hippocampus, such as reduction in hippocampal volume and neuronal and microglial density (Oliveira et al., 2015). Pascual et al. (2011) report that 5 months of EtOH exposure followed by a 15-day withdrawal period in mice, promotes long-term anxiety-like behavior associated with alcohol-induced inflammatory damage with epigenetic modifications mediated by TLR4. Furthermore, the adolescence period is also characterized by the rapid maturation of brain systems mediating reward and by changes in the secretion of stress-related hormones, events that might participate in increasing anxiety, and the initiation pattern of EtOH and drug consumption (Guerri & Pascual, 2010).

Besides, there is a large and growing evidence suggesting a strong relationship between stress, anxiety, and AUDs (Silberman et al., 2009). The transition to EtOH dependence involves not only the neural circuits associated with the reward process but also the dysregulation of circuits that mediate behavioral responses to stressors. In the stress system, the signaling molecule corticotropin-releasing factor (CRF) activate the hypothalamic–pituitary–adrenal (HPA) axis and produces high anxiety-like states (Gilpin & Koob, 2008). Moreover, Chu, Koob, Cole, Zorrilla, and Roberts (2007) reported that increased intake in EtOH-dependent animals could be effectively reduced by treatments with CRF1-R (receptor) antagonists and attenuate withdrawal-associated anxiety. Therefore, preclinical or clinical studies have reported that heavy or binge drinking of EtOH displays a high level of anxiety, regardless of age or gender.

Depression

Depression is the most common psychiatric disorder in patients with substance-use disorders such as EtOH abuse (Stevenson et al., 2009; Yang et al., 2014). As previously mentioned, the abrupt cessation of chronic EtOH intake is related to negative affective state, anxiety, and depression (Bertotto et al., 2006; De Witte et al., 2003; Gilpin & Koob, 2008).

Clinical studies show that the development of depressive and addictive disorders among adolescents are considered even higher than that observed in adults (Sullivan, Fiellin, & O'Connor, 2005). The development of depression during the chronic EtOH withdrawal may be attributed to the cyclic pattern of euphoria after EtOH consumption followed by the dysphoric mood that accompanies the withdrawal period (Briones & Woods, 2013).

Several reports consistently have shown that adolescent limbic system is susceptibility to alcohol-induced memory impairment and mood-altering properties (Crews, Braun, Hoplight, Switzer, & Knapp, 2000). However, the aetiological mechanism is not well understood. Briones and Woods (2013) reported that the depression-like symptoms seen during the withdrawal period following chronic binge EtOH exposure possibly are related to reduced brain-derived neurotrophic factor (BDNF) signaling in the hippocampus that consequently alter neural plasticity.

Even in adults, chronic EtOH consumption, during the withdrawal period or not, can display depression-like behavior. In the withdrawal period, the depressive behavior was associated with the reduction in number of proliferating neural progenitor cells and immature neurons in the dentate gyrus (DG) of the hippocampus in mice. Such results indicate that ethanol withdrawal–induced depression is associated with reductions in hippocampal neurogenesis (Rehm et al., 2015; Stevenson et al., 2009).

Motor Deficits

The motor cortex has been recognized to play an important role in fine motor control and independent movements of reach and hold, sensorimotor integration, and movement of cognitive motor of higher order (Teixeira et al., 2014). Besides, it has been reported that the motor system is planned and controlled by the cortex, under the supervision of the control unit, which comprises the basal ganglia and cerebellum. Such CNS structures are able to adequately correct the movement carried out by resending the information to cortical structures through the motor feedback loop (Handley, Medcalf, Hellier, & Dutta, 2009).

All these structures as cortex motor, cerebellum, and basal ganglia are affected by EtOH exposure, and it is well known that EtOH produces marked impairments in balance and motor coordination. In fact, it has been highlighted that EtOH intake provokes structural changes in these CNS regions related to motor function. In the cerebellum, the occurrence of alcohol-induced atrophy related to Purkinje cells loss is well defined (Duckett & Schoedler, 1991; Sullivan, 2003). Contradictorily, in the basal ganglia and motor cortex, anatomical changes are more difficult to observe following ethanol consumption. However, morphological changes have been extensively described (Maia et al., 2009; Teixeira et al., 2014).

As reported to other features of EtOH-induced impairment, the motor function also depends on the period and time of exposure. In fact, adolescent rats are less vulnerable to EtOH-induced motor impairments when compared with adults (Van Skike et al., 2010; White et al., 2002), which might enable adolescents to drink larger amounts of alcohol with no perception of the toxicological effects. However, although the young cerebellum may be less sensitive to EtOH than adults, excessive alcohol consumption at early age may lead to cerebellar long-term adverse effects (Forbes, Cooze, Malone, French, & Weber, 2013).

In this sense, in the tests of loss of righting reflex (LORR), Little, Kuhn, Wilson, and Swartzwelder (1996) found that adolescent rats recovered the righting reflex (i.e., recover the normal upright position) earlier than adult, despite that BAC of adolescent rats was higher than alcohol exposure of adult rats. Little and colleagues study indicated that adolescence period is less sensitive to the sedative effects of ethanol than adulthood.

Nevertheless, long-term alcohol-induced motor impairment occurs after chronic exposure, independent of age. Pascual et al. (2007) show that intermittent EtOH intake during the adolescence stage (25–38 days old) affects the motor function demonstrated by impaired performance on rotarod and beam walking tests, despite the lower sensitivity to sedative effects described by Little et al. study. However, these motor effects did not persist after 3 weeks of abstinence, in the adulthood (61 days old) (Little et al., 1996).

In any case, the principal motor deficits related to chronic alcohol consumption are the cerebellar ataxia, balance, gait, muscle strength impairments, and motor incoordination (Kumar et al., 2009; Yang et al., 2014).

Furthermore, bradykinesia can occur after EtOH consumption. This movement disorder is associated with abnormal functioning of the intrinsic circuitry of the basal ganglia, probably altered by injury and cell death of dopaminergic terminals, suggesting the striatal and cerebellar disorders (Handley et al., 2009).

Cognition

The hippocampus plays a crucial role in learning and memory processes (Deng, Aimone, & Gage, 2010; Pascual et al., 2011, 2007). It is well reported that hippocampus is one of the brain regions most vulnerable to EtOH-induced neurotoxicity (Ward et al., 2009; Yang et al., 2014). The CA1 hippocampal region is related to memory consolidation process and CA3 to retrieval of memories (Ward et al., 2009).

The memories can be classified according to the function, content, duration, nature, and motivation. Depending on the type of memory, different CNS structures can be mobilized (Cippitelli et al., 2010; Izquierdo et al., 1998). For example, spatial and nonspatial memories are dependent on entorhinal cortex and hippocampus (Cippitelli et al., 2010) while implicit memory (i.e., habituation), which is a kind of content classification, requires the striatum (Izquierdo et al., 1998). Regardless the type of memory, the process of memory consolidation is associated with molecular events in the hippocampal CA1 region involved in long-term potentiation (LTP), and it also requires equivalent events to occur with different timings in the basolateral amygdala and the entorhinal, as well as parietal and cingulate cortex (Izquierdo et al., 1998, 2006; Squire & Cohen, 1984).

Human epidemiological studies have demonstrated that adult alcoholics often suffer from cognitive dysfunction, including short-term and long-term spatial memories (Sullivan & Pfefferbaum, 2005). Studies about EtOH consumption perinatally showed impairment in cognitive functions related to implicit memory procedures and short-term memory in proles (Maia et al., 2010, 2009). Remarkably, memory problems are among the most common dysfunctions in adolescents with AUD, and these effects have been mainly associated with abnormalities in the brain response to a spatial working memory task (Guerri & Pascual, 2010).

It is noteworthy that it is not clear if acute EtOH exposure displays a greater, lesser, or similar impairments in the spatial learning and memory between adolescents and adult rats (Chin et al., 2011). Although some reports indicate that acute EtOH-induced learning impairment is similar between adolescents and adults (Chin et al., 2011; Rajachandran, Spear, & Spear, 1993), other cognitive studies claim that adolescents present greater deficits in the spatial learning on the Morris Water maze task than adults (Markwiese, Acheson, Levin, Wilson, & Swartzwelder, 1998; McKinzie et al., 1996). Thus, these last studies indicate that the sensitivity to the cognitive-impairing effects of acute EtOH may be more pronounced in younger than in older organisms (Yang et al., 2014).

Contradictorily to acute EtOH consumption effects, adolescent rats are more sensitive than adults to the disruptive effects to chronic EtOH intoxication in cognitive processes (Guerri & Pascual, 2010). In addition, there is evidence that some of the cognitive effects induced by repeated EtOH treatment in adolescent rats, such as learning impairments, might persist into adulthood (Cippitelli et al., 2010; Pascual et al., 2011, 2007). Besides, it has been reported that chronic or intermittent EtOH exposure promote both spatial and nonspatial memory impairments (novel object recognition) in adolescent and adult EtOH-treated rodents (Cippitelli et al., 2010; Pascual et al., 2011, 2007; Zhao et al., 2013).

Nevertheless, the possible recovery during abstinence remains unclear. Zhao et al. (2013) reported the recovery in the spatial and nonspatial memory. However, Cippitelli et al. (2010) founded the reversibility in the object recognition but not in the spatial memory impairment after alcohol abstinence. In addition to spatial memory, long-term memory is also affected by the alcohol-induced cognitive dysfunctions (Pascual et al., 2011). Notably, the dose, timing, and duration of EtOH exposure are determinants to the recovery of hippocampal-dependent memory impairment after EtOH abstinence (Cippitelli et al., 2010).

PERSPECTIVES AND FINAL COMMENTS

During 2000s, chronic alcohol consumption has become a public health problem. Much more than social problems, EtOH misuse has been reported by numerous researcher groups as being responsible for CNS alterations. Among the central disorders identified, anatomical, morphological, neurochemical, and behavioral damages have been identified. Due to its widespread brain harmful effects, alcohol exposure displays neuronal cell death by several mechanisms as inflammatory, neurotoxicity, and oxidative stress process. In addition to neuronal structure changes, EtOH intake can display psychiatric disorders as anxiety and depression, as well as motor and cognitive deficits.

The extent of the damage depends on the period and timing of exposure allied to the pattern of drinking. The new perspective focuses on the molecular signaling affected by alcohol addiction and the discovery of therapeutic drugs that could recover from or minimize the damage among alcoholics.

References

Acevedo, M. B., Nizhnikov, M. E., Molina, J. C., & Pautassi, R. M. (2014). Relationship between ethanol-induced activity and anxiolysis in the open field, elevated plus maze, light-dark box, and ethanol intake in adolescent rats. *Behavioural Brain Research, 265*, 203–215.

Adams, C. L., Short, J. L., & Lawrence, A. J. (2010). Cue-conditioned alcohol seeking in rats following abstinence: Involvement of metabotropic glutamate 5 receptors. *British Journal of Pharmacology, 159*, 534–542.

Alikunju, S., Abdul Muneer, P. M., Zhang, Y., Szlachetka, A. M., & Haorah, J. (2011). The inflammatory footprints of alcohol-induced oxidative damage in neurovascular components. *Brain, Behavior, and Immunity, 25*(Suppl. 1), S129–S136.

Aragon, C. M. G., Rogan, F., & Amit, Z. (1992). Ethanol metabolism in rat brain homogenates by a catalase-H_2O_2 system. *Biochemical Pharmacology, 44*, 93–98.

Bannach, S. V., Teixeira, F. B., Fernandes, L. M. P., Ferreira, R. O., Santana, L. N., Fontes-Júnior, E. A., ... Lima, R. R. (2015). Alveolar bone loss induced by chronic ethanol consumption from adolescence to adulthood in Wistar rats. *Indian Journal of Experimental Biology, 53*, 93–97.

Beenstock, J., Adams, J., & White, M. (2011). The association between time perspective and alcohol consumption in university students: Cross-sectional study. *European Journal of Public Health, 21*, 438–443.

Bertotto, M. E., Bustos, S. G., Molina, V. A., & Martijena, I. D. (2006). Influence of ethanol withdrawal on fear memory: Effect of D-cycloserine. *Neuroscience, 142*, 979–990.

Bhattacharjee, A. K., Nagashima, T., Kondoh, T., & Tamaki, N. (2001). The effects of the Na^+/Ca^{++} exchange blocker on osmotic blood–brain barrier disruption. *Brain Research, 900*, 157–162.

Boutros, N., Semenova, S., & Markou, A. (2014). Adolescent intermittent ethanol exposure diminishes anhedonia during ethanol withdrawal in adulthood. *European Neuropsychopharmacology, 24*, 856–864.

Brien, J. F., Loomis, C. W., Tranmer, J., & McGrath, M. (1983). Disposition of ethanol in human maternal venous blood and amniotic fluid. *American Journal of Obstetrics and Gynecology, 146*, 181–186.

Briones, T. L., & Woods, J. (2013). Chronic binge-like alcohol consumption in adolescence causes depression-like symptoms possibly mediated by the effects of BDNF on neurogenesis. *Neuroscience, 254*, 324–334.

Brown, S. A., & Tapert, S. F. (2004). Adolescence and the trajectory of alcohol use: Basic to clinical studies. *Annals of New York Academy of Sciences, 1021*, 234–244.

Camarini, R., & Pautassi, R. M. (2016). Behavioral sensitization to ethanol: Neural basis and factors that influence its acquisition and expression. *Brain Research Bulletin, 125*, 53–78.

Casillas-Espinosa, P. M., Powell, K. L., & O'Brien, T. J. (2012). Regulators of synaptic transmission: Roles in the pathogenesis and treatment of epilepsy. *Epilepsia, 53*, 41–58.

Cederbaum, A. I. (2012). Alcohol metabolism. *Clinics in Liver Disease, 16*, 667–685.

Chen, G., Ma, C., Bower, K. A., Shi, X., Ke, Z., & Luo, J. (2008). Ethanol promotes endoplasmic reticulum stress-induced neuronal death: Involvement of oxidative stress. *Journal of Neuroscience Research, 86*, 937–946.

Chin, V. S., Van Skike, C. E., Berry, R. B., Kirk, R. E., Diaz-Granados, J., & Matthews, D. B. (2011). Effect of acute ethanol and acute allopregnanolone on spatial memory in adolescent and adult rats. *Alcohol, 45*, 473–483.

Chu, K., Koob, G. F., Cole, M., Zorrilla, E. P., & Roberts, A. J. (2007). Dependence-induced increases in ethanol self-administration in mice are blocked by the CRF1 receptor antagonist antalarmin and by CRF1 receptor knockout. *Pharmacology, Biochemistry, and Behavior, 86*, 813–821.

Cippitelli, A., Zook, M., Bell, L., Damadzic, R., Eskay, R. L., Schwandt, M., & Heilig, M. (2010). Reversibility of object recognition but not spatial memory impairment following binge-like alcohol exposure in rats. *Neurobiology of Learning and Memory, 94*, 538–546.

Cohen-Kerem, R., & Koren, G. (2003). Antioxidants and fetal protection against ethanol teratogenicity. I. Review of the experimental data and implications to humans. *Neurotoxicology and Teratology, 25*, 1–9.

Crego, A., Rodriguez-Holguín, S., Parada, M., Mota, N., Corral, M., & Cadaveira, F. (2010). Reduced anterior prefrontal cortex activation in young binge drinkers during a visual working memory task. *Drug and Alcohol Dependence, 109*, 45–56.

Crews, F. T., Braun, C. J., Hoplight, B., Switzer, R. C., & Knapp, D. J. (2000). Binge ethanol consumption causes differential brain damage in young adolescent rats compared with adult rats. *Alcoholism, Clinical and Experimental Research, 24*, 1712–1723.

Das, S. K., & Vasudevan, D. M. (2007). Alcohol-induced oxidative stress. *Life Sciences, 81*, 177–187.

De Witte, P., Pinto, E., Ansseau, M., & Verbanck, P. (2003). Alcohol withdrawal: From animal research to clinical issues. *Neuroscience and Biobehavioral Reviews, 27*, 189–197.

Deng, W., Aimone, J. B., & Gage, F. H. (2010). New neurons and new memories: How does adult hippocampal neurogenesis affect learning and memory. *Nature Reviews Neuroscience, 11*, 339–350.

Dodd, P. R., Beckmann, A. M., Davidson, M. S., & Wilce, P. A. (2000). Glutamate-mediated transmission, alcohol, and alcoholism. *Neurochemistry International, 37*, 509–533.

Duckett, S., & Schoedler, S. (1991). Nutritional disorders and alcoholism. In S. Duckett (Ed.), *The pathology of the aging human nervous system*. Philadelphia: Lea & Febiger.

Duryee, M. J., Willis, M. S., Freeman, T. L., Kuszynski, C. A., Tuma, D. J., Klassen, L. W., & Thiele, G. M. (2004). Mechanisms of alcohol liver damage: Aldehydes, scavenger receptors, and autoimmunity. *Frontiers in Bioscience, 9*, 3145–3155.

Fernandes, L. M. P., Lima, R. R., Monteiro, M. C., Gomes-Leal, W., & Maia, C. S. F. (2014). Alcohol in pregnancy and its effects on the central nervous system. In L. V. Berhardt (Ed.), *Advances in medicine and biology, 74* pp. 25–42).

Fernandes, L. M., Teixeira, F. B., Alves-Junior, S. M., Pinheiro Jde, J., Maia, C. S., & Lima, R. R. (2015). Immunohistochemical changes and atrophy after chronic ethanol intoxication in rat salivary glands. *Histology and Histopathology, 30*, 1069–1078.

Ferreira, M. P., & Willoughby, D. (2008). Alcohol consumption: The good, the bad, and the indifferent. *Applied Physiology, Nutrition, and Metabolism, 33*, 12–20.

Fiala, M., Liu, Q. N., Sayre, J., Pop, V., Brahmandam, V., Graves, M. C., & Vinters, H. V. (2002). Cyclooxygenase-2-positive macrophages infiltrate the Alzheimer's disease brain and damage the blood–brain barrier. *European Journal of Clinical Investigation, 32*, 360–371.

Forbes, A., Cooze, J., Malone, C., French, V., & Weber, J. T. (2013). Effects of intermittent binge alcohol exposure on long-term motor function in young rats. *Alcohol, 47*, 95–102.

Fraser, A. G. (1997). Pharmacokinetic interactions between alcohol and other drugs. *Clinical Pharmacokinetics, 33*, 79–90.

Galicia-Moreno, M., & Gutiérrez-Reyes, G. (2014). The role of oxidative stress in the development of alcoholic liver disease. *Revista de Gastroenterología de México, 79*, 135–144.

García-Moreno, L. M., & Cimadevilla, J. M. (2012). Acute and chronic ethanol intake: Effects on spatial and non-spatial memory in rats. *Alcohol, 46*, 757–762.

Gilpin, N. W., & Koob, G. F. (2008). Neurobiology of alcohol dependence. *Alcohol Research & Health, 31*, 185–195.

Goullé, J. P., & Guerbet, M. (2015). Pharmacokinetics, metabolism, and analytical methods of ethanol. *Annales Pharmaceutiques Françaises, 73*, 313–322.

Graham, K., Wilsnack, R., Dawson, D., & Vogeltanz, N. (1998). Should alcohol consumption measures be adjusted for gender differences? *Addiction, 93*, 1137–1147.

Guerri, C., & Pascual, M. (2010). Mechanisms involved in the neurotoxic, cognitive, and neurobehavioral effects of alcohol consumption during adolescence. *Alcohol, 44*, 15–26.

Haastrup, M. B., Pottegård, A., & Damkier, P. (2014). Alcohol and breastfeeding. *Basic & Clinical Pharmacology & Toxicology, 114*, 168–173.

Hampton, M. B., & Orrenius, S. (1998). Redox regulation of apoptotic cell death in the immune system. *Toxicology Letters, 103*, 355–358.

Handler, J. A., & Thurman, R. G. (1990). Redox interactions between catalase and alcohol dehydrogenase pathways of ethanol metabolism in the perfused rat liver. *Journal of Biological Chemistry, 265*, 1510–1515.

Handley, A., Medcalf, P., Hellier, K., & Dutta, D. (2009). Movement disorders after stroke. *Age and Ageing, 38*, 260–266.

Haorah, J., Heilman, D., Knipe, B., Chrastil, J., Leibhart, J., Ghorpade, A., ... Persidsky, Y. (2005). Ethanol-induced activation of myosin light chain kinase leads to dysfunction of tight junctions and blood–brain barrier compromise. *Alcoholism, Clinical and Experimental Research, 29*, 999–1009.

Haorah, J., Knipe, B., Gorantla, S., Zheng, J., & Persidsky, Y. (2007). Alcohol-induced blood–brain barrier dysfunction is mediated via inositol 1,4,5-triphosphate receptor IP3R-gated intracellular calcium release. *Journal of Neurochemistry, 100*, 324–336.

Haorah, J., Knipe, B., Leibhart, J., Ghorpade, A., & Persidsky, Y. (2005). Alcohol-induced oxidative stress in brain endothelial cells causes blood–brain barrier dysfunction. *Journal of Leukocyte Biology, 78*, 1223–1232.

Haorah, J., Ramirez, S. H., Floreani, N., Gorantla, S., Morsey, B., & Persidsky, Y. (2008). Mechanism of alcohol-induced oxidative stress and neuronal injury. *Free Radical Biology and Medicine, 45*, 1542–1550.

Haorah, J., Ramirez, S. H., Schall, K., Smith, D., Pandya, R., & Persidsky, Y. (2007). Oxidative stress activates protein tyrosine kinase and matrix metalloproteinases leading to blood–brain barrier dysfunction. *Journal of Neurochemistry, 101*, 566–576.

Haorah, J., Schall, K., Ramirez, S. H., & Persidsky, Y. (2008). Activation of protein tyrosine kinases and matrix metalloproteinases causes blood–brain barrier injury: Novel mechanism for neurodegeneration associated with alcohol abuse. *Glia, 56*, 78–88.

Harper, C. (1998). The neuropathology of alcohol-specific brain damage, or does alcohol damage the brain? *Journal of Neuropathology and Experimental Neurology, 57*, 101–110.

Harper, C. (2009). The neuropathology of alcohol-related brain damage. *Alcohol and Alcoholism, 44*, 136–140.

Hawkins, B. T., & Davis, T. P. (2005). The blood–brain barrier/neurovascular unit in health and disease. *Pharmacological Reviews, 57*, 173–185.

Heilig, M., Goldman, D., Berrettini, W., & O'Brien, C. P. (2011). Pharmacogenetic approaches to the treatment of alcohol addiction. *Nature Reviews Neuroscience, 12*, 670–684.

Holford, N. H. G. (1987). Clinical pharmacokinetics of ethanol. *Clinical Pharmacokinetics, 13*, 273–292.

Idanpaan-Heikkila, J., Jouppila, P., Akerblom, H. K., Isoaho, R., Kauppila, E., & Koivisto, M. (1972). Elimination and metabolic effects of ethanol in mother, fetus, and newborn infant. *American Journal of Obstetrics and Gynecology, 112*, 387–393.

Ikonomidou, C., Stefovska, V., & Turski, L. (2000). Neuronal death enhanced by *N*-methyl-D-aspartate antagonists. *Proceedings of the National Academy of Sciences of United States of America, 97*, 12885–12890.

INPAD, Instituto Nacional de Ciência e Tecnologia para Políticas Públicas de Álcool e outras Drogas. (2013). *II Levantamento Nacional de Álcool e Drogas (LENAD)*. http://inpad.org.br/wpcontent/uploads/2013/04/LENAD_PressRelease_Alcohol.pdf.

Izquierdo, I., Barros, D. M., Mello e Souza, T., de Souza, M. M., Izquierdo, L. A., & Medina, J. H. (1998). Mechanisms for memory types differ. *Nature, 393*, 635–636.

Izquierdo, I., Bevilaqua, L. R., Rossato, J. I., Bonini, J. S., Medina, J. H., & Cammarota, M. (2006). Different molecular cascades in different sites of the brain control memory consolidation. *Trends in Neurosciences, 29*, 496–505.

Jacobus, J., & Tapert, S. F. (2013). Neurotoxic effects of alcohol in adolescence. *Annual Review of Clinical Psychology, 9*, 703–721.

Ji, C. (2012). Mechanisms of alcohol-induced endoplasmic reticulum stress and organ injuries. *Biochemistry Research International*. http://dx.doi.org/10.1155/2012/216450.

Johnston, L. D., O'Malley, P. M., Bachman, J. G., & Schulenberg, J. E. (2009). Monitoring the future national results on adolescent drug use: Overview of key findings, 2008. In *Abuse* (p. 09). Bethesda, MD: National Institute on Drug Abuse. NIH Publication.

Johnston, L. D., O'Malley, P. M., Miech, R. A., Bachman, J. G., & Schulenberg, J. E. (2014). Monitoring the future national survey results on drug use: Overview of key findings on adolescent drug use. In *Abuse* (p. 38). Bethesda, MD: National Institute on Drug Abuse. NIH Publication.

Jung, M. E., & Metzger, D. B. (2010). Alcohol withdrawal and brain injuries: Beyond classical mechanisms. *Molecules, 15*, 4984–5011.

Kent, W. (2012). The pharmacokinetics of alcohol in healthy adults. *WebmedCentral Pharmacology, 3*, WMC003291.

Kroenke, C. D., Flory, G. S., Park, B., Shaw, J., Rau, A. R., & Grant, K. A. (2013). Chronic ethanol (EtOH) consumption differentially alters gray and white matter EtOH methyl 1H magnetic resonance intensity in the primate brain. *Alcoholism, Clinical and Experimental Research, 37*, 1325–1332.

Kumari, M., & Ticku, M. K. (2000). Regulation of NMDA receptors by ethanol. *Progress in Drug Research, 54*, 152–189.

Kumar, A., LaVoie, H. A., DiPette, D. J., & Singh, U. S. (2013). Ethanol neurotoxicity in the developing cerebellum: Underlying mechanisms and implications. *Brain Science, 3*, 941–963.

Kumar, S., Porcu, P., Werner, D. F., Matthews, D. B., Diaz-Granados, J. L., Helfand, R. S., & Morrow, A. L. (2009). The role of $GABA_A$ receptors in the acute and chronic effects of ethanol: A decade of progress. *Psychopharmacology, 205*, 529–564.

Kuperman, S., Chan, G., Kramer, J. R., Bierut, L., Bucholz, K. K., Fox, L., ... Schuckit, M. A. (2005). Relationship of age of first drink to child behavioral problems and family psychopathology. *Alcoholism, Clinical and Experimental Research, 29*, 1869–1876.

Lieber, C. S. (1991). Hepatic metabolic and toxic effects of ethanol: 1991 uptake. *Alcoholism, Clinical and Experimental Research, 15*, 573–592.

Little, P. J., Kuhn, C. M., Wilson, W. A., & Swartzwelder, H. S. (1996). Differential effects of ethanol in adolescent and adult rats. *Alcoholism, Clinical and Experimental Research, 20*, 1346–1351.

Maia, C. S. F., Ferreira, V. M., Diniz, J. S., Carneiro, F. P., de Sousa, J. B., da Costa, E. T., & Tomaz, C. (2010). Inhibitory avoidance acquisition in adult rats exposed to a combination of ethanol and methylmercury during central nervous system development. *Behavioural Brain Research, 211*, 191–197.

Maia, C. S., Lucena, G. M., Corrêa, P. B., Serra, R. B., Matos, R. W., Menezes, F. C., ... Ferreira, V. M. (2009). Interference of ethanol and methylmercury in the developing central nervous system. *Neurotoxicology, 30*, 23–30.

Mann, K., Agartz, I., Harper, C., Shoaf, S., Rawlings, R. R., Momenan, R., ... Heinz, A. (2001). Neuroimaging in alcoholism: Ethanol and brain damage. *Alcoholism, Clinical and Experimental Research, 25*, 104S–109S.

Mantle, D., & Preedy, V. R. (1999). Free radicals as mediators of alcohol toxicity. *Adverse Drug Reactions and Toxicological Reviews, 18*, 235–252.

Manzo-Avalos, S., & Saavedra-Molina, A. (2010). Cellular and mitochondrial effects of alcohol consumption. *International Journal of Environmental Research and Public Health, 7*, 4281–4304.

Marek, E., & Kraft, W. K. (2014). Ethanol pharmacokinetics in neonates and infants. *Current Therapeutic Research, Clinical and Experimental, 76*, 90–97.

Markwiese, B. J., Acheson, S. K., Levin, E. D., Wilson, W. A., & Swartzwelder, H. S. (1998). Differential effects of ethanol on memory in adolescent and adult rats. *Alcoholism, Clinical and Experimental Research, 22*, 416–421.

Marshall, A. W., Kingstone, D., Boss, M., & Morgan, M. Y. (1983). Ethanol elimination in males and females: Relationship to menstrual cycle and body composition. *Hepatology, 3*, 701–706.

Maudens, K. E., Patteet, L., Van Nuijs, A. L., Van Broekhoven, C., Covaci, A., & Neels, H. (2014). The influence of the body mass index (BMI) on the volume of distribution of ethanol. *Forensic Science International, 243*, 74–78.

McClain, J. A., Morris, S. A., Marshall, S. A., & Nixon, K. (2014). Ectopic hippocampal neurogenesis in adolescent male rats following alcohol dependence. *Addiction Biology, 19*, 687–699.

McKinzie, D. L., Eha, R., Murphy, J. M., McBride, W. J., Lumeng, L., & Li, T. K. (1996). Effects of taste aversion training on the acquisition of alcohol drinking in adolescent P and HAD rat lines. *Alcoholism, Clinical and Experimental Research, 20*, 682–687.

Melis, M., Spiga, S., & Diana, M. (2005). The dopamine hypothesis of drug addiction: Hypodopaminergic state. *International Review of Neurobiology, 63*, 101–154.

Meloni, J. N., & Laranjeira, R. (2004). Custo Social e de Saúde do Consumo do Álcool. *Revista Brasileira Psiquiátrica, 26*, 7–10.

Miller, M. W. (2006). Effect of prenatal exposure to ethanol on glutamate and GABA immunoreactivity in macaque somatosensory and motor cortices: Critical timing of exposure. *Neuroscience, 138*, 97–107.

Mirand, A. L., & Welte, J. W. (1994). Total body water adjustment of mean alcohol intakes. *Journal of Substance Abuse, 6*, 419–425.

Möykkynen, T., & Korpi, E. R. (2012). Acute effects of ethanol on glutamate receptors. *Basic & Clinical Pharmacology & Toxicology, 111*, 4–13.

Nagy, J. (2004). The NR2B subtype of NMDA receptor: A potential target for the treatment of alcohol dependence. *Current Drug Targets. CNS and Neurological Disorders, 3*, 169–179.

Nakamura, K., Iwahashi, K., Furukawa, A., Ameno, K., Kinoshita, H., Ijiri, I., ... Mori, N. (2003). Acetaldehyde adducts in the brain of alcoholics. *Archives of Toxicology, 77*, 591–593.

Nakamura, K., Iwahashi, K., Itoh, M., Ameno, K., Ijiri, I., Takeuchi, Y., & Suwaki, H. (2000). Immunohistochemical study on acetaldehyde adducts in alcohol-fed mice. *Alcoholism, Clinical and Experimental Research, 24*, 93S–96S.

Nava-Ocampo, A. A., Velazquez-Armenta, Y., Brien, J. F., & Koren, G. (2004). Elimination kinetics of ethanol in pregnant women. *Reproductive Toxicology, 18*, 613–617.

NIAAA, National Institute on Alcohol Abuse, and Alcoholism. (2004). *National Institute of Alcohol Abuse and Alcoholism Council approves definition of binge drinking*. Bethesda http://pubs.niaaa.nih.gov/publications.

Norberg, A., Jones, A. W., Hahn, R. G., & Gabrielsson, J. L. (2003). Role of variability in explaining ethanol pharmacokinetics: Research and forensic applications. *Clinical Pharmacokinetics, 42*, 1–31.

Oliveira, G. B., Fontes, E. A., Jr., Carvalho, S., Silva, J. B., Fernandes, L. M. P., Oliveira, M. C. S. P., ... Maia, C. S. F. (2014). Minocycline mitigates motor impairments and cortical neuronal loss induced by focal ischemia in rats chronically exposed to ethanol during adolescence. *Brain Research, 1561*, 23–34.

Oliveira, A. C., Pereira, M. C., Santana, L. N., Fernandes, R. M., Teixeira, F. B., Oliveira, G. B., ... Maia, C. S. F. (2015). Chronic ethanol exposure during adolescence through early adulthood in female rats induces emotional and memory deficits associated with morphological and molecular alterations in hippocampus. *Journal of Psychopharmacology, 29*, 712–724.

Pascual, M., Baliño, P., Alfonso-Loeches, S., Aragón, C. M., & Guerri, C. (2011). Impact of TLR4 on behavioral and cognitive dysfunctions associated with alcohol-induced neuroinflammatory damage. *Brain, Behavior, and Immunity, 25*(Suppl. 1), S80–S91.

Pascual, M., Blanco, A. M., Cauli, O., Miñarro, J., & Guerri, C. (2007). Intermittent ethanol exposure induces inflammatory brain damage and causes long-term behavioural alterations in adolescent rats. *The European Journal of Neuroscience, 25*, 541–550.

Paton, A. (2005). Alcohol in the body. *BMJ, 330*, 85–87.

Petit, G., Kornreich, C., Verbanck, P., & Campanella, S. (2013). Gender differences in reactivity to alcohol cues in binge drinkers: A preliminary assessment of event-related potentials. *Psychiatry Research, 209*, 494–503.

Pickens, C. L., & Calu, D. J. (2011). Alcohol reward, dopamine depletion, and GDNF. *The Journal of Neuroscience, 31*, 14833–14834.

Pohorecky, L. A., & Brick, J. (1988). Pharmacology of ethanol. *Pharmacology & Therapeutics, 36*, 335–427.

Rajachandran, L., Spear, N. E., & Spear, L. P. (1993). Effects of the combined administration of the 5-HT3 antagonist MDL 72222 and ethanol on conditioning in the periadolescent and adult rat. *Pharmacology, Biochemistry, and Behavior, 46*, 535–542.

Ramchandani, V. A., Bosron, W. F., & Li, T. K. (2001). Research advances in ethanol metabolism. *Pathologie Biologie (Paris), 49*, 676–682.

Rehm, J., Allamani, A., Aubin, H. J., Della Vedova, R., Elekes, Z., Frick, U., ... Wojnar, M. (2015). People with alcohol use disorders in specialized care in eight different European countries. *Alcohol and Alcoholism, 50*, 310–318.

Rehm, J., Mathers, C., Popova, S., Thavorncharoensap, M., Teerawattananon, Y., & Patra, J. (2009). Global burden of disease and injury and economic cost attributable to alcohol use and alcohol-use disorders. *Lancet, 373*, 2223–2233.

Rodd-Henricks, Z. A., Melendez, R. I., Zaffaroni, A., Goldstein, A., McBride, W. J., & Li, T. K. (2002). The reinforcing effects of acetaldehyde in the posterior ventral tegmental area of alcohol-preferring rats. *Pharmacology, Biochemistry, and Behavior, 72*, 55–64.

Rubin, L. L., & Staddon, J. M. (1999). The cell biology of the blood–brain barrier. *Annual Review of Neuroscience, 22*, 11–28.

Sanchis, C., & Aragón, C. M. G. (2007). ¿Qué bebemos cuando bebemos? El papel del acetaldehído en el consumo de alcohol. *Adicciones, 19*, 5–11.

Schmitt, G., Aderjan, R., Keller, T., & Wu, M. (1995). Ethyl glucuronide: An unusual ethanol metabolite in humans: Synthesis, analytical data, and determination in serum and urine. *Journal of Analytical Toxicology, 19*, 91–94.

Schuckit, M. A. (2009). Alcohol-use disorders. *Lancet, 373*, 492–501.

Sergent, O., Griffon, B., Cillard, P., & Cillard, J. (2001). Alcohol and oxidative stress. *Pathologie Biologie (Paris), 49*, 689–695.

Silberman, Y., Bajo, M., Chappell, A. M., Christian, D. T., Cruz, M., Diaz, M. R., ... Weiner, J. L. (2009). Neurobiological mechanisms contributing to alcohol-stress-anxiety interactions. *Alcohol, 43*, 509–519.

Skala, K., & Walter, H. (2013). Adolescence and Alcohol: A review of the literature. *Neuropsychiatrie, 27*, 202–211.

Squire, L. R., & Cohen, N. J. (1984). Human memory and amnesia. In G. Lynch, J. L. McGaugh, & N. M. Weinberger (Eds.), *Neurobiology of learning and memory* (pp. 3–64). New York: Guilford Press.

Stevenson, J. R., Schroeder, J. P., Nixon, K., Besheer, J., Crews, F. T., & Hodge, C. W. (2009). Abstinence following alcohol drinking produces depression-like behavior and reduced hippocampal neurogenesis in mice. *Neuropsychopharmacology, 34*, 1209–1222.

Sullivan, E. V. (2003). Compromised pontocerebellar and cerebellothalamocortical systems: Speculations on their contributions to cognitive and motor impairment in nonamnesic alcoholism. *Alcoholism, Clinical and Experimental Research, 27*, 1409–1419.

Sullivan, L. E., Fiellin, D. A., & O'Connor, P. G. (2005). The prevalence and impact of alcohol problems in major depression: A systematic review. *The American Journal of Medicine, 118*, 330–341.

Sullivan, E. V., & Pfefferbaum, A. (2005). Neurocircuitry in alcoholism: A substrate of disruption and repair. *Psychopharmacology, 180*, 583–594.

Teixeira, F. B., Santana, L. N. S., Bezerra, F. R., De Carvalho, S., Fontes-Júnior, E. A., Prediger, R. D., ... Lima, R. R. (2014). Chronic ethanol exposure during adolescence in rats induces motor impairments and cerebral cortex damage associated with oxidative stress. *PloS One, 26*, e101074.

Van Skike, C. E., Botta, P., Chin, V. S., Tokunaga, S., McDaniel, J. M., Venard, J., ... Matthews, D. B. (2010). Behavioral effects of ethanol in cerebellum are age dependent: Potential system and molecular mechanisms. *Alcoholism, Clinical and Experimental Research, 34*, 2070–2080.

Vengeliene, V., Bilbao, A., Molander, A., & Spanagel, R. (2008). Neuropharmacology of alcohol addiction. *British Journal of Pharmacology, 154*, 299–315.

Volkow, N. D., Wang, G. J., Telang, F., Fowler, J. S., Logan, J., Jayne, M., ... Wong, C. (2007). Profound decreases in dopamine release in striatum in detoxified alcoholics: Possible orbitofrontal involvement. *Journal of Neuroscience, 27*, 12700–12706.

Vonghia, L., Leggio, L., Ferrulli, A., Bertini, M., Gasbarrini, G., & Addolorato, G. (2008). Acute alcohol intoxication. *European Journal of Internal Medicine, 19*, 561–567.

Ward, R. J., Colivicchi, M. A., Allen, R., Schol, F., Lallemand, F., de Witte, P., ... Dexter, D. (2009). Neuro-inflammation induced in the hippocampus of 'binge drinking' rats may be mediated by elevated extracellular glutamate content. *Journal of Neurochemistry, 111*, 1119–1128.

Waring, P. (2005). Redox active calcium ion channels and cell death. *Archives of Biochemistry and Biophysics, 434*, 33–42.

White, A. M., Truesdale, M. C., Bae, J. G., Ahmad, S., Wilson, W. A., Best, P. J., & Swartzwelder, H. S. (2002). Differential effects of ethanol on motor coordination in adolescent and adult rats. *Pharmacology, Biochemistry, and Behavior, 73*, 673–677.

WHO, World Health Organization. (2010). *Global strategy to reduce the harmful use of alcohol 2010*. Geneva http://www.who.int/substance_abuse/msbalcstragegy.pdf.

WHO, World Health Organization. (2014). *Global status report on alcohol and health*. Geneva http://apps.who.int/iris/bitstream/10665/112736/1/9789240692763_eng.pdf.

Willis, M. S., Thiele, G. M., Tuma, D. J., & Klassen, L. W. (2003). T cell proliferative responses to malondialdehyde-acetaldehyde haptenated protein are scavenger receptor mediated. *International Immunopharmacology, 3*, 1381–1399.

Witt, E. D. (2010). .Research on alcohol and adolescent brain development: Opportunities and future directions. *Alcohol, 44*, 119–124.

Wu, D., & Cederbaum, A. I. (2003). Alcohol, oxidative stress, and free radical damage. *Alcohol Research and Health, 27*, 277–284.

Yang, J. Y., Xue, X., Tian, H., Wang, X. X., Dong, Y. X., Wang, F., ... Wu, C. F. (2014). Role of microglia in ethanol-induced neurodegenerative disease: Pathological and behavioral dysfunction at different developmental stages. *Pharmacology & Therapeutics, 144*, 321–337.

Young, C., Straiko, M. M., Johnson, S. A., Creeley, C., & Olney, J. W. (2008). Ethanol causes and lithium prevents neuroapoptosis and suppression of pERK in the infant mouse brain. *Neurobiology of Disease, 31*, 355–360.

Zakhari, S. (2006). Overview: How is alcohol metabolized by the body? *Alcohol Research and Health, 29*, 245–254.

Zeigler, D. W., Wang, C. C., Yoast, R. A., Dickinson, B. D., McCaffree, M. A., Robinowitz, C. B., & Sterling, M. L. (2005). The neurocognitive effects of alcohol on adolescents and college students. *Preventive Medicine, 40*, 23–32.

Zhao, Y. N., Wang, F., Fan, Y. X., Ping, G. F., Yang, J. Y., & Wu, C. F. (2013). Activated microglia are implicated in cognitive deficits, neuronal death, and successful recovery following intermittent ethanol exposure. *Behavioural Brain Research, 236*, 270–282.

Zhuo, M. (2009). Plasticity of NMDA receptor NR2B subunit in memory and chronic pain. *Molecular Brain, 2*, 2–4.

Zimatkin, S. M. (1991). Histochemical study of aldehyde dehydrogenase in the rat CNS. *Journal of Neurochemistry, 56*, 1–11.

Zimatkin, S. M., & Deitrich, R. A. (1997). Ethanol metabolism in the brain. *Addiction Biology, 2*, 387–400.

Zimatkin, S. M., Pronko, S. P., Vasiliou, V., Gonzalez, F. J., & Deitrich, R. A. (2006). Enzymatic mechanisms of ethanol oxidation in the brain. *Alcoholism, Clinical and Experimental Research, 30*, 1500–1505.

CHAPTER

21

Functionally Relevant Brain Alterations in Polysubstance Users: Differences to Monosubstance Users, Study Challenges, and Implications for Treatment

D.J. Meyerhoff

University of California San Francisco, San Francisco, CA, United States

INTRODUCTION

The realization that the use of multiple psychoactive substances by the same individual is a common way of abusing substances is not new. Survey data during the early 1990s report on the rising rates of polysubstance use (PSU) in the general population (see review in Rounsaville, Petry, & Carroll, 2003), and multiple substance abuse in clinical treatment trials was discussed already in 1991 (Kosten, 1991). PSU in its disordered form (polysubstance use disorder, PSUD) is defined as the simultaneous or concurrent use of more than one addictive substance. Common drug combinations (or "typologies" of PSU) are alcohol and tobacco products with and without marijuana (cannabis) use, different amphetamines with and without cigarette use, the comorbid use of alcohol and stimulants such as cocaine or amphetamines with and without cigarettes and marijuana consumption, or the co-consumption of heroin and cocaine ("speedballing") (e.g., Quek et al., 2013). Depending on the reasons for which these substances are combined, they are either consumed simultaneously (i.e., co-ingested at the same time) or concurrently over a given time period (but not necessarily at the same time, as they may be taken sequentially). Substances are often used to self-medicate negative mood states or negative affect, and are frequently combined either to enhance another's psychoactive properties or to counterbalance another's real or perceived actions. PSUD is a pervasive health problem worldwide that constitutes an enormous individual and societal burden, but has no consistently effective pharmacological or behavioral treatment (see, e.g., Amato et al., 2011). An important reason for this lack of beneficial treatment may be that the complex biological or functional correlates of PSU on the brain have rarely been studied explicitly, so that any specific effects of certain typologies of PSU are unknown; this has made it extremely difficult to devise scientifically grounded and evidence-based treatment targeted at PSU or even specific typologies of PSU.

By contrast, the use of individual substances (monosubstance use) and their effects on the brain have been studied extensively, and national research institutes have been formed in the United States based on the most common types of substances abused. Many decades of research into the effects and correlates of monosubstance use on the animal and human brain has taught us what we know today about addiction. With the advent of brain imaging methodology, imaging of brain structure, metabolism, and function has become commonplace, and studies have related quantity and frequency of specific substance use to results of quantitative neuroimaging and neurocognitive assessments with varying degrees of success. Most extensively studied have been alcohol and stimulants and, more recently, also tobacco and marijuana, and this research has critically informed the development of pharmacological and behavioral treatment of monosubstance use disorders.

It is acknowledged that the combined use of several substances is difficult to study properly with the exact same scientific methods and rigor as the use of a single

Addictive Substances and Neurological Disease
http://dx.doi.org/10.1016/B978-0-12-805373-7.00021-9

substance; the corresponding issues and challenges have been reviewed (e.g., Connor, Gullo, White, & Kelly, 2014). They include but are not limited to diagnostic challenges, the need for additional control groups, the complexities of outcome measurements, data interpretation, and of the PSU population itself. For these reasons, relatively few peer-reviewed research reports have focused specifically on PSU, so that little is known about the neuroadaptations and cognitive deficits related to concurrent chronic misuse of illicit and licit psychoactive substances, their changes with abstinence, and their influence on the ability of individuals with PSU to engage in adaptive, pro-health behavior. However, studying the consequences and behavioral correlates of one of the most common forms of substance use cannot be neglected any longer if we strive to design evidence-based treatment methods for an ever-increasing number of substance users who urgently need effective treatment (Hohmann & Shear, 2002). We cannot simply assume that the brain and its function in polysubstance users are affected similarly to the brain and function in monosubstance users, just perhaps simply amplified because of the greater cumulative substance use. In other words, generalizing brain research and treatment findings from studies of alcoholics or stimulant users to polysubstance users does not do justice to the real needs of the complex PSU population. Emerging evidence demonstrates that PSUD affects the human brain in different ways than monosubstance use disorders. We dare say that unless we know more about the specific neurobiological and functional correlates of common PSU typologies, we cannot intelligently inform more effective treatment of such disorders.

The inclusion in clinical research trials of individuals who abuse several drugs has previously been argued (Kenna, Nielsen, Mello, Schiesl, & Swift, 2007; Rounsaville et al., 2003) and counterargued (O'Brien & Lynch, 2003), but it appears that surprisingly little research has specifically included individuals with comorbid substance use disorders (SUDs) in neuroimaging or neuropsychological research, let alone substance use treatment trials (but see Kenna et al., 2007; Strain, 2003). Therefore, there are neither FDA-approved treatments nor treatment guidelines for treatment-seeking individuals with PSU, and effective interventions are urgently needed (e.g., Kedia, Sell, & Relyea, 2007; Verdejo-Garcia, Lopez-Torrecillas, Gimenez, & Perez-Garcia, 2004; Yucel & Lubman, 2007). However, the complexity of PSU (e.g., the different typologies of PSU and their common psychiatric comorbidities) makes studying, understanding, and treating PSU a formidable challenge (see, e.g., Connor et al., 2014; Strain, 2003). This challenge, though, cannot be avoided if we are true to our promises as medical professionals to provide better and more effective treatment for this exceedingly common pattern/form of chronic substance abuse.

In the following, we review the prevalence of PSU in the United States and Europe and briefly survey the extant literature of some monosubstance use effects on the human brain, as revealed by neuropsychological and magnetic resonance (MR)-based neuroimaging studies; we then continue to review in more detail neuropsychological and neuroimaging research that specifically focuses on polysubstance users vis-a-vis healthy control populations and typical monosubstance users. This research provides evidence that PSU affects the human brain in different ways than single drug use; these differential brain effects are likely related to the worse health consequences generally observed in poly- than in monosubstance users, and they warrant development of different targeted treatment approaches. The reader may note that this chapter does not cover the medical and nonmedical use of prescription drugs and their common substance use comorbidities, and that the review of the human neuroimaging studies in this chapter do not cover the vast and ever-expanding field of functional magnetic resonance imaging (fMRI) studies.

PREVALENCE OF PSU

Data from the National Epidemiologic Survey on Alcohol and Related Conditions (NESARC), which has surveyed probability samples of noninstitutionalized US adults 18 years of age or older, indicate that the proportion of individuals with comorbid substance use over those with alcohol use disorders (AUDs) has increased since mid-1990s. According to the 2014 National Survey in the United States (SAMSHA http://www.samhsa.gov/disorders/substance-use), more people aged 12 or older were current illicit drug users in 2014 than anytime between 2002 and 2013. Of the about 21.5 million people aged 12 or older with an SUD in the past year, 17.0 million had an AUD, 7.1 million an illicit drug use disorder, and 2.6 million had both an AUD and an illicit drug use disorder (i.e., they were polysubstance users). Although the percentage of people aged 12 or older with an SUD decreased between 2002 and 2014, absolute numbers increased. The rise in illicit drug use from before 2009 to 2014 may primarily reflect increases by adults aged 26 or older and secondarily increases in illicit drug use among young adults (18–25 years). Reasons for increases nationally and worldwide of both illicit and licit drug use are likely the greater availability of any drugs, including nonmedical use of prescription drugs and tobacco and, particularly in the developing countries, relatively more disposable income and low education.

Among the young, specific personality traits may contribute to greater willingness to experiment with new substances, whereas among the older adults (and adolescent and young adults with major depressive disorder) substances are commonly abused to deal with negative affect. Worldwide prevalence and current patterns of and reasons for comorbid substance abuse have been discussed in detail (Connor et al., 2014).

With the abuse of a single substance becoming less common (except perhaps among those initiating drug use), substance use patterns have also changed among treatment seeking populations. In the United States, within the past 10 years, at least half the individuals under substance abuse treatment are polysubstance users, who abuse two, three, or more substances simultaneously and/or concurrently (Hanlon, Beveridge, & Porrino, 2013; Medina, Shear, Schafer, Armstrong, & Dyer, 2004; national admissions data at www.samhsa.gov/data/DASIS/teds05/TEDSAd2k5web.pdf); 56% of treatment seekers in 2012 sought help for an addiction to more than one substance (National Center on Addiction and Substance Abuse, Columbia University). In Europe, 62% of cocaine and 85% of cannabis users seeking treatment reported PSU (European Monitoring Centre for Drugs and Drug Addiction, EMCDDA, Annual Report 2009). Further, cigarette smoking is at least three times as prevalent among polysubstance users as in the general US population (Patkar et al., 2006; Stark & Campbell, 1993), with higher smoking rates in polysubstance users than in monosubstance users (Richter, Ahluwalia, Mosier, Nazir, & Ahluwalia, 2002). Not surprisingly, compared to monosubstance users, polysubstance users are at an increased risk for greater toxicity from cumulative substance use and/or potential interactions between abused substances, higher risks of overdose, and major chronic health problems. Alcohol abusers who also used illicit drugs concurrently over the past year had greater drinking severity and comorbid psychopathology, such as major depression, anxiety, and personality disorders (Connor et al., 2013, 2014; Moss, Goldstein, Chen, & Yi, 2015; Schmidt, Pennington, Cardoos, Durazzo, & Meyerhoff, 2017). The greater prevalence of these comorbid conditions among polysubstance users is also the likely reason for the even higher prevalence of PSU among treatment seekers compared to the general adult population. Correspondingly, within 1 year of treatment, 40–60% of patients treated for PSU relapse (McLellan, Lewis, O'Brien, & Kleber, 2000), a rate considered to be higher than for most monosubstance users (except tobacco). Yet, clinical assessment, diagnosis, and treatment of current substance users focus on individual drug-specific SUDs rather than comorbid SUDs found in polysubstance users.

OVERVIEW OF COGNITION IN MONOSUBSTANCE USERS

In order to understand the effects of substance abuse on brain biology and neurocognition and to contribute evidence-based information for better treatment approaches for individual SUDs, many years of clinical research has focused on the abuse of common individual substances such as alcohol or psychostimulants. For the vast majority of these studies, additional (comorbid) drug use was either excluded from study, treated as a nuisance variable and/or simply acknowledged as a study limitation. In brief, this research has demonstrated that monoabuse of or dependence on alcohol, cocaine, methamphetamine, or cannabis is associated with cognitive dysfunction (Hester & Garavan, 2004; Kalechstein, Newton, & Green, 2003; Kubler, Murphy, & Garavan, 2005; Lundqvist, 2005; McKetin & Mattick, 1997; Oscar-Berman, 1997, 2000; Salo et al., 2002; Simon, Dacey, Glynn, Rawson, & Ling, 2004; Solowij et al., 2002; Verdejo-Garcia et al., 2012, 2004), primarily in learning/memory (Barber, Panikkar, & McKeith, 2001; Lundqvist, 2005; Oscar-Berman, 2000; Pope, Gruber, Hudson, Huestis, & Yurgelun-Todd, 2001; Pope & Yurgelun-Todd, 1996; Simon et al., 2000; Solowij et al., 2002; Volkow, Chang, Wang, Fowler, Franceschi, et al., 2001; Volkow, Chang, Wang, Fowler, Leonido-Yee, et al., 2001), working memory (Lundqvist, 2005; Oscar-Berman, 2000), and executive skills (Barber et al., 2001; Hermann et al., 2007; Hester & Garavan, 2004; Kim et al., 2005; Moeller et al., 2005; Oscar-Berman, 2000; Pope et al., 2001; Pope & Yurgelun-Todd, 1996; Salo et al., 2002; Simon et al., 2000; Solowij et al., 2002). Recent reviews of AUD patients describe deficits related to working memory, visuospatial functions, inhibition, and executive-based functions such as mental flexibility, problem solving, divided attention (Bernardin, Maheut-Bosser, & Paille, 2014), and cognitive control (Wilcox, Dekonenko, Mayer, Bogenschutz, & Turner, 2014). Deficits in inhibitory control are greater in actively drinking alcoholics than controls (Moody, Franck, Hatz, & Bickel, 2016; Vuchinich & Simpson, 1998) and they predict relapse to drinking (Rupp et al., 2016). In addition, chronic cigarette smoking, ubiquitous among substance users, relates to cognitive deficits in both nonclinical a1nd clinical samples (e.g., Almeida et al., 2008; Brody, 2006; Durazzo, Insel, Weiner, & The Alzheimer's Disease Neuroimaging Initiative A, 2012; Durazzo & Meyerhoff, 2007; Durazzo, Mon, Gazdzinski, & Meyerhoff, 2011; Durazzo, Rothlind, Gazdzinski, Banys, & Meyerhoff, 2006; Gazdzinski et al., 2006; Glass et al., 2006; Morales, Lee, Hellemann, O'Neill, & London, 2012; Pennington et al., 2013). Greater smoking severity in AUD patients predicted worse executive function (Glass et al., 2009), and smoking alcoholics performed worse than nonsmoking alcoholics on domains of auditory–verbal learning and memory, processing speed, cognitive efficiency, and working memory at 1 week and 4 weeks of abstinence (Durazzo, Rothlind et al., 2006; Pennington et al., 2013). Fernandez-Serrano, Perez-Garcia, and Verdejo-Garcia (2011) in an impressively extensive review outlined specific versus generalized effects of drugs of abuse, indicating that all substances studied are commonly associated with neuropsychological deficits in episodic memory, emotional processing, and aspects of executive functions. However, specific substances appear to be relatively robustly related to specific cognitive and self-regulatory dysfunctions [e.g., alcohol and stimulants to impulsivity and cognitive flexibility, alcohol and MDMA (3,4-methylenedioxymethamphetamine, or ecstasy) to spatial processing, perceptual speed, and selective attention, and cannabis and methamphetamine to prospective memory deficits].

There are some data on the changes of cognitive function in substance users during abstinence. However, as retention in treatment is a common problem in substance use research studies (primarily of psychostimulant users), this body of data is not very large. In AUD, neurocognitive functions recover at least partially during sustained abstinence (for review see Bernardin et al., 2014). We and others showed substantial cognitive recovery with abstinence from alcohol over 1 and 6–9 months of abstinence that was hindered by comorbid cigarette smoking (Durazzo, Gazdzinski, & Meyerhoff, 2007; Durazzo, Meyerhoff, & Nixon, 2010; Pennington et al., 2013). Specifically, smoking was shown to significantly hinder recovery of visuospatial learning and processing speed in AUD (Durazzo, Pennington, Schmidt, & Meyerhoff, 2014; Pennington et al., 2013). Cognitive performance was largely unchanged over 1 month in abstinent methamphetamine patients (Simon, Dean, Cordova, Monterosso, & London, 2010), but cognition improved to some degree over 1 year of abstinence (Iudicello et al., 2010). Of the very few longitudinal studies in adult monodrug abusers (Fernandez-Serrano et al., 2011), none of them used task-based measures of self-regulation.

OVERVIEW OF NEUROIMAGING IN MONOSUBSTANCE USERS

MR-based neuroimaging studies of individuals chronically consuming licit or illicit substances (alcohol, cocaine, amphetamines, marijuana, or tobacco), that is, monosubstance users, revealed correlates of cognitive dysfunction including regional brain atrophy (Adalsteinsson, Sullivan, Mayer, & Pfefferbaum, 2006; Bae et al., 2006; Bartzokis et al., 2000; Block et al., 2000; Durazzo, Tosun et al., 2011; Franklin et al., 2002; Goldstein & Volkow, 2011; Liu, Matochik, Cadet, & London, 1998; Matochik, Eldreth, Cadet, & Bolla, 2005; Matochik, London, Eldreth, Cadet, & Bolla, 2003; Pfefferbaum, Adalsteinsson, & Sullivan, 2006; Pfefferbaum & Sullivan, 2002; Rando et al., 2011; Sullivan, 2000; Sullivan & Pfefferbaum, 2005; Wang, Xu, Qian, Shen, & Zhang, 2015; Wrase et al., 2008; Xiao et al., 2015). Brain atrophy could be driven by loss or damage to cells, dendrites, and/or synapses or by less functionally relevant changes to interstitial space or relative water distributions. Other neuroimaging methods help assess functional changes or neurobiological mechanisms underlying such structural change in monosubstance users. They include low cerebral blood flow (Chang et al., 2002; Ernst, Chang, Leonido-Yee, & Speck, 2000; Ernst, Chang, Oropilla, Gustavson, & Speck, 2000; Gansler et al., 2000; Heinz, Beck, Grusser, Grace, & Wrase, 2009; Hwang et al., 2006; Nicolas et al., 1993; Oishi, Mochizuki, & Shikata, 1999; Rogers, Meyer, Shaw, & Mortel, 1983; Tunving, Thulin, Risberg, & Warkentin, 1986; Volkow, Mullani, Gould, Adler, & Krajewski, 1988) and altered regional brain glucose metabolism (Eldreth, Matochik, Cadet, & Bolla, 2004; Kim et al., 2005; Volkow, Chang, Wang, Fowler, Franceschi, et al., 2001; Volkow, Chang, Wang, Fowler, Leonido-Yee, et al., 2001; Volkow et al., 1993; Volkow, Hitzemann, Wang, Fowler, Burr, et al., 1992; Volkow, Hitzemann, Wang, Fowler, Wolf, et al., 1992; Volkow et al., 1988, 1994). Furthermore, proton MR spectroscopy (MRS) revealed metabolic abnormalities in chronic substance users that are consistent with abnormal markers of neuronal integrity (via lower concentrations of regional N-acetylaspartate, NAA), cell membrane turnover/synthesis (via altered levels of choline-containing metabolites, Cho), and gliosis (via altered levels of myo-inositol, mI) (Durazzo, Gazdzinski, Banys, & Meyerhoff, 2004; Durazzo, Gazdzinski, Yeh, & Meyerhoff, 2008; Durazzo, Pathak, Gazdzinski, Mon, & Meyerhoff, 2010; Ernst, Chang, Leonido-Yee, et al., 2000; Ernst, Chang, Oropilla, et al., 2000; Fein, Meyerhoff, & Weiner, 1995; Hermann et al., 2007; Ke et al., 2004; Meyerhoff et al., 2004; Meyerhoff, MacKay, Weiner, & Fein, 1994; Nordahl et al., 2005; Parks et al., 2002; Sailasuta, Abulseoud, Hernandez, Haghani, & Ross, 2010) [reviewed by (Meyerhoff, Durazzo, & Ende, 2013)]. These studies have indicated that different substances similarly alter neuronal integrity, energy metabolism, membrane turnover, neurotransmission (glutamate, GABA), and inflammatory processes (Licata & Renshaw, 2010; Yang et al., 2009), albeit via different neural pathways. Nevertheless, neuroimaging of abusers of different individual substances describes alterations in largely similar brain regions, primarily in the prefrontal cortex (PFC), the underlying white matter, basal ganglia, and thalami (Bae et al., 2006;

Dom, Sabbe, Hulstijn, & van den Brink, 2005; Eldreth et al., 2004; Ernst, Chang, Leonido-Yee, et al., 2000; Ernst, Chang, Oropilla, et al., 2000; Franklin et al., 2002; Hermann et al., 2007; Holman et al., 1991; Matochik et al., 2005, 2003; Nordahl et al., 2005; Sorg et al., 2012; Sullivan, 2000; Sullivan, Rosenbloom, & Pfefferbaum, 2000; Thompson et al., 2004; Volkow, Hitzemann, Wang, Fowler, Burr, et al., 1992; Volkow, Hitzemann, Wang, Fowler, Wolf, et al., 1992; Wang et al., 2015; Weber et al., 1993), all regions critically involved in cognitive control, drug-seeking behavior, and reward.

Longitudinal neuroimaging studies of treatment-seeking monosubstance users during abstinence from an illicit drug are rare, likely because of poor subject retention (e.g., Hanlon et al., 2013; Mackey & Paulus, 2013; Morales et al., 2012; Salo & Fassbender, 2012). Therefore, it is unclear to what degree the observed structural and metabolic brain injuries in illicit drug users are reversible with abstinence. Using longitudinal proton MRS, a reduced glutamate level in anterior cingulate cortex (ACC) of methamphetamine abusers at <1 month of abstinence (which was related to drug craving) was shown to recover over 5 months of abstinence (Ernst & Chang, 2008). However, prefrontal NAA and Cho in another study of patients with intravenous methamphetamine dependence did not change significantly over just 1 month of abstinence (Yoon et al., 2010), whereas gray matter volume did recover in similar individuals (Morales et al., 2012). Longitudinal neuroimaging studies of abstinent AUD patients have provided the greatest evidence for beneficial neuroplasticity during sobriety (see recent reviews by Buhler & Mann, 2011; Durazzo, Mon, Gazdzinski, Yeh, & Meyerhoff, 2015; Meyerhoff, 2014; Zahr, 2014). This cellular and subcellular neuroplasticity during abstinence that leads to measurable volume changes is thought to represent the reversal of processes associated with atrophy, such as (re)growth of cells (mostly glia, but also neurons in some brain regions), dendrites, and synapses (e.g., Anderson, 2011).

REVIEW OF COGNITION IN POLYSUBSTANCE USERS

Cognition in Polysubstance Users Relative to Drug-Free Controls

In polysubstance users, the degree of clinically relevant cognitive deficits has been debated. We found cognitive deficits in both cocaine-dependent and cocaine-plus-alcohol-dependent individuals at 3 months of abstinence (Di Sclafani et al., 1998). Decision making was still impaired in polysubstance users abstinent for 8 months (Verdejo-Garcia, Rivas-Perez, Vilar-Lopez, & Perez-Garcia, 2007), and a review suggests that psychostimulant-related deficits on episodic memory, planning, and cognitive flexibility in polysubstance users can persist for several years of abstinence (Fernandez-Serrano et al., 2011). The consequences of persistent cognitive deficits on relapse risk have been discussed (Verdejo-Garcia et al., 2012, 2007). Lower executive function in polysubstance users is related to the amount of cocaine and cannabis consumed (Fernandez-Serrano, Perez-Garcia, Perales, & Verdejo-Garcia, 2010; Fernandez-Serrano, Perez-Garcia, Schmidt Rio-Valle, & Verdejo-Garcia, 2010). Compared to nonsmoking, drug-free controls, polysubstance users had significantly greater age-related losses in processing speed, general intelligence, global cognition, and cognitive efficiency, indicating detrimental cumulative effects of the abuse of substances on cognition over lifetime (Meyerhoff et al., unpublished).

Cognition in PSU as a Function of Smoking Status

One common factor across different SUDs and within PSU populations is chronic cigarette smoking, which is particularly common in substance abusers (see Prevalence of PSU section). An ever-increasing number of studies since the early 2000s have indicated measurable injury to brain function and biology from chronic use of tobacco products. Chronic cigarette smoking has significant effects on cognition in healthy controls (Durazzo, Meyerhoff, & Nixon 2011), in samples of treatment-seeking alcoholics (Durazzo, Fryer et al., 2010; Durazzo, Rothlind et al., 2006), and in nontreatment-seeking hazardous drinkers (Durazzo, Rothlind et al., 2007; Durazzo, Rothlind, Weiner, & Meyerhoff, 2005). Typically, smoking individuals showed worse visuospatial memory and greater motor impulsivity than their nonsmoking counterparts. In these studies, the degree of functional deficits varies considerably among substance users investigated, ranging from weak if any deficits to clinically significant cognitive impairment defined as greater than two times the standard deviation of a healthy control sample. But even modest neuropsychological deficits in substance users can impact quality-of-life and relapse risk (Goldstein et al., 2004). Demonstrating functionally significant and clinically relevant decrements of cognition in smoking versus nonsmoking substance users warrants inclusion of smoking cessation in substance abuse treatment, will further motivate smoking cessation, increase treatment seeking for all forms of substance use, and inform targeted behavioral and pharmacological therapy for comorbid tobacco and other substance use disorders. Smoking is a modifiable health risk with greater annual mortality than SUD and AUD combined, and we cannot afford to ignore it further in substance abuse research and treatment.

Tobacco use and potentially other, newer forms of nicotine delivery ought to be considered common substance use comorbidities in PSU similar to abuse of alcohol and stimulants.

Inhibitory Control in PSU

In addition to assessing traditional domains of cognitive function, greater sensitivity to neuropsychological deficits may be attained through tasks that target specific cognitive/emotional dysfunction (Goldstein et al., 2004) and simulate real-life decision making, risk taking, and impulsivity. These measures of inhibitory control in PSU are clinically relevant, as they relate to both substance seeking and the ability to achieve and maintain abstinence: high impulsivity and novelty/sensation seeking increase risks for substance abuse (Conway, Swendsen, Rounsaville, & Merikangas, 2002) and poor treatment outcomes (Loree, Lundahl, & Ledgerwood, 2015); impulsive and compulsive substance seeking relates to dysregulated consumption in the face of repeated adverse biopsychosocial consequences, and it is associated with deficient attention, information processing, self-monitoring, problem solving/reasoning, decision making, inhibition, and anticipation of consequences of current behavior (George & Koob, 2010; Heatherton & Wagner, 2011). Deficits in self-regulation or inhibitory control should be targeted in substance abuse treatment approaches.

In children with disruptive behavior disorders and healthy controls, aspects of self-regulation are subserved by "top-down" prefrontal cortical regions (Goto et al., 2010; Hill et al., 2009; Wright, Feczko, Dickerson, & Williams, 2007). Similarly, neurobiological injury in fronto-striatal brain regions of polysubstance users (e.g., George & Koob, 2010; Heatherton & Wagner, 2011; Muller, Weijers, Boning, & Wiesbeck, 2008; Verdejo-Garcia et al., 2012; Volkow, Wang, Fowler, Tomasi, & Telang, 2011) is also likely related to deficits in self-regulation (e.g., Crews & Boettiger, 2009; Durazzo & Meyerhoff, 2007; Kalivas, 2009; Volkow et al., 2011). In alcohol and other substance abusers, these specific inhibitory control deficits have been linked to abnormal neural activation and morphometry in fronto-striatal brain systems including the PFC (Connolly, Foxe, Nierenberg, Shpaner, & Garavan, 2012; Ersche et al., 2012). Given the relative paucity of studies in this area, it is unclear if these neuropsychological phenotypes are a preexisting vulnerability to high impulsivity and sensation seeking that increase the risk for substance abuse (Conway et al., 2002) or a consequence of abuse (Ersche et al., 2012; Verdejo-Garcia, Lawrence, & Clark, 2008); if the latter, neuropsychological deficits may (at least partially) recover with sustained abstinence.

Improving our understanding of different aspects of inhibitory control deficits in polysubstance users is key to better treatment. Current treatments such as cognitive behavioral treatment (CBT) of cognitive control over craving and cognitive enhancement are available that work through enhancing PFC/executive function/inhibition/self-control (Sofuoglu, DeVito, Waters, & Carroll, 2013; Vocci, 2008; Xu et al., 2010), and perhaps also decrease impulsivity, even before the development of addictive use (Volkow & Baler, 2012). This review supports the view that task-based measures of decision making, risk taking, and impulsivity should augment traditional tests of executive skills, processing speed, and learning and memory in research studies of neuropsychology and behavior; they increase the clinical specificity and sensitivity of cognitive assessments (Potenza & de Wit, 2010; Stephens et al., 2010), and—when linked to relevant neurobiology—these self-regulatory/inhibitory measures can yield specific and urgently needed targets for both behavioral and pharmacological treatment of PSUD.

To sum up this section, some deficits relative to healthy controls are reported in nearly all neuropsychological studies of substance users, including PSU individuals; this relates to both the fraction of substance users showing deficits and the number of cognitive domains affected. Intact neurocognition and inhibitory control are critical for addiction treatment efficacy, retention (Aharonovich et al., 2006; Passetti, Clark, Mehta, Joyce, & King, 2008; Streeter et al., 2008), and maintenance of abstinence during treatment (de Wit, 2009; Rupp et al., 2016) and as such constitute an important target for substance abuse treatment.

Cognition in Polysubstance Users Relative to Monosubstance Users

A 2015 review suggests that improved patient care may depend on the better characterization of neuropsychological phenotypes associated with the chronic abuse of different drugs, as only this will inform the development of better treatment interventions with the goal of improving cognitive functions in subgroups of users (Cadet & Bisagno, 2015). However, science-based evidence for differential brain effects that would inform such targeted treatment is lacking. Only few investigators have tried to specifically contrast differences of neuropsychological performance between different substance using groups, or—specifically for the purpose of this chapter—differences between polysubstance users and monosubstance users (Fernandez-Serrano et al., 2011; Schulte et al., 2014). Some investigators reported on a wider range of and more severe and long-lasting deficits in polysubstance users compared to monosubstance users, and many on different aspects of memory (Horner, 1997; Selby & Azrin, 1998;

TABLE 21.1 Demographic, Clinical, Behavioral, and Substance Use Variables for PSU and AUD Groups Contributing to the Author's Recent PSU Research (Mean ± Standard Deviation)

Group	PSU	AUD
Number of participants (female)	36 (2)*	69 (17)*
Age at baseline (years)	46.3 ± 10.1	47.8 ± 10.8
Education (years)	12.6 ± 1.3*	14.5 ± 2.3*
Cigarette smoker (%)	69	57
Abstinent from any drugs (days)	≥30 ± 8	≥29 ± 10
Race/ethnicity (%)		
African–American	58*	10*
Asian	0	1
Caucasian	25*	71*
Latino	11	13
Native American/Aleutian	3	4
Polynesian/Pacific Islander	3	0
Other	0	1
Hypertension (%)	19	28
Hepatitis C positive (%)	19*	6*
Being on prescribed psychoactive medication (%)	25*	62*
Mood disorder (%)	19	25
Substance-induced mood disorder (%)	8	9
Family history of alcohol problems (%)	83	86
American National Adult Reading Test, AMNART	105.3 ± 9.1*	115.3 ± 9.1*
Beck Depression Inventory, BDI	11.2 ± 7.9	12.3 ± 8.3
State–Trait Anxiety Inventory, STAI: State	34.6 ± 11.0	35.2 ± 11.3
State–Trait Anxiety Inventory, STAI: Trait	34.8 ± 11.0	35.0 ± 11.2
Barratt Impulsivity Scale (BIS-11) total impulsivity	68.2 ± 10.5*	64.5 ± 10.3*
BIS-11 nonplanning (cognitive control)	27.6 ± 4.1*	25.4 ± 5.0*
BIS-11 attentional	16.8 ± 4.2	16.6 ± 4.3
BIS-11 motor	23.8 ± 4.7	22.6 ± 3.8
Fagerstrom tolerance test for nicotine dependence (FTND) total	4.4 ± 1.4	4.0 ± 1.6
FTND lifetime highest	15.6 ± 6.5	16.7 ± 8.1
FTND daily smoking (years)	23.2 ± 12.6	25.3 ± 10.3
1-year average drinks/month	233 ± 232	302 ± 173
Lifetime average drinks/month	211 ± 191	182 ± 108
Lifetime years drinking	31.2 ± 9.9	30.8 ± 11.1
Years, heavy drinking[a]	19.2 ± 10.7	18.5 ± 10.7
Onset, heavy drinking[a] (age)	22.1 ± 8.9	25.5 ± 8.8
Years, cocaine use ($n = 27$)	18.4 ± 10.6	n/a
Onset, cocaine use (age)	22.0 ± 7.8	n/a
Years, amphetamine use ($n = 12$)	12.8 ± 8.1	n/a
Onset, amphetamine use (age)	37.3 ± 11.8	n/a
Years, marijuana use ($n = 7$)	26.9 ± 13.4	n/a
Onset, marijuana use (age)	15.6 ± 7.5	n/a
Years, opioid use ($n = 5$)	6.3 ± 5.1	n/a
Onset, opioid use (age)	26.2 ± 13.7	n/a

[a]*Heavy drinking defined as >100 alcoholic drinks/month in men and >80 alcoholic drinks/month in women.*
* *Significantly different at* $p \leq .05$.
Schmidt, T. P., Pennington, D. L., Cardoos, S. L., Durazzo, T. C., & Meyerhoff, D. J. (2017). Neurocognition and inhibitory control in polysubstance use disorders: Comparison with alcohol use disorders and changes with abstinence. Journal of Clinical and Experimental Neuropsychology, *39*, 22–34.

Verdejo-Garcia et al., 2004, 2007), presumably from synergistic/additive neurotoxicity of concurrent or simultaneous substance use. Treatment-seeking polysubstance users perform worse than those presenting with AUD on executive functions such as verbal fluency, working memory, planning, and multitasking (Fernandez-Serrano, Perez-Garcia, Perales, et al., 2010; Fernandez-Serrano, Perez-Garcia, Schmidt Rio-Valle, et al., 2010). In earlier reports, usually polysubstance users do not uniformly perform worse on traditional cognitive tests than AUD or stimulant-only groups (e.g., Gonzalez et al., 2004; Lawton-Craddock, Nixon, & Tivis, 2003; Nixon, Paul, & Phillips, 1998).

We also compared measures of cognitive performance and inhibitory control in relatively large samples of treatment-seeking individuals with PSU and "pure" AUD, both at 1 month of abstinence (Schmidt et al., 2017). All 36 PSU patients were dependent on alcohol (just as our AUD sample) and had comorbid SUDs (cocaine 75%, amphetamines 33%, marijuana 19%, and opioids 14%). Both groups were matched well on age (PSU 46 years, AUD 48 years), lifetime drinking and smoking histories, mood scores (depressive and anxiety symptomatologies), as well as rates of mood disorders, hypertension, family histories of alcohol problems, and cigarette smoking. The PSU group started drinking alcohol regularly at age 15 (2 years earlier than the AUD group) and heavily at age 22 (4 years earlier than AUD), and both groups started smoking cigarettes daily at age 23. Polysubstance users with other SUDs started using marijuana at age 16, cocaine at age 22, opioids at age 26 and amphetamines at age 37. The PSU participants were African-American (58%), Caucasian (25%), Latino (11%), and other (6%), whereas the 69 AUD participants were primarily Caucasian (71%), followed by Latino (13%), African-American (10%), and other (6%). The high percentage of African-American in the PSU group clearly distinguishes this sample from that of most AUD samples described in the literature. Table 21.1 has a summary of the demographic, clinical, behavioral, and basic substance use measures of our treatment-seeking PSU cohort compared to those of a typical treatment-seeking AUD cohort that form the basis for the following comparisons.

After covarying for premorbid verbal intelligence (American National Adult Reading Test, AMNART, which was lower in polysubstance users than in those with AUD), polysubstance users showed numerically lower scores than those with AUD in all tested cognitive domains. Polysubstance users performed significantly worse than AUD patients on auditory–verbal memory, auditory–verbal learning, and intelligence. Polysubstance users also tended to be worse than AUD patients in decision making (Iowa Gambling Task, IGT) but not in risk taking (Balloon Analogue Risk Task, BART). Cigarette smoking status did not affect these group differences. Polysubstance users were more impulsive than those with AUD (Barratt Impulsiveness Scale 11, BIS-11, total and nonplanning impulsivity, a measure of cognitive control), and motor impulsivity was higher in smoking polysubstance users compared to both smoking and nonsmoking individuals with AUD. For both PSU and AUD individuals, worse performance on global cognition, cognitive efficiency, general intelligence, and visuospatial skills are related to more lifetime years of drinking. Within the PSU group, higher self-reported impulsivity and worse performance on executive function and fine motor skills were associated with greater cumulative lifetime substance use and/or earlier onset age of substance use, suggesting that the measured deficits are a consequence of substance use, and not necessarily premorbid.

The specific neurocognitive and inhibitory control measures that differentiate PSU and AUD individuals may provide helpful insights into the specific clinical needs of the understudied (Connor et al., 2014), yet highly prevalent population of polysubstance users in substance use treatment centers today. At the very least, demonstrating differences between these groups of substance users should make it obvious that the different groups may require more tailored treatment approaches to improve treatment outcome.

Review of Neuropsychological Changes During Abstinence From PSU

In abstinent PSU, cross-sectional studies either did (Fernandez-Serrano et al., 2011; Verdejo-Garcia et al., 2004) or did not reveal correlations of abstinence duration with cognitive dysfunction (Medina et al., 2004). Although this suggests recovery from cognitive dysfunction, these cross-sectional studies cannot determine within-subject change or causality. Very few longitudinal studies have explicitly investigated changes in neurocognition or inhibitory control in abstinent PSU individuals (Fernandez-Serrano et al., 2011). Individuals with comorbid alcohol and cocaine use disorders demonstrated significant improvements on measures of immediate memory over 6 months of abstinence (Fein, Di Sclafani, & Meyerhoff, 2002), while another study described improvements in verbal short-term memory over 3 to 4 months of abstinence from multiple substances (Block, Erwin, & Ghoneim, 2002). The additive detrimental effects of concurrent cocaine and alcohol dependence persisted over 1 month of abstinence (Bolla, Funderburk, & Cadet, 2000). Additional use of other substances in people with AUD was shown to hamper neurocognitive recovery (Schulte et al., 2014). In a series of studies, we showed that comorbid cigarette smoking hinders cognitive recovery in treatment-seeking abstinent alcoholics (Durazzo, Gazdzinski et al., 2007; Durazzo, Meyerhoff et al., 2010; Pennington et al., 2013).

We measured the change of neuropsychological functioning between 1 and 4 months of abstinence in those polysubstance users who are described in the preceding paragraph cross-sectionally (Schmidt et al., 2017). Neurocognitive functions in abstinent polysubstance users improved significantly in the domains of general intelligence, cognitive efficiency, and executive function, working memory, and visuospatial skills, while weaker improvements were observed for global cognition and processing speed. Abstinent polysubstance users did not change significantly in the domains of fine motor skills and learning and memory (both auditory–verbal and visuospatial). The lack of significant changes in these domains were related to significant time-by-smoking status interactions, in which only nonsmokers increased on fine motor skills whereas only smokers improved on visuospatial memory. Risk-taking (BART) scores increased significantly with abstinence in these polysubstance users, the decision-making (IGT) scores did not change (i.e., they remained low during abstinence), and self-reported total (including motor and nonplanning) impulsivity (BIS-11) decreased with abstinence. Thus, polysubstance users recovered quite effectively over 3 months of continued abstinence, despite decades of substance abuse and repeated treatment attempts.

It is informative to compare the successful abstainers to those polysubstance users who were either shown or were presumed to have relapsed within the 4 months of the follow-up window. The relapsed polysubstance users had more lifetime years of cocaine use than the abstainers (24 vs. 15 years), and they performed significantly worse at treatment entry than the abstainers on cognitive efficiency, processing speed, and visuospatial learning. These subgroups did not differ significantly on years of education, AMNART, tobacco use severity, and proportions of smokers or of family members with problem drinking, or the proportion of individuals taking a prescribed psychoactive medication. This analysis can inform the identification of potential markers of relapse risk. Overall, our analyses demonstrate that deficits in important cognitive domains and high self-reported impulsivity/disinhibition in polysubstance users are not premorbid and can improve with abstinence. The demonstration of significant recovery of critical brain functions in polysubstance users is motivating to treatment seekers and should help reduce stigma and modify public perception of addiction. The fact that such functional recovery occurs after decades of substance abuse, often after repeated attempts at becoming sober, suggests effective neuroplasticity in severe PSU; there appears to exist a critical window of opportunity for supporting interventions with plasticity-based cognitive remediation methods. The corresponding improvement of function (and neurobiology—see the following) in polysubstance users will likely promote long-term abstinence in these individuals, similar to demonstrations of an association between better treatment response and longitudinal cognitive recovery in AUD (Bates, Buckman, & Nguyen, 2013).

Underlying these clinically relevant neuropsychological deficits and their changes with abstinence from substances are neurobiological abnormalities and their changes during abstinence. These are reviewed in the following.

REVIEW OF NEUROIMAGING IN POLYSUBSTANCE USERS

Neuroimaging in Polysubstance Users Relative to Drug-Free Controls

Although PSU is prevalent among today's substance users and most current treatment center clients present with PSU, very little research has been designated to specifically study potentially altered neurobiology in polysubstance users. An early structural MRI study of 2-week-abstinent polysubstance users (with cocaine as substance of choice, followed by cannabis and alcohol) reported prefrontal gray matter atrophy (Liu et al., 1998), a finding also common to AUD. Another study found reduced white matter volumes only in the frontal cortex of polysubstance users compared to matched controls (Schlaepfer et al., 2006), whereas others reported gray matter volume reductions in the orbitofrontal cortex (OFC) (Tanabe et al., 2009) and cortical thinning of the insula in long-term abstinent female PSU patients (Tanabe et al., 2013). Cortical gray matter volumes were also reduced in the dorsolateral PFC, the inferior frontal gyrus, and the cerebellum of male adolescents with substance and conduct problems (Dalwani et al., 2011). Similarly, a 2016 voxel-based morphometry study reported on widespread gray matter volume reductions in polysubstance users, which were related to the duration of PSU (Noyan et al., 2016).

Only two small studies used other than structural MR methods to evaluate the brain in polysubstance users. Phosphorus MR spectroscopy (^{31}P MRS) of short-term abstinent cocaine-dependent polysubstance users compared to healthy controls showed cerebral high-energy phosphate and phospholipid metabolite abnormalities of the central white matter in the absence of significant brain atrophy (Christensen et al., 1996; MacKay, Meyerhoff, Dillon, Weiner, & Fein, 1993), suggesting central myelin and cell membrane injury associated with PSU. The characterization of aberrant neurobiology in PSUD can identify new biomarkers of brain injury mechanisms and their understudied clinical/behavioral correlates.

Neuroimaging in Polysubstance Users Relative to Monosubstance Users

Review of the Literature

Although not borne out by the limited number of neuroimaging studies that compare polysubstance users to drug-free controls, a basic perception among the general population and health care providers is that the neurobiological abnormalities associated with concurrent abuse of multiple substances are likely greater than those related to the abuse of a single substance. This is understandable if we consider injury from substance use to be cumulative or additive. However, very few formal comparisons have been made that would support such supposition. One of those studies relates to the measurement of the microstructural integrity of the brain's white matter and indicates progressively greater white matter injury, including demyelination, with the consumption of a larger number of different substances (alcohol, cocaine, and marijuana) (Kaag et al., 2016). However, most other studies that compare brain health between mono- and polysubstance users do not necessarily support more widespread or severe injury associated with cumulative substance use: the effects of PSU appear to be more complex than that, with interactions of drug effects on neurobiology and cognition not being unusual. For example, cigarette smoking among methamphetamine abusers is associated with significant gray matter volume loss (Morales et al., 2012), and separate and interactive effects of co-occurring cocaine and alcohol dependence on brain structure and metabolite levels measured by MR were described (Morales et al., 2012; O'Neill, Cardenas, & Meyerhoff, 2001). Short-term-abstinent polysubstance users (with cocaine as their substance of choice plus significant cannabis and alcohol abuse) had *higher* normalized glucose metabolism than controls in frontotemporal cortex, including OFC (Stapleton et al., 1995), indicating increased energy metabolism. However, 1-week-abstinent individuals with both methamphetamine and cannabis use disorders had *lower* metabolism in temporal cortex, right hippocampus, and ventral striatum than individuals abusing methamphetamine only (Voytek et al., 2005). Comorbid alcohol- and cocaine-dependent individuals had less white matter than individuals with alcohol dependence alone (Bjork, Grant, & Hommer, 2003). And while PSU and AUD individuals in this study had similar cortical gray matter loss after correction for different alcohol consumption, polysubstance users—in contrast to AUD individuals—had no volume loss in the striatum. This observation is consistent with enlargements of the striatum reported in another study of PSU by the same group (Grodin, Lin, Durkee, Hommer, & Momenan, 2013), in studies of amphetamine abusers, some of whom also abused alcohol (Berman, O'Neill, Fears, Bartzokis, & London, 2008), and in corresponding reports of methamphetamine users (Thompson et al., 2004) and young recreational marijuana users (Gilman et al., 2014). A possible explanation for these different volume findings in the striatum of AUD versus stimulant-abusing PSU patients are potential anti-inflammatory/neuroprotective effects of cannabinoids, including dendritic arborization, or neuroinflammation from stimulant use (perhaps related to gliosis or astrocytosis) that counteract primary atrophic effects related to alcohol abuse.

An alternative explanation for anomalies in striatal and prefrontal morphology in stimulant users has been discussed and relates to the possible presence of personality disorder symptomatology, which is often comorbid with stimulant use disorders (e.g., antisocial, borderline or obsessive–compulsive disorder) (Payer et al., 2015). Alternatively, morphological alterations of the caudate may be compensatory to deficiencies in the dorsolateral PFC (see Review of Our Own Studies section), a prefrontal structure highly connected to the caudate.

Review of Our Own Studies

Our own early cross-sectional studies of PSU effects on brain structure and metabolite levels also suggested primary injury to frontal brain regions with corresponding functional correlates: volume loss of gray matter in the PFC of alcohol-dependent cocaine addicts at 6 weeks of abstinence correlated with executive dysfunction (Fein et al., 2002) and persisted for 1–3 years of abstinence (Fein et al., 2002; O'Neill et al., 2001). The frontal gray matter atrophy was greater in the comorbid cocaine-and-alcohol-dependent sample than in monosubstance users; in largely the same subjects, we found similarly low NAA in frontal gray matter (suggesting neuronal injury) and high Cho (suggesting gliosis, high membrane turnover) in frontal white matter of both cocaine using groups at 4–5 months of abstinence (Meyerhoff et al., 1999; O'Neill et al., 2001). These neurobiological abnormalities may constitute the potential biological substrates for the significant cognitive deficits detected in both cocaine-dependent and comorbid cocaine-and-alcohol-dependent samples at 3 months of abstinence (Di Sclafani et al., 1998).

In our neuroimaging studies of treatment-seeking individuals with PSUD, our anatomical focus was on the brain regions most impacted in all forms of SUDs. These form a part of a network of discrete, overlapping anterior frontal cortical–subcortical circuits collectively referred to as brain reward/executive oversight system (BREOS) (Haber & Knutson, 2010; Makris, Gasic et al., 2008; Makris, Oscar-Berman et al., 2008; Meyerhoff, Durazzo, & Ende, 2011; Meyerhoff et al., 2013; Verdejo-Garcia & Bechara, 2009). This extensive brain network forms the biological substrate for motivation/drive, reward/saliency, and inhibitory control/executive function and is crucial to the development and persistence of all addictive disorders (e.g., Koob & Volkow, 2009; Volkow, Wang, Fowler, & Telang, 2008; Volkow et al., 2011). Different SUDs have been linked to enduring changes in neuronal and glial tissues of BREOS regions that subserve "top-down" inhibitory control/executive skills (Alexander & Stuss, 2000; Gazzaley & D'Esposito, 2007; George & Koob, 2010; Volkow, Wang, Fowler, & Tomasi, 2012). Consequently, BREOS abnormalities have been related to deficits in executive skills, learning and memory, working memory, processing speed, and visuospatial skills (Durazzo & Meyerhoff, 2007; Gazzaley & D'Esposito, 2007), which in turn negatively affect a person's ability to engage in planned, adaptive, pro-health behavior that is critical for long-term abstinence (Connolly et al., 2012; Verdejo-Garcia et al., 2004).

We analyzed cross-sectional neuroimaging data (various MR-based modalities, but no functional MRI) from a subset of the 36 PSU subjects described in the foregoing (see Table 21.1 above), who were abstinent from substances (except tobacco) for about 1 month, and compared them to the corresponding neuroimaging data from 1-month-abstinent AUD patients. According to DSM-IV criteria, all AUD and all PSU individuals were alcohol dependent; in addition, 62% of the PSU individuals were dependent on both cocaine and alcohol (including two with marijuana dependence), 14% were dependent on methamphetamines and alcohol (including one with marijuana dependence), 14% were dependent on all four substances, and 11% were opioid and alcohol dependent. Polysubstance users had similar lifetime alcohol consumption than the "pure" AUD sample (monosubstance users) ($\sim$230 standard alcoholic drinks/month over $\sim$33 years of drinking); 68% of all PSU and 59% of all AUD participants were current cigarette smokers with lifetime smoking histories and prevalence that were also comparable across the subsets of substance users. Smoking duration was similar across PSU and AUD individuals, but polysubstance users smoked fewer cigarettes per day over lifetime than AUD individuals (11 ± 7 vs. 18 ± 7 daily cigarettes); nevertheless, polysubstance users had a similar Fagerstrom score than AUD individuals (4.2 ± 1.7 vs. 4.8 ± 1.6), indicating similar low-to-medium smoking severity, but comparable to our control group (15 ± 10 cigarettes per day, Fagerstrom score 4.8 ± 1.6). To isolate the effects of poly- versus monosubstance use, all statistical analyses covaried for age, AMNART, BDI, and smoking status (in all comparisons including controls), and for alcohol use in addition when comparing PSU and AUD only.

MRI Studies of Brain Structure: T1-weighted brain MRIs from 19 1-month-abstinent polysubstance users, 40 1-month-abstinent AUD individuals, and 27 non/light-drinking controls were segmented automatically into lobar gray matter, white matter, and cerebrospinal fluid volumes and normalized to intracranial volume (Mon et al., 2014). Despite similar lifetime drinking and smoking histories and comorbid stimulant abuse, the polysubstance users had *larger* white matter volumes than the AUD patients in all major lobes; polysubstance users also had *larger* frontal and parietal white matter volumes than the controls. Parietal white matter volume in polysubstance users correlated positively with prior year cocaine use, which suggests a substance-related volume expansion as opposed to white matter hypertrophy predisposing to substance abuse. Such white matter hypertrophy has been described also in methamphetamine dependence (Thompson et al., 2004), heavy cannabis users (Matochik et al., 2005), and cigarette smokers (Gazdzinski et al., 2005). Notably and in contrast to AUD patients with similar lifetime drinking severities, polysubstance users had more sulcal cerebrospinal fluid volumes and less cortical and subcortical atrophy than AUD patients; this observation is reminiscent of the effects of potential neuroinflammation associated with stimulant use described earlier for the striatum. Thus, it is possible that widespread cerebral gliosis masks gross gray matter atrophy in alcohol-dependent PSU (Mackey & Paulus, 2013).

In a follow-up MRI study of a similar patient cohort (Pennington et al., 2015), parcellated morphological data (regional volumes, surface areas, and cortical thickness) were obtained from T1-weighted brain images for dorsal PFC, ACC, OFC, and insula. Polysubstance users had smaller left OFC and right dorsal PFC volumes and surface areas than controls, but they did not differ significantly from AUD patients on these measures. Instead, polysubstance users had thinner right ACC and left dorsal PFC than AUD patients. No significant relationships between morphometry and quantity/frequency/duration of substance, alcohol, or tobacco use were observed. Polysubstance users had focal frontal cortical atrophy, and only PSU exhibited distinct relationships of smaller dorsal PFC, OFC, and ACC morphometrics to poorer cognitive efficiency, executive function, intelligence, processing speed, and inhibitory control (higher impulsivity and risk taking), functions that have been related to relapse. Overall, these data suggest premorbid effects on frontal cortical morphometry in both AUD and PSU patients, but with stronger functional ramifications in PSU than AUD patients. These neuroimaging abnormalities may serve as PSU biomarkers and as potential targets for pharmacological and behavioral PSU-specific treatment aimed at decreasing the high relapse rates in PSU. Structural brain differences between AUD and PSU patients, together with potentially related neuropsychological differences between the groups, may demand different treatment approaches to mitigate functionally relevant brain abnormalities and improve substance use behavior.

Within the PSU group, we investigated the effects of chronic cigarette smoking by comparing 20 smokers to 10 nonsmokers who were all abstinent from substances for 1 month and who did not differ significantly in cocaine or alcohol use quantities and frequencies. The smoking polysubstance users had trends to thinner cortices than the nonsmoking polysubstance users in dorsolateral PFC and OFC, consistent with significant smoking-related thinning in AUD patients and light-drinking controls (Durazzo, Mon, Gazdzinski, & Meyerhoff, 2013). In addition, smoking polysubstance users showed greater age-related volume losses in parietal gray matter and white matter than nonsmoking polysubstance users, consistent with findings in AUD patients (Gazdzinski et al., 2005). These smoking effects suggest that offering smoking cessation in PSU treatment may help alleviate smoking-related neurobiological deficits in this population, with the prospect of improving cognition and consequently treatment outcome.

MRI Studies of Cerebral Blood Flow: Cerebral blood flow in cocaine-dependent individuals is both decreased and increased in different cortical brain regions, suggesting compensatory energy-consuming activities within the brain. We used arterial spin labeling perfusion at 4 T to evaluate regional cerebral blood flow noninvasively in many cortical regions identified on co-registered parcellated structural MR images (Abé, Mon, Durazzo, et al., 2013; Abé, Mon, Hoefer, et al., 2013; Murray, Durazzo, Mon, Schmidt, & Meyerhoff, 2015); 20 PSU and 26 AUD patients at 1 month of abstinence showed reduced cortical blood flow in anterior and posterior brain regions. Regional cortical perfusion was almost always numerically *higher* in PSU than in AUD group (except in OFC where both groups had similar reductions). It is possible that interactions between cocaine and alcohol consumed concurrently may increase vasoactive effects in polysubstance users (McCance-Katz, Kosten, & Jatlow, 1998) and/or that postulated gliosis in polysubstance users may increase regional blood flow (Gottschalk & Kosten, 2002) seen in alcohol-dependent PSU versus AUD patients, as observed in an arterial spin labeling study (Murray et al., 2015). In both groups, hypoperfusion in the BREOS correlated with greater drinking severity, poorer cognitive performance, worse decision making (IGT), greater impulsivity (BIS), and greater risk taking (BART), but the patterns of perfusion–function relationships were different among AUD and PSU groups. In contrast to AUD, subcortical perfusion in PSU group was higher in smokers than in nonsmokers. Thus, because of potential interactions between substances consumed by polysubstance users, chronic smoking appears to have different effects on subcortical vasculature/tissue in PSU and AUD individuals, highlighting yet again that the greater cumulative substance consumption in polysubstance users than in monosubstance users is not always and directly related to more severe brain injury in polysubstance users.

Magnetic Resonance Spectroscopy (MRS) Studies of Brain Metabolites: We also measured and compared brain metabolite concentrations in the PSU, AUD, and control participants described in the foregoing (Abé, Mon, Durazzo, et al., 2013; Abé, Mon, Hoefer et al., 2013). Using single-volume proton (^{1}H) MRS at 4 T in 28 polysubstance users relative to 40 patients with AUD we detected significantly lower NAA, Cr, Cho, and mI in the dorsolateral PFC and a strong trend to lower GABA in the ACC (Abé, Mon, Durazzo, et al., 2013), whereas the corresponding metabolite levels in 1-month-abstinent AUD were similar to those in 16 drug-free controls (see also Mon, Durazzo, & Meyerhoff, 2012). Within polysubstance users, lower dorsolateral PFC GABA related to more cocaine use, lower dorsolateral PFC NAA related to worse visuospatial and working memory, and lower NAA and Cho in the ACC was associated with poorer visuospatial learning and processing speed. The greater and functionally significant metabolic injury in PSU than in AUD patients at 1 month of abstinence was confirmed in our proton MR spectroscopic imaging (^{1}H MRSI) study of similar subjects (Abé, Mon, Hoefer, et al., 2013). While the 1-month-abstinent AUD had normal metabolite concentrations which had recovered from reduced levels measured in AUD earlier in abstinence (Durazzo et al., 2004; Durazzo, Gazdzinski, Banys, & Meyerhoff, 2006), the 1-month-abstinent polysubstance users had metabolic abnormalities (primarily higher mI, indicative of astrocytic alterations, perhaps gliosis) in temporal gray matter, cerebellar vermis, and lenticular nuclei. Less NAA in lobar gray matter correlated with higher cocaine use quantities in PSU, but not with any quantitative measures of alcohol use. Also in PSU, reduced cortical gray matter NAA was related to slower cognitive processing speed.

In sum, the greater and more persistent metabolic injury in PSU than in AUD patients at 1 month of abstinence (who exhibit similar lifetime drinking and smoking histories) are related to illicit substance use and cognitive deficits. These PSU findings, together with our earlier PSU studies (Fein et al., 2002; Meyerhoff et al., 1999) and reports in the monosubstance use literature (Hanlon, Dufault, Wesley, & Porrino, 2011), suggest greater neuronal injury/dysfunction primarily in the dorsolateral PFC, a brain area critical to inhibitory control, and altered (subcortical) glial function or myelination, perhaps related to neuroinflammation in PSU. While most of our frontal gray matter NAA and GABA results are similar to findings in monosubstance users (cocaine, methamphetamine, or alcohol; e.g., Ke et al., 2004; Licata & Renshaw, 2010), our Cho, Cr, and mI results in polysubstance users are unique. They suggest strong effects of abstinence duration (see following heading) on metabolite levels in polysubstance users, and the findings are consistent with differential cognitive control demands subserved by ACC and dorsolateral PFC in different stages of abstinence (Connolly et al., 2012), consistent with substance interactions, and with effects of PSU on metabolite levels (i.e., greater cumulative oxidative stress, and glial and/or inflammatory changes). Overall, the neurobiological alterations in polysubstance users may represent specific PSU biomarkers and potential targets for improved behavioral and pharmacological treatment of this ever-increasing yet understudied population.

Relapse Prediction in Substance Users

Sociodemographic, clinical, neuropsychological, and physiological factors predict substance abuse treatment success (e.g., Dean et al., 2009; Durazzo, Gazdzinski et al., 2008; Fisher, Elias, & Ritz, 1998; Reske & Paulus, 2008). Specifically, intact "top-down" control of emotions and behavior relates to successful drug abstinence (Connolly et al., 2012; Volkow et al., 2011). The consequences of persistent cognitive deficits on relapse behavior in cocaine dependence and PSUD have been discussed (Verdejo-Garcia et al., 2012, 2007), but not studied explicitly. We and others found that treatment-seeking AUD patients have abnormal brain structure (Beck et al., 2012; Cardenas et al., 2011; Durazzo, Gazdzinski et al., 2008; Durazzo, Tosun et al., 2011; Rando et al., 2011; Wrase et al., 2008) and metabolite levels (Durazzo, Gazdzinski et al., 2008; Durazzo, Pathak et al., 2010; Parks et al., 2002) in "top-down" BREOS regions at entry into alcoholism treatment (baseline) that differentiate future relapsers from abstainers (Durazzo, Rothlind, Gazdzinski, & Meyerhoff, 2008) (see also Research Society on Alcoholism 2007 symposium). Volume, surface areas, and cortical thickness of various top-down BREOS regions in AUD are differentially sensitive to relapse and post-treatment alcohol use (Durazzo, Tosun et al., 2011).

In contrast to our AUD cohort, our 16 PSU individuals who relapsed within 4 months after treatment entry compared to the 20 polysubstance users who stayed abstinent had *thicker* ACC and trends to thicker OFC at 1 month of abstinence (unpublished); these thicker cortices may reflect increased glial volume (suggesting gliosis and/or neuroinflammation) specific to PSU or more active use of ACC/OFC function (Lyoo et al., 2011). These abnormalities are clinically relevant, as structural and functional PFC deficits are postulated to impair the ability to control future drinking (Duka et al., 2011). Studies in cocaine dependence showed that high mu-opioid receptor binding in the brain reward system predicts treatment outcome (Ghitza et al., 2010), and that white matter integrity measured by diffusion tensor imaging relates to abstinence duration within 8 weeks of cocaine-dependence treatment (Xu et al., 2010); furthermore, a functional MRI study of cognitive control in ex-smokers suggested increased PFC activity (Nestor, McCabe, Jones, Clancy, & Garavan, 2011). We postulate that underlying these structural abnormalities in polysubstance users are alterations of metabolite levels and/or blood flow associated with oxidative stress and inflammation that influence relapse risk. In addition, measures of self-regulation relate to relapse behavior, including decision making (Evren, Durkaya, Evren, Dalbudak, & Cetin, 2012; Hanson, Luciana, & Sullwold, 2008), post-treatment anger expression (Patterson, Kerrin, Wileyto, & Lerman, 2008), and negative emotionality (Fisher et al., 1998). Consistent with this, greater anger frequency/intensity in our 1-month-abstinent polysubstance users related to smaller BREOS volume (unpublished), which in turn predicted post-treatment drinking in AUD patients (Durazzo, Tosun et al., 2011).

To summarize this section, different MR outcome and behavioral measures in PSU relate to treatment outcome and/or relapse propensity, and these findings, if confirmed in further studies, suggest targets for more efficient pharmacological and behavioral treatment of PSUD. Treatment providers and patients will benefit substantially from identifying those polysubstance users who are especially vulnerable to relapse, as this will help select a specific treatment course early in recovery to improve overall outcome. Further, if comorbid smoking is shown to compound the brain abnormalities in polysubstance users, to impede brain recovery or to promote relapse, this will bolster clinical practice to provide specific smoking cessation support at the inception of substance abuse treatment (Prochaska, Delucchi, & Hall, 2004).

Review of Neuroimaging Changes During Abstinence From PSU

There is a clear lack of truly longitudinal MR studies of abstinent mono- and polysubstance users, likely because of poor subject retention (e.g., Hanlon et al., 2013; Mackey & Paulus, 2013; Morales et al., 2012; Salo & Fassbender, 2012). However, studying longitudinal change in abstinent polysubstance users is extremely valuable. If effective neuroplasticity after removal of the chronic insult can be demonstrated in PSU, it implies (1) the existence of a critical window of opportunity for effective intervention (e.g., plasticity-based cognitive remediation); (2) that critical aspects of cognitive dysfunction and their neurobiological correlates are not stable premorbid/risk factors for abuse but rather consequences of abuse that can be altered with discontinuing abuse; and (3) that relapse risk is reduced with effective recovery of brain biology and function.

Identifying any underlying mechanisms of such improvements can inform interventions that enhance specific recovery processes (e.g., strengthen prefrontal connectivity or employ GABAergic therapy for better inhibitory control).

In our preliminary neuroimaging studies of 20 PSU patients between 1 and 4 months of abstinence (unpublished), we found some evidence for widespread volumetric change, including an increase in left caudate volume, a decrease in right superior temporal lobe volume, and trends to decreasing insula and entorhinal cortex volumes. Given that cross-sectional abnormalities in polysubstance users involved both smaller and larger regional tissue volumes, observing both volume increases and decreases with successful abstinence is not surprising. However, none of the volumetric changes correlated meaningfully with cognitive improvements. Preliminary metabolic changes in 15 of these abstinent polysubstance users indicate normalization of metabolite levels by 4 months of abstinence. Specifically, GABA and NAA in the ACC and NAA and Cho in the dorsolateral PFC increased, while ACC mI and glutamate, and mI in the parieto-occipital gray matter decreased. In contrast to structural changes, some of these metabolite level changes correlated with change in cognitive domain scores (e.g., NAA increases in the ACC and dorsolateral PFC were associated with improvements in visuospatial learning and working memory, respectively).

Replicating these promising preliminary observations of improvements to brain health in abstinent polysubstance users is important for motivation to treat as well as for public perception of substance abuse and drug policy. Public knowledge of clinically relevant brain recovery with abstinence will help move us away from the prevailing public mindset of "your brain on drugs" to one more helpfully described as "your brain in recovery."

REVIEW SUMMARY AND OUTLOOK

With the wider availability of illicit drugs today than in the past, the simultaneous or concurrent abuse of several addictive substances has soared. "Pure" monosubstance users are rare among treatment seekers today, and PSU is the current clinical reality. Polysubstance users comprise the largest and fastest growing group of treatment seekers in the developed world today, with greater than 50% prevalence in many regions. As such, PSUDs with its relatively high comorbid psychopathology and its correspondingly strong detrimental effects on brain biology and function constitute a major health problem. Nevertheless, individuals with PSUD have rarely been the expressed target of scientific research, let alone clinical treatment studies. It can be argued that the corresponding lack of information on clinically and functionally significant health effects in PSUD contributes to a large extent to the lack of effective treatment and poor treatment outcome among individuals with PSUD. A better understanding of the neurobiological and neuropsychological correlates of maladaptive abuse of several substances, of adaptive brain plasticity during successful abstinence, and of critical predictors of all-too-common relapse in PSUD can contribute urgently needed new knowledge to the design of novel neurobehavioral and pharmacological interventions. In addition, from a more immediate and practical point of view, such information can inform treatment providers of the specific cerebral and cognitive dysfunction associated with PSUD for better appreciation of the specific challenges individuals with PSU face in attempting to reduce their substance use.

Almost all research on addiction has focused on individuals ostensibly abusing a single drug, most commonly alcohol, cocaine, amphetamines, marijuana, heroin, or tobacco. Understandable from a scientific point of view (scientists like to isolate their research target), these studies have yielded a wealth of information on addiction and on the brain changes associated with a given drug. They contributed to the development of interventions that successfully reduce alcohol consumption or tobacco use, to name a few. Regrettably though, this approach has also done little to advance effective treatment of PSUD.

In this chapter, we reviewed the relatively small body of peer-reviewed research on the particular and specific brain alterations that are associated with PSUD vis-a-vis those related to monosubstance use disorders. This chapter demonstrates neuropsychological deficits in polysubstance users who at first glance appear to be largely similar in nature to those of single drug users, but are often also more severe and long-lasting in polysubstance users. Careful comparisons have also revealed differences between individuals with PSUD and AUD in different aspects of learning and memory, various measures of inhibitory control, and intelligence, which all have been related to relapse risk in single drug studies. Underlying these differences in brain function are neurobiological abnormalities (brain structure and metabolism) that differ in location, spatial extent, nature of injury, and persistence during abstinence from abused substances. We also showed that chronic tobacco use in PSUD and AUD populations and controls affects neurobiology and function, both cross-sectionally and longitudinally, in regions critical to executive control/oversight and impulse behavior; thus, smoking in these populations may be behaviorally relevant. Preliminary serial assessments within the first few months of abstinence from all drugs but tobacco also demonstrate effective recovery from several cognitive, self-regulatory, and neurobiological deficits in polysubstance users. The studies that have contrasted single drug users and PSU individuals suggest that different treatment approaches may be indicated for poly- and monosubstance users, with strategies that leverage our knowledge of the unique differences between different typologies of substance users. It would be a fallacy to attempt to build PSU treatment development solely on what we have learned from studies of "pure" SUDs. This chapter shows that individuals with PSU experience neuropsychological and neurobiological abnormalities, which are sufficiently different from those of monosubstance users that we cannot simply generalize from treating AUD or heroin addiction to the more complex PSU populations.

The reviewed PSU studies have limitations in common with most single drug studies, primarily in that cross-sectional designs cannot infer causality and that the findings can be confounded by premorbid vulnerability factors, development and aging, gender, severity of substance use, duration of abstinence, lifestyle, and comorbid psychopathology. But PSU studies also have their unique challenges. Compared to participants in single drug studies, many PSU study populations may differ in ethnic composition and comorbid psychopathology, while all are by definition more heterogeneous in regard to types, quantities, and patterns of substances abused over lifetime; this poses unique challenges regarding recruitment of study participants and appropriate control groups, diagnostic assessment, reproducibility, generalizability, and, ultimately, design of clinical trials. Studying, understanding, and treating PSUD is a formidable challenge from scientific and methodological viewpoints, and it requires adhering to the best scientific principles. As daunting as the task may appear, however, we cannot be allowed to shy away from the challenge as we have to be true to our promises as medical professionals to provide better and more effective treatment to an exceedingly common form of substance use.

Recent developments in the United States promise to accelerate and improve PSU research: National Institute of Drug Abuse (NIDA) has made the development of new interventions including pharmacology and cognitive remediation for PSUD a research priority (Vocci, 2008; Wexler, 2011). More recently, several NIH (National Institutes of Health) institutes have joined forces to face the challenges and tackle common substance use comorbidities including the growing PSU epidemic. In its new 2016–20 Strategic Plan, NIDA prioritizes research that incorporates real-world complexities including PSU. Our research communities have to apply the same scientific rigor we have in the past to study other diseases and provide sound practice-based evidence for what works to reduce all forms of substance use in our society. Most and above all, we have to strive to better understand the correlates of substance use in the very same population(s) we seek to treat. Going forward in this endeavor, we ignore clinical reality at our peril.

Acknowledgments

This review was supported by NIH AA10788 and DA039903.

References

Abé, C., Mon, A., Durazzo, T. C., Pennington, D. L., Schmidt, T. P., & Meyerhoff, D. J. (2013). Polysubstance and alcohol dependence: Unique abnormalities of magnetic resonance-derived brain metabolite levels. *Drug and Alcohol Dependence, 130*, 30–37.

Abé, C., Mon, A., Hoefer, M. E., Durazzo, T. C., Pennington, D. L., Schmidt, T. P., & Meyerhoff, D. J. (2013). Metabolic abnormalities in lobar and subcortical brain regions of abstinent polysubstance users: Magnetic resonance spectroscopic imaging. *Alcohol and Alcoholism, 48*, 543–551.

Adalsteinsson, E., Sullivan, E. V., Mayer, D., & Pfefferbaum, A. (December 2006). In vivo quantification of ethanol kinetics in rat brain. *Neuropsychopharmacology, 31*(12), 2683–2691. PubMed PMID: 16407891.

Aharonovich, E., Hasin, D. S., Brooks, A. C., Liu, X., Bisaga, A., & Nunes, E. V. (2006). Cognitive deficits predict low treatment retention in cocaine dependent patients. *Drug and Alcohol Dependence, 81*, 313–322.

Alexander, M. P., & Stuss, D. T. (2000). Disorders of frontal lobe functioning. *Seminars in Neurology, 20*, 427–437.

Almeida, O. P., Garrido, G. J., Lautenschlager, N. T., Hulse, G. K., Jamrozik, K., & Flicker, L. (2008). Smoking is associated with reduced cortical regional gray matter density in brain regions associated with incipient Alzheimer disease. *The American Journal of Geriatric Psychiatry, 16*, 92–98.

Amato, L., Davoli, M., Vecchi, S., Ali, R., Farrell, M., Faggiano, F., … Chengzheng, Z. (2011). Cochrane systematic reviews in the field of addiction: What's there and what should be. *Drug and Alcohol Dependence, 113*, 96–103.

Anderson, B. J. (2011). Plasticity of gray matter volume: The cellular and synaptic plasticity that underlies volumetric change. *Developmental Psychobiology, 53*, 456–465.

Bae, S. C., Lyoo, I. K., Sung, Y. H., Yoo, J., Chung, A., Yoon, S. J., … Renshaw, P. F. (2006). Increased white matter hyperintensities in male methamphetamine abusers. *Drug and Alcohol Dependence, 81*, 83–88.

Barber, R., Panikkar, A., & McKeith, I. G. (2001). Dementia with Lewy bodies: Diagnosis and management. *International Journal of Geriatric Psychiatry, 16*(Suppl. 1), S12–S18.

Bartzokis, G., Beckson, M., Lu, P. H., Edwards, N., Rapoport, R., Wiseman, E., & Bridge, P. (2000). Age-related brain volume reductions in amphetamine and cocaine addicts and normal controls: Implications for addiction research. *Psychiatry Research, 98*, 93–102.

Bates, M. E., Buckman, J. F., & Nguyen, T. T. (2013). A role for cognitive rehabilitation in increasing the effectiveness of treatment for alcohol use disorders. *Neuropsychology Review, 23*, 27–47.

Beck, A., Wustenberg, T., Genauck, A., Wrase, J., Schlagenhauf, F., Smolka, M. N., … Heinz, A. (2012). Effect of brain structure, brain function, and brain connectivity on relapse in alcohol-dependent patients. *Archives of General Psychiatry, 69*, 842–852.

Berman, S., O'Neill, J., Fears, S., Bartzokis, G., & London, E. D. (2008). Abuse of amphetamines and structural abnormalities in the brain. *Annals of the New York Academy of Sciences, 1141*, 195–220.

Bernardin, F., Maheut-Bosser, A., & Paille, F. (2014). Cognitive impairments in alcohol-dependent subjects. *Frontiers in Psychiatry, 5*, 78.

Bjork, J. M., Grant, S. J., & Hommer, D. W. (2003). Cross-sectional volumetric analysis of brain atrophy in alcohol dependence: Effects of drinking history and comorbid substance use disorder. *The American Journal of Psychiatry, 160*, 2038–2045.

Block, R. I., Erwin, W. J., & Ghoneim, M. M. (2002). Chronic drug use and cognitive impairments. *Pharmacology, Biochemistry, and Behavior, 73*, 491–504.

Block, R. I., O'Leary, D. S., Ehrhardt, J. C., Augustinack, J. C., Ghoneim, M. M., Arndt, S., & Hall, J. A. (2000). Effects of frequent marijuana use on brain tissue volume and composition. *Neuroreport, 11*, 491–496.

Bolla, K. I., Funderburk, F. R., & Cadet, J. L. (2000). Differential effects of cocaine and cocaine alcohol on neurocognitive performance. *Neurology, 54*, 2285–2292.

Brody, A. L. (August 2006). Functional brain imaging of tobacco use and dependence. *Journal of Psychiatric Research, 40*(5), 404–418. Review. PubMed PMID: 15979645; PubMed Central PMCID: PMC2876087.

Buhler, M., & Mann, K. (2011). Alcohol and the human brain: A systematic review of different neuroimaging methods. *Alcoholism, Clinical and Experimental Research, 35*, 1771–1793.

Cadet, J. L., & Bisagno, V. (2015). Neuropsychological consequences of chronic drug use: Relevance to treatment approaches. *Frontiers in Psychiatry, 6*, 189.

Cardenas, V. A., Durazzo, T. C., Gazdzinski, S., Mon, A., Studholme, C., & Meyerhoff, D. J. (2011). Brain morphology at entry into treatment for alcohol dependence is related to relapse propensity. *Biological Psychiatry, 70*, 561–567.

Chang, L., Ernst, T., Speck, O., Patel, H., DeSilva, M., Leonido-Yee, M., & Miller, E. N. (2002). Perfusion MRI and computerized cognitive test abnormalities in abstinent methamphetamine users. *Psychiatry Research, 114*, 65–79.

Christensen, J. D., Kaufman, M. J., Levin, J. M., Mendelson, J. H., Holman, B. L., Cohen, B. M., & Renshaw, P. F. (1996). Abnormal cerebral metabolism in polydrug abusers during early withdrawal: A 31P MR spectroscopy study. *Magnetic Resonance in Medicine, 35*, 658–663.

Connolly, C. G., Foxe, J. J., Nierenberg, J., Shpaner, M., & Garavan, H. (2012). The neurobiology of cognitive control in successful cocaine abstinence. *Drug and Alcohol Dependence, 121*, 45–53.

Connor, J. P., Gullo, M. J., Chan, G., Young, R. M., Hall, W. D., & Feeney, G. F. (2013). Polysubstance use in cannabis users referred for treatment: Drug use profiles, psychiatric comorbidity and cannabis-related beliefs. *Frontiers in Psychiatry, 4*, 79.

Connor, J. P., Gullo, M. J., White, A., & Kelly, A. B. (2014). Polysubstance use: Diagnostic challenges, patterns of use and health. *Current Opinion in Psychiatry, 27*, 269–275.

Conway, K. P., Swendsen, J. D., Rounsaville, B. J., & Merikangas, K. R. (2002). Personality, drug of choice, and comorbid psychopathology among substance abusers. *Drug and Alcohol Dependence, 65*, 225–234.

Crews, F. T., & Boettiger, C. A. (2009). Impulsivity, frontal lobes and risk for addiction. *Pharmacology, Biochemistry, and Behavior, 93*, 237–247.

Dalwani, M., Sakai, J. T., Mikulich-Gilbertson, S. K., Tanabe, J., Raymond, K., McWilliams, S. K., ... Crowley, T. J. (2011). Reduced cortical gray matter volume in male adolescents with substance and conduct problems. *Drug and Alcohol Dependence, 118*, 295–305.

Dean, A. C., London, E. D., Sugar, C. A., Kitchen, C. M., Swanson, A. N., Heinzerling, K. G., ... Shoptaw, S. (2009). Predicting adherence to treatment for methamphetamine dependence from neuropsychological and drug use variables. *Drug and Alcohol Dependence, 105*, 48–55.

Di Sclafani, V., Bloomer, C., Clark, H., Norman, D., Hannauer, D., & Fein, G. (1998). Abstinent chronic cocaine and cocaine/alcohol abusers evidence normal hippocampal volumes on MRI despite cognitive impairments. *Addiction Biology, 3*, 261–270.

Dom, G., Sabbe, B., Hulstijn, W., & van den Brink, W. (2005). Substance use disorders and the orbitofrontal cortex: Systematic review of behavioral decision-making and neuroimaging studies. *British Journal of Psychiatry, 187*, 209–220.

Duka, T., Trick, L., Nikolaou, K., Gray, M. A., Kempton, M. J., Williams, H., ... Stephens, D. N. (2011). Unique brain areas associated with abstinence control are damaged in multiply detoxified alcoholics. *Biological Psychiatry, 70*, 545–552.

Durazzo, T. C., Fryer, S. L., Rothlind, J. C., Vertinski, M., Gazdzinski, S., Mon, A., & Meyerhoff, D. J. (2010). Measures of learning, memory and processing speed accurately predict smoking status in short-term abstinent treatment-seeking alcohol-dependent individuals. *Alcohol and Alcoholism, 45*, 507–513.

Durazzo, T. C., Gazdzinski, S., Banys, P., & Meyerhoff, D. J. (2004). Cigarette smoking exacerbates chronic alcohol-induced brain damage: A preliminary metabolite imaging study. *Alcoholism, Clinical and Experimental Research, 28*, 1849–1860.

Durazzo, T. C., Gazdzinski, S., Banys, P., & Meyerhoff, D. J. (2006). Brain metabolite concentrations and neurocognition during short-term recovery from alcohol dependence: Preliminary evidence of the effects of concurrent chronic cigarette smoking. *Alcoholism, Clinical and Experimental Research, 30*, 539–551.

Durazzo, T. C., Gazdzinski, S., & Meyerhoff, D. J. (2007). The neurobiological and neurocognitive consequences of chronic cigarette smoking in alcohol use disorders. *Alcohol and Alcoholism, 42*, 174–185.

Durazzo, T. C., Gazdzinski, S., Yeh, P. H., & Meyerhoff, D. J. (2008). Combined neuroimaging, neurocognitive and psychiatric factors to predict alcohol consumption following treatment for alcohol dependence. *Alcohol and Alcoholism, 43*, 683–691.

Durazzo, T., Insel, P. S., Weiner, M. W., & The Alzheimer's Disease Neuroimaging Initiative, A. (2012). Greater regional brain atrophy rate in healthy elders with a history of cigarette smoking. *Alzheimer's & Dementia, 8*, 513–519.

Durazzo, T. C., & Meyerhoff, D. J. (2007). Neurobiological and neurocognitive effects of chronic cigarette smoking and alcoholism. *Frontiers in Bioscience, 12*, 4079–4100.

Durazzo, T. C., Meyerhoff, D. J., & Nixon, S. J. (2010). Chronic cigarette smoking: Implications for neurocognition and brain neurobiology. *International Journal of Environmental Research and Public Health, 7*, 3760–3791.

Durazzo, T. C., Meyerhoff, D. J., & Nixon, S. J. (2011). A comprehensive assessment of neurocognition in middle-aged chronic cigarette smokers. *Drug and Alcohol Dependence, 122*, 105–111.

Durazzo, T. C., Mon, A., Gazdzinski, S., & Meyerhoff, D. J. (2011). Chronic cigarette smoking in alcohol dependence: Associations with cortical thickness and N-acetylaspartate levels in the extended brain reward system. *Addiction Biology, 18*, 379–391.

Durazzo, T. C., Mon, A., Gazdzinski, S., & Meyerhoff, D. J. (2013). Chronic cigarette smoking in alcohol dependence: Associations with cortical thickness and N-acetylaspartate levels in the extended brain reward system. *Addiction Biology, 18*, 379–391.

Durazzo, T. C., Mon, A., Gazdzinski, S., Yeh, P. H., & Meyerhoff, D. J. (2015). Serial longitudinal magnetic resonance imaging data indicate non-linear regional gray matter volume recovery in abstinent alcohol-dependent individuals. *Addiction Biology, 20*, 956–967.

Durazzo, T. C., Pathak, V., Gazdzinski, S., Mon, A., & Meyerhoff, D. J. (2010). Metabolite levels in the brain reward pathway discriminate those who remain abstinent from those who resume hazardous alcohol consumption after treatment for alcohol dependence. *Journal of Studies on Alcohol and Drugs, 71*, 278–289.

Durazzo, T. C., Pennington, D. L., Schmidt, T. P., & Meyerhoff, D. J. (2014). Effects of cigarette smoking history on neurocognitive recovery over 8 months of abstinence in alcohol-dependent individuals. *Alcoholism, Clinical and Experimental Research, 38*, 2816–2825.

Durazzo, T. C., Rothlind, J. C., Cardenas, V. A., Studholme, C., Weiner, M. W., & Meyerhoff, D. J. (2007). Chronic cigarette smoking and heavy drinking in human immunodeficiency virus: Consequences for neurocognition and brain morphology. *Alcohol, 41*, 489–501.

Durazzo, T. C., Rothlind, J. C., Gazdzinski, S., Banys, P., & Meyerhoff, D. J. (2006). A comparison of neurocognitive function in nonsmoking and chronically smoking short-term abstinent alcoholics. *Alcohol, 39*, 1–11.

Durazzo, T. C., Rothlind, J. C., Gazdzinski, S., & Meyerhoff, D. J. (2008). The relationships of sociodemographic factors, medical, psychiatric, and substance-misuse co-morbidities to neurocognition in short-term abstinent alcohol-dependent individuals. *Alcohol, 42*, 439–449.

Durazzo, T. C., Rothlind, J. C., Weiner, M. W., & Meyerhoff, D. J. (2005). Effects of chronic cigarette smoking on neuropsycholigical test performance in heavy social drinkers. In *28th annual meeting of the research society on alcoholism* (Vol. 29, p. 67A). Santa Barbara, CA: ACER.

Durazzo, T. C., Tosun, D., Buckley, S., Gazdzinski, S., Mon, A., Fryer, S. L., & Meyerhoff, D. J. (2011). Cortical thickness, surface area, and volume of the brain reward system in alcohol dependence: Relationships to relapse and extended abstinence. *Alcoholism, Clinical and Experimental Research, 35*, 1187–1200.

Eldreth, D. A., Matochik, J. A., Cadet, J. L., & Bolla, K. I. (2004). Abnormal brain activity in prefrontal brain regions in abstinent marijuana users. *NeuroImage, 23*, 914–920.

Ernst, T., & Chang, L. (2008). Adaptation of brain glutamate plus glutamine during abstinence from chronic methamphetamine use. *Journal of Neuroimmune Pharmacology, 3*, 165–172.

Ernst, T., Chang, L., Leonido-Yee, M., & Speck, O. (2000). Evidence for long-term neurotoxicity associated with methamphetamine abuse: A 1H MRS study. *Neurology, 54*, 1344–1349.

Ernst, T., Chang, L., Oropilla, G., Gustavson, A., & Speck, O. (2000). Cerebral perfusion abnormalities in abstinent cocaine abusers: A perfusion MRI and SPECT study. *Psychiatry Research, 99*, 63–74.

Ersche, K. D., Jones, P. S., Williams, G. B., Turton, A. J., Robbins, T. W., & Bullmore, E. T. (2012). Abnormal brain structure implicated in stimulant drug addiction. *Science, 335*, 601–604.

Evren, C., Durkaya, M., Evren, B., Dalbudak, E., & Cetin, R. (January 2012). Relationship of relapse with impulsivity, novelty seeking and craving in male alcohol-dependent inpatients. *Drug and Alcohol Review, 31*(1), 81–90. http://dx.doi.org/10.1111/j.1465-3362.2011.00303.x. PubMed PMID: 21450046.

Fein, G., Di Sclafani, V., & Meyerhoff, D. J. (2002). Prefrontal cortical volume reduction associated with frontal cortex function deficit in 6-week abstinent crack-cocaine dependent men. *Drug and Alcohol Dependence, 68*, 87–93.

Fein, G., Meyerhoff, D. J., & Weiner, M. W. (1995). Magnetic resonance spectroscopy of the brain in alcohol abuse. *Alcohol Health & Research World, 19*, 306–314.

Fernandez-Serrano, M. J., Perez-Garcia, M., Perales, J. C., & Verdejo-Garcia, A. (2010). Prevalence of executive dysfunction in cocaine, heroin and alcohol users enrolled in therapeutic communities. *European Journal of Pharmacology, 626*, 104–112.

Fernandez-Serrano, M. J., Perez-Garcia, M., Schmidt Rio-Valle, J., & Verdejo-Garcia, A. (2010). Neuropsychological consequences of alcohol and drug abuse on different components of executive functions. *Journal of Psychopharmacology, 24*, 1317–1332.

Fernandez-Serrano, M. J., Perez-Garcia, M., & Verdejo-Garcia, A. (2011). What are the specific vs. generalized effects of drugs of abuse on neuropsychological performance? *Neuroscience and Biobehavioral Reviews, 35*, 377–406.

Fisher, L. A., Elias, J. W., & Ritz, K. (1998). Predicting relapse to substance abuse as a function of personality dimensions. *Alcoholism, Clinical and Experimental Research, 22*, 1041–1047.

Franklin, T. R., Acton, P. D., Maldjian, J. A., Gray, J. D., Croft, J. R., Dackis, C. A., … Childress, A. R. (2002). Decreased gray matter concentration in the insular, orbitofrontal, cingulate, and temporal cortices of cocaine patients. *Biological Psychiatry, 51*, 134–142.

Gansler, D. A., Harris, G. J., Oscar-Berman, M., Streeter, C., Lewis, R. F., Ahmed, I., & Achong, D. (2000). Hypoperfusion of inferior frontal brain regions in abstinent alcoholics: A pilot SPECT study. *Journal of Studies on Alcohol, 61*, 32–37.

Gazdzinski, S., Durazzo, T., Jahng, G. H., Ezekiel, F., Banys, P., & Meyerhoff, D. (2006). Effects of chronic alcohol dependence and chronic cigarette smoking on cerebral perfusion: A preliminary magnetic resonance study. *Alcoholism, Clinical and Experimental Research, 30*, 947–958.

Gazdzinski, S., Durazzo, T. C., Studholme, C., Song, E., Banys, P., & Meyerhoff, D. J. (2005). Quantitative brain MRI in alcohol dependence: Preliminary evidence for effects of concurrent chronic cigarette smoking on regional brain volumes. *Alcoholism, Clinical and Experimental Research, 29*, 1484–1495.

Gazzaley, A., & D'Esposito, M. (2007). Unifying prefrontal cortex function: Executive control, neural networks, and top-down modulation. In B. L. Miller (Ed.), *The human frontal lobes: Functions and disorders* (2nd ed., pp. 187–206). New York: The Guilford Press.

George, O., & Koob, G. F. (2010). Individual differences in prefrontal cortex function and the transition from drug use to drug dependence. *Neuroscience and Biobehavioral Reviews, 35*, 232–247.

Ghitza, U. E., Preston, K. L., Epstein, D. H., Kuwabara, H., Endres, C. J., Bencherif, B., … Gorelick, D. A. (2010). Brain mu-opioid receptor binding predicts treatment outcome in cocaine-abusing outpatients. *Biological Psychiatry, 68*, 697–703.

Gilman, J. M., Kuster, J. K., Lee, S., Lee, M. J., Kim, B. W., Makris, N., … Breiter, H. C. (2014). Cannabis use is quantitatively associated with nucleus accumbens and amygdala abnormalities in young adult recreational users. *Journal of Neuroscience, 34*, 5529–5538.

Glass, J. M., Adams, K. M., Nigg, J. T., Wong, M. M., Puttler, L. I., Buu, A., … Zucker, R. A. (April 28, 2006). Smoking is associated with neurocognitive deficits in alcoholism. *Drug and Alcohol Dependence, 82*(2), 119–126.

Glass, J. M., Buu, A., Adams, K. M., Nigg, J. T., Puttler, L. I., Jester, J. M., & Zucker, R. A. (2009). Effects of alcoholism severity and smoking on executive neurocognitive function. *Addiction, 104*, 38–48.

Goldstein, R. Z., Leskovjan, A. C., Hoff, A. L., Hitzemann, R., Bashan, F., Khalsa, S. S., … Volkow, N. D. (2004). Severity of neuropsychological impairment in cocaine and alcohol addiction: Association with metabolism in the prefrontal cortex. *Neuropsychologia, 42*, 1447–1458.

Goldstein, R. Z., & Volkow, N. D. (2011). Dysfunction of the prefrontal cortex in addiction: Neuroimaging findings and clinical implications. *Nature Reviews. Neuroscience, 12*, 652–669.

Gonzalez, R., Rippeth, J. D., Carey, C. L., Heaton, R. K., Moore, D. J., Schweinsburg, B. C., … Grant, I. (2004). Neurocognitive performance of methamphetamine users discordant for history of marijuana exposure. *Drug and Alcohol Dependence, 76*, 181–190.

Goto, N., Yoshimura, R., Moriya, J., Kakeda, S., Hayashi, K., Ueda, N., … Nakamura, J. (2010). Critical examination of a correlation between brain gamma-aminobutyric acid (GABA) concentrations and a personality trait of extroversion in healthy volunteers as measured by a 3 Tesla proton magnetic resonance spectroscopy study. *Psychiatry Research, 182*, 53–57.

Gottschalk, P., & Kosten, T. (2002). Cerebral perfusion defects in combined cocaine and alcohol dependence. *Drug and Alcohol Dependence, 68*, 95.

Grodin, E. N., Lin, H., Durkee, C. A., Hommer, D. W., & Momenan, R. (2013). Deficits in cortical, diencephalic and midbrain gray matter in alcoholism measured by VBM: Effects of co-morbid substance abuse. *NeuroImage. Clinical, 2*, 469–476.

Haber, S. N., & Knutson, B. (2010). The reward circuit: Linking primate anatomy and human imaging. *Neuropsychopharmacology, 35*, 4–26.

Hanlon, C. A., Beveridge, T. J., & Porrino, L. J. (November 2013). Recovering from cocaine: Insights from clinical and preclinical investigations. *Neuroscience and Biobehavioral Reviews, 37*(9 Pt A), 2037–2046.

Hanlon, C. A., Dufault, D. L., Wesley, M. J., & Porrino, L. J. (2011). Elevated gray and white matter densities in cocaine abstainers compared to current users. *Psychopharmacology, 218*, 681–692.

Hanson, K. L., Luciana, M., & Sullwold, K. (2008). Reward-related decision-making deficits and elevated impulsivity among MDMA and other drug users. *Drug and Alcohol Dependence, 96*, 99–110.

Heatherton, T. F., & Wagner, D. D. (2011). Cognitive neuroscience of self-regulation failure. *Trends in Cognitive Sciences, 15*, 132–139.

Heinz, A., Beck, A., Grusser, S. M., Grace, A. A., & Wrase, J. (2009). Identifying the neural circuitry of alcohol craving and relapse vulnerability. *Addiction Biology, 14*, 108–118.

Hermann, D., Sartorius, A., Welzel, H., Walter, S., Skopp, G., Ende, G., & Mann, K. (2007). Dorsolateral prefrontal cortex N-acetylaspartate/total creatine (NAA/tCr) loss in male recreational cannabis users. *Biological Psychiatry, 61*, 1281–1289.

Hester, R., & Garavan, H. (2004). Executive dysfunction in cocaine addiction: Evidence for discordant frontal, cingulate, and cerebellar activity. *Journal of Neuroscience, 24*, 11017–11022.

Hill, S. Y., Wang, S., Kostelnik, B., Carter, H., Holmes, B., McDermott, M., … Keshavan, M. S. (2009). Disruption of orbitofrontal cortex laterality in offspring from multiplex alcohol dependence families. *Biological Psychiatry, 65*, 129–136.

Hohmann, A. A., & Shear, M. K. (2002). Community-based intervention research: Coping with the "noise" of real life in study design. *The American Journal of Psychiatry, 159*, 201–207.

Holman, B. L., Carvalho, P. A., Mendelson, J., Teoh, S. K., Nardin, R., Hallgring, E., … Johnson, K. A. (1991). Brain perfusion is abnormal in cocaine-dependent polydrug users: A study using technetium-99m-HMPAO and ASPECT. *Journal of Nuclear Medicine, 32*, 1206–1210.

Horner, M. D. (1997). Cognitive functioning in alcoholic patients with and without cocaine dependence. *Archives of Clinical Neuropsychology, 12*, 667–676.

Hwang, J., Lyoo, I. K., Kim, S. J., Sung, Y. H., Bae, S., Cho, S. N., … Renshaw, P. F. (2006). Decreased cerebral blood flow of the right anterior cingulate cortex in long-term and short-term abstinent methamphetamine users. *Drug and Alcohol Dependence, 82*, 177–181.

Iudicello, J. E., Woods, S. P., Vigil, O., Scott, J. C., Cherner, M., Heaton, R. K., … Grant, I. (2010). Longer term improvement in neurocognitive functioning and affective distress among methamphetamine users who achieve stable abstinence. *Journal of Clinical and Experimental Neuropsychology, 32*, 704–718.

Kaag, A. M., van Wingen, G. A., Caan, M. W., Homberg, J. R., van den Brink, W., & Reneman, L. (February 10, 2016). White matter alterations in cocaine users are negatively related to the number of additionally (ab)used substances. *Addiction Biology.* http://dx.doi.org/10.1111/adb.12375. [Epub ahead of print] PubMed PMID: 26860848.

Kalechstein, A. D., Newton, T. F., & Green, M. (2003). Methamphetamine dependence is associated with neurocognitive impairment in the initial phases of abstinence. *The Journal of Neuropsychiatry and Clinical Neurosciences, 15*, 215–220.

Kalivas, P. W. (2009). The glutamate homeostasis hypothesis of addiction. *Nature Reviews. Neuroscience, 10*, 561–572.

Kedia, S., Sell, M. A., & Relyea, G. (2007). Mono- versus polydrug abuse patterns among publicly funded clients. *Substance Abuse Treatment, Prevention, and Policy, 2*, 33.

Kenna, G. A., Nielsen, D. M., Mello, P., Schiesl, A., & Swift, R. M. (2007). Pharmacotherapy of dual substance abuse and dependence. *CNS Drugs, 21*, 213–237.

Ke, Y., Streeter, C. C., Nassar, L. E., Sarid-Segal, O., Hennen, J., Yurgelun-Todd, D. A., … Renshaw, P. F. (2004). Frontal lobe GABA levels in cocaine dependence: A two-dimensional, J-resolved magnetic resonance spectroscopy study. *Psychiatry Research, 130*, 283–293.

Kim, S. J., Lyoo, I. K., Hwang, J., Sung, Y. H., Lee, H. Y., Lee, D. S., … Renshaw, P. F. (2005). Frontal glucose hypometabolism in abstinent methamphetamine users. *Neuropsychopharmacology, 30*, 1383–1391.

Koob, G. F., & Volkow, N. (2009). Neurocircuitry of addiction. *Neuropsychopharmacology Reviews, 35*, 217–238.

Kosten, T. R. (1991). Client issues in drug abuse treatment: Addressing multiple drug abuse. *NIDA Research Monograph, 106*, 136–151.

Kubler, A., Murphy, K., & Garavan, H. (2005). Cocaine dependence and attention switching within and between verbal and visuospatial working memory. *The European Journal of Neuroscience, 21*, 1984–1992.

Lawton-Craddock, A., Nixon, S. J., & Tivis, R. (2003). Cognitive efficiency in stimulant abusers with and without alcohol dependence. *Alcoholism, Clinical and Experimental Research, 27*, 457–464.

Licata, S. C., & Renshaw, P. F. (2010). Neurochemistry of drug action: Insights from proton magnetic resonance spectroscopic imaging and their relevance to addiction. *Annals of the New York Academy of Sciences, 1187*, 148–171.

Liu, X., Matochik, J. A., Cadet, J. L., & London, E. D. (1998). Smaller volume of prefrontal lobe in polysubstance abusers: A magnetic resonance imaging study. *Neuropsychopharmacology, 18*, 243–252.

Loree, A. M., Lundahl, L. H., & Ledgerwood, D. M. (2015). Impulsivity as a predictor of treatment outcome in substance use disorders: Review and synthesis. *Drug and Alcohol Review, 34*, 119–134.

Lundqvist, T. (2005). Cognitive consequences of cannabis use: Comparison with abuse of stimulants and heroin with regard to attention, memory and executive functions. *Pharmacology, Biochemistry, and Behavior, 81*, 319–330.

Lyoo, I. K., Kim, J. E., Yoon, S. J., Hwang, J., Bae, S., & Kim, D. J. (2011). The neurobiological role of the dorsolateral prefrontal cortex in recovery from trauma. Longitudinal brain imaging study among survivors of the South Korean subway disaster. *Archives of General Psychiatry, 68*, 701–713.

MacKay, S., Meyerhoff, D. J., Dillon, W. P., Weiner, M. W., & Fein, G. (1993). Alteration of brain phospholipid metabolites in cocaine-dependent polysubstance abusers. *Biological Psychiatry, 34*, 261–264.

Mackey, S., & Paulus, M. (2013). Are there volumetric brain differences associated with the use of cocaine and amphetamine-type stimulants? *Neuroscience and Biobehavioral Reviews, 37*, 300–316.

Makris, N., Gasic, G. P., Kennedy, D. N., Hodge, S. M., Kaiser, J. R., Lee, M. J., … Breiter, H. C. (2008). Cortical thickness abnormalities in cocaine addiction—a reflection of both drug use and a pre-existing disposition to drug abuse? *Neuron, 60*, 174–188.

Makris, N., Oscar-Berman, M., Jaffin, S. K., Hodge, S. M., Kennedy, D. N., Caviness, V. S., … Harris, G. J. (2008). Decreased volume of the brain reward system in alcoholism. *Biological Psychiatry, 64*, 192–202.

Matochik, J. A., Eldreth, D. A., Cadet, J. L., & Bolla, K. I. (2005). Altered brain tissue composition in heavy marijuana users. *Drug and Alcohol Dependence, 77*, 23–30.

Matochik, J. A., London, E. D., Eldreth, D. A., Cadet, J. L., & Bolla, K. I. (2003). Frontal cortical tissue composition in abstinent cocaine abusers: A magnetic resonance imaging study. *NeuroImage, 19*, 1095–1102.

McCance-Katz, E. F., Kosten, T. R., & Jatlow, P. (1998). Concurrent use of cocaine and alcohol is more potent and potentially more toxic than use of either alone—A multiple-dose study. *Biological Psychiatry, 44*, 250—259.

McKetin, R., & Mattick, R. P. (1997). Attention and memory in illicit amphetamine users. *Drug and Alcohol Dependence, 48*, 235—242.

McLellan, A. T., Lewis, D. C., O'Brien, C. P., & Kleber, H. D. (2000). Drug dependence, a chronic medical illness: Implications for treatment, insurance, and outcomes evaluation. *JAMA, 284*, 1689—1695.

Medina, K. L., Shear, P. K., Schafer, J., Armstrong, T. G., & Dyer, P. (2004). Cognitive functioning and length of abstinence in polysubstance dependent men. *Archives of Clinical Neuropsychology, 19*, 245—258.

Meyerhoff, D. J. (2014). Brain proton magnetic resonance spectroscopy of alcohol use disorders. *Handbook of Clinical Neurology, 125*, 313—337.

Meyerhoff, D. J., Bloomer, C., Salas, G., Schuff, N., Norman, D., Weiner, M. W., & Fein, G. (1999). Cortical metabolite in abstinent cocaine and cocaine/alcohol dependent subjects: An in-vivo proton magnetic resonance spectroscopic imaging study. *Addiction Biology, 4*, 405—419.

Meyerhoff, D., Blumenfeld, R., Truran, D., Lindgren, J., Flenniken, D., Cardenas, V., ... Weiner, H. (2004). Effects of heavy drinking, binge drinking, and family history of alcoholism on regional brain metabolites. *Alcoholism, Clinical and Experimental Research, 28*, 650—661.

Meyerhoff, D. J., Durazzo, T. C., & Ende, G. (2011). Chronic alcohol consumption, abstinence and relapse: Brain proton magnetic resonance spectroscopy studies in animals and humans. *Current Topics in Behavioral Neurosciences, 13*, 511—540.

Meyerhoff, D. J., Durazzo, T. C., & Ende, G. (2013). Chronic alcohol consumption, abstinence and relapse: Brain proton magnetic resonance spectroscopy studies in animals and humans. In *Behavioral neurobiology of alcohol addiction* (pp. 511—540). Springer.

Meyerhoff, D. J., MacKay, S., Weiner, M. W., & Fein, G. (1994). Reduced phospholipid resonances in chronic alcoholics. In *Research society on alcoholism meeting*.

Moeller, F. G., Hasan, K. M., Steinberg, J. L., Kramer, L. A., Dougherty, D. M., Santos, R. M., ... Narayana, P. A. (2005). Reduced anterior corpus callosum white matter integrity is related to increased impulsivity and reduced discriminability in cocaine-dependent subjects: Diffusion tensor imaging. *Neuropsychopharmacology, 30*, 610—617.

Mon, A., Durazzo, T. C., Abe, C., Gazdzinski, S., Pennington, D., Schmidt, T., & Meyerhoff, D. J. (2014). Structural brain differences in alcohol-dependent individuals with and without comorbid substance dependence. *Drug and Alcohol Dependence, 144*, 170—177.

Mon, A., Durazzo, T., & Meyerhoff, D. J. (2012). Glutamate, GABA, and other cortical metabolite concentrations during early abstinence from alcohol and their associations with neurocognitive changes. *Drug and Alcohol Dependence, 125*, 27—36.

Moody, L., Franck, C., Hatz, L., & Bickel, W. K. (February 2016). Impulsivity and polysubstance use: A systematic comparison of delay discounting in mono-, dual-, and trisubstance use. *Experimental and Clinical Psychopharmacology, 24*(1), 30—37.

Morales, A. M., Lee, B., Hellemann, G., O'Neill, J., & London, E. D. (2012). Gray-matter volume in methamphetamine dependence: Cigarette smoking and changes with abstinence from methamphetamine. *Drug and Alcohol Dependence, 125*, 230—238.

Moss, H. B., Goldstein, R. B., Chen, C. M., & Yi, H. Y. (2015). Patterns of use of other drugs among those with alcohol dependence: Associations with drinking behavior and psychopathology. *Addictive Behaviors, 50*, 192—198.

Muller, S. E., Weijers, H. G., Boning, J., & Wiesbeck, G. A. (2008). Personality traits predict treatment outcome in alcohol-dependent patients. *Neuropsychobiology, 57*, 159—164.

Murray, D. E., Durazzo, T. C., Mon, A., Schmidt, T. P., & Meyerhoff, D. J. (2015). Brain perfusion in polysubstance users: Relationship to substance and tobacco use, cognition, and self-regulation. *Drug and Alcohol Dependence, 150*, 120—128.

Nestor, L., McCabe, E., Jones, J., Clancy, L., & Garavan, H. (2011). Differences in "bottom-up" and "top-down" neural activity in current and former cigarette smokers: Evidence for neural substrates which may promote nicotine abstinence through increased cognitive control. *NeuroImage, 56*, 2258—2275.

Nicolas, J. M., Catafau, A. M., Estruch, R., Lomena, F. J., Salamero, M., Herranz, R., ... Urbano-Marquez, A. (1993). Regional cerebral blood flow-SPECT in chronic alcoholism: Relation to neuropsychological testing. *Journal of Nuclear Medicine, 34*, 1452—1459.

Nixon, S. J., Paul, R., & Phillips, M. (1998). Cognitive efficiency in alcoholics and polysubstance abusers. *Alcoholism, Clinical and Experimental Research, 22*, 1414—1420.

Nordahl, T. E., Salo, R., Natsuaki, Y., Galloway, G. P., Waters, C., Moore, C. D., ... Buonocore, M. H. (2005). Methamphetamine users in sustained abstinence: A proton magnetic resonance spectroscopy study. *Archives of General Psychiatry, 62*, 444—452.

Noyan, C. O., Kose, S., Nurmedov, S., Metin, B., Darcin, A. E., & Dilbaz, N. (2016). Volumetric brain abnormalities in polysubstance use disorder patients. *Neuropsychiatric Disease and Treatment*, 1355—1363.

O'Brien, C. P., & Lynch, K. G. (2003). Can we design and replicate clinical trials with a multiple drug focus. *Drug and Alcohol Dependence, 70*, 135—137.

O'Neill, J., Cardenas, V. A., & Meyerhoff, D. J. (2001). Separate and interactive effects of cocaine and alcohol dependence on brain structures and metabolites: Quantitative MRI and proton MR spectroscopic imaging. *Addiction Biology, 6*, 347—361.

Oishi, M., Mochizuki, Y., & Shikata, E. (1999). Corpus callosum atrophy and cerebral blood flow in chronic alcoholics. *Journal of the Neurological Sciences, 162*, 51—55.

Oscar-Berman, M. (1997). Impairments of brain and behavior: The neurological effects of alcohol. *Alcohol Health and Research World, 21*, 65—75.

Oscar-Berman, M. (2000). NIAAA research monograph no. 34: Neuropsychological vulnerabilities in chronic alcoholism. In *Review of NIAAA's neuroscience and behavioral research portfolio* (pp. 437—472). Bethesda, MD: NIAAA.

Parks, M. H., Dawant, B. M., Riddle, W. R., Hartmann, S. L., Dietrich, M. S., Nickel, M. K., ... Martin, P. R. (2002). Longitudinal brain metabolic characterization of chronic alcoholics with proton magnetic resonance spectroscopy. *Alcoholism, Clinical and Experimental Research, 26*, 1368—1380.

Passetti, F., Clark, L., Mehta, M. A., Joyce, E., & King, M. (2008). Neuropsychological predictors of clinical outcome in opiate addiction. *Drug and Alcohol Dependence, 94*, 82—91.

Patkar, A. A., Mannelli, P., Peindl, K., Murray, H. W., Meier, B., & Leone, F. T. (2006). Changes in tobacco smoking following treatment for cocaine dependence. *The American Journal of Drug and Alcohol Abuse, 32*, 135—148.

Patterson, F., Kerrin, K., Wileyto, E. P., & Lerman, C. (2008). Increase in anger symptoms after smoking cessation predicts relapse. *Drug and Alcohol Dependence, 95*, 173—176.

Payer, D. E., Park, M. T., Kish, S. J., Kolla, N. J., Lerch, J. P., Boileau, I., & Chakravarty, M. M. (2015). Personality disorder symptomatology is associated with anomalies in striatal and prefrontal morphology. *Frontiers in Human Neuroscience, 9*, 472.

Pennington, D. L., Durazzo, T. C., Schmidt, T. P., Abe, C., Mon, A., & Meyerhoff, D. J. (2015). Alcohol use disorder with and without stimulant use: Brain morphometry and its associations with cigarette smoking, cognition, and inhibitory control. *PLoS One, 10*, e0122505.

Pennington, D. L., Durazzo, T. C., Schmidt, T. P., Mon, A., Abe, C., & Meyerhoff, D. J. (2013). The effects of chronic cigarette smoking on cognitive recovery during early abstinence from alcohol. *Alcoholism, Clinical and Experimental Research, 37*, 1220–1227.

Pfefferbaum, A., Adalsteinsson, E., & Sullivan, E. V. (July 2006). Dysmorphology and microstructural degradation of the corpus callosum: Interaction of age and alcoholism. *Neurobiology of Aging, 27*(7), 994–1009.

Pfefferbaum, A., & Sullivan, E. V. (2002). Microstructural but not macrostructural disruption of white matter in women with chronic alcoholism. *NeuroImage, 15*, 708–718.

Pope, H. G., Jr., Gruber, A. J., Hudson, J. I., Huestis, M. A., & Yurgelun-Todd, D. (2001). Neuropsychological performance in long-term cannabis users. *Archives of General Psychiatry, 58*, 909–915.

Pope, H. G., Jr., & Yurgelun-Todd, D. (1996). The residual cognitive effects of heavy marijuana use in college students. *JAMA, 275*, 521–527.

Potenza, M. N., & de Wit, H. (2010). Control yourself: Alcohol and impulsivity. *Alcoholism, Clinical and Experimental Research, 34*, 1303–1305.

Prochaska, J. J., Delucchi, K., & Hall, S. M. (2004). A meta-analysis of smoking cessation interventions with individuals in substance abuse treatment or recovery. *Journal of Consulting and Clinical Psychology, 72*, 1144–1156.

Quek, L. H., Chan, G. C., White, A., Connor, J. P., Baker, P. J., Saunders, J. B., & Kelly, A. B. (2013). Concurrent and simultaneous polydrug use: Latent class analysis of an Australian nationally representative sample of young adults. *Frontiers in Public Health, 1*, 61.

Rando, K., Hong, K. I., Bhagwagar, Z., Li, C. S., Bergquist, K., Guarnaccia, J., & Sinha, R. (February 2011). Association of frontal and posterior cortical gray matter volume with time to alcohol relapse: A prospective study. *The American Journal of Psychiatry, 168*(2), 183–192.

Reske, M., & Paulus, M. P. (2008). Predicting treatment outcome in stimulant dependence. *Annals of the New York Academy of Sciences, 1141*, 270–283.

Richter, K. P., Ahluwalia, H. K., Mosier, M. C., Nazir, N., & Ahluwalia, J. S. (2002). A population-based study of cigarette smoking among illicit drug users in the United States. *Addiction, 97*, 861–869.

Rogers, R. L., Meyer, J. S., Shaw, T. G., & Mortel, K. F. (1983). Reductions in regional cerebral blood flow associated with chronic consumption of alcohol. *Journal of the American Geriatrics Society, 31*, 540–543.

Rounsaville, B. J., Petry, N. M., & Carroll, K. M. (2003). Single versus multiple drug focus in substance abuse clinical trials research. *Drug and Alcohol Dependence, 70*, 117–125.

Rupp, C. I., Beck, J. K., Heinz, A., Kemmler, G., Manz, S., Tempel, K., & Fleischhacker, W. W. (2016). Impulsivity and alcohol dependence treatment completion: Is there a neurocognitive risk factor at treatment entry? *Alcoholism, Clinical and Experimental Research, 40*, 152–160.

Sailasuta, N., Abulseoud, O., Hernandez, M., Haghani, P., & Ross, B. D. (2010). Metabolic abnormalities in abstinent methamphetamine dependent subjects. *Substance Abuse, 2010*, 9–20.

Salo, R., & Fassbender, C. (2012). Structural, functional and spectroscopic MRI studies of methamphetamine addiction. *Current Topics in Behavioral Neurosciences, 11*, 321–364.

Salo, R., Nordahl, T. E., Possin, K., Leamon, M., Gibson, D. R., Galloway, G. P., ... Sullivan, E. V. (2002). Preliminary evidence of reduced cognitive inhibition in methamphetamine-dependent individuals. *Psychiatry Research, 111*, 65–74.

Schlaepfer, T. E., Lancaster, E., Heidbreder, R., Strain, E. C., Kosel, M., Fisch, H. U., & Pearlson, G. D. (2006). Decreased frontal white-matter volume in chronic substance abuse. *International Journal of Neuropsychopharmacology, 9*, 147–153.

Schmidt, T. P., Pennington, D. L., Cardoos, S. L., Durazzo, T. C., & Meyerhoff, D. J. (2017). Neurocognition and inhibitory control in polysubstance use disorders: Comparison with alcohol use disorders and changes with abstinence. *Journal of Clinical and Experimental Neuropsychology, 39*, 22–34.

Schulte, M. H., Cousijn, J., den Uyl, T. E., Goudriaan, A. E., van den Brink, W., Veltman, D. J., ... Wiers, R. W. (2014). Recovery of neurocognitive functions following sustained abstinence after substance dependence and implications for treatment. *Clinical Psychology Review, 34*, 531–550.

Selby, M. J., & Azrin, R. L. (1998). Neuropsychological functioning in drug abusers. *Drug and Alcohol Dependence, 50*, 39–45.

Simon, S. L., Dacey, J., Glynn, S., Rawson, R., & Ling, W. (2004). The effect of relapse on cognition in abstinent methamphetamine abusers. *Journal of Substance Abuse Treatment, 27*, 59–66.

Simon, S. L., Dean, A. C., Cordova, X., Monterosso, J. R., & London, E. D. (2010). Methamphetamine dependence and neuropsychological functioning: Evaluating change during early abstinence. *Journal of Studies on Alcohol and Drugs, 71*, 335–344.

Simon, S. L., Domier, C., Carnell, J., Brethen, P., Rawson, R., & Ling, W. (2000). Cognitive impairment in individuals currently using methamphetamine. *The American Journal on Addictions, 9*, 222–231.

Sofuoglu, M., DeVito, E. E., Waters, A. J., & Carroll, K. M. (2013). Cognitive enhancement as a treatment for drug addictions. *Neuropharmacology, 64*, 452–463.

Solowij, N., Stephens, R. S., Roffman, R. A., Babor, T., Kadden, R., Miller, M., ... Vendetti, J. (2002). Cognitive functioning of long-term heavy cannabis users seeking treatment. *JAMA, 287*, 1123–1131.

Sorg, S. F., Taylor, M. J., Alhassoon, O. M., Gongvatana, A., Theilmann, R. J., Frank, L. R., & Grant, I. (February 1, 2012). Frontal white matter integrity predictors of adult alcohol treatment outcome. *Biological Psychiatry, 71*(3), 262–268.

Stapleton, J. M., Morgan, M. J., Phillips, R. L., Wong, D. F., Yung, B. C., Shaya, E. K., ... London, E. D. (1995). Cerebral glucose utilization in polysubstance abuse. *Neuropsychopharmacology, 13*, 21–31.

Stark, M. J., & Campbell, B. K. (1993). Drug use and cigarette smoking in applicants for drug abuse treatment. *Journal of Substance Abuse, 5*, 175–181.

Stephens, D. N., Duka, T., Crombag, H. S., Cunningham, C. L., Heilig, M., & Crabbe, J. C. (2010). Reward sensitivity: Issues of measurement, and achieving consilience between human and animal phenotypes. *Addiction Biology, 15*, 145–168.

Strain, E. C. (2003). Single versus multiple drug focus in substance abuse clinical trials research: The devil is in the details. *Drug and Alcohol Dependence, 70*, 131–134.

Streeter, C. C., Terhune, D. B., Whitfield, T. H., Gruber, S., Sarid-Segal, O., Silveri, M. M., ... Yurgelun-Todd, D. A. (2008). Performance on the Stroop predicts treatment compliance in cocaine-dependent individuals. *Neuropsychopharmacology, 33*, 827–836.

Sullivan, E. V. (2000). NIAAA Research Monograph No. 34: Human brain vulnerability to alcoholism: Evidence from neuroimaging studies. In A. Noronha, M. Eckardt, & K. Warren (Eds.), *Review of NIAAA's neuroscience and behavioral research portfolio* (pp. 473–508). Bethesda, MD: National Institute on Alcohol Abuse and Alcoholism.

Sullivan, E. V., & Pfefferbaum, A. (2005). Neurocircuitry in alcoholism: A substrate of disruption and repair. *Psychopharmacology, 180*, 583–594.

Sullivan, E. V., Rosenbloom, M. J., & Pfefferbaum, A. (2000). Brain vulnerability to alcoholism: Evidence from neuroimaging studies. In *NIAAA*.

Tanabe, J., Tregellas, J. R., Dalwani, M., Thompson, L., Owens, E., Crowley, T., & Banich, M. (2009). Medial orbitofrontal cortex gray matter is reduced in abstinent substance-dependent individuals. *Biological Psychiatry, 65*, 160–164.

Tanabe, J., York, P., Krmpotich, T., Miller, D., Dalwani, M., Sakai, J. T., … Rojas, D. C. (2013). Insula and orbitofrontal cortical morphology in substance dependence is modulated by sex. *AJNR. American Journal of Neuroradiology, 34*, 1150–1156.

Thompson, P. M., Hayashi, K. M., Simon, S. L., Geaga, J. A., Hong, M. S., Sui, Y., … London, E. D. (2004). Structural abnormalities in the brains of human subjects who use methamphetamine. *Journal of Neuroscience, 24*, 6028–6036.

Tunving, K., Thulin, S. O., Risberg, J., & Warkentin, S. (1986). Regional cerebral blood flow in long-term heavy cannabis use. *Psychiatry Research, 17*, 15–21.

Verdejo-Garcia, A., & Bechara, A. (2009). A somatic marker theory of addiction. *Neuropharmacology, 56*(Suppl. 1), 48–62.

Verdejo-Garcia, A., Betanzos-Espinosa, P., Lozano, O. M., Vergara-Moragues, E., Gonzalez-Saiz, F., Fernandez-Calderon, F., … Perez-Garcia, M. (April 1, 2012). Self-regulation and treatment retention in cocaine dependent individuals: A longitudinal study. *Drug and Alcohol Dependence, 122*(1–2), 142–148.

Verdejo-Garcia, A., Lawrence, A. J., & Clark, L. (2008). Impulsivity as a vulnerability marker for substance-use disorders: Review of findings from high-risk research, problem gamblers and genetic association studies. *Neuroscience and Biobehavioral Reviews, 32*, 777–810.

Verdejo-Garcia, A., Lopez-Torrecillas, F., Gimenez, C. O., & Perez-Garcia, M. (2004). Clinical implications and methodological challenges in the study of the neuropsychological correlates of cannabis, stimulant, and opioid abuse. *Neuropsychology Review, 14*, 1–41.

Verdejo-Garcia, A., Rivas-Perez, C., Vilar-Lopez, R., & Perez-Garcia, M. (2007). Strategic self-regulation, decision-making and emotion processing in poly-substance abusers in their first year of abstinence. *Drug and Alcohol Dependence, 86*, 139–146.

Vocci, F. J. (2008). Cognitive remediation in the treatment of stimulant abuse disorders: A research agenda. *Experimental and Clinical Psychopharmacology, 16*, 484–497.

Volkow, N. D., & Baler, R. D. (2012). Neuroscience. To stop or not to stop? *Science, 335*, 546–548.

Volkow, N. D., Chang, L., Wang, G. J., Fowler, J. S., Franceschi, D., Sedler, M. J., … Logan, J. (2001). Higher cortical and lower subcortical metabolism in detoxified methamphetamine abusers. *The American Journal of Psychiatry, 158*, 383–389.

Volkow, N. D., Chang, L., Wang, G. J., Fowler, J. S., Leonido-Yee, M., Franceschi, D., … Miller, E. N. (2001). Association of dopamine transporter reduction with psychomotor impairment in methamphetamine abusers. *The American Journal of Psychiatry, 158*, 377–382.

Volkow, N. D., Fowler, J. S., Wang, G. J., Hitzemann, R., Logan, J., Schlyer, D. J., … Wolf, A. P. (1993). Decreased dopamine D2 receptor availability is associated with reduced frontal metabolism in cocaine abusers. *Synapse, 14*, 169–177.

Volkow, N. D., Hitzemann, R., Wang, G. J., Fowler, J. S., Burr, G., Pascani, K., … Wolf, A. P. (1992). Decreased brain metabolism in neurologically intact healthy alcoholics. *The American Journal of Psychiatry, 149*, 1016–1022.

Volkow, N. D., Hitzemann, R., Wang, G. J., Fowler, J. S., Wolf, A. P., Dewey, S. L., & Handlesman, L. (1992). Long-term frontal brain metabolic changes in cocaine abusers. *Synapse, 11*, 184–190.

Volkow, N. D., Mullani, N., Gould, K. L., Adler, S., & Krajewski, K. (1988). Cerebral blood flow in chronic cocaine users: A study with positron emission tomography. *The British Journal of Psychiatry, 152*, 641–648.

Volkow, N. D., Wang, G. J., Fowler, J. S., & Telang, F. (2008). Overlapping neuronal circuits in addiction and obesity: Evidence of systems pathology. *Philosophical Transactions of the Royal Society of London. Series B, Biological Sciences, 363*, 3191–3200.

Volkow, N. D., Wang, G. J., Fowler, J. S., & Tomasi, D. (2012). Addiction circuitry in the human brain. *Annual Review of Pharmacology and Toxicology, 52*, 321–336.

Volkow, N. D., Wang, G.-J., Fowler, J. S., Tomasi, D., & Telang, F. (September 13, 2011). Addiction: Beyond dopamine reward circuitry. *Proceedings of the National Academy of Sciences of the United States of America, 108*(37), 15037–15042.

Volkow, N. D., Wang, G. J., Hitzemann, R., Fowler, J. S., Overall, J. E., Burr, G., & Wolf, A. P. (1994). Recovery of brain glucose metabolism in detoxified alcoholics. *The American Journal of Psychiatry, 151*, 178–183.

Voytek, B., Berman, S. M., Hassid, B. D., Simon, S. L., Mandelkern, M. A., Brody, A. L., … London, E. D. (2005). Differences in regional brain metabolism associated with marijuana abuse in methamphetamine abusers. *Synapse, 57*, 113–115.

Vuchinich, R. E., & Simpson, C. A. (1998). Hyperbolic temporal discounting in social drinkers and problem drinkers. *Experimental and Clinical Psychopharmacology, 6*, 292–305.

Wang, C., Xu, X., Qian, W., Shen, Z., & Zhang, M. (2015). Altered human brain anatomy in chronic smokers: A review of magnetic resonance imaging studies. *Neurological Sciences, 36*, 497–504.

Weber, D. A., Franceschi, D., Ivanovic, M., Atkins, H. L., Cabahug, C., Wong, C. T., & Susskind, H. (1993). SPECT and planar brain imaging in crack abuse: Iodine-123-iodoamphetamine uptake and localization. *Journal of Nuclear Medicine, 34*, 899–907.

Wexler, B. E. (2011). Computerized cognitive remediation treatment for substance abuse disorders. *Biological Psychiatry, 69*, 197–198.

Wilcox, C. E., Dekonenko, C. J., Mayer, A. R., Bogenschutz, M. P., & Turner, J. A. (2014). Cognitive control in alcohol use disorder: Deficits and clinical relevance. *Reviews in the Neurosciences, 25*, 1–24.

de Wit, H. (2009). Impulsivity as a determinant and consequence of drug use: A review of underlying processes. *Addiction Biology, 14*, 22–31.

Wrase, J., Makris, N., Braus, D., Mann, K., Smolka, M., Kennedy, D., … Heinz, A. (2008). Amygdala volume associated with alcohol abuse relapse and craving. *American Journal of Psychiatry, 165*, 1179–1184.

Wright, C. I., Feczko, E., Dickerson, B., & Williams, D. (2007). Neuroanatomical correlates of personality in the elderly. *NeuroImage, 35*, 263–272.

Xiao, P., Dai, Z., Zhong, J., Zhu, Y., Shi, H., & Pan, P. (2015). Regional gray matter deficits in alcohol dependence: A meta-analysis of voxel-based morphometry studies. *Drug and Alcohol Dependence, 153*, 22–28.

Xu, J., DeVito, E. E., Worhunsky, P. D., Carroll, K. M., Rounsaville, B. J., & Potenza, M. N. (2010). White matter integrity is associated with treatment outcome measures in cocaine dependence. *Neuropsychopharmacology, 35*, 1541–1549.

Yang, S., Salmeron, B. J., Ross, T. J., Xi, Z. X., Stein, E. A., & Yang, Y. (2009). Lower glutamate levels in rostral anterior cingulate of chronic cocaine users – A (1)H-MRS study using TE-averaged PRESS at 3 T with an optimized quantification strategy. *Psychiatry Research, 174*, 171–176.

Yoon, S. J., Lyoo, I. K., Kim, H. J., Kim, T. S., Sung, Y. H., Kim, N., … Renshaw, P. F. (2010). Neurochemical alterations in methamphetamine-dependent patients treated with cytidine-5′-diphosphate choline: A longitudinal proton magnetic resonance spectroscopy study. *Neuropsychopharmacology, 35*, 1165–1173.

Yucel, M., & Lubman, D. I. (2007). Neurocognitive and neuroimaging evidence of behavioural dysregulation in human drug addiction: Implications for diagnosis, treatment and prevention. *Drug and Alcohol Review, 26*, 33–39.

Zahr, N. M. (2014). Structural and microstructral imaging of the brain in alcohol use disorders. *Handbook of Clinical Neurology, 125*, 275–290.

CHAPTER

22

Deep Brain Stimulation: A Possible Therapeutic Technique for Treating Refractory Alcohol and Drug Addiction Behaviors

S.R. Hauser[1], *J.A. Wilden*[2], *V. Batra*[3], *Z.A. Rodd*[1]

[1]Indiana University School of Medicine, Indianapolis, IN, United States; [2]Willis-Knighton Health Systems, Shreveport, LA, United States; [3]Louisiana State University Health Sciences Center, Shreveport, LA, United States

INTRODUCTION

Addiction is probably the most common psychiatric disorder in the world. Incorporating all forms of addiction (drugs, sex, eating, gambling, and others) a significantly large portion of the world population has an addiction. In general, there are no effective pharmacological or therapeutic treatments for addiction. Therefore, Alcoholics Anonymous (AA), for example, is considered the standard behavioral treatment for alcoholism, and similar programs are used for other drugs of abuse (e.g., Narcotic Anonymous). About 50% of individuals leave AA within the first 6 months, and 95% leave after the first year. Vaillant (2005) concluded that there were no significant differences between untreated alcoholics and AA participants. In Project Match, all cognitive behavioral therapies had highly variable/short-lived effectiveness in promoting and maintaining abstinence, especially for those with severe alcoholism, and did not differ from untreated patients (Cutler & Fishbain, 2005). Methadone is widely thought to be a pharmacological substitute for opioid addictions. The failure of methadone therapy is clearly indicated in the 2011/2012 Crime Survey of England and Wales which revealed that the majority of methadone patients simultaneously use other drugs of abuse/alcohol including heroin. For cocaine, there is no United States Food and Drug Administration (FDA) approved pharmacological treatment and behavioral therapies have little, no, or harmful effects (Fischer et al., 2015). Given the lack of effective pharmacological or therapeutic treatment and the great need for successful treatment for addiction, the use of a surgical treatment for drug and alcohol abuse could be seen as the last attempt to save an individual's life.

Deep Brain Stimulation

Deep brain stimulation (DBS) is a surgical technique that has been used to treat neurological disorders for over 50 years. Although electrical stimulation of the brain has been investigated as early as 1870s, the history of chronically stimulating targeted brain areas for treating neurological disorders began to arise during 1960s (Perlmutter & Mink, 2006). By 1970s there were reports on patients being therapeutically treated with chronic stimulation for movement, pain, or epilepsy disorders (Perlmutter & Mink, 2006). During the late 1980s to early 1990s combining the techniques of implanting deep brain electrodes connected via wires to an implantable subcutaneous external pacemaker (called neurostimulator/battery pack) under skin of the upper chest resulted in the modern-day use of long-term chronic DBS (Benabid et al., 1996, 1991; Benabid, Pollak, Louveau, Henry, & de Rougemont, 1987; Perlmutter & Mink, 2006). Currently, DBS utilizes stereotactic surgery and neuroimaging techniques to implant a micro-electrode into the brain connected to a subcutaneous neurostimulator that controls the electrical pulses sent to the targeted brain area (McIntyre, Savasta, Kerkerian-Le Goff, & Vitek, 2004). Chronic high-frequency stimulation (HFS; 120–180 Hz) is commonly used for the therapeutic DBS.

Addictive Substances and Neurological Disease
http://dx.doi.org/10.1016/B978-0-12-805373-7.00022-0

While the mechanisms underlying the effectiveness of DBS are still poorly understood, it has been reported that deep brain HFS appears to inhibit neuronal impulses as well as excite other neurons at their axon terminals (Benabid, 2015). This dual effect may be due to the fact that DBS is not restricted to the targeted brain area and can induce changes in the surrounding electrical fields, which have a variable effect on nearby regions, depending on the neuronal composition (Benabid, 2015). In addition, DBS can increase neurotransmission [i.e., dopamine (DA), serotonin (5-HT), and gamma-aminobutyric acid (GABA)] and sustain changes in the firing of neurons (McIntyre et al., 2004).

DBS is approved by the FDA to treat essentially tremors, Parkinson's disease, dystonia, and obsessive–compulsive disorders (OCDs) that are treatment resistant to pharmacological and/or behavioral therapies (McIntyre et al., 2004). Clinical and preclinical researchers are broadening their investigations of possible therapeutic effects of DBS in various brain targets to other neuropsychiatric disorders (i.e., depression and Tourette's syndrome) (Kringelbach et al., 2007; Yadid, Gispan, & Lax, 2013). Recently, there has been emerging interest in examining DBS effects on drug addiction behaviors.

Drug Addiction Disorders

All forms of addiction are characterized as a chronically relapsing condition that is typified by continuing to use a drug despite negative consequences of that use. Chronic use of drugs of abuse can cause or exacerbate health problems, increase mortality rates, cause social problems at home, work, or school, increase illegal activities, and cost society billions of dollars each year. Millions of individuals meet Diagnostic and Statistical Manual of Mental Disorders-V (DSM-V) criteria for addiction to drugs of abuse, and substance abuse is indisputably a brain disease. Drugs of abuse can alter brain functioning and lead to long-lasting neuroadaptive changes in the brain that can impair cognition and behavioral self-control and increase drug reward and craving leading to recurrent relapses (cf. Everitt et al., 2008). It is likely the plethora of changes produced by chronic drug use that has inhibited the development of successful pharmacotherapeutics for the treatment of addiction and why the more general action of DBS could be effective.

The targets of interest for DBS treatment are either neural areas within or those that act upon the mesocorticolimbic (MCL) DA system which consists of DA neurons that originate in the ventral tegmental area (VTA) projecting to nucleus accumbens (Acb), the amygdala, and the frontal cortex (Feltenstein & See, 2008). The MCL-DA is involved in the reinforcement of drugs of abuse, drug craving, motivation, learning, memory, and movement (Feltenstein & See, 2008). It is the primary system studied to get better understanding of the underlying neural substrates and mechanisms that are involved in mediating drug addiction behaviors. It is key to highlight that downstream structures, including the Acb, innervate the VTA as a feedback loop and may be able to alter the function of the MCL reward pathway. The aim of this chapter is to focus on the possible application of using DBS for drug addictions such as alcohol, cocaine, methamphetamine, nicotine, and heroin.

EFFECTS OF DEEP BRAIN STIMULATION ON ALCOHOL ADDICTION BEHAVIORS

Alcoholism is chronic and recurring illness that leads to significant mental and physical consequence, and it is the third leading preventable cause of death in the world (Lim et al., 2012). It has been reported that about 17 million people in the United States have been diagnosed with alcohol addiction (National Institute on Alcohol Abuse and Alcoholism, 2016). It is estimated that 45–75% of treated alcoholics will relapse within 3 years (Anton et al., 2006; Bottlender & Soyka, 2005; Finney & Moos, 1991), therefore there is a need for more effective therapies.

Effects of DBS of the Nucleus Accumbens on Alcohol Consumption in Rodents

There have been a limited number of preclinical studies that have examined the effects of the DBS of the Acb on the ethanol (EtOH) consumptions in rodent models. The Acb is divided into two major subregions, known as the core (AcbCore) and shell (AcbSh). The AcbSh has been implicated in the rewarding effects of various drugs that increase local DA release (Di Chiara & Imperato, 1988), and it mediates EtOH- seeking behaviors (Chaudhri, Sahuque, & Janak, 2009; Hauser et al., 2015). As for the AcbCore, it mediates cue-induced drug-seeking behaviors (Bossert, Poles, Wihbey, Koya, & Shaham, 2007; Chaudhri, Sahuque, Cone, & Janak, 2008; Fuchs, Evans, Parker, & See, 2004) and the incentive value of reward-conditioned stimuli (Ito, Dalley, Howes, Robbins, & Everitt, 2000; Ito, Robbins, Everitt, 2004), but it does not appear to have a primary role in the reinforcement of EtOH (Ding, Ingraham, Rodd, & McBride, 2015; Engleman et al., 2009). Interestingly, DBS (130 Hz; 30 μA) in naive animals can increase DA and serotonin release and lower DA/serotonin turnover in the AcbSh (Sesia et al., 2010). In contrast, AcbCore DA release was not altered by DBS (130 Hz or 120 Hz) at currents of 30 μA (Sesia et al., 2010) or 300–400 μA (van Dijk, Klompmakers, Feenstra, &

Denys, 2012). However, DBS of the AcbCore can increase DA and serotonin levels in medial prefrontal cortex as well DA and noradrenaline in orbital frontal cortex (van Dijk, Mason, Klompmakers, Feenstra, & Denys, 2011). Overall, these findings provide evidence that Acb is a suitable target for investigating the effects of DBS on addiction behaviors.

Knapp, Tozier, Pak, Ciraulo, and Kornetsky (2009) first investigated the effects of DBS of the Acb on EtOH intake in Long–Evans rats. The DBS frequency was 160 Hz, and several electrical current intensities were administered bilaterally in the AcbSh or the AcbCore 5 min before and during the daily 30 min limited access to oral EtOH (Knapp et al., 2009). DBS of the AcbSh and AcbCore significantly reduced EtOH intake but only at 150 μA. This study provided evidence that DBS of Acb may be an effective tool in attenuating alcohol consumption and established that 150 μA was the minimum effective dose (Knapp et al., 2009).

Other studies have examined DBS of AcbSh in the selective genetically bred alcohol-preferring (P) rat, a line that is well characterized both behaviorally and neurobiologically as an animal model of alcoholism (McBride & Li, 1998; Murphy et al., 2002). In the P rat, the AcbSh, but not the AcbCore, supports the reinforcements of EtOH (Engleman et al., 2009), and there are differences in the sensitivity and response to EtOH of the AcbSh between P and their progenitor stock of Wistar rats (Engleman et al., 2009). These findings make the P rat a valuable tool to investigate alcohol addiction behaviors. In a Henderson et al. (2010) study, DBS (140–150 Hz; and 200 μA) decreased EtOH consumption by about 30% in seven of nine P rats, while one showed no effects from DBS, and one actually showed increased EtOH intake during DBS. The authors' findings also indicated that DBS of Acb was effective in reducing EtOH intake by about 50% of baseline levels during a 24-h drinking session following 4–6 weeks of abstinence, thus suggesting that DBS of Acb may be beneficial for relapse prevention (Henderson et al., 2010).

A more recent study extended the findings of DBS of the AcbSh in P rats. DBS can mimic lesioning effects, and research has shown that microinjection of GABA agonists baclofen (Bac; $GABA_B$) + muscimol (Mus; $GABA_A$) into a discrete brain region can produce a rapid but temporary reduction in local neuronal activity without inhibiting passing fiber tracts (Martin & Ghez, 1999; McFarland & Kalivas, 2001; Rocha & Kalivas, 2010). A mixture of GABA agonists acts like a reversible lesion. Using microinjection techniques, Wilden et al. (2014) demonstrated that inactivation of the AcbSh reduced maintenance operant EtOH self-administration in P rats. In contrast, inactivation of the AcbSh did not alter acquisition to operant EtOH self-administration (Wilden et al., 2014), suggesting that neuroadaptations following chronic EtOH drinking may need to occur in order for intervention to be effective (Wilden et al., 2014). A second study examined the effects of DBS into the AcbSh on the maintenance of EtOH self-administration in P rats. DBS (100 Hz) of AcbSh had a transient effect on decreasing chronic EtOH intake, reducing EtOH intake on days 2 and 3 of 5 DBS treatment days (Wilden et al., 2014). In contrast, 200 μA DBS of AcbSh attenuated EtOH drinking on days 3, 4, 5, and 6 of 6 DBS treatment days; EtOH intake did not return to baseline until DBS was terminated (Wilden et al., 2014). Collectively, these studies provide evidence that DBS of Acb may be a viable therapy for refractory alcoholism.

Effects of DBS of the Nucleus Accumbens on Alcohol Consumption Craving in Humans

The first clinical report to suggest that DBS of the Acb may be viable treatment for alcohol addiction was the case report by Kuhn et al. (2007). In this report, DBS (130 Hz; 3 V) of Acb was used to help a patient suffering from refractory anxiety disorder. This patient also fulfilled the DSM-IV diagnostic criteria for alcohol dependency and had high pathological score on WHO Alcohol Use Disorder Identification Test (Kuhn et al., 2007). DBS of the Acb proved to be only moderately effective for the patient's primary disorder (anxiety); however, after months of continuous DBS treatment the patient's alcohol drinking began to decrease. The patient's average of 10 or more drinks per day was reduced to 1 to 2 drinks per day with more alcohol-free days (Kuhn et al., 2007). In a second case report by this group, the patient's primary disorder was alcohol addiction. This patient's alcohol consumption deceased after 8 months of DBS of Acb and he was completely alcohol free after 1 year. In addition, his alcohol dependency and craving scores for alcohol dependence fell below the pathological level, suggesting a normalization of that patient's addictive behavior and craving (Kuhn et al., 2011).

The transition from moderate alcohol consumption to excessive alcohol consumption has been hypothesized to be based upon a "loss of control," with reports suggesting that the development and course of alcohol use and dependence are complicated by heightened impulsivity (Dom, Hulstijn, & Sabbe, 2006; Miller, 1991). Impulse control and decision making are thought to be mediated by Acb, orbitofrontal cortex, and anterior mid-cingulate cortex (aMCC) (Kuhn et al., 2011). Kuhn et al. (2011) found that DBS of the Acb can improve impulse control. The patient's impulse control and decision making were accessed using an electrophysiological marker of error processing which is linked to function of the aMCC (Kuhn et al., 2011). When DBS was turned on, the patient's performance improved but the reverses happen when DBS was turned off. The authors' report provided further evidence that DBS of the Acb has a positive effect on addiction via normalization of craving associated with aMCC dysfunction. DBS (130 Hz; 3.5 V) of the Acb has also been shown to induce slower and less risky choices compared to turning DBS off, suggesting that DBS may attenuate impulsive, riskier, and less controlled behaviors (Heldmann et al., 2012). Taken together, these two case reports suggest that effectiveness of DBS of the Acb may be due in part to its ability to improve behavioral and impulse control through activating other cortical areas.

Recently, a pilot study was conducted to investigate the effects of DBS (130 Hz; 3.5 or 4.5 V) of the Acb on five patients who met DSM-IV criteria for alcohol dependence, and the following criteria: (1) 10-year history of alcoholism, (2) long-term inpatient therapy for at least 6 months, (3) unsuccessful attempts of forensic therapy for 2 years, and (4) pharmacological therapy was ineffective (Müller et al., 2016, 2009; Voges, Müller, Bogerts, Münte, & Heinze, 2013). Follow-up showed that patients 1 and 2 were able to stop drinking, did not experience any cravings, and remained abstinent at 18 months (Müller et al., 2009) and 4 years (Voges et al., 2013). Nicotine consumption was also reduced in patient 2, which suggest that DBS may also be effective for co-abuse (Müller et al., 2009). The follow-up showed that patient 1 was still abstinent (Müller et al., 2016) and still did not experience any cravings for alcohol. Contact was lost with patient 2, therefore the authors were unable to carry out an 8-year follow-up (Müller et al., 2016). There was improvement in the drinking behavior of patients 3, 4, and 5, and their alcohol craving did vanish; however, these patients did experience relapses (Müller et al., 2016, 2009; Voges et al., 2013). All three of these patients claimed that the relapses were due to negative stress (Müller et al., 2016; Voges et al., 2013). Collectively, this pilot trial demonstrated that DBS of the Acb appears to consistently eliminate craving for alcohol but eliminated craving does not seem to be enough to maintain abstinence. These findings show the complexity of treating alcohol addictions because negative emotional events can act as a trigger to cause relapse even though the patient does not crave alcohol, thus providing further evidence that stress can make patients vulnerable to relapse.

EFFECTS OF DEEP BRAIN STIMULATION ON PSYCHOSTIMULANTS ADDICTION BEHAVIORS

Cocaine and Nucleus Accumbens

Cocaine is an addictive psychostimulant, and as per the National Survey on Drug Use and Health (NSDUH) in 2008, about 1.9 million people were current cocaine users in the United States (NSDUH, 2009). It was also estimated that about 1.4 million Americans met the DSM criteria for cocaine dependence or abuse of cocaine in 2008 (NSDUH, 2009). Like all drugs of abuse the reinforcing effects of cocaine are primarily mediated by enhanced DA transmission within the MCL-DA circuitry. Animals will self-administer cocaine directly into the AcbSh but not the AcbCore (Katner et al., 2011). Cocaine-seeking studies have reported that inactivation of the AcbCore, but not the AcbSh, with GABA agonists Bac+Mus and lidocaine, is involved in mediating priming-induced reinstatement of cocaine seeking (McFarland & Kalivas, 2001). A D_1 receptor antagonist into the AcbSh reduced context- and drug-primed cocaine-seeking behaviors (Anderson, Bari, & Pierce, 2003; Bachtell, Whisler, Karanian, & Self, 2005) but did not alter cue-induced cocaine seeking (Anderson et al., 2003). Activation of D_1 receptors with an agonist into the AcbSh and AcbCore has been shown to reinstate cocaine seeking (Bachtell et al., 2005; Schmidt & Pierce, 2006). Collectively, these studies provide evidence that the Acb is involved in cocaine reinforcement and "craving-like" behaviors, therefore some DBS studies chose the Acb as a target to examine.

Intravenous (i.v.) operant self-administration techniques have been used to examine the effects of DBS on cocaine intake and cocaine-seeking behaviors. Vassoler et al. (2008) examined the effects of bilateral DBS at frequency of 160 Hz into the AcbSh on drug-primed reinstatement of cocaine seeking. The authors' findings showed that priming doses of cocaine reinstated cocaine seeking and that DBS at 150 μA in the AcbSh attenuated the reinstatement of cocaine seeking (Vassoler et al., 2008). In contrast, DBS into AcbSh did not alter the reinstatement of food seeking nor did DBS into dorsal striatum alter cocaine seeking. These findings were replicated in a subsequent study, and the new results revealed that DBS into the AcbCore did not alter drug-primed reinstatement of cocaine seeking, thus extending the previous findings that DBS ameliorating effects may be specific to the AcbSh (Vassoler et al., 2008, 2013).

However, similar to the McFarland and Kalivas (2001) study, Vassoler and colleagues indicated that inactivation of the AcbCore, but not the AcbSh, with Bac+Mus or lidocaine attenuated cocaine seeking (Vassoler et al., 2013). These results suggested that the DBS effect did not appear to be due to the inactivation of the target nuclei since inactivating the AcbSh failed to mimic the effects of DBS (Vassoler et al., 2013). Further investigation of the neural circuitry showed the effects of DBS increased c-Fos immunoreactivity locally at Acb and distally in the infralimbic prefrontal cortex (IL). Bac+Mus microinjected into anterior cingulate (AC), prelimbic prefrontal cortex (PL), or IL before cocaine reinstatement reduced drug-primed reinstatement of cocaine seeking (Vassoler et al., 2013). The authors suggested that DBS may be producing antidromic activation of afferent structures (Vassoler et al., 2013). Another subsequent study from this group indicated that DBS of AcbSh also attenuates cue-induced cocaine- and sucrose-seeking behaviors (Guercio, Schmidt, & Pierce, 2015). However, this nonspecific effect suggests that DBS of AcbSh may not be an effective treatment for cue-induced cocaine seeking because of the general reduction of both drug and natural reward seeking.

It is hypothesized that time points beyond the acute withdrawal phase of drugs abuse may increase the vulnerability to cue- or context-induced craving behaviors (Gawin & Kleber, 1986). Hamilton, Lee, and Canales (2015) examine the long-term efficacy of unilateral DBS of the Acb during abstinence. During 14 consecutive days of abstinence period the animals received 30 min of unilateral DBS of Acb at either low-frequency stimulation (LFS, 20 Hz) or HFS (160 Hz) (Hamilton et al., 2015). DBS was discontinued from abstinence days 15 to 30. Cocaine seeking was expressed at every abstinence time period with the most robust occurring at day 15. Interestingly, both the LF and HF of DBS reduced primed-induced cocaine seeking after 15 days of abstinence but not at day 1 or 30 (Hamilton et al., 2015). These findings showed that DBS stimulation needed at least 2 weeks of daily administration to be effective because it was not effective on day 1, and when it was discontinued the effects of DBS were no longer apparent on the last day of abstinence. The authors suggested that daily unilateral DBS of the Acb at low and high frequency attenuated cocaine relapse after 15 days of abstinence, with therapeutic-like effects seemingly diminishing after DBS discontinuation (Hamilton et al., 2015).

Cocaine and Lateral Habenula

The lateral habenula (LHb) is another emerging target for DBS on cocaine addiction behaviors. Studies have shown that the LHb is activated by stimulus showing no reward or aversive event; therefore, it plays an important role in encoding information for negative-reward signals and learning from/avoiding aversive experiences in future (Fakhoury & Lopez, 2014). The LHb has glutamatergic projections that act on DAnergic midbrain structures such as the VTA and rostro-medial tegmentum (RMTg), as well as serotonergic raphe nuclei (Fakhoury & Lopez, 2014). Research has shown that the activation of neurons within the LHb decreases DAergic neurons activity, whereas inhibition of neurons in the LHb results in activation of DAnergic neurons (Fakhoury & Lopez, 2014). The LHb can inhibit DAnergic cell firing in VTA by activating RMTg GABAergic neurons, leading to reductions in motivation and reward (Fakhoury & Lopez, 2014). Neuronal activity in LHb can be increased by cocaine and cocaine-related cues (James, Charnley, Flynn, Smith, & Dayas, 2011; Mahler & Aston-Jones, 2012).

DBS research has investigated LHb involvement in ongoing maintenance, extinction, and drug seeking for cocaine. DBS at high frequencies is the most effective treatment for neurological disorders including addiction behaviors, and these effects may be similar to ablation/lesioning techniques. Interestingly, lesioning the LHb does not alter ongoing cocaine self-administration and increased cocaine-seeking behavior by delaying the extinction response (Friedman et al., 2010). Moreover, HFS (100 Hz) had no effect at all on cocaine responses while LFS (10 Hz) increased cocaine responses. Findings of Friedman et al. (2010) demonstrated that a specific combination of alternating low (10 Hz; 200 μA) and high (100 Hz; 200 μA) frequencies produced the optimal effects for DBS on cocaine addiction behaviors. Their findings showed that combined DBS in the LHb, administered 15 min during the session, attenuated cocaine self-administration and priming-induced cocaine seeking. In addition, the combined DBS accelerated extinction for all cocaine groups (Friedman et al., 2010). As mentioned, the LHb has glutamatergic projections that act on DAnergic midbrain structures such as the VTA and RMTg as well as serotonergic structures in the raphe nuclei (Fakhoury & Lopez, 2014). In the same study, Friedman and colleagues examined how DBS in the LHb would affect glutamatergic and GABAergic proteins in the VTA. Their results revealed that cocaine self-administration enhanced VTA protein levels of the glutamatergic receptor NR1 subunit of the N-methyl-D-aspartate (NMDA) receptor, the GluR1 subunit of the α-Amino-3-hydroxy-5-methyl-4-isoxazolepropionic Acid (AMPA) receptor, and the scaffolding protein PSD95, but had no effect on the GABAergic $GABA_A\beta$ receptor (Friedman et al., 2010). Moreover, DBS treatment in the LHb, either alone or with cocaine, returned NR1, GluR1, and PSD95 to baseline levels in VTA (Friedman et al., 2010). Collectively, these findings suggested that the effect of LHb DBS on cocaine reinforcement may be due to attenuation of the cocaine-induced increases in glutamatergic input to the VTA (Friedman et al., 2010). However, it is worthwhile to note that the integrity of the fiber connection between the LHb and the VTA must be intact for neuromodulation to exert its effects, thus rendering DBS questionable for heavy cocaine users with degeneration of the major LHb efferent tract to the VTA, the fasciculus retroflexus (Lax et al., 2013).

Cocaine and Subthalamic Nucleus

The FDA has already approved DBS of subthalamic nucleus (STN) for Parkinson's disease, and the STN is also a possible target for cocaine addiction. Preclinical research has demonstrated that bilaterally lesioning the STN appears to have opposite effects on the motivation for food and cocaine. Lesioning the STN increases the reward saliency for food reward, but decreases the reward saliency for cocaine (Baunez, Dias, Cador, & Amalric, 2005). Conditioned place preference (CPP) was also enhanced for food while decreased for cocaine following lesioning the STN (Baunez et al., 2005). In addition, inactivating the STN with muscimol has a similar effect as lesioning the STN on progressive ratio (PR) performance for food reinforcement (Baunez et al., 2005). Rouaud et al. (2010) demonstrated that DBS (130 Hz; 50–130 μA) of the STN did increase rats' motivation to work for one sucrose pellet, whereas the opposite effect was observed for the rats' willingness to work for cocaine. These findings suggest that DBS of STN may reduce motivation for drug rewards while enhancing the motivation for natural rewards (Rouaud et al., 2010).

Methamphetamine and Nucleus Accumbens

Methamphetamine is another highly addictive psychostimulant that is abused by about 12 million people in their life with 1.2 million people using it in the past year and about 440,000 of those identified as past-month users [Substance Abuse and Mental Health Services Administration (SAMHSA), 2013]. There is no known treatment, and rates of relapse are extremely high. An initial study examining the effects of DBS on methamphetamine self-administration in rats has been performed (Batra, Guerin, Goeders, & Wilden, 2016). The results suggest that DBS into the AcbSh is effective at reducing i.v. methamphetamine self-administration. It is noteworthy that this chapter describes electrical stimulation therapy that is delivered intermittently, not continuously, and is temporally and spatially separate from the drug-use environment, which more closely approximates what may be possible in human methamphetamine addicts. In other words, not only traditional DBS, but also newer noninvasive neuromodulatory techniques like transcranial magnetic stimulation (TMS), which is delivered in daily 15–30 min outpatient sessions, may have a role in treating psychostimulant addiction.

EFFECTS OF DEEP BRAIN STIMULATION ON NICOTINE ADDICTION BEHAVIORS

Nicotine and Nucleus Accumbens

Tobacco use is considered to be the most preventable cause of disease and death in the United States [US Department of Health and Human Services (DHHS), 2014]. About 69.6 million Americans aged 12 or older reported current use of tobacco (NSDUH, 2011). It has been estimated that smoking causes more than 400,000 deaths and over 157 billion dollars are spent each year for health costs associated with tobacco (US DHHS, 2014). Nicotine is the addictive agent in tobacco products that causes and reinforces the repetitive use of tobacco. Nicotine affects the brain by interacting with nicotinic acetylcholine receptors (nAChRs) that are distributed throughout the CNS. The MCL-DA is activated by nicotine. It can enhance DA neurotransmission by increasing the firing rate of DA neurons (Grenhoff, Aston-Jones, & Svensson, 1986) leading to an increase in extracellular levels of DA in the Acb (Ferrari, Le, Picciotto, Changeux, & Zoli, 2002; Nisell, Nomikos, & Svensson, 1994). The Acb is a possible target for DBS for nicotine addiction behaviors. In one patient who received DBS (185 Hz) of the Acb for severe refractory OCD, not only did OCD symptoms disappear after 10 months but also the patient stopped smoking and remained abstinent without effort, craving, and withdrawal symptoms (Mantione, van de Brink, Schuurman, & Denys, 2010). In a pilot study of 10 patients with nicotine dependency, DBS of the Acb was able to reduce smoking in 3 patients (Kuhn et al., 2009). While both of these reports have limitations, specifically small sample size and effect (only 30% patients quit smoking), they do suggest, respectively, that Acb may be an area of interest for effects of DBS.

Nicotine and Granular Insular Cortex

The insula is involved in conscious emotional feelings through its role in the representation of bodily (interoceptive) state (cf. Naqvi, Gaznick, Tranel, & Bechara, 2014). Nicotine does produce bodily sensations, and it was proposed that both the central and peripheral effects of nicotine are involved in the conscious pleasure produced by smoking (cf. Naqvi & Bechara, 2010). In addition, imaging studies have shown that the insula is activated by drug-associated cues (cf. Naqvi et al., 2014). In 2007, Naqvi and colleagues reported that patients with brain damage to the insula underwent a disruption of smoking addiction, characterized by the ability to quit smoking easily, immediately, without relapse, and without persistence of the urge to smoke. This study proposed that the insula may be an important neural substrate involved in smoking addiction (Naqvi, Rudrauf, Damasio, & Bechara, 2007). Naqvi et al. (2007) considered cue-induced craving as an emotional response and postulated that if the insula was generally involved in emotional feelings, then insula lesions would disrupt emotional feelings of cue-induced cravings and reduce the likelihood of relapse.

Preclinical studies using the GABA agonist inactivation technique provided evidence for this theory. Inactivation of the granular insula resulted in decreases in i.v. nicotine self-administration, reduced break points for PR schedules of reinforcement, and prevented priming- and cue-induced nicotine seeking in rats. Food self-administration and food seeking were not altered by the inactivation of the granular insula, thus suggesting that disruption of granular insula may be specific to drug craving and not natural reward craving (Forget, Pushparaj, & Le Foll, 2010). Similar effects were also observed with bilateral DBS (130 Hz, 50 μA) of the granular insula (Pushparaj et al., 2013). Findings of Pushparaj et al. (2013) indicated that DBS of the insular region significantly attenuated nicotine self-administration and PR schedules of reinforcement as well as reduced priming- and cue-induced nicotine seeking. This clinical and preclinical work provides support that insula may be a possible DBS therapeutic target for nicotine addiction behaviors, however, further research is warranted.

EFFECTS OF DEEP BRAIN STIMULATION ON HEROIN ADDICTION BEHAVIORS

Heroin use in the United States has reached epidemic levels in some regions. About 2.5 million Americans currently are addicted to prescription opioids and/or heroin (ASAM, 2016). Clinical case reports have shown that DBS of the Acb may be effective for heroin addiction. A case report on 24-year-old heroin addict showed that DBS (145 Hz) of the Acb resulted in a decrease of smoking after 40 days of treatment, an improvement in cognitive, depression, and anxiety scores after 3 months and the patient stopped abusing heroin completely after the DBS procedure (Zhou, Xu, & Jiang, 2011). After 2.5 years the patient requested that the DBS be removed and a 6-year follow-up showed that the patient was drug free even after 3.5 years following removal of the device with no relapse. In a pilot trial, DBS (130–140 Hz; 4.5 or 5.0 V, respectively) of Acb was used to treat two patients who met the DSM-IV criteria for heroin dependence (Kuhn et al., 2014). DBS reduced drug craving to the point that patients were tapered completely off levomethadone (a substitution treatment for opioid addiction), and this did not result in an increase in craving for heroin or levomethadone (Kuhn et al., 2014). The patients remained mostly abstinent (one single incident of heroin use reported) of heroin and levomethadone use and their psychiatric symptoms were ameliorated. However, the patients still reported using other drugs occasionally out of boredom or due to stress, even though they did not have cravings for those drugs (Kuhn et al., 2014). These cases suggest that DBS stimulation in the Acb may be beneficial for treatment-resistant heroin addicts, however, further investigations are warranted.

Previous studies have indicated that lesioning the AcbCore can severely impair heroin acquisition (Hutcheson, Parkinson, Robbins, & Everitt, 2001). Primed- and cue-induced heroin seeking can increase the glutamate release in the AcbCore, and AMPA antagonist into AcbCore can attenuate glutamate release and drug-primed heroin seeking (LaLumiere & Kalivas, 2008). In addition, antagonism of D1 receptors in the AcbCore can reduce primed- and cue-induced heroin seeking (Bossert et al., 2007; LaLumiere & Kalivas, 2008). In 2013 Guo and colleagues investigated long-term effects of DBS in AcbCore on cue- and drug-primed induced heroin reinstatement of drug seeking. DBS (130 Hz; 0, 75, or 150 μA) was administered bilaterally or unilaterally into the AcbCore for 60 min a day for 7 days during abstinence. Bilateral DBS in the AcbCore was most effective because both 75 and 150 μA stimulation reduced cue- and heroin-primed induced reinstatement of drug seeking, while it did not have effect on locomotor activity, spatial learning, or memory-retention capabilities. Unilateral DBS stimulation into AcbCore was only effective at 150 μA for both attenuated cue- and heroin-induced drug seeking. In addition, this study indicated that bilateral stimulation increased the transcription factor phosphorylated cAMP response element–binding protein (pCREB) in AcbCore while reducing delta FosB. cAMP response element–binding protein (CREB) and delta FosB have been implicated in the rewarding and reinforcement effects of drugs abuse (Berke & Hyman, 2000; Nestler, 2004). Previous studies have shown that upregulation of CREB in Acb by drugs of abuse mediates tolerance to the reinforcing effects of drugs while delta FosB accumulates only after chronic drug abuse and remains elevated weeks after withdrawal, and the prolonged exposure to delta FosB increases the sensitivity to the rewarding effects of drugs of abuse (McClung & Nestler, 2003). Collectively, these findings suggest that DBS stimulation into Acb may be an effective therapeutic option for preventing relapse to heroin addiction by altering abnormal protein production (Guo et al., 2013).

CONCLUSION

Accumulating preclinical and clinical evidence indicate that DBS may be an effective therapeutic tool on drug intake and drug craving. More importantly, research is starting to establish the biological consequences of DBS treatment in the MCL-DA pathway. Understanding the effects of DBS within this system may elucidate the biological basis of the effectiveness of

DBS for the treatment of substance abuse. This information could be critical for the development of efficacious pharmacotherapeutics, and novel pharmacological tools to increase the effectiveness of DBS treatment, and/or other potentially less invasive neuromodulatory techniques, such as TMS, for the treatment of addiction. Currently, we are at the cusp of understanding the effectiveness of DBS for the treatment of drug and alcohol abuse, and the application of that powerful tool for the treatment of a common infliction.

References

American Society of Addiction Medicine. (2016). *Opioid addiction 2016 facts & figures*. http://www.asam.org/docs/default-source/advocacy/opioid-addiction-disease-facts-figures.

Anderson, S. M., Bari, A. A., & Pierce, R. C. (2003). Administration of the D1-like dopamine receptor antagonist SCH-23390 into the medial nucleus accumbens shell attenuates cocaine priming-induced reinstatement of drug-seeking behavior in rats. *Psychopharmacology, 168*, 132–138.

Anton, R. F., O'Malley, S. S., Ciraulo, D. A., Cisler, R. A., Couper, D., Donovan, D. M., ... Zweben, A. (2006). Combined pharmacotherapies and behavioral interventions for alcohol dependence: The COMBINE study: A randomized controlled trial. *JAMA, 295*, 2003–2017.

Bachtell, R. K., Whisler, K., Karanian, D., & Self, D. W. (2005). Effects of intra-nucleus accumbens shell administration of dopamine agonists and antagonists on cocaine-taking and cocaine-seeking behaviors in the rat. *Psychopharmacology, 183*, 41–53.

Batra, V., Guerin, G. F., Goeders, N. E., & Wilden, J. A. (2016). A general method for evaluating deep brain stimulation effects on intravenous methamphetamine self-administration. *Journal of Visualized Experiments, 107*.

Baunez, C., Dias, C., Cador, M., & Amalric, M. (2005). The subthalamic nucleus exerts opposite control on cocaine and 'natural' rewards. *Nature Neuroscience, 8*, 484–489.

Benabid, A. L. (2015). Neuroscience: Spotlight on deep-brain stimulation. *Nature, 519*, 299–300.

Benabid, A. L., Pollak, P., Gao, D., Hoffmann, D., Limousin, P., Gay, E., ... Benazzouz, A. (1996). Chronic electrical stimulation of the ventralis intermedius nucleus of the thalamus as a treatment of movement disorders. *Journal of Neurosurgery, 84*, 203–214.

Benabid, A. L., Pollak, P., Gervason, C., Hoffmann, D., Gao, D. M., Hommel, M., ... de Rougemont, J. (1991). Long-term suppression of tremor by chronic stimulation of the ventral intermediate thalamic nucleus. *Lancet, 337*, 403–406.

Benabid, A. L., Pollak, P., Louveau, A., Henry, S., & de Rougemont, J. (1987). Combined (thalamotomy and stimulation) stereotactic surgery of the VIM thalamic nucleus for bilateral Parkinson disease. *Applied Neurophysiology, 50*, 344–346.

Berke, J. D., & Hyman, S. E. (2000). Addiction, dopamine, and the molecular mechanisms of memory. *Neuron, 25*, 515–532.

Bossert, J. M., Poles, G. C., Wihbey, K. A., Koya, E., & Shaham, Y. (2007). Differential effects of blockade of dopamine D1-family receptors in nucleus accumbens core or shell on reinstatement of heroin seeking induced by contextual and discrete cues. *The Journal of Neuroscience, 27*, 12655–12663.

Bottlender, M., & Soyka, M. (2005). Outpatient alcoholism treatment: Predictors of outcome after 3 years. *Drug and Alcohol Dependence, 80*, 83–89.

Chaudhri, N., Sahuque, L. L., Cone, J. J., & Janak, P. H. (2008). Reinstated ethanol-seeking in rats is modulated by environmental contextual and requires the nucleus accumbens core. *The European Journal of Neuroscience, 28*, 2288–2298.

Chaudhri, N., Sahuque, L. L., & Janak, P. H. (2009). Ethanol seeking triggered by environmental contextual is attenuated by blocking dopamine D1 receptors in the nucleus accumbens core and shell in rats. *Psychopharmacology, 207*, 303–314.

Cutler, R. B., & Fishbain, D. A. (2005). Are alcoholism treatments effective? The Project MATCH data. *BMC Public Health, 5*, 75.

Di Chiara, G., & Imperato, A. (1988). Drugs abused by humans preferentially increase synaptic dopamine concentrations in the mesolimbic system of freely moving rats. *Proceedings of the National Academy of Sciences of the United States of America, 85*, 5274–5278.

van Dijk, A., Klompmakers, A. A., Feenstra, M. G., & Denys, D. (2012). Deep brain stimulation of the accumbens increases dopamine, serotonin, and noradrenaline in the prefrontal cortex. *Journal of Neurochemistry, 123*, 897–903.

van Dijk, A., Mason, O., Klompmakers, A. A., Feenstra, M. G., & Denys, D. (2011). Unilateral deep brain stimulation in the nucleus accumbens core does not affect local monoamine release. *Journal of Neuroscience Methods, 202*, 113–118.

Ding, Z. M., Ingraham, C. M., Rodd, Z. A., & McBride, W. J. (2015). The reinforcing effects of ethanol within the posterior ventral tegmental area depend on dopamine neurotransmission to forebrain cortico limbic systems. *Addiction Biology, 20*, 458–468.

Dom, G., Hulstijn, W., & Sabbe, B. (2006). Differences in impulsivity and sensation seeking between early- and late-onset alcoholics. *Addictive Behaviors, 31*, 298–308.

Engleman, E. A., Ding, Z. M., Oster, S. M., Toalston, J. E., Bell, R. L., Murphy, J. M., ... Rodd, Z. A. (2009). Ethanol is self-administered into the nucleus accumbens shell, but not the core: Evidence of genetic sensitivity. *Alcoholism, Clinical and Experimental Research, 33*, 2162–2171.

Everitt, B. J., Belin, D., Economidou, D., Pelloux, Y., Dalley, J. W., & Robbins, T. W. (2008). Review. Neural mechanisms underlying the vulnerability to develop compulsive drug-seeking habits and addiction. *Philosophical Transactions of the Royal Society of London. Series B, Biological Sciences, 363*, 3125–3135.

Fakhoury, M., & Lopez, S. D. (2014). The role of habenula in motivation and reward. *Advances in Neuroscience, 2014*, 862048. http://dx.doi.org/10.1155/2014/862048.

Feltenstein, M. W., & See, R. E. (2008). The neurocircuitry of addiction: An overview. *British Journal of Pharmacology, 154*, 261–274.

Ferrari, R., Le, N. N., Picciotto, M. R., Changeux, J. P., & Zoli, M. (2002). Acute and long-term changes in the mesolimbic dopamine pathway after systemic or local single nicotine injections. *The European Journal of Neuroscience, 15*, 1810–1818.

Finney, J. W., & Moos, R. H. (1991). The long-term course of treated alcoholism: I. Mortality, relapse and remission rates and comparisons with community controls. *Journal of Studies on Alcohol, 52*, 44–54.

Fischer, B., Blanken, P., Da Silveira, D., Gallassi, A., Goldner, E. M., Rehm, J., ... Wood, E. (2015). Effectiveness of secondary prevention and treatment interventions for crack cocaine abuse: A comprehensive narrative overview of English-language studies. *International Journal on Drug Policy, 26*, 352–363.

Forget, B., Pushparaj, A., & Le Foll, B. (2010). Granular insular cortex inactivation as a novel therapeutic strategy for nicotine addiction. *Biological Psychiatry, 68*, 265–271.

Friedman, A., Lax, E., Dikshtein, Y., Abraham, L., Flaumenhaft, Y., Sudai, E., ... Yadid, G. (2010). Electrical stimulation of the lateral habenula produces enduring inhibitory effect on cocaine seeking behavior. *Neuropharmacology, 59*, 452–459.

Fuchs, R. A., Evans, K. A., Parker, M. C., & See, R. E. (2004). Differential involvement of the core and shell subregions of the nucleus accumbens in conditioned cue-induced reinstatement of cocaine seeking in rats. *Psychopharmacology, 176*, 459–465.

Gawin, F. H., & Kleber, H. D. (1986). Abstinence symptomatology and psychiatric diagnosis in cocaine abusers. Clinical observations. *Archives of General Psychiatry, 43*, 107–113.

Grenhoff, J., Aston-Jones, G., & Svensson, T. H. (1986). Nicotinic effects on the firing pattern of midbrain dopamine neurons. *Acta Physiologica Scandinavica, 128*, 351–358.

Guercio, L. A., Schmidt, H. D., & Pierce, R. C. (2015). Deep brain stimulation of the nucleus accumbens shell attenuates cue-induced reinstatement of both cocaine and sucrose seeking in rats. *Behavioural Brain Research, 281*, 125–130.

Guo, L., Zhou, H., Wang, R., Xu, J., Zhou, W., Zhang, F., ... Jiang, J. (2013). DBS of nucleus accumbens on heroin seeking behaviors in self-administering rats. *Drug and Alcohol Dependence, 129*, 70–81.

Hamilton, J., Lee, J., & Canales, J. J. (2015). Chronic unilateral stimulation of the nucleus accumbens at high or low frequencies attenuates relapse to cocaine seeking in an animal model. *Brain Stimulation, 8*, 57–63.

Hauser, S. R., Deehan, G. A., Jr., Dhaher, R., Knight, C. P., Wilden, J. A., McBride, W. J., & Rodd, Z. A. (2015). D1 receptors in the nucleus accumbens-shell, but not the core, are involved in mediating ethanol-seeking behavior of alcohol-preferring (P) rats. *Neuroscience, 295*, 243–251.

Heldmann, M., Berding, G., Voges, J., Bogerts, B., Galazky, I., Müller, U., ... Münte, T. F. (2012). Deep brain stimulation of nucleus accumbens region in alcoholism affects reward processing. *PLoS One, 7*, e36572.

Henderson, M. B., Green, A. I., Bradford, P. S., Chau, D. T., Roberts, D. W., & Leiter, J. C. (2010). Deep brain stimulation of the nucleus accumbens reduces alcohol intake in alcohol-preferring rats. *Neurosurgical Focus, 29*, E12.

Hutcheson, D. M., Parkinson, J. A., Robbins, T. W., & Everitt, B. J. (2001). The effects of nucleus accumbens core and shell lesions on intravenous heroin self-administration and the acquisition of drug-seeking behaviour under a second-order schedule of heroin reinforcement. *Psychopharmacology, 153*, 464–472.

Ito, R., Dalley, J. W., Howes, S. R., Robbins, T. W., & Everitt, B. J. (2000). Dissociation in conditioned dopamine release in the nucleus accumbens core and shell in response to cocaine cues and during cocaine-seeking behavior in rats. *The Journal of Neuroscience, 20*, 7489–7495.

Ito, R., Robbins, T. W., & Everitt, B. J. (2004). Differential control over cocaine-seeking behavior by nucleus accumbens core and shell. *Nature Neuroscience, 7*, 389–397.

James, M. H., Charnley, J. L., Flynn, J. R., Smith, D. W., & Dayas, C. V. (2011). Propensity to 'relapse' following exposure to cocaine cues is associated with the recruitment of specific thalamic and epithalamic nuclei. *Neuroscience, 199*, 235–242.

Katner, S. N., Oster, S. M., Ding, Z. M., Deehan, G. A., Jr., Toalston, J. E., Hauser, S. R., ... Rodd, Z. A. (2011). Alcohol-preferring (P) rats are more sensitive than Wistar rats to the reinforcing effects of cocaine self-administered directly into the nucleus accumbens shell. *Pharmacology, Biochemistry, and Behavior, 99*, 688–695.

Knapp, C. M., Tozier, L., Pak, A., Ciraulo, D. A., & Kornetsky, C. (2009). Deep brain stimulation of the nucleus accumbens reduces ethanol consumption in rats. *Pharmacology, Biochemistry, and Behavior, 92*, 474–479.

Kringelbach, M. L., Jenkinson, N., Green, A. L., Owen, S. L., Hansen, P. C., Cornelissen, P. L., ... Aziz, T. Z. (2007). Deep brain stimulation for chronic pain investigated with magnetoencephalography. *Neuroreport, 18*, 223–228.

Kuhn, J., Bauer, R., Pohl, S., Lenartz, D., Huff, W., Kim, E. H., ... Sturm, V. (2009). Observations on unaided smoking cessation after deep brain stimulation of the nucleus accumbens. *European Addiction Research, 15*, 196–201.

Kuhn, J., Gründler, T. O., Bauer, R., Huff, W., Fischer, A. G., Lenartz, D., ... Sturm, V. (2011). Successful deep brain stimulation of the nucleus accumbens in severe alcohol dependence is associated with changed performance monitoring. *Addiction Biology, 16*, 620–623.

Kuhn, J., Lenartz, D., Huff, W., Lee, S., Koulousakis, A., Klosterkoetter, J., & Sturm, V. (2007). Remission of alcohol dependency following deep brain stimulation of the nucleus accumbens: Valuable therapeutic implications? *Journal of Neurology, Neurosurgery, Psychiatry, 78*, 1152–1153.

Kuhn, J., Möller, M., Treppmann, J. F., Bartsch, C., Lenartz, D., Gruendler, T. O., ... Sturm, V. (2014). Deep brain stimulation of the nucleus accumbens and its usefulness in severe opioid addiction. *Molecular Psychiatry, 19*, 145–146.

LaLumiere, R. T., & Kalivas, P. W. (2008). Glutamate release in the nucleus accumbens core is necessary for heroin seeking. *The Journal of Neuroscience, 28*, 3170–3177.

Lax, E., Friedman, A., Croitoru, O., Sudai, E., Ben-Moshe, H., Redlus, L., ... Yadid, G. (2013). Neurodegeneration of lateral habenula efferent fibers after intermittent cocaine administration: Implications for deep brain stimulation. *Neuropharmacology, 75*, 246–254.

Lim, S. S., Vos, T., Flaxman, A. D., Danaei, G., Shibuya, K., Adair-Rohani, H., ... Memish, Z. A. (2012). A comparative risk assessment of burden of disease and injury attributable to 67 risk factors and risk factor clusters in 21 regions, 1990–2010: A systematic analysis for the Global Burden of Disease Study. *Lancet, 380*, 2224–2260, 2012.

Mahler, S. V., & Aston-Jones, G. S. (2012). Fos activation of selective afferents to ventral tegmental area during cue-induced reinstatement of cocaine seeking in rats. *The Journal of Neuroscience, 32*, 13309–13326. ocd.

Mantione, M., van de Brink, W., Schuurman, P. R., & Denys, D. (2010). Smoking cessation and weight loss after chronic deep brain stimulation of the nucleus accumbens: Therapeutic and research implications: Case report. *Neurosurgery, 66*, E218. discussion E218.

Martin, J. H., & Ghez, C. (1999). Pharmacological inactivation in the analysis of the central control of movement. *Journal of Neuroscience Methods, 86*, 145–159.

McBride, W. J., & Li, T. K. (1998). Animal models of alcoholism: Neurobiology of high alcohol-drinking behavior in rodents. *Critical Reviews in Neurobiology, 12*, 339–369.

McClung, C. A., & Nestler, E. J. (2003). Regulation of gene expression and cocaine reward by CREB and DeltaFosB. *Nature Neuroscience, 6*, 1208–1215.

McFarland, K., & Kalivas, P. W. (2001). The circuitry mediating cocaine induced reinstatement of drug-seeking behavior. *The Journal of Neuroscience, 21*, 8655–8663.

McIntyre, C. C., Savasta, M., Kerkerian-Le Goff, L., & Vitek, J. L. (2004). Uncovering the mechanism(s) of action of deep brain stimulation: Activation, inhibition, or both. *Clinical Neurophysiology, 115*, 1239–1248.

Miller, L. (1991). Predicting relapse and recovery in alcoholism and addiction: Neuropsychology, personality, and cognitive style. *Journal of Substance Abuse Treatment, 8*, 277–291.

Müller, U. J., Sturm, V., Voges, J., Heinze, H. J., Galazky, I., Büntjen, L., ... Bogerts, B. (2016). Nucleus accumbens deep brain stimulation for alcohol addiction – Safety and clinical long-term results of a pilot trial. *Pharmacopsychiatry, 49*, 170–173.

Müller, U. J., Sturm, V., Voges, J., Heinze, H. J., Galazky, I., Heldmann, M., ... Bogerts, B. (2009). Successful treatment of chronic resistant alcoholism by deep brain stimulation of nucleus accumbens: First experience with three cases. *Pharmacopsychiatry, 42*, 288–291.

Murphy, J. M., Stewart, R. B., Bell, R. L., Badia-Elder, N. E., Carr, L. G., McBride, W. J., ... Li, T. K. (2002). Phenotypic and genotypic characterization of the Indiana University rat lines selectively bred for high and low alcohol preference. *Behavior Genetics, 32*, 363–388.

Naqvi, N. H., & Bechara, A. (2010). The insula and drug addiction: An interoceptive view of pleasure, urges, and decision-making. *Brain Structure & Function, 214*, 435–450.

Naqvi, N. H., Gaznick, N., Tranel, D., & Bechara, A. (2014). The insula: A critical neural substrate for craving and drug seeking under conflict and risk. *Annals of the New York Academy of Sciences, 1316*, 53–70.

Naqvi, N. H., Rudrauf, D., Damasio, H., & Bechara, A. (2007). Damage to the insula disrupts addiction to cigarette smoking. *Science, 315*, 531–534.

National Institute on Alcohol Abuse, and Alcoholism (NIAAA). (2016). *Alcohol facts and statistics*. www.niaaa.nih.gov/alcohol-health/overview-alcohol-consumption/alcohol-facts-and-statistics.

Nestler, E. J. (2004). Molecular mechanisms of drug addiction. *Neuropharmacology, 47*(Suppl. 1), 24–32.

Nisell, M., Nomikos, G. G., & Svensson, T. H. (1994). Systemic nicotine-induced dopamine release in the rat nucleus accumbens is regulated by nicotinic receptors in the ventral tegmental area. *Synapse, 16*, 36–44.

Perlmutter, J. S., & Mink, J. W. (2006). Deep brain stimulation. *Annual Review of Neuroscience, 29*, 229–257.

Pushparaj, A., Hamani, C., Yu, W., Shin, D. S., Kang, B., Nobrega, J. N., & Le Foll, B. (2013). Electrical stimulation of the insular region attenuates nicotine-taking and nicotine-seeking behaviors. *Neuropsychopharmacology, 38*, 690–698.

Rocha, A., & Kalivas, P. W. (2010). Role of the prefrontal cortex and nucleus accumbens in reinstating methamphetamine seeking. *The European Journal of Neuroscience, 31*, 903–909.

Rouaud, T., Lardeux, S., Panayotis, N., Paleressompoulle, D., Cador, M., & Baunez, C. (2010). Reducing the desire for cocaine with subthalamic nucleus deep brain stimulation. *Proceedings of the National Academy of Sciences of the United States of America, 107*, 1196–1200.

Schmidt, H. D., & Pierce, R. C. (2006). Cooperative activation of D1-like and D2-like dopamine receptors in the nucleus accumbens shell is required for the reinstatement of cocaine-seeking behavior in the rat. *Neuroscience, 142*, 451–461.

Sesia, T., Bulthuis, V., Tan, S., Lim, L. W., Vlamings, R., Blokland, A., … Temel, Y. (2010). Deep brain stimulation of the nucleus accumbens shell increases impulsive behavior and tissue levels of dopamine and serotonin. *Experimental Neurology, 225*, 302–309.

Substance Abuse, and Mental Health Services Administration. (2009). *Office of applied studies. Results from the 2008 national survey on drug use and health: National findings, NSDUH Series H-36, HHS Pub. No. SMA 09–4443*. Rockville, MD: SAMHSA.

Substance Abuse, and Mental Health Services Administration. (2011). *Results from the 2010 national survey on drug use and health: Summary of national findings, NSDUH Series H-41, HHS Publication No. (SMA) 11–4658*. Rockville, MD: Substance Abuse and Mental Health Services Administration.

Substance Abuse, and Mental Health Services Administration. (2013). *Results from the 2012 national survey on drug use and health: Summary of national findings*. Rockville, MD: Substance Abuse and Mental Health Services Administration.

U.S. Department of Health, and Human Services. (2014). *The health consequences of Smoking—50 Years of progress: A report of the surgeon general*. Atlanta: U.S. Department of Health and Human Services, Centers for Disease Control and Prevention, National Center for Chronic Disease Prevention and Health Promotion, Office on Smoking and Health.

Vaillant, G. E. (2005). Alcoholics anonymous: Cult or cure? *The Australian and New Zealand Journal of Psychiatry, 39*, 431–436.

Vassoler, F. M., Schmidt, H. D., Gerard, M. E., Famous, K. R., Ciraulo, D. A., Kornetsky, C., … Pierce, R. C. (2008). Deep brain stimulation of the nucleus accumbens shell attenuates cocaine priming-induced reinstatement of drug seeking in rats. *The Journal of Neuroscience, 28*, 8735–8739.

Vassoler, F. M., White, S. L., Hopkins, T. J., Guercio, L. A., Espallergues, J., Berton, O., … Pierce, R. C. (2013). Deep brain stimulation of the nucleus accumbens shell attenuates cocaine reinstatement through local and antidromic activation. *The Journal of Neuroscience, 33*, 14446–14454.

Voges, J., Müller, U., Bogerts, B., Münte, T., & Heinze, H. J. (2013). Deep brain stimulation surgery for alcohol addiction. *World Neurosurgery, 80*, S28.e21–S28.e31.

Wilden, J. A., Qing, K. Y., Hauser, S. R., McBride, W. J., Irazoqui, P. P., & Rodd, Z. A. (2014). Reduced ethanol consumption by alcohol-preferring (P) rats following pharmacological silencing and deep brain stimulation of the nucleus accumbens shell. *Journal of Neurosurgery, 120*, 997–1005.

Yadid, G., Gispan, I., & Lax, E. (2013). Lateral habenula deep brain stimulation for personalized treatment of drug addiction. *Frontiers in Human Neuroscience, 7*, 806.

Zhou, H., Xu, J., & Jiang, J. (2011). Deep brain stimulation of nucleus accumbens on heroin-seeking behaviors: A case report. *Biological Psychiatry, 69*, e41–e42.

PART III

TOBACCO SMOKING IN NEUROMODULATION

CHAPTER

23

Understanding the Roles of Genetic and Environmental Influences on the Neurobiology of Nicotine Use

E. Prom-Wormley, G. Langi, J. Clifford, J. Real

Virginia Commonwealth University, Richmond, VA, United States

INTRODUCTION

The prevalence of current smoking in American adults has remained relatively stable over the past decade, decreasing from 28.0% to 25.2% over 2005–13 (Agaku, King, Dube, & Centers for Disease Control and Prevention, 2014). Most likely associated with policy-level action to decrease the rates of smoking initiation (SI) in adolescents, this level of success has not yet translated into increased numbers of successful quit attempts or improved population health (Goren, Annunziata, Schnoll, & Suaya, 2014). Smoking remains the leading cause of preventable death around the world, resulting in about 440,000 deaths annually in the United States (Centers for Disease Control and Prevention, 2002) and over 4 million deaths globally every year (Jackson, Muldoon, Biasi, & Damaj, 2015; Le Foll & Goldberg, 2009). Studies suggest that successful quitting is associated with reductions in mortality rates to those of nonsmokers for those who quit by age 35 (Samet, 1990; Jha et al., 2013). However, successful smoking abstinence, defined as a former smoker not engaging in any smoking for 1 year or more (Velicer, Prochaska, Rossi, & Snow, 1992), is difficult to achieve. For example, 80% of quit attempts result in relapse (Baker et al., 2007; Hendricks, Prochaska, Humfleet, & Hall, 2008; Kozlowski, Porter, Orleans, Pope, & Heatherton, 1994; West, 2005). In addition, although more than 70% of smokers would like to quit, about 44% engage in a quit attempt and only 4–7% of adults who attempt to quit smoking are successful (Fiore et al., 2008). It is possible that those who continue to smoke, despite societal disincentives and wider availability of pharmacological treatment, require individualized approaches to increase the possibility of successful quit attempts (Boardman et al., 2011). Developing such approaches requires an understanding of the biological pathways shared between nicotine dependence (ND) and the process of smoking cessation, most of which are likely to occur within various regions of the brain.

The relatively stable rates of current smoking as well as the difficulty associated with successful abstinence are thought to be due to etiological pathways related to ND and smoking cessation, defined as the process by which an individual proceeds from being a smoker to becoming a nonsmoker (Centers for Disease Control and Prevention, 2002). Historically, ND was defined by the Diagnostic and Statistical Manual, fourth edition (DSM-IV) as a maladaptive pattern of nicotine use leading to clinically significant impairment or distress. It is characterized according to four criteria: (1) tolerance, (2) withdrawal, (3) consuming nicotine in larger amounts than originally anticipated, and (4) difficulty quitting despite a desire to do so (American Psychiatric Association, 2000, 2013). ND has been recently renamed "tobacco use disorder," and its definition updated in the DSM-V to meet just three criteria by removing the DSM-IV category of "difficulty quitting" and adding an item related to craving within the "nicotine consumption in larger amounts than originally anticipated" criterion. The DSM-V definition is relatively new and used in few published studies thus far in the fields of genetic epidemiology or brain morphometry. However, there is substantial overlap with the DSM-IV definition; therefore, the remainder of this chapter focuses on the concept of "nicotine dependence" in order to provide a review that reflects the majority of prior research results.

Understanding the neurobiological mechanisms that are involved in maintaining smoking behaviors,

Addictive Substances and Neurological Disease
http://dx.doi.org/10.1016/B978-0-12-805373-7.00023-2

particularly ND and withdrawal, may improve the likelihood of successful smoking abstinence. Structural neuroimaging studies of genetically informative samples have demonstrated evidence for genetic and environmental influences on the neurobiological mechanisms that underpin smoking behaviors. This growing body of research is providing a rich understanding of smoking-related alterations in the brain, as well as potential avenues by which to address and improve outcomes related to smoking cessation.

THE BRAIN DISEASE MODEL OF NICOTINE DEPENDENCE

Addiction and therefore substance use disorders such as ND may be conceptualized and treated as an acquired disease of the brain. Volkow, Koob, and McLellan (2016) summarize three stages of addiction: binge and intoxication, withdrawal and negative affect, and preoccupation and anticipation (craving). Each stage is associated with the activation of specific neurobiological circuits (Volkow & Baler, 2014): (1) reward/saliency, (2) memory/learning-conditioning/habits, (3) inhibitory control/executive function, (4) motivation/drive, (5) introspection, and (6) aversion. Furthermore, each stage involves different brain regions. As an example, the prefrontal cortex is involved in salience attribution [the process of making objects, such as a cigarette, attention grabbing (Jensen & Kapur, 2009)], inhibitory control (the circuitry to prevent inappropriate behaviors), and motivation circuitry (encouraging efforts to achieve a goal), while the anterior cingulate cortex is involved in inhibitory control (Volkow & Baler, 2014). Since disruption of these two regions is associated with compulsivity and impulsivity (Fineberg et al., 2010), reductions in the size of these regions are expected to be involved in compulsive smoking behaviors such as those that occur in the preoccupation and anticipation stage. Genetic epidemiological studies using measures of brain structure have further strengthened this model and provide insight into potential prevention and treatment approaches.

ELUCIDATING THE NEUROBIOLOGY OF NICOTINE DEPENDENCE USING STRUCTURAL MAGNETIC RESONANCE IMAGING

Understanding how structural integrity of brain regions reflects etiology is expected to inform an understanding of the neurobiology of ND, since the measure of structures is based on its constituent composition of neurons, glial cells, and myelinated axons, which are all important in the processing of information. The most common measure of brain morphometry is volume, an indicator of the amount and size of neurons, dendritic processes, and glial cells. Subcortical regions are consistently measured as volume while cortical volume is the product of cerebral cortical thickness and surface area. Cortical surface area and cortical thickness quantify different aspects of structural development. The radial hypothesis of cortical development (Rakic, 1988, 1995, 2007) suggests that neurons in the cerebral cortex are first organized into ontogenetic columns that run perpendicular to the surface of the brain (Mountcastle, 1997; Panizzon et al., 2009). Additionally, the radial hypothesis suggests that cortical surface area is driven by the number of columns and that cortical thickness is influenced by the number of cells within a column (Rakic, 1988). At the global level, there is no significant genetic covariance shared between the surface area and thickness (Panizzon et al., 2009). Moreover, studies of interindividual variation in adult brain size report that the majority of differences in cortical gray matter (GM) volume are due to differences in cortical surface area rather than cortical thickness (Im et al., 2008; Pakkenberg & Gundersen, 1997). Therefore, surface area and cortical thickness are distinct features of cortical structure, may reflect different features of cortical function, and their study provides insight beyond knowledge gained from cortical volume.

The study of brain morphometry relies on data derived from structural magnetic resonance imaging (MRI), which allows for the safe in vivo examination of cerebral atrophy as well as structural integrity with high contrast. An MRI machine produces a static magnetic field using a strong magnet and electromagnetic radiation in the form of a radio frequency pulse. The magnetic field aligns protons in a fashion either with or against the field. Simultaneously, the radio frequency pulse excites and synchronizes photons within tissue to rotate by a 90- or 180-degree angle. The MRI signals produced from the excited tissue vary across tissue types, that is, GM versus white matter (WM), as a result of differential electromagnetic activity. Once the radio frequency source is switched off, the magnetic vector returns to its resting state. The rate at which protons relax from a higher to a lower energy state and the rate at which protons become unsynchronized cause a signal that reflects measurable properties and is affected by tissue type. Receiver coils detect the location and strength of the electromagnetic waves and produce data representing tissue-specific images (Berger, 2002).

Recent technological advancements such as automated digital image processing as well as improved processing speed for the computation of digital neuroimaging data have resulted in the rapid generation of large quantities of MRI-based data. MRI-produced data use such automated approaches and algorithms to create a voxel-based image of brain structure (Paus, 2010). A voxel represents the smallest volumetric three-dimensional digital representation of an image and is analogous to pixels in two-dimensional digital images (Gao, Huth, Lescroart, & Gallant, 2015). Typical computer-automated methods for image processing normalize MRI images to the same space, and partition the normalized image into GM, WM, and cerebrospinal fluid on the basis of voxel intensities (Ashburner & Friston, 2000; Dennis & Thompson, 2013; Mechelli, Price, Friston, & Ashburner, 2005).

Neuroimaging Studies of Smokers Highlight Brain Structures Involved in Addiction

Smoking is consistently associated with widespread decreases in structural size of brain regions involved in the addiction process. For example, case-control studies using structural MRI have reported significant volumetric decreases in smokers' brains compared to those of nonsmokers. Brody et al. (2004) used hand-drawn regions of interest to compare nicotine-dependent adult smokers (21–65 years of age) with nonsmokers and identified GM volume decreases in left dorsal anterior cingulate cortex, left dorsolateral prefrontal cortex, and bilateral ventrolateral prefrontal cortex in the smokers. Other case-control studies have used voxel-based morphometry and FreeSurfer, an open source brain imaging software (http://surfer.nmr.mgh.harvard.edu), to analyze MRI images, and have reported similar findings in the anterior cingulate cortex (Gallinat et al., 2006; Liao, Tang, Liu, Chen, & Hao, 2012), prefrontal cortex (Gallinat et al., 2006; Hanlon et al., 2016; Liao et al., 2012; Morales, Lee, Hellemann, O'Neill, & London, 2012), and orbitofrontal cortex (Gallinat et al., 2006), with additional regions affected such as temporal gyrus (Gallinat et al., 2006), lingual gyrus (Gallinat et al., 2006; Hanlon et al., 2016), parahippocampal gyrus (Gallinat et al., 2006; Hanlon et al., 2016), thalamus (Franklin et al., 2014; Gallinat et al., 2006; Hanlon et al., 2016; Liao et al., 2012), cerebellum (Franklin et al., 2014; Kühn et al., 2012; Yu, Zhao, & Lu, 2011), amygdala (Hanlon et al., 2016), and putamen (Hanlon et al., 2016). In addition, some studies demonstrated decreased cortical thickness in smokers' brain structures, such as orbitofrontal cortex (Kühn, Schubert, & Gallinat, 2010; Li et al., 2015), anterior cingulate cortex, temporal gyrus, parahippocampus, and insula (Li et al., 2015).

Cross-sectional studies of community-based samples have replicated earlier findings of decreased GM volume in the prefrontal cortex, anterior cingulate cortex, insula, and temporal gyrus (Fritz et al., 2014), with additional discoveries of reduced olfactory gyrus volume (Fritz et al., 2014) and reduced total cerebral brain volume ratio in men only (Seshadri et al., 2004). However, few community-based studies have found no significant relationship between in striatal volume and smoking status (i.e., smokers vs. nonsmokers), perhaps due to a less stringent definition of smoking status than the case-control studies (Das, Cherbuin, Anstey, Sachdev, & Easteal, 2012; Taki et al., 2004).

Although prior studies reported widespread decreases in GM, they also indicated increases in GM volume and density in smokers, or in some cases no significant associations between brain structure/size and smoking. For example, increased GM volume was observed in some parts of the caudate (Li et al., 2015), putamen, parahippocampus (Franklin et al., 2014), lingual gyrus, and occipital cortex (Hanlon et al., 2016). Further, a cross-sectional study of community-based sample reported increased GM density in fusiform and temporal subgyrus regions of female smokers (Chen, Wen, Anstey, & Sachdev, 2006). Similarly, studies on smokers' WM volume have generated conflicting results, with some detecting no significant differences in WM volume between smokers and nonsmokers (Gallinat et al., 2006; Paul et al., 2008), and others reporting decreased WM volume in the corpus callosum (Choi et al., 2010) and increased WM volume in the prefrontal cortex, Rolandic operculum, temporal lobe (Fritz et al., 2014), anterior cingulate cortex, cingulate cortex, and putamen (Yu et al., 2011) of smokers.

GM consists of neural cell bodies, neuropil (unmyelinated axons, dendrites, and glial cells), and synapses, while WM has myelinated axons and commissures. The GM makes up the cerebral cortex (the highly folded outer layer of the cerebrum), which has sensory, motor, and association (for speech and decision-making) regions (Yeo et al., 2011). GM loss is common in numerous mental disorders, including substance addiction, resulting in decreased executive functioning and emotional problems (Goldstein & Volkow, 2011; Goodkind et al., 2015). The WM comprises a much larger proportion of the brain compared to GM and serves to connect distant brain regions (Fields, 2008). Disruption of myelin in WM might impair cognitive ability, mood, and the ability for organized thinking due to slower signal conduction (Fields, 2008). The decreases in GM and WM volume that are typically associated with smoking therefore suggest that alterations in the neuronal connections between regions are involved in the processing of smoking-related information (i.e., nicotine concentration in a delivery device, nicotine delivery device used, and situations that increase the likelihood of smoking).

Two meta-analyses provide further confirmation for regional GM changes according to smoking status, demonstrating robust GM decreases in the anterior cingulate cortex (Pan et al., 2013) and prefrontal cortex (Zhong et al., 2016), but a GM increase in the lingual cortex (Zhong et al., 2016). In addition to volumetric differences in specific structures (see Table 23.1), neuroimaging studies have also revealed relatively small overall changes in the size of a region as indicated by moderate to weak associations with smoking status (Brody et al., 2004; Choi et al., 2010; Fritz et al., 2014; Gallinat et al., 2006; Hanlon et al., 2016; Liao et al., 2012; Morales et al., 2012). It is possible that these small changes reflect dynamic structural alterations and are likely to mirror functional processing in order to address environmental demands placed on the brain (May, 2007). For example, although cigarette smoking has neurotoxic properties (Ferrea & Winterer, 2009), chronic exposure to nicotine is likely to result in neuroadaptation (Menossi et al., 2013) rather than severe dysfunction. These structural changes likely reflect drug-related neuroplastic changes in the brain (Seo & Sinha, 2015). Further, these structural and functional changes may occur as a response to environmental stimuli during nicotine intake (Sale, Berardi, & Maffei, 2014).

Consistent widespread changes in volume occur during development across brain structures in both men and women (see Batouli, Trollor, Wen, & Sachdev, 2014 for a detailed review) and as such the development of the brain is likely to be dynamic throughout the life course. Smoking and later ND may exacerbate these changes, and neuroimaging studies of smokers across the stages of development throughout life may highlight structures that are particularly important in the ND pathway. For example, one study compared brain structure volumes of early (20- to 29-year-old) and established (30- to 49-year-old) smokers to age-matched nonsmoking subjects (Hanlon et al., 2016). Both early and established smokers compared to nonsmoking

TABLE 23.1 Structural Changes in Total Brain Structure and Cerebellar and Subcortical Gray (GM) and White Matter (WM) Volume Associated With Smoking

Location	Change in Volume	GM/WM	N	Study Design	Reference(s)
B-cerebellum	Decrease	GM	32	Case-control	Yu et al. (2011)
R-cerebellum	Decrease	GM	55	Case-control	Kühn et al. (2012)
Total cerebral brain	Decrease	–	1841	Cross-sectional	Seshadri et al. (2004)
Total brain	Decrease	–	315	Cross-sectional	Das et al. (2012)
Ventricular volume	Decrease	–	315	Cross-sectional	Das et al. (2012)
B-amygdala	Decrease	GM	118	Case-control	Hanlon et al. (2016)
L-anterior cingulate cortex	Decrease	GM	88	Case-control	Liao et al. (2012)
L-anterior cingulate cortex	Decrease	GM	36	Case-control	Brody et al. (2004)
L-anterior cingulate cortex	Increase	WM	32	Case-control	Yu et al. (2011)
R-anterior cingulate cortex	Decrease	GM	974	Cross-sectional	Fritz et al. (2014)
R-caudate	Increase	Both	49	Case-control	Li et al. (2015)
R-cingulate gyrus	Decrease	GM	45	Case-control	Gallinat et al. (2006)
L-posterior cingulate	Decrease	GM	45	Case-control	Gallinat et al. (2006)
B-corpus callosum	Decrease	WM	58	Case-control	Choi et al. (2010)
L-middle cingulate	Increase	WM	32	Case-control	Yu et al. (2011)
R-putamen	Decrease	GM	118	Case-control	Hanlon et al. (2016)
B-putamen	Increase	WM	32	Case-control	Yu et al. (2011)
L-thalamus	Decrease	GM	118	Case-control	Hanlon et al. (2016)
L-thalamus	Decrease	GM	88	Case-control	Liao et al. (2012)
R-thalamus	Decrease	GM	45	Case-control	Gallinat et al. (2006)

B, bilateral; *L*, left; *R*, right.

controls had significant reductions in GM volume of the insula, amygdala, medial and orbital frontal cortex, thalamus, frontal pole, and putamen. Significant increases in occipital cortex and lingual gyrus were also identified. However, established smokers had an overall lower GM volume than did younger smokers and nonsmokers. These findings suggest that although all smokers experience reductions in brain volume over time, smoking may affect brain structure within the first few years of SI (Hanlon et al., 2016). Such studies of smokers across the lifespan are expected to identify the neurobiological networks pivotal to development of the various stages of ND and related symptom severity.

In summary, results from structural MRI studies highlight the role of several brain regions in the etiology of ND (Tables 23.1 and 23.2) and support the brain disease model for addiction, which may be a result of neuroadaptation. Nicotine-dependent individuals consume nicotine products in large quantities and build tolerance

TABLE 23.2 Structural Changes in Cortical Gray (GM) and White Matter (WM) Volume Associated With Smoking

Hemisphere and Lobe	Region	Change in Volume	GM/WM	N	Study Design	Reference(s)
L-frontal	Medial frontal cortex	Decrease	GM	118	Case-control	Hanlon et al. (2016)
L-frontal	Medial frontal cortex	Decrease	GM	88	Case-control	Liao et al. (2012)
L-frontal	Orbital frontal cortex	Decrease	GM	118	Case-control	Hanlon et al. (2016)
R-frontal	Frontal pole	Decrease	GM	118	Case-control	Hanlon et al. (2016)
L-frontal	Inferior frontal gyrus	Decrease	GM	45	Case-control	Gallinat et al. (2006)
L-frontal	Medial frontal gyrus	Decrease	GM	45	Case-control	Gallinat et al. (2006)
R-frontal	Orbitofrontal cortex	Decrease	GM	43	Case-control	Morales et al. (2012)
L-frontal	Dorsolateral prefrontal cortex	Decrease	GM	36	Case-control	Brody et al. (2004)
B-frontal	Dorsolateral prefrontal cortex	Decrease	GM	974	Cross-sectional	Fritz et al. (2014)
B-frontal	Dorsolateral prefrontal cortex	Increase	WM	974	Cross-sectional	Fritz et al. (2014)
B-frontal	Dorsomedial prefrontal cortex	Decrease	GM	974	Cross-sectional	Fritz et al. (2014)
B-frontal	Ventrolateral prefrontal cortex	Decrease	GM	36	Case-control	Brody et al. (2004)
B-frontal	Ventrolateral prefrontal cortex	Decrease	GM	974	Cross-sectional	Fritz et al. (2014)
B-frontal	Ventromedial prefrontal cortex	Decrease	GM	974	Cross-sectional	Fritz et al. (2014)
R-frontal	Supplementary motor area	Increase	WM	974	Cross-sectional	Fritz et al. (2014)
R-frontal	Olfactory gyrus	Decrease	GM	974	Cross-sectional	Fritz et al. (2014)
B-occipital	Cuneus	Decrease	GM	45	Case-control	Gallinat et al. (2006)
R-occipital	Lingual gyrus	Increase	GM	118	Case-control	Hanlon et al. (2016)
B-occipital	Lingual gyrus	Decrease	GM	45	Case-control	Gallinat et al. (2006)
L-occipital	Occipital cortex	Increase	GM	118	Case-control	Hanlon et al. (2016)
L-parietal	Postcentral gyrus	Decrease	GM	45	Case-control	Gallinat et al. (2006)
L-temporal	Superior temporal gyrus	Decrease	GM	45	Case-control	Gallinat et al. (2006)
L-temporal	Fusiform gyrus	Decrease	GM	45	Case-control	Gallinat et al. (2006)
L-temporal	Parahippocampal gyrus	Decrease	GM	45	Case-control	Gallinat et al. (2006)
B-temporal	Parahippocampal gyrus	Decrease	GM	118	Case-control	Hanlon et al. (2016)
R-temporal	Temporal lobe	Increase	WM	974	Cross-sectional	Fritz et al. (2014)
B-temporal	Insula	Decrease	GM	118	Case-control	Hanlon et al. (2016)
L-temporal	Insula	Decrease	GM	974	Cross-sectional	Fritz et al. (2014)
R-temporal	Rolandic operculum	Increase	WM	974	Cross-sectional	Fritz et al. (2014)

B, bilateral; *L*, left; *R*, right.

to the substance (American Psychiatric Association, 2013). Numerous structural MRI studies have shown significant decreases in brain volume across cortical and subcortical regions with robust findings associated with the prefrontal cortex and anterior cingulate cortex. These dynamic structural alterations are of small magnitude, which suggests neuroplasticity as a response to mediate the neurotoxic effects of chronic nicotine exposure and consumption.

THE GENETIC EPIDEMIOLOGY OF SMOKING AND BRAIN STRUCTURE

Genetically informative study designs can be used to build upon knowledge developed in prior neuroimaging studies by estimating the degree to which smoking occurs in families, estimating the magnitude of genetic and environmental influences on smoking, and describing the mechanisms by which genetic and environmental effects function to increase risk for smoking. These studies typically use data collected from related individuals such as twins and their family members, adopted twins reared apart, and multigenerational families. Additional approaches such as candidate gene association studies and genome-wide association studies (GWAS) build on results from twin and family studies to provide a more detailed view of specific genetic factors involved in the etiology of an outcome (here, ND or brain morphology, for example). Finally, study designs testing the influence of epigenetic processes explore the direct differences that shape how genes are expressed. Each of these approaches uses their unique strengths to paint a picture of the genetic architecture underlying many phenotypes and diseases.

TWIN AND FAMILY STUDIES

Family and twin studies estimate the degree to which outcomes/phenotypes cluster in families and as such provide an estimate of the relative contribution of genetic and environmental factors. Family studies do so by estimating "familiality," or the degree to which a trait is associated within families. Such a trait is expected to be due to some genetic and environmental influences shared among family members. Many studies examine first-degree relatives, whereas others use a multigenerational study design spanning three or more generations of relatives. However, while family studies of first-degree relatives can offer early clues to genetic influences on a trait or behavior, familiality alone cannot distinguish the unique contributions of genetic and environmental influences.

Twin study designs can estimate both genetic and environmental contributions to an outcome. Studies of twins reared in the same home use data from monozygotic (MZ) twins, who result from a single fertilized egg and as such inherit the same genetic material, as well as dizygotic (DZ) twins who result from two different eggs and share, on average, 50% of their genetic material. If MZ pairs are more similar for a trait than are DZ pairs, then the trait is expected to be subject to some genetic influences. If DZ pairs are equally similar (or even more so) for a trait than are MZ pairs, then environmental factors are expected to play a substantial role. To determine which type of environmental factor matters more—shared or unique—researchers can assess the correlation between MZ pairs (Neale, 2009).

Twin studies use MZ and DZ twin-pair variances and covariances to estimate the proportion of total phenotypic variance of a trait that is due to additive genetic, shared environmental, and unique environmental influences. Additive genetic effects (A) refer to the additive effects of alleles at every locus, common environmental effects (C) are effects, which are common to both twins such as the family environment, and unique environmental effects (E) refer to effects not shared by members of the twin pair, which in turn make the twins less similar. Unique environmental effects might include the uniqueness of peer influences, where twins can have different friends, and include measurement error (Cherny, 2009).

Twin and Family Studies of Brain Structure

Twin and family studies of brain structure may be most useful in studying neurobiological pathways related to ND since there are more twin and family studies that have produced results for these measures compared to other neuroimaging technologies (i.e., functional MRI, diffusion tensor imaging, or blood oxygen level–dependent signals) (Jansen, Mous, White, Posthuma, & Polderman, 2015). Consequently, to date, structural MRI provides the richest evidence regarding the genetic influences on the neurobiology of smoking, although this is likely to change as more data become available.

Twin and family studies generally suggest that all measures of brain structure are fairly heritable. Family studies indicate moderate to high heritability (i.e., the proportion of the total variance due to additive genetic influence) of cortical structures, with A accounting for 19–78% of the total variance for volumetric measures. These heritability estimates remain moderate for measures of average cortical thickness (A = 21–83%) as well as surface area (A = 25–82%) (DeStefano et al., 2009; Winkler et al., 2010). Meta-analyses of twin studies in 2010s (Blokland, de Zubicaray, McMahon, & Wright,

2012; Strike et al., 2015) reported that most brain structures were highly heritable with little contribution from common environmental factors (C = 0.0–28.3%, but accounting for less than 20% of the total variance in most subjects). Cortical structures showed high heritability for GM volume (A = 49.9–76.5%) and WM volume (A = 62.3–84%) measures, while cortical thickness was of moderate to high heritability (A = 21.6–76.1%) except for a small number of regions (4 of 62) with low heritability (A = 0.4–19.6%). Similarly, subcortical volumes were highly heritable, with genetic factors accounting for 45.5–81.6% of the total variance (Blokland et al., 2012; Strike et al., 2015).

Across the life course, most brain structures remain under a high degree of genetic influence with dynamic heritability estimates from childhood to adulthood (Batouli et al., 2014; Jansen et al., 2015). In neonates, for example, high heritability has been reported, with genetic influences accounting for 42–74% of the total variance for GM and WM volumetric measures, except in the corpus callosum and cerebellum where the heritability was 4% and 17%, respectively (Gilmore et al., 2010). Furthermore, in children and adolescents (5–19 years of age), moderate to high heritability was reported for GM and WM volumetric (A = 41–91%) and density (A = 82–83%) measures, subcortical measures (A = 26–95%) as well as cortical thickness (A = 33–72%) (van Leeuwen et al., 2009; Lenroot et al., 2009; Peper et al., 2009; van Soelen et al., 2013; Wallace et al., 2006; Yoon, Fahim, Perusse, & Evans, 2010; Yoon, Perusse, Lee, & Evans, 2011). Significant age–heritability interactions have also been reported in children—increasing WM volume heritability with age but the opposite for GM volume (Wallace et al., 2006) and increasing cortical thickness heritability with age in late-maturing regions (Lenroot et al., 2009). Significant genetic factor influences on brain structure appear to be maintained in adults (19–69 years of age), as moderate to high heritability was also observed for GM and WM volumetric (A = 66–87%), density (A = 76–83%), and cortical thickness measures (A = 20–78%) (Baaré et al., 2001; Hulshoff Pol et al., 2006; Kremen et al., 2010; Panizzon et al., 2009; Posthuma et al., 2003; Winkler et al., 2010), whereas surface area and subcortical measures displayed a wider range of heritability (A = 3–94% and A = 0–84%, respectively) (Chen et al., 2012; DeStefano et al., 2009; Eyler et al., 2011a, 2011b; Panizzon et al., 2009; Scamvougeras, Kigar, Jones, Weinberger, & Witelson, 2003; Wright, Sham, Murray, Weinberger, & Bullmore, 2002). Studies of older adults (68–78 years of age) have reported similar findings of moderate to high heritability for volumetric (A = 27–85%) and surface area (A = 22–85%) measures (Carmelli et al., 1998; Geschwind, Miller, DeCarli, & Carmelli, 2002; Pfefferbaum, Sullivan, & Carmelli, 2001; Pfefferbaum, Sullivan, Swan, & Carmelli, 2000; Sullivan, Pfefferbaum, Swan, & Carmelli, 2001).

Longitudinal twin studies have demonstrated the influence of genetic factors on normal developmental brain structure changes, including GM and WM density and volume, as we age (Batouli et al., 2014). In children, researchers observed changes in heritability measures for cortical thickness with age (50% at 5 years, 38–45% at 9 years, 45–53% at 12 years, and 90% at 17 years) (Schmitt et al., 2014; van Soelen et al., 2012), with significant yet different genetic contributions to cortical thickness change between ages 9 and 12 (A = 49–50%) (van Soelen et al., 2012). Genetic factors have also been shown to be important for volumetric brain changes (A = 19–45%, except for GM and WM volume changes), but unlike those for cortical thickness measures, the genetic correlations of brain volumes between ages 9 and 12 were close to 1 (van Soelen et al., 2013). Studies in adults mirror those in children, showing significant genetic influences for cortical thickness change (A = 28–56%) and volumes (A = 29–33%), with some exceptions, such as low to nonsignificant heritability for volumetric changes in cerebrum GM (Brans et al., 2010; Brouwer et al., 2014). However, the sole longitudinal study thus far on volumetric changes in older adults demonstrated a very small genetic influence (A = 7%) for volumetric change between baseline and follow-up 4 years later (Lessov-Schlaggar et al., 2012), and additional research in this age group is needed (Jansen et al., 2015).

In contrast to studies on the effect of age on heritability of brain structures, few twin studies have tested for sex differences. Such mechanisms can be tested using data from same-sex as well as opposite-sex twin pairs. Chiang et al. (2011) reported significantly greater genetic influences on WM development in males (A = 60–80%) than in females (A = 20–60%). However, a study by Swagerman, Brouwer, de Geus, Hulshoff Pol, and Boomsma (2014) reported no significant sex differences in subcortical GM volumes. Although sex differences may provide insight into the role of sex-specific neurobiological mechanisms on brain structure (Dennison et al., 2013; Koolschijn & Crone, 2013; Lenroot & Giedd, 2010), knowledge in this area is currently limited.

Genetic and Environmental Factors Shared Between Brain Structures

Twin and family studies have also identified genetic and environmental factors that are shared between brain structures in order to understand their patterning. Several phenotypic correlations have been detected across subcortical as well as cortical regions, and it is possible that these correlations highlight neurobiological pathways with regions that function together via genetic or environmental factors that are shared between structures (Strike et al., 2015). For example, Eyler et al. (2011a, 2011b) reported high genetic correlations between specific subcortical regions, which could be partitioned into four sets of genetic factors. These genetic factors explained relationships between the subcortical structures and specifically reflected (1) a basal ganglia/thalamic factor that consisted of the putamen, pallidum, caudate, and thalamus; (2) a limbic factor, consisting of hippocampus and amygdala; (3) a ventricular factor consisting of all the ventricular measures included in the subcortical area; and (4) a factor consisting of bilateral accumbens. Consequently, these subcortical results indicate that multiple sets of genes are responsible for the relationships between subcortical structures. A similar factor structure was not detected for common environmental influences (Eyler et al., 2011a, 2011b). However, the genetic factor structure was consistent in a sample of younger adults (Rentería et al., 2014), and as such the study of genetic and environmental influences shared between structures may provide a framework to consider networks of brain regions that function together to promote specific behaviors.

Twin and Family Studies of Smoking

The stages by which smoking behaviors are measured include: (1) SI, when a never-smoking individual attempts smoking for the first time, a necessary step to other stages of smoking or ND; (2) smoking persistence (SP), regular/heavy tobacco use which is an intermediate step to becoming nicotine dependent; and (3) ND. The SI is generally measured as having attempted smoking at any time during the life course (study participants are asked, "Have you ever smoked a cigarette? If so, how many?") or as the age at which experimentation occurred ("How old were you when you first started to smoke?"). There are few standards for measuring SP, but it is typically measured as the maximum number of cigarettes smoked in the past 24 h, current smoking (i.e., "Do you currently smoke?" or "Have you smoked in the past 30 days"), and average number of packs of cigarettes smoked in a year (pack-years). ND is typically measured either through instruments developed to reflect symptomatology highlighted in the most recent versions of the DSM diagnosis of ND (American Psychiatric Association, 1994, 2013) or as a sum score of symptoms detailed in the Fägerstrom Test for Nicotine Dependence (Heatherton, Kozlowski, Frecker, & Fagerström, 1991). Some studies use smoking amount (number of cigarettes/packs smoked per day) as a proxy measure for ND (Rose, Broms, & Korhonen, 2009).

Family studies indicate clustering within families and suggest genetic and environmental contributions in the development of smoking. Siblings of habitual-smoking probands were 77% (relative risk = 1.77) more likely to engage in habitual smoking themselves. Agrawal and Lynskey (2008) looked at seven twin studies from five countries and found that heritability for ND ranged from 30% to 72% with no significant common environmental influences. Genetic factors seem to play a greater role; environmental influences drop dramatically from early adulthood to mid-adulthood (Agrawal & Lynskey, 2008).

Twin studies across stages of smoking behavior indicate increasing influence of additive genetic effects (A) as ND develops. The heritability of SI is moderate, with A accounting for about 11–65% of the total variance of SI and common shared environmental influences (C) contributing a substantial proportion (7–75%) (Fowler et al., 2007; Hopfer, Crowley, & Hewitt, 2003; Huizink et al., 2010; Kendler, Schmitt, Aggen, & Prescott, 2008; Rose et al., 2009; Sartor et al., 2009), and no evidence for sex differences in adolescence (Rhee et al., 2003). Genetic influences increase for SP (A = 39–80%) with little to no common environmental influences evident (C = 0–30%) (Koopmans, Slutske, Heath, Neale, & Boomsma, 1999; Rhee et al., 2003; Rose et al., 2009). Interestingly, SP is subject to age and sex differences, as a meta-analysis concluded that genetic influences were more important for SP in adult males (A = 0.59, C = 8%) (Li, Cheng, Ma, & Swan, 2003), while a twin-adoption study of adolescence reported common environments to be more important for SP in adolescent males (A = 6%, C = 43%) (Rhee et al., 2003). As for SP, additive genetic influences were important for ND (A = 40–86%) with common environmental influences absent (C = 0%) in adults and adolescence (Koopmans et al., 1999; Rose et al., 2009), although one study of 17-year-old twins showed significant common environmental influences for ND (C = 37%). Subsequently, twin studies have highlighted the importance of genetic heritability to the development of ND and SP as opposed to SI, and these estimates remained consistent across different definitions of SP and ND.

Twin studies also suggest moderate overlap of genetic influences across stages of smoking severity. Some twin studies suggest moderate to high genetic influence overlap between SI and SP (correlation of 0.17–0.87) (Broms, Silventoinen, Madden, Heath, & Kaprio, 2006; Fowler et al., 2007; Huizink et al., 2010; Morley et al., 2007), while some earlier studies show a smaller overlap (correlation of 0–0.056) (Hardie, Moss, & Lynch, 2006; Heath, Martin, Lynskey, Todorov, & Madden, 2002). This suggests that to some extent, etiological similarities exist between SI and the development of ND through common genetic and environmental factors shared between these two stages (Fowler et al., 2007).

For SI, longitudinal studies of adolescent twins indicate increasing genetic influences from early adolescence into adulthood. A study of Dutch twins reported that genetic influences were not significant for young adolescents (12–14 year olds), but significant for older adolescents (17–25 year olds), with heritability of 0.66 for boys and 0.33 for girls (Koopmans, van Doornen, & Boomsma, 1997). A three-wave study in Australian twins reported increasing heritability over time: 0.15 at wave 1, 0.20 at wave 2, and 0.35 at wave 3, accompanied by a decrease in common environmental factors. This study also demonstrated that a proportion of the heritability for SI might reflect an indirect influence of genetic influences on choice of peers (White, Hopper, Wearing, & Hill, 2003).

UNDERSTANDING THE NEUROBIOLOGICAL PATHWAYS INVOLVED IN NICOTINE DEPENDENCE

Few twin/family studies have resolved the degree to which genetic and environmental influences between smoking and brain structure are shared. A cross-sectional study of middle-aged male twin pairs detected widespread, negative phenotypic correlations between cigarette pack-years and several cortical as well as subcortical brain structures. These phenotypic correlations were found to be the result of genetic and unique environmental factors that are shared between cigarette pack-years and subcortical volume as well as cortical volume and surface area. Regions for which associations were due to significant genetic covariance included right lingual gyrus, right caudal anterior cingulate cortex, left cuneus, left cerebellum cortex, right pars opercularis, and right precuneus, which are expected to be involved in the control circuitry as defined by the brain disease model. Similarly, regions for which associations were due to significant unique environmental covariance included left posterior cingulate cortex, right rostral middle frontal gyrus, left pars orbitalis, right inferior temporal gyrus, left accumbens, left rostral middle frontal gyrus, left pallidum, and left middle temporal gyrus, which are expected to be involved in learning-conditioning circuitry. However, after adjustment for multiple testing, the results were no longer significant, emphasizing the need for additional study on the pathways shared between brain structure and smoking in larger samples (Prom-Wormley et al., 2015).

Results from twin and family studies should be evaluated while noting some limitations. First, sample size, particularly for studies of brain structure size may be low and as such power to detect significant genetic variance may be limited. Neuroimaging studies are expensive to conduct and become more expensive when ascertaining related individuals. Similarly, the racial/ethnic composition of many twin and family studies may not be generalizable to all populations since many studies use samples from European nations or have a high proportion of participants of European racial/ethnic background. Second, some brain structures are easier to measure than others as a function of structure size, leading to variability in estimates of genetic and environmental influences. Therefore, measurement error is a concern, particularly for brain structure. Measurement error is also a concern for measures of smoking because most measures of smoking rely on respondent self-report that may be subject to recall bias. Further, self-report measures of smoking are typically measured as binary or ordinal outcomes and may not completely capture the complete underlying liability for engaging in specific smoking-related behaviors.

Results from basic twin models such as those summarized in this chapter are subject to several assumptions, including: (1) MZ and DZ pairs are exposed to the shared environment to the same extent (i.e., equal environments assumption); (2) heritability estimates in twin studies only reflects additive genetic influences; (3) influence due to nonadditive genetic sources of variance are not estimated including gene–environment interaction, gene–environment correlation, epistasis, or epigenetic mechanisms; (4) twin studies assume random mating and as such individuals are as likely to choose mates that are different from themselves as they are to choose mates that are phenotypically or genotypically similar to themselves; (5) variation due to the mutual influences between twins are not estimated (i.e., cooperation and competition within twin-pair interactions).

In summary, twin and family studies indicate substantial heritability for brain structure and smoking behavior. Additive genetic factors influence the development of most brain structures with minimal contribution from common environmental factors. Consistent associations between smoking and brain structures have been discovered with higher genetic components in grouped regions with similar functionalities, confirming the involvement of genetic factors in neurobiological pathways critical for a healthy brain. In smoking behaviors, there is evidence for shared genetic and environmental influences across the different stages of ND (SI and SP). Moderate additive genetic effects with substantial common environment contributions were reported for SI, but genetic influences become increasingly important as one progresses to becoming nicotine dependent, with some overlap of these factors between these two stages, and as smokers age. To some degree, the genetic and environmental influences between smoking and brain structure are shared and may be region specific. Interestingly, brain regions with significant genetic and environmental covariance with smoking are involved in the control and memory circuitry of the brain disease model. Yet, it is unclear from twin and family studies if these substantial genetic influences are due to the same genes or if they are merely indicative of an overarching latent genetic factor encompassing all genetic influences. These results also highlight additional areas for future study to address the limited knowledge on the influence of sex on the genetic and environmental contributions on brain structure and smoking, as well as the narrow understanding of how genetic influences are involved in brain structure and smoking develops across the life course.

GENE-BASED GENETIC EPIDEMIOLOGICAL STUDY DESIGNS

Results from twin and family studies have encouraged additional research focused on understanding the role of specific genetic influences on biological pathways involved in the neurocircuitry of ND. Consequently, additional genetically informed study designs have focused on understanding the role of specific genetic factors in the etiology of ND and smoking cessation. Results from genetic association and epigenetic studies of smoking and ND have been central in accomplishing this goal, and it is expected that these study designs will continue to produce results that will lead to improved treatment of ND and prevention of smoking-related outcomes.

CANDIDATE GENE ASSOCIATION STUDIES

Candidate gene association (CGA) studies test whether specific genetic markers, or specific locations in the DNA that differ among individuals, are significantly associated with an outcome of interest in individuals affected with an outcome compared to those who are not affected. One assumption of CGA studies is that genetic markers will display variation among cases and controls for a given phenotype. Another assumption is that the genetic markers used in CGA studies are expected to reflect gene regions with established functional relevance as determined from prior molecular genetic, animal model, and/or human studies. For example, markers located within the coding or regulatory regions of genes should affect gene expression or function/activity of the encoded protein. When the functional relevance of gene is known a priori, results from CGA studies can be clinically appealing, particularly for the development of personalized medicine applications (Peters, Rodin, de Boer, & Maitland-van der Zee, 2010).

Candidate Gene Studies of Smoking

Several genes have been identified in prior CGA studies of smoking and ND that are involved in (1) synaptic cholinergic neurotransmitter function (e.g., genes encoding for neuronal acetylcholine receptor (AChR) subunits, such as *CHRNA*2 *CHRNA4*, and *CHRNB2*); (2) synaptic dopaminergic neurotransmitter function (e.g., genes responsible for dopamine receptor function, such as *DRD1*, *DRD2*, and *DRD4*, as well as dopamine metabolism genes like *DBH* which encodes for dopamine β-hydroxylase) pathways; (3) GABAergic neurotransmitter function (e.g., genes responsible for genes encoding for neuronal GABA receptor subunits, such as *GABBR2*, *GABRA2*, and *GABRA4*); (4) serotonergic neurotransmitter function (e.g., genes encoding serotonin receptor function such as *HTR3A* and HTR5A); (5) glutaminergic neurotransmitter function (i.e., genes that encode for glutamate receptor function, such as *GRIN3A* and *GRIN2B*); (6) nicotine metabolism (e.g., *CYP2A6* and *CYP2B6*); (7) cellular transport of neurotransmitters across the synaptic cleft (e.g., *SLC6A4*); and (8) signal transduction via the mitogen-activated protein kinase signaling pathway (e.g., *BDNF* which encodes brain-derived neurotrophic factor and *NTRK2* which encodes neurotrophic tyrosine kinase receptor type 2) (Wang & Li, 2010; Yang & Li, 2016).

CGA results highlight the complexity involved in the development of ND. This is likely to consist of genetic variants involved in nicotine metabolism and reflects the degree to which nicotine is available for use in the brain as well as the processing of nicotine throughout the brain via neurobiological mechanisms related to the use and uptake of nicotine through the neurotransmitter systems. Of the neurotransmitter pathways that have been implicated in CGA studies, details regarding the neural substrates in the dopaminergic and serotonergic neurotransmitter pathways have received greatest attention. Dopaminergic pathway activation occurs via mesocorticolimbic projections (Biasi & Dani, 2011), and is involved in nicotine-related behaviors and moods via mesolimbic (i.e., from ventral tegmental area to the ventral striatum and nucleus accumbens) and mesocortical pathways (i.e., from the ventral tegmentum area to the prefrontal cortex and anterior cingulate cortex). The mesolimbic pathway is involved in producing feelings of euphoria and pleasure due to reward motivation (Dichter, Damiano, & Allen, 2012). The mesocortical pathway affects cognitive behavior (Jasinska, Zorick, Brody, & Stein, 2014). The serotonergic pathway connects serotonin production in the raphe nucleus through projections that innervate most of the rest of the brain through diffuse projections (Berger, Gray, & Roth, 2009). The serotonergic pathway is involved in addiction as well as a variety of behaviors related to the processing environments that are likely to be involved in ND [e.g., mood, memory, anger, fear, and stress responses (Berger, Gray, & Roth, 2009)].

GENOME-WIDE ASSOCIATION STUDIES

Genome-wide association studies (GWAS) test for genetic associations using several thousand single nucleotide polymorphisms (SNPs) located across the genome. An SNP represents one type of genetic marker that refers to a change in a base pair (nucleotide). GWAS is based on the concept of linkage disequilibrium (LD), defined as a process of nonrandom segregation of SNPs (Hirschhorn & Daly, 2005; McCarthy & Hirschhorn, 2008). Markers that are physically located close to one another will segregate together and as a result be observed to be in stronger LD with one another as quantified by stronger correlations (i.e., correlations > 0.8). Consequently, the alleles of nongenotyped markers in high LD with the genotyped markers can be probabilistically inferred via imputation (Marchini & Howie, 2010). Imputation predicts genomic sequences of areas that were not previously genotyped using genomic reference panels such as HapMap and 1000 Genomes. GWAS can be expanded to include data from multiple datasets in order to increase power to detect significant associations and as such GWAS meta-analysis or the pooling of GWAS data is becoming increasingly utilized (for reviews see Dudbridge & Gusnanto, 2008; Evangelou & Ioannidis, 2013; Medland, Jahanshad, Neale, & Thompson, 2014).

Genome-Wide Association Studies of Smoking Behavior

The chromosome 15 nicotinic acetylcholine receptor (nAChR) gene cluster *CHRNA5-A3-B4* is most consistently associated with ND (measured as daily average smoking quantity) by GWAS (Berrettini et al., 2008; Bierut et al., 2008; Saccone et al., 2007; Thorgeirsson & Stefansson, 2008). The genes in this cluster are often co-expressed and co-regulated and genetic association results alongside functional information of these variants may provide insight into the complexity of ND. For example, the most consistently associated SNP within this cluster is a functional missense variant (rs16969968) that results in a nonsynonymous substitution of aspartic acid to asparagine at position 398 (D398N), which may be involved in addiction severity as reported by studies of humans as well as mice (Hong et al., 2010; Morel et al., 2014; Tammimaki et al., 2012). The D398N substitution produces decreased acetylcholine-evoked function in α5 nicotinic receptors, which in turn decreases sensitivity to the rewarding effects of nicotine (Bierut et al., 2008; George et al., 2012). Additionally, the *CHRNA3* variant, rs578776 lowers activation of the anterior cingulate cortex and decreases its function to the thalamus pathway (Hong et al., 2010; Nees et al., 2013). It is expected that this reduced function is involved with feedback on recent nicotine exposure, and may be involved in craving or withdrawal processes (Brunzell, Stafford, & Dixon, 2015). Consequently, variants in the *CHRNA5-A3-B4* cluster may be responsible for different, although interrelated aspects of ND. To date, there are no known studies of the relationship between *CHRNA5-A3-B4* variants and brain structure in studies of smoking.

Meta-analyses using multiple GWAS samples have also identified associations between genetic variants involved in nicotine-mediated function in the brain and specific smoking behaviors, including (1) SI with *BDNF* (a gene encoding brain-derived neurotrophic factor, a protein involved in regulation of neuronal survival and behavior-related plasticity); (2) smoking quantity with *CHRNB3 and CHRNA6* (genes encoding nAChR subunits involved in nicotine-induced dopamine release); (3) nicotine metabolism with the nicotine-metabolizing enzymes, *CYP2A6* and *CYP2B6*; and (4) smoking cessation with *DBH* (a gene encoding dopamine β-hydroxylase, an enzyme in synaptic vesicles that catalyzes the conversion of dopamine to norepinephrine) (Thorgeirsson et al., 2010; Tobacco and Genetics Consortium, 2010).

Genetic Association Studies of Brain Structure

Candidate gene association studies of brain structure have been largely underpowered and have led to considerable inconsistency. Three candidate genes have received most attention in the literature: *BDNF*, *COMT* (a gene encoding catechol-O-methyltransferase, a dopaminergic system enzyme responsible for the regulation of dopamine metabolism), and *APOE* (a gene encoding apolipoprotein E, a protein involved in lipid metabolism and injury repair in the brain). *BDNF* and *COMT* have been identified as promising candidate genes in genetic association studies of smoking. However, these markers have not been consistently replicated in CGA studies of GWAS of brain morphology for any of these markers.

Table 23.3 summarizes the location and function of genetic variants reported from GWAS of brain structure. Two key points should be noted. First, three of the five SNPs are intronic, meaning that they are in the areas between protein-coding regions (exons). These SNPs are in areas that were previously thought to be irrelevant and may provide evidence for the importance of regulatory mechanisms (e.g., methylation and acetylation, see the following section). Second, many of the

TABLE 23.3 GWAS-Identified Variants Associated With Brain Structure

Structure	Reference Number	Gene	Function	Reference(s)
Intracranial volume	rs10784502	*HMGA2* (intronic)	Encodes an architectural protein	Stein et al. (2012)
	rs17689882	*CRHR1* (intronic)	Encodes a corticotrophin-releasing hormone receptor	Hibar et al. (2015)
Visual cortical surface (occipital lobe surface area)	rs2618516	*GPCPD1* (intronic)	Encodes a protein that hydrolyzes glycerophosphocholine	Bakken et al. (2012)
Hippocampal volume	rs7294919	Intergenic	Unknown	Stein et al. (2012)
	rs145212527	*FBLN2*	Tissue organization and neuron differentiation	Hibar, Medland, et al. (2013)
Lentiform nucleus volume	rs1795240	Intergenic	Unknown	Hibar, Stein, et al. (2013)

genes, such as *HMGA2* that encodes for an architectural protein, have functions that are of an organizational nature with respect to overall brain structure. These genes are important for the developing brain and organization of neuronal tissues, which may be impacted by substance use.

THE IMPACT OF GENOME-WIDE SIGNIFICANT VARIANTS FOR SMOKING ON BRAIN STRUCTURE

Genetic association studies of either smoking or brain structure continue to grow in number. However, to the best of our knowledge, there are no genetic association studies yet published that test for the influence of genetic markers on the relationship between smoking and brain structure, in contrast to other psychiatric disorders. Indeed, numerous case-control and some cross-sectional studies have reported associations between genetic variants and structural brain changes for schizophrenia and bipolar disorder (as reviewed by Gurung & Prata, 2015), as well as major depressive disorder (as reviewed by Ueda et al., 2016).

Additional follow-up testing the functional role of the variants identified through genetic association studies in the *CHRNA5-A3-B4* cluster is necessary but has been challenging. In general, nAChRs are expressed throughout the brain with the most common expression occurring in the dorsal striatum, thalamus, amygdala, ventral tegmental area, cortex, hippocampus, and basal ganglia (Clarke & Pert, 1985; Séguéla , Wadiche, Dineley-Miller, Dani, & Patrick, 1993; Zoli, Lena, Picciotto, & Changeux, 1998). There are 12 types of neuronal AChR subunits (α2–10 and β2–4), which combine together as a pentameric transmembrane cation channel. The exact composition of the subunits in a given channel make neuronal nAChRs differentially responsive to various compounds. Further, the localization of receptors in the brain also results in specificity and complexity of cholinergic signaling. The α3-β4*nAChRs (where * refers to assembly with other subunits) have enriched expression in the habenula, with a small subset of these receptors containing the α5 subunit (Baldwin, Alanis, & Salas, 2011; Grady et al., 2009; Scholze, Koth, Orr-Urtreger, & Huck, 2012; Shih et al., 2014). The habenula and interpeduncular nucleus pathway regulates the mesolimbic system and has been demonstrated to be critical for nicotine withdrawal symptoms in mice. However, knowledge on the function of the habenula in humans has been slow to emerge because it is very small and only accounts for only few voxels in automated neuroimaging methods. The voxel-based morphometry method itself poses difficulties in obtaining high-quality images of subcortical abnormalities, with image noise further complicating differentiation of the small habenula. However, a protocol for manual tracing of the habenula using high-resolution T1-weighted structural images has been successfully used in early 2010 to study volumetric changes of the habenula in bipolar and posttraumatic stress disorder patients (Lawson, Drevets, & Roiser, 2013). This method may also facilitate exploration of the association of genetic variants in the *CHRNA5-A3-B4* cluster with structural changes in the habenula.

LIMITATIONS OF GENETIC ASSOCIATION STUDIES

Although significant genetic associations for ND have been reported in several variants, particularly in dopamine and serotonin pathway genes, these results suffer from a high degree of inconsistency. For example, the direction of these associations are not consistent (i.e., genetic variants either increasing or decreasing smoking liability) and only a handful of associations have been replicated (Lerman & Berrettini, 2003; Quaak, van

Schayck, Knaapen, & van Schooten, 2009). Similar inconsistencies have been identified in genetic association studies of brain morphology (for review, see Strike et al., 2015). CGA studies in particular are subject to false positive (i.e., detection of a significant association when it actually does not exist) and false negative (i.e., nondetection of a significant association where one actually exists) associations. This results from (1) samples sizes that are too small to detect significant associations of polygenic traits where each marker produces a small magnitude of influence (i.e., odds ratios less than 1.3 or resulting in a 30% increase risk in affected individuals compared to those who are unaffected) (Marjoram, Zubair, & Nuzhdin, 2014); (2) misclassification due to differences in the measurement of smoking behaviors (i.e., SI vs. number of cigarettes per day vs. a diagnosis of ND) or genotyping error; (3) lack of adjustment for racial/ethnic genetic heterogeneity (i.e., the distribution of genetic variants may vary as a result of sample ancestry instead of the presence of a significant association); and (4) inadequate adjustment of statistical significance, particularly for novel CGA study results (e.g., $p = 1 \times 10^{-5}$) compared to results identified from prior GWAS that have also have demonstrated functional significance ($p = .05$) (Strike et al., 2015). Consequently, testing and reporting the replication of results from CGA studies is strongly encouraged (Munafo & Gage, 2013). Further, as results from GWAS become more abundant, CGA studies can be used for follow-up on candidate genes of interest that are also supported by additional literature.

Classification of smoking behaviors in GWAS using multiple samples is likely to decrease the power to detect significant associations. Meta-analysis of GWAS suffers from sample heterogeneity and as such studies are often forced to use measures of smoking that allow for the greatest overlap across studies. For example, studies rely on self-reported measures of ND, SI, or cigarettes per day, and the method by which these measures are collected are not currently universally standardized during data collection. Biological measures such as serum cotinine levels (as a measure of smoking quantity) may provide a more accurate method of characterizing smokers and increase the power to detect significant associations (Munafò et al., 2012). For example, the influence of the risk allele in rs1051730 of *CHRNA3* has a higher effect size for cotinine level ($\beta = 0.30$, $p = 2.23 \times 10^{-6}$) than the number of cigarettes consumed per day ($\beta = 0.13$, $p = .034$) (Keskitalo et al., 2009).

EPIGENETIC STUDIES

As of 2016, SNPs that have been identified in genetic association studies only account for a small proportion of the expected heritability on smoking, as estimated by twin studies (Ware & Munafò, 2015). The remaining genetic variance may result from epigenetic mechanisms—defined as dynamic changes in gene expression that may result from exposure to any number of environmental factors. Epigenetic mechanisms refer to regulatory processes affecting the genome, such as DNA methylation, acetylation of histones, and regulation of messenger RNA (mRNA) expression (Egger, Liang, Aparicio, & Jones, 2004). Methylation of a portion of a gene will result in a reduction in gene expression with subsequent alteration in protein function (Grewal & Rice, 2004). Acetylation will expose portions of a gene to transcription factors, potentially increasing gene expression and altering protein function in a different way (Kouzarides, 2007). Changes in mRNA expression cause changes in protein function as a result of either altered levels of the protein or altered protein function/activity.

DNA METHYLATION AND SMOKING BEHAVIOR

Among the different types of epigenetic mechanisms, DNA methylation is the most studied epigenetic process in smoking behavior research (as reviewed by Lee & Pausova, 2013). DNA methylation refers to a process where a methyl group is enzymatically added to cytosines in CpG dinucleotides (CpGs) in the DNA sequence. CpGs only occur in about 1% of locations within the human genome and most CpGs (70%) remain methylated. Generally, methylation inhibits DNA transcription, especially if it occurs in promoters. However, methylation can also enhance DNA transcription when methylation occurs in the body of the gene. Thus, the process of DNA methylation allows cells to rapidly respond to environmental stimuli.

Epigenetic studies of peripheral blood cells have consistently reported associations between chronic smoking and decreased DNA methylation of CpG sites in three genes. Most current studies use DNA methylation microarrays (e.g., the Illumina Infinium Human Methylation 450K BeadChip) and have identified numerous CpG dinucleotides differentially methylated between current smokers and never smokers, with 4% (62 of 1460 CpG sites) discovered in three or more studies (Gao, Jia, Zhang, Breitling, & Brenner, 2015). The most frequently reported sites were cg03636183 in the second exon of coagulation factor II receptor-like 3 gene (*F2RL3*), which is important for cell signaling and platelet activation, cg05575921 in the third intron of the aryl hydrocarbon receptor repressor gene (*AHRR*), which is a tumor suppressor and might be important for cigarette toxin metabolism, and cg19859270 in the first exon of the G-protein-coupled receptor 15 gene (*GPR15*), which is linked with chronic inflammation (Gao, Jia, et al., 2015).

A limitation of the methylation study is the difficulty to ensure the ability to detect a true association since it is possible that unmeasured confounding factors (i.e., race/ethnicity, cigarette use prior to sample collection, age, or sex) may bias results. The MZ co-twin study design may become a powerful design for methylation research since it uses twin pairs that are genetically identical and only one member of the pair has been exposed to an environment of interest (i.e., smoking) which allows for a perfect match in order to decrease variance from several potentially unmeasured covariates. A few studies have used this study design, and of these studies there are associations between previously reported genes related to cigarette toxin/smoke metabolism (e.g., *AHRR* and *F2RL3*) (Allione et al., 2015).

Another possible limitation of this study design is that DNA methylation is likely to be tissue specific and as such, lung and brain tissue would be most relevant for studies of smoking. However, most studies use peripheral lymphocytes, buccal cells, or saliva to examine differences in DNA methylation patterns between smokers and nonsmokers, because such cells/body fluids are easy to sample. Nevertheless, there is growing evidence that easily accessible tissues can act as surrogates for tissues. Moreover, DNA methylation patterns between peripheral blood and saliva have been reported to be comparable (Teschendorff et al., 2015; Thompson et al., 2013). Further, DNA methylation patterns in saliva were about 3% more likely to agree with those in brain regions than those in blood (Smith et al., 2015). Thus, comparison of DNA methylation patterns in such "proxy" tissues, particularly saliva, may be reliable for epigenetic research protocols examining smoking behaviors. However, it is possible that methylation studies of smoking behaviors and ND in particular may be better studied using brain samples rather than proxy samples.

In summary, current methylation studies of smoking likely highlight biological pathways in the inflammatory system and those related to cigarette toxin metabolism, which are likely to address the processing of cigarette smoke. These processes may be involved in nicotine availability to the brain, resulting in dependence. However, additional study is required since the initial methylation study results do not support genetic variants in pathways identified from prior genetic association results. Further, it is possible that future studies of smoking behavior and ND may benefit from the study of brain samples in addition to proxy samples such as saliva.

EPIGENETIC STUDIES OF BRAIN MORPHOLOGY

Epigenetic studies of brain morphology are currently limited to results from animal studies. Epigenetic studies in human populations will require additional technological advances, including increased computing power and improved bioinformatics approaches, to handle the immense amount of data flowing from this type of work (Robinson & Nestler, 2011). Additionally, current studies of epigenetic mechanisms are limited to individual genes, rather than the regulation of whole systems of genes in a biological pathway. Nevertheless, discoveries from animal studies are providing some initial insights. For example, Belichenko, Belichenko, Li, Mobley, and Francke (2008) reported morphological differences in knockout mice with reduced expression of the methyl CpG binding protein 2 gene (*MeCP2*) against their wild-type littermates. Specifically, MeCP2 deficiency was associated with a decrease in cortical, hippocampal, and cerebellar volume (Belichenko et al., 2008). The MeCP2 protein is key to mature nerve function and potentially silencing other genes (Chahrour et al., 2008). Additionally, animal studies have indicated the importance of epigenetic mechanisms in the brain development of mice during puberty (Morrison, Rodgers, Morgan, & Bale, 2014). DNA methylation, histone acetylation, and microRNAs (miRNAs) have all been implicated in the inhibition of time-sensitive hormonal changes that affect certain genes during puberty—for example, the gonadotropin-releasing hormone gene *GnRH* (Morrison et al., 2014). Humans have brain structures similar to mice, and it is possible that epigenetic mechanisms play an important role in human brain maturation (and by extension all aspects of brain morphology) and may be important for sex-specific etiologies related to ND.

FUTURE DIRECTIONS

Understanding the role of genetic and environmental factors on the etiology of ND, an acquired disease of the brain, continues to produce many insights as well as challenges. Results from genetic epidemiology studies of smoking and brain structure including twin/family studies, genetic association studies, and epigenetic studies have highlighted important structures and processes highlighted within the brain disease model and as more data become available, such knowledge is expected to assist in the development of more effective treatment and therapy to treat ND resulting in greater rates of smoking cessation.

Epigenetic and whole genome sequencing (WGS) studies are likely to be at the forefront of the next wave of genetic research as these methods offer the capability to investigate structural and rare variation (Medland et al., 2014). The future of this exciting area of research is expected to capture a substantial proportion of genetic variation that is currently poorly understood. Epigenetic data in combination with whole genome data as well as brain morphometry will allow for deeper interrogation of the genome and the neurobiology of ND. Expectations of the brain disease model can be tested, following the assumption that substance use will result in several biological changes including brain morphology particularly in regions with strong roles in all neurotransmitter pathways, and in particular the dopaminergic pathway.

It is also possible that different types of genetic data will provide different insights into the neurobiology of ND. For example, methylation data may provide insight into acute changes associated with smoking exposure. In contrast, genetic association, WGS, and brain morphology data may address questions regarding long-term changes or of fundamental biological pathways involved in ND.

Current studies remain focused on the search for a single gene or group of genes responsible for these phenotypes. However, environmental influences continue to be important in the maintenance of ND, and as such gene–environment interactions need to be considered in future investigations. For example, how do specific genetic and environmental risk factors function together to decrease the likelihood of successful smoking abstinence? Lastly, the scope of genetic epidemiological questions has been limited to smoking behaviors of the individual. Few, if any studies have analyzed the role of genetic and environmental influences on brain structure and exposure to second-hand smoke.

Questions regarding neurobiological pathways of polydrug use versus ND is one of increasing importance. Twin studies have identified a significant proportion of genetic influence on addiction to several illicit substances, which could be attributed to a common genetic factor (Young, Rhee, Stallings, Corley, & Hewitt, 2006). Further, alcohol and nicotine have some overlapping genetic and environmental influences (Agrawal et al., 2012), and as such the study of the genetic and environmental contributions shared between brain structure and polydrug use may provide valuable insight into comorbid nicotine use as well as addiction in general. Taken as a whole, these results also suggest shared genetic pathways that are involved in the brain disease model of addiction and that there are likely some features of this model that will reflect a specific etiology for ND. The future of ND research is exciting as research collaborations continue to grow and as more data become available alongside additional statistical methods to analyze these data. Consequently, genetic epidemiological studies on brain morphology and the etiology of nicotine are expected to continue producing promising leads for clinical and public health success against ND.

References

Agaku, I. T., King, B. A., Dube, S. R., & Centers for Disease Control and Prevention. (2014). Current cigarette smoking among adults – United States, 2005–2012. *MMWR. Morbidity and Mortality Weekly Report, 63*(2), 29–34.

Agrawal, A., & Lynskey, M. T. (2008). Are there genetic influences on addiction: Evidence from family, adoption and twin studies. *Addiction, 103*(7), 1069–1081. http://dx.doi.org/10.1111/j.1360-0443.2008.02213.x.

Agrawal, A., Verweij, K. J., Gillespie, N. A., Heath, A. C., Lessov-Schlaggar, C. N., Martin, N. G., ... Lynskey, M. T. (2012). The genetics of addiction-a translational perspective. *Translational Psychiatry, 2*, e140. http://dx.doi.org/10.1038/tp.2012.54.

Allione, A., Marcon, F., Fiorito, G., Guarrera, S., Siniscalchi, E., Zijno, A., ... Matullo, G. (2015). Novel epigenetic changes unveiled by monozygotic twins discordant for smoking habits. *PLoS One, 10*(6), e0128265. http://dx.doi.org/10.1371/journal.pone.0128265.

American Psychiatric Association. (2000). *Diagnostic and statistical manual of mental disorders: DSM-IV* (4th ed.). Washington, DC: American Psychiatric Association.

American Psychiatric Association. (2013). *Diagnostic and statistical manual of mental disorders* (5th ed.). Arlington, VA: American Psychiatric Publishing.

APA. (1994). *Diagnostic and statistical manual of mental disorder* (4th ed.). Washington, DC: American Psychiatric Association.

Ashburner, J., & Friston, K. J. (2000). Voxel-based morphometry—the methods. *NeuroImage, 11*(6 Pt 1), 805–821. http://doi.org/10.1006/nimg.2000.0582.

Baaré, W. F., Hulshoff Pol, H. E., Boomsma, D. I., Posthuma, D., de Geus, E. J., Schnack, H. G., & Kahn, R. S. (2001). Quantitative genetic modeling of variation in human brain morphology. *Cerebral Cortex (New York, N.Y. : 1991), 11*(9), 816–824.

Baker, T. B., Piper, M. E., McCarthy, D. E., Bolt, D. M., Smith, S. S., Kim, S. Y., ... Toll, B. A. (2007). Time to first cigarette in the morning as an index of ability to quit smoking: Implications for nicotine dependence. *Nicotine & Tobacco Research, 9*(Suppl. 4), S555–S570.

Bakken, T. E., Roddey, J. C., Djurovic, S., Akshoomoff, N., Amaral, D. G., Bloss, C. S., ... Carlson, H. (2012). Association of common genetic variants in GPCPD1 with scaling of visual cortical surface area in humans. *Proceedings of the National Academy of Sciences of the United States of America, 109*(10), 3985–3990.

Baldwin, P. R., Alanis, R., & Salas, R. (2011). The role of the habenula in nicotine addiction. *Journal of Addiction Research & Therapy, S1*(2). http://doi.org/10.4172/2155-6105.S1-002.

Batouli, S. A. H., Trollor, J. N., Wen, W., & Sachdev, P. S. (2014). The heritability of volumes of brain structures and its relationship to age: A review of twin and family studies. *Ageing Research Reviews, 13*, 1–9. http://doi.org/10.1016/j.arr.2013.10.003.

Belichenko, N. P., Belichenko, P. V., Li, H. H., Mobley, W. C., & Francke, U. (2008). Comparative study of brain morphology in Mecp2 mutant mouse models of Rhett syndrome. *Journal of Comparative Neurology, 508*(1), 184–195.

Berger, A. (2002). Magnetic resonance imaging. *BMJ (Clinical Research Ed.), 324*(7328), 35.

Berger, M., Gray, J. A., & Roth, B. L. (2009). The expanded biology of serotonin. *Annual Review of Medicine, 60*, 355–366. http://dx.doi.org/10.1146/annurev.med.60.042307.110802.

Berrettini, W., Yuan, X., Tozzi, F., Song, K., Francks, C., Chilcoat, H., … Mooser, V. (2008). Alpha-5/alpha-3 nicotinic receptor subunit alleles increase risk for heavy smoking. *Molecular Psychiatry, 13*(4), 368–373. http://dx.doi.org/10.1038/sj.mp.4002154.

Biasi, M. D., & Dani, J. A. (2011). Reward, addiction, withdrawal to nicotine. *Annual Review of Neuroscience, 34*(1), 105–130. http://dx.doi.org/10.1146/annurev-neuro-061010-113734.

Bierut, L. J., Stitzel, J. A., Wang, J. C., Hinrichs, A. L., Grucza, R. A., Xuei, X., … Goate, A. M. (2008). Variants in nicotinic receptors and risk for nicotine dependence. *The American Journal of Psychiatry, 165*(9), 1163–1171. http://dx.doi.org/10.1176/appi.ajp.2008.07111711.

Blokland, G. A. M., de Zubicaray, G. I., McMahon, K. L., & Wright, M. J. (2012). Genetic and environmental influences on neuroimaging phenotypes: A meta-analytical perspective on twin imaging studies. *Twin Research and Human Genetics, 15*(3), 351–371. http://doi.org/10.1017/thg.2012.11.

Boardman, J. D., Blalock, C. L., Pampel, F. C., Hatemi, P. K., Heath, A. C., & Eaves, L. J. (2011). Population composition, public policy, and the genetics of smoking. *Demography, 48*(4), 1517–1533.

Brans, R. G. H., Kahn, R. S., Schnack, H. G., van Baal, G. C. M., Posthuma, D., van Haren, N. E. M., … Hulshoff Pol, H. E. (2010). Brain plasticity and intellectual ability are influenced by shared genes. *The Journal of Neuroscience, 30*(16), 5519–5524. http://doi.org/10.1523/JNEUROSCI.5841-09.2010.

Brody, A. L., Mandelkern, M. A., Jarvik, M. E., Lee, G. S., Smith, E. C., Huang, J. C., … London, E. D. (2004). Differences between smokers and nonsmokers in regional gray matter volumes and densities. *Biological Psychiatry, 55*(1), 77–84. http://doi.org/10.1016/S0006-3223(03)00610-3.

Broms, U., Silventoinen, K., Madden, P. A. F., Heath, A. C., & Kaprio, J. (2006). Genetic architecture of smoking behavior: A study of Finnish adult twins. *Twin Research and Human Genetics, 9*(1), 64–72. http://doi.org/10.1375/183242706776403046.

Brouwer, R. M., Hedman, A. M., van Haren, N. E. M., Schnack, H. G., Brans, R. G. H., Smit, D. J. A., … Hulshoff Pol, H. E. (2014). Heritability of brain volume change and its relation to intelligence. *NeuroImage, 100*, 676–683. http://doi.org/10.1016/j.neuroimage.2014.04.072.

Brunzell, D. H., Stafford, A. M., & Dixon, C. I. (2015). Nicotinic receptor contributions to smoking: Insights from human studies and animal models. *Current Addiction Reports, 2*(1), 33–46. http://dx.doi.org/10.1007/s40429-015-0042-2.

Carmelli, D., DeCarli, C., Swan, G. E., Jack, L. M., Reed, T., Wolf, P. A., & Miller, B. L. (1998). Evidence for genetic variance in white matter hyperintensity volume in normal elderly male twins. *Stroke, 29*(6), 1177–1181.

Centers for Disease Control and Prevention. (2002). Annual smoking-attributable mortality, years of potential life lost, and economic costs—United States, 1995–1999. *MMWR. Morbidity and Mortality Weekly Report, 51*(14), 300–303.

Chahrour, M., Jung, S. Y., Shaw, C., Zhou, X., Wong, S. T., Qin, J., & Zoghbi, H. Y. (2008). MeCP2, a key contributor to neurological disease, activates and represses transcription. *Science, 320*(5880), 1224–1229. http://dx.doi.org/10.1126/science.1153252.

Chen, C.-H., Gutierrez, E. D., Thompson, W., Panizzon, M. S., Jernigan, T. L., Eyler, L. T., … Dale, A. M. (2012). Hierarchical genetic organization of human cortical surface area. *Science (New York, N.Y.), 335*(6076), 1634–1636. http://doi.org/10.1126/science.1215330.

Chen, X., Wen, W., Anstey, K. J., & Sachdev, P. S. (2006). Effects of cerebrovascular risk factors on gray matter volume in adults aged 60–64 years: A voxel-based morphometric study. *Psychiatry Research, 147*(2–3), 105–114. http://doi.org/10.1016/j.pscychresns.2006.01.009.

Cherny, S. S. (2009). QTL methodology in behavior genetics. In *Handbook of behavior genetics* (pp. 35–45). New York, NY: Springer. http://doi.org/10.1007/978-0-387-76727-7_3.

Chiang, M.-C., McMahon, K. L., de Zubicaray, G. I., Martin, N. G., Hickie, I., Toga, A. W., … Thompson, P. M. (2011). Genetics of white matter development: A DTI study of 705 twins and their siblings aged 12 to 29. *NeuroImage, 54*(3), 2308–2317. http://doi.org/10.1016/j.neuroimage.2010.10.015.

Choi, M.-H., Lee, S.-J., Yang, J.-W., Kim, J.-H., Choi, J.-S., Park, J. Y., … Chung, S.-C. (2010). Difference between smokers and non-smokers in the corpus callosum volume. *Neuroscience Letters, 485*(1), 71–73. http://doi.org/10.1016/j.neulet.2010.08.066.

Clarke, P. B., & Pert, A. (1985). Autoradiographic evidence for nicotine receptors on nigrostriatal and mesolimbic dopaminergic neurons. *Brain Research, 348*(2), 355–358.

Das, D., Cherbuin, N., Anstey, K. J., Sachdev, P. S., & Easteal, S. (2012). Lifetime cigarette smoking is associated with striatal volume measures. *Addiction Biology, 17*(4), 817–825. http://doi.org/10.1111/j.1369-1600.2010.00301.x.

Dennis, E. L., & Thompson, P. M. (2013). Typical and atypical brain development: A review of neuroimaging studies. *Dialogues in Clinical Neuroscience, 15*(3), 359–384.

Dennison, M., Whittle, S., Yücel, M., Vijayakumar, N., Kline, A., Simmons, J., & Allen, N. B. (2013). Mapping subcortical brain maturation during adolescence: Evidence of hemisphere- and sex-specific longitudinal changes. *Developmental Science, 16*(5), 772–791. http://doi.org/10.1111/desc.12057.

DeStefano, A. L., Seshadri, S., Beiser, A., Atwood, L. D., Massaro, J. M., Au, R., … DeCarli, C. (2009). Bivariate heritability of total and regional brain volumes: The Framingham Study. *Alzheimer Disease and Associated Disorders, 23*(3), 218–223. http://doi.org/10.1097/WAD.0b013e31819cadd8.

Dichter, G. S., Damiano, C. A., & Allen, J. A. (2012). Reward circuitry dysfunction in psychiatric and neurodevelopmental disorders and genetic syndromes: Animal models and clinical findings. *Journal of Neurodevelopmental Disorders, 4*(1), 19. http://dx.doi.org/10.1186/1866-1955-4-19.

Dudbridge, F., & Gusnanto, A. (2008). Estimation of significance thresholds for genomewide association scans. *Genetic Epidemiology, 32*(3), 227–234. http://dx.doi.org/10.1002/gepi.20297.

Egger, G., Liang, G., Aparicio, A., & Jones, P. A. (2004). Epigentics in human disease and prospects for epigenetic therapy. *Nature, 429*(6990), 257–463.

Evangelou, E., & Ioannidis, J. P. (2013). Meta-analysis methods for genome-wide association studies and beyond. *Nature Reviews Genetics, 14*(6), 379–389. http://dx.doi.org/10.1038/nrg3472.

Eyler, L. T., Prom-Wormley, E., Fennema-Notestine, C., Panizzon, M. S., Neale, M. C., Jernigan, T. L., … Kremen, W. S. (2011). Genetic patterns of correlation among subcortical volumes in humans: Results from a magnetic resonance imaging twin study. *Human Brain Mapping, 32*(4), 641–653. http://doi.org/10.1002/hbm.21054.

Eyler, L. T., Prom-Wormley, E., Panizzon, M. S., Kaup, A. R., Fennema-Notestine, C., Neale, M. C., … Kremen, W. S. (2011). Genetic and environmental contributions to regional cortical surface area in humans: A magnetic resonance imaging twin study. *Cerebral Cortex (New York, N.Y.: 1991), 21*(10), 2313–2321. http://doi.org/10.1093/cercor/bhr013.

Ferrea, S., & Winterer, G. (2009). Neuroprotective and neurotoxic effects of nicotine. *Pharmacopsychiatry, 42*(6), 255–265. http://doi.org/10.1055/s-0029-1224138.

Fields, R. D. (2008). White matter in learning, cognition and psychiatric disorders. *Trends in Neurosciences, 31*(7), 361–370. http://doi.org/10.1016/j.tins.2008.04.001.

Fineberg, N. A., Potenza, M. N., Chamberlain, S. R., Berlin, H. A., Menzies, L., Bechara, A., … Hollander, E. (2010). Probing compulsive and impulsive behaviors, from animal models to endophenotypes: A narrative review. *Neuropsychopharmacology, 35*(3), 591–604. http://doi.org/10.1038/npp.2009.185.

Fiore, M. C., Jaén, C. R., Baker, T. B., Bailey, W. C., Benowitz, N. L., Curry, S. J., ... Leitzke, C. (2008). A clinical practice guideline for treating tobacco use and dependence: 2008 update. A U.S. Public Health Service report. *American Journal of Preventive Medicine, 35*(2), 158–176. http://dx.doi.org/10.1016/j.amepre.2008.04.009.

Fowler, T., Lifford, K., Shelton, K., Rice, F., Thapar, A., Neale, M. C., ... van den Bree, M. B. M. (2007). Exploring the relationship between genetic and environmental influences on initiation and progression of substance use. *Addiction (Abingdon, England), 102*(3), 413–422. http://doi.org/10.1111/j.1360-0443.2006.01694.x.

Franklin, T. R., Wetherill, R. R., Jagannathan, K., Johnson, B., Mumma, J., Hager, N., ... Childress, A. R. (2014). The effects of chronic cigarette smoking on gray matter volume: Influence of sex. *PLoS One, 9*(8), e104102. http://doi.org/10.1371/journal.pone.0104102.

Fritz, H.-C., Wittfeld, K., Schmidt, C. O., Domin, M., Grabe, H. J., Hegenscheid, K., ... Lotze, M. (2014). Current smoking and reduced gray matter volume-a voxel-based morphometry study. *Neuropsychopharmacology, 39*(11), 2594–2600. http://doi.org/10.1038/npp.2014.112.

Gallinat, J., Meisenzahl, E., Jacobsen, L. K., Kalus, P., Bierbrauer, J., Kienast, T., ... Staedtgen, M. (2006). Smoking and structural brain deficits: A volumetric MR investigation. *The European Journal of Neuroscience, 24*(6), 1744–1750. http://doi.org/10.1111/j.1460-9568.2006.05050.x.

Gao, J. S., Huth, A. G., Lescroart, M. D., & Gallant, J. L. (2015). Pycortex: An interactive surface visualizer for fMRI. *Frontiers in Neuroinformatics, 9*, 23. http://dx.doi.org/10.33.89/fninf.2015.00023.

Gao, X., Jia, M., Zhang, Y., Breitling, L. P., & Brenner, H. (2015). DNA methylation changes of whole blood cells in response to active smoking exposure in adults: A systematic review of DNA methylation studies. *Clinical Epigenetics, 7*, 113. http://doi.org/10.1186/s13148-015-0148-3.

George, A. A., Lucero, L. M., Damaj, M. I., Lukas, R. J., Chen, X., & Whiteaker, P. (2012). Function of human alpha3beta4alpha5 nicotinic acetylcholine receptors is reduced by the alpha5(D398N) variant. *The Journal of Biological Chemistry, 287*(30), 25151–25162. http://dx.doi.org/10.1074/jbc.M112.379339.

Geschwind, D. H., Miller, B. L., DeCarli, C., & Carmelli, D. (2002). Heritability of lobar brain volumes in twins supports genetic models of cerebral laterality and handedness. *Proceedings of the National Academy of Sciences of the United States of America, 99*(5), 3176–3181. http://doi.org/10.1073/pnas.052494999.

Gilmore, J. H., Schmitt, J. E., Knickmeyer, R. C., Smith, J. K., Lin, W., Styner, M., ... Neale, M. C. (2010). Genetic and environmental contributions to neonatal brain structure: A twin study. *Human Brain Mapping, 31*(8), 1174–1182. http://doi.org/10.1002/hbm.20926.

Goldstein, R. Z., & Volkow, N. D. (2011). Dysfunction of the prefrontal cortex in addiction: Neuroimaging findings and clinical implications. *Nature Reviews. Neuroscience, 12*(11), 652–669. http://doi.org/10.1038/nrn3119.

Goodkind, M., Eickhoff, S. B., Oathes, D. J., Jiang, Y., Chang, A., Jones-Hagata, L. B., ... Etkin, A. (2015). Identification of a common neurobiological substrate for mental illness. *JAMA Psychiatry, 72*(4), 305–315. http://doi.org/10.1001/jamapsychiatry.2014.2206.

Goren, A., Annunziata, K., Schnoll, R. A., & Suaya, J. A. (2014). Smoking cessation and attempted cessation among adults in the United States. *PLoS One, 9*(3), e93014.

Grady, S. R., Moretti, M., Zoli, M., Marks, M. J., Zanardi, A., Pucci, L., ... Gotti, C. (2009). Rodent habenulo-interpeduncular pathway expresses a large variety of uncommon nAChR subtypes, but only the alpha3beta4* and alpha3beta3beta4* subtypes mediate acetylcholine release. *The Journal of Neuroscience: The Official Journal of the Society for Neuroscience, 29*(7), 2272–2282. http://dx.doi.org/10.1523/jneurosci.5121-08.2009.

Grewal, S. I., & Rice, J. C. (2004). Regulation of heterochromatin by histone methylation and small RNAs. *Current Opinion in Cell Biology, 16*(3), 230–238. http://dx.doi.org/10.1016/j.ceb.2004.04.002.

Gurung, R., & Prata, D. P. (2015). What is the impact of genome-wide supported risk variants for schizophrenia and bipolar disorder on brain structure and function? A systematic review. *Psychological Medicine, 45*(12), 2461–2480. http://doi.org/10.1017/S0033291715000537.

Hanlon, C. A., Owens, M. M., Joseph, J. E., Zhu, X., George, M. S., Brady, K. T., & Hartwell, K. J. (2016). Lower subcortical gray matter volume in both younger smokers and established smokers relative to non-smokers. *Addiction Biology, 21*(1), 185–195. http://doi.org/10.1111/adb.12171.

Hardie, T. L., Moss, H. B., & Lynch, K. G. (2006). Genetic correlations between smoking initiation and smoking behaviors in a twin sample. *Addictive Behaviors, 31*(11), 2030–2037. http://doi.org/10.1016/j.addbeh.2006.02.010.

Heath, A. C., Martin, N. G., Lynskey, M. T., Todorov, A. A., & Madden, P. A. F. (2002). Estimating two-stage models for genetic influences on alcohol, tobacco or drug use initiation and dependence vulnerability in twin and family data. *Twin Research, 5*(2), 113–124. http://doi.org/10.1375/1369052022983.

Heatherton, T. F., Kozlowski, L. T., Frecker, R. C., & Fagerström, K. O. (1991). The Fagerström test for nicotine dependence: A revision of the Fagerström Tolerance Questionnaire. *British Journal of Addiction, 86*(9), 1119–1127.

Hendricks, P. S., Prochaska, J. J., Humfleet, G. L., & Hall, S. M. (2008). Evaluating the validities of different DSM-IV-based conceptual constructs of tobacco dependence. *Addiction, 103*, 1215–1223.

Hibar, D. P., Medland, S. E., Stein, J. L., Kim, S., Shen, L., Saykin, A. J., & Thompson, P. M. (2013). Genetic clustering on the hippocampal surface for genome-wide association studies. *Medical Image Computing and Computer-assisted Interventions, 16*(Pt2), 690–697.

Hibar, D. P., Stein, J. L., Renteria, M. E., Arias-Vasquez, A., Desrivieres, S., Jahanshad, N., & Medland, S. E. (2015). Common genetic variants influence human subcortical brain structures. *Nature, 520*(7546), 224–229.

Hibar, D. P., Stein, J. L., Ryles, A. B., Kohannim, O., Jahanshad, N., Medland, S. E., & Thompson, P. M. (2013). Genome-wide association identifies genetic variants associated with lentiform nucleus volume in N = 1345 young and elderly subjects. *Brain Imaging and Behavior, 7*(2), 102–115.

Hirschhorn, J. N., & Daly, M. J. (2005). Genome-wide association studies for common diseases and complex traits. *Nature Reviews Genetics, 6*(2), 95–108. http://dx.doi.org/10.1038/nrg1521.

Hong, L. E., Hodgkinson, C. A., Yang, Y., Sampath, H., Ross, T. J., Buchholz, B., ... Stein, E. A. (2010). A genetically modulated, intrinsic cingulate circuit supports human nicotine addiction. *Proceedings of the National Academy of Sciences of the United States of America, 107*(30), 13509–13514. http://dx.doi.org/10.1073/pnas.1004745107.

Hopfer, C. J., Crowley, T. J., & Hewitt, J. K. (2003). Review of twin and adoption studies of adolescent substance use. *Journal of the American Academy of Child and Adolescent Psychiatry, 42*(6), 710–719.

Huizink, A. C., Levälahti, E., Korhonen, T., Dick, D. M., Pulkkinen, L., Rose, R. J., & Kaprio, J. (2010). Tobacco, cannabis, and other illicit drug use among Finnish adolescent twins: Causal relationship or correlated liabilities? *Journal of Studies on Alcohol and Drugs, 71*(1), 5–14.

Hulshoff Pol, H. E., Schnack, H. G., Posthuma, D., Mandl, R. C. W., Baaré, W. F., van Oel, C., ... Kahn, R. S. (2006). Genetic contributions to human brain morphology and intelligence. *The Journal of Neuroscience, 26*(40), 10235–10242. http://doi.org/10.1523/JNEUROSCI.1312-06.2006.

Im, K., Lee, J. M., Lyttelton, O., Kim, S. H., Evans, A. C., & Kim, S. I. (2008). Brain size and cortical structure in the adult human brain. *Cerebral Cortex, 18*(9), 2181–2191. http://dx.doi.org/10.1093/cercor/bhm244.

Jackson, K., Muldoon, P., Biasi, M. D., & Damaj, M. (2015). New mechanisms and perspectives in nicotine withdrawal. *Neuropharmacology, 96*, 223–234. http://dx.doi.org/10.1016/j.neuropharm.2014.11.009.

Jansen, A. G., Mous, S. E., White, T., Posthuma, D., & Polderman, T. J. C. (2015). What twin studies tell us about the heritability of brain development, morphology, and function: A review. *Neuropsychology Review, 25*(1), 27–46. http://doi.org/10.1007/s11065-015-9278-9.

Jasinska, A. J., Zorick, T., Brody, A. L., & Stein, E. A. (2014). Dual role of nicotine in addiction and cognition: A review of neuroimaging studies in humans. *Neuropharmacology, 84*, 111–122. http://dx.doi.org/10.1016/j.neuropharm.2013.02.015.

Jensen, J., & Kapur, S. (2009). Salience and psychosis: Moving from theory to practise. *Psychological Medicine, 39*(2), 197–198. http://doi.org/10.1017/S0033291708003899.

Jha, P., Ramasundarahettige, C., Landsman, V., Rostron, B., Thun, M., Anderson, R. N., ... Peto, R. (2013). 21st-century hazards of smoking and benefits of cessation in the United States. *The New England Journal of Medicine, 368*(4), 341–350. http://dx.doi.org/10.1056/NEJMsa1211128.

Kendler, K. S., Schmitt, E., Aggen, S. H., & Prescott, C. A. (2008). Genetic and environmental influences on alcohol, caffeine, cannabis, and nicotine use from early adolescence to middle adulthood. *Archives of General Psychiatry, 65*(6), 674–682.

Keskitalo, K., Broms, U., Heliövaara, M., Ripatti, S., Surakka, I., Perola, M., ... Kaprio, J. (2009). Association of serum cotinine level with a cluster of three nicotinic acetylcholine receptor genes (CHRNA3/CHRNA5/CHRNB4) on chromosome 15. *Human Molecular Genetics, 18*(20), 4007–4012. http://doi.org/10.1093/hmg/ddp322.

Koolschijn, P. C. M. P., & Crone, E. A. (2013). Sex differences and structural brain maturation from childhood to early adulthood. *Developmental Cognitive Neuroscience, 5*, 106–118. http://doi.org/10.1016/j.dcn.2013.02.003.

Koopmans, J. R., Slutske, W. S., Heath, A. C., Neale, M. C., & Boomsma, D. I. (1999). The genetics of smoking initiation and quantity smoked in Dutch adolescent and young adult twins. *Behavior Genetics, 29*(6), 383–393.

Koopmans, J. R., van Doornen, L. J., & Boomsma, D. I. (1997). Association between alcohol use and smoking in adolescent and young adult twins: A bivariate genetic analysis. *Alcoholism, Clinical and Experimental Research, 21*(3), 537–546.

Kouzarides, T. (2007). Chromatin modifications and their function. *Cell, 128*(4), 693–705. http://dx.doi.org/10.1016/j.cell.2007.02.005.

Kozlowski, L. T., Porter, C. Q., Orleans, C. T., Pope, M. A., & Heatherton, T. (1994). Predicting smoking cessation with self-reported measures of nicotine dependence: FTQ, FTND, and HSI. *Drug and Alcohol Dependence, 34*, 211–216.

Kremen, W. S., Prom-Wormley, E., Panizzon, M. S., Eyler, L. T., Fischl, B., Neale, M. C., ... Fennema-Notestine, C. (2010). Genetic and environmental influences on the size of specific brain regions in midlife: The VETSA MRI study. *NeuroImage, 49*(2), 1213–1223. http://doi.org/10.1016/j.neuroimage.2009.09.043.

Kühn, S., Romanowski, A., Schilling, C., Mobascher, A., Warbrick, T., Winterer, G., & Gallinat, J. (2012). Brain grey matter deficits in smokers: Focus on the cerebellum. *Brain Structure & Function, 217*(2), 517–522. http://doi.org/10.1007/s00429-011-0346-5.

Kühn, S., Schubert, F., & Gallinat, J. (2010). Reduced thickness of medial orbitofrontal cortex in smokers. *Biological Psychiatry, 68*(11), 1061–1065. http://doi.org/10.1016/j.biopsych.2010.08.004.

Lawson, R. P., Drevets, W. C., & Roiser, J. P. (2013). Defining the habenula in human neuroimaging studies. *NeuroImage, 64*, 722–727. http://doi.org/10.1016/j.neuroimage.2012.08.076.

Le Foll, B., & Goldberg, S. R. (2009). Effects of nicotine in experimental animals and humans: An update on addictive properties. *Handbook of Experimental Pharmacology (192)*, 335–367. doi:10.1007/978-3-540-69248-5_12.

Lee, K. W. K., & Pausova, Z. (2013). Cigarette smoking and DNA methylation. *Frontiers in Genetics, 4*(132). http://doi.org/10.3389/fgene.2013.00132.

van Leeuwen, M., Peper, J. S., van den Berg, S. M., Brouwer, R. M., Hulshoff Pol, H. E., Kahn, R. S., & Boomsma, D. I. (2009). A genetic analysis of brain volumes and IQ in children. *Intelligence, 37*(2), 181–191. http://doi.org/10.1016/j.intell.2008.10.005.

Lenroot, R. K., & Giedd, J. N. (2010). Sex differences in the adolescent brain. *Brain and Cognition, 72*(1), 46–55. http://doi.org/10.1016/j.bandc.2009.10.008.

Lenroot, R. K., Schmitt, J. E., Ordaz, S. J., Wallace, G. L., Neale, M. C., Lerch, J. P., ... Giedd, J. N. (2009). Differences in genetic and environmental influences on the human cerebral cortex associated with development during childhood and adolescence. *Human Brain Mapping, 30*(1), 163–174. http://doi.org/10.1002/hbm.20494.

Lerman, C., & Berrettini, W. (2003). Elucidating the role of genetic factors in smoking behavior and nicotine dependence. *American Journal of Medical Genetics. Part B, Neuropsychiatric Genetics, 118B*(1), 48–54. http://doi.org/10.1002/ajmg.b.10003.

Lessov-Schlaggar, C. N., Hardin, J., DeCarli, C., Krasnow, R. E., Reed, T., Wolf, P. A., ... Carmelli, D. (2012). Longitudinal genetic analysis of brain volumes in normal elderly male twins. *Neurobiology of Aging, 33*(4), 636–644. http://doi.org/10.1016/j.neurobiolaging.2010.06.002.

Li, M. D., Cheng, R., Ma, J. Z., & Swan, G. E. (2003). A meta-analysis of estimated genetic and environmental effects on smoking behavior in male and female adult twins. *Addiction (Abingdon, England), 98*(1), 23–31.

Li, Y., Yuan, K., Cai, C., Feng, D., Yin, J., Bi, Y., ... Tian, J. (2015). Reduced frontal cortical thickness and increased caudate volume within fronto-striatal circuits in young adult smokers. *Drug and Alcohol Dependence, 151*, 211–219. http://doi.org/10.1016/j.drugalcdep.2015.03.023.

Liao, Y., Tang, J., Liu, T., Chen, X., & Hao, W. (2012). Differences between smokers and non-smokers in regional gray matter volumes: A voxel-based morphometry study. *Addiction Biology, 17*(6), 977–980. http://doi.org/10.1111/j.1369-1600.2010.00250.x.

Marchini, J., & Howie, B. (2010). Genotype imputation for genome-wide association studies. *Nature Reviews Genetics, 11*(7), 499–511. http://dx.doi.org/10.1038/nrg2796.

Marjoram, P., Zubair, A., & Nuzhdin, S. V. (2014). Post-GWAS: Where next? More samples, more SNPs or more biology? *Heredity, 112*(1), 79–88. http://dx.doi.org/10.1038/hdy.2013.52.

May, A. (2007). Neuroimaging: Visualising the brain in pain. *Neurological Sciences, 28*(S2). http://dx.doi.org/10.1007/s10072-007-0760-x.

McCarthy, M. I., & Hirschhorn, J. N. (2008). Genome-wide association studies: Potential next steps on a genetic journey. *Human Molecular Genetics, 17*(R2), R156–R165. http://dx.doi.org/10.1093/hmg/ddn289.

Medland, S. E., Jahanshad, N., Neale, B. M., & Thompson, P. M. (2014). Whole-genome analyses of whole-brain data: Working within an expanded search space. *Nature Neuroscience, 17*(6), 791–800. http://dx.doi.org/10.1038/nn.3718.

Mechelli, A., Price, C. J., Friston, K. J., & Ashburner, J. (June 1, 2005). Voxel-based morphometry of the human brain: Methods and applications. *Current Medical Imaging Reviews, 1*(1), 1–9. Bentham Science Publ Ltd. Retrieved from http://discovery.ucl.ac.uk/136300/.

Menossi, H. S., Goudriaan, A. E., de Azevedo-Marques Périco, C., Nicastri, S., de Andrade, A. G., D'Elia, G., ... Castaldelli-Maia, J. M. (2013). Neural bases of pharmacological treatment of nicotine dependence – Insights from functional brain imaging: A systematic review. *CNS Drugs, 27*(11), 921–941. http://doi.org/10.1007/s40263-013-0092-8.

Morales, A. M., Lee, B., Hellemann, G., O'Neill, J., & London, E. D. (2012). Gray-matter volume in methamphetamine dependence: Cigarette smoking and changes with abstinence from methamphetamine. *Drug and Alcohol Dependence, 125*(3), 230–238. http://doi.org/10.1016/j.drugalcdep.2012.02.017.

Morel, C., Fattore, L., Pons, S., Hay, Y. A., Marti, F., Lambolez, B., … Faure, P. (2014). Nicotine consumption is regulated by a human polymorphism in dopamine neurons. *Molecular Psychiatry, 19*(8), 930–936. http://dx.doi.org/10.1038/mp.2013.158.

Morley, K. I., Lynskey, M. T., Madden, P. A. F., Treloar, S. A., Heath, A. C., & Martin, N. G. (2007). Exploring the inter-relationship of smoking age-at-onset, cigarette consumption and smoking persistence: Genes or environment? *Psychological Medicine, 37*(9), 1357–1367. http://doi.org/10.1017/S0033291707000748.

Morrison, K. E., Rodgers, A. B., Morgan, C. P., & Bale, T. L. (2014). Epigenetic mechanisms in pubertal brain maturation. *Neuroscience, 264*, 17–24.

Mountcastle, V. (1997). The columnar organization of the neocortex. *Brain: A Journal of Neurology, 120*(4), 701–722. http://dx.doi.org/10.1093/brain/120.4.701.

Munafo, M. R., & Gage, S. H. (2013). Improving the reliability and reporting of genetic association studies. *Drug and Alcohol Dependence, 132*(3), 411–413. http://dx.doi.org/10.1016/j.drugalcdep.2013.03.023.

Munafò, M. R., Timofeeva, M. N., Morris, R. W., Prieto-Merino, D., Sattar, N., Brennan, P., … Davey Smith, G. (2012). Association between genetic variants on chromosome 15q25 locus and objective measures of tobacco exposure. *Journal of the National Cancer Institute, 104*(10), 740–748. http://doi.org/10.1093/jnci/djs191.

Neale, M. C. (2009). Biometrical models in behavioral genetics. In *Handbook of behavior genetics* (pp. 15–33). Springer.

Nees, F., Witt, S. H., Lourdusamy, A., Vollstadt-Klein, S., Steiner, S., Poustka, L., … IMAGEN Consortium. (2013). Genetic risk for nicotine dependence in the cholinergic system and activation of the brain reward system in healthy adolescents. *Neuropsychopharmacology: Official Publication of the American College of Neuropsychopharmacology, 38*(11), 2081–2089. http://dx.doi.org/10.1038/npp.2013.131.

Pakkenberg, B., & Gundersen, H. J. (1997). Neocortical neuron number in humans: Effect of sex and age. *The Journal of Comparative Neurology, 384*(2), 312–320. http://dx.doi.org/10.1002/(sici)1096-9861(19970728)384:23.0.co;2-k.

Pan, P., Shi, H., Zhong, J., Xiao, P., Shen, Y., Wu, L., … He, G. (2013). Chronic smoking and brain gray matter changes: Evidence from meta-analysis of voxel-based morphometry studies. *Neurological Sciences, 34*(6), 813–817. http://doi.org/10.1007/s10072-012-1256-x.

Panizzon, M. S., Fennema-Notestine, C., Eyler, L. T., Jernigan, T. L., Prom-Wormley, E., Neale, M., & Kremen, W. S. (2009). Distinct genetic influences on cortical surface area and cortical thickness. *Cerebral Cortex, 19*(11), 2728–2735. http://dx.doi.org/10.1093/cercor/bhp026.

Paul, R. H., Grieve, S. M., Niaura, R., David, S. P., Laidlaw, D. H., Cohen, R., … Gordon, E. (2008). Chronic cigarette smoking and the microstructural integrity of white matter in healthy adults: A diffusion tensor imaging study. *Nicotine & Tobacco Research, 10*(1), 137–147. http://doi.org/10.1080/14622200701767829.

Paus, T. (2010). Growth of white matter in the adolescent brain: Myelin or axon? *Brain and Cognition, 72*(1), 26–35. http://dx.doi.org/10.1016/j.bandc.2009.06.002.

Peper, J. S., Schnack, H. G., Brouwer, R. M., Van Baal, G. C. M., Pjetri, E., Székely, E., … Hulshoff Pol, H. E. (2009). Heritability of regional and global brain structure at the onset of puberty: A magnetic resonance imaging study in 9-year-old twin pairs. *Human Brain Mapping, 30*(7), 2184–2196. http://doi.org/10.1002/hbm.20660.

Peters, B. J. M., Rodin, A. S., de Boer, A., & Maitland-van der Zee, A.-H. (2010). Methodological and statistical issues in pharmacogenomics. *The Journal of Pharmacy and Pharmacology, 62*(2), 161–166. http://doi.org/10.1211/jpp.62.02.0002.

Pfefferbaum, A., Sullivan, E. V., & Carmelli, D. (2001). Genetic regulation of regional microstructure of the corpus callosum in late life. *Neuroreport, 12*(8), 1677–1681.

Pfefferbaum, A., Sullivan, E. V., Swan, G. E., & Carmelli, D. (2000). Brain structure in men remains highly heritable in the seventh and eighth decades of life. *Neurobiology of Aging, 21*(1), 63–74.

Posthuma, D., Baaré, W. F. C., Hulshoff Pol, H. E., Kahn, R. S., Boomsma, D. I., & De Geus, E. J. C. (2003). Genetic correlations between brain volumes and the WAIS-III dimensions of verbal comprehension, working memory, perceptual organization, and processing speed. *Twin Research, 6*(02), 131–139. http://doi.org/10.1375/twin.6.2.131.

Prom-Wormley, E., Maes, H. H. M., Schmitt, J. E., Panizzon, M. S., Xian, H., Eyler, L. T., … Neale, M. C. (2015). Genetic and environmental contributions to the relationships between brain structure and average lifetime cigarette use. *Behavior Genetics, 45*(2), 157–170. http://doi.org/10.1007/s10519-014-9704-4.

Quaak, M., van Schayck, C. P., Knaapen, A. M., & van Schooten, F. J. (2009). Implications of gene-drug interactions in smoking cessation for improving the prevention of chronic degenerative diseases. *Mutation Research, 667*(1–2), 44–57. http://doi.org/10.1016/j.mrfmmm.2008.10.015.

Rakic, P. (1988). Specification of cerebral cortical areas. *Science, 241*(4862), 170–176.

Rakic, P. (1995). Radial versus tangential migration of neuronal clones in the developing cerebral cortex. *Proceedings of the National Academy of Sciences of the United States of America, 92*(25), 11323–11327.

Rakic, P. (2007). The radial edifice of cortical architecture: From neuronal silhouettes to genetic engineering. *Brain Research Reviews, 55*(2), 204–219. http://doi.org/10.1016/j.brainresrev.2007.02.010.

Rentería, M. E., Hansell, N. K., Strike, L. T., McMahon, K. L., de Zubicaray, G. I., Hickie, I. B., … Wright, M. J. (2014). Genetic architecture of subcortical brain regions: Common and region-specific genetic contributions. *Genes, Brain, and Behavior, 13*(8), 821–830. http://dx.doi.org/10.1111/gbb.12177.

Rhee, S. H., Hewitt, J. K., Young, S. E., Corley, R. P., Crowley, T. J., & Stallings, M. C. (2003). Genetic and environmental influences on substance initiation, use, and problem use in adolescents. *Archives of General Psychiatry, 60*(12), 1256–1264. http://doi.org/10.1001/archpsyc.60.12.1256.

Robinson, A. J., & Nestler, E. J. (2011). Transcriptional and epigenetic mechanism of addiction. *Nature Reviews Neuroscience, 12*, 623–637.

Rose, R. J., Broms, U., Korhonen, T., Dick, D. M., & Kapiro, J. (2009). Genetics of smoking behavior. In Y. K. Kim (Ed.), *Handbook of behavior genetics* (pp. 411–432). New York, NY: Springer. Retrieved from http://link.springer.com/10.1007/978-0-387-76727-7_28.

Saccone, S. F., Hinrichs, A. L., Saccone, N. L., Chase, G. A., Konvicka, K., Madden, P. A. F., … Bierut, L. J. (2007). Cholinergic nicotinic receptor genes implicated in a nicotine dependence association study targeting 348 candidate genes with 3713 SNPs. *Human Molecular Genetics, 16*(1), 36–49. http://doi.org/10.1093/hmg/ddl438.

Sale, A., Berardi, N., & Maffei, L. (2014). Environment and brain plasticity: Towards an endogenous pharmacotherapy. *Physiological Reviews, 94*(1), 189–234. http://doi.org/10.1152/physrev.00036.2012.

Samet, J. M. (1990). The 1990 report of the Surgeon General: The health benefits of smoking cessation. *The American Review of Respiratory Disease, 142*(5), 993–994. http://dx.doi.org/10.1164/ajrccm/142.5.993.

Sartor, C. E., Agrawal, A., Lynskey, M. T., Bucholz, K. K., Madden, P. A., & Heath, A. C. (2009). Common genetic influences on the timing of first use for alcohol, cigarettes, and cannabis in young African-American women. *Drug and Alcohol Dependence, 102*(1–3), 49–55.

Scamvougeras, A., Kigar, D. L., Jones, D., Weinberger, D. R., & Witelson, S. F. (2003). Size of the human corpus callosum is genetically determined: An MRI study in mono and dizygotic twins. *Neuroscience Letters, 338*(2), 91–94.

Schmitt, J. E., Neale, M. C., Fassassi, B., Perez, J., Lenroot, R. K., Wells, E. M., & Giedd, J. N. (2014). The dynamic role of genetics on cortical patterning during childhood and adolescence. *Proceedings of the National Academy of Sciences of the United States of America, 111*(18), 6774–6779. http://doi.org/10.1073/pnas.1311630111.

Scholze, P., Koth, G., Orr-Urtreger, A., & Huck, S. (2012). Subunit composition of alpha5-containing nicotinic receptors in the rodent habenula. *Journal of Neurochemistry, 121*(4), 551–560. http://dx.doi.org/10.1111/j.1471-4159.2012.07714.x.

Séguéla, P., Wadiche, J., Dineley-Miller, K., Dani, J. A., & Patrick, J. W. (1993). Molecular cloning, functional properties, and distribution of rat brain alpha 7: A nicotinic cation channel highly permeable to calcium. *The Journal of neuroscience: The Official Journal of the Society for Neuroscience, 13*(2), 596–604.

Seo, D., & Sinha, R. (2015). Neuroplasticity and predictors of alcohol recovery. *Alcohol Research, 37*(1), 143–152.

Seshadri, S., DeStefano, A. L., Au, R., Massaro, J. M., Beiser, A. S., Kelly-Hayes, M., & Wolf, P. A. (2007). Genetic correlates of brain aging on MRI and cognitive test measures: A genome-wide association and linkage analysis in the Framingham study. *BMC Medical Genetics, 8*(Suppl. 1), S15.

Seshadri, S., Wolf, P. A., Beiser, A., Elias, M. F., Au, R., Kase, C. S., … DeCarli, C. (2004). Stroke risk profile, brain volume, and cognitive function: The Framingham Offspring Study. *Neurology, 63*(9), 1591–1599.

Shih, P. Y., Engle, S. E., Oh, G., Deshpande, P., Puskar, N. L., Lester, H. A., & Drenan, R. M. (2014). Differential expression and function of nicotinic acetylcholine receptors in subdivisions of medial habenula. *The Journal of Neuroscience: The Official Journal of the Society for Neuroscience, 34*(29), 9789–9802. http://dx.doi.org/10.1523/jneurosci.0476-14.2014.

Smith, A. K., Kilaru, V., Klengel, T., Mercer, K. B., Bradley, B., Conneely, K. N., … Binder, E. B. (2015). DNA extracted from saliva for methylation studies of psychiatric traits: Evidence tissue specificity and relatedness to brain. *American Journal of Medical Genetics. Part B, Neuropsychiatric Genetics, 168B*(1), 36–44. http://doi.org/10.1002/ajmg.b.32278.

van Soelen, I. L. C., Brouwer, R. M., van Baal, G. C. M., Schnack, H. G., Peper, J. S., Chen, L., … Hulshoff Pol, H. E. (2013). Heritability of volumetric brain changes and height in children entering puberty. *Human Brain Mapping, 34*(3), 713–725. http://doi.org/10.1002/hbm.21468.

van Soelen, I. L. C., Brouwer, R. M., van Baal, G. C. M., Schnack, H. G., Peper, J. S., Collins, D. L., … Hulshoff Pol, H. E. (2012). Genetic influences on thinning of the cerebral cortex during development. *NeuroImage, 59*(4), 3871–3880. http://doi.org/10.1016/j.neuroimage.2011.11.044.

Stein, J. L., Medland, S. E., Vasquez, A. A., Hibar, D. P., Senstad, R. E., Winkler, A. M., & Thompson, P. M. (2012). Identification of common variants associated with human hippocampal and intracranial volume. *Nature Genetics, 44*(5), 552–561.

Strike, L. T., Couvy-Duchesne, B., Hansell, N. K., Cuellar-Partida, G., Medland, S. E., & Wright, M. J. (2015). Genetics and brain morphology. *Neuropsychology Review, 25*(1), 63–96. http://doi.org/10.1007/s11065-015-9281-1.

Sullivan, E. V., Pfefferbaum, A., Swan, G. E., & Carmelli, D. (2001). Heritability of hippocampal size in elderly twin men: Equivalent influence from genes and environment. *Hippocampus, 11*(6), 754–762. http://doi.org/10.1002/hipo.1091.

Swagerman, S. C., Brouwer, R. M., de Geus, E. J. C., Hulshoff Pol, H. E., & Boomsma, D. I. (2014). Development and heritability of subcortical brain volumes at ages 9 and 12. *Genes, Brain, and Behavior, 13*(8), 733–742. http://doi.org/10.1111/gbb.12182.

Taki, Y., Goto, R., Evans, A., Zijdenbos, A., Neelin, P., Lerch, J., … Fukuda, H. (2004). Voxel-based morphometry of human brain with age and cerebrovascular risk factors. *Neurobiology of Aging, 25*(4), 455–463. http://doi.org/10.1016/j.neurobiolaging.2003.09.002.

Tammimaki, A., Herder, P., Li, P., Esch, C., Laughlin, J. R., Akk, G., & Stitzel, J. A. (2012). Impact of human D398N single nucleotide polymorphism on intracellular calcium response mediated by alpha3beta4alpha5 nicotinic acetylcholine receptors. *Neuropharmacology, 63*(6), 1002–1011. http://dx.doi.org/10.1016/j.neuropharm.2012.07.022.

Teschendorff, A. E., Yang, Z., Wong, A., Pipinikas, C. P., Jiao, Y., Jones, A., … Widschwendter, M. (2015). Correlation of smoking-associated DNA methylation changes in buccal cells with DNA methylation changes in epithelial cancer. *JAMA Oncology, 1*(4), 476–485. http://doi.org/10.1001/jamaoncol.2015.1053.

Thompson, T. M., Sharfi, D., Lee, M., Yrigollen, C. M., Naumova, O. Y., & Grigorenko, E. L. (2013). Comparison of whole-genome DNA methylation patterns in whole blood, saliva, and lymphoblastoid cell lines. *Behavior Genetics, 43*(2), 168–176. http://doi.org/10.1007/s10519-012-9579-1.

Thorgeirsson, T. E., Gudbjartsson, D. F., Surakka, I., Vink, J. M., Amin, N., Geller, F., … Stefansson, K. (2010). Sequence variants at CHRNB3-CHRNA6 and CYP2A6 affect smoking behavior. *Nature Genetics, 42*(5), 448–453. http://doi.org/10.1038/ng.573.

Thorgeirsson, T. E., & Stefansson, K. (2008). Genetics of smoking behavior and its consequences: The role of nicotinic acetylcholine receptors. *Biological Psychiatry, 64*(11), 919–921. http://dx.doi.org/10.1016/j.biopsych.2008.09.010.

Tobacco and Genetics Consortium. (2010). Genome-wide meta-analyses identify multiple loci associated with smoking behavior. *Nature Genetics, 42*(5), 441–447. http://doi.org/10.1038/ng.571.

Ueda, I., Kakeda, S., Watanabe, K., Yoshimura, R., Kishi, T., Abe, O., … Korogi, Y. (2016). Relationship between G1287A of the NET gene polymorphisms and brain volume in major depressive disorder: A voxel-based MRI study. *PLoS One, 11*(3), e0150712. http://doi.org/10.1371/journal.pone.0150712.

Velicer, W. F., Prochaska, J. O., Rossi, J. S., & Snow, M. G. (1992). Assessing outcome in smoking cessation studies. *Psychological Bulletin, 111*(1), 23–41.

Volkow, N. D., & Baler, R. D. (2014). Addiction science: Uncovering neurobiological complexity. *Neuropharmacology, 76*(Pt B), 235–249. http://doi.org/10.1016/j.neuropharm.2013.05.007.

Volkow, N. D., Koob, G. F., & McLellan, A. T. (2016). Neurobiologic advances from the brain disease model of addiction. *The New England Journal of Medicine, 374*(4), 363–371. http://doi.org/10.1056/NEJMra1511480.

Wallace, G. L., Eric Schmitt, J., Lenroot, R., Viding, E., Ordaz, S., Rosenthal, M. A., … Giedd, J. N. (2006). A pediatric twin study of brain morphometry. *Journal of Child Psychology and Psychiatry, and Allied Disciplines, 47*(10), 987–993. http://doi.org/10.1111/j.1469-7610.2006.01676.x.

Wang, J., & Li, M. D. (2010). Common and unique biological pathways associated with smoking initiation/progression, nicotine dependence, and smoking cessation. *Neuropsychopharmacology: Official Publication of the American College of Neuropsychopharmacology, 35*(3), 702–719. http://dx.doi.org/10.1038/npp.2009.178.

Ware, J. J., & Munafò, M. R. (2015). Genetics of smoking behaviour. *Current Topics in Behavioral Neurosciences, 23*, 19–36. http://doi.org/10.1007/978-3-319-13665-3_2.

West, R. (2005). Defining and assessing nicotine dependence in humans. In G. Bock, & J. Goode (Eds.), *Understanding nicotine and tobacco addiction: Novartis foundation symposium, No 275* (pp. 36–58). Chichester, UK: Wiley.

White, V. M., Hopper, J. L., Wearing, A. J., & Hill, D. J. (2003). The role of genes in tobacco smoking during adolescence and young adulthood: A multivariate behaviour genetic investigation. *Addiction, 98*(8), 1087–1100.

Winkler, A. M., Kochunov, P., Blangero, J., Almasy, L., Zilles, K., Fox, P. T., & Glahn, D. C. (2010). Cortical thickness or grey matter volume? The importance of selecting the phenotype for imaging genetics studies. *NeuroImage, 53*(3), 1135–1146. http://doi.org/10.1016/j.neuroimage.2009.12.028.

Wright, I. C., Sham, P., Murray, R. M., Weinberger, D. R., & Bullmore, E. T. (2002). Genetic contributions to regional variability in human brain structure: Methods and preliminary results. *NeuroImage, 17*(1), 256–271. http://doi.org/10.1006/nimg.2002.1163.

Yang, J., & Li, M. D. (2016). Converging findings from linkage and association analyses on susceptibility genes for smoking and other addictions. *Molecular Psychiatry, 21*(8), 992–1008. http://dx.doi.org/10.1038/mp.2016.67.

Yeo, B. T. T., Krienen, F. M., Sepulcre, J., Sabuncu, M. R., Lashkari, D., Hollinshead, M., … Buckner, R. L. (2011). The organization of the human cerebral cortex estimated by intrinsic functional connectivity. *Journal of Neurophysiology, 106*(3), 1125–1165. http://doi.org/10.1152/jn.00338.2011.

Yoon, U., Fahim, C., Perusse, D., & Evans, A. C. (2010). Lateralized genetic and environmental influences on human brain morphology of 8-year-old twins. *NeuroImage, 53*(3), 1117–1125. http://doi.org/10.1016/j.neuroimage.2010.01.007.

Yoon, U., Perusse, D., Lee, J.-M., & Evans, A. C. (2011). Genetic and environmental influences on structural variability of the brain in pediatric twin: Deformation based morphometry. *Neuroscience Letters, 493*(1–2), 8–13. http://doi.org/10.1016/j.neulet.2011.01.070.

Young, S. E., Rhee, S. H., Stallings, M. C., Corley, R. P., & Hewitt, J. K. (2006). Genetic and environmental vulnerabilities underlying adolescent substance use and problem use: General or specific? *Behavior Genetics, 36*(4), 603–615. http://dx.doi.org/10.1007/s10519-006-9066-7.

Yu, R., Zhao, L., & Lu, L. (2011). Regional grey and white matter changes in heavy male smokers. *PLoS One, 6*(11), e27440. http://doi.org/10.1371/journal.pone.0027440.

Zhong, J., Shi, H., Shen, Y., Dai, Z., Zhu, Y., Ma, H., & Sheng, L. (2016). Voxelwise meta-analysis of gray matter anomalies in chronic cigarette smokers. *Behavioural Brain Research, 311*, 39–45. http://doi.org/10.1016/j.bbr.2016.05.016.

Zoli, M., Lena, C., Picciotto, M. R., & Changeux, J. P. (1998). Identification of four classes of brain nicotinic receptors using beta2 mutant mice. *The Journal of Neuroscience: The Official Journal of the Society for Neuroscience, 18*(12), 4461–4472.

Further Reading

Franke, B., Stein, J. L., Ripke, S., Anttila, V., Hibar, D. P., van Hulzen, K. J., … Sullivan, P. F. (2016). Genetic influences on schizophrenia and subcortical brain volumes: Large-scale proof of concept. *Nature Neuroscience, 19*(3), 420–431. http://dx.doi.org/10.1038/nn.4228.

van Geuns, R. J., Wielopolski, P. A., de Bruin, H. G., Rensing, B. J., van Ooijen, P. M., Hulshoff, M., … de Feyter, P. J. (1999). Basic principles of magnetic resonance imaging. *Progress in Cardiovascular Diseases, 42*(2), 149–156.

Schmaal, L., Hibar, D. P., Samann, P. G., Hall, G. B., Baune, B. T., Jahanshad, N., … Veltman, D. J. (2016). Cortical abnormalities in adults and adolescents with major depression based on brain scans from 20 cohorts worldwide in the ENIGMA Major Depressive Disorder Working Group. *Molecular Psychiatry.* http://dx.doi.org/10.1038/mp.2016.60.

Schmaal, L., Veltman, D. J., van Erp, T. G., Samann, P. G., Frodl, T., Jahanshad, N., … Hibar, D. P. (2016). Subcortical brain alterations in major depressive disorder: Findings from the ENIGMA Major Depressive Disorder working group. *Molecular Psychiatry, 21*(6), 806–812. http://dx.doi.org/10.1038/mp.2015.69.

Thompson, P. M., Andreassen, O. A., Arias-Vasquez, A., Bearden, C. E., Boedhoe, P. S., Brouwer, R. M., … ENIGMA Consortium. (2017). ENIGMA and the individual: Predicting factors that affect the brain in 35 countries worldwide. *NeuroImage, 145*(Pt B), 389–408. http://dx.doi.org/10.1016/j.neuroimage.2015.11.057.

Thompson, P. M., Stein, J. L., Medland, S. E., Hibar, D. P., Vasquez, A. A., Renteria, M. E., … Alzheimer's Disease Neuroimaging Initiative, EPIGEN Consortium, IMAGEN Consortium, Sagunay Youth Study (SYS) Group. (2014). The ENIGMA Consortium: Large-scale collaborative analyses of neuroimaging and genetic data. *Brain Imaging and Behavior, 8*(2), 153–182. http://dx.doi.org/10.1007/s11682-013-9269-5.

Vink, J. M., Jansen, R., Brooks, A., Willemsen, G., van Grootheest, G., de Geus, E., … Boomsma, D. I. (2015). Differential gene expression patterns between smokers and non-smokers: Cause or consequence? *Addiction Biology.* http://doi.org/10.1111/adb.12322.

World Health Organization. (2009). *Global health risks: Mortality and burden of disease attributable to selected major risks.* Retrieved from WHO http://www.who.int/healthinfo/global_burden_disease/GlobalHealthRisks_report_full.pdf.

World Health Organization. (2011). *Who report on the global tobacco epidemic, 2011: Warning about the dangers of tobacco.* Retrieved from WHO http://apps.who.int/iris/bitstream/10665/44616/1/9789240687813_eng.pdf.

C H A P T E R

24

Tobacco Smoke and Nicotine: Neurotoxicity in Brain Development

L.H. Lobo Torres[1], *R.C. Tamborelli Garcia*[2], *R. Camarini*[3], *T. Marcourakis*[3]

[1]Federal University of Alfenas, Alfenas, Brazil; [2]Federal University of São Paulo (UNIFESP), Diadema, Brazil; [3]University of São Paulo, São Paulo, Brazil

INTRODUCTION

Exposure to certain environmental agents during prenatal and early postnatal periods may be harmful to brain development. Smoking during pregnancy and the exposure of children to passive smoke have been associated with neurobehavioral disorders, such as learning impairments. Dysfunctions of the cholinergic systems may be involved in these disorders (Slotkin et al., 2004). Among the constituents of tobacco smoke that are potentially toxic to brain development are nicotine and carbon monoxide. The possible mechanisms of harm that is caused by prenatal tobacco exposure include hypoxia that is produced by carbon monoxide, which increases carboxyhemoglobin and nicotine-induced vasoconstriction, consequently decreasing vascularization of the placenta and fetus (Gressens, Laudenbach, & Marret, 2003). A correlation was found between the amount of gliosis in the brain stem and maternal cigarette smoking during pregnancy in childhood victims of sudden infant death syndrome (SIDS; Storm, Nylander, & Saugstad, 1999).

During brain development and the maturation of neural circuits, the effects of drugs of abuse can be more deleterious than in adult brains. Our focus in this chapter is on studies of the mechanisms of action of nicotine and tobacco smoke on the development of the central nervous system (CNS).

DEVELOPMENT OF THE CENTRAL NERVOUS SYSTEM

CNS tissue originates from the ectoderm during the embryonic period, with the formation of the neural tube, neurogenesis, and the migration of cells that form the forebrain, midbrain, and hindbrain. During fetal periods (gestational week 9 to birth), such processes as histogenesis and organogenesis occur. These processes depend on progenitor cell proliferation, migration, differentiation, synaptogenesis, apoptosis, and myelination (Houston, Herting, & Sowell, 2014).

The period of brain development is a critical period in which endogenous and exogenous factors can affect the orchestrated sequence of this phase. Several studies have suggested that changes in the normal brain development process may result in clinical disorders, such as epilepsy, schizophrenia, and autism (Rice & Barone, 2000). Exposure to xenobiotics during this critical period can result in neurotoxicity. Alcohol and tobacco are licit drugs that are used among pregnant women worldwide. The consumption of alcohol during the prenatal period is associated with perinatal death and fetal alcohol syndrome. Exposure to tobacco smoke is related to a higher incidence of SIDS. Among illicit drugs, prenatal cocaine exposure can lead to impairments in neurotransmitter signaling and significant peripheral effects. Marijuana is associated with deleterious effects on executive function, learning, and memory (Campolongo et al., 2007; Scott-Goodwin, Puerto, & Moreno, 2016).

Importantly, the immature brain has special features that may contribute to its vulnerability. The immature brain requires high cerebral blood flow (40 mL/min/100 g at birth, 80 mL/min/100 g from 2 to 10 years of age, and 45 mL/min/100 g during adulthood). The response of the immature brain to insults from free radicals is different from the adult brain because the immature brain's antioxidant systems are not yet fully developed (Blomgren & Hagberg, 2006). To better

Addictive Substances and Neurological Disease
http://dx.doi.org/10.1016/B978-0-12-805373-7.00024-4

understand the ways in which xenobiotics may interfere with brain development, the sequence of events that characterize this period of development needs to be known.

Synaptogenesis and Synaptic Plasticity

The formation of synapses is essential for the brain to become functional. The synaptogenic process requires (1) cell–cell interactions that occur in the early stages of development and depend on genetic information, (2) the refinement of recently developed synaptic connections, which requires proper neuronal activity and involves the elimination of a large number of synapses and growth and reinforcement of surviving synapses, and (3) the regulation of synaptic efficacy, which is experience dependent and occurs over the individual's entire lifespan (Kandel, Schwartz, & Jessell, 2013, Chapters 12 and 55).

Synaptogenesis is characterized by biochemical and morphological changes in pre- and postsynaptic elements, which are related to neurotransmitter release. Most neurotransmitters are stored and released by synaptic vesicles in the presynaptic terminal. This release requires a set of proteins that are responsible for synaptic vesicle exocytosis. Synapsins are peripheral proteins that are located in the vesicular membrane that bind to components of the cytoskeleton, such as actin. The phosphorylation of synapsin releases synaptic vesicles of the cytoskeleton, allowing the migration of neurotransmitter vesicles (Hosaka, Hammer, & Südhof, 1999). Soluble *N*-ethylmaleimide sensitive–factor attachment protein receptors (SNAREs) comprise a large protein superfamily that includes synaptobrevin, syntaxin, and SNAP-25, which catalyze the fusion of synaptic vesicles with the plasma membrane (Kandel et al., 2013, Chapters 12 and 55). Synaptophysins (i.e., a group of non-SNAREs) are abundant proteins of synaptic vesicles that bind synaptobrevin, forming the synaptophysin–synaptobrevin complex. Previous studies suggested that synaptophysin modulates the availability of synaptobrevin, thereby regulating synaptic vesicle exocytosis (Becher et al., 1999). Postsynaptic scaffolding proteins, such as postsynaptic density-95 (PSD95), are abundant components of the PSD, which is a marker of mature synapses (Toni et al., 2007).

Synaptic activity is important for the modulation of PSD95 turnover and regulation of the production and responsiveness of neurotrophins factors, such as brain-derived neurotrophic factor (BDNF). BDNF is related to synaptic transmission and long-term potentiation, playing an important role in brain plasticity, which requires a change of structure and function of neural circuits in response to stimuli. Brain plasticity is associated with several processes, such as brain development, learning, and memory (McAllister, Katz, & Lo, 1996).

Learning and memory require a specific set of neural processes. The cholinergic system is essential for the development of cognitive functions, and its influence on memory has been well described (for review, see Deiana, Platt, & Riedel, 2011).

Development of the Cholinergic System

Acetylcholine (ACh) is a neurotransmitter that plays a crucial role in the peripheral and central nervous systems and is considered primordial for cortical plasticity during brain development (Robertson et al., 1998). In addition to its involvement in cognitive functions, ACh participates in other functions, such as motivation, arousal, and attention.

ACh is the endogenous agonist of nicotinic acetylcholine receptors (nAChRs), which are a diverse family of pentameric transmembrane proteins that allow the entry of cations, such as Na^+ and Ca^{2+} (Gotti, Zoli, & Clementi, 2006). Nine α subunits (α_2 to α_{10}) and three β subunits (β_2 to β_4) of nAChRs have been identified in the mammalian brain. Heteromeric $\alpha_4\beta_2$ and $\alpha_3\beta_4$ subtypes and the homomeric α_7 subtype are the most predominant subtypes in the human brain (Benowitz, 2009).

Acetylcholine receptors (AChRs) are largely distributed with transient up-regulation in diverse brain regions during the development of the fetal nervous system and are involved in several processes, such as proliferation, differentiation, synaptogenesis, axonal pathfinding, and neurotransmitter release. Atluri et al. (2001) found that functional nAChRs are detectable on embryonic day 10 (E10) in mice, and the α_3, α_4, and α_7 subunits were expressed from E10 until birth. Cholinergic innervation dynamically matures in the cerebral cortex in rodents. Endogenous ACh or exogenous ACh agonists, such as nicotine, can interfere with synaptic signaling (Mechawar & Descarries, 2001).

Myelination

Myelin is a lipid-protein membrane that is spirally wound in a multilamellar sheath that is formed by a high lipid content and proteins. The myelin sheath is related to motor, sensory, and cognitive development and is the most recent evolution of the nervous system among the human species (Van der Knaap et al., 1991). In fact, there has been an increase in the white matter/gray matter ratio, with a consequent increase in complexity of the nervous system (Frahm, Stephan, & Stephan, 1982). Brain myelination generally follows a caudal-to-rostral gradient, although the opposite direction, rostral to caudal, is observed in the spinal cord (Baumann & Pham-Dinh, 2001). Myelination occurs mainly between the second trimester of gestation and the early years of postnatal life in humans. In rodents, myelination starts at birth in the spinal cord and is completed in almost all regions of the brain around postnatal days (PD) 45–60 (Doretto et al., 2011).

Oligodendrocytes are essential for the myelination process. Among glial cells, oligodendrocytes are responsible for producing brain myelin. The myelin sheath is a stable and resistant structure. During early stages of the myelination process (i.e., synthesis and deposition of the myelin); however, it is vulnerable to external influences. Oligodendrocyte maturation requires intrinsic and extrinsic factors, such as growth factors, transcription factors, and extracellular matrix proteins, that spatiotemporally regulate the myelination process (Bercury & Macklin, 2015). Oligodendrocyte progenitor cells (OPCs) are highly migratory and proliferative cells that express platelet-derived growth factor alpha receptors (PDGFαRs) and the proteoglycan NG2 (Butts, Houde, & Mehmet, 2008). The transcription factors Olig1 and Olig2 regulate PDGFαR expression, which is a marker for OPCs and stimulates the proliferation, motility, and survival of OPCs. Olig1 is a basic helix–loop–helix transcription factor that is also a marker of the early stages of myelination because it is expressed in OPCs and premyelinating oligodendrocytes (Bercury & Macklin, 2015).

Mature, myelinating oligodendrocytes are responsible for forming the myelin sheath around axons and express myelin basic protein (MBP), proteolipid protein (PLP), and myelin-associated glycoprotein (MAG). A single mature oligodendrocyte can produce myelin sheaths in many different segments of axons in different neurons (Hardy & Friedrich, 1996).

Advances in imaging techniques have contributed to studies on the dynamic process of myelin sheath biosynthesis (Snaidero et al., 2014; Sobottka, Ziegler, Kaech, Becher, & Goebels, 2011). Myelin biogenesis requires a system of cytoplasmic channels that are present during the early stages of myelination in mice, allowing communication from the outside to the inside of the myelin sheath under development. Snaidero et al. (2014) found abundant myelin sheaths with cytoplasmic channels on PD10, whereas they were practically extinct on PD60, suggesting a gradual decrease in these cytoplasmic channels at later stages of development and highlighting their importance during early development of the nervous system.

EFFECT OF NICOTINE ON DEVELOPMENT OF THE CENTRAL NERVOUS SYSTEM

Nicotine is a natural compound that is found in tobacco leaves. Its content differs among several manufactured products. Cigarette tobacco, oral snuff, and pipe tobacco have similar nicotine content, whereas only half of this average nicotine content is found in cigars and chewing tobacco (Jacob, Yu, Shulgin, & Benowitz, 1999). After absorption and distribution, nicotine quickly reaches the brain and diffuses into brain tissue.

The $\alpha_4\beta_2$ nAChR subtype is the most abundant nicotinic receptor subtype in the human brain. It mediates the reinforcing properties of nicotine. Picciotto et al. (1998) reported that β_2 knockout mice exhibited decreases in nicotine self-administration and nicotine-induced dopamine release in the ventral striatum.

Nicotine and nAChRs have been found to be crucial for learning and memory processes, especially during brain development, which occurs during early postnatal periods in several species, referred to as a "brain growth spurt" (first 3–4 weeks after birth in rodents; Miao et al., 1998). The involvement of homomeric α_7 nAChRs in learning and memory was described, in which an α_7 nAChR agonist improved working memory in rats with lesions of the hippocampal region (Levin, Bettegowda, Blosser, & Gordon, 1999). Despite its crucial role in learning-dependent reinforcement, the overstimulation of nAChRs by nicotine may lead to several harmful developmental effects, depending on the pharmacological properties and localization of nAChRs (Dwyer, Broide, & Leslie, 2008).

Prenatal nicotine exposure can decrease the proliferation of neuronal progenitors in the ventricular and subventricular zones of the embryonic neural tube, mainly through α_7 nAChR activation, by decreasing glutamatergic neurons in the medial prefrontal cortex (Aoyama et al., 2016). Mice that were prenatally exposed to nicotine exhibited behavioral impairment in attention in adulthood, which was ameliorated by the *N*-methyl-D-aspartate (NMDA) partial agonist D-cycloserine, suggesting that this exposure may cause cognitive deficits in offspring via alterations in NMDA receptors. Because of its neuroteratogenic effect, nicotine changes the expression of genes that are involved in both differentiation and apoptosis, causing mitotic arrest and cell death (Trauth, Seidler, & Slotkin, 2000).

Berger, Gage, and Vijayaraghavan (1998) found that low doses of nicotine activated one subclass of AChRs that bind the snake venom toxin α-bungarotoxin and are responsible for neurotransmitter modulation in the brain. This activation induces the tumor suppressor protein p53, a cell cycle–related protein, leading nicotine-exposed hippocampal progenitor cells to apoptosis. Interestingly, this apoptotic effect also depends on the ability of cells to tolerate changes in intracellular calcium levels, in which undifferentiated cells are unable to buffer it. Indeed, calbindin expression, a calcium buffering system, is only present after the differentiation process, protecting differentiated cells from calcium cytotoxicity (Mattson, Rychlik, Chu, & Christakos, 1991).

The apoptotic effects of nicotine have been described in vitro and in vivo during brain development (Abreu-Villaca, Seidler, & Slotkin, 2004; Qiao, Seidler, Violin, & Slotkin, 2003). Moreover, although the transitory expression of c-*fos* is correlated with short-term cell stimulation, constitutive c-*fos* overexpression elicits apoptosis, even in healthy cells, and this can be found after prenatal nicotine exposure (Trauth et al., 2000). Chronic nicotine administration decreased hippocampal neurogenesis, whereas the differentiation of neuronal progenitor cells was less affected (Shingo & Kito, 2005). Zhu et al. (2012) reported that mice that were prenatally exposed to nicotine presented a decrease in the volume of the cingulate cortex, which may be associated with a reduction of cell number. Similar decreases in cingulate cortex volume are observed in attention deficit/hyperactivity disorder (ADHD; Makris et al., 2010).

Another important issue regarding the effects of nicotine on CNS development is early postnatal nicotine exposure. Maternal nicotine exposure in animals is related to cognitive deficits in offspring. Miao et al. (1998) studied the chronic effects of nicotine exposure during different neonatal periods on the developmental regulation of both the expression of mRNAs that encode nAChR subunits and the number of nAChRs that are labeled by different nicotinic ligands in the brain. Their results showed that the second postnatal week was the most susceptible to the effects of nicotine, with an upregulation of nAChRs during PD1–21 in the rat cortex, striatum, hippocampus, thalamus, and brain stem. Rats that were exposed to nicotine through lactation during the early postnatal period had impaired CA1-dependent memory in the hippocampus in adolescence (Chen, Nakauchi, Su, Tanimoto, & Sumikawa, 2016).

Evidence suggests that α_2* nAChR-mediated responses (the asterisk indicates the possible presence of other subunits in the receptor) in CA1 oriens-lacunosum moleculare cells are lost in adolescent rats that are maternally exposed to nicotine, changing the controlled sensory information flow. Moreover, coadministration of the nonselective nAChR antagonist mecamylamine prevented the effects of nicotine, suggesting that such effects are mediated by nAChR activation. These findings suggest that changes in α_2* nAChRs may be the mechanism involved to the maternal nicotine-induced impairments in hippocampus-dependent memory.

Nicotine can be excreted via maternal milk and may cause long-lasting cognitive deficits in offspring. Mouse pups that were exposed to nicotine through this route presented long-term, rather than short-term, hippocampus-dependent memory impairment in adolescence (Nakauchi et al., 2015). The authors further showed that postnatal nicotine exposure modulated the activity of the CA1 subfield of the hippocampus.

Another issue is nicotine exposure during adolescence. The majority of tobacco smokers initiate tobacco use during this period. Strong evidence indicates that synaptogenesis, apoptosis, and synaptic programming continue during adolescence. Trauth et al. (2000) showed that male and female rats that were exposed to nicotine on PD30–47.5 presented both cell loss and a change in cell size in the hippocampus, cerebral cortex, and midbrain. These findings may be associated with a gender-specific increase in p53 expression in both the cerebral cortex and hippocampus. Male rats exhibited an increase in p53 expression in the cerebral cortex during a nicotine infusion on PD45, whereas females presented this increase on PD60. No differences in c-*fos* expression were observed among male and female rats in adolescence, in contrast to prenatal nicotine exposure.

To summarize, these data suggest that the adolescent brain is more tolerant to nicotine-induced developmental neurotoxicity compared with the fetal brain. However, we cannot exclude the possibility that nicotine-induced neurotoxicity extends into the adolescent period. Such cytotoxicity depends on the gene expression of p53 and c-*fos*, cell number, and cell morphology. Together with changes in synaptic function, these may be important factors that underlie nicotine addiction and long-term neurotoxic consequences.

TOBACCO SMOKING AFFECTS THE DEVELOPMENT OF THE CENTRAL NERVOUS SYSTEM

The adverse outcomes of prenatal active or passive exposure to tobacco smoke on fetal development have been demonstrated in several clinical and animal studies (for review, see Pagani, 2014). Passive smoking, secondhand smoke, or environmental tobacco smoke (ETS) is the inhalation of tobacco smoke in the air, which comprises 15–20% of mainstream smoke (i.e., the smoke exhaled by the smoker) and 80–85% of sidestream smoke (i.e., smoke that comes from the burning tip of a cigarette). Sidestream smoke is 10-times more toxic than mainstream smoke because it includes a high concentration of pollutants. The composition of tobacco smoke has around 8000 substances, such as nicotine, carbon monoxide, benzene, butane, toluene, hydrogen cyanide, ammonia, polycyclic aromatic hydrocarbons, and heavy metals, such as cadmium, lead, and arsenic. Worldwide, according to Öberg, Jaakkola, Woodward, Peruga, and Prüss-Ustün (2011), around 40% of children, 35% of women, and 33% of men are exposed, especially in homes or cars, to indoor ETS. A direct effect of this exposure is an increase in the rate of headaches in preadolescent children who are exposed to maternal tabagism (Arruda, Guidetti, Galli, Albuquerque, & Bigal, 2011).

Evidence suggests that nicotine may contribute to teratogenic effects in children who are prenatally exposed to tobacco smoke. To verify the relationship between maternal smoking and nonsyndromic orofacial clefts, a meta-analysis was performed by Sabbagh et al. (2015). The evaluation of the 14 children showed a 1.5-fold increase in the risk of nonsyndromic orofacial clefts with maternal smoking. Moreover, a decrease in the retinal nerve fiber layer was observed in children who were exposed to tobacco smoke during pregnancy (Pueyo et al., 2011).

Many authors have related prenatal exposure to smoking to SIDS, which is considered to result from a reduction of normal respiration and arousal during sleep. Exposure to tobacco smoke is believed to be a risk factor for SIDS because of the effect of nicotine on nAChRs in the brain stem. Lavezzi, Mecchia, and Matturri (2012) evaluated the correlation between prenatal smoking and unexplained fetal and infant deaths. They showed that fetal and infant victims of sudden death had more hypoplasia, a deficiency of vascularization, and reactive gliosis in the area postrema compared with age-matched controls. They argued that substances in tobacco smoke can affect neurons in this area and disturb centers that are involved in the control of vital functions.

Machaalani, Say, and Waters (2011) investigated brain stem nuclei that are responsible for the control of respiration and arousal in infant SIDS victims. The comparison between SIDS and non-SIDS infants revealed a decrease in α_7 and β_2 nAChRs in specific brain stem nuclei. The effect of tobacco smoke was detected among the SIDS group, in which an association was found between exposure and the increase in α_7 and β_2 nAChRs in brain stem nuclei. The authors concluded that tobacco smoke exposure may cause a loss of the regulation of those nAChR subunits in SIDS infants.

Neurobehavioral disorders, such as poor working-memory performance and learning disabilities, including ADHD, and structural and neurobiological changes were reported to result from maternal smoking (Slotkin, Pinkerton, & Seidler, 2006; Slotkin, Seidler, & Spindel 2011).

The effect of prenatal exposure to tobacco smoke was evaluated in 12-year-old children who were asked to perform an inhibitory control and attention test (Bennett et al., 2009) and a working memory task (Bennett et al., 2013) using event-related functional magnetic resonance imaging (fMRI). In the first study, the analysis showed that more brain regions were activated in tobacco-exposed children during the task. The regions that were activated included left frontal, right occipital, bilateral temporal, and parietal regions. The authors argued that the activation of multiple brain regions may indicate that children who are exposed to maternal smoking have a less mature brain (Bennett et al., 2009). In the second study, greater activation was observed in inferior parietal areas in prenatally tobacco-exposed children during a working-memory task, whereas the unexposed group presented greater activation in bilateral frontal areas. The frontal region is increasingly used in the transition from childhood to adolescence, and the activation of other areas in tobacco-exposed children may suggest a delay in the maturation of their frontal area.

Although Boucher et al. (2014) did not find a relationship between prenatal tobacco smoke exposure and behavioral performance using a visual inhibition paradigm in 186 children (11 years old), they found an association between prenatal tobacco smoke exposure and smaller amplitudes on components that underlie response inhibition. The authors concluded that this impaired response may suggest neurophysiological deficits that may be involved in ADHD-like symptoms that are seen in children who are prenatally exposed to maternal smoking.

A very interesting 25-year prospective brain imaging study examined the long-term consequences of prenatal smoking (Holz et al., 2014). Brain volume was assessed by fMRI in 178 young adults, and morphometric data, lifetime ADHD symptoms, and novelty seeking were evaluated. The subjects were followed since birth and evaluated at 25 years of age in an inhibitory paradigm. The group that was exposed to maternal smoking had more ADHD symptoms and psychosocial adversity during their lifespan. Although no differences were seen in inhibitory task performance between the two groups, a group effect was found at the functional level. fMRI revealed a reduction of activation in areas that are related to attention and response inhibition, including the anterior cingulate cortex (ACC) and inferior frontal gyrus (IFG), in subjects who were exposed to prenatal smoking. ADHD has been associated with high levels of novelty seeking, and the authors concluded that there was an inverse correlation between diminished activity in the ACC and IFG, and ADHD symptoms and novelty seeking. These impairments at the functional level may reflect dysregulation of the attention process, which can result in a personality that is more susceptible to ADHD.

Similar to human studies, prenatal exposure to tobacco smoke in animals is also associated with harmful outcomes. Yochum et al. (2014) investigated neurobehavioral development in mice that were prenatally exposed to mainstream tobacco smoke. The offspring were evaluated at 4 weeks (adolescence) and 4–6 months (adulthood) of age. Only male mice exhibited greater locomotor activity and aggression compared with nonexposed mice. They also exhibited a decrease in dopamine and serotonin in the striatum and cortex and a reduction of BDNF. These results are in agreement with the prevalence of ADHD in boys that may persist from childhood into adulthood (Pagani, 2014).

The long-term consequences of early postnatal exposure to tobacco smoke were investigated by Torres, Garcia, et al. (2015) and Torres, Annoni, et al. (2015). Mice were exposed to a mixture of mainstream and sidestream tobacco smoke twice daily from PD3 to PD14. Myelination, reference, and working memory, as well as synaptic proteins (synapsin, synaptophysin, and PSD95), and BDNF were assessed from late infancy to early adulthood. Mice that were exposed to tobacco smoke during the early postnatal period presented worse performance in the spatial reference memory task in infancy, adolescence, and adulthood. They also exhibited disruption of the maintenance of spatial working memory in both infancy and adolescence. These results were consistent with synaptic components, in which exposed mice exhibited decreases in synaptic proteins and BDNF levels in the hippocampus at all three ages evaluated. A lower percentage of myelinated fibers was found in the optic nerve in infant mice that were exposed to ETS. With regard to proteins that involved in myelination, the following was reported: a decrease in Olig1 levels in the cerebellum and brain stem, and an increase in MBP levels in the cerebellum at infancy; a decrease in MBP levels in the telencephalon and brain stem in adolescence; and a decrease in MBP levels in the cerebellum and diencephalon in adulthood. These two studies demonstrate that ETS exposure during a critical phase of development affects several processes that may be irreversible even after a long exposure-free period.

A decrease in neurogenesis in the hippocampal dentate gyrus was reported by Bruijnzeel et al. (2011). Adolescent rats were exposed to tobacco smoke for 14 days, and an increase in gliogenesis was observed, together with fewer dividing progenitor cells and surviving cells. The decrease in neurogenesis may be attributable to an increase in the susceptibility of immature neurons and deficits in differentiation and apoptosis. The mechanisms that underlie these consequences of tobacco smoke exposure involve increases in glutamate release, glucocorticoids, and apoptotic cells.

To evaluate the oxidant effects of cigarette smoke in the brain, La Maestra et al. (2011) exposed mice to different concentrations of cigarette smoke for 4 weeks. They observed dose-dependent increases in DNA damage and malondialdehyde (MDA), a lipid peroxidation marker, with an increase in the phosphorylation of tau at serine 199. Interestingly, DNA damage recovered 1 week after smoking cessation. Lobo Torres et al. (2012) also evaluated oxidative stress and lipid peroxidation in mice that were exposed to cigarette smoke for 14 days beginning on PD5. The results showed a disturbance of antioxidant enzymes that depended on both the brain region and the time of euthanasia (immediately or 3 h after the last exposure). However, no changes were seen in MDA and 3-nitrotyrosine. Altogether, these studies revealed alterations in oxidative stress parameters and tau phosphorylation, as well as DNA damage that were caused by tobacco smoke during a critical period of brain development.

The consequences of sidestream tobacco smoke (at levels that did not inhibit fetal growth) during pregnancy on gene regulation were investigated by Mukhopadhyay, Horn, Greene, and Pisano (2010). The sidestream tobacco smoke group exhibited alterations in the expression of 61 genes in the fetal hippocampus (25 up-regulated genes and 36 down-regulated genes). The functions of the up-regulated genes were related to ion transport; lipid metabolism; neurotransmission, neurogenesis, and synaptogenesis; platelet aggregation; protein metabolism/modification; and signal transduction and transcription/translation. The functions of the down-regulated genes were related to apoptosis; cell cycle; cellular organization; ion transport; lipid metabolism; metabolism; neurotransmission, neurogenesis, and synaptogenesis; platelet aggregation; protein metabolism/modification; metabolism; signal transduction; and transcription/translation and transport. Neal et al. (2014) used a similar approach and exposed mice from gestational day 1 to PD21, a period that corresponds to preimplantation through the third trimester of pregnancy. Their results indicated the impact of exposure on many processes (e.g., glycolysis, oxidative phosphorylation, and fatty acid metabolism) and pathways that are important for the development of the nervous system.

Although exposure to tobacco smoke during lactation has been reported to be associated with lower body weight and retroperitoneal fat mass, Santos-Silva et al. (2013) found that these offspring show obesity parameters in adulthood, such as hyperphagia, higher central adiposity, dyslipidemia, and hyperglycemia. Pinheiro et al. (2015) reported that rats that were exposed to tobacco smoke by maternal milk had a preference for fat in adulthood, probably because of changes in the dopaminergic reward pathway.

In conclusion, these studies provide evidence that, in addition to causing respiratory problems, smoking during pregnancy causes deleterious neural effects on the offspring brain. Thus, improving smoking cessation programs for women during pregnancy is of fundamental importance.

References

Abreu-Villaca, Y., Seidler, F. J., & Slotkin, T. A. (2004). Does prenatal nicotine exposure sensitize the brain to nicotine-induced neurotoxicity in adolescence? *Neuropsychopharmacology, 29*, 1440–1450.

Aoyama, Y., Toriumi, K., Mouri, A., Hattori, T., Ueda, E., Shimato, A., ... Yamada, K. (2016). Prenatal nicotine exposure impairs the proliferation of neuronal progenitors, leading to fewer glutamatergic neurons in the medial prefrontal cortex. *Neuropsychopharmacology, 41*, 578–589.

Arruda, M. A., Guidetti, V., Galli, F., Albuquerque, R. C. A., & Bigal, M. E. (2011). Prenatal exposure to tobacco and alcohol are associated with chronic daily headaches at childhood. A population-based study. *Arquivos de Neuro-Psiquiatria, 69*, 27–33.

Atluri, P., Fleck, M. W., Shen, Q., Mah, S. J., Stadfelt, D., Barnes, W., ... Schneider, A. S. (2001). Functional nicotinic acetylcholine receptor expression in stem and progenitor cells of the early embryonic mouse cerebral cortex. *Developmental Biology, 240*, 143–156.

Baumann, N., & Pham-Dinh, D. (2001). Biology of oligodendrocyte and myelin in the mammalian central nervous system. *Physiological Reviews, 81*, 871–927.

Becher, A., Drenckhahn, A., Pahner, I., Margittai, M., Jahn, R., & Ahnert-Hilger, G. (1999). The synaptophysin-synaptobrevin complex: A hallmark of synaptic vesicle maturation. *Journal of Neuroscience, 19*, 1922–1931.

Bennett, D. S., Mohamed, F. B., Carmody, D. P., Bendersky, M., Patel, S., Khorrami, M., ... Lewis, M. (2009). Response inhibition among early adolescents prenatally exposed to tobacco: An fMRI study. *Neurotoxicology and Teratology, 31*, 283–290.

Bennett, D. S., Mohamed, F. B., Carmody, D. P., Malik, M., Faro, S. H., & Lewis, M. (2013). Prenatal tobacco exposure predicts differential brain function during working memory in early adolescence: A preliminary investigation. *Brain Imaging and Behavior, 7*, 49–59.

Benowitz, N. L. (2009). Pharmacology of nicotine: Addiction, smoking-induced disease, and therapeutics. *Annual Review of Pharmacology and Toxicology, 49*, 57–71.

Bercury, K. K., & Macklin, W. B. (2015). Dynamics and mechanisms of CNS myelination. *Developmental Cell, 32*, 447–558.

Berger, F., Gage, F. H., & Vijayaraghavan, S. (1998). Nicotinic receptor-induced apoptotic cell death of hippocampal progenitor cells. *Journal of Neuroscience, 18*, 6871–6881.

Blomgren, K., & Hagberg, H. (2006). Free radicals, mitochondria, and hypoxia-ischemia in the developing brain. *Free Radical Biology and Medicine, 40*, 388–397.

Boucher, O., Jacobson, J. L., Burden, M. J., Dewailly, E., Jacobson, S. W., & Muckle, G. (2014). Prenatal tobacco exposure and response inhibition in school-aged children: An event-related potential study. *Neurotoxicology and Teratology, 44*, 81–88.

Bruijnzeel, A. W., Bauzo, R. M., Munikoti, V., Rodrick, G. B., Yamada, H., Fornal, C. A., ... Jacobs, B. L. (2011). Tobacco smoke diminishes neurogenesis and promotes gliogenesis in the dentate gyrus of adolescent rats. *Brain Research, 1413*, 32–42.

Butts, B. D., Houde, C., & Mehmet, H. (2008). Maturation-dependent sensitivity of oligodendrocyte lineage cells to apoptosis: Implications for normal development and disease. *Cell Death and Differentiation, 15*, 1178–1186.

Campolongo, P., Trezza, V., Cassano, T., Gaetani, S., Morgese, M. G., Ubaldi, M., ... Cuomo, V. (2007). Perinatal exposure to delta-9-tetrahydrocannabinol causes enduring cognitive deficits associated with alteration of cortical gene expression and neurotransmission in rats. *Addiction Biology, 12*, 485–495.

Chen, K., Nakauchi, S., Su, H., Tanimoto, S., & Sumikawa, K. (2016). Early postnatal nicotine exposure disrupts the α2* nicotinic acetylcholine receptor-mediated control of oriens-lacunosum moleculare cells during adolescence in rats. *Neuropharmacology, 101*, 57–67.

Deiana, S., Platt, B., & Riedel, G. (2011). The cholinergic system and spatial learning. *Behavioural Brain Research, 221*, 389–411.

Doretto, S., Malerba, M., Ramos, M., Ikrar, T., Kinoshita, C., De Mei, C., ... Borrelli, E. (2011). Oligodendrocytes as regulators of neuronal networks during early postnatal development. *PLoS One, 6*(5), e19849.

Dwyer, J. B., Broide, R. S., & Leslie, F. M. (2008). Nicotine and brain development. *Birth Defects Research. Part C, Embryo Today, 84*, 30–44.

Frahm, H. D., Stephan, H., & Stephan, M. (1982). Comparison of brain structure volumes in Insectivora and Primates. I. Neocortex. *Journal für Hirnforschung, 23*, 375–389.

Gotti, C., Zoli, M., & Clementi, F. (2006). Brain nicotinic acetylcholine receptors: Native subtypes and their relevance. *Trends in Pharmacological Sciences, 27*, 482–491.

Gressens, P., Laudenbach, V., & Marret, S. (2003). Les mécanismes d'action du tabac sur le cerveau en developpement. *Journal de Gynécologie Obstétrique et Biologie de la Reproduction (Paris), 32*(Suppl. 1), 1S30–1S32.

Hardy, R. J., & Friedrich, V. L., Jr. (1996). Progressive remodeling of the oligodendrocyte process arbor during myelinogenesis. *Developmental Neuroscience, 18*, 243–254.

Holz, N. E., Boecker, R., Baumeister, S., Hohm, E., Zohsel, K., Buchmann, A. F., ... Laucht, M. (2014). Effect of prenatal exposure to tobacco smoke on inhibitory control neuroimaging results from a 25-year prospective study. *JAMA Psychiatry, 71*, 786–796.

Hosaka, M., Hammer, R. E., & Südhof, T. C. (1999). A phospho-switch controls the dynamic association of synapsins with synaptic vesicles. *Neuron, 24*, 377–387.

Houston, S. M., Herting, M. M., & Sowell, E. R. (2014). The neurobiology of childhood structural brain development: Conception through adulthood. *Current Topics in Behavioral Neurosciences, 16*, 3–17.

Jacob, P., 3rd, Yu, L., Shulgin, A. T., & Benowitz, N. L. (1999). Minor tobacco alkaloids as biomarkers for tobacco use: Comparison of users of cigarettes, smokeless tobacco, cigars and pipes. *American Journal of Public Health, 89*, 731–736.

Kandel, E. R., Schwartz, J. H., & Jessell, T. M. (2013). *Principles of neural science* (5th ed.). United States of America: McGraw-Hill Companies, Inc.

La Maestra, S., Kisby, G. E., Micale, R. T., Johnson, J., Kow, Y. W., Bao, G., ... De Flora, S. (2011). Cigarette smoke induces DNA damage and alters base-excision repair and Tau levels in the brain of neonatal mice. *Toxicological Sciences, 123*, 471–479.

Lavezzi, A. M., Mecchia, D., & Matturri, L. (2012). Neuropathology of the area postrema in sudden intrauterine and infant death syndromes related to tobacco smoke exposure. *Autonomic Neuroscience: Basic & Clinical, 166*, 29–34.

Levin, E. D., Bettegowda, C., Blosser, J., & Gordon, J. (1999). AR-R17779, and alpha7 nicotinic agonist, improves learning and memory in rats. *Behavioral Pharmacology, 10*, 675–680.

Lobo Torres, L. H., Moreira, W. L., Tamborelli Garcia, R. C., Annoni, R., Nicoletti Carvalho, A. L., Teixeira, S. A., ... Marcourakis, T. (2012). Environmental tobacco smoke induces oxidative stress in distinct brain regions of infant mice. *Journal of Toxicology and Environmental Health A, 75*, 971–980.

Machaalani, R., Say, M., & Waters, K. A. (2011). Effects of cigarette smoke exposure on nicotinic acetylcholine receptor subunits α7 and β2 in the sudden infant death syndrome (SIDS) brainstem. *Toxicology and Applied Pharmacology, 257*, 396–404.

Makris, N., Seidman, L. J., Valera, E. M., Biederman, J., Monuteaux, M. C., Kennedy, D. N., & Faraone, S. V. (2010). Anterior cingulate volumetric alterations in treatment-naive adults with ADHD: A pilot study. *Journal of Attention Disorders, 13*, 407–413.

Mattson, M. P., Rychlik, B., Chu, C., & Christakos, S. (1991). Evidence for calcium-reducing and excitoprotective roles for the calcium-binding protein calbindin D28K in cultured hippocampal neurons. *Neuron, 6*, 41–51.

McAllister, A. K., Katz, L. C., & Lo, D. C. (1996). Neurotrophin regulation of cortical dendritic growth requires activity. *Neuron, 17*, 1057–1064.

Mechawar, N., & Descarries, L. (2001). The cholinergic innervation develops early and rapidly in the rat cerebral cortex: A quantitative immunocytochemical study. *Neuroscience, 108*, 555–567.

Miao, H., Liu, C., Bishop, K., Gong, Z. H., Nordberg, A., & Zhang, X. (1998). Nicotine exposure during a critical period of development leads to persistent changes in nicotinic acetylcholine receptors of adult rat brain. *Journal of Neurochemistry, 70*, 752–762.

Mukhopadhyay, P., Horn, K. H., Greene, R. M., & Pisano, M. M. (2010). Prenatal exposure to environmental tobacco smoke alters gene expression in the developing murine hippocampus. *Reproductive Toxicology, 29*, 164–175.

Nakauchi, S., Malvaez, M., Su, H., Kleeman, E., Dang, R., Wood, M. A., & Sumikawa, K. (2015). Early postnatal nicotine exposure causes hippocampus-dependent memory impairments in adolescent mice: Association with altered nicotinic cholinergic modulation of LTP, but not impaired LTP. *Neurobiology of Learning and Memory, 118*, 178–188.

Neal, R. E., Chen, J., Jagadapillai, R., Jang, H. J., Abomoelak, B., Brock, G., ... Pisano, M. M. (2014). Developmental cigarette smoke exposure: Hippocampus proteome and metabolome profiles in low birth weight pups. *Toxicology, 317*, 40–49.

Öberg, M., Jaakkola, M. S., Woodward, A., Peruga, A., & Prüss-Ustün, A. (2011). World-wide burden of disease from exposure to second-hand smoke: A retrospective analysis of data from 192 countries. *Lancet, 377*, 139–146.

Pagani, L. S. (2014). Environmental tobacco smoke exposure and brain development: The case of attention deficit/hyperactivity disorder. *Neuroscience and Biobehavioral Reviews, 44*, 195–205.

Picciotto, M. R., Zoli, M., Rimondini, R., Lena, C., Marubio, M., Pich, E. M., ... Changeux, J. P. (1998). Acetylcholine receptors containing the beta2 subunit are involved in the reinforcing properties of nicotine. *Nature, 391*, 173–177.

Pinheiro, C. R., Moura, E. G., Manhães, A. C., Fraga, M. C., Claudio-Neto, S., Abreu-Villaça, Y., ... Lisboa, P. C. (2015). Concurrent maternal and pup postnatal tobacco smoke exposure in Wistar rats changes food preference and dopaminergic reward system parameters in the adult male offspring. *Neuroscience, 301*, 178–192.

Pueyo, V., Güerri, N., Oros, D., Valle, S., Tuquet, H., González, I., ... Pablo, L. E. (2011). Effects of smoking during pregnancy on the optic nerve neurodevelopment. *Early Human Development, 87*, 331–334.

Qiao, D., Seidler, F. J., Violin, J. D., & Slotkin, T. A. (2003). Nicotine is a developmental neurotoxicant and neuroprotectant: Stage-selective inhibition of DNA synthesis coincident with shielding from effects of chlorpyrifos. *Brain Research. Developmental Brain Research, 147*, 183–190.

Rice, D., & Barone, S., Jr. (2000). Critical periods of vulnerability for the developing nervous system: Evidence from humans and animal models. *Environmental Health Perspectives, 108*, 511–533.

Robertson, R. T., Gallardo, K. A., Claytor, K. J., Ha, D. H., Ku, K. H., Yu, B. P., ... Leslie, F. M. (1998). Neonatal treatment with 192 IgG-saporin produces long-term forebrain cholinergic deficits and reduces dendritic branching and spine density of neocortical pyramidal neurons. *Cerebral Cortex, 8*, 142–155.

Sabbagh, H. J., Hassan, M. H. A., Innes, N. P. T., Elkodary, H. M., Little, J., & Mossey, P. A. (2015). Passive smoking in the etiology of non-syndromic orofacial clefts: A systematic review and meta-analysis. *PLoS One, 10*(3), e0116963.

Santos-Silva, A. P., Oliveira, E., Pinheiro, C. R., Santana, A. C., Nascimento-Saba, C. C., Abreu-Villaça, Y., ... Lisboa, P. C. (2013). Endocrine effects of tobacco smoke exposure during lactation in weaned and adult male offspring. *Journal of Endocrinology, 218*, 13–24.

Scott-Goodwin, A. C., Puerto, M., & Moreno, I. (2016). Toxic effects of prenatal exposure to alcohol, tobacco and other drugs. *Reproductive Toxicology, 61*, 120–130.

Shingo, A. S., & Kito, S. (2005). Effects of nicotine on neurogenesis and plasticity of hippocampal neurons. *Journal of Neural Transmission, 112*, 1475–1478.

Slotkin, T. A. (2004). Cholinergic systems in brain development and disruption by neurotoxicants: Nicotine, environmental tobacco smoke, organophosphates. *Toxicology and Applied Pharmacology, 198*, 132–151.

Slotkin, T. A., Pinkerton, K. E., & Seidler, F. J. (2006). Perinatal environmental tobacco smoke exposure in rhesus monkeys: Critical periods and regional selectivity for effects on brain cell development and lipid peroxidation. *Environmental Health Perspectives, 114*, 34–39.

Slotkin, T. A., Seidler, F. J., & Spindel, E. R. (2011). Prenatal nicotine exposure in rhesus monkeys compromises development of brainstem and cardiac monoamine pathways involved in perinatal adaptation and sudden infant death syndrome: Amelioration by vitamin C. *Neurotoxicology and Teratology, 33*, 431–434.

Snaidero, N., Möbius, W., Czopka, T., Hekking, L. H., Mathisen, C., Verkleij, D., ... Simons, M. (2014). Myelin membrane wrapping of CNS axons by PI(3,4,5)P3-dependent polarized growth at the inner tongue. *Cell, 156*, 277–290.

Sobottka, B., Ziegler, U., Kaech, A., Becher, B., & Goebels, N. (2011). CNS live imaging reveals a new mechanism of myelination: The liquid croissant model. *Glia, 59*, 1841–1849.

Storm, H., Nylander, G., & Saugstad, O. D. (1999). The amount of brainstem gliosis in sudden infant death syndrome (SIDS) victims correlates with maternal cigarette smoking during pregnancy. *Acta Paediatrica, 88*, 13–18.

Toni, N., Teng, E. M., Bushong, E. A., Aimone, J. B., Zhao, C., Consiglio, A., ... Gage, F. H. (2007). Synapse formation on neurons born in the adult hippocampus. *Nature Neuroscience, 10*, 727–734.

Torres, L. H., Annoni, R., Balestrin, N. T., Coleto, P. L., Duro, S. O., Garcia, R. C., ... Marcourakis, T. (2015). Environmental tobacco smoke in the early postnatal period induces impairment in brain myelination. *Archives of Toxicology, 89*, 2051–2058.

Torres, L. H., Garcia, R. C., Blois, A. M., Dati, L. M., Durão, A. C., Alves, A. S., ... Marcourakis, T. (2015). Exposure of neonatal mice to tobacco smoke disturbs synaptic proteins and spatial learning and memory from late infancy to early adulthood. *PLoS One, 10*(8), e0136399.

Trauth, J. A., Seidler, F. J., & Slotkin, T. A. (2000). An animal model of adolescent nicotine exposure: Effects on gene expression and macromolecular constituents in rat brain regions. *Brain Research, 867*, 29–39.

Van der Knaap, M. S., Valk, J., Bakker, C. J., Schooneveld, M., Faber, J. A., Willemse, J., & Gooskens, R. H. (1991). Myelination as an expression of the functional maturity of the brain. *Developmental Medicine and Child Neurology, 33*, 849–857.

Yochum, C., Doherty-Lyon, S., Hoffman, C., Hossain, M. M., Zellikoff, J. T., & Richardson, J. R. (2014). Prenatal cigarette smoke exposure causes hyperactivity and aggressive behavior: Role of altered catecholamines and BDNF. *Experimental Neurolology, 254*, 145–152.

Zhu, J., Zhang, X., Xu, Y., Spencer, T. J., Biederman, J., & Bhide, P. G. (2012). Prenatal nicotine exposure mouse model showing hyperactivity, reduced cingulate cortex volume, reduced dopamine turnover, and responsiveness to oral methylphenidate treatment. *Journal of Neuroscience, 32*, 9410–9418.

CHAPTER

25

Paradise Lost: A New Paradigm for Explaining the Interaction Between Neural and Psychological Changes in Nicotine Addiction Patients

T. Isomura[1], *T. Murai*[2], M. *Kano*[3]

[1]Reset Behavior Research Group, Nagoya, Japan; [2]Kyoto University, Kyoto, Japan; [3]Shin-Nakagawa Hospital, Yokohama, Japan

INTRODUCTION

Tobacco is easily obtained and can be purchased at an affordable price. Therefore, smoking is somewhat common from the age as early as adolescence. However, since nicotine is a substance that causes addiction, many smokers continue smoking, even after suffering from serious health problems. In this chapter, we review major hypotheses explaining the development of nicotine addiction, including the reward deficiency syndrome hypothesis, the sensitization theory of addiction, and the integrative neurodevelopmental model. We then describe and expand our own hypothesis, the Paradise Lost Theory (PLT) of Addiction.

Studies of Neural Bases

Most studies investigating neural bases of nicotine dependence have focused on brain reward circuits. Studies assessing animal models have revealed overactivation, and long-lasting sensitization, within reward circuits during the self-administration of nicotine (DiFranza & Wellman, 2007; Robinson & Berridge, 2003; Robinson & Kolb, 1997). Studies on humans using fMRI also show hyperresponsivity in reward circuits (i.e., the ventral striatum, cingulate cortex, and orbitofrontal and ventromedial prefrontal cortices) in the presence of anticipatory tobacco stimuli (David et al., 2005; Diekhof, Falkai, & Gruber, 2008). These findings corroborate the sensitization theory of addiction, which hypothesizes that hyper- and everlasting activation within the neural reward circuitry facilitates nicotine dependence (Robinson & Berridge, 2003).

Conversely, for nondrug stimuli (such as food and money), hyporesponsivity within reward circuits is observed (Garavan et al., 2000; Goldstein et al., 2007; Wrase et al., 2007). Based on these findings, a hyporeactive dopaminergic neural system suggests a preexisting susceptibility to addiction, which leads to the "Reward Deficiency Syndrome Hypothesis" (Blum, Cull, & Hommer, 1996). From this viewpoint, drug-taking behavior is considered an opponent process or self-medication. These two hypotheses seem at odds, with one arguing for a hyperresponsivity, while the other presumes a hyporesponsivity, as cause. To address this controversy, a neurodevelopmental model has been proposed, which places the root cause of addiction on hypersensitivity. However, the model still tries to account for both aspects of dopaminergic neural responsivity: hyper toward drug cues and hypo toward nondrug cues (Leyton & Vezina, 2014).

This model also suggests that activation level in reward circuits peaks during adolescence, which could be predictive of increased risk and novelty-seeking behaviors (Cohen et al., 2010). Therefore, the model supposes that smoking that starts in adolescence is the consequence of increased reward-related neural activation. The model further hypothesizes that among adolescents who start smoking, and especially for individuals whose reward circuits are particularly

http://dx.doi.org/10.1016/B978-0-12-805373-7.00025-6

hypersensitive, neural reactivity to tobacco-related cues develops faster and more extensively through conditioning and drug-induced sensitization. As a result, neural responsivity exceeds what is observed with nondrug cues (Leyton & Vezina, 2014). Although mechanisms underlying nicotine addiction are still not fully known, these three hypotheses have a common assumption: preexisting susceptibility is the root cause of addiction (Table 25.1). Thus, be it hyper- or hyposensitivity, if a preexisting abnormality within the reward circuit is the cause, are individuals with a more normal reward system safe from nicotine dependence, even if they smoke? Or does smoking still pose a risk for addiction for an individual with even a moderate range of dopaminergic neural activity?

We believe that adolescents with a healthy reward system are still at risk for becoming addicted to tobacco, if they begin to smoke. Once these children develop the combination of nicotine-induced neural alterations and various cognitive fallacies, a strong tendency to continue smoking could result (PLT of Addiction) (Isomura, Suzuki, & Murai, 2014) (Table 25.1).

NEURAL EVENTS DURING THE DEVELOPMENT OF NICOTINE ADDICTION

Increased Dopamine Responsivity and Reward Seeking Prior to Drug Use During Adolescence

Dopaminergic activity in the brain's reward system peaks during adolescence and is associated with increased risk and novelty-seeking behavior (including drug use) (Cohen et al., 2010). Leyton and Vezina (2014) reviewed several studies and concluded that increased reward circuit reactivity could predispose adolescents to engage in drug use. Additionally, levels of dopamine responsiveness can predict individual differences in externalizing behaviors. For example, heightened striatal responses to images of food and sex have been linked to weight gain and sexual activity, respectively (Demos, Heatherton, & Kelley, 2012). This association is thought to be causal; for instance, when reward circuits are weakened by dietary tyrosine/phenylalanine depletion, behavioral tendencies toward rewarding stimuli are diminished (Bjork, Grant, Chen, & Hommer, 2014).

What Happens After Smoking Initiation?

It is unclear how sensitivity within reward circuits changes after a lifetime nonsmoker begins smoking. However, indirect observations are available. For example, Bühler et al. (2010) compared nondependent smokers (who smoked less than one cigarette a day) and dependent smokers on a tobacco exposure cue task. When a tobacco-related cue was presented, both groups showed similar activation in the ventral striatum. However, in response to a monetary reward, nondependent smokers showed greater activation than dependent smokers (Bühler et al., 2010). One limitation of this result is that it is unlikely that those who smoke only one cigarette a day maintain such a low smoking frequency. Most will likely increase their tobacco consumption. Therefore, assuming that casual smokers observed in this study increased their smoking frequency (thus becoming dependent smokers), ventral striatal responses to nontobacco reward should decrease as the number of cigarettes smoked per day increases.

Peters et al. (2011) conducted a study on adolescents and demonstrated that reduction in reward responsivity to nondrug cues (chocolate candies) starts at a very early smoking stage. Furthermore, activation within the ventral striatum to nontobacco cues was inversely correlated with the number of cigarettes smoked. Strikingly, this reduction was observed very shortly after the participants began smoking (only after a total of 10 cigarettes smoked over their lifetime) (Peters et al., 2011). Taken together, evidence suggests that adolescents demonstrate enhanced sensitivity within the brain's reward circuits, even before they begin smoking. After smoking has commenced, responsivity to nondrug reward stimuli decreases significantly.

Questions arise as to the underlying mechanism for selective reductions in nondrug-related cue responses. Leyton and Vezina (2014) proposed the following explanation based on their neurodevelopmental model, accounting for individual differences in reward circuitry. In other words, individuals with extremely hyperactive reward circuits (among those who start smoking) develop rapid and extensive elevation within those circuits through conditioning and sensitization. This exceeds the neural responses enacted toward nondrug stimuli, which can inhibit nondrug-related behaviors (Leyton & Vezina, 2014). However, will 10 or fewer cigarettes provide enough conditioning and sensitization to exceed reward responses to nondrug cues, leading to decreased responsivity toward nondrug rewards? PLT hypothesizes that nicotine directly decreases reward circuit responsivity toward nondrug cues, from almost immediately after a person begins smoking (Isomura et al., 2014). This theory helps provide a plausible explanation of various aspects of addiction development, including neural, cognitive, and behavioral events. Furthermore, it is possible that even a healthy adolescent is at risk for acquiring nicotine dependence.

TABLE 25.1 Comparison of Major Hypotheses on Development of Nicotine Addiction and the Paradise Lost Theory of Addiction

Feature	**Opponent Process/ Reward Deficiency**	**Incentive Sensitization**	**Integrative Neurodevelopmental Model**	**Paradise Lost Theory**
Positive reinforcement	No	Yes	Yes	Yes
Negative reinforcement	Yes	No	No	Yes
Hyperactive incentive salience	No	Yes	Yes	Yes
Hypoactive incentive salience	Yes	No	Yes	Yes
Preexisting susceptibility	Yes	Yes	Yes	No
COGNITIVE FALLACIES				
Caused by experiencing the pleasure of tobacco	No	No	No	Yes
Caused by Paradise Lost state	No	No	No	Yes
INTERVENTION STRATEGIES				
Prevention	?	?	Yes	Yes
Treat high dopamine responses	No	Yes	Yes	Yes
Treat low dopamine responses	Yes	No	Yes	Yes
Redirect attentional biases	No	Yes	Yes	Yes
Reset cognitive fallacies	No	No	No	Yes

A few items are adapted from Leyton, M., & Vezina, P. (2014). Dopamine ups and downs in vulnerability to addictions: A neurodevelopmental model. Trends in Pharmacological Sciences, *35(6), 268–276.*

FOUR CHARACTERISTICS OF CIGARETTE SMOKING THAT CAUSE ADDICTION

The following sections review research based on PLT. As previously mentioned, fMRI findings suggest that as the number of cigarettes smoked over one's lifetime increases, neural responses to nontobacco cues decreases, and hyperresponsivity to tobacco cues emerges. We consider these changes as being directly induced by nicotine's pharmacological effects, indicating a biological basis for addiction. We propose four mechanisms for explaining how these changes occur: (1) Cognitive fallacies induced through smoking. (2) Self-reproductive rewards provided by smoking. (3) High upward quantitative endurability of cigarette smoking. (4) Long-lasting sensitization due to repeated smoking. All four factors are influenced by trait variables; however, even individuals without any typical risk factors can acquire a nicotine addiction. Among the aforementioned factors, cognitive fallacies could be a key addiction facilitator.

Cognitive Fallacies

According to a report from the Surgeon General (2010) [Centers for Disease Control and Prevention (US), 2010, Chap. 4], "Positive reinforcement from nicotine may play a more significant role in the initiation of smoking, and negative reinforcement, particularly relief from withdrawal, is an important contributor to the persistence of smoking and relapse." Hence, we believe that cognitive fallacies play a crucial role in the both phases as follows:

1. Cognitive fallacies during the first "pleasure of tobacco" experience: the point of no return.

   When an individual experiences smoking for the first time, the taste is noxious, coughing is induced, and headaches are often experienced. However, it does not take long before smoking becomes enjoyable. Nearly half of individuals who experiment with smoking experience the "pleasure of tobacco" after less than 20 cigarettes smoked; nearly 80% have this experience after less than 40 cigarettes (Isomura et al., 2014). Since reward circuits become blunted at very early smoking stages (i.e., after less than 10 cigarettes), adolescents hardly become aware of nicotine-induced alterations in their neural circuitry, particularly in terms of diminished feelings toward nontobacco rewards. However, while individuals may be "thoughtlessly" experimenting with smoking, neural responsivity to nondrug reward begins to precipitously decline under the influence of nicotine. If an adolescent stops smoking after trying only few cigarettes, these neural consequences tend not to emerge. However, if she or he continues experimenting, the fallacious "pleasure of tobacco" experience may take hold.

   Neural (Knott, 1997) and pharmacological treatment studies (Foulds, 2006) suggest that the "pleasure of tobacco" can be attributed to the relief felt from avoiding nicotine withdrawal. However, adolescents are not able to recognize this as being a withdrawal relief because the human sensory system is more sensitive to rapid and salient (e.g., immediate resolution of nicotine withdrawal after puffing a cigarette) rather than slow and deliberate (gradual emergence of nicotine withdrawal after puffing a previous cigarette) changes.

   As a result, adolescents recognize this post-smoking hedonic experience as "genuine pleasure." In other words, nicotine imposes itself upon a "blind spot" in the human sensory system, inducing this cognitive fallacy among smokers.

   We consider that whether an individual experiences the "pleasure of tobacco" is critical for whether she or he becomes a habitual smoker. Additionally, we assume not only that this experience is a gateway to addiction, it is also at this moment that the individual encounters a "point of no return." At the moment an adolescent experiences his or her first salient "pleasure of tobacco," the cognitive fallacy that smoking is providing genuine pleasure manifests. At the same time, the hedonic experience that emerges immediately after smoking works as a reinforcer, leading to repetitive smoking.

   Once the pleasure of tobacco has been perceived, individuals who had been smoking only during social occasions (e.g., borrowing their friends' cigarettes) begin to buy cigarettes themselves. From here, rapid progressions in neural, cognitive, and behavioral alterations toward the Paradise Lost state begin. Repeated nicotine use causes the reward system to be less responsive to nondrug cues. At the same time, neural reactivity within the ventral striatum in response to tobacco is established, which further accelerates smoking behavior. In this way, adolescents begin to smoke habitually, leading toward addiction. In fact, in one of our studies, 53 of 115 students experienced the "pleasure of smoking"; only one of these students remained a nonsmoker. The rest were either current or former smokers (Isomura et al., 2014).

2. Cognitive fallacies regarding Paradise Lost state.

   Although an individual may cross the line of no return without realizing it, the period soon after first experiencing the "pleasure of tobacco" might be the only pleasurable period for a long-term smoker. This could be because most individuals think they have control over their smoking at the start. The idea that neural alterations can occur after a fairly trivial number of cigarettes does not even occur to them. Thus, these individuals may just think that they are having fun and will stop before the behavior gets serious. However, just like it is difficult for young smokers to notice the impact of smoking on their physical body (e.g., lung functioning), the ill effects of nicotine on the brain are even less noticeable. That is, in reality, repeated smoking decreases reward reactivity toward nondrug rewards (including food and money). This results in diminished sensitivity toward stimuli/behaviors that were previously enjoyed prior to smoking (a Paradise Lost state).

Next, we provide some examples of a Paradise Lost state. Generally speaking, an instance where an individual feels like smoking after meals could be accounted for by nicotine depletion while eating and/or conditioning to the environment. Here, a Paradise Lost state could be impactful. For instance, if not for decreased sensitivity to nondrug rewards due to chronic nicotine influences, the individual should feel satisfied after eating (especially when eating a favorite food). However, since smokers' reward circuits are weakened, they are not able to fully feel pleasant, which leaves them feeling that something is missing. In order to compensate, smokers may decide to smoke a cigarette in order to forcibly stimulate dopaminergic neurons.

Let us take another example, such as the experiences one has after climbing a mountain. Even after getting to the top, feeling accomplished, and seeing such a great view, many smokers wish to smoke. From the viewpoint of a nonsmoker, it is difficult to understand why smokers need to smoke, given the beautiful view and the fresh/clean air. Yet, for smokers with a weakened reward system, this environment does not produce full satisfaction. It could even be the case that the more gorgeous the view and climbing experience, the more a smoker wishes to smoke; this is because smokers might feel they are missing out on something based on their blunted reward sensitivity, leading them to smoke a cigarette.

It is important to note that smokers are unable to recognize the effects of nicotine on their numbed sensitivity to intrinsic rewards. Since this effect takes place gradually, smokers may initially feel a vague, uncomfortable feeling when they are without nicotine. Then, the process proceeds to the next stage whereby smokers are unable to fully enjoy pleasurable stimuli or events without smoking. In this way, the transition from an intact state to a Paradise Lost state develops slowly over time. Therefore, smokers are unable to understand what is happening internally; all they may recognize is that it is difficult for them to feel pleasure toward nonsmoking activities. This feeling of anhedonia might be deepened when nicotine depletion decreases dopaminergic activity. In fact, anhedonia is a key symptom of nicotine withdrawal, in addition to negative affect, concentration difficulties, and cravings (Cook et al., 2015).

From the perspective of a smoker experiencing anhedonia, who is unaware that this feeling is likely caused by nicotine, it is understandable that they find smoking soothing and comfortable. It is natural that the more severe the anhedonia, the more an individual will crave a cigarette, and quitting becomes very difficult. Additionally, the experience of relieving anhedonia by smoking (even if only temporarily) establishes a new cognitive distortion whereby smoking is the only pastime they can enjoy, and quitting would be detrimental to their pleasure and happiness. Therefore, PLT focuses on the inability to feel happiness (anhedonia), along with the experience of several negative emotions, including anxiety and irritation. Thus, the more severe the Paradise Lost state, the more the cravings, manifesting cognitive distortions that prioritize the short-term benefits of smoking a cigarette over the long-term dangers (health and monetary costs).

Research has shown that cigarette cravings are stronger among individuals whose reward circuits are blunted (Peechatka, Whitton, Farmer, Pizzagalli, & Janes, 2015). Leventhal, Waters, Kahler, Ray, and Sussman (2009) examined underlying psychological factors related to smoking and demonstrated that smokers who reported severe anhedonia were more likely to immediately start smoking again once they tried to quit. Furthermore, quitting success rates were lower compared with a group of smokers displaying less anhedonia. The authors also revealed that more severe anhedonia during abstinence led to stronger cigarette cravings (Leventhal et al., 2009). Smokers who experience intense anhedonia place more value on the short-term benefits of cigarette smoking (Leventhal et al., 2014). This is compatible with PLT. Conversely, no significant difference was detected between individuals high and low in anhedonia in terms of smoking history and number of cigarettes smoked (Leventhal et al., 2009). Thus, there might be additional factors that create susceptibility toward developing anhedonia (Fig. 25.1).

Self-Reproductive Reward

Another mechanism for maintaining striatal responses to drug-related cues is the self-reproductive

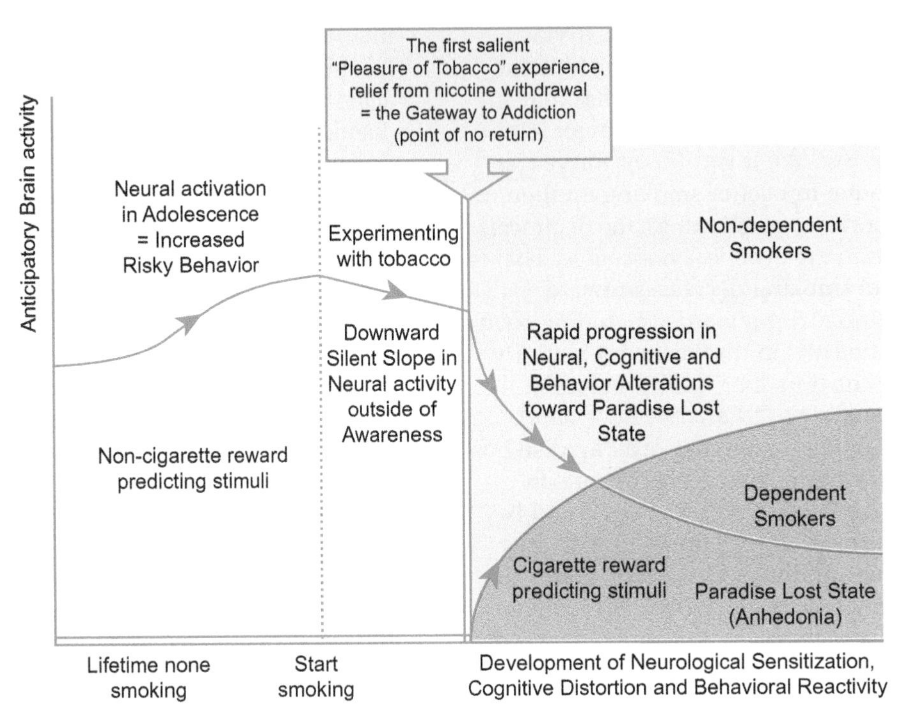

FIGURE 25.1 Hypothetical anticipatory brain activity curve in the course of addiction development according to the Paradise Lost theory.

nature of reward associated with cigarette smoking. Regarding the role dopamine neurons play in learning through incentive salience (Schultz, Dayan, & Montague, 1997), reward from avoiding nicotine withdrawal via smoking has a unique self-reproductive feature (Isomura et al., 2014). Regarding ordinary rewards, if the same reward is repeatedly presented, it can start to lose its novelty and become less salient. Thus, an old and familiar reward becomes less attractive, and one's focus moves to a different target, which is likely new and more salient. However, for rewards that result from avoiding withdrawal, repeated smoking can actually strengthen physical dependence, which creates even stronger withdrawal symptoms. Stronger withdrawal symptoms mean a greater hedonic impact from the relief of that withdrawal. In this way, the reward of avoiding nicotine withdrawal does not lose its hedonic impact and remains endlessly salient. Furthermore, many addicts feel shameful about their smoking and/or suffer from a sense of guilt. They often continue to use drugs in order to escape from these negative feelings, which perpetuate more feelings of shame and/or guilt (Flanagan, 2013). Therefore, not only the physical reward of avoiding withdrawal but also the psychological reward of escaping from negative affect is self-reproductive. Thus, the endless and repeated hedonic feelings experienced immediately after smoking cause strong operant conditioning and cognitive distortion (Isomura et al., 2014).

Quantitative Endurability of Cigarette Smoking

Sustained neural activity to tobacco cues, and decreased ventral striatal responsivity to nondrug cues, might be partly due to the peculiar nature of smoking rewards. In other words, smoking quantity can be easily multiplied as dependence proceeds. However, each individual likely has an upper limit as to the number of cigarettes smoked per day. For instance, only few people smoke more than 50 cigarettes a day due to noticeable damage to the respiratory system. However, it is quite common that an individual who was initially satisfied with just few cigarettes a day will gradually increase to 20 to 30 cigarettes a day. Thus, for some people, daily nicotine quantity provided by smoking can increase 10-fold or more from the time an individual started smoking. Here, even though the amount of cigarettes smoked increases, most smokers can endure. This is because the discomfort from excessive smoking is not unbearable and they can continue smoking. In short, cigarette smoking has large and upward "quantitative endurability."

This endurability allows smokers to increase the number of cigarettes for maintaining their hedonic experience, even when they find it difficult to feel pleasure on account of their nicotine tolerance. In this way, striatal responsivity to tobacco cues remains active, while neural responses to nondrug cues decreases. We refer to the extent that a behavior can be pervasively enacted without experiencing unbearable aversive effects as "quantitative endurability." The size of this endurability is likely to be associated with whether the behavior is addictive. For a behavior with small quantitative endurability, when tolerance is induced through repetition, behavioral frequency decreases as the experience becomes less pleasurable and more uncomfortable. However, a more endurable behavior allows individuals to continue doing it, because an individual can increase the behavior in order to keep their hedonic experience without substantial unbearable/unpleasant effects. However, even though the behavioral amount has increased, tolerance reemerges. Nevertheless, it is possible to increase the amount again. In the case of smoking, an individual is first satisfied with few cigarettes a day. This then proceeds toward 10 to 20 a day, and then finally habitual smoking throughout the day.

Before cigarettes were invented, tobacco was mainly consumed via cigars or pipes. It is likely that several people became addicted to nicotine in this manner. However, smoking a cigar is not as effective as smoking a cigarette in terms of nicotine delivery to the brain. When smoking a cigar, nicotine needs to be absorbed through a slower passage across the buccal mucosa (the membrane that lines the mouth) (Travell, 1940). Additionally, compared with the acrid and strong-tasting cigar smoke that is in the alkaline range, cigarette smoke is relatively acidic and more pleasant to smoke (Slade, 1993). In other words, the tobacco industry successfully developed a new nicotine delivery system with greater quantitative endurability. The invention of the cigarette also coincided with tobacco's transformation into a product of mass consumption, resulting in the pandemic of nicotine addiction. The tobacco industry has been continuing an effort to increase tobacco's quantitative endurability, including the production of filtered and mentholated cigarettes. Although speculative, as compared with a cigar, a modern cigarette is able to smoothly deliver a larger quantity of nicotine across a wide range of individuals.

The important point here is that a behavior with larger quantitative endurability can generate stronger stimulation within neural circuits. In regard to cigarette smoking, this amplification is more than 10-fold. More importantly, for greater amount of stimuli provided to the reward system, a more severe response reduction toward nondrug cues ensues, resulting in a more serious Paradise Lost state. Therefore, compared to behaviors with low quantitative endurability, a behavior with high endurability is more likely to induce a more severe Paradise Lost state, leading to stronger cravings and more extensive cognitive fallacies. Additionally, increased smoking frequency results in sensitization within reward circuits. Furthermore, other addictions beyond nicotine have high quantitative endurability, including alcohol, cocaine, and gambling. Thus, the concept of quantitative endurability sheds light on our understanding regarding which individuals have an affinity for specific addictive behaviors. For example, individuals with an inherently weak and sensitive respiratory system are less likely to become smokers due to low quantitative endurability. However, other addictive behaviors may result within other domains so long as the quantitative endurability is high.

Long-Lasting Sensitization

Sensitization within reward circuits is the consequence of repeated smoking resulting from a combination of embedded cognitive fallacies, self-reproductive reward, and high quantitative endurability. This sensitization involves eternal structural changes within neural synapses, which may underlie the final stages of an addiction (Robinson & Kolb, 1997). Hence, the neural mechanisms are geared toward oriented conditioning, whereby relapses can occur (even after long-term abstinence) with the reintroduction of just one cigarette (Everitt, 2014).

DRUG OR TRAIT?

Type of Drug Delivery Has Different Influences on Cognition

Clinical morphine use among terminally ill individuals provides a good example as to the critical role of cognition in addiction development. Slow acting and sustained release of morphine is prescribed to relieve pain. Interestingly, those who receive this treatment are often able to handle large doses of morphine and not become addicted. More specifically, while a physical dependence may emerge, a psychological dependence is less likely (Chapman & Hill, 1989; Portenoy, 1990; Savage et al., 2003). Now, due to a physical dependence, strong withdrawal symptoms can appear. However, drug cravings are not very common: most individuals do not desire any more morphine than is needed to ease pain (Portenoy, 1990; Savage et al., 2003). It is because as sustained morphine release does not create a rapid increase of morphine concentration in the blood, euphoric states are not elicited. Additionally, due to morphine's slow-acting effects, it is difficult for patients to recognize the association between drug intake and its pain relief properties.

This example clearly shows that physical drug dependence does not necessarily lead to an addiction. Rather, each individual's cognitive associations with a drug may be more important for producing an addiction. Therefore, we do not simply think that taking nicotine into the body causes neural alterations and addiction. We suggest that the combination of nicotine-induced neural changes and cognitive fallacies drive people to continue smoking until they become addicted (Isomura et al., 2014). This can be exemplified by examining nicotine replacement therapy. While it is true that even nicotine gums and patches can manifest an addiction, these delivery methods have very low rates of addictive outcomes (Etter, 2007; West et al., 2000). As compared to a cigarette, the concentration of nicotine in the blood slowly increases toward a gentle peak. Here, it is difficult for individuals to vividly feel any increase of nicotine in the blood, as a more substantial period of time is needed until drug effects appear (Schneider, Olmstead, Franzon, & Lunell, 2001). Therefore, even though a drug effect can be noticed, individuals merely consider this a mitigation of nicotine withdrawal. Being able to feel any sort of euphoric effect from nicotine replacement products is exceptionally rare.

Cognitive Fallacies Facilitate Addiction

In contrast, although the "pleasurable effect of nicotine" arises from the resolution of nicotine withdrawal, smokers recognize smoking as a genuinely pleasant activity for two reasons. First, a cigarette's effects appear immediately after smoking. Second, the knowledge of forgoing nicotine withdrawal is difficult to comprehend, especially when an individual experienced "the pleasure of tobacco" when initially experimenting. Later, as the number of cigarettes smoked increases, smokers eventually recognize feelings of nicotine withdrawal. However, even when detecting a sense of nicotine depletion, this is not always correctly identified as nicotine withdrawal. Smokers often confuse nicotine withdrawal with other subtle, negative sensations (e.g., consequences of an oral fixation, needing something to occupy one's hands, or feeling bored, insecure, or stressed). Therefore, smokers tend to consider the effects of smoking as purely positive whereby tobacco provides oral gratification and helps with stress. In short, when considering the rates of addiction from cigarette smoking and nicotine replacement products, the critical factor is the cognitions related to the object (e.g., cigarette or nicotine patch) that is providing specific pleasurable or euphoric effects.

Addictive Risks for Cigarette Smoking Among Healthy Adolescents

If cognitive appraisals of tobacco are important, it is quite dangerous, even for an individual with intact reward circuits, to begin smoking. Just through repeated smoking experience, nicotine invariably decreases reactivity within neural reward systems, resulting in the "pleasure of tobacco" and generation of cognitive fallacies. There are considerable individual differences in reward system activity among individuals when they start smoking. For instance, an individual with elevated dopaminergic activity is more likely to engage in risk and reward-seeking behavior, increasing the probability of smoking initiation until the "pleasure of tobacco" has been experienced. On the other hand, an individual with a hyporeactive reward system might reach a Paradise Lost state sooner and develop an addiction earlier than an individual with intact reward circuitry (Fig. 25.2).

Nevertheless, no matter what a person's inherent activity level is, repeated smoking will weaken processing within the reward system, leading to cognitive distortions triggered by the "pleasure of tobacco." This can result in combinatorial changes within neurons and thought processes that keep people smoking until they become addicted. In other words, after initial exposure, even an intact reward system is not always enough to stave off addiction.

IMPLICATIONS FOR INTERVENTION

Prevention Is Most Important

The genetic traits associated with nicotine addiction are not clear. It is our presumption that regardless of genetic factors, any individual can become addicted to tobacco once the "pleasure of tobacco" has been experienced. Then, once addicted to nicotine, treatment is very difficult. Therefore, the most important issue is how to prevent adolescents from even starting to smoke.

Various methods have been attempted in regard to interventions. From a PLT framework, the importance of education regarding the physical and psychological mechanisms underlying nicotine addiction has been recently suggested. Given the lack of individuals' awareness regarding nicotine influences on the brain, smokers are under the illusion that cigarettes are needed to cope with stress. This way of thinking helps justify smoking as useful, smart, and culturally acceptable. This perception not only spreads among smokers but to nonsmokers as well.

These psychosocially influenced ways of thinking are referred to as social nicotine dependence. This is evaluated using the Kano Test of Social Nicotine Dependence (KTSND) (Otani et al., 2009; Yoshii et al., 2006). The KTSND is comprised of 10 items, and the maximum possible score is 30. Higher scores are associated with increased justification and admiration for smoking. Nonsmokers have been shown to display

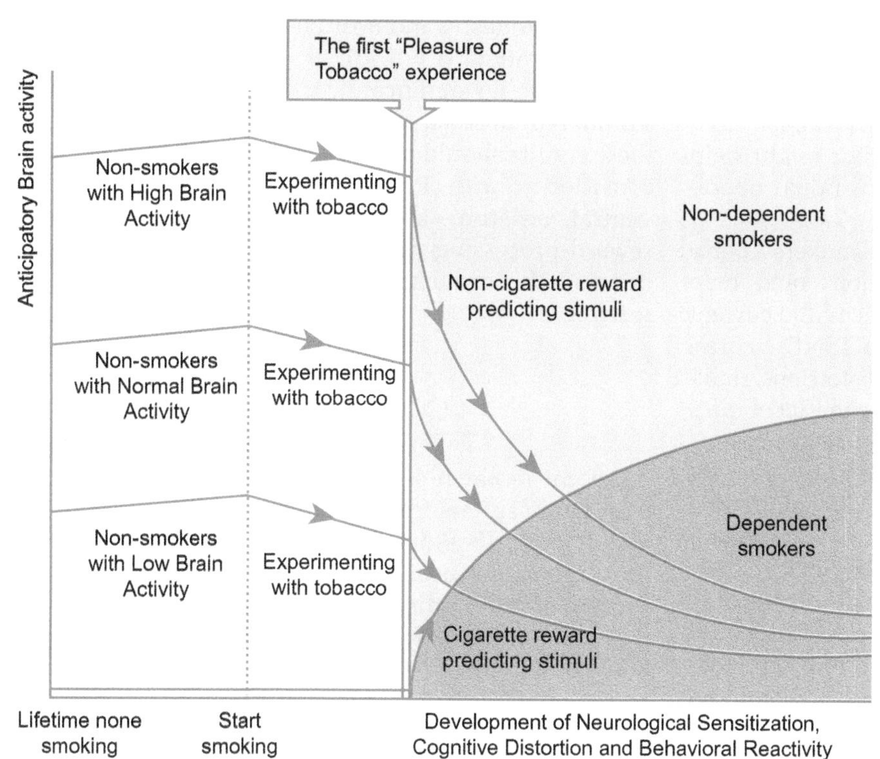

FIGURE 25.2 Drug or trait? Addictive risks for cigarette smoking among healthy adolescents. No matter what a person's inherent brain activity is, repeated smoking will weaken the reward system, leading to cognitive distortions triggered by the "pleasure of tobacco." This combinatorial change within neurons and thought processes can keep people smoking until they become addicted. An individual with elevated dopaminergic activity is more likely to engage in risk and reward-seeking behavior, increasing the probability of smoking initiation. An individual with a hyporeactive reward system might reach a Paradise Lost state sooner and develop an addiction earlier than an individual with intact reward circuitry. Even for an individual with intact reward circuits, it is dangerous to begin smoking.

rather low KTSND scores. Current smokers tend to have the highest scores, while ex-smokers also show relatively high scores. Among nonsmokers, those who previously experimented with smoking report slightly higher scores than those who never smoked (Otani et al., 2009).

One longitudinal study revealed information regarding predictive cognitive factors related to future smoking (Kitada, Amagai, Oura, Tanigushi, & Kano, 2011). That is, among nonsmoking college students, those with high KTSND scores tended to become smokers within 2 years. This was especially the case for students who answered negatively to Item 1 and affirmatively to Item 7 (Item 1: odds ratio = 2.1, $p < .05$; Item 7: odds ratio = 2.0, $p < .05$). Item 1 stated, "Smoking itself is a disease." Item 7 stated, "Tobacco helps relieve stress." Therefore, targeted education regarding these specific appraisals is necessary in order to diminish these socially distorted concepts.

Interventions for Current Smokers

PLT does not take a single-factor view of addiction in terms of either hyper- or hypo-mesolimbic dopamine activation. However, the theory does provide a new point of view regarding the relationship between neurological and psychological changes due to an addiction. Therefore, the theory offers some novel starting points for intervention. Neurologically, smokers' reward circuits are hyperreactive to tobacco cues and hyporeactive to nontobacco cues. Therefore, ideal interventions aim to decrease hypersensitivity to drug rewards while increasing reactivity to nondrug rewards. Many different neural systems are involved in addiction. Thus, the hyperactivity toward drug cues and hypoactivity to nondrug cues depend on different neural networks (Janes et al., 2012). Thus, pharmacological interventions, with at least two targets (hyper and hypo), may be quite beneficial.

The attenuation of cravings after exercise or during self-expanding activities has also been reported (Janse Van Rensburg, Taylor, Benattayallah, & Hodgson, 2012). These effects could have neural network antecedents, such as within cue-reactivity regions in the temporo-parietal junction or reward-processing regions within the frontal cortex (Xu, Aron, Westmaas, Wang, & Sweet, 2014). Thus, certain behavioral interventions could be promising.

Several psychological interventions have also been implemented. For instance, individuals' speech that favors the status quo during conversations (e.g., pro-smoking utterances), elevates reward circuit activity. However, when making pro-quitting utterances, reward activation diminishes considerably (Feldstein Ewing, Filbey, Sabbineni, Chandler, & Hutchison, 2011). Compared with smokers who maintain a status quo (i.e., continue smoking), smokers who feel strong ambivalence toward smoking show attenuated cigarette-related activation in brain areas linked to reward processing, motivation, and attention (i.e., rostral anterior cingulate, medial prefrontal cortex, caudate nucleus, and visual cortex) (Wilson, Creswell, Sayette, & Fiez, 2012). Therefore, psychological interventions that reduce smokers' justification for smoking might help directly decrease cravings through various neural alterations.

This suggests that attempts to resolve smokers' cognitive fallacies via psychological education could help buttress attempts to quit. Compared with current smokers, ex-smokers display lower KTSND scores, which could be due to weak cognitive distortions, making it more likely that they would quit (or the act of quitting produced declines in KTSND scores). In any case, if it is possible to reset smokers' cognitive fallacies, individuals may be more motivated to actually quit smoking. In addition, realizing that difficulties with feeling happy outside of smoking are due to neurological alterations (and that smoking cessation could help recover blunted neural sensitivity to various rewards) could help generate a motivation to quit. While a few self-help books have tried to pursue this tactic, more intense evaluations are needed (Isomura, Rexford, Fukuzawa, & Fukuzawa, 2011).

Expanding Into Other Addictions

Understanding nicotine addiction based on PLT may be applicable to other addictions. For example, an individual who is addicted to gambling and uses his or her winnings to buy gifts for family members (i.e., social reward was obtained from doing something nice for loved ones) may no longer do so once the reward has diminished during the emergence of an addiction. This change is the most common and essential feature for addiction onset. Thus, attempts to apply PLT to the development of gambling addiction have now begun (Isomura & Inagaki, 2016).

PLT does not necessarily define addiction as an inherent trait. While trait influences are important during the development of an addiction, we propose a dynamic interaction between neural alterations and cognitive fallacies as an addiction's root cause. Therefore, although recovery from sensitization is difficult, there is certainly a possibility of recovering from a Paradise Lost state. In other words, psychological education regarding the mechanisms underlying addiction can be encouraging.

Future research is needed to assess individuals' ability to recover from a Paradise Lost state. However, with regard to gambling addictions, Tsurumi et al. (2014) observed a significant negative correlation between activation in the insula and illness duration ($r = -0.654$, $p = .001$). Activation within this region was also modestly and positively correlated with abstinence duration ($r = 0.401$, $p = .058$) (Tsurumi et al., 2014). Since no significant correlations between illness duration or abstinence and striatal activation emerged, these results should be carefully interpreted in terms of consistency with PLT. However, if the insula and ventral striatum are important components of a reward-processing network, these results are compatible with the idea that abstinence induces partial neural recovery.

CONCLUSIONS

Recent research estimates that 47% of the US adult population suffers from maladaptive signs of an addictive disorder (Sussman, Lisha, & Griffiths, 2011). Thus, more research is needed to verify the PLT. Additionally, further understanding as to the relationship between neurological and psychological changes during addiction should provide improved targets of prevention and intervention for treating addiction.

References

Bjork, J. M., Grant, S. J., Chen, G., & Hommer, D. W. (2014). Dietary tyrosine/phenylalanine depletion effects on behavioral and brain signatures of human motivational processing. *Neuropsychopharmacology, 39*(3), 595–604.

Blum, K., Cull, J. G., & Hommer, D. W. (1996). Reward deficiency syndrome. *The American Scientist, 84*, 132–145.

Bühler, M., Vollstädt-Klein, S., Kobiella, A., Budde, H., Reed, L. J., Braus, D. F., ... Smolka, M. N. (2010). Nicotine dependence is characterized by disordered reward processing in a network driving motivation. *Biological Psychiatry, 67*, 745–752.

Centers for Disease Control and Prevention (US); National Center for Chronic Disease Prevention and Health Promotion (US); Office on Smoking and Health (US). (2010). *How tobacco smoke causes disease: The biology and behavioral basis for smoking-attributable disease. A report of the surgeon general*. Atlanta (GA): Centers for Disease Control and Prevention (US).

Chapman, C. R., & Hill, H. F. (April 15, 1989). Prolonged morphine self-administration and addiction liability. Evaluation of two theories in a bone marrow transplant unit. *Cancer, 63*(8), 1636–1644.

Cohen, J. R., Asarnow, R. F., Sabb, F. W., Bilder, R. M., Bookheimer, S. Y., Knowlton, B. J., & Poldrack, R. A. (2010). A unique adolescent response to reward prediction errors. *Nature Neuroscience, 13*, 669–671.

Cook, J. W., Piper, M. E., Leventhal, A. M., Schlam, T. R., Fiore, M. C., & Baker, T. B. (2015). Anhedonia as a component of the tobacco withdrawal syndrome. *Journal of Abnormal Psychology, 124*(1), 215–225.

David, S. P., Munafó, M. R., Johansen-Berg, H., Smith, S. M., Rogers, R. D., Matthews, P. M., & Walton, R. T. (2005). Ventral striatum/nucleus accumbens activation to smoking-related pictorial cues in smokers and nonsmokers: A functional magnetic resonance imaging study. *Biological Psychiatry, 58*, 488–494.

Demos, K. E., Heatherton, T. F., & Kelley, W. M. (2012). Individual differences in nucleus accumbens activity to food and sexual images predict weight gain and sexual behavior. *The Journal of Neuroscience, 32*(16), 5549–5552.

Diekhof, E. K., Falkai, P., & Gruber, O. (2008). Functional neuroimaging of reward processing and decision-making: A review of aberrant motivational and affective processing in addiction and mood disorders. *Brain Research Reviews, 59*, 164–184.

DiFranza, J. R., & Wellman, R. J. (2007). Sensitization to nicotine: How the animal literature might inform future human research. *Nicotine and Tobacco Research, 9*(1), 9–20.

Etter, J. F. (2007). Addiction to the nicotine gum in never smokers. *BMC Public Health, 17*(7), 159.

Everitt, B. J. (2014). Neural and psychological mechanisms underlying compulsive drug seeking habits and drug memories—indications for novel treatments of addiction. *The European Journal of Neuroscience, 40*(1), 2163–2182.

Feldstein Ewing, S. W., Filbey, F. M., Sabbineni, A., Chandler, L. D., & Hutchison, K. E. (2011). How psychosocial alcohol interventions work: A preliminary look at what fMRI can tell us. *Alcoholism, Clinical and Experimental Research, 35*(4), 643–651.

Flanagan, O. (2013). The shame of addiction. *Frontiers in Psychiatry, 4*, 120.

Foulds, J. (2006). The neurobiological basis for partial agonist treatment of nicotine dependence: Varenicline. *International Journal of Clinical Practice, 60*, 571–576.

Garavan, H., Pankiewicz, J., Bloom, A., Cho, J., Sperry, L., Ross, T. J., ... Stein, E. A. (2000). Cue-induced cocaine craving: Neuroanatomical specificity for drug users and drug stimuli. *American Journal of Psychiatry, 157*, 1789–1798.

Goldstein, R. Z., Alia-Klein, N., Tomasi, D., Zhang, L., Cottone, L. A., Maloney, T., ... Volkow, N. D. (2007). Is decreased prefrontal cortical sensitivity to monetary reward associated with impaired motivation and self-control in cocaine addiction? *The American Journal of Psychiatry, 164*(1), 43–51.

Isomura, T., & Inagaki, K. (2016). The Index of Positive Perceptions Toward Pachinko and Slot (PPPS): A new instrument for evaluation of socio-psychological aspects of problematic gambling. *Cutting-Edge Research on Health and Behavior, 1*(1), 1–16.

Isomura, T., Rexford, E., Fukuzawa, H., & Fukuzawa, R. (2011). *How to reset your smoking* (Kindle Ed.). Amazon Digital Services LLC.

Isomura, T., Suzuki, J., & Murai, T. (2014). Paradise Lost: The relationships between neurological and psychological changes in nicotine-dependent patients. *Addiction Research and Theory, 22*(2), 158–165.

Janes, A. C., Smoller, J. W., David, S. P., Frederick, B. D., Haddad, S., Basu, A., ... Kaufman, M. J. (2012). Association between CHRNA5 genetic variation at rs16969968 and brain reactivity to smoking images in nicotine dependent women. *Drug and Alcohol Dependence, 120*(1–3), 7–13.

Janse Van Rensburg, K., Taylor, A., Benattayallah, A., & Hodgson, T. (2012). The effects of exercise on cigarette cravings and brain activation in response to smoking-related images. *Psychopharmacology, 221*(4), 659–666.

Kitada, M., Amagai, K., Oura, A., Tanigushi, H., & Kano, M. (2011). Effects of attitude to the Kano test for social dependence: KTSND and towards smoke-free regulation to undertake smoking behavior on never smokers; prospective cohort study among university students. *Japanese Journal of Tobacco Control, 6*(6), 98–107.

Knott, V. J. (1977). EEG Alpha correlates of non-smokers, smokers, and smoking deprivation. *Psychophysiology, 14*, 150–156.

Leventhal, A. M., Trujillo, M., Ameringer, K. J., Tidey, J. W., Sussman, S., & Kahler, C. W. (2014). Anhedonia and the relative reward value of drug and nondrug reinforcers in cigarette smokers. *Journal of Abnormal Psychology, 123*(2), 375–386.

Leventhal, A. M., Waters, A. J., Kahler, C. W., Ray, L. A., & Sussman, S. (2009). Relations between anhedonia and smoking motivation. *Nicotine and Tobacco Research, 11*(9), 1047–1054.

Leyton, M., & Vezina, P. (2014). Dopamine ups and downs in vulnerability to addictions: A neurodevelopmental model. *Trends in Pharmacological Sciences, 35*(6), 268–276.

Otani, T., Yoshii, C., Kano, M., Kitada, M., Inagaki, K., Kurioka, N., ... Koyama, H. (2009). Validity and reliability of Kano test for social nicotine dependence. *Annals of Epidemiology, 19*(11), 815–822.

Peechatka, A. L., Whitton, A. E., Farmer, S. L., Pizzagalli, D. A., & Janes, A. C. (2015). Cigarette craving is associated with blunted reward processing in nicotine-dependent smokers. *Drug and Alcohol Dependence, 155*, 202–207.

Peters, J., Bromberg, U., Schneider, S., Brassen, S., Menz, M., Banaschewski, T., ... IMAGEN Consortium. (2011). Lower ventral striatal activation during reward anticipation in adolescent smokers. *The American Journal of Psychiatry, 168*(5), 540–549.

Portenoy, R. K. (1990). Chronic opioid therapy in nonmalignant pain. *Journal of Pain and Symptom Management, 5*(Suppl. 1), S46–S62.

Robinson, T. E., & Berridge, K. C. (2003). Addiction. *Annual Review of Psychology, 54*, 25–53.

Robinson, T. E., & Kolb, B. (1997). Persistent structural modifications in nucleus accumbens and prefrontal cortex neurons produced by previous experience with amphetamine. *The Journal of Neuroscience, 17*(21), 8491–8497.

Savage, S. R., Joranson, D. E., Covington, E. C., Schnoll, S. H., Heit, H. A., & Gilson, A. M. (2003). Definitions related to the medical use of opioids: Evolution towards universal agreement. *Journal of Pain and Symptom Management, 26*(1), 655–667.

Schneider, N. G., Olmstead, R. E., Franzon, M. A., & Lunell, E. (2001). The nicotine inhaler: Clinical pharmacokinetics and comparison with other nicotine treatments. *Clinical Pharmacokinetics, 40*(9), 661–684.

Schultz, W., Dayan, P., & Montague, P. R. (1997). A neural substrate of prediction and reward. *Science, 275*, 1593–1596.

Slade, J. (1993). Nicotine delivery devices. In C. T. Orleans, & J. Slade (Eds.), *Nicotine addiction: Principles and management* (pp. 3–23). New York: Oxford University Press.

Sussman, S., Lisha, N., & Griffiths, M. (2011). Prevalence of the addictions: A problem of the majority or the minority? *Evaluation & the Health Professions, 34*(1), 3–56.

Travell, J. (1940). The influence of the hydrogen ion concentration on the absorption of alkaloids from the stomach. *The Journal of Pharmacology and Experimental Therapeutics, 69*, 23–33.

Tsurumi, K., Kawada, R., Yokoyama, N., Sugihara, G., Sawamoto, N., Aso, T., ... Takahashi, H. (2014). Insular activation during reward anticipation reflects duration of illness in abstinent pathological gamblers. *Frontiers in Psychology, 5*, 1013.

West, R., Hajek, P., Foulds, J., Nilsson, F., May, S., & Meadows, A. (2000). A comparison of the abuse liability and dependence potential of nicotine patch, gum, spray and inhaler. *Psychopharmacology, 149*(3), 198–202.

Wilson, S. J., Creswell, K. G., Sayette, M. A., & Fiez, J. A. (2012). Ambivalence about smoking and cue-elicited neural activity in quitting-motivated smokers faced with an opportunity to smoke. *Addictive Behaviors, 38*(2), 1541–1549.

Wrase, J., Schlagenhauf, F., Kienast, T., Würstenberg, T., Bermpohl, F., Kahnt, T., ... Heinz, A. (2007). Dysfunction of reward processing correlate with alcohol craving in detoxified alcoholics. *NeuroImage, 35*, 787–794.

Xu, X., Aron, A., Westmaas, J. L., Wang, J., & Sweet, L. H. (April 11, 2014). An fMRI study of nicotine-deprived smokers' reactivity to smoking cues during novel/exciting activity. *PLoS One, 9*(4), e94598.

Yoshii, C., Kano, M., Isomura, T., Kunitomo, F., Aizawa, M., Harada, H., ... Kido, M. (March 1, 2006). Innovative questionnaire examining psychological nicotine dependence, "The Kano test for social nicotine dependence (KTSND)". *Journal of UOEH, 28*(1), 45–55.

CHAPTER

26

Interactions of Alcohol and Nicotine: CNS Sites and Contributions to Their Co-abuse

W.J. McBride

Indiana University School of Medicine, Indianapolis, IN, United States

INTRODUCTION

Alcohol drinking and smoking are serious health problems, with about 50 million individuals in the United States smoking cigarettes and almost 20 million being alcohol dependent or abusing alcohol (Falk, Yi, & Hiller-Sturmhofel, 2006). In individuals diagnosed with alcohol dependency, the rate of smoking is estimated to be between 80% and 97% (Gulliver et al., 1995; Hughes, 1996; Hurt et al., 1994; John, Hill, Rumpf, Hapke, & Meyer, 2003; John, Mayer, et al., 2003), and has remained constant while the overall smoking rate has decreased (John, Hill, et al., 2003; John, Mayer, et al., 2003). The severity of nicotine dependency is linked to more severe levels of alcohol relapse (Abrams et al., 1992; Gulliver et al., 1995) and impairs the likelihood that an individual with alcohol dependency will succeed in becoming abstinent if they continue to smoke during this period (Daeppen et al., 2000; Gulliver et al., 1995; Sobell, Sobell, & Kozolowski, 1995). In general, individuals that co-abuse ethanol and nicotine have a worse clinical outcome than individuals who use only one of the drugs (Lajtha & Sershen, 2010; Weinberger, Platt, Jiang, & Goodwin, 2015). Furthermore, the combination of smoking and drinking produces significant structural and neuro-cognitive effects in the brain (Durazzo, Cardenas, Studholme, Weiner, & Meyerhoff, 2007; Durazzo et al., 2014; Durazzo, Pennington, Schmidt, Mon, & Meyerhoff, 2013). Therefore, there is a need to better understand mechanisms within the central nervous system (CNS) underlying the interactions of ethanol and nicotine that contribute to their co-abuse.

One factor contributing to the co-abuse of alcohol and nicotine may be the counteracting adverse effects that nicotine may have on alcohol. For example, nicotine may decrease alcohol-induced cognitive impairment and reduce the intoxicating effects of alcohol (reviewed in Hurley, Taylor, & Tizabi, 2012).

There have been few studies examining mechanisms underlying the interactions of ethanol and nicotine. One study (Clark & Little, 2004) examined interactions of ethanol and nicotine in tissue slices containing the ventral tegmental area (VTA). A second study (Doyon, Dong, et al., 2013) used acute intraperitoneal (i.p.) nicotine and intravenous (i.v.) experimenter-administered ethanol to examine their interactions on dopamine (DA) release in the nucleus accumbens (Acb). Other studies examined acute interactions of ethanol and nicotine on flash-evoked potentials in the visual cortex and superior colliculus (Hetzler & Martin, 2006), and the acute effects of ethanol and nicotine on conditioned taste aversion (Rinker et al., 2008). In addition to few studies examining the acute interactions of ethanol and nicotine, even less attention has been given to the CNS effects of the chronic interactions of the self-administration of ethanol and nicotine that could contribute to their co-abuse.

ANIMAL MODELS OF CO-ABUSE OF ETHANOL AND NICOTINE

One of the difficulties of studying CNS mechanisms underlying the co-abuse of ethanol and nicotine is having animal models that self-administer both drugs to abuse levels. Although smoking is the most common way to self-administer nicotine by humans, in animals, the most common method is i.v. self-administration (Corrigall & Coen, 1989).

One laboratory (Le, Funk, Lo, & Coen, 2014; Le et al., 2010) reported that Wistar rats could be trained to co-self-administer nicotine and ethanol in the same operant session. These investigators reported ethanol intakes of

Addictive Substances and Neurological Disease
http://dx.doi.org/10.1016/B978-0-12-805373-7.00026-8

0.9–1.4 g/kg and nicotine intakes of 0.9–1.7 mg/kg. These levels of intake would be expected to produce blood ethanol concentrations (BECs) of around 80 mg % and blood nicotine levels 15–25 ng/mL. Blood alcohol concentrations (BACs) of 80 mg% would qualify as binge levels of intake and blood nicotine levels around 25 ng/mL would qualify as abuse levels (Benowitz, 1997; Hatsukami, Pickens, Svikis, & Hughes, 1988).

Another study (Hauser et al., 2012) developed an oral operant procedure for the co-self-administration of ethanol and nicotine. This procedure required adding the nicotine to a single (15%) or multiple (10%, 20%, and 30%) ethanol solutions. Operant responding in 1-h sessions produced BECs of about 80 mg% and blood nicotine levels of about 50 ng/ml. The alcohol levels qualify as hazardous binge drinking and the nicotine levels exceed abuse levels (Benowitz, 1997; Hatsukami et al., 1988). The oral consumption of nicotine has some face validity. Nicotine-laced water (Nico Water), nicotine infused fruit drink (Platinum Products), beer brewed with tobacco (NicoShot), and a 15% nicotine energy drink (Liquid Smoking) have all been successfully introduced into the market.

Although not ideal, the two animal procedures offer an experimental approach toward examining brain mechanisms involved in the co-abuse of alcohol and nicotine, and also examining the negative consequences of their co-abuse.

IMPACT OF NICOTINE ON ALCOHOL DRINKING, SEEKING, AND RELAPSE

Several studies have documented that repeated pretreatment with nicotine increases home-cage ethanol intake (Blomqvist, Ericson, Johnson, Engel, & Soderpalm, 1996; Olausson, Ericson, Lof, & Engel, 2001; Potthoff, Ellison, & Nelson, 1983), as well as operant ethanol self-administration (Clark, Lindgren, Brooks, Watson, & Little, 2001; Le, Corrigall, Harding, Juzytsch, & Li, 2000; Le, Wang, Harding, Juzytsch, & Shaham, 2003), supporting the idea that nicotine is an effective stimulus for promoting and maintaining alcohol drinking. Repeated daily nicotine injections also promoted the development of compulsive alcohol drinking in alcohol-dependent rats (Leao et al., 2015).

Alcohol-seeking behavior was markedly enhanced under the reinstatement of responding model, when testing was done the day after the last extinction session (Le et al., 2003), or with the Pavlovian Spontaneous Recovery model, when testing was done 2 weeks after extinction training (Hauser et al., 2011). These results suggest that nicotine activates neuronal circuits triggering alcohol craving, and support the idea that continued smoking while attempting to stop drinking will make it more difficult to maintain abstinence.

One of the aforementioned studies (Hauser et al., 2011) also examined the effects of nicotine on relapse drinking of female alcohol-preferring (P) rats and reported that, following a 4- to 5-week alcohol-abstinence period, treatment with 0.3 or 1.0 mg/kg nicotine, about 4 h prior to the operant session, increased responding for ethanol 1.6-fold. These results are in general agreement with another study, which reported that repeated treatment with similar doses of nicotine increased responding on the ethanol lever by male Wistar rats after a 5-day deprivation period (Lopez-Moreno et al., 2004). Overall, these results suggest that nicotine activates neuronal circuits that promote relapse drinking, and provide additional support for the idea that preventing relapse drinking will be more difficult if the individual continues smoking while undergoing treatment for alcohol abuse.

COMMON RECEPTORS FOR THE INTERACTIONS OF ALCOHOL AND NICOTINE

Neuronal nicotinic cholinergic (nACh) and serotonin-3 (5-HT_3) receptors are members of the Cys-loop ligand-gated ion channel superfamily: 5-HT_3 and nACh receptors share up to 30% sequence homology (Mascia, Trudell, & Harris, 2000; Peters et al., 2006). Thus far, nACh receptors have been a focus toward understanding molecular mechanisms underlying the co-use of ethanol and nicotine (reviewed in Doyon, Thomas, Ostroumov, Dong, & Dani, 2013; Feduccia, Chatterjee, & Bartlett, 2012; Hendrickson, Guildford, & Tapper, 2013).

Ethanol enhances the excitatory action of serotonin (5-HT) at 5-HT_3 receptors (Lovinger & White, 1991; Sung, Engel, Allan, & Lovinger, 2000; Yu et al., 1996); evidence supports the involvement of this receptor in promoting the stimulating actions of ethanol on DA release within the mesolimbic system, and in mediating ethanol self-administration (Campbell, Kohl, & McBride, 1996; Rodd et al., 2010; Rodd-Henricks et al., 2003). Nicotine binds with higher affinity to the 5-HT_3 receptor compared to any nACh receptor (Breitinger, Geetha, & Hess, 2001; Gurley & Lanthorn, 1998; Jackson & Yakel, 1995). However, it appears that nicotine antagonizes the actions of 5-HT at 5-HT_3 receptors (Breitinger et al., 2001; Gurley & Lanthorn, 1998). These findings would suggest that nicotine would inhibit the actions of ethanol at the 5-HT_3 receptor. However, the interactions of ethanol and nicotine at the 5-HT_3 receptors have not been studied.

Ethanol has been reported to act at certain nACh receptors (Borghese, Henderson, Bleck, Trudell, & arris, 2003; Xiao et al., 2013; Yu et al., 1996; Zuo, Nagata, Yeh, & Narahashi, 2004). Ethanol at concentrations of

10–100 mM inhibited nicotine responses at the α7 subunit in *Xenopus* oocytes (Yu et al., 1996). Another study (Aistrup, Marszalec, & Narahashi, 1999) reported enhancement of α-bungarotoxin-insensitive ACh currents by 30–100 mM ethanol in cultured rat cortical neurons, possibly reflecting actions at α4β2 subunits in these neurons. In addition, ethanol appears to reduce chronic nicotine-induced nACh receptor channel desensitization at α4β2 nACh receptors (Marszalec, Aistrup, & Narahashi, 1999). Other studies (Zuo et al., 2001, 2004) reported that high concentrations (100 and 300 mM) of ethanol potentiated ACh-induced currents in α4β2 subunits expressed in human embryonic kidney cells. Ethanol (200 mM) potentiated the action of ACh at α2β4 subunits expressed in *Xenopus* oocytes (Borghese et al., 2003). Unfortunately, ethanol concentrations of 100 mM (450 mg%) and higher are in the toxic range and well beyond blood levels (100–150 mg% or 20–30 mM) that are considered intoxicating, and more physiologically relevant.

The involvement of α4 subunits in mediating the effects of ethanol were examined in α4 knockout (KO) and knock-in (KI) mice (Liu et al., 2013). Using in vitro slice preparations, these investigators reported that ethanol activation of DA neurons was significantly reduced in KO mice, whereas preparations from KI mice were more sensitive to the stimulating actions of ethanol. The depressant effect of ethanol at nACh receptors in primary cultured superior cervical ganglion has been proposed to be due to its action at α7 subunits (Xiao et al., 2013). Overall, these results suggest that ethanol can inhibit the action of ACh at α7 subunits but may potentiate the actions of ACh at different combinations of αβ subunits, possibly including α4β2 subunits, and oppose nicotine-induced receptor channel desensitization.

Immunocytochemical studies reported expression of 5-HT_3 receptors in cell bodies and dendrites in forebrain regions (Morales, Battenberg, & Bloom, 1998), as well as on axons in the Acb (Ricci, Stellar, & Todtenkopf, 2004). In addition, there is evidence that nACh and 5-HT_3 receptors can be colocalized on the same striatal nerve terminals (Dougherty & Nichols, 2009; Nayak, Ronde, Spier, Lummis, & Nichols, 2000) with colocalization of the 5-HT_3 receptor with the α4 subunit, but not with the α3 and/or α5 subunits (Nayak et al., 2000). These results suggest nerve terminals expressing both nACh and 5-HT_3 receptors may be sites where the actions of ethanol and nicotine could converge.

CNS SITES FOR THE INTERACTIONS OF ALCOHOL AND NICOTINE

There is evidence that ethanol and nicotine individually have actions at multiple CNS sites (see reviews by Doyon, Thomas, et al., 2013; Pistillo, Clementi, Zoli, & Gotti, 2015). In addition to acute effects, it is important to study not just chronic effects but chronic effects of self-administration of both ethanol and nicotine. However, to understand their co-abuse, it may be necessary to identify CNS sites where both drugs have their effects. Moreover, it would be important to determine within these common sites whether ethanol and nicotine interact additively or synergistically, and that their interactions could potentially influence their co-abuse. The VTA and other regions within the cortico-limbic system have received the most attention as possible CNS sites mediating the co-abuse of alcohol and nicotine.

INTERACTIONS WITHIN THE VTA

The VTA is a major brain site in the neuro-circuitry of addiction (Koob & Volkow, 2010). The VTA receives its major cholinergic inputs from the lateral dorsal tegmentum and pedunculopontine tegmentum, and contains nACh receptors with different combinations of subunits, that is, α4β2, α4α5β2, α6β3β2, α6α4β3β2, α3β2, and α7 (reviewed in Hendrickson et al., 2013; Pistillo et al., 2015). Within the VTA, the α4β2 subtype is expressed in both DA and GABA neurons, whereas the α6β2* subtype is mainly expressed on DA neurons (Fig. 26.1). The α7-homomeric nACh receptor is expressed in about half of the DA neurons and glutamatergic afferents in the VTA (reviewed in Pistillo et al., 2015).

The activity of DA neurons within the VTA of rats is involved in the i.v. self-administration of nicotine (Corrigall, Coen, & Adamson, 1994) and the oral self-administration of ethanol (Hodge, Haraguchi, Chappelle, & Samson, 1996; Hodge, Haraguchi, Erickson, & Samson, 1993; Rodd et al., 2010). Furthermore, the posterior VTA is a neuroanatomical site supporting the rewarding actions of nicotine (Hauser, Bracken, et al., 2014; Ikemoto, Qin, & Liu, 2006) and ethanol (Rodd et al., 2004; Rodd-Henricks, McKinzie, Crile, Murphy, & McBride, 2000) in the rat. In addition, activation of DA neurons (Ikemoto et al., 2006; Rodd et al., 2004) and 5-HT_3 receptors (Hauser, Bracken, et al., 2014; Rodd-Henricks et al., 2003) are involved in mediating the rewarding actions of ethanol and nicotine within the posterior VTA. The VTA also appears to be a neuroanatomical site supporting the rewarding actions of nicotine in the mouse, possibly involving α4β2 and α7 subunits (Besson et al., 2012; David, Besson, Changeux, Granon, & Cazala, 2006; Exley et al., 2011; Molles et al., 2006).

The interactions of low concentrations of ethanol on the firing rate of VTA DA neurons were determined in slice preparations from the mouse (Clark & Little, 2004). These investigators reported that at certain

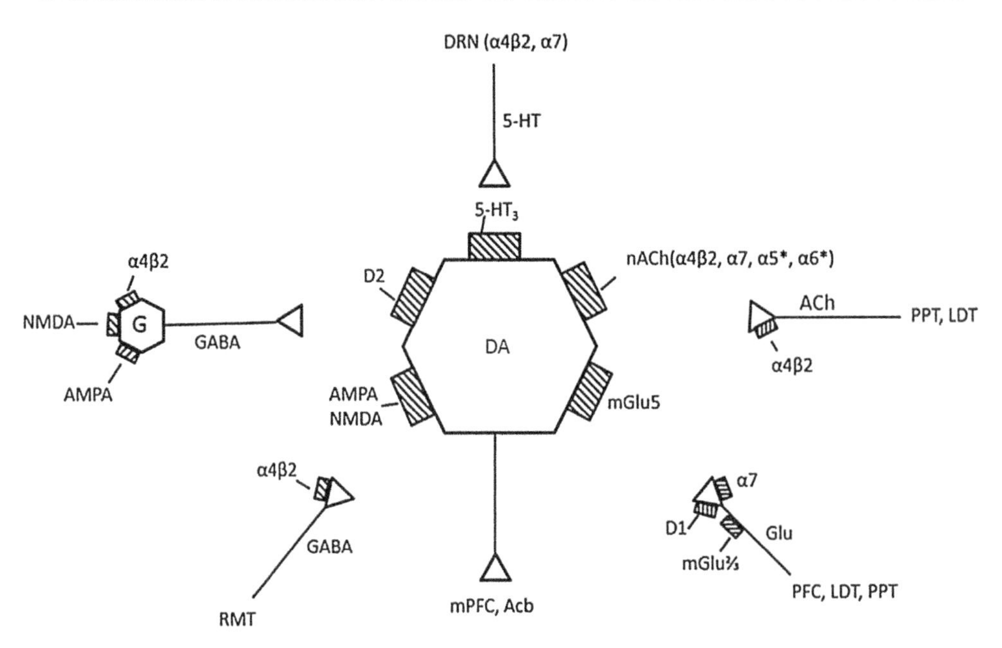

FIGURE 26.1 Simplified diagram of neural circuits within the ventral tegmental area (VTA) showing potential cellular sites where ethanol and nicotine could interact. The posterior VTA is a CNS subregion where ethanol and nicotine produce rewarding effects and ethanol and nicotine can interact synergistically to produce rewarding effects. Ethanol could be potentiating the actions of serotonin (5-HT) at 5-HT_3 receptors and acetylcholine (ACh) or nicotine at certain nicotinic cholinergic (nACh) receptors, for example, $\alpha4\beta2$; thus, the interactions of ethanol and nicotine at 5-HT_3 and $\alpha4\beta2$ receptors could be mediating their synergistic rewarding interactions within the posterior VTA. Dopamine (DA) neurons receive multiple inputs from a number of sources some of which are: (1) 5-HT inputs from the dorsal raphe nucleus (DRN); (2) ACh inputs from the pedunculopontine tegmentum (PPT) and lateral dorsal tegmentum (LDT); (3) GABA inputs from the rostromedial tegmentum (RMT) and other regions; and (4) glutamatergic (Glu) inputs from prefrontal cortex (PFC), LDT, and PPT, as well as other regions. In addition, GABA interneurons (G) also play a major role in regulating DA neuronal activity; nACh receptors are involved in regulating the activity of ACh (e.g., $\alpha4\beta2$), GABA (e.g., $\alpha4\beta2$), and Glu (e.g., $\alpha7$) inputs to the VTA, as well as local GABA interneuronal activity. 5-HT_3 and nACh (e.g., $\alpha4\beta2$, $\alpha7$, $\alpha5^*$, and $\alpha6^*$) receptors, located on DA neurons as well as on afferents to the posterior VTA, may be involved in regulating the co-abuse of ethanol and nicotine. DRN 5-HT neuronal activity appears to be regulated by $\alpha4\beta2$ and $\alpha7$ nACh receptors.

concentrations these drugs interacted synergistically to enhance VTA DA neuronal activity. Another study (Tizabi, Copeland, Louis, & Taylor, 2002) reported that the combination of systemic ethanol and local nicotine administration into the VTA produced a greater effect on DA release in the Acb than either drug alone. This effect was observed at low doses, but not at higher doses, of both drugs, and appeared to be additive rather than synergistic. In addition, several studies supported the involvement of the $\alpha6$ subunit contributing to ethanol activation of VTA DA neurons (Larsson, Jerlhag, Svensson, Soderpalm, & Engel, 2004; Liu, Zhao-Shea, McIntosh, & Tapper, 2015) and to the cooperative interaction of ethanol and nicotine to enhance α-amino-3-hydroxy-5-methyl-4-isoxazolepropionic acid (AMPA) receptor function in the VTA (Engle, McIntosh, & Drenan, 2015).

The effects of interactions of ethanol and nicotine were studied in midbrain tissue slices from Long–Evans rats (Doyon, Dong, et al., 2013). These investigators reported that prior in vivo treatment with 0.07 mg/kg i.v. nicotine 15 h before preparation of the tissue slices inhibited the stimulation of VTA DA neurons by 50 mM (225 mg%) ethanol. Blocking the stimulatory effects of ethanol appeared to be due to increased GABAergic transmission on VTA DA neurons (Doyon, Dong, et al., 2013). These results do not support an additive or synergistic interaction between nicotine and ethanol. However, the measurements conducted 15 h after the administration of nicotine may reflect neuronal changes associated with withdrawal. Furthermore, the effects of nicotine in this study involved stress hormone signaling. Therefore, it appears that the timing of exposure between nicotine and ethanol has a significant influence on the effects of their interactions.

It is important to consider that the self-administration of ethanol and nicotine may produce different effects than those observed with acute experimenter-administered drugs or conducting in vitro tests. The intracranial self-administration (ICSA) technique was used to examine the interactions of the rewarding effects of ethanol and nicotine within the posterior VTA (Truitt et al., 2015). The results of this study indicated that P rats would self-infuse a mixture of 50 mg% ethanol plus 1 μM nicotine into the posterior VTA; individually these concentrations were not self-infused. Overall, the results support the posterior VTA as a subregion where ethanol and nicotine can interact.

Nicotinic cholinergic receptors within the VTA regulate voluntary ethanol intake (Ericson, Blomqvist, Engel, & Soderpalm, 1998), the reinforcing properties of ethanol-associated conditioned cues (Lof et al., 2007), and context-induced ethanol-seeking behavior (Hauser, Deehan, et al., 2014). Mecamylamine, a nonselective nACh receptor antagonist, inhibited ethanol intake (Ericson et al., 1998), the reinforcing effects of ethanol-associated conditioned cues (Lof et al., 2007), and nicotine-enhanced ethanol-seeking behavior (Hauser, Deehan, et al., 2014), when administered into the VTA, supporting the idea of involvement of nACh receptors in these behaviors. However, mecamylamine may also act at 5-HT_3 receptors (Drisdel, Sharp, Henderson, Hales, & Green, 2008). Overall, the data support the involvement of the VTA in initiating and promoting the use, co-use, and co-abuse of ethanol and nicotine, and possible roles for $\alpha4\beta2$ subunits, $\alpha7$ subunits, and 5-HT_3 receptors in mediating the actions of nicotine and interactions of nicotine with ethanol.

Sensitization to the stimulating effects of ethanol may be an important factor underlying its rewarding actions and the promotion of alcohol abuse. Sensitization to the local DA-stimulating effects of ethanol developed in the posterior VTA of female Wistar rats after repeated local pretreatment with 200 mg% ethanol (Ding, Rodd, Engleman, & McBride, 2009). Similarly, repeated exposure of the posterior VTA to nicotine produced sensitization to the local DA-stimulating effects of ethanol within the posterior VTA (Ding et al., 2012). The DA-stimulating effects of ethanol (Ding et al., 2009) and nicotine (Zhao-Shea et al., 2011), as well as the rewarding effects of ethanol (Rodd-Henricks et al., 2000) and nicotine (Hauser, Bracken, et al., 2014; Ikemoto et al., 2006), were observed in the posterior but not the anterior VTA. These results provide additional support for the posterior VTA being an important CNS subregion mediating the interactions of ethanol and nicotine that support their co-abuse. Within the posterior VTA, the activation of nACh receptors containing $\alpha4$ and $\alpha6$ subunits may be involved in the stimulating/rewarding effects of nicotine (Exley et al., 2011; Zhao-Shea et al., 2011) and activation of 5-HT_3 receptors may be modulating the stimulating/rewarding effects of ethanol (Ding et al., 2011; Rodd-Henricks et al., 2003). Furthermore, these studies illustrate that only the posterior part of the VTA may be involved in the co-abuse of ethanol and nicotine.

INTERACTIONS WITHIN THE NUCLEUS ACCUMBENS

The nucleus accumbens (Acb) contains cholinergic interneurons, which comprise only a small percent (2–5%) of the total neurons; the Acb does not appear to receive cholinergic inputs from other regions (see Meredith, Blank, & Groenewegen, 1989; Pistillo et al., 2015). The Acb receives major DA inputs from the VTA (Fig. 26.2), and is a major part of the neurocircuits involved in alcohol and drug addiction (Koob & Volkow, 2010). Cholinergic interneurons play a critical role in regulating DA release within the Acb (Cachope et al., 2012; Threlfell et al., 2012); $\alpha6\beta2$ subunits appear to have a major role in triggering the release of DA in the Acb (Exley, Clements, Hartung, McIntosh, & Cragg, 2008). Furthermore, it appears that DA projections to the shell portion of the Acb, and not the core portion, are involved in processing the feedforward reward information (Ikemoto, Glazier, Murphy, & McBride, 1997). The i.v. self-administration of nicotine produced a preferential increase in the extracellular concentrations of DA in the shell compared to the core (Lecca et al., 2006); the reinforcing effects of i.v. nicotine appear to be regulated by $\alpha6\beta2$ subunits within the Acb-shell (Brunzell, Boschen, Hendrick, Beardsley, & McIntosh, 2010). In the rodent, $\alpha6\beta2\beta3^*$ and $\alpha4\beta2^*$ subtypes appear to be the main nACh receptor subtypes regulating the release of DA in the Acb, whereas mainly $\alpha7$ subunits appear to be localized on cortical and thalamic glutamatergic afferents (reviewed in Pistillo et al., 2015; Quik, Perez, & Grady, 2011).

Ethanol self-administration increased DA release in the Acb of P and Wistar rats (Gonzales & Weiss, 1998; Melendez et al., 2002; Weiss, Lorang, Bloom, & Koob, 1993; Weiss et al.,1996); the microinjection of D2 receptor antagonists into the Acb reduced ethanol self-administration (Rassnick, Pulvirenti, & Koob, 1992; Samson, Hodge, Tolliver, & Haraguchi, 1993). These studies provide support for the involvement of DA systems within the Acb in mediating ethanol self-administration. Unfortunately, these studies did not delineate the effects of the shell versus the core.

The Acb-shell appears to be a neuroanatomical site mediating the rewarding actions of both ethanol and nicotine. Ethanol (75–300 mg%) is self-infused directly into the Acb-shell, but not into the Acb-core (Engleman et al., 2009) by P rats. The Acb-shell also mediates the rewarding actions of nicotine (Deehan et al., 2015).

The Acb-shell might be a neuroanatomical site where the interactions of ethanol and nicotine could occur and promote their co-abuse. Using the oral ethanol–nicotine self-administration procedure (Hauser et al., 2012), the effects of self-administration of water, nicotine alone, ethanol alone, and ethanol plus nicotine on the rewarding actions of nicotine within the Acb-shell were determined (Deehan et al., 2015). Comparisons among the water, nicotine, and ethanol groups indicated no significant differences among any of these three groups with regard to the self-infusion of nicotine into

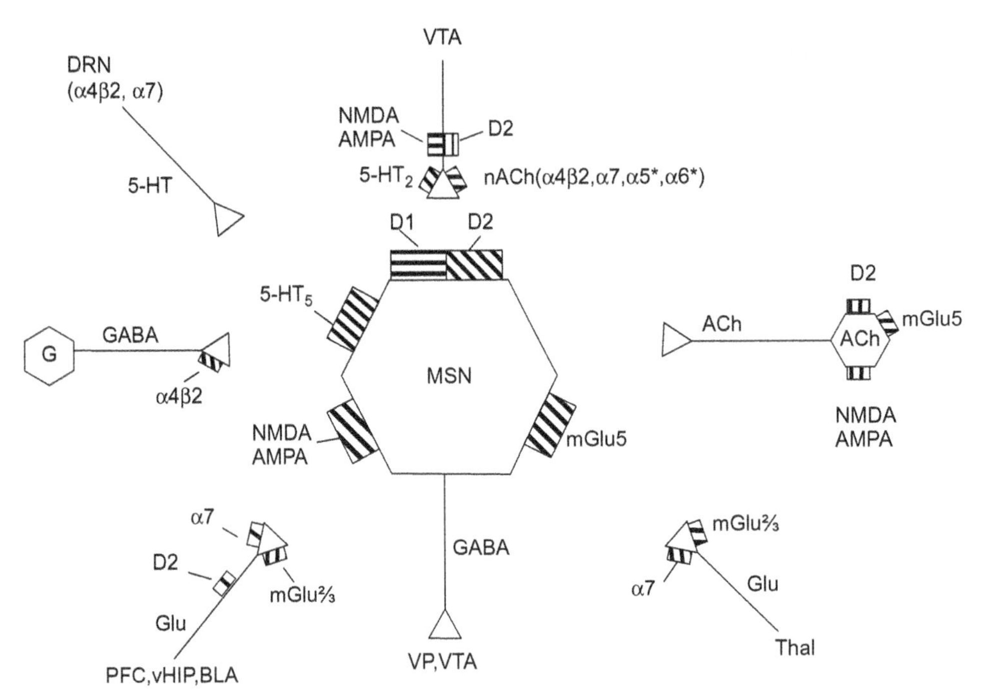

FIGURE 26.2 Simplified diagram of neural circuits within the nucleus accumbens (Acb) showing potential cellular sites where ethanol and nicotine could interact. The Acb-shell is a CNS subregion where both ethanol and nicotine produce rewarding effects, and a site where the co-abuse of ethanol and nicotine markedly increase the rewarding actions of nicotine. The Acb does not receive acetylcholine (ACh) inputs from other CNS regions, but this subregion contains large cholinergic tonically active interneurons (ACh). These ACh interneurons are activated by glutamate (Glu) projections from the prefrontal cortex (PFC) and thalamus (Thal). The Acb also receives major Glu inputs from the ventral hippocampus (vHIP) and basolateral amygdala (BLA). ACh interneurons, mainly via α4β2 or α7 nACh receptors, activate GABA interneurons, and dopamine (DA) and Glu inputs, which regulate the activity of medium spiny neurons (MSNs). The MSNs can be classified in two specific subtypes: those expressing D1 receptors and those expressing D2 receptors. The activity of MSNs is regulated by DA, Glu, and GABA inputs. MSNs exhibit little if any expression of 5-HT_3 or nACh receptors. The DA input from the VTA is regulated by nACh receptors (α4β2, α5*, and α6*), ionotropic Glu receptors (NMDA and AMPA), D_2 auto-receptors, and 5-HT_3 receptors. The dorsal raphe nucleus (DRN) 5-HT neurons are regulated by α4β2 and α7 nACh receptors. Ethanol could be potentiating the actions of 5-HT at 5-HT_3 receptors and ACh (or nicotine) at certain nACh receptors, for example, α4β2 on VTA DA projections to the Acb-shell; these interactions could be mediating their rewarding interactions. The increased sensitivity and response of the Acb-shell to the rewarding actions of nicotine following chronic co-abuse of ethanol and nicotine could be due to up-regulation of certain nACh receptor subunits, and/or alteration in the excitatory–inhibitory balance produced by co-abuse which makes the Acb-shell more sensitive and responsive to nicotine. The major GABA projections of the MSNs in the Acb-shell are to the ventral pallidum (VP) and negative feedback to the VTA.

the Acb-shell. However, the group that consumed both ethanol and nicotine, self-infused lower doses of nicotine and received fourfold more self-infusions of nicotine compared to the other three groups (Deehan et al., 2015). These results suggest that the co-abuse of ethanol and nicotine produced unique effects within the Acb-shell on the rewarding actions of nicotine, which could not be accounted for by examining the effects of ethanol and nicotine individually. Since the individual intakes and the co-abuse intakes of ethanol and nicotine were not significantly different across groups, the marked enhancement of nicotine self-infusions is not likely due to differences in intake. Furthermore, since the marked enhancement of nicotine self-infusion was observed after a 2-week drug-free period, the results suggest that co-abuse of ethanol and nicotine may have produced enduring neuronal alterations within the Acb-shell that maintained and/or enhanced the rewarding actions of nicotine. Taken together, these findings support the Acb-shell as another neuroanatomical site to study neuronal mechanisms underlying the interactions of ethanol and nicotine that contribute to their co-abuse. Moreover, the results of this study provide additional evidence that treating the co-abuse of ethanol and nicotine will likely require the development of unique strategies, and that individuals who co-abuse ethanol and nicotine will be more difficult to treat.

INTERACTIONS WITHIN THE MEDIAL PREFRONTAL CORTEX

The medial prefrontal cortex (mPFC) is an important area to examine for the co-abuse of ethanol and nicotine. The mPFC is involved in drug-seeking behavior and is a major brain region involved in alcohol and drug abuse that has major glutamatergic projections to the VTA and Acb. Glutamate systems have been implicated in drug addiction (George et al., 2012; Kalivas & Volkow, 2005; Olive, Cleva, Kalivas, & Malcolm, 2012), including nicotine (Parikh, Ji, Decker, & Sarter, 2010; Wang, Chen, & Sharp, 2008; Wang, Chen, Steketee, & Sharp, 2007) and ethanol (Holmes et al., 2012; Trantham-Davidson et al., 2014).

The principle source of cholinergic input to cortical structures of the rat originates in the nucleus basalis (Mesulam, Mufson, Wainer, & Levey, 1983; Pistillo et al., 2015), although there is evidence for intrinsic cholinergic interneurons in the cortex (Von Engelhardt, Eliava, Meyer, Rozov, & Monyer, 2007). Cortical cholinergic afferents originating from the basal forebrain play an important role in performance of tasks to assess attentional function (Muir, Page, Sirinathsinghji, Robbins, & Everitt, 1993). The mPFC receives dense cholinergic innervation and modulation of its pyramidal cells and GABA interneurons in layers V and VI by nACh receptors, that is, α7, α4β2*, and α4β2α5* (reviewed in Bloem, Poorthuis, & Mansvelder, 2014). Within the prefrontal cortex, DA and glutamate release are regulated by α4β2* or α7 nACh receptors (Fig. 26.3). Pyramidal neurons in layers II–III do not express nACh receptors, and the majority of pyramidal neurons in layer V express α7 receptors, whereas the majority of pyramidal cells in layer VI express mainly α4β2* receptors (reviewed in Pistillo et al., 2015).

The mPFC of the rat consists of four main subdivisions: the medial agranular (AGm), the anterior cingulate (AC), the prelimbic (PL), and the infralimbic (IL) cortices (reviewed in Vertes, 2004). The AGm and AC cortices have been implicated in various motor behaviors. The IL cortex appears to be involved in visceral/autonomic activity, for example, respiration, heart rate, blood pressure, and so on. The PL cortex appears to be involved in cognitive processes and goal-directed behaviors. Some studies suggest that the PL and IL cortex may have opposite effects on drug-seeking behavior (Peters, Kalivas, & Quirk, 2009) and regulation of spatial contextual control over appetitive learning (Ashwell & Ito, 2014). The PL cortex is important for initiating cocaine seeking, whereas activating the IL cortex inhibits cocaine seeking (Peters, Lalumiere, & Kalivas,

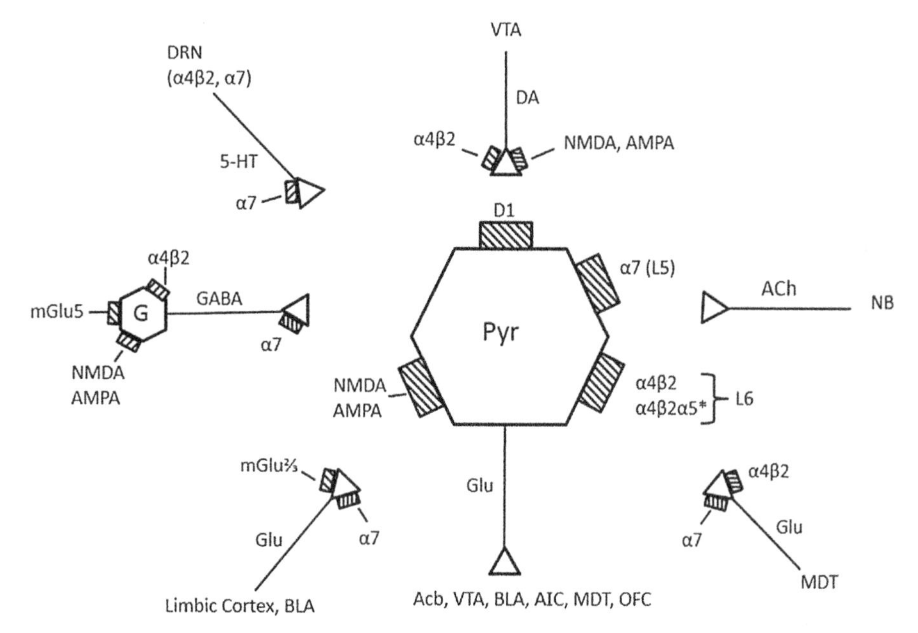

FIGURE 26.3 Simplified diagram of neural circuits within the mPFC showing potential cellular sites where ethanol and nicotine could interact. The mPFC is a CNS region where chronic self-administration of ethanol and nicotine produces unique effects on glutamate transmission. Pyramidal (Pyr) neurons in layer VI (L6) mainly express α4β2 and α4β2α5 nicotinic cholinergic (nACh) receptors, whereas those in layer V (L5) mainly express α7 nACh receptors. Acetylcholine (ACh) projections originate in the nucleus basalis (NB). GABA interneurons in L5 are regulated by α7 and α4β2 nACh receptors, whereas those in L6 are regulated mainly by α4β2 nACh receptors. There are two types of GABA neurons in the mPFC—fast-spiking interneurons that inhibit Pyr neurons and nonfast-spiking interneurons that inhibit fast-spiking GABA interneurons. The mPFC receives significant (1) glutamatergic (Glu) inputs from the limbic cortex, for example, agranular insular cortex (AIC) and orbital frontal cortex (OFC), as well as from the basolateral amygdala (BLA) and medial dorsal thalamus (MDT); (2) major dopamine (DA) inputs from the VTA; and (3) major serotonin (5-HT) inputs from the dorsal raphe nucleus (DRN). These inputs are regulated in part by nACh receptors, primarily those containing α4β2 or α7 subunits. The Pyr neurons in the pre-limbic (PL) and infra-limbic (IL) subregions of the mPFC project to overlapping as well as different regions, for example, (1) nucleus accumbens (Acb) shell and core, BLA, AIC, and DRN for the PL cortex; (2) basomedial amygdala, substantia innominata, and bed nucleus of the stria terminalis for the IL cortex; and (3) MDT, OFC, VTA, and caudate–putamen for both subregions. Within the mPFC, postsynaptic Glu actions are mediated by AMPA, NMDA, and metabotropic glutamate 5 (mGlu5) receptors; mGlu2/3 auto-receptors regulate the release of Glu from some Glu terminals.

2008). Activation of mPFC glutamatergic projections to the Acb-core appears to be necessary to sustain compulsive alcohol drinking in rats (Seif et al., 2013). The PL cortex and Acb-shell appear to be involved in mediating conditioned nicotine-seeking behavior (Kodas, Cohen, Louis, & Griebel, 2007). Nicotine, via activation of α4β2 receptors, regulates extracellular concentrations of glutamate in the PL cortex (Gioanni et al., 1999). In addition, the IL cortex may be involved in maintaining nicotine self-administration (Caille, Guillem, Cador, Manzoni, & Georges, 2009). Taken together, these findings indicate complex roles for the PL and IL cortices in drug self-administration and drug-seeking behaviors, some of which may be in the opposite direction. The involvement of the mPFC in mediating nicotine self-administration and alcohol intake support a role for this region in the co-abuse of ethanol and nicotine.

Quantitative microdialysis data (Deehan et al., 2015) indicated that the co-abuse of ethanol and nicotine, compared to water control, ethanol-alone, and nicotine-alone groups, produced an elevation in the extracellular levels of glutamate without altering clearance within the mPFC, suggesting increased glutamate transmission developed or persisted during the 2-week drug-free period. The equivalent intake of nicotine alone markedly elevated glutamate clearance without significantly altering extracellular levels of glutamate, suggesting up-regulation of glutamate transporter systems. The co-abuse of alcohol with nicotine prevented the development of these compensatory transporter systems (possibly through inhibition of transcription–translational processes). The elevated levels of extracellular glutamate may not be due to the effects of ethanol alone, since results in the posterior VTA and Acb-shell indicate that elevated glutamate transmission produced by chronic alcohol drinking of P rats dissipated after 2 weeks of deprivation (Ding et al., 2013), but may be due to the stimulating actions of nicotine. Overall, these results suggest that chronic co-abuse of ethanol and nicotine produced unique neuronal alterations within the mPFC that elevated glutamate transmission.

In conclusion, the findings support the idea that ethanol and nicotine may interact within the mPFC to produce unique effects on glutamate transmission. However, it is not clear whether these effects are occurring on glutamate afferents and/or intrinsic glutamate cells.

GENETIC ASSOCIATIONS BETWEEN NICOTINE AND ALCOHOL DEPENDENCE

The high proportion of individuals who abuse both nicotine and alcohol (Falk et al., 2006; John, Hill, et al., 2003; John, Meyer, et al., 2003) suggests that common genetic factors may be contributing to their vulnerability to abuse multiple drugs. A male–male twin study reported a high correlation for the co-occurrence of nicotine and alcohol dependence (True et al., 1999), supporting the idea of common genetic factors contributing to vulnerability to abuse both drugs. Another study identified an area on human chromosome 2 containing a locus for the common genetic vulnerability to abuse both alcohol and nicotine (Bierut et al., 2004). The A1 allele of the DA D2 receptor appears to be associated with higher levels of alcohol consumption and smoking compared to individuals having the A2/A2 genotype (Connor et al., 2007). In addition, genetic variation in the gene cluster containing α5, α4, and β4 subunits appears to be a determinant of early alcohol and tobacco initiation in young adults (Schlaepfer et al., 2008).

Animal studies demonstrated a relationship between sensitivity to the reinforcing actions of nicotine and high alcohol preference. Comparison of the i.v. self-administration of nicotine between P and NP (non-preferring) rats indicated that the P rats were more sensitive to the reinforcing effects of nicotine and administered more self-infusions of nicotine (Le et al., 2006). In addition, this study demonstrated nicotine-primed reinstatement of responding by P rats but not by NP rats. These results support the idea that selective breeding for high alcohol preference is associated with vulnerability to abuse nicotine.

A second study reported that the P rats were more sensitive to the rewarding actions of nicotine and received more self-infusions of nicotine than Wistar rats in the posterior VTA (Hauser, Bracken, et al., 2014). Similar differences were observed for the self-infusion of ethanol into the posterior VTA between P and Wistar rats (Rodd et al., 2004). Taken together, these results suggest that there may be common genetic factors contributing to vulnerability to abuse both alcohol and nicotine, and that the posterior VTA may be one neuroanatomical site influenced by common genetic factors.

CONCLUSIONS

Alcohol drinking and smoking are serious health problems and often occur together (John, Hill, et al., 2003; John, Meyer, et al., 2003). Individuals who co-abuse alcohol and nicotine are difficult to treat (Lajtha & Sershen, 2010; Weinberger et al., 2015). Findings from human (Bierut et al., 2004; Connor et al., 2007) and animal (Hauser, Bracken, et al., 2014; Le et al., 2006) studies provide support for common genetic factors contributing to vulnerability to abuse both alcohol and nicotine. Other findings indicate that nicotine can enhance ethanol self-administration, alcohol-seeking behavior, and relapse drinking (Hauser et al., 2011; Le et al., 2000, 2003), supporting the clinical data that co-abuse may be more difficult to treat if the individual undergoing treatment for alcohol abuse continues to smoke. Common receptors for ethanol and nicotine include 5-HT_3 and nACh receptors containing the $\alpha 4$ and $\alpha 7$ subunits (Jackson & Yakel, 1995; Liu et al., 2013; Yu et al., 1996). The posterior VTA (Truitt et al., 2015), Acb-shell (Deehan et al., 2015), and mPFC (Deehan et al., 2015) are three CNS sites where ethanol and nicotine can interact; sites where ethanol and nicotine interact may contribute to promoting their co-abuse. Thus far, limited studies have been conducted to examine neurobiological mechanisms underlying the co-abuse of ethanol and nicotine. The co-abuse of ethanol and nicotine may produce unique neuronal alterations that cannot be observed with their individual abuse (Deehan et al., 2015). The development of suitable animal models for the co-abuse of alcohol and nicotine will greatly aid in generating findings on the neurobiological basis for the co-addiction of alcohol and nicotine. A better understanding of the complex neurobiological mechanisms underlying the co-abuse of alcohol and nicotine is important for developing appropriate strategies for treating their co-abuse.

References

Abrams, D. B., Rohsenow, D. J., Niaura, R. S., Pedraza, M., Longbaugh, R., Beatties, M. C., ... Monti, P. M. (1992). Smoking and treatment outcome for alcoholics: Effects on coping skills, urge to drink and drinking rates. *Behavioral Therapeutics, 23*, 283–297.

Aistrup, G. L., Marszalec, W., & Narahashi, T. (1999). Ethanol modulation of nicotinic acetylcholine receptor currents in cultured cortical neurons. *Molecular Pharmacology, 55*, 39–49.

Ashwell, R., & Ito, R. (2014). Excitotoxic lesions of the infralimbic, but not prelimbic cortex facilitate reversal of appetitive discrimination context conditioning: The role of the infralimbic cortex in context generalization. *Frontiers in Behavioral Neuroscience, 28*. http://dx.doi.org/10.3389/fnbeh.2014.00063.

Benowitz, N. L. (1997). Systemic absorption and effects of nicotine from smokeless tobacco. *Advances in Dental Research, 11*, 336–341.

Besson, M., David, V., Baudonnat, M., Cazala, P., Guilloux, J. P., Reperant, C., ... Granon, S. (2012). Alpha7-nicotinic receptors modulate nicotine-induced reinforcement and extracellular dopamine outflow in the mesolimbic system in mice. *Psychopharmacology, 220*, 1–14.

Bierut, L. J., Rice, J. P., Goate, A., Hinrichs, A. L., Saccone, N. L., Foroud, T., ... Reich, T. (2004). A genomic scan for habitual smoking in families of alcoholics: Common and specific genetic factors in substance dependence. *American Journal Medical Genetics. Part A, 124A*, 19–27.

Bloem, B., Poorthuis, R. B., & Mansvelder, H. D. (2014). Cholinergic modulation of the medial prefrontal cortex: The role of nicotinic receptors in attention and regulation of neuronal activity. *Frontiers Neural Circuits, 8*, 1–16.

Blomqvist, O., Ericson, M., Johnson, D. H., Engel, J. A., & Soderpalm, B. (1996). Voluntary ethanol intake in the rat: Effects of nicotinic acetylcholine receptor blockade or subchronic nicotine treatment. *European Journal of Pharmacology, 314*, 257–267.

Borghese, C. M., Henderson, L. A., Bleck, V., Trudell, J. R., & Harris, R. A. (2003). Sites of excitatory and inhibitory actions of alcohols on neuronal alpha 2 beta 4 nicotinic acetylcholine receptors. *Journal Pharmacology & Experimental Therapeutic, 307*, 45–52.

Breitinger, H.-G. A., Geetha, N., & Hess, G. P. (2001). Inhibition of the serotonin 5-HT_3 receptor by nicotine, cocaine and fluoxetine investigated by rapid chemical kinetic techniques. *Biochemistry, 40*, 8419–8429.

Brunzell, D. H., Boschen, K. E., Hendrick, E. S., Beardsley, P. M., & McIntosh, J. M. (2010). Alpha-conotoxin MII-sensitive nicotinic acetylcholine receptors in the nucleus accumbens shell regulate progressive ratio responding maintained by nicotine. *Neuropsychopharmacology: Official Publication of the American College of Neuropsychopharmacology, 35*, 665–673.

Cachope, R., Mateo, Y., Mathur, B. N., Irving, J., Wang, H.-L., Morales, M., ... Cheer, J. F. (2012). Selective activation of cholinergic interneurons enhances accumbal phasic dopamine release: Setting the tone for reward processing. *Cell Reports, 2*, 33–41.

Caille, S., Guillem, K., Cador, M., Manzoni, O., & Georges, F. (2009). Voluntary nicotine consumption triggers in vivo potentiation of cortical excitatory drives to midbrain dopaminergic neurons. *Journal of Neuroscience, 29*, 10410–10415.

Campbell, A. D., Kohl, R. R., & McBride, W. J. (1996). Serotonin-3 receptor and ethanol-stimulated somatodendritic dopamine release. *Alcohol, 13*, 569–574.

Clark, A., Lindgren, S., Brooks, S. P., Watson, W. P., & Little, H. J. (2001). Chronic infusion of nicotine can increase operant self-administration of alcohol. *Neuropharmacology, 41*, 108–117.

Clark, A., & Little, H. J. (2004). Interaction between low concentrations of ethanol and nicotine on firing rate of ventral tegmental dopamine neurons. *Drug and Alcohol Dependence, 75*, 199–206.

Connor, J. P., Young, R. M., Lawford, B. R., Saunders, J. B., Ritchie, T. L., & Noble, E. P. (2007). Heavy nicotine and alcohol use in alcohol dependence is associated with D2 dopamine receptor (DRD2) polymorphism. *Addiction Behavior, 32*, 310–319.

Corrigall, W. A., & Coen, K. M. (1989). Nicotine maintains robust self-administration in rats on a limited-access schedule. *Psychopharmacology, 99*, 473–478.

Corrigall, W. A., Coen, K. M., & Adamson, K. L. (1994). Self-administered nicotine activates the mesolimbic dopamine system through the ventral tegmental area. *Brain Research, 653*, 278–284.

Daeppen, J. B., Smith, T. L., Danko, G. P., Gordon, L., Landi, N. A., Nurnberger, J. I., Jr., ... Schuckit, M. A. (2000). Clinical correlates of cigarette smoking and nicotine dependence in alcohol-dependent men and women. The Collaborative Study Group on the Genetics of Alcoholism. *Alcohol Alcoholism, 35*, 171–175.

David, V., Besson, M., Changeux, J. P., Granon, S., & Cazala, P. (2006). Reinforcing effects of nicotine microinjections into the ventral tegmental area of mice: Dependence on cholinergic nicotinic and dopaminergic D1 receptors. *Neuropharmacology, 50*, 1030–1040.

Deehan, G. A., Jr., Hauser, S. R., Waeiss, R. A., Knight, C. P., Toalston, J. E., Truitt, W. A., ... Rodd, Z. A. (2015). Co-administration of ethanol and nicotine: The enduring alterations in the rewarding properties of nicotine and glutamate activity within the mesocorticolimbic system of female alcohol-preferring (P) rats. *Psychopharmacology, 232*, 4293–4302.

Ding, Z. M., Katner, S. N., Rodd, Z. A., Truitt, W., Hauser, S. R., Deehan, G. A., Jr., ... McBride, W. J. (2012). Repeated exposure of the posterior ventral tegmental area to nicotine increases the sensitivity of local dopamine neurons to the stimulating effects of ethanol. *Alcohol, 46*, 217–223.

Ding, Z. M., Oster, S. M., Hall, S. R., Engleman, E. A., Hauser, S. R., McBride, W. J., & Rodd, Z. A. (2011). The stimulating effects of ethanol on ventral tegmental area dopamine neurons projecting to the ventral pallidum and medial prefrontal cortex in female Wistar rats: Regional difference and involvement of serotonin-3 receptors. *Psychopharmacology, 216*, 245–255.

Ding, Z. M., Rodd, Z. A., Engleman, E. A., Bailey, J. A., Lahiri, D. K., & McBride, W. J. (2013). Alcohol drinking and deprivation alter basal extracellular glutamate concentrations and clearance in the mesolimbic system of alcohol-preferring (P) rats. *Addiction Biology, 18*, 297–306.

Ding, Z. M., Rodd, Z. A., Engleman, E. A., & McBride, W. J. (2009). Sensitization of ventral tegmental area dopamine neurons to the stimulating effects of ethanol. *Alcoholism, Clinical & Experimental Research, 33*, 1571–1581.

Dougherty, J. J., & Nichols, R. A. (2009). Cross-regulation between colocalized nicotinic acetylcholine and 5-HT_3 serotonin receptors on presynaptic nerve terminals. *Acta Pharmacologia Sinica, 30*, 788–794.

Doyon, W. M., Dong, Y., Ostroumov, A., Thomas, A. M., Zhang, T. A., & Dani, J. A. (2013). Nicotine decreases ethanol-induced dopamine signaling and increases self-administration via stress hormones. *Neuron, 79*, 530–540.

Doyon, W. M., Thomas, A. M., Ostroumov, A., Dong, Y., & Dani, J. A. (2013). Potential substrates for nicotine and alcohol interactions: A focus on the mesocorticolimbic dopamine system. *Biochemical Pharmacology, 86*, 1181–1193.

Drisdel, R. C., Sharp, D., Henderson, T., Hales, T. G., & Green, W. N. (2008). High affinity binding of epibatidine to serotonin type 3 receptors. *Journal Biological Chemistry, 283*, 9659–9665.

Durazzo, T. C., Cardenas, V. A., Studholme, C., Weiner, M. W., & Meyerhoff, D. J. (2007). Non-treatment-seeking heavy drinkers: Effects of chronic cigarette smoking on brain structure. *Drug and Alcohol Dependence, 87*, 76–82.

Durazzo, T. C., Mattsson, N., Weiner, M. W., Korecka, M., Trojanowski, J. Q., & Shaw, L. M. (2014). History of cigarette smoking in cognitively-normal elders is associated with elevated cerebrospinal fluid biomarkers of oxidative stress. *Drug and Alcohol Dependence, 142*, 262–268.

Durazzo, T. C., Pennington, D. L., Schmidt, T. P., Mon, A. C., & Meyerhoff, D. J. (2013). Neurocognition in 1-month-abstinent treatment-seeking alcohol-dependent individuals: Interactive effects of age and chronic cigarette smoking. *Alcoholism, Clinical & Experimental Research, 37*, 1794–1803.

Engle, S. E., McIntosh, J. M., & Drenan, R. M. (2015). Nicotinic and ethanol cooperate to enhance ventral tegmental area AMPA receptor function via α6-containing nicotinic receptors. *Neuropharmacology, 91*, 13–22.

Engleman, E. A., Ding, Z.-M., Oster, S. M., Toalston, J. E., Bell, R. L., Murphy, J. M., ... Rodd, Z. A. (2009). Ethanol is self-administered into the nucleus accumbens shell, but not the core: Evidence of genetic sensitivity. *Alcoholism, Clinical & Experimental Research, 12*, 2162–2171.

Ericson, M., Blomqvist, O., Engel, J. A., & Soderpalm, B. (1998). Voluntary ethanol intake in the rat and the associated accumbal dopamine overflow are blocked by ventral tegmental mecamylamine. *European Journal of Pharmacology, 358*, 189–196.

Exley, R., Clements, M. A., Hartung, H., McIntosh, J. M., & Cragg, S. J. (2008). α6-containing nicotinic acetylcholine receptors dominate the nicotine control of dopamine neurotransmission in nucleus accumbens. *Neuropsychopharmacology: Official Publication of the American College of Neuropsychopharmacology, 33*, 2158–2166.

Exley, R., Maubourguet, N., David, V., Eddine, R., Evard, A., Pons, S., ... Faure, P. (2011). Distinct contributions of nicotinic acetylcholine receptor subunit alpha4 and subunit alpha6 to the reinforcing effects of nicotine. *Proceedings of the National Academy of Sciences of the United States of America, 108*, 7577–7582.

Falk, D. E., Yi, H. Y., & Hiller-Sturmhofel, S. (2006). An epidemiologic analysis of co-occurring alcohol and tobacco use and disorders: Findings from the National Epidemiologic Survey on Alcohol and Related Conditions. *Alcohol Research Health, 29*, 162–171.

Feduccia, A. A., Chatterjee, S., & Bartlett, S. E. (2012). Neuronal nicotinic acetylcholine receptors: Neuroplastic changes underlying alcohol and nicotine addictions. *Frontier Molecular Neuroscience, 5*, 1–18.

George, O., Sanders, C., Freiling, J., Grigoryan, E., Vu, S., Allen, C. D., ... Koob, G. F. (2012). Recruitment of medial prefrontal cortex neurons during alcohol withdrawal predicts cognitive impairment and excessive alcohol drinking. *Proceedings of the National Academy of Sciences of the United States of America, 109*, 18156–18161.

Gioanni, Y., Rougeot, C., Clarke, P. B., Lepouse, C., Thierry, A. M., & Vidal, C. (1999). Nicotinic receptors in the rat prefrontal cortex: Increase in glutamate release and facilitation of mediodorsal thalamocortical transmission. *European Journal Neuroscience, 11*, 18–30.

Gonzales, R. A., & Weiss, F. (1998). Suppression of ethanol-reinforced behavior by naltrexone is associated with attenuation of the ethanol-induced increase in dialysate dopamine levels in the nucleus accumbens. *Journal of Neuroscience, 18*, 10663–10671.

Gulliver, S. B., Rohsenow, D. J., Colby, S. M., Dey, A. N., Abrams, D. B., Niaura, R. S., & Monti, P. M. (1995). Interrelationship of smoking and alcohol dependence, use and urges to use. *Journal of Studies on Alcohol, 56*, 202–206.

Gurley, D. A., & Lanthorn, T. H. (1998). Nicotinic agonists competitively antagonize serotonin at mouse 5-HT_3 receptors expressed in *Xenopus* oocytes. *Neuroscience Letters, 247*, 107–110.

Hatsukami, D. K., Pickens, R. W., Svikis, D. S., & Hughes, J. R. (1988). Smoking topography and nicotine blood levels. *Addiction Behavior, 13*, 91–95.

Hauser, S. R., Bracken, A. L., Deehan, G. A., Jr., Toalston, J. E., Ding, Z.-M., Truitt, W. A., ... Rodd, Z. A. (2014). Selective breeding for high alcohol preference increases the sensitivity of the posterior VTA to the reinforcing effects of nicotine. *Addiction Biology, 19*, 800–811.

Hauser, S. R., Deehan, G. A., Jr., Toalston, J. E., Bell, R. L., McBride, W. J., & Rodd, Z. A. (2014). Enhanced alcohol-seeking behavior by nicotine in the posterior ventral tegmental area of female alcohol-preferring (P) rats: Modulation by serotonin-3 and nicotinic cholinergic receptors. *Psychopharmacology, 231*, 3745–3755.

Hauser, S. R., Getachew, B., Oster, S. M., Dhaher, R., Ding, Z.-M., Bell, R. L., ... Rodd, Z. A. (2011). Nicotine modulates alcohol-seeking and relapse by alcohol-preferring (P) rats in a time dependent manner. *Alcoholism, Clinical & Experimental Research, 35*, 43–54.

Hauser, S. R., Katner, S. N., Deehan, G. A., Jr., Ding, Z.-M., Toalston, J. E., Scott, B. J., ... Rodd, Z. A. (2012). Development of an oral operant nicotine/ethanol co-use model in alcohol-preferring (P) rats. *Alcoholism, Clinical & Experimental Research, 36*, 1963–1972.

Hendrickson, L. M., Guildford, M. J., & Tapper, A. R. (2013). Neuronal nicotinic acetylcholine receptors: Common molecular substrates of nicotine and alcohol dependence. *Frontier Psychiatry, 4*, 1–16.

Hetzler, B. E., & Martin, E. I. (2006). Nicotine–ethanol interactions in flash-evoked potentials and behavior of Long-Evans rats. *Pharmacology Biochemistry & Behavior, 83*, 76–89.

Hodge, C. W., Haraguchi, M., Chappelle, A. M., & Samson, H. H. (1996). Effects of ventral tegmental microinjections of the $GABA_A$ agonist muscimol on self-administration of ethanol and sucrose. *Pharmacology Biochemistry & Behavior, 53*, 971–977.

Hodge, C. W., Haraguchi, M., Erickson, H., & Samson, H. H. (1993). Ventral tegmental microinjections of quinpirole decrease ethanol and sucrose-reinforced responding. *Alcoholism, Clinical & Experimental Research, 17*, 370–375.

Holmes, A., Firzgerald, P. J., MacPherson, K. P., DeBrouse, L., Colacicco, G., Flynn, S. M., ... Camp, M. (2012). Chronic alcohol remodels prefrontal neurons and disrupts NMDAR-mediated fear extinction encoding. *Nature Neuroscience, 15*, 1359–1361.

Hughes, J. R. (1996). Treating smokers with current or past alcohol dependence. *American Journal of Health & Behavior, 20*, 286–290.

Hurley, L. L., Taylor, R. E., & Tizabi, Y. (2012). Positive and negative effects of alcohol and nicotine and their interactions: A mechanistic review. *Neurotoxicology Research, 21*, 57–69.

Hurt, R. D., Eberman, K. M., Croghan, I. T., Offord, K. P., Davis, L. J., Jr., Morse, R. M., ... Bruce, B. K. (1994). Nicotine dependence treatment during inpatient treatment for other addictions: A prospective intervention trial. *Alcoholism, Clinical & Experimental Research, 18*, 867–872.

Ikemoto, S., Glazier, B. S., Murphy, J. M., & McBride, W. J. (1997). Role of dopamine D-1 and D-2 receptors in the nucleus accumbens in mediating reward. *Journal Neuroscience, 17*, 8580–8585.

Ikemoto, S., Qin, M., & Liu, Z.-H. (2006). Primary reinforcing effects of nicotine are triggered from multiple regions both inside and outside the ventral tegmental area. *Journal Neuroscience, 26*, 723–730.

Jackson, M. B., & Yakel, J. L. (1995). The 5-HT_3 receptor channel. *Annual Reviews of Physiology, 57*, 447–468.

John, U., Hill, A., Rumpf, H. J., Hapke, U., & Meyer, C. (2003). Alcohol high risk drinking, abuse, and dependence among tobacco smoking medical care patients and the general population. *Drug and Alcohol Dependence, 64*, 233–241.

John, U., Meyer, C., Rumpf, H. J., Schumann, A., Thyrian, J. R., & Hapke, U. (2003). Strength of the relationship between tobacco smoking, nicotine dependence and the severity of alcohol dependence syndrome criteria in a population-based sample. *Alcohol Alcoholism, 38*, 606–612.

Kalivas, P. W., & Volkow, N. D. (2005). The neural basis of addiction: A pathology of motivation and choice. *American Journal of Psychiatry, 162*, 1403–1413.

Kodas, E., Cohen, C., Louis, C., & Griebel, G. (2007). Cortico-limbic circuitry for conditioned nicotine-seeking behavior in rats involves endocannabinoid signaling. *Psychopharmacology, 194*, 161–171.

Koob, G. F., & Volkow, N. D. (2010). Neurocircuitry of addiction. *Neuropsychopharmacology: Official Publication of the American College of Neuropsychopharmacology, 35*, 217–238.

Lajtha, A., & Sershen, H. (2010). Nicotine: Alcohol reward interactions. *Neurochemical Research, 35*, 1248–1258.

Larsson, A., Jerlhag, E., Svensson, L., Soderpalm, B., & Engel, J. A. (2004). Is an alpha-conotoxin MII-sensitive mechanism involved in the neurochemical, stimulatory, and rewarding effects of ethanol? *Alcohol, 34*, 239–250.

Le, A. D., Corrigall, W. A., Harding, J. W., Juzytsch, W., & Li, T.-K. (2000). Involvement of nicotinic receptors in alcohol self-administration. *Alcoholism, Clinical and Experimental Research, 24*, 155–163.

Le, A. D., Funk, D., Lo, S., & Coen, K. (2014). Operant self-administration of alcohol and nicotine in a preclinical model of co-abuse. *Psychopharmacology, 231*, 4019–4029.

Le, A. D., Li, Z., Funk, D., Shram, M., Li, T.-K., & Shaham, Y. (2006). Increased vulnerability to nicotine self-administration and relapse in alcohol-naïve offspring of rats selectively bred for high alcohol intake. *The Journal of Neuroscience, 26*, 1872–1879.

Le, A. D., Lo, S., Harding, S., Juzytsch, W., Marinelli, P. W., & Funk, D. (2010). Co-administration of intravenous nicotine and oral alcohol in rats. *Psychopharmacology, 208*, 475–486.

Le, A. D., Wang, A., Harding, S., Juzytsch, W., & Shaham, Y. (2003). Nicotine increases alcohol self-administration and reinstates alcohol seeking in rats. *Psychopharmacology, 168*, 216–226.

Leao, R. M., Cruz, F. C., Vendruscolo, L. F., De Guglielmo, G., Logrip, M. L., Planeta, C. S., ... George, O. (2015). Chronic nicotine activates stress/reward-related brain regions and facilitates the transition to compulsive alcohol drinking. *Journal Neuroscience, 35*, 6241–6253.

Lecca, D., Cacciapaglia, F., Valentini, V., Gronli, J., Spiga, S., & Di Chiara, G. (2006). Preferential increase of extracellular dopamine in the rat nucleus accumbens shell as compared to that in the core during acquisition and maintenance of intravenous nicotine self-administration. *Psychopharmacology, 184*, 435–446.

Liu, L., Hendrickson, L. M., Guildford, M. J., Zhao-Shea, R., Gardner, P. D., & Tapper, A. R. (2013). Nicotinic acetylcholine receptors containing the α4 subunit modulate alcohol reward. *Biological Psychiatry, 73*, 738–746.

Liu, L., Zhao-Shea, R., McIntosh, J. M., & Tapper, A. R. (2015). Nicotinic acetylcholine receptors containing the α6 subunit contribute to ethanol activation of ventral tegmental area dopaminergic neurons. *Biochemical Pharmacology, 86*, 1194–1200.

Lof, E., Olausson, P., deBejczy, A., Stomberg, R., McIntosh, J. M., Taylor, J. R., & Soderpalm, B. (2007). Nicotinic acetylcholine receptors in the ventral tegmental area mediate the dopamine activating and reinforcing properties of ethanol cues. *Psychopharmacology, 195*, 333–343.

Lopez-Moreno, J. A., Trigo-Diaz, J. M., Rodriguez de Fonseca, F., Gonzalez Cuevas, G., Gomez se Heras, R., Crespo Galan, I., & Navarro, M. (2004). Nicotine in alcohol deprivation increases alcohol operant self-administration during reinstatement. *Neuropharmacology, 47*, 1036–1044.

Lovinger, D. M., & White, G. (1991). Ethanol potentiation of 5-hydroxtryptamine-3 receptor-mediated ion current in neuroblastoma cells and isolated adult mammalian neurons. *Molecular Pharmacology, 40*, 263–270.

Marszalec, W., Aistrup, G. L., & Narahashi, T. (1999). Ethanol-nicotine interactions at alpha-bungarotoxin-insensitive nicotinic acetylcholine receptors in rat cortical neurons. *Alcoholism, Clinical & Experimental Research, 23*, 439–445.

Mascia, M. P., Trudell, J. R., & Harris, R. A. (2000). Specific binding sites for alcohols and anesthetics on ligand-gated ion channels. *Proceedings of the National Academy of Sciences of the United States of America, 97*, 9305–9310.

Melendez, R. I., Rodd-Henricks, Z. A., Engleman, E. A., Li, T.-K., McBride, W. J., & Murphy, J. M. (2002). Microdialysis of dopamine in the nucleus accumbens of alcohol-preferring (P) rats during anticipation and operant self-administration of ethanol. *Alcoholism, Clinical & Experimental Research, 26*, 318–325.

Meredith, G. E., Blank, B., & Groenewegen, H. J. (1989). The distribution and compartmental organization of the cholinergic neurons in nucleus accumbens of the rat. *Neuroscience, 31*, 327–345.

Mesulam, M. M., Mufson, E. J., Wainer, B. H., & Levey, A. I. (1983). Central cholinergic pathways in the rat: An overview based on an alternative nomenclature (Ch1–Ch6). *Neuroscience, 10*, 1185–1201.

Molles, B. E., Maskos, U., Pons, S., Besson, M., Guiard, P., Guilloux, J. P., ... Changeux, J. P. (2006). Targeted in vivo expression of nicotinic acetylcholine receptors in mouse brain using lentiviral expression vectors. *Journal Molecular Neuroscience, 30*, 105–106.

Morales, M., Battenberg, E., & Bloom, F. E. (1998). Distribution of neurons expressing immunoreactivity for the 5-HT_3 receptor subtype in the rat brain and spinal cord. *Journal of Comparative Neurology, 385*, 385–401.

Muir, J. L., Page, K. J., Sirinathsinghji, D. J. S., Robbins, T. W., & Everitt, B. J. (1993). Excitotoxic lesions of basal forebrain cholinergic neurons: Effects on learning, memory and attention. *Behavioural Brain Research, 57*, 123–131.

Nayak, S. V., Ronde, P., Spier, A. D., Lummis, S. C. R., & Nichols, R. A. (2000). Nicotinic receptors co-localize with 5-HT_3 serotonin receptors on striatal nerve terminals. *Neuropharmacology, 39*, 2681–2690.

Olausson, P., Ericson, M., Lof, E., & Engel, J. A. (2001). Nicotine-induced behavioral disinhibition and ethanol preference correlate after repeated nicotine treatment. *European Journal of Pharmacology, 417*, 117–123.

Olive, M. F., Cleva, R. M., Kalivas, P. W., & Malcolm, R. J. (2012). Glutamatergic medications for the treatment of drug and behavioral addictions. *Pharmacology Biochemistry & Behavior, 100*, 801–810.

Parikh, V., Ji, J., Decker, M. W., & Sarter, M. (2010). Prefrontal beta2 subunit-containing and alpha7 nicotinic acetylcholine receptors differentially control glutamatergic and cholinergic signaling. *Journal Neuroscience, 30*, 3518–3530.

Peters, J. A., Carland, J. E., Cooper, M. A., Livesey, M. R., Deeb, T. Z., Hales, T. G., & Lambert, J. J. (2006). Novel structural determinants of single-channel conductance in nicotinic acetylcholine and 5-hydroxytryptamine type-3 receptors. *Biochemical Society Transactions, 34*, 882–886.

Peters, J., Kalivas, P. W., & Quirk, G. I. (2009). Extinction circuits for fear and addiction overlap in prefrontal cortex. *Learn & Memory, 16*, 279–288.

Peters, J., Lalumiere, R. T., & Kalivas, P. W. (2008). Infralimbic prefrontal cortex is responsible for inhibiting cocaine seeking in extinguished rats. *Journal of Neuroscience, 28*, 6046–6053.

Pistillo, F., Clementi, F., Zoli, M., & Gotti, C. (2015). Nicotinic, glutamatergic and dopaminergic synaptic transmission and plasticity in the mesocorticolimbic system: Focus on nicotine effects. *Progress in Neurobiology, 124*, 1–27.

Potthoff, A. D., Ellison, G., & Nelson, L. (1983). Ethanol intake increases during continuous administration of amphetamine and nicotine, but not several other drugs. *Pharmacology Biochemistry & Behavior, 18*, 489–493.

Quik, M., Perez, X. A., & Grady, S. R. (2011). Role of α6 nicotinic receptors in CNS dopaminergic function: Relevance to addiction and neurological disorders. *Biochemical Pharmacology, 82*, 873–882.

Rassnick, S., Pulvirenti, L., & Koob, G. F. (1992). Oral ethanol self-administration in rats is reduced by the administration of dopamine and glutamate receptor antagonists into the nucleus accumbens. *Psychopharmacology, 109*, 92–98.

Ricci, L. A., Stellar, J. R., & Todtenkopf, M. S. (2004). Subregion-specific down-regulation of 5-HT_3 immunoreactivity in the nucleus accumbens shell during the induction of cocaine sensitization. *Pharmacology Biochemistry & Behavior, 77*, 415–422.

Rinker, J. A., Busse, G. D., Roma, P. G., Chen, S. A., Barr, C. S., & Riley, A. L. (2008). The effects of nicotine on ethanol-induced conditioned taste aversions in Long-Evans rats. *Psychopharmacology, 197*, 409–419.

Rodd, Z. A., Bell, R. L., Oster, S. M., Toalston, J. E., Pommer, T. J., McBride, W. J., & Murphy, J. M. (2010). Serotonin-3 receptors in the posterior ventral tegmental area regulate ethanol self-administration of alcohol-preferring (P) rats. *Alcohol, 44*, 245–255.

Rodd, Z. A., Melendez, R. I., Bell, R. L., Kuc, K. A., Zhang, Y., Murphy, J. M., & McBride, W. J. (2004). Intracranial self-administration of ethanol within the ventral tegmental area of male Wistar rats: Evidence for involvement of dopamine neurons. *Journal of Neuroscience, 24*, 1050–1057.

Rodd-Henricks, Z. A., McKinzie, D. L., Crile, R. S., Murphy, J. M., & McBride, W. J. (2000). Regional heterogeneity for the intracranial self-administration of ethanol within the ventral tegmental area of female Wistar rats. *Psychopharmacology, 149*, 217–224.

Rodd-Henricks, Z. A., McKinzie, D. L., Melendez, R. I., Berry, N., Murphy, J. M., & McBride, W. J. (2003). Effects of serotonin-3 receptor antagonists on the intracranial self-administration of ethanol within the posterior VTA of Wistar rats. *Psychopharmacology, 165*, 252–259.

Samson, H. H., Hodge, C. W., Tolliver, G. A., & Haraguchi, M. (1993). Effect of dopamine agonists and antagonists on ethanol-reinforced behavior: The involvement of the nucleus accumbens. *Brain Research Bulletin, 30*, 133–141.

Schlaepfer, I. R., Hoft, N. R., Collins, A. C., Corley, R. P., Hewitt, J. K., Hopfer, C. J., ... Ehringer, M. A. (2008). The CHRNA5/A3/B4 gene cluster variability as an important determinant of early alcohol and tobacco initiation in young adults. *Biological Psychiatry, 63*, 1039–1046.

Seif, T., Chang, S.-J., Simms, J. A., Gibb, S. L., Dadgar, J., Chen, B. T., ... Hopf, F. W. (2013). Cortical activation of accumbens hyperpolarization-active NMDARs mediates aversion-resistant alcohol intake. *Nature Neuroscience, 16*, 1094–1100.

Sobell, M. B., Sobell, L. C., & Kozolowski, L. T. (1995). Dual recoveries from alcohol and smoking problems. In J. B. Fertig, & J. S. Allen (Eds.), *Alcohol and tobacco: From basic science to clinical practice*. Bethesda, MD: NIAAA.

Sung, K. W., Engel, S. R., Allan, A. M., & Lovinger, D. M. (2000). 5-HT_3 receptor function and potentiation by alcohols in frontal cortex neurons from transgenic mice over expressing the receptor. *Neuropharmacology, 39*, 2346–2351.

Threlfell, S., Lalic, T., Platt, N. J., Jennings, K. A., Deisseroth, K., & Cragg, S. J. (2012). Striatal dopamine release is triggered by synchronizes activity in cholinergic interneurons. *Neuron, 75*, 58–64.

Tizabi, Y., Copeland, R. L., Louis, V. A., & Taylor, R. E. (2002). Effects of combined systemic alcohol and central nicotine administration into ventral tegmental area on dopamine release in the nucleus accumbens. *Alcoholism, Clinical & Experimental Research, 26*, 394–399.

Trantham-Davidson, H., Burnett, E. J., Gass, J. T., Lopez, M. F., Mulholland, P. J., Centanni, S. W., ... Chandler, L. J. (2014). Chronic alcohol disrupts dopamine receptor activity and the cognitive function of the medial prefrontal cortex. *Journal of Neuroscience, 34*, 3706–3718.

True, W. R., Xian, H., Scherrer, J. F., Madden, P. A., Bucholz, K. K., Heath, A. C., ... Tsuang, M. (1999). Common genetic vulnerability for nicotine and alcohol dependence in men. *Archives General Psychiatry, 56*, 655–661.

Truitt, W. A., Hauser, S. R., Deehan, G. A., Jr., Toalston, J. E., Wilden, J. A., Bell, R. L., ... Rodd, Z. A. (2015). Ethanol and nicotine interaction within the posterior ventral tegmental area in male and female alcohol-preferring rats: Evidence of synergy and differential gene activation in the nucleus accumbens shell. *Psychopharmacology, 232*, 639–649.

Vertes, R. P. (2004). Differential projections of the infralimbic and prelimbic cortex in the rat. *Synapse, 51*, 32–58.

Von Engelhardt, J., Eliava, M., Meyer, A. H., Rozov, A., & Monyer, H. (2007). Functional characterization of intrinsic cholinergic interneurons in the cortex. *Journal of Neuroscience, 27*, 5633–5642.

Wang, F., Chen, H., & Sharp, B. M. (2008). Neuroadaptive changes in the mesocortical glutamatergic system during chronic nicotine self-administration and after extinction in rats. *Journal of Neurochemistry, 106*, 943–956.

Wang, F., Chen, H., Steketee, J. D., & Sharp, B. M. (2007). Upregulation of ionotropic glutamate receptor subunits within specific mesocorticolimbic regions during chronic nicotine self-administration. *Neuropsychopharmacology: Official Publication of the American College of Neuropsychopharmacology, 32*, 103–109.

Weinberger, A. H., Platt, J., Jiang, B., & Goodwin, R. D. (2015). Cigarette smoking and risk of alcohol use relapse among adults in recovery from alcohol use disorders. *Alcoholism, Clinical & Experimental Research, 39*, 1989–1996.

Weiss, F., Lorang, M. T., Bloom, F. E., & Koob, G. F. (1993). Oral alcohol self-administration stimulates dopamine release in the rat nucleus accumbens: Genetic and motivational determinants. *Journal Pharmacology & Experimental Therapeutics, 267*, 250–258.

Weiss, F., Parsons, L. H., Schulteis, G., Hyytia, P., Lorang, M. T., Bloom, F. E., & Koob, G. F. (1996). Ethanol self-administration restores withdrawal-associated deficiencies in accumbal dopamine and 5-hydroxytryptamine release in dependent rats. *Journal of Neuroscience, 16*, 3474–3485.

Xiao, Z., Zhu, F., Lu, Z., Pan, S., Liang, J., Zheng, J., & Liu, Z. (2013). Modulation of nicotinic acetylcholine receptors in native sympathetic neurons by ethanol. *International Journal of Physiology Pathophysiology & Pharmacology, 5*, 161–168.

Yu, D., Zhang, L., Eisele, J.-L., Bertrand, D., Changeux, J.-P., & Weight, F. F. (1996). Ethanol inhibition of nicotinic acetylcholine type α7 receptors involves the amino-terminal domain of the receptor. *Molecular Pharmacology, 50*, 1010–1016.

Zhao-Shea, R., Liu, L., Soll, L. G., Improgo, M. R., Meyers, E. E., McIntosh, J. M., ... Tapper, A. R. (2011). Nicotine-mediated activation of dopaminergic neurons in distinct regions of the ventral tegmental area. *Neuropsychopharmacology: Official Publication of the American College of Neuropsychopharmacology, 36*, 1021–1032.

Zuo, Y., Aistrup, G. L., Marszalec, W., Gillespie, A., Chavez-Noriega, L. E., Yeh, J. Z., & Narahashi, T. (2001). Dual action of n-alcohols on neuronal nicotinic acetylcholine receptors. *Molecular Pharmacology, 60*, 700–711.

Zuo, Y., Nagata, K., Yeh, J. Z., & Narahashi, T. (2004). Single-channel analyses of ethanol modulation of neuronal nicotinic acetylcholine receptors. *Alcoholism, Clinical & Experimental Research, 28*, 688–696.

CHAPTER

27

Role of Basal Forebrain in Nicotine Alcohol Co-abuse

R. Sharma, P. Sahota, M.M. Thakkar

HSTMV Hospital, University of Missouri, Columbia, MO, United States

INTRODUCTION

As early as 1860, Reverend George Trask wrote "Do you know of one drunkard that does not use tobacco?" (Trask, 1860). Thus, "smokers drink and drinkers smoke" is not only a popular expression, but also an unfortunate well-documented fact. Strong epidemiological, clinical, and laboratory evidence indicates that smoking is a major contributing factor for the development of alcoholism. According to the Substance Abuse and Mental Health Services Administration (SAMHSA, 2005), the prevalence of smoking is about three times higher in alcoholics than in general population. According to DSM-IV (Diagnostic and Statistical Manual of Mental Disorders, 4th edition), more than 8% of adults in United States are alcohol dependent, of whom, over 85% are addicted to nicotine, the primary component of tobacco and the primary cause of tobacco addiction. Why do alcohol and nicotine dependencies show such high comorbidity? Is it because nicotine and alcohol are readily and legally available? Or is it because smoking and drinking are socially accepted without any social stigma? Or is there a neurobiological basis?

It has been suggested that one major reason why people use drugs, such as alcohol, is for recreation and/or to enjoy euphoric/pleasurable sensations. However, accompanying negative "aversive" side-effects limit euphoric/pleasurable sensations and hence "spoil the fun" (Chatterjee & Bartlett, 2010; Koob, 2000; Robinson & Berridge, 2001). For example, alcohol consumption promotes sleepiness and causes attention deficits and cognitive impairments, which limits euphoric pleasures and recreational experiences. If a stimulant like nicotine is consumed (or smoked), along with alcohol, negative side effects are reduced while positive effects are enhanced resulting in longer drinking sessions and/or binge drinking. This alcohol misuse/abuse can have serious consequences including increased risk toward the development of alcohol-use disorders (Harrison, Desai, & McKee, 2008; Li, Volkow, Baler, & Egli, 2007; McKee & Weinberger, 2013).

Studies conducted in rodents also demonstrate an intimate relationship between alcohol and nicotine use. For example, rodents exposed to nicotine increase alcohol consumption; the effect is dose and treatment duration dependent (Hauser et al., 2012; Le, Wang, Harding, Juzytsch, & Shaham, 2003; Locker, Marks, Kamens, & Klein, 2015; Lopez-Moreno et al., 2004; Smith, Horan, Gaskin, & Amit, 1999). Systemic or central administration of nicotine receptor, partial agonist or antagonist, reduces alcohol consumption and/or preference (Bito-Onon, Simms, Chatterjee, Holgate, & Bartlett, 2011; Blomqvist, Ericson, Johnson, Engel, & Soderpalm, 1996; Steensland, Simms, Holgate, Richards, & Bartlett, 2007). Thus, although nicotine is believed to be one of the greatest risk factors for the development of alcoholism, little is known about underlying neurobiological mechanisms responsible for nicotine and alcohol co-abuse (Davis & de Fiebre, 2006).

It has been suggested that combined nicotine and alcohol exposure activates the mesolimbic dopamine reward system to enhance positive effects and/or "pleasurable sensations" (Clark & Little, 2004; Tizabi, Bai, Copeland, & Taylor, 2007; Tizabi, Copeland, Louis, & Taylor, 2002). However, it is yet unclear as to how and where nicotine acts to suppress/reduce the negative or sedative effects of alcohol and increase alcohol consumption. In order to understand the mechanism of how nicotine suppresses the sedative effects of alcohol, it is important to understand the anatomical substrates mediating sleepiness following alcohol consumption.

Addictive Substances and Neurological Disease
http://dx.doi.org/10.1016/B978-0-12-805373-7.00027-X

This chapter begins with a brief description of sleep–wakefulness and neuronal substrates mediating the sleep-promoting effects of alcohol followed by our research describing how and where nicotine acts to suppress alcohol-induced sleepiness. Next, it provides a brief description of the reward centers responsible for mediating the positive effects of alcohol followed by our recent attempts to elucidate the neuronal mediators of alcohol and nicotine co-use.

NEURONAL MEDIATORS OF SLEEP–WAKEFULNESS

Behavioral arousal or wakefulness is a manifestation of increased activity in the cortex also known as cortical desynchronization or cortical activation (Thakkar, 2011; Thakkar, Sharma, & Sahota, 2015). Cortical activation is controlled by an ensemble of multiple neuronal systems residing in several brain regions including the brain stem, hypothalamus, and basal forebrain (BF) (Brown, Basheer, McKenna, Strecker, & McCarley, 2012). Among these systems, the BF is pivotal because it represents the final node in the ventral extra-thalamic relay, integrating ascending and descending inputs from the brain stem and the forebrain to control cortical activation along with other brain functions including sleep, cognition, attention, and memory (Thakkar et al., 2015). The BF is a heterogeneous collection of structures located close to the medial and ventral surfaces of the cerebral hemispheres (Woolf, 1991). The nuclei that make up the BF region include the medial septum/vertical limb of the diagonal band, horizontal limb of the diagonal band, and substantia innominata region (Brown et al., 2012). Although several neuronal phenotypes are present in the BF, the cholinergic, GABAergic, and the glutamatergic neurons are critical for promoting cortical activation and behavioral arousal (Gritti, Mainville, Mancia, & Jones, 1997; Manns, Alonso, & Jones, 2003; Zaborszky, van den Pol, & Gyengesi, 2012; Zant et al., 2016).

Sleep is broadly divided into two distinct *states*: nonrapid eye movement (NREM) sleep and rapid eye movement (REM) sleep. While NREM sleep is characterized by the presence of high-amplitude, low-frequency EEG and low muscle activity, REM sleep is characterized by low-amplitude, high-frequency EEG and complete absence of muscle activity.

Transition from wakefulness to sleep requires the inhibition of the BF. A major source of inhibition involves Process S or the homeostasis-mediated regulation of NREM sleep (Thakkar et al., 2015). The physiological regulation of mammalian sleep is controlled by Process S (homeostasis process), Process C (circadian process), and their interaction (Borbely, 1982). Process C, controlled by the suprachiasmatic nucleus, provides a circadian alerting signal and is independent of sleep and wakefulness. Process S manifests as sleep pressure that increases during wakefulness and declines during sleep (Brown et al., 2012; Edgar, Dement, & Fuller, 1993). Strong and consistent evidence exists to suggest that purine nucleotide adenosine (AD) and the BF are key mediators of Process S (Brown et al., 2012; Porkka-Heiskanen et al., 1997; Thakkar, Delgiacco, Strecker, & McCarley, 2003; Thakkar, Winston, & McCarley, 2003).

AD is a by-product of energy (adenosine triphosphate, ATP) metabolism and thus provides a link between energy metabolism and neuronal activity. During wakefulness, energy (ATP) usage is high in wake-promoting systems, especially in the BF, due to increased neuronal firing, synaptic activity, and synaptic potentiation. This wakefulness-induced increase in energy usage is reflected in increased accumulation of extracellular AD that corresponds to increased accumulation of sleep pressure. The longer the period of wakefulness, the greater the accumulation of AD and sleep pressure, the longer it takes for sleep pressure to dissipate during sleep (Porkka-Heiskanen, 2013).

Experimentally, sleep pressure is examined by sleep deprivation followed by recovery sleep. During sleep deprivation, extracellular AD increases, most markedly, in the wake-promoting BF region where it acts via AD A_1 receptors (A1R) to inhibit wake-promoting neurons and facilitate the transition from wakefulness to NREM sleep (Thakkar, Delgiacco, et al., 2003; Thakkar, Winston, et al., 2003). Interestingly, AD is also implicated as a key mediator of neuronal responses to alcohol (Dunwiddie & Masino, 2001; Hack & Christie, 2003; Newton & Messing, 2006). Does AD mediate alcohol-induced sleepiness?

NEURONAL MEDIATORS OF ALCOHOL-INDUCED SLEEPINESS

In vitro studies have shown that acute alcohol exposure inhibits an NBTI-sensitive equilibrative nucleoside transporter 1 (ENT1) resulting in extracellular accumulation of AD. In contrast, chronic alcohol exposure causes down-regulation of ENT1 expression (Krauss, Ghirnikar, Diamond, & Gordon, 1993; Nagy, Diamond, Casso, Franklin, & Gordon, 1990). The ENT1-null mice show decreased AD tone and increased alcohol consumption coupled with reduced hypnotic (loss of righting reflex) and ataxic responses to alcohol. While treatment with an A1R agonist reduces alcohol consumption in ENT1-null mice, $A_{2A}R$ agonist is ineffective (Choi et al., 2004). Finally, AD levels are significantly increased in fetal cortex following maternal alcohol consumption (Watson et al., 1999). Furthermore, there is strong and convincing evidence demonstrating the importance of AD and its A1R in mediating the effects of alcohol on ataxia and other behaviors, including anxiety, tremors, and seizures, during alcohol withdrawal (Batista, Prediger, Morato, & Takahashi, 2005; Concas et al., 1996; Dar, 1993, 1997, 2001; Kaplan, Bharmal, Leite-Morris, & Adams, 1999; Phan, Gray, & Nyce, 1997; Prediger, da Silva, Batista, Bittencourt, & Takahashi, 2006; Saeed, 2006).

Several clinical studies also support the role of AD in alcohol-induced behaviors including sleepiness. Caffeine, a nonspecific AD receptor antagonist, offsets alcohol's debilitating and performance-impairment effects while preventing sleepiness and improving alertness (Drake, Roehrs, Turner, Scofield, & Roth, 2003; Franks, Hagedorn, Hensley, Hensley, & Starmer, 1975; Hasenfratz, Bunge, Dal, & Battig, 1993; Kerr, Sherwood, & Hindmarch, 1991; Liguori & Robinson, 2001).

Thus, strong clinical and preclinical studies suggest that AD mediates several neuronal/behavioral effects of alcohol including alcohol consumption. Since AD is the homeostatic regulator of sleep, it is very likely that AD may be the mediator of sleepiness following alcohol consumption. Indeed, our laboratory was the first to demonstrate that local infusion of alcohol, dose dependently, increases extracellular AD in the BF in freely behaving rats (Sharma, Engemann, Sahota, & Thakkar, 2010b). Furthermore, systemic alcohol administration, at the onset of wake period, inhibits and/or reduces the activation of wake-promoting neurons in the BF resulting in the suppression of cortical desynchronization and promotion of NREM sleep. However, the NREM sleep-promoting effect was attenuated if A1R was blocked in the BF (Thakkar, Engemann, Sharma, & Sahota, 2010). These studies suggest that alcohol via AD inhibits the BF wake-promoting neurons to promote NREM sleep and suppress wakefulness. Does nicotine act via the BF to suppress the sleep-promoting effects of alcohol?

We first provide a brief description of nicotine acetylcholine receptors (nAChRs) followed by the nAChR subtypes implicated in alcohol and nicotine co-use. Next, we present our finding describing the role of nicotine, via its action on the BF, in attenuating sleepiness following alcohol consumption.

NICOTINE ACTS VIA BF TO SUPPRESS NEGATIVE EFFECTS OF ALCOHOL

Nicotine is a ligand at nAChRs with agonistic properties. nAChRs are ligand-gated ion channels consisting of unique combinations from a family of at least 17 ($\alpha1-\alpha10$, $\beta1-\beta4$, γ, δ, and ε) similar, but distinct, subunits. While most nAChRs exist as hetero-pentamers containing two (or more) different kinds of subunits (e.g., $\alpha4\beta2$ nAChRs), nAChRs containing $\alpha7$ subunit exists as homo-pentamers (Wu & Lukas, 2011). The heteromeric $\alpha_4\beta_2$ and the homomeric $\alpha7$ receptors are among the most abundant nAChRs in the brain (Millar & Gotti, 2009; Wu & Lukas, 2011).

There is strong evidence implicating $\alpha_4\beta_2$ and $\alpha7$ nAChR subtypes in nicotine and alcohol addiction (Tapper et al., 2004; Tuesta, Fowler, & Kenny, 2011; Wu & Lukas, 2011). For example, mice with constitutive deletion of $\alpha7$ nAChRs display reduced alcohol consumption. Mecamylamine, a central nicotinic receptor antagonist, reduces alcohol consumption in rats, whereas in normal healthy humans, mecamylamine attenuates positive/euphoric effects of alcohol along with a reduction in self-reported desire to consume additional alcohol (Blomqvist et al., 1996; Chi & de Wit, 2003; Young, Mahler, Chi, & de Wit, 2005). The FDA-approved drug for smoking cessation, varenicline, is a full agonist of $\alpha7$ and a partial agonist of $\alpha4\beta2$ nAChR subtypes. Varenicline significantly reduces alcohol cravings and consumption in human and animals by its action on $\alpha4\beta2$ nAChR subtype (Chatterjee & Bartlett, 2010; Hendrickson, Zhao-Shea, Pang, Gardner, & Tapper, 2010; McKee et al., 2009; Steensland et al., 2007).

Interestingly, the two major subtypes implicated in nicotine and alcohol addiction, $\alpha4\beta2$ and $\alpha7$, are predominant nAChRs expressed in the BF. Furthermore, several studies (in vivo and vitro) have demonstrated that nAChRs agonists including nicotine activate the BF to promote cortical activation and behavioral arousal (Biton et al., 2007; Jaehne et al., 2014; Khateb et al., 1997; Nair et al., 2015; Salin-Pascual, Moro-Lopez, Gonzalez-Sanchez, & Blanco-Centurion, 1999; Thomsen, Hay-Schmidt, Hansen, & Mikkelsen, 2010).

Since nicotine acts via the BF to promote wakefulness and, as described in the foregoing, alcohol acts via the BF to promote sleep, we hypothesized that nicotine infusion in the BF may attenuate alcohol-induced sleepiness.

To test our hypothesis, adult male Sprague–Dawley rats were instrumented to record electroencephalogram (EEG) and electromyogram (EMG) as previously described (Sharma, Bradshaw, Sahota, & Thakkar, 2014; Sharma, Engemann, Sahota, & Thakkar, 2010a; Thakkar, Engemann, Sharma, Mohan, & Sahota, 2010). On the day of treatment, rats were divided into following groups: (1) ASCF+W: rats were bilaterally microinjected with artificial cerebrospinal fluid (ACSF; 500 nL/side) into the BF followed by intragastric water (10 mL/kg) administration. (2) ACSF+EtOH: rats were bilaterally microinjected with ACSF into the BF followed by intragastric alcohol (3 g/kg) administration. (3) NiC+W: nicotine (11 ng nicotine/500 nL/side) was bilaterally microinjected into the BF followed by intragastric water (10 mL/kg) administration. (4) NiC+EtOH: nicotine

(11 ng nicotine/500 nL/side) was bilaterally microinjected into the BF followed by intragastric alcohol (3 g/kg) administration. On completion, electrographic recording of sleep–wakefulness was begun and continued for 6 h.

All treatments were performed 30 min before the onset of sleep (light) period to mimic the temporal pattern of nicotine and alcohol consumption in humans (Chandra, Shiffman, Scharf, Dang, & Shadel, 2007; Dawson, 1996). The dose of nicotine was chosen based on consumption of nicotine by an average smoker (Nashmi et al., 2003). The electrographic recording of sleep–wakefulness was begun immediately after the completion of infusion and intragastric treatments, coinciding with the onset of sleep (light) period, and continued for 6 h. The results obtained from this study are described in Fig. 27.1.

Intragastric administration of alcohol produced a strong sleep-promoting effect as was evident by a significant decrease in the amount of time to fall asleep (sleep latency; *Panel A*) and a significant increase in the amount of time spent in NREM sleep (*Panel B*). However, local infusion of nicotine in the BF blocked alcohol-induced sleepiness [for details see (Sharma, Lodhi, Sahota, & Thakkar, 2015)].

These results suggest that nicotine via the BF suppresses alcohol-induced sleepiness. Although this study did not examine the effects of nicotine in the BF on alcohol-induced attention and cognition deficits, based on importance of the BF in attention and cognition control, it is very likely that nicotine acts via the BF to mitigate several negative effects of alcohol including attention and cognitive deficits. Does nicotine act via the BF to enhance the rewarding effects of alcohol?

Before we address this question, we would like to provide a very brief overview about the brain centers that mediate positive/rewarding effects of alcohol in the brain [for detailed reviews, see (Ostroumov, Thomas, Dani, & Doyon, 2015; Soderpalm & Ericson, 2013; Tabakoff & Hoffman, 2013)] followed by describing our recent findings that implicate the BF in mediating the rewarding effects of alcohol and nicotine co-use.

NEUROANATOMICAL SUBSTRATES MEDIATING THE REWARDING EFFECTS OF ALCOHOL

Reward promotes a state of happiness/positivity that leads to motivation and reinforcement of "drug use." The discovery of "reward centers" in the brain began with the observation that rats willingly and repeatedly chose to self-stimulate particular areas of the brain with electricity. This led to the discovery of dopaminergic ventral tegmental area (VTA) (Nutt, Lingford-Hughes, Erritzoe, & Stokes, 2015; Wise, 2008). Further research led to the discovery of mesolimbic dopaminergic projections and nucleus accumbens (NAc) as the terminal site for the processing of pleasurable sensations resulting in the "dopamine hedonia" or "dopamine pleasure" hypothesis (Wise, 1980, 2008). Conversely, the "dopamine pleasure hypothesis" postulated that reduction of dopamine neurotransmission causes loss of pleasure as evident by ability of dopamine receptors antagonists (neuroleptics) in reducing the reinforcing properties of stimulants (Wise & Bozarth, 1987). Since 1970s the mesolimbic dopamine pathway has been at the center stage and implicated to serve as a "common

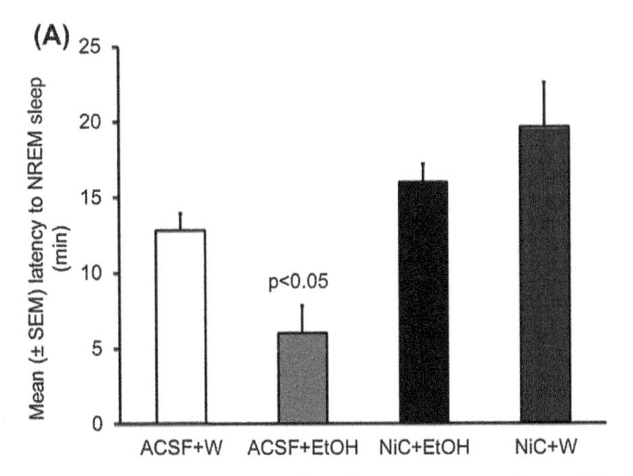

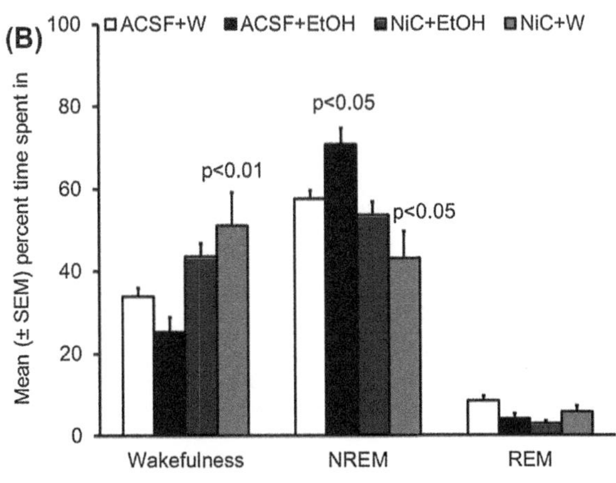

FIGURE 27.1 Nicotine via BF inhibits sleep-promoting effects of alcohol [for details see (Sharma et al., 2015)]. *Panel A*: Rats exposed to alcohol (ACSF+EtOH) displayed a significant reduction in sleep onset latency as compared to controls (ACSF+W). This effect was normalized in rats pretreated with nicotine in the BF (NiC+EtOH). Sleep onset latency remained unaffected in rats with BF nicotine and systemic water administration (NiC+W group). *Panel B*: As compared to controls (ACSF+W), rats exposed to alcohol (ACSF+EtOH) displayed a significant increase in NREM sleep while suppressing wakefulness. This effect was normalized in rats pretreated with nicotine in the BF (NiC+EtOH). Rats exposed to only nicotine infusion in the BF (NiC+W) displayed a significant increase in wakefulness while suppressing NREM sleep. REM sleep remained unaffected.

neural currency" for pleasure and reward. However, in 2000s or so, the role of dopamine in addiction, especially its role in "pleasure" has been questioned [for details see (Berridge & Kringelbach, 2015; Nutt et al., 2015)]. While, recent studies implicate multiple neuronal substrates in mediating positive/rewarding effects, the role of glutamate, especially the corticostraital glutamatergic circuitry, has been gaining immense popularity in alcohol research.

The glutamatergic neurons of the medial prefrontal cortex (mPFC) and their projections to the ventral striatum (nucleus accumbens, NAc) are the major component of the corticostraital pathway implicated in alcohol addiction. The mPFC is typically referred to as the structures located along the medial wall of PFC. These structures are often grouped as dorsal (precentral cortex and anterior cingulate cortex) and ventral (prelimbic, infralimbic, and ventral orbital cortices) subregions. The prelimbic area sends strong glutamatergic projections to the NAc core and is necessary for the acquisition of goal-directed behavior. The infralimbic PFC sends extensive projections to the NAc shell, as well as the amygdala, and is critical for the expression of habitual behavior. The mPFC receives several inputs for multiple brain regions including the BF [for more details see (Barker et al., 2015; Heidbreder & Groenewegen, 2003; Hoover & Vertes, 2007; Moorman, James, McGlinchey, & Aston-Jones, 2015; Vertes, 2004, 2006)].

Alcohol was long thought to exert its actions on the brain solely via potentiation of GABAergic transmission and/or changes in plasma membrane fluidity. However, subsequent studies highlight the importance of glutamate in mediating the rewarding effects of alcohol (Gass & Olive, 2008; Krystal, Petrakis, Mason, Trevisan, & D'Souza, 2003).

Acute alcohol exposure has a dose-dependent effect on extracellular levels of glutamate in the NAc. While systemic administration of low dose (0.5–1 g/kg) increases glutamate levels, high dose (2 g/kg or more) reduces it (Moghaddam & Bolinao, 1994; Selim & Bradberry, 1996; Yan, Reith, Yan, & Jobe, 1998). In contrast, repeated alcohol exposure increases extracellular glutamate in NAc (Kapasova & Szumlinski, 2008; Melendez, Hicks, Cagle, & Kalivas, 2005). Alcohol intake–induced increase in NAc glutamate promotes continued excessive alcohol consumption resulting in hyperglutamatergic, hyperexcitable state (Griffin, Haun, Hazelbaker, Ramachandra, & Becker, 2014; Rao, Bell, Engleman, & Sari, 2015). Administration of NMDA receptor antagonists attenuate positive effects of alcohol and alcohol consumption in rodents (Bienkowski, Koros, Kostowski, & Danysz, 1999; Boyce-Rustay & Cunningham, 2004; Broadbent, Kampmueller, & Koonse, 2003; Camarini, Frussa-Filho, Monteiro, & Calil, 2000; Escher, Call, Blaha, & Mittleman, 2006; Kotlinska, Biala, Rafalski, Bochenski, & Danysz, 2004; McMillen, Joyner, Parmar, Tyer, & Williams, 2004; Piasecki et al., 1998; Shelton & Balster, 1997). Acamprosate, a clinically effective FDA-approved drug for the treatment of alcoholism, mediates its effect via glutamatergic mechanism (D'Souza, 2015). Topiramate, another glutamatergic drug, is often used off-label in the treatment of alcoholism. This drug displays greater efficacy than naltrexone (an FDA-approved opioid antagonist) and reduces alcohol consumption in both humans and rodents (Baltieri, Daro, Ribeiro, & de Andrade, 2008; Holmes, Spanagel, & Krystal, 2013). Finally, studies reported in 2015 suggest that glutamate transporters can serve as targets for attenuating the rewarding effects of alcohol (Rao et al., 2015). Thus, much evidence exists indicating the importance of glutamatergic corticostraital pathway as the mediator of rewarding effects of alcohol.

NICOTINE ACTS VIA BF TO ENHANCE THE REWARDING EFFECTS OF ALCOHOL

The BF sends strong cholinergic and noncholinergic projections to the cortex, and these projections have a vital role in promoting cortical activation. In addition, the BF is the sole provider of the cholinergic inputs to the cortex. Thus, it is not surprising that the BF has strong interactions with glutamatergic systems in the cortex. Acetylcholine released from the cholinergic inputs from the BF have a modulatory role at cortical glutamatergic synapses (Metherate, Tremblay, & Dykes, 1987; Tremblay, Warren, & Dykes, 1990). Lesions of the BF cholinergic neurons impair cortical glutamatergic transmission and alter the cortical expression of glutamate transporter and metabotropic glutamate receptor (Garrett, Kim, Wilson, & Wellman, 2006; Reine, Samuel, Nieoullon, & Kerkerian-Le Goff, 1992). Activation of presynaptic α7 nAChRs on cortical glutamatergic terminals increases extracellular glutamate (Huang et al., 2014; Lambe, Picciotto, & Aghajanian, 2003). Studies published in mid-2010s implicate the BF in reward-associated behaviors and motivational salience (Devore, Pender-Morris, Dean, Smith, & Linster, 2016; Lin, Brown, Hussain Shuler, Petersen, & Kepecs, 2015; Raver & Lin, 2015). Based on the studies described in the foregoing, we hypothesized that nicotine may act via the BF to enhance the rewarding effects of alcohol. Two experiments were performed to test our hypothesis.

The first experiment was designed to examine the activation of NAc in Sprague–Dawley rats. Since neuronal activation is indicated by induction of c-Fos, activation of NAc was examined by monitoring c-Fos expression with immunohistochemistry (IH) (Sharma et al., 2010a; Thakkar, Engemann, Sharma, & Sahota, 2010). Both the BF and NAc are implicated in sleep regulation (Qiu et al., 2012; Thakkar, Winston, et al., 2003). Therefore, to prevent any confounds, all experiments were performed at the onset of dark period when rats are maximally active.

Rats were surgically implanted with bilateral guide cannulas targeted toward the BF regions. On the day of treatment, rats were divided in four groups: (1) ASCF+W: rats were bilaterally microinjected with artificial cerebrospinal fluid (ACSF; 500 nL/side) into the BF followed by intragastric water (10 mL/kg) administration. (2) ACSF+EtOH: rats were bilaterally microinjected with ACSF into the BF followed by intragastric alcohol (3 g/kg) administration. (3) NiC+W: nicotine (11 ng nicotine/500 nL/side) was bilaterally microinjected into the BF followed by intragastric water (10 mL/kg) administration. (4) NiC+EtOH: Nicotine (11 ng nicotine/500 nL/side) was bilaterally microinjected into the BF followed by intragastric alcohol (3 g/kg) administration. Subsequently, rats were left undisturbed for 2 h and then euthanized, brains removed, and processed for c-Fos immunohistochemistry in the

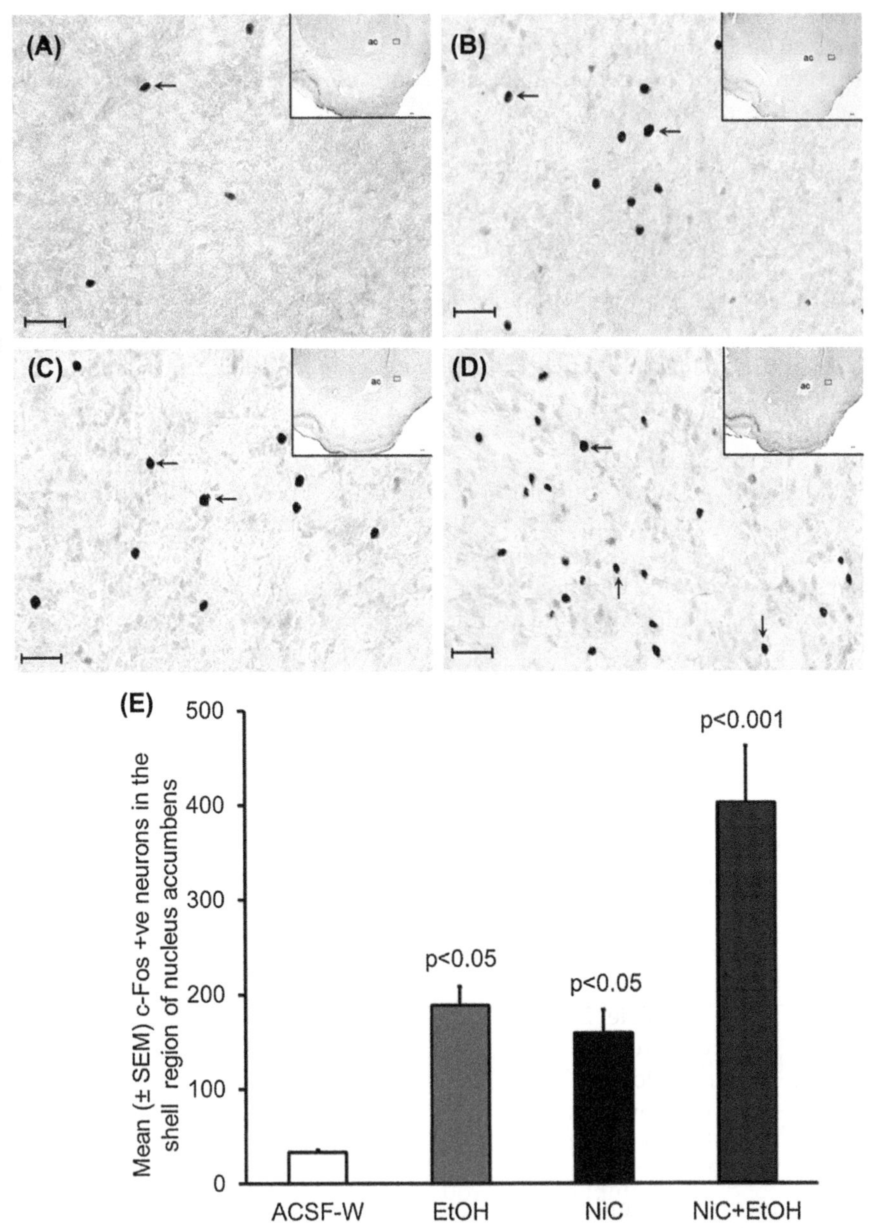

FIGURE 27.2 Nicotine via BF enhances alcohol-induced activation of nucleus accumbens [for details see (Sharma, Dumontier, et al., 2014)]. Representative high-magnification photomicrograph depicting c-Fos + ve nuclei (*arrows*) in the nucleus accumbens (NAc) shell region of rats in ACSF + W (*Panel A*), ACSF+EtOH (*Panel B*), NiC+W (*Panel C*), and NiC+EtOH (*Panel D*) groups. The graphical representation of the results is described in *Panel E*. Calibration bar: 30 μm. The boxes in the insets describe the localization of the photomicrograph in the NAc shell. Calibration bar: 100 μm. *ac*, anterior commissure; *AcbC*, core region of NAc.

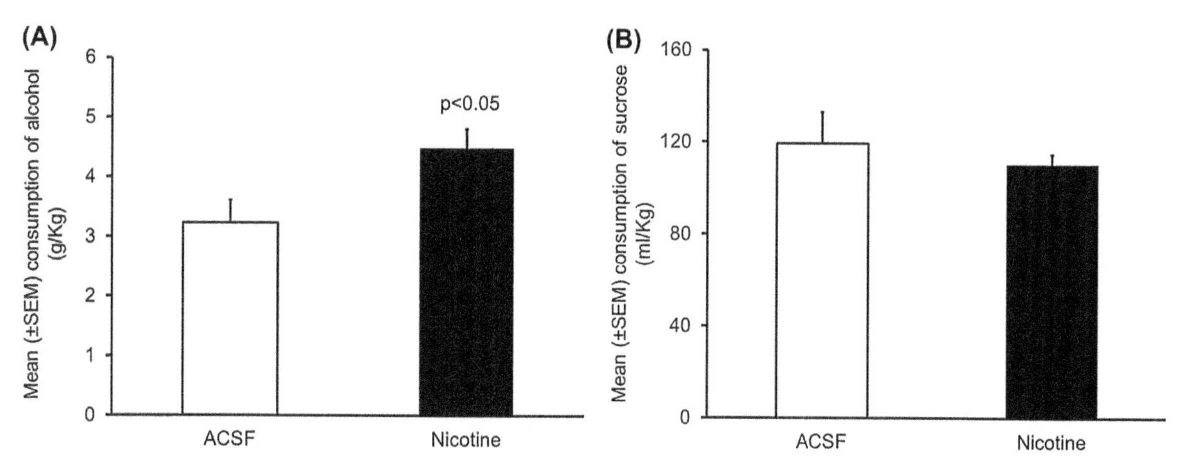

FIGURE 27.3 Nicotine via BF enhances alcohol consumption (*Panel A*) without affecting sucrose consumption (*Panel B*). ACSF, artificial cerebrospinal fluid [for details see (Sharma et al., 2014b)].

NAc and Nissl staining to verify microinjection sites in the BF.

Our results are described in Fig. 27.2. Nicotine infusion in the BF combined with systemic alcohol more than doubled c-Fos expression in the shell region of NAc (*Panel E*) but not in the core region of NAc [for details see (Sharma, Dumontier, deRoode, Sahota, & Thakkar, 2014)]. These results suggest that nicotine via its action on BF neurons activates the NAc. Since rewarding or reinforcing properties of alcohol and/or nicotine are associated with the activation of NAc, we suggest that the nicotine via BF enhances the positive/rewarding effects of alcohol (Koob & Volkow, 2010; Pontieri, Tanda, Orzi, & Di, 1996).

If nicotine via the BF suppresses negative effects of alcohol while enhancing rewarding effects then nicotine should act via the BF to increase alcohol consumption. To test this hypothesis, we used C57BL/6J mice as our animal model and the "drinking-in-the-dark" paradigm to examine the effect of nicotine infusion in the BF on binge alcohol consumption. The C57BL/6J mice voluntary consumed huge amount of alcohol in relatively short time for post-ingestive intoxicating effects, and not simply for taste or caloric fulfillment (Sharma, Sahota, & Thakkar, 2014a; Sprow & Thiele, 2012). The "drinking-in-the-dark" is a relatively simple and extensively used method to examine binge alcohol consumption in mice (Rhodes, Best, Belknap, Finn, & Crabbe, 2005; Sprow & Thiele, 2012).

Mice were stereotaxically implanted with bilateral stainless steel guide cannula above the BF. Following postoperative recovery, mice were voluntarily allowed to self-administer alcohol (20%) for 2 h daily for 3 days. On day 4, 1 h before the onset of alcohol exposure, mice were randomly divided into two groups: The ACSF group was bilaterally infused with ACSF (100 nL/side). The Nic group was bilaterally infused with nicotine (1.6 ng of nicotine/100 nL/side) into the BF [for details see (Sharma, Sahota, & Thakkar, 2014b)].

In order to validate the specificity of the effect on alcohol consumption, we examined the effects of nicotine infusion in the BF on sucrose (10%) consumption in separate groups (ACSF and Nicotine) of mice.

The results of our study are described in Fig. 27.3. As compared to control (ACSF) perfusion, nicotine perfusion in the BF significantly increased the amount of alcohol consumed (*Panel A*). However, nicotine infusion had no effect on sucrose consumption (*Panel B*) [for details see (Sharma et al., 2014b)]. The results of this study suggest that nicotine infusion into the BF enhances alcohol consumption.

SUMMARY

Overall, our studies have highlighted the importance of the BF in mediating the effects of nicotine, not only to reduce the negative/aversive effects, but also in enhancing the rewarding effects of alcohol. Although the detailed mechanisms of how BF interacts with the reward center is a "work-in-progress," we suggest that alcohol-induced inhibition of the BF results in cortical synchronization and sleepiness. However, when nicotine is consumed with alcohol, nicotine acts via BF to promote cortical activation and activation of the cortex activates the NAc, resulting in the enhancement of rewarding effects.

Acknowledgments

This work was supported by the Harry S. Truman Memorial Veterans Hospital, and funding from the National Institute of Alcohol Abuse and Alcoholism. We would also like to acknowledge the help of Abhilasha Sharma in the preparation of this manuscript.

References

Baltieri, D. A., Daro, F. R., Ribeiro, P. L., & de Andrade, A. G. (2008). Comparing topiramate with naltrexone in the treatment of alcohol dependence. *Addiction, 103*(12), 2035–2044. http://dx.doi.org/10.1111/j.1360-0443.2008.02355.x.

Barker, J. M., Corbit, L. H., Robinson, D. L., Gremel, C. M., Gonzales, R. A., & Chandler, L. J. (2015). Corticostriatal circuitry and habitual ethanol seeking. *Alcohol, 49*(8), 817–824. http://dx.doi.org/10.1016/j.alcohol.2015.03.003.

Batista, L. C., Prediger, R. D., Morato, G. S., & Takahashi, R. N. (2005). Blockade of adenosine and dopamine receptors inhibits the development of rapid tolerance to ethanol in mice. *Psychopharmacology (Berl), 181*(4), 714–721.

Berridge, K. C., & Kringelbach, M. L. (2015). Pleasure systems in the brain. *Neuron, 86*(3), 646–664. http://dx.doi.org/10.1016/j.neuron.2015.02.018.

Bienkowski, P., Koros, E., Kostowski, W., & Danysz, W. (1999). Effects of *N*-methyl-D-aspartate receptor antagonists on reinforced and nonreinforced responding for ethanol in rats. *Alcohol, 18*(2–3), 131–137.

Bito-Onon, J. J., Simms, J. A., Chatterjee, S., Holgate, J., & Bartlett, S. E. (2011). Varenicline, a partial agonist at neuronal nicotinic acetylcholine receptors, reduces nicotine-induced increases in 20% ethanol operant self-administration in Sprague-Dawley rats. *Addiction Biology, 16*(3), 440–449. http://dx.doi.org/10.1111/j.1369-1600.2010.00309.x.

Biton, B., Bergis, O. E., Galli, F., Nedelec, A., Lochead, A. W., Jegham, S., & Scatton, B. (2007). SSR180711, a novel selective alpha7 nicotinic receptor partial agonist: (1) Binding and functional profile. *Neuropsychopharmacology, 32*(1), 1–16. http://dx.doi.org/10.1038/sj.npp.1301189.

Blomqvist, O., Ericson, M., Johnson, D. H., Engel, J. A., & Soderpalm, B. (1996). Voluntary ethanol intake in the rat: Effects of nicotinic acetylcholine receptor blockade or subchronic nicotine treatment. *European Journal of Pharmacology, 314*(3), 257–267.

Borbely, A. A. (1982). A two process model of sleep regulation. *Human Neurobiology, 1*, 195–204.

Boyce-Rustay, J. M., & Cunningham, C. L. (2004). The role of NMDA receptor binding sites in ethanol place conditioning. *Behavioral Neuroscience, 118*(4), 822–834. http://dx.doi.org/10.1037/0735-7044.118.4.822.

Broadbent, J., Kampmueller, K. M., & Koonse, S. A. (2003). Expression of behavioral sensitization to ethanol by DBA/2J mice: The role of NMDA and non-NMDA glutamate receptors. *Psychopharmacology (Berl), 167*(3), 225–234. http://dx.doi.org/10.1007/s00213-003-1404-3.

Brown, R. E., Basheer, R., McKenna, J. T., Strecker, R. E., & McCarley, R. W. (2012). Control of sleep and wakefulness. *Physiological Reviews, 92*(3), 1087–1187.

Camarini, R., Frussa-Filho, R., Monteiro, M. G., & Calil, H. M. (2000). MK-801 blocks the development of behavioral sensitization to the ethanol. *Alcoholism, Clinical and Experimental Research, 24*(3), 285–290.

Chandra, S., Shiffman, S., Scharf, D. M., Dang, Q., & Shadel, W. G. (2007). Daily smoking patterns, their determinants, and implications for quitting. *Experimental and Clinical Psychopharmacology, 15*(1), 67–80. http://dx.doi.org/10.1037/1064-1297.15.1.67.

Chatterjee, S., & Bartlett, S. E. (2010). Neuronal nicotinic acetylcholine receptors as pharmacotherapeutic targets for the treatment of alcohol use disorders. *CNS & Neurological Disorders Drug Targets, 9*(1), 60–76.

Chi, H., & de Wit, H. (2003). Mecamylamine attenuates the subjective stimulant-like effects of alcohol in social drinkers. *Alcoholism, Clinical and Experimental Research, 27*(5), 780–786. http://dx.doi.org/10.1097/01.ALC.0000065435.12068.24.

Choi, D. S., Cascini, M. G., Mailliard, W., Young, H., Paredes, P., McMahon, T., & Messing, R. O. (2004). The type 1 equilibrative nucleoside transporter regulates ethanol intoxication and preference. *Nature Neuroscience, 7*(8), 855–861. http://dx.doi.org/10.1038/nn1288.

Clark, A., & Little, H. J. (2004). Interactions between low concentrations of ethanol and nicotine on firing rate of ventral tegmental dopamine neurones. *Drug and Alcohol Dependence, 75*(2), 199–206.

Concas, A., Mascia, M. P., Cuccheddu, T., Floris, S., Mostallino, M. C., Perra, C., & Biggio, G. (1996). Chronic ethanol intoxication enhances [3H]CCPA binding and does not reduce A1 adenosine receptor function in rat cerebellum. *Pharmacology, Biochemistry, and Behavior, 53*(2), 249–255.

D'Souza, M. S. (2015). Glutamatergic transmission in drug reward: Implications for drug addiction. *Frontiers in Neuroscience, 9*, 404. http://dx.doi.org/10.3389/fnins.2015.00404.

Dar, M. S. (1993). Brain adenosinergic modulation of acute ethanol-induced motor impairment. *Alcohol and Alcoholism Supplement, 2*, 425–429.

Dar, M. S. (1997). Mouse cerebellar adenosinergic modulation of ethanol-induced motor incoordination: Possible involvement of cAMP. *Brain Research, 749*(2), 263–274.

Dar, M. S. (2001). Modulation of ethanol-induced motor incoordination by mouse striatal A(1) adenosinergic receptor. *Brain Research Bulletin, 55*(4), 513–520.

Davis, T. J., & de Fiebre, C. M. (2006). Alcohol's actions on neuronal nicotinic acetylcholine receptors. *Alcohol Research and Health, 29*(3), 179–185.

Dawson, D. A. (1996). Temporal drinking patterns and variation in social consequences. *Addiction, 91*(11), 1623–1635.

Devore, S., Pender-Morris, N., Dean, O., Smith, D., & Linster, C. (2016). Basal forebrain dynamics during nonassociative and associative olfactory learning. *Journal of Neurophysiology, 115*(1), 423–433. http://dx.doi.org/10.1152/jn.00572.2015.

Drake, C. L., Roehrs, T., Turner, L., Scofield, H. M., & Roth, T. (2003). Caffeine reversal of ethanol effects on the multiple sleep latency test, memory, and psychomotor performance. *Neuropsychopharmacology, 28*(2), 371–378.

Dunwiddie, T. V., & Masino, S. A. (2001). The role and regulation of adenosine in the central nervous system. *Annual Review of Neuroscience, 24*, 31–55.

Edgar, D. M., Dement, W. C., & Fuller, C. A. (1993). Effect of SCN lesions on sleep in squirrel monkeys: Evidence for opponent processes in sleep-wake regulation. *The Journal of Neuroscience, 13*(3), 1065–1079.

Escher, T., Call, S. B., Blaha, C. D., & Mittleman, G. (2006). Behavioral effects of aminoadamantane class NMDA receptor antagonists on schedule-induced alcohol and self-administration of water in mice. *Psychopharmacology (Berl), 187*(4), 424–434. http://dx.doi.org/10.1007/s00213-006-0465-5.

Franks, H. M., Hagedorn, H., Hensley, V. R., Hensley, W. J., & Starmer, G. A. (1975). The effect of caffeine on human performance, alone and in combination with ethanol. *Psychopharmacologia, 45*(2), 177–181.

Garrett, J. E., Kim, I., Wilson, R. E., & Wellman, C. L. (2006). Effect of *N*-methyl-D-aspartate receptor blockade on plasticity of frontal cortex after cholinergic deafferentation in rat. *Neuroscience, 140*(1), 57–66. http://dx.doi.org/10.1016/j.neuroscience.2006.01.029.

Gass, J. T., & Olive, M. F. (2008). Glutamatergic substrates of drug addiction and alcoholism. *Biochemical Pharmacology, 75*(1), 218–265. http://dx.doi.org/10.1016/j.bcp.2007.06.039.

Griffin, W. C., 3rd, Haun, H. L., Hazelbaker, C. L., Ramachandra, V. S., & Becker, H. C. (2014). Increased extracellular glutamate in the nucleus accumbens promotes excessive ethanol drinking in ethanol dependent mice. *Neuropsychopharmacology, 39*(3), 707–717. http://dx.doi.org/10.1038/npp.2013.256.

Gritti, I., Mainville, L., Mancia, M., & Jones, B. E. (1997). GABAergic and other noncholinergic basal forebrain neurons, together with cholinergic neurons, project to the mesocortex and isocortex in the rat. *Journal of Comparative Neurology, 383*(2), 163–177.

Hack, S. P., & Christie, M. J. (2003). Adaptations in adenosine signaling in drug dependence: Therapeutic implications. *Critical Reviews in Neurobiology, 15*(3–4), 235–274.

Harrison, E. L., Desai, R. A., & McKee, S. A. (2008). Nondaily smoking and alcohol use, hazardous drinking, and alcohol diagnoses among young adults: Findings from the NESARC. *Alcoholism, Clinical and Experimental Research, 32*(12), 2081–2087. http://dx.doi.org/10.1111/j.1530-0277.2008.00796.x.

Hasenfratz, M., Bunge, A., Dal, P. G., & Battig, K. (1993). Antagonistic effects of caffeine and alcohol on mental performance parameters. *Pharmacology, Biochemistry, and Behavior, 46*(2), 463–465.

Hauser, S. R., Getachew, B., Oster, S. M., Dhaher, R., Ding, Z. M., Bell, R. L., & Rodd, Z. A. (2012). Nicotine modulates alcohol-seeking and relapse by alcohol-preferring (P) rats in a time-dependent manner. *Alcoholism, Clinical and Experimental Research, 36*(1), 43–54. http://dx.doi.org/10.1111/j.1530-0277.2011.01579.x.

Heidbreder, C. A., & Groenewegen, H. J. (2003). The medial prefrontal cortex in the rat: Evidence for a dorso-ventral distinction based upon functional and anatomical characteristics. *Neuroscience and Biobehavioral Reviews, 27*(6), 555–579. http://dx.doi.org/10.1016/j.neubiorev.2003.09.003.

Hendrickson, L. M., Zhao-Shea, R., Pang, X., Gardner, P. D., & Tapper, A. R. (2010). Activation of alpha4* nAChRs is necessary and sufficient for varenicline-induced reduction of alcohol consumption. *The Journal of Neuroscience, 30*(30), 10169–10176. http://dx.doi.org/10.1523/jneurosci.2601-10.2010.

Holmes, A., Spanagel, R., & Krystal, J. H. (2013). Glutamatergic targets for new alcohol medications. *Psychopharmacology (Berl), 229*(3), 539–554. http://dx.doi.org/10.1007/s00213-013-3226-2.

Hoover, W. B., & Vertes, R. P. (2007). Anatomical analysis of afferent projections to the medial prefrontal cortex in the rat. *Brain Structure & Function, 212*(2), 149–179.

Huang, M., Felix, A. R., Flood, D. G., Bhuvaneswaran, C., Hilt, D., Koenig, G., & Meltzer, H. Y. (2014). The novel alpha7 nicotinic acetylcholine receptor agonist EVP-6124 enhances dopamine, acetylcholine, and glutamate efflux in rat cortex and nucleus accumbens. *Psychopharmacology (Berl), 231*(23), 4541–4551. http://dx.doi.org/10.1007/s00213-014-3596-0.

Jaehne, A., Unbehaun, T., Feige, B., Herr, S., Appel, A., & Riemann, D. (2014). The influence of 8 and 16 mg nicotine patches on sleep in healthy non-smokers. *Pharmacopsychiatry, 47*(2), 73–78. http://dx.doi.org/10.1055/s-0034-1371867.

Kapasova, Z., & Szumlinski, K. K. (2008). Strain differences in alcohol-induced neurochemical plasticity: A role for accumbens glutamate in alcohol intake. *Alcoholism, Clinical and Experimental Research, 32*(4), 617–631. http://dx.doi.org/10.1111/j.1530-0277.2008.00620.x.

Kaplan, G. B., Bharmal, N. H., Leite-Morris, K. A., & Adams, W. R. (1999). Role of adenosine A1 and A2A receptors in the alcohol withdrawal syndrome. *Alcohol, 19*(2), 157–162.

Kerr, J. S., Sherwood, N., & Hindmarch, I. (1991). Separate and combined effects of the social drugs on psychomotor performance. *Psychopharmacology (Berl), 104*(1), 113–119.

Khateb, A., Fort, P., Williams, S., Serafin, M., Jones, B. E., & Muhlethaler, M. (1997). Modulation of cholinergic nucleus basalis neurons by acetylcholine and *N*-methyl-D-aspartate. *Neuroscience, 81*(1), 47–55.

Koob, G. F. (2000). Animal models of craving for ethanol. *Addiction, 95*(Suppl. 2), S73–S81.

Koob, G. F., & Volkow, N. D. (2010). Neurocircuitry of addiction. *Neuropsychopharmacology, 35*(1), 217–238.

Kotlinska, J., Biala, G., Rafalski, P., Bochenski, M., & Danysz, W. (2004). Effect of neramexane on ethanol dependence and reinforcement. *European Journal of Pharmacology, 503*(1–3), 95–98. http://dx.doi.org/10.1016/j.ejphar.2004.09.036.

Krauss, S. W., Ghirnikar, R. B., Diamond, I., & Gordon, A. S. (1993). Inhibition of adenosine uptake by ethanol is specific for one class of nucleoside transporters. *Molecular Pharmacology, 44*(5), 1021–1026.

Krystal, J. H., Petrakis, I. L., Mason, G., Trevisan, L., & D'Souza, D. C. (2003). *N*-methyl-D-aspartate glutamate receptors and alcoholism: Reward, dependence, treatment, and vulnerability. *Pharmacology & Therapeutics, 99*(1), 79–94. http://dx.doi.org/10.1016/s0163-7258(03)00054-8.

Lambe, E. K., Picciotto, M. R., & Aghajanian, G. K. (2003). Nicotine induces glutamate release from thalamocortical terminals in prefrontal cortex. *Neuropsychopharmacology, 28*(2), 216–225. http://dx.doi.org/10.1038/sj.npp.1300032.

Le, A. D., Wang, A., Harding, S., Juzytsch, W., & Shaham, Y. (2003). Nicotine increases alcohol self-administration and reinstates alcohol seeking in rats. *Psychopharmacology (Berl), 168*(1–2), 216–221.

Liguori, A., & Robinson, J. H. (2001). Caffeine antagonism of alcohol-induced driving impairment. *Drug and Alcohol Dependence, 63*(2), 123–129.

Lin, S. C., Brown, R. E., Hussain Shuler, M. G., Petersen, C. C., & Kepecs, A. (2015). Optogenetic dissection of the basal forebrain neuromodulatory control of cortical activation, plasticity, and cognition. *The Journal of Neuroscience, 35*(41), 13896–13903. http://dx.doi.org/10.1523/JNEUROSCI.2590-15.2015.

Li, T. K., Volkow, N. D., Baler, R. D., & Egli, M. (2007). The biological bases of nicotine and alcohol co-addiction. *Biological Psychiatry, 61*(1), 1–3. http://dx.doi.org/10.1016/j.biopsych.2006.11.004.

Locker, A. R., Marks, M. J., Kamens, H. M., & Klein, L. C. (2015). Exposure to nicotine increases nicotinic acetylcholine receptor density in the reward pathway and binge ethanol consumption in C57BL/6J adolescent female mice. *Brain Research Bulletin.* http://dx.doi.org/10.1016/j.brainresbull.2015.09.009.

Lopez-Moreno, J. A., Trigo-Diaz, J. M., Rodriguez de Fonseca, F., Gonzalez Cuevas, G., Gomez de Heras, R., Crespo Galan, I., & Navarro, M. (2004). Nicotine in alcohol deprivation increases alcohol operant self-administration during reinstatement. *Neuropharmacology, 47*(7), 1036–1044. http://dx.doi.org/10.1016/j.neuropharm.2004.08.002.

Manns, I. D., Alonso, A., & Jones, B. E. (2003). Rhythmically discharging basal forebrain units comprise cholinergic, GABAergic, and putative glutamatergic cells. *Journal of Neurophysiology, 89*(2), 1057–1066.

McKee, S. A., Harrison, E. L., O'Malley, S. S., Krishnan-Sarin, S., Shi, J., Tetrault, J. M., & Balchunas, E. (2009). Varenicline reduces alcohol self-administration in heavy-drinking smokers. *Biological Psychiatry, 66*(2), 185–190. http://dx.doi.org/10.1016/j.biopsych.2009.01.029.

McKee, S. A., & Weinberger, A. H. (2013). How can we use our knowledge of alcohol-tobacco interactions to reduce alcohol use? *Annual Review of Clinical Psychology, 9*, 649–674. http://dx.doi.org/10.1146/annurev-clinpsy-050212-185549.

McMillen, B. A., Joyner, P. W., Parmar, C. A., Tyer, W. E., & Williams, H. L. (2004). Effects of NMDA glutamate receptor antagonist drugs on the volitional consumption of ethanol by a genetic drinking rat. *Brain Research Bulletin, 64*(3), 279–284. http://dx.doi.org/10.1016/j.brainresbull.2004.08.001.

Melendez, R. I., Hicks, M. P., Cagle, S. S., & Kalivas, P. W. (2005). Ethanol exposure decreases glutamate uptake in the nucleus accumbens. *Alcoholism, Clinical & Experimental Research, 29*(3), 326–333. http://dx.doi.org/10.1097/01.alc.0000156086.65665.4d.

Metherate, R., Tremblay, N., & Dykes, R. W. (1987). Acetylcholine permits long-term enhancement of neuronal responsiveness in cat primary somatosensory cortex. *Neuroscience, 22*(1), 75–81.

Millar, N. S., & Gotti, C. (2009). Diversity of vertebrate nicotinic acetylcholine receptors. *Neuropharmacology, 56*(1), 237–246. http://dx.doi.org/10.1016/j.neuropharm.2008.07.041.

Moghaddam, B., & Bolinao, M. L. (1994). Biphasic effect of ethanol on extracellular accumulation of glutamate in the hippocampus and the nucleus accumbens. *Neuroscience Letters, 178*(1), 99–102.

Moorman, D. E., James, M. H., McGlinchey, E. M., & Aston-Jones, G. (2015). Differential roles of medial prefrontal subregions in the regulation of drug seeking. *Brain Research, 1628*(Pt A), 130–146. http://dx.doi.org/10.1016/j.brainres.2014.12.024.

Nagy, L. E., Diamond, I., Casso, D. J., Franklin, C., & Gordon, A. S. (1990). Ethanol increases extracellular adenosine by inhibiting adenosine uptake via the nucleoside transporter. *Journal of Biological Chemistry, 265*(4), 1946–1951.

Nair, A., Sharma, A., Rice, I., Sahota, P., Sharma, R., Murugesan, S., & Thakkar, M. (2015). Nicotine promotes wakefulness via activation of wake-promoting cholinergic neurons of the basal forebrain. In *Society for neuroscience abstract, neuroscience meeting planner*. Washington, DC: Society for Neuroscience, 2015. Online.

Nashmi, R., Dickinson, M. E., McKinney, S., Jareb, M., Labarca, C., Fraser, S. E., & Lester, H. A. (2003). Assembly of alpha4beta2 nicotinic acetylcholine receptors assessed with functional fluorescently labeled subunits: Effects of localization, trafficking, and nicotine-induced upregulation in clonal mammalian cells and in cultured midbrain neurons. *The Journal of Neuroscience, 23*(37), 11554–11567.

Newton, P. M., & Messing, R. O. (2006). Intracellular signaling pathways that regulate behavioral responses to ethanol. *Pharmacology & Therapeutics, 109*(1–2), 227–237.

Nutt, D. J., Lingford-Hughes, A., Erritzoe, D., & Stokes, P. R. (2015). The dopamine theory of addiction: 40 Years of highs and lows. *Nature Reviews. Neuroscience, 16*(5), 305–312. http://dx.doi.org/10.1038/nrn3939.

Ostroumov, A., Thomas, A. M., Dani, J. A., & Doyon, W. M. (2015). Cigarettes and alcohol: The influence of nicotine on operant alcohol self-administration and the mesolimbic dopamine system. *Biochemical Pharmacology, 97*(4), 550–557. http://dx.doi.org/10.1016/j.bcp.2015.07.038.

Phan, T. A., Gray, A. M., & Nyce, J. W. (1997). Intrastriatal adenosine A1 receptor antisense oligodeoxynucleotide blocks ethanol-induced motor incoordination. *European Journal of Pharmacology, 323*(2–3), R5–R7.

Piasecki, J., Koros, E., Dyr, W., Kostowski, W., Danysz, W., & Bienkowski, P. (1998). Ethanol-reinforced behaviour in the rat: Effects of uncompetitive NMDA receptor antagonist, memantine. *European Journal of Pharmacology, 354*(2–3), 135–143.

Pontieri, F. E., Tanda, G., Orzi, F., & Di, C. G. (1996). Effects of nicotine on the nucleus accumbens and similarity to those of addictive drugs. *Nature, 382*(6588), 255–257.

Porkka-Heiskanen, T. (2013). Sleep homeostasis. *Current Opinion in Neurobiology, 23*(5), 799–805.

Porkka-Heiskanen, T., Strecker, R. E., Thakkar, M., Bjorkum, A. A., Greene, R. W., & McCarley, R. W. (1997). Adenosine: A mediator of the sleep-inducing effects of prolonged wakefulness. *Science, 276*(5316), 1265–1268.

Prediger, R. D., da Silva, G. E., Batista, L. C., Bittencourt, A. L., & Takahashi, R. N. (2006). Activation of adenosine A1 receptors reduces anxiety-like behavior during acute ethanol withdrawal (hangover) in mice. *Neuropsychopharmacology, 31*(10), 2210–2220.

Qiu, M. H., Liu, W., Qu, W. M., Urade, Y., Lu, J., & Huang, Z. L. (2012). The role of nucleus accumbens core/shell in sleep-wake regulation and their involvement in modafinil-induced arousal. *PLoS ONE, 7*(9), e45471.

Rao, P. S., Bell, R. L., Engleman, E. A., & Sari, Y. (2015). Targeting glutamate uptake to treat alcohol use disorders. *Frontiers in Neuroscience, 9*, 144. http://dx.doi.org/10.3389/fnins.2015.00144.

Raver, S. M., & Lin, S. C. (2015). Basal forebrain motivational salience signal enhances cortical processing and decision speed. *Frontiers in Behavioral Neuroscience, 9*, 277. http://dx.doi.org/10.3389/fnbeh.2015.00277.

Reine, G., Samuel, D., Nieoullon, A., & Kerkerian-Le Goff, L. (1992). Effects of lesion of the cholinergic basal forebrain nuclei on the activity of glutamatergic and GABAergic systems in the rat frontal cortex and hippocampus. *Journal of Neural Transmission. General Section, 87*(3), 175–192.

Rhodes, J. S., Best, K., Belknap, J. K., Finn, D. A., & Crabbe, J. C. (2005). Evaluation of a simple model of ethanol drinking to intoxication in C57BL/6J mice. *Physiology & Behavior, 84*(1), 53–63.

Robinson, T. E., & Berridge, K. C. (2001). Incentive-sensitization and addiction. *Addiction, 96*(1), 103–114.

Saeed, D. M. (2006). Co-modulation of acute ethanol-induced motor impairment by mouse cerebellar adenosinergic A1 and $GABA_A$ receptor systems. *Brain Research Bulletin, 71*(1–3), 287–295.

Salin-Pascual, R. J., Moro-Lopez, M. L., Gonzalez-Sanchez, H., & Blanco-Centurion, C. (1999). Changes in sleep after acute and repeated administration of nicotine in the rat. *Psychopharmacology, 145*(2), 133–138.

SAMHSA. (2005). *Results from the 2005 national survey on drug use and health: National findings*. Retrieved from http://www.samhsa.gov/data/nsduh/2k5nsduh/2k5results.htm.

Selim, M., & Bradberry, C. W. (1996). Effect of ethanol on extracellular 5-HT and glutamate in the nucleus accumbens and prefrontal cortex: Comparison between the Lewis and Fischer 344 rat strains. *Brain Research, 716*(1–2), 157–164. http://dx.doi.org/10.1016/0006-8993(95)01385-7.

Sharma, R., Bradshaw, K., Sahota, P., & Thakkar, M. M. (2014). Acute binge alcohol administration reverses sleep-wake cycle in Sprague Dawley rats. *Alcoholism, Clinical and Experimental Research, 38*(7), 1941–1946.

Sharma, R., Dumontier, S., deRoode, D., Sahota, P., & Thakkar, M. M. (2014). Nicotine infusion in the wake-promoting basal forebrain enhances alcohol-induced activation of nucleus accumbens. *Alcoholism, Clinical and Experimental Research, 38*(10), 2590–2596.

Sharma, R., Engemann, S., Sahota, P., & Thakkar, M. M. (2010a). Role of adenosine and wake-promoting basal forebrain in insomnia and associated sleep disruptions caused by ethanol dependence. *Journal of Neurochemistry, 115*(3), 782–794.

Sharma, R., Engemann, S. C., Sahota, P., & Thakkar, M. M. (2010b). Effects of ethanol on extracellular levels of adenosine in the basal forebrain: An in vivo microdialysis study in freely behaving rats. *Alcoholism, Clinical and Experimental Research, 34*(5), 813–818.

Sharma, R., Lodhi, S., Sahota, P., & Thakkar, M. M. (2015). Nicotine administration in the wake-promoting basal forebrain attenuates sleep-promoting effects of alcohol. *Journal of Neurochemistry, 135*(2), 323–331.

Sharma, R., Sahota, P., & Thakkar, M. M. (2014a). Rapid tolerance development to the NREM sleep promoting effect of alcohol. *Sleep, 37*(4), 821–824.

Sharma, R., Sahota, P., & Thakkar, M. M. (2014b). Nicotine administration in the cholinergic basal forebrain increases alcohol consumption in C57BL/6J mice. *Alcoholism, Clinical and Experimental Research, 38*(5), 1315–1320.

Shelton, K. L., & Balster, R. L. (1997). Effects of gamma-aminobutyric acid agonists and *N*-methyl-D-aspartate antagonists on a multiple schedule of ethanol and saccharin self-administration in rats. *The Journal of Pharmacology and Experimental Therapeutics, 280*(3), 1250–1260.

Smith, B. R., Horan, J. T., Gaskin, S., & Amit, Z. (1999). Exposure to nicotine enhances acquisition of ethanol drinking by laboratory rats in a limited access paradigm. *Psychopharmacology, 142*(4), 408–412.

Soderpalm, B., & Ericson, M. (2013). Neurocircuitry involved in the development of alcohol addiction: The dopamine system and its access points. *Current Topics in Behavioral Neurosciences, 13*, 127–161. http://dx.doi.org/10.1007/7854_2011_170.

Sprow, G. M., & Thiele, T. E. (2012). The neurobiology of binge-like ethanol drinking: Evidence from rodent models. *Physiology & Behavior, 106*(3), 325–331.

Steensland, P., Simms, J. A., Holgate, J., Richards, J. K., & Bartlett, S. E. (2007). Varenicline, an alpha4beta2 nicotinic acetylcholine receptor partial agonist, selectively decreases ethanol consumption and seeking. *Proceedings of the National Academy of Sciences of the United States of America, 104*(30), 12518–12523.

Tabakoff, B., & Hoffman, P. L. (2013). The neurobiology of alcohol consumption and alcoholism: An integrative history. *Pharmacology, Biochemistry, and Behavior, 113*, 20–37. http://dx.doi.org/10.1016/j.pbb.2013.10.009.

Tapper, A. R., McKinney, S. L., Nashmi, R., Schwarz, J., Deshpande, P., Labarca, C., & Lester, H. A. (2004). Nicotine activation of alpha4* receptors: Sufficient for reward, tolerance, and sensitization. *Science, 306*(5698), 1029–1032.

Thakkar, M. M. (2011). Histamine in the regulation of wakefulness. *Sleep Medicine Reviews, 15*(1), 65–74.

Thakkar, M. M., Delgiacco, R. A., Strecker, R. E., & McCarley, R. W. (2003). Adenosinergic inhibition of basal forebrain wakefulness-active neurons: A simultaneous unit recording and microdialysis study in freely behaving cats. *Neuroscience, 22*(4), 1107–1113.

Thakkar, M. M., Engemann, S. C., Sharma, R., Mohan, R. R., & Sahota, P. (2010). Sleep-wakefulness in alcohol preferring and non-preferring rats following binge alcohol administration. *Neuroscience, 170*(1), 22–27.

Thakkar, M. M., Engemann, S. C., Sharma, R., & Sahota, P. (2010). Role of wake-promoting basal forebrain and adenosinergic mechanisms in sleep-promoting effects of ethanol. *Alcoholism, Clinical and Experimental Research, 34*(6), 997–1005.

Thakkar, M. M., Sharma, R., & Sahota, P. (2015). Alcohol disrupts sleep homeostasis. *Alcohol, 49*(4), 299–310.

Thakkar, M. M., Winston, S., & McCarley, R. W. (2003). A1 receptor and adenosinergic homeostatic regulation of sleep-wakefulness: Effects of antisense to the A1 receptor in the cholinergic Basal forebrain. *The Journal of Neuroscience, 23*(10), 4278–4287.

Thomsen, M. S., Hay-Schmidt, A., Hansen, H. H., & Mikkelsen, J. D. (2010). Distinct neural pathways mediate alpha7 nicotinic acetylcholine receptor-dependent activation of the forebrain. *Cerebral Cortex, 20*(9), 2092–2102. http://dx.doi.org/10.1093/cercor/bhp283.

Tizabi, Y., Bai, L., Copeland, R. L., Jr., & Taylor, R. E. (2007). Combined effects of systemic alcohol and nicotine on dopamine release in the nucleus accumbens shell. *Alcohol and Alcoholism, 42*(5), 413–416. http://dx.doi.org/10.1093/alcalc/agm057.

Tizabi, Y., Copeland, R. L., Jr., Louis, V. A., & Taylor, R. E. (2002). Effects of combined systemic alcohol and central nicotine administration into ventral tegmental area on dopamine release in the nucleus accumbens. *Alcoholism, Clinical and Experimental Research, 26*(3), 394–399.

Trask, G. (1860). *Letters on tobacco, for American Lads*. Fitchburg, MA: Trask.

Tremblay, N., Warren, R. A., & Dykes, R. W. (1990). Electrophysiological studies of acetylcholine and the role of the basal forebrain in the somatosensory cortex of the cat. I. Cortical neurons excited by glutamate. *Journal of Neurophysiology, 64*(4), 1199–1211.

Tuesta, L. M., Fowler, C. D., & Kenny, P. J. (2011). Recent advances in understanding nicotinic receptor signaling mechanisms that regulate drug self-administration behavior. *Biochemical Pharmacology, 82*(8), 984–995. http://dx.doi.org/10.1016/j.bcp.2011.06.026.

Vertes, R. P. (2004). Differential projections of the infralimbic and prelimbic cortex in the rat. *Synapse, 51*(1), 32–58. http://dx.doi.org/10.1002/syn.10279.

Vertes, R. P. (2006). Interactions among the medial prefrontal cortex, hippocampus and midline thalamus in emotional and cognitive processing in the rat. *Neuroscience, 142*(1), 1–20. http://dx.doi.org/10.1016/j.neuroscience.2006.06.027.

Watson, C. S., White, S. E., Homan, J. H., Kimura, K. A., Brien, J. F., Fraher, L., & Bocking, A. D. (1999). Increased cerebral extracellular adenosine and decreased PGE2 during ethanol-induced inhibition of FBM. *Journal of Applied Physiology, 86*(4), 1410–1420.

Wise, R. A. (1980). The dopamine synapse and the notion of 'pleasure centers' in the brain. *Trends in Neurosciences, 3*.

Wise, R. A. (2008). Dopamine and reward: The anhedonia hypothesis 30 years on. *Neurotoxicity Research, 14*(2–3), 169–183. http://dx.doi.org/10.1007/bf03033808.

Wise, R. A., & Bozarth, M. A. (1987). A psychomotor stimulant theory of addiction. *Psychological Review, 94*(4), 469–492.

Woolf, N. J. (1991). Cholinergic systems in mammalian brain and spinal cord. *Progress in Neurobiology, 37*(6), 475–524.

Wu, J., & Lukas, R. J. (2011). Naturally-expressed nicotinic acetylcholine receptor subtypes. *Biochemical Pharmacology, 82*(8), 800–807. http://dx.doi.org/10.1016/j.bcp.2011.07.067.

Yan, Q. S., Reith, M. E., Yan, S. G., & Jobe, P. C. (1998). Effect of systemic ethanol on basal and stimulated glutamate releases in the nucleus accumbens of freely moving Sprague-Dawley rats: A microdialysis study. *Neuroscience Letters, 258*(1), 29–32.

Young, E. M., Mahler, S., Chi, H., & de Wit, H. (2005). Mecamylamine and ethanol preference in healthy volunteers. *Alcoholism, Clinical & Experimental Research, 29*(1), 58–65. http://dx.doi.org/10.1097/01.alc.0000150007.34702.16.

Zaborszky, L., van den Pol, A., & Gyengesi, E. (2012). The basal forebrain cholinergic projection system in mice. In C. Watson, G. Paxinos, & L. Puelles (Eds.), *The mouse nervous system* (pp. 684–718). San Diego: Academic Press.

Zant, J. C., Kim, T., Prokai, L., Szarka, S., McNally, J., McKenna, J. T., & Basheer, R. (2016). Cholinergic neurons in the basal forebrain promote wakefulness by actions on neighboring non-cholinergic neurons: An opto-dialysis study. *The Journal of Neuroscience, 36*(6), 2057–2067. http://dx.doi.org/10.1523/JNEUROSCI.3318-15.2016.

CHAPTER

28

Chronic and Acute Nicotine Exposure Versus Placebo in Smokers and Nonsmokers: A Systematic Review of Resting-State fMRI Studies

S.J. Brooks[1], *J. Ipser*[1], *D.J. Stein*[1,2]

[1]University of Cape Town, Cape Town, South Africa; [2]MRC Unit on Anxiety & Stress Disorders, Cape Town, South Africa

BACKGROUND

Resting-state networks provide useful information about the natural mode of functioning in the healthy and disordered brain, independent of stimulus-biased responses. Resting-state functional magnetic resonance imaging (RS-fMRI) was pioneered by Biswal, Yetkin, Haughton, and Hyde (1995) after observations that spontaneous low-level synchronous fluctuations (<0.1 Hz) in blood oxygen level-dependent (BOLD) signal occurred independently of task stimulation. The default mode network (DMN) was the first resting-state network identified, using positron emission tomography (PET) by Raichle et al. (2001) and confirmed with fMRI by Greicius, Krasnow, Reiss, and Menon (2003). The DMN involves the medial prefrontal cortex (mPFC), the posterior cingulate/precuneus, superior temporal cortex, hippocampus, and the inferior parietal cortex and is also known by alternative nomenclature, such as the task-negative network (TNN), or the hippocampal–cortex memory system (Vincent, Kahn, Snyder, Raichle, & Buckner, 2008). The DMN is recruited during nongoal-oriented cognitive activity, introspective rumination, low-level arousal, homeostatic- and self-regulation, such as during "daydreaming," and supporting autobiographical, internally focused, declarative, episodic memory retrieval.

The DMN is deactivated antagonistically and anticorrelated with the executive control network (ECN), a frontoparietal system involving the dorsolateral prefrontal cortex (dlPFC) and inferior parietal cortex (Barkhof, Haller, & Rombouts, 2014) (see Fig. 28.1). The ECN or task-positive network underlies executive functioning such as working memory, goal-oriented cognition, and impulse control and is connected via reciprocal and parallel segregated pathways to different locations within the dorsal striatum (Barkhof et al., 2014). Although the ECN can be distinguished anatomically from the dorsal attention network (DAN), which lies intermediate to the ECN and DMN in the frontal and parietal cortices, and is therefore well-placed to integrate these two networks (Vincent et al., 2008), most RS-fMRI literature on the ECN does not make this distinction. In addition to the DMN and ECN, there are many other major canonical RS networks that are frequently identified in the literature (for review, see Barkhof et al., 2014). These networks include the salience network (SN: frontal cortex, anterior cingulate, and anterior insular cortex circuitry), dorsal attention network (DAN: insular cortex and posterior parietal), auditory (temporal cortex), sensorimotor network (SMN: striatal and parietal cortex), and visual network (VN: occipital cortex). The insular cortex is commonly involved in the salience and dorsal attention RS circuits (Cohen, 2014), both of which likely influence the activation of other RS networks by shifting attention to internal or external stimulation, for example (Lerman et al., 2014). In terms of nicotine addiction, damage to the insular cortex has been associated with the cessation of smoking and reduced craving (Naqvi, Gaznick, Tranel, & Bechara, 2014), and a review of RS studies of those

Addictive Substances and Neurological Disease
http://dx.doi.org/10.1016/B978-0-12-805373-7.00028-1

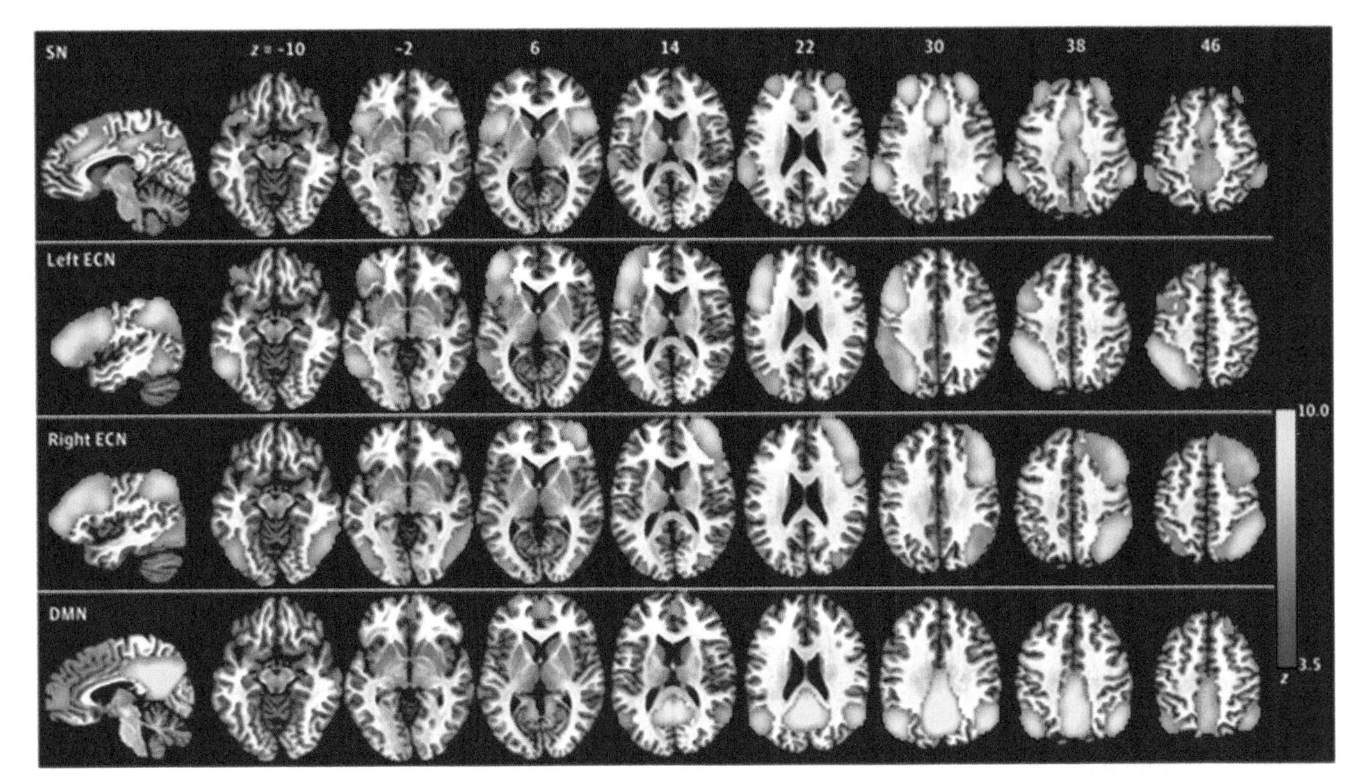

FIGURE 28.1 Acute nicotine exposure is associated with enhanced synchronicity in the limbic connectivity, such that the salience network (SN) and executive control network (ECN) may temporarily improve attention in the short-term, and anticorrelate with default mode network (DMN). *From Lerman, C., Gu, H., Loughead, J., Ruparel, K., Yang, Y., & Stein, E. A. (2014). Large-scale brain network coupling predicts acute nicotine abstinence effects on craving and cognitive function.* JAMA Psychiatry, *71(5), 523–530. http://dx.doi.org/10.1001/jamapsychiatry.2013.4091. Permissions granted from JAMA Psychiatry.*

with nicotine dependence implicates an ACC–insula neural model (Sutherland, McHugh, Pariyadath, & Stein, 2012) (see Fig. 28.2). Furthermore, the inherent efficacy of intrinsic cognitive control networks in their RS is pertinent to the study of nicotine addiction due to the persistence of nicotine-consumption-dependence and -withdrawal effects (Jasinska et al., 2014).

Brain imaging technologies provide a noninvasive method for examining the effects of pharmacological challenge on RS networks in the brain, using nicotine for example. The use of RS-fMRI has helped the advancement of nicotine addiction research (McClernon & Gilbert, 2004). RS-fMRI uses correlations between a proxy measure of neuronal activity in voxels situated in regions of interest (including the whole brain) to make inferences about brain connectivity when participants are engaged in a nonspecific task requiring minimal cognitive resources (i.e., a "resting" condition). Common approaches to analyze RS-fMRI include methods such as seed-based correlations, independent components analyses (ICAs), and graph theory to examine patterns of functional connectivity between nodes, or vertices (e.g., random and lattice). Modern advances in the methods of analysis of RS-fMRI include structural equation modeling (SEM) and measures to

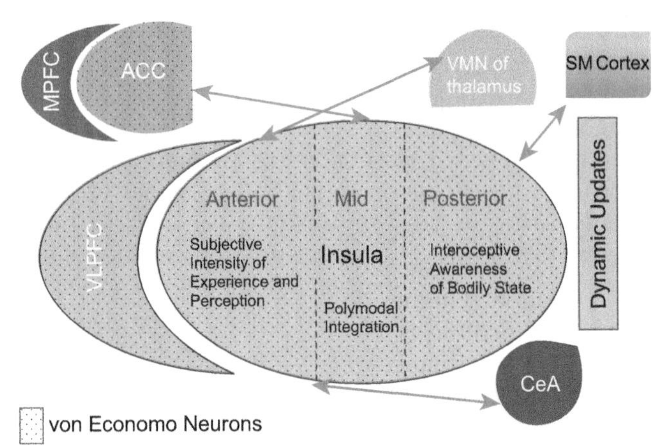

FIGURE 28.2 The insula. Representation of the insula and its cytoarchitectural similarities with the anterior cingulate cortex (ACC) and the ventrolateral prefrontal cortex (VLPFC). Illustrated above are the main connectivities involved in dynamic updates between the following: (1) the posterior insula and the sensory motor (SM) cortex; and (2) the anterior insula and both the central nucleus of the amygdala(CeA) and the ventro-medial nucleus (VMN) of the thalamus. The mid-insula is also connected to the thalamus and is intermediary in translating the somatosensory stimuli to the anterior region where they are perceived and further evaluated in conjunction with the ACC and the medial prefrontal cortex (mPFC). *From Pavuluri, M., & May, A. (2015). I feel, therefore, I am: The insula and its role in human emotion, cognition and the sensory-motor system.* AIMS Neuroscience, *2(1), 18–27. http://dx.doi.org/10.3934/Neuroscience.2015.1.18. Permissions granted by the authors and also the editor of AIMS Neuroscience by email communication.*

assess the direction of functional connectivity, such as Granger's causality (Vincent et al., 2008). However, RS-fMRI studies of the effects of nicotine on brain function have so far not used these modern analytic procedures, and have rather relied on traditional methods.

Reviews of RS-fMRI in those who consume nicotine highlight connectivity patterns at rest that may underlie alterations in cognitive control, with these patterns associated with addiction liability, cognitive enhancement, and the development of neuroimaging-based biomarkers for addiction (Barkhof et al., 2014; Fedota & Stein, 2015; Jasinska, Zorick, Brody, & Stein, 2014; Lu & Stein, 2014; Mihov & Hurlemann, 2012; Sutherland et al., 2012). For example, acute nicotine administration may boost cognitive performance via activation of the SN that suppresses the task-negative, internally driven DMN, while temporarily enhancing the intrinsic connectivity of the ECN, or task-positive network to focus attention on external goals (Lerman et al., 2014; Sutherland et al., 2012). This is consistent with Volkow's work on stimulant use and the effects on the brain, where drug cues are suggested to capture the salience network, thus reducing attention toward naturally rewarding stimuli (Goldstein & Volkow, 2011).

Conversely, for chronic smokers, only few RS-fMRI studies reported consistent alteration of brain function during rest, and so the effects of chronic smoking on resting brain function are not clear (Cole et al., 2010; Wang et al., 2007). Additionally, interpretation of RS-fMRI data among abstinent chronic smokers may be confounded by psychological symptoms following acute nicotine withdrawal, including craving, anger, irritability, frustration, anxiety, difficulty concentrating, restlessness, depression, increased appetite, insomnia, and impatience (Hughes, 2007). Notably, a hallmark mental feature of nicotine withdrawal after chronic use, alongside these psychological effects, is reduced concentration (Hughes, 2007), whereas acute administration of nicotine is linked to temporarily enhanced cognitive performance in smokers (Heishman, Kleykamp, & Singleton, 2010).

Despite advances in our understanding of how nicotine affects the brain, reviews suggest that it would be simplistic to posit at this stage a single profile of neural response to nicotine intake and that potential network interactions should be taken into account (Fedota & Stein, 2015). Instead, changes in brain function are dependent on various factors, including duration of use, with chronic and acute exposure associated with different neural signatures (Hong et al., 2009). This is to be expected, given the pronounced reductions in the volumes of certain brain regions, such as the bilateral PFC and ACC volume, in chronic smokers versus those who have never smoked, consistent with deficits in working memory (Brody et al., 2004; Evans & Drobes, 2009; Gallinat et al., 2006; Kühn, Schubert, & Gallinat, 2010; Myers, Taylor, Moolchan, & Heishman, 2008). Differences in intrinsic functional connectivity in relation to chronic versus acute nicotine exposure are also observed, although no study and only one review (Fedota & Stein, 2015) has compared acute (state) versus chronic (trait) nicotine intake, although unlike our review these authors focused less on the methods of measuring RS networks. For instance, although RS-fMRI studies of chronic and acute exposure frequently focus on the effects of nicotine on connectivity of the ACC and insula (Sutherland et al., 2012), the specific regions within these structures in which perturbations of connectivity are observed appear to depend on whether acute or dependence-related effects of nicotine are being investigated (Hong et al., 2009), as well as which statistical method is used.

Against the background of heterogeneous data analysis methods for RS-fMRI studies into nicotine consumption a meta-analysis in this case was not possible. Therefore, here we attempt to provide a comprehensive, systematic review of RS-fMRI studies which differs from the most recent, informative review by Fedota and Stein (2015) in five main ways: (1) we provide a more in-depth description of a broader range of RS-fMRI methods (see the following) whereas by their own admission the scope of the previous review did not include an extensive methodological overview; (2) we only describe data from healthy subjects who smoke (including nonaffected first-degree relatives of individuals with schizophrenia), whereas Fedota et al. include brain activation of psychiatric patients who also smoke; (3) we only focus on fMRI methods to increase the homogeneity of reported findings, whereas Fedota et al. also describe, for example, arterial spin labeling (ASL) and PET; (4) we update the review to include four more studies: in chronic (our $n = 13$ vs. their $n = 12$) and acute (our $n = 10$ vs. their $n = 7$); and finally (5) the previous review introduces a sophisticated heuristic framework to characterize neurobiological markers of addiction, whereas we attempt to simplify the field and the methodology for a general audience to appreciate chronic versus acute consumption of nicotine and patterns of neural activation at rest. Thus, here we review: (1) chronic nicotine exposure versus those who have never smoked on resting neural function in otherwise healthy adults and (2) acute nicotine exposure versus placebo in current smokers. We chose to focus on RS-fMRI studies (and not, for example, perfusion fMRI such as ASL) in an attempt to reduce the heterogeneity of methods and thus hone specific RS networks associated with a variation in blood oxygenation following nicotine consumption.

METHODS

Searching

Inclusion/Exclusion Criteria

Two types of studies employing RS-fMRI were considered potentially eligible for inclusion in this review: (1) those investigating the effects of current smoking in chronic smokers compared to individuals who have never smoked (henceforth referred to as chronic studies) and (2) studies assessing the effect of an acute administration of nicotine compared to placebo in individuals who currently smoke or with prior history of smoking (referred to as acute studies). PubMed, MEDLINE, and ScienceDirect were searched for eligible studies up to December 2015. Moreover, manual searches of the reference lists of eligible study reports were performed. Search terms included: (resting state) AND (fMRI) AND (smoking); OR (resting state), AND (fMRI) AND (nicotine); OR (fMRI) AND (nicotine); OR (fMRI) AND (smoking). Studies were eligible for inclusion if they reported RS functional connectivity analyses of fMRI performed on adults in English language peer-reviewed journals. Studies and/or data collected from psychiatric populations were excluded (e.g., schizophrenia). However, if publications focusing on psychiatric populations also provided data on smokers and nonsmokers among their healthy controls these were examined for potential eligibility for inclusion.

For details of searching procedure see Supplementary Fig. 28.3 PRISMA diagram.

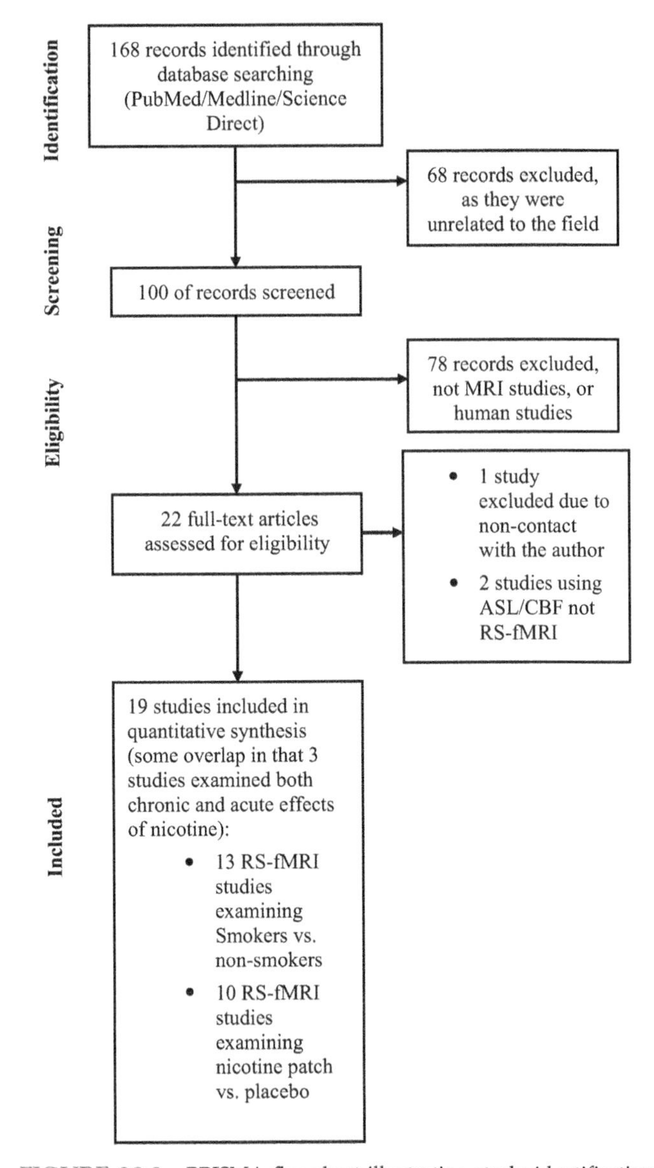

FIGURE 28.3 PRISMA flowchart illustrating study identification, exclusion, and inclusion in meta-analysis.

RESULTS

See Table 28.1 for details of included studies in the review for this chapter.

Chronic Studies (n = 13)

It should be noted that the authors of the included studies examining chronic smokers versus those who had purportedly never smoked could not be entirely certain that self-reported nonsmokers had never smoked in their lifetime. Nevertheless, all studies included in this review did state that at the time of the scan, and at least 1 year prior to the study the control subjects had not smoked.

Two studies examining the effects of chronic cigarette use on RS connectivity used seed-based methods located in regions across the DMN and ECN, namely in the dlPFC, dorsal medial prefrontal cortex (dmPFC), dorsal anterior cingulate cortex/cingulate cortex (dACC), rostral anterior cingulate cortex (rACC), occipital cortex, and insula/operculum (Moran, Sampath, Kochunov, & Hong, 2012; Moran, Sampath, Stein, & Hong, 2012; Zhang et al., 2011). Seed-based methods analyze connectivity patterns between a predetermined brain region and other distal regions. Connectivity patterns are based on inferences of synchronicity in the BOLD time courses for the respective regions. DMN connectivity increases intrinsically during introspective internally focused thought (e.g., self-referential processing), but connectivity between the regions of the DMN decreases during extrinsic cognitive tasks that demand attention on externally derived stimuli, which rather utilizes the ECN. Zhang et al. (2011) placed seeds in six regions across the DMN and ECN and compared 18 male and female smokers against 18 age- and gender-matched controls, while Moran, Sampath, Kochunov, et al. (2012) and Moran, Sampath, Stein, et al. (2012) positioned a seed in the dACC in a

TABLE 28.1 Resting-State fMRI Studies (in Alphabetical Order) Included in the Review

	Study Name:	Nicotine Intake (FTND > 6)	Control Condition:	Total N, Sex, Age Range	Seed Regions	Analysis	Main Results (Smokers Compared to Nonsmokers)
(A) CHRONIC SMOKERS VERSUS NONSMOKERS (N = 13)							
1	Breckel et al. (2013)	16.3 cigarettes per day for at least 2 years, last cigarette smoked 2 h before scan	Never smoked	35:16 f, 19 m, 19–44 years 18 CS, 17 NS	Functional network integration (global efficiency and clustering)	Graph theory, (nodes/vertices)	$\leftrightarrow$ left insula $\leftrightarrow$ middle frontal gyrus
2	Chu et al. (2014)	Smoked 10 or more cigarettes per day during the previous year, had smoked for more than 5 years, had no period of smoking abstinence longer than 3 months in the past year, and met ICD 10 criteria for nicotine dependence were eligible for the study	Never smoked	39 m, Chinese Han, 23–40 years 20 CS, 19 NS	Quantitative measurements of network of low-frequency oscillations, correcting for physiological noise in nonspecific brain areas (fALFF)	Fractional Amplitude of Low-Frequency Fluctuation (fALFF)	Smokers > nonsmokers $\uparrow$ fALFF in the left middle occipital gyrus, left limbic lobe, and left cerebellum posterior lobe $\downarrow$ fALFF in the right middle frontal gyrus, right superior temporal gyrus, right extra nuclear, left postcentral gyrus, and left cerebellum anterior lobe. Heavy smokers > light smokers $\uparrow$ fALFF in the right superior temporal gyrus, right precentral gyrus, and right occipital lobe/cuneus $\downarrow$ fALFF in the right/left limbic lobe/cingulate gyrus, right/left frontal lobe/subgyral, and right/left cerebellum posterior lobe
3	Janes et al. (2012)	At least 10 cigarettes per day in the last 6 months; "allowed to smoke until shortly before the study visit"	13 Never smoked; two smoked <2 packs of cigarettes in their lifetime about 30 years previously, and one smoked <10 packs of cigarettes 14 years previously	29:29 f (age range not given) 13 CS, 16 NS	Subcortical limbic network Medial prefrontal cortex network Left frontoparietal network Right frontoparietal network	Independent Components Analysis (ICA), and dual regression approach	$\uparrow$ coupling between left frontoparietal and medial prefrontal networks $\uparrow$ subcortical limbic network amplitude

Continued

TABLE 28.1 Resting-State fMRI Studies (in Alphabetical Order) Included in the Review—cont'd

	Study Name:	Nicotine Intake (FTND > 6)	Control Condition:	Total N, Sex, Age Range	Seed Regions	Analysis	Main Results (Smokers Compared to Nonsmokers)
4	Huang et al. (2014)	Smoked at least daily for the past year and have a lifetime history of smoking >100 cigarettes	Smoked no more than two cigarettes in lifetime, and none in the year prior to scan	21 CS; "both genders"; 18–39 years	Anterior cingulate	ICA, and seed-based approach	↑ functional connectivity between the anterior cingulate cortex and the precuneus, caudate, putamen, and frontal cortex
5	Lin et al. (2015)	20 cigarettes per day for at least 5 years, no period of abstinence for more than 3 months; no special instruction regarding whether they should or should not smoke before the study, but no withdrawal symptoms before the study	Smoked no more than five cigarettes in their lifetime	64:12 f, 52 m; 33–58 years 31 CS (heavy) versus 33 NS	45 nodes of anatomical interest, repeated for each hemisphere, correlational matrices across the 90 regions	Area under the curve (AUC) measures of topographic networks, graph network analysis (nodes/vertices)	↓ global efficiency ↑ local efficiency and clustering coefficients and greater path length ↓ nodal global efficiency mainly in brain regions within the default mode network (DMN) ↑ nodal local efficiency in the visual-related regions Association between the altered network metrics and the duration of cigarette use or the severity of nicotine dependence
6	Moran, Sampath, Kochunov, et al. (2012) and Moran, Sampath, Stein, et al. (2012)	First-degree smoking relatives of individuals diagnosed with schizophrenia: 19.5 cigarettes per day; 17.5 packs per year; smoke 1 cigarette prior to scan but not 30 min before scan	Never smoked	65 (gender not given) 18–58 years 37 CS, 28 NS	dACC	Seed-based analysis	↓dACC—right limbic cluster (vs. parahippocampal, amygdalar, and posterior insular regions) resting connectivity strength that correlated with nicotine addiction severity
7	Pariyadath et al. (2014)	Scored at least 6 on the Fagerstrom test, smoking ad libitum before the scan (no details of number of cigarettes and number of years of smoking)	Never smoked	42, 20 f, 22 m, 28–48 years	16 RS networks, including: sensorimotor, auditory, thalamus/caudate, ECN (mPFC, dlPFC, DS, midbrain), DMN (mPFC, cerebellum, cuneus), visual, frontal, cerebellum, frontoparietal, higher	Support vector machine (SVM)-based classification applied to resting-state networks	Within-network functional connectivity measures better predict smoking status than between-network connectivity (the representativeness of each individual node with respect to its parent network). Connectivity measures

					order network (dlPFC, parietal)		between the executive control network and frontoparietal network predict smoking status
8	Tang et al. (2012)	10 or more cigarettes per day during the previous year; no abstinence longer than 3 months	Never smoked	89:18 f, 71 m 19–39 years 45 CS, 44 NS	A measure of connectivity synchronization between brain regions	Whole-brain Regional Homogeneity (Wb ReHo)	↓ ReHo in the right inferior frontal cortex ↑ ReHo in the left superior parietal lobe
9	Wang et al. (2007)	No smoking at least 12 h before the scan	Never smoked	42 m, 19–28 years 22 CS 20 NS	T-tests of GBC between groups	Global brain connectivity (GBC)	State of abstinence smokers > controls ↑ GBC in the insula and superior frontal gyrus
10	Weiland et al. (2015)	12.2 cigarettes per day on average	Never smoked	Total n = 650 452 CS, age 31.4 (276 m, 176 f) 198 NS, age 30.1 years (117 m, 81 f)	Dorsal DMN and the right and left executive control networks (RECN and LECN)	ICA within DMN and ECNs, group differences in network connectivity strength using predefined algorithms	↓ network strength in smokers than nonsmokers in the left ECN and DMN ↓ lower connectivity than nonsmokers associated with key network hubs: the dorsolateral prefrontal cortex (dlPFC), and parietal nodes within ECNs. Further, ECN connectivity strength was negatively associated with pack years of cigarette use
11	Wu et al. (2015)	37.26 cigarettes per day on average; smoking for 25.42 years, Fagerstrom test score 8.87	Never smoked	Total n = 68 34 CS, age 46.29 years, 24 m, 7 f 34 NS, age 46.91 years, 28 m, 5 f	Voxel-by-voxel intrinsic brain activity	ReHo with the resting-state fMRI data analysis toolkit (REST) (Song et al., 2011)	Compared with nonsmokers, heavy smokers showed decreased ReHo primarily in brain regions associated with the default-mode, frontoparietal attention, and inhibitory control networks; heavy smokers showed increased ReHo predominately in regions related to motor planning
12	Yu et al. (2013)	At least 15 cigarettes per day for at least 10 years	Never smoked	32 m 39–47 years 16 CS, 16 NS	A measure of connectivity synchronization between brain regions	Regional Homogeneity with ROIs (ROI ReHo)	↓ ReHo in medial and lateral PFC ↑ ReHo in insula and posterior cingulate cortex

Continued

TABLE 28.1 Resting-State fMRI Studies (in Alphabetical Order) Included in the Review—cont'd

	Study Name:	Nicotine Intake (FTND > 6)	Control Condition:	Total N, Sex, Age Range	Seed Regions	Analysis	Main Results (Smokers Compared to Nonsmokers)
13	Zhang et al. (2011)	Average 21 cigarettes per day for average 12 years; no cigarette for at least 2 h before the scan	Never smoked	44:21 f, 23 m, 21–50 years For resting-state analyses a subset was examined: 18 CS, 18 NS	Six regions that were more activated in smokers compared to nonsmokers during prior cue-induced fMRI: dlPFC, dorsal medial prefrontal cortex (dmPFC), dorsal anterior cingulate cortex/cingulate cortex, rostral anterior cingulate cortex (rACC), occipital cortex, and insula/operculum.	Seed-based resting-state functional connectivity (rsFC), derived from cue-induced activations in smokers	rsFC strength between rACC and dlPFC was positively correlated with the cue-elicited activity in dlPFC rsFC strength between dlPFC and dmPFC was positively correlated with the cue-elicited activity in dmPFC, while rsFC strength between dmPFC and insula/operculum was negatively correlated with the cue-elicited activity in both dmPFC and insula/operculum
(B) ACUTE NICOTINE VERSUS ABSTINENCE (N = 10)							
1	Cole et al. (2010)	4-mg nicotine patch; current smokers who abstain during scan or prior smokers Double-blind, placebo controlled, cross-over design	Placebo patch	17:4 f, 13 m, 19–49 years 17 CS	ECN/DMN	Seed-based functional connectivity	↑ inverse coupling between ECN and DMN; improvements in withdrawal symptoms negatively correlated with altered functional connectivity within the DMN, and connectivity between the ECN and regions implicated in reward processing
2	Ding & Lee (2013)	>5 years, >15 cigarettes/day. Scored at least 6 on the Fagerstrom test; smoked before the scan until satisfied	Repeat scan in smokers, but after abstaining for 12 h prior to scan (to evoke nicotine withdrawal)	21 CS m, 22–30 years	Initial ICA to obtain maps for DMN, executive-control network (ECN), and salience network (SN); secondary analyses with GCA	ICA and Grainger Causality Analysis (GCA)	↓ effective connectivity from SN to DMN after smoking ↑ from ECN/DMN to SN was after smoking A paired t-test on ICA spatial patterns revealed functional connectivity variation in regions such as the insula, parahippocampus, precuneus, anterior cingulate cortex, supplementary motor

							area, and ventromedial/dlPFC These regions were later selected as the regions of interest (ROIs), and their effective connectivity was investigated using GCA ↑ insula effective connectivity with the other ROIs; while in smoking satiety, the parahippocampus had the enhanced interarea effective connectivity
3	Janes et al. (2014)	Current dependent smokers, average 7 pack years of smoking use (number of packs of cigarettes smoked/day × years as a smoker) ≥10 cigarettes per day for the last 6 months	1-h abstinence	17:8 m, 9 f, 20–30 years 17 CS	OMPFC resting-state network, including the sgACC extending into the ventral striatum	Seed-based functional connectivity and correlation with craving	↑ coupling with greater craving between the OMPFC network and other cortical, limbic, striatal, and visceromotor brain regions
4	Hong et al. (2009)	21–35 mg single dose nicotine patch; current smokers ≥10 cigarettes per day, who abstain during the scan or prior smokers; double-blind place controlled, randomized, crossover design. Two scans between 5 and 14 days apart. Abstain from smoking for 2 h prior to scan and 4.5 h prior to patch administration	Placebo patch	19:14 m, 5 f, 18–50 years 19 CS	LdACC, LvPCC, LdPCC, LsACC, RMCC, RvPCC, RdPCC	Seed-based functional connectivity and correlation with craving	↑ dACC-dorsal striatal functional connectivity circuits correlated positively with severity of nicotine addiction (not modified by nicotine patch administration) ↑ cingulate-neocortical functional connectivity enhanced by acute nicotine administration
5	Huang et al. (2014)	Smoked at least daily for the past year and have a lifetime history of smoking >100 cigarettes.	11 h of abstinence	21 CS; both genders, 18–39 years	Anterior cingulate	ICA, and seed-based approach	↑ connectivity between the anterior cingulate cortex and the precuneus, insula, orbital frontal gyrus, superior frontal gyrus, posterior cingulate

Continued

TABLE 28.1 Resting-State fMRI Studies (in Alphabetical Order) Included in the Review—cont'd

	Study Name:	Nicotine Intake (FTND > 6)	Control Condition:	Total N, Sex, Age Range	Seed Regions	Analysis	Main Results (Smokers Compared to Nonsmokers)
							cortex, superior temporal, and inferior temporal lobe ↑ strength of connectivity with intensity of craving between the anterior cingulate cortex and the precuneus, insula, caudate, putamen, middle cingulate gyrus, and precentral gyrus
6	Lerman et al. (2014)	Smoking satiety in smokers, who had cigarette 20 min before scan. Smokers having ≥10 cigarettes per day for ≥ 6 months	24 h of abstinence	37, 19–61 years 37 CS	Strength of coupling between DMN, ECN, and SN	Resource Allocation Index (RAI)	↓ RAI in the abstinent compared with the smoking satiety states suggesting weaker inhibition between the default mode and salience networks ↓ abstinence-induced cravings to smoke and less suppression of default mode activity (vmPFC, PCC)
7	Moran, Sampath, Kochunov, et al. (2012) and Moran, Sampath, Stein, et al. (2012)	21–35-mg nicotine patch; current smokers who abstain during the scan or prior smokers	Placebo patch	24 CS: ns, 18–50 years	L pIns, R pIns, L dACC L daCC R dACC	Seed-based functional connectivity and correlation with craving	↓ correlations between ↑ smoking severity and rsFC between insula, dACC, and striatum
8	Tanabe, et al. (2011)	7-mg nicotine patch; all subjects were nicotine-free for at least 3 years prior to the study, and so were prior, not current smokers	Placebo patch	19 CS: 11 m, 8 f Average age: 30 years	DMN	Group ICA was performed	↓ coupling in the default network in the posterior cingulate cortex, precuneus, paracentral lobule, and medial orbitofrontal cortex
9	Wang et al. (2014)	Smokers allowed to smoke two cigarettes before the scan (SOS)	No smoking for 12 h prior to scan (SOA)	42 CS m, 19–28 years	T-tests of GBC between groups	Whole-brain Regional Homogeneity (Wb ReHo)	State of satisfaction smokers > state of abstinence ↓ GBC in several regions of the DMN, as compared with smokers in the SOA condition

10 Wylie et al. (2012)	All subjects were nonsmokers, 10 had never smoked, while 5 had minimal previous tobacco use. No subject had a lifetime use of more than 100 cigarettes and all were nicotine-free for at least 3 years prior to the study. Subjects were scanned before and 90 min after receiving a 7-mg transdermal nicotine patch	Placebo patch	15 CS: 9 m, 6 f Average age: 29 years 15 NS	DMN, ECN	Network topology using graph theory (nodes/vertices) based on previously published ROIs	No effect on average connectivity was observed ↑ local efficiency ↑ global efficiency ↑in the regional efficiency of limbic and paralimbic areas

Summary of smoking versus nonsmoking RS-fMRI studies:
Increased coupling between frontoparietal (precuneus) networks, increased limbic (striatum, cerebellum) network amplitude. Increased connectivity between the insula and superior frontal gyrus and posterior cingulate. Increased local efficiency and clustering coefficients and greater path length and amplitude frequency in visual networks. Reduced dACC—right limbic cluster (VS, parahippocampal, amygdalar, and posterior insular). Reduced global efficiency within the DMN, reduced frequency amplitude in a frontotemporal network. Within-network functional connectivity measures better predict smoking status than between-network connectivity. Reduced regional connectivity within the right inferior frontal cortex but increased regional connectivity in the left superior parietal lobe. Connectivity measures between the executive control network and frontoparietal network predict smoking status and nicotine dependence. *Main circuits implicated*: Frontoparietal increases and increased DMN in general with reduced efficiency. Reduced connectivity between anterior cingulate and limbic regions. Greater connectivity in DMN and reduced ECN.

Summary of acute nicotine versus abstinence RS-fMRI studies:
Increased inverse coupling between ECN and DMN. Increased coupling with greater craving between the OMPFC network and other cortical and limbic brain regions. Increased cingulate–neocortical functional connectivity enhanced by acute nicotine administration. Increased connectivity between the anterior cingulate cortex and the precuneus, insula, frontal cortex, posterior cingulate cortex, and temporal lobe. Weaker inhibition between the default mode and salience networks in smoking abstinence versus smoking satisfaction. Higher smoking severity and lower rsFC between insula, dACC, and striatum after nicotine. *Main circuits implicated*: Cingulate cortex and connections to the parietal and limbic regions, greater connectivity in ECNs.

CS, chronic smoker; *ECN*, executive control network; *f*, female; *GBC*, global brain connectivity; *l*, left; *m*, male; *n*, number of participants; *NS*, nonsmoker; *ns*, not specified; *OMPFC*, orbitomedial prefrontal cortex; *r*, right; *ReHo*, regional homogeneity (functional coherence between neighboring voxels); *ROI*, region of interest analysis; *WB*, whole-brain analysis; ↔, equal connectivity between two brain regions; ↑, increase in connectivity; ↓, decrease in connectivity.

comparison of 37 chronic male and female smokers (who were first-degree relatives of individuals with schizophrenia) to 28 who had never smoked (Moran, Sampath, Kochunov, et al., 2012; Moran, Sampath, Stein, et al., 2012). Zhang et al. found that functional connectivity between the dmPFC and insula was negatively correlated with nicotine addiction severity, and also increased functional connectivity between rACC, dlPFC, and dmPFC. Moran et al. found decreased functional connectivity between the dACC and ventral striatum, parahippocampal, amygdalar, and posterior insula regions, with more greatly decreased connectivity between these regions associated with nicotine addiction severity.

Three studies used ICAs to examine nodes of network activity across the whole brain (Huang et al., 2014; Janes, Nickerson, Frederick Bde, & Kaufman, 2012; Weiland, Sabbineni, Calhoun, Welsh, & Hutchison, 2015). ICA is a data-driven, model-free method of identifying spatially orthogonal networks that have been applied to intrinsic connectivity fMRI data (Beckmann, DeLuca, Devlin, & Smith, 2005). Although this method extracts spatially orthogonal networks, it can be used to interrogate connectivity between networks. Using ICA, Janes et al. (2012) did not identify connectivity differences but rather greater activation amplitude between left frontoparietal and mPFC, as well as greater subcortical limbic network amplitude in chronic smokers versus those who had never smoked. Amplitude, as opposed to functional connectivity illustrates collective strength of activation rather than the correlation coefficient of functional association between specific regions within a given network. Similarly, Huang et al. compared smokers (who were in withdrawal for 11 h) to nonsmokers with ICA and found stronger functional connectivity in the smokers between the ACC and the bilateral precuneus, left inferior and middle temporal cortex, bilateral middle frontal cortex, right caudate and putamen, and left inferior parietal cortex. However, care must be taken in interpretation as the strength of connectivity and increased amplitude of activation in current smokers versus nonsmokers, between the ACC and the regions described foregoing could be attributable to the withdrawal state (e.g., conflict monitoring, aversion sensitivity, and so on). Finally, Weiland et al. (2015) used ICA to examine the dorsal DMN and left/right ECN in chronic heavy smokers versus those who have never smoked. They found reduced network strength between the left ECN and the DMN, as well as lower connectivity in network hubs within the ECN, including the dlPFC, and parietal nodes. Finally, they reported that ECN connectivity strength was negatively correlated with duration of smoking.

Three studies used an approach called Regional Homogeneity (ReHo) to examine intrinsic connectivity between voxels and time series data derived from the BOLD signal (Tang et al., 2012; Wu, Yang, Zhu, & Lin, 2015; Yu et al., 2013). ReHo is a method that enables the investigation of local brain functional connectivity on an individual subject basis (Zhang et al., 2011). ReHo measures correlations between the BOLD time series for a specified voxel and neighboring voxels within a prespecified radius, with Kendall's coefficient of concordance (KCC) most frequently used as an estimate of connectivity. Tang et al. (2012) used a whole-brain approach and examined 45 chronic male and female smokers against 44 who had never smoked, and found decreased ReHo in the right inferior frontal cortex and increased ReHo in the left superior parietal cortex in smokers. Yu et al. (2013) examined 16 male smokers and 16 male nonsmokers using a regional brain approach within the frontoparietal network (FPN), and found that smokers compared to nonsmokers had reduced ReHo in medial and lateral PFC and increased ReHo in the insula and posterior cingulate cortex. Finally, Wu et al. (2015) used ReHo to compare network synchronicity between smokers and nonsmokers. The study revealed reduced synchronicity in brain regions associated with the DM, frontoparietal attention, and inhibitory control networks. Furthermore, heavy smokers showed increased ReHo predominately in regions related to motor planning.

Two studies used graph theory to examine the difference between RS connectivity in chronic smokers versus those who have never smoked (Breckel, Thiel, & Giessing, 2013; Lin, Wu, Zhu, & Lei, 2015). Graph theory assesses network structure in terms of regions of interest, or nodes, as well as connections between the nodes, known as edges. In the context of RS-fMRI, the edges consist of correlations between BOLD time courses in the respective nodes that exceed a threshold, typically arbitrarily set to Pearson's $R = 0.2$ or 0.3. Breckel et al. (2013) examined 18 male and female smokers versus 17 nonsmoking age- and gender-matched controls, and found that there was no difference in the intrinsic connectivity between the left insula and middle frontal cortex when minimally deprived chronic smokers were compared with those who have never smoked. They suggest that changes observed in network connectivity in psychiatric populations (e.g., schizophrenia) who also chronically smoke may be attributable to the psychiatric condition rather than to changes associated with chronic nicotine consumption. Also, using graph theory, Lin et al. (2015) compared 31 heavy male and female smokers to 33 nonsmokers and examined nodal connections between 90 regions of interest across the whole brain (Lin et al., 2015). They used nonparametric permutation testing to examine group differences in connections between these regions, and regression analyses to relate differences in nodal connections to severity of nicotine addiction scores. They found that compared to those who have never smoked, heavy smokers showed lower global efficiency (lower synchronous functional connectivity coupling across the whole-brain/larger networks), and higher local efficiency (higher synchronous functional connectivity coupling within local/smaller brain networks). They also demonstrated higher local clustering coefficients (areas of locally increased synchronous functional connectivity) and greater path length (functional connectivity spreading to more distal brain regions), all indicative, according to the authors, of reduced overall efficiency of RS networks in chronic smokers compared to those who have never smoked. Furthermore, heavy smokers showed decreased nodal global efficiency within the DMN, and increased nodal local efficiency in the visual-related regions. Decreased nodal global efficiency within the DMN might be indicative of diminished self-regulation (Mel'nikov et al., 2014), whereas increased nodal local efficiency might be associated with cognitive biases toward substance-related cues in those who are substance dependent (Bueichekú et al., 2015).

The remaining three chronic studies used different analysis approaches. Pariyadath et al. used a machine learning approach, known as support vector machine (SVM), using the AdaBoost algorithm (Freund & Schapire, 1997) to examine network connectivity in 42 smokers versus nonsmokers (Pariyadath, Stein, & Ross, 2014). This algorithm trains the SVM classifier (e.g., smoker vs. nonsmoker) via an iterative process on a weighted set of samples, where the weights are determined by the accuracy of the classifier for smokers versus nonsmokers on a previous iteration. The resulting classification occurs via a linear combination of individual classifiers, where each SVM classifier is weighted by its performance accuracy. In this way, AdaBoost builds a collection of nonlinear classifiers from a weighted combination of multiple linear SVM classifiers (Pariyadath et al., 2014). Their analyses used 16 regions of interest derived from a previous ICA including: sensorimotor, auditory, thalamus/caudate, ECN (mPFC, dlPFC, dorsal striatum, and midbrain), DMN (mPFC, cerebellum, and cuneus), visual, frontal, cerebellum, frontoparietal, and higher-order (dlPFC and parietal) network. They showed that increased connectivity measures within the executive control and FPNs were particularly informative in predicting smoking status. Specifically, their FPN incorporated the left angular gyrus (DMN/ECN), left middle frontal cortex (DMN), left middle temporal cortex, left cingulate cortex (DMN/ECN and SN), right supramarginal cortex, right middle frontal cortex (DMN and SN), right middle temporal cortex, and right medial PFC (DMN). It is of note that some of the frontoparietal regions they use overlap between the DMN, SN, and ECN (see suggestions in the preceding parentheses). Pariyadath and colleagues finally conducted post hoc t-tests to reveal group differences in network connectivity, showing significantly lower connectivity in chronic smokers compared to never-smoked controls in two "higher-order networks." One higher-order network they identified consisted of the left precuneus, bilateral inferior parietal cortex, bilateral middle frontal cortex, bilateral temporal cortex, left occipital cortex, and left cingulate cortex. The second higher-order network that appeared to be reduced in smokers compared to controls included the bilateral precuneus, left lateral occipital cortex, right posterior cingulate cortex, right middle PFC, left lingual gyrus, right culmen, and left posterior cingulate cortex.

Chu et al. used fractional Amplitude of Low-Frequency Fluctuation (fALFF) to measure the strength of correlation between low-frequency oscillations, correcting for physiological noise in nonspecific brain areas (Chu et al., 2014) in 20 Chinese Han male chronic smokers and 19 matched controls who had never smoked. fALFF is a relative measure of the proportion of the amplitude of intrinsic BOLD T2* signal across the entire frequency range that is contributed by low frequencies (typically 0.01−0.1 Hz) (Zou et al., 2008). Smokers compared to nonsmokers had higher fALFF, or in other words, higher correlations in low-frequency oscillations between the left middle occipital cortex, a large area encompassing the left limbic region (e.g., striatum and hippocampal−amygdala complex) and left posterior cerebellum, but lower fALFF in the right medial PFC, right superior temporal cortex, right extra nuclear region (insular cortex, striatum, and amygdala), left postcentral cortex, and left cerebellum anterior lobe. In additional analyses, Chu et al. compared light-to-heavy smokers, and revealed that heavy smokers had increased fALFF in the right superior temporal cortex, right precentral primary motor cortex, and right occipital cortex/cuneus, and reduced fALFF in a large area encompassing the bilateral limbic region/cingulate gyrus, bilateral PFC, and bilateral posterior cerebellum. The physiological relevance of higher or lower fALFF (correlations of low-frequency oscillations in brain activation) may be indicative of increased or decreased neural communication or synchronicity between these regions. In other words, increased fALFF in limbic regions in smokers compared to nonsmokers may indicate greater neural communication in sensory and reward regions associated with craving for nicotine. Conversely, reduced fALFF in heavy versus light smokers in limbic regions may indicate reduced neural communication that is temporarily stimulated by the consumption of nicotine. Also, reduced fALFF in prefrontal regions may be indicative of executive function deficits that may ordinarily help to regulate nicotine craving and consumption.

Finally, Wang et al. (2014) compared 22 chronic male 12-h abstinent smokers prior to the scan, to 20 males who have never smoked. The authors compared global brain connectivity (GBC) across the two groups (Wang et al., 2014). GBC uses a specific algorithm as defined by others (Buckner et al., 2009; Tomasi & Volkow, 2010, 2011, 2012). In brief, GBC measures the connectivity between each voxel in the whole brain and creates an estimate of global connectivity. The time course of each voxel from each participant is correlated with every other voxel. Thus, a matrix of Pearson's correlation coefficients is obtained. Subsequently, the number of voxel connections for each voxel is counted within a threshold to compute the GBC map. The vertex degree is calculated as the number of adjacent links using an undirected and weighted adjacency matrix. The map is finally transformed to Fisher Z values, so that maps across participants can be averaged and compared (Wang et al., 2014). Compared to nonsmokers, smokers had greater GBC in the left insula and left superior frontal cortex, and craving scores negatively correlated with left superior frontal cortex activation but not left insula.

In summary, intrinsic connectivity between the dACC, dmPFC, and within a broad area collectively referred to as the limbic region (e.g., striatum, amygdala, and hippocampus) associated with both the DMN and salience networks appears to be reduced in chronic smokers compared to nonsmokers. Furthermore, nicotine dependence severity appears to negatively correlate with intrinsic connectivity within these regions. Moreover, reduced connectivity in superior frontal cortex in chronic smokers is associated with greater nicotine dependence. In contrast to the reduced connectivity observed between networks in chronic smokers compared to nonsmokers, chronic smoking is associated with increased within-network connectivity, particularly with respect to the DMN (e.g., left frontoparietal/precuneus and mPFC networks), as well as the SN, and may be indicative of heightened sense of self-regulation need in the presence of craving for nicotine. Additionally, localized estimates of connectivity are increased in the left superior parietal cortex, insula, and posterior cingulate cortex (regions associated with the DMN and SN), and reduced in the right inferior, medial, and lateral PFC (associated with the ECN) in chronic smokers compared to nonsmokers. However, others show no increased connectivity between the insula cortex and mPFC in relation to nicotine addiction. Graph theory methods indicate reduced overall global efficiency of resting-state networks in chronic smokers compared to those who have never smoked. However, local efficiency in smaller networks (as opposed to global efficiency across the whole brain) such as those incorporating limbic structures extending to the insula and ACC is also increased in smokers versus nonsmokers (e.g., Huang et al., 2014; Janes et al., 2012) or with no difference to nonsmokers (Breckel et al., 2013). While this may appear to be counterintuitive, local increases observed in smokers may still not pass the connectivity threshold of global efficiency within neural networks observed in nonsmoking controls.

Furthermore, heavy smokers compared to light smokers show decreased nodal global efficiency within the DMN, and increased nodal local efficiency in the visual-related regions, indicative of an inverted U-shape relationship between functional connectivity in light-to-heavy smoking. Other machine-learning approaches show reduced connectivity in chronic, compared to nonsmokers in frontoparietal regions associated with both the DMN and ECN (left precuneus, bilateral inferior parietal cortex, bilateral mPFC, and ACC), and also temporal and occipital cortices. Thus, the effects of chronic smoking on RS connectivity are variable, but chronic smoking compared to nonsmoking appears to broadly influence frontoparietal and limbic connectivity associated with the DMN and SN.

Acute Studies (n = 10)

Four studies examined the effects of acute nicotine administration versus abstinence using seed-based analyses. Cole et al. (2010) used seeds in the ECN and DMN and examined 17 smokers who received a 4-mg nicotine patch versus placebo and found increased inverse coupling between ECN and DMN. Improvements in withdrawal symptoms negatively correlated with altered functional connectivity within the DMN, and connectivity between the ECN and regions implicated in reward processing. Hong et al. (2009) also compared functional connectivity when smokers were given a 21–35-mg nicotine patch versus placebo, with seeds in the anterior and posterior cingulate regions. They found increased dACC–dorsal striatal functional connectivity circuits that correlated positively with severity of nicotine addiction, and increased cingulate–neocortical functional connectivity enhanced by acute nicotine administration. Moran, Sampath, Kochunov, et al. (2012) and Moran, Sampath, Stein, et al. (2012) also gave smokers a 21–35-mg nicotine patch versus placebo, and placed seeds in the bilateral posterior insula and bilateral dorsal ACC. They found increased functional connectivity in line with greater smoking severity and rsFC among insula, dACC, and striatum. Janes, Farmer, Frederick, Nickerson, and Lukas (2014) compared 17 smokers before and after a 1-h abstinence period using a seed-based approach based in the orbitomedial prefrontal cortex (OMPFC) and subgenual anterior cingulate cortex (sgACC) and found increased coupling with greater craving between the OMPFC–sgACC network and other cortical, limbic, striatal, and viscera–motor brain regions.

Three acute studies used ICAs: one to probe the effects of smoking as usual before the scan versus an 11-h period of abstinence in 21 current smokers (Huang et al., 2014); second after the administration of a 7-mg nicotine patch versus placebo in 19 current smokers (Tanabe et al., 2011); and the third used ICA with Grainger Causality Analysis (GCA) to measure effective connectivity (directional analysis) within and between regional networks (Ding & Lee, 2013) in smokers in withdrawal versus satiation. Huang et al., after identifying spatially orthogonal networks, placed seeds in the ACC and DMN, respectively, and found increased connectivity between the ACC and the precuneus, insula, orbital frontal gyrus, superior frontal gyrus, posterior cingulate cortex, superior temporal, and inferior temporal lobe in smokers versus the abstinent period. They also found increased strength of connectivity with intensity of craving between the ACC and the precuneus, insula, caudate, putamen, middle cingulate gyrus, and precentral gyrus. Tanabe et al. found decreased coupling in the DMN after nicotine patch administration, specifically between the posterior cingulate cortex, precuneus, paracentral lobule, and medial orbitofrontal cortex. Ding et al. reported that there was reduced effective connectivity from SN to DMN and increased connectivity from ECN/DMN to the SN after smoking satiation compared to withdrawal. Furthermore, the authors reported increased connectivity from the insula to the parahippocampus, precuneus, ACC, SMA, and ventromedial/dlPFC after smoking satiation.

The remaining three acute studies used varying methods. Lerman et al. (2014) used a Resource Allocation Index (RAI) to examine the strength of coupling between DMN, ECN, and SN when comparing smoking satiety versus 24 h of abstinence. RAI is a composite quantitative network association index integrating the SN-ECN (positive) correlation and the SN-DMN (negative) correlation, to represent nicotine-satiety versus nicotine-withdrawal response, respectively, proposed by some (e.g., Sutherland et al., 2012). Specifically, the SN is suggested to toggle neural activity bias between the DMN and ECN in order to shift attention toward external (nicotine satiety) or internal (nicotine withdrawal) stimulation. Using Group Independent Component Activation (GICA) maps as spatial predictors for each participant's four-dimensional data, a linear regression generated a time course for each component. As a measure of cross-network coupling, the authors calculated correlation coefficients (CC) between component time courses derived from the SN, ECN, and the DMN ($CC_{SN,ECN}$ and $CC_{SN,DMN}$). To assess the actions of the SN on DMN and ECN in the smoking and abstinence states with a single value, the authors defined a composite network association index as $m = zSN,ECN - zSN,DMN = f(CC_{SN,ECN}) - f(CC_{SN,DMN})$, where $f(CC) = ½ \ln (1 + CC/1 - CC)$ and m refers to the RAI (Lerman et al., 2014). With this approach, Lerman et al. found reduced RAI in the abstinent compared to the smoking satiety state, suggesting weaker inhibition between the DMN and SN (negative network). The authors suggest that weaker inhibition of the DMN–SN is the driving force of abstinence-induced lower RAI and that weaker network connectivity in the ECN–SN contributes to the urge to smoke. This is in line with less neural activation suppression in the DMN–SN RAI during a working memory task, and consistent with observations that nicotine addiction is associated with working memory deficits (Evans & Drobes, 2009; Myers et al., 2008).

In another study, Wang et al. (2014) permitted 42 male smokers to smoke 2 cigarettes before the scan, and compared functional connectivity using ReHo whole-brain measures to a condition where the smokers were not permitted to smoke for 12 h prior to the scan. They found reduced global ReHo in several regions of the DMN in the smoking, as compared with smokers in the abstinence condition. Finally, Wylie, Rojas, Tanabe, Martin, and Tregellas (2012) used network topology using graph theory (nodes/vertices) based on previously published ROIs in the DMN and ECN in nonsmokers who received a 7-mg transdermal nicotine patch versus placebo. They found no effect on average connectivity, but an increase in local regional network efficiency in the limbic and paralimbic areas.

In summary, and in contrast to studies examining the effects of chronic smoking on RS networks, acute nicotine exposure is associated with elevated functional connectivity in the ECN and SN, combined with reduced intrinsic connectivity in regions of the DMN. Similarly, reduced feelings of craving and withdrawal appear to be associated with reduced DMN and greater ECN connectivity with limbic regions, suggesting greater executive control of reward networks. Moreover, nicotine craving is associated with greater functional connectivity in regions of the SN (e.g., cingulate cortex and insula) and DMN following nicotine patch administration.

DISCUSSION

We present a systematic review of RS-fMRI studies that examines the effects of chronic smoking versus the acute administration of nicotine on functional connectivity of brain networks, with more focus on RS-fMRI methods as compared to other reviews that have included other imaging modalities (e.g., Fedota & Stein, 2015). Additionally, here we only synthesize data derived from smoking and nonsmoking participants in otherwise good psychiatric health, whereas other reviews have included patient populations who smoke (e.g., those with schizophrenia). Considering the differences between this and previous reviews, we have attempted to provide a homogeneous account of the methods used and the reported effects of acute versus chronic nicotine exposure on RS neural networks.

Acute administration of nicotine offers a brief "spike" in nicotine exposure versus no exposure, and is, on the whole, associated with enhanced connectivity within and between the ECN and SN, as well as diminished DMN connectivity (see Fig. 28.1, adapted from Lerman et al., 2014). Chronic administration of nicotine in smokers versus nonsmokers, however, shows differential effects on RS networks; some studies report increased DMN and reduced ECN, while others demonstrate increased ECN and SN connectivity. There was some indication that severity of nicotine craving, dependence, and withdrawal effects in chronic smokers are associated with reduced connectivity within PFC networks, although other studies show increased frontostriatal and limbic connectivity at rest in conjunction with nicotine-dependence severity. Our systematic review adds to the extant reviews on RS networks in those with nicotine dependence. For example, one review provides a neural model of RS networks in those with nicotine dependence linking functional connectivity within the DMN and ECN with activation of the ACC and the insula (Sutherland et al., 2012). In 2015 review by Fedota and Stein on acute versus chronic nicotine effects on RS networks in the brain provided a sophisticated heuristic of the most implicated neural networks and biomarkers for addiction, involving, broadly, the DMN, ECN, and SN, and more specifically the ACC and insula. This most recent review emphasizes, as is the case here, that acute nicotine exposure appears to heighten activation of attention networks, whereas chronic nicotine exposure dampens global brain processes and synchronicity within executive functioning networks, while potentially heightening DMN activation associated with arousal, somatosensation, and addiction.

Our review, which has also furnished the reader with a methodological overview of RS-fMRI studies in the hope to better illuminate how RS networks are derived, suggests that acute administration of nicotine is most consistent compared to the chronic effects of nicotine on rsFC. This may be due in part, to a greater level of experimental control achieved over extenuating variables in pharmacological challenge studies (e.g., nicotine patch vs. placebo). For example, other factors may influence RS networks in chronic smokers, including: frequency and type of smoking behavior, brand of cigarette and levels of nicotine, tar and carbon monoxide, individual metabolism and lung health, as well as periods of abstinence during the chronic smoking period. Acute administration of nicotine, on the other hand, offers an "odd-ball event" in nicotine exposure versus no exposure, although the administration of nicotine via a patch or lozenge is undoubtedly delayed in its potency than intravenously, for example. However, caution must be exercised in interpreting the findings of acute nicotine studies as the vehicle type (e.g., cigarette, lozenge, and patch) has varying impact on the entry in to the nervous system and thus can have differing effects on brain processes.

The findings from investigations of the chronic effects of nicotine consumption in smokers versus nonsmokers while less consistent than acute studies seem in general, to implicate alterations in frontoparietal and limbic connectivity, especially involving the PFC, insula, and ACC. However, the precise nature of these changes, as well as the causal relationship between symptoms of dependence/craving and disruptions of these networks is currently unclear, although weak functional connectivity between the superior frontal gyrus and the right insula is suggested to correspond to nicotine addiction and dysfunctional executive control (Fedota et al., 2016). Furthermore, some studies reported no differences in functional connectivity between chronic smokers and nonsmokers (e.g., Breckel et al., 2013), while the findings of other chronic nicotine consumption studies were confounded by the study of otherwise healthy first-degree relatives of individuals with schizophrenia (Moran, Sampath, Kochunov, et al., 2012; Moran, Sampath, Stein, et al., 2012). The effects of chronic smoking on RS neural connectivity are likely influenced by multilevel factors, not just in smoking behavior and other lifestyle choices, but also differences in physiology (Piasecki, Trela, Hedeker, & Mermelstein, 2014). Additionally, nicotine intake contributes to oxidative stress, which can compromise vasculature architecture (Zanetti et al., 2014). Thus, it might be that varying levels of chronic oxidative stress lead to variation in the effects of chronic smoking on RS neural connectivity. Given that nicotine at first acts on peripheral sensory receptors before more slowly entering neural systems, future RS studies must also consider other such physiological factors (Kiyatkin, 2014).

In terms of the effects of acute administration of nicotine in comparison to placebo or abstinence on RS brain activity in smokers, more experimental control over extraneous variables can be achieved, not least because of the controlled administration of nicotine, but also because these studies often use a repeated measures design, using a placebo or abstinence-controlled condition within the same participant. For example, acute administration of nicotine may temporarily improve cognitive and behavioral performance by rather enhancing the ECN–SN and deactivating the DMN–SN (Cole et al., 2010; Lerman et al., 2014). Additionally, acute nicotine administration may reduce functional connectivity between distal regions of the frontolimbic circuitry, but increase connectivity between local networks within the PFC—to temporarily improve executive function (Ding & Lee, 2013; Hong et al., 2009; Huang et al., 2014; Janes et al., 2014). Enhancement of PFC connectivity by acute nicotine administration seems to particularly center on the ACC and connections to parietal cortex at rest (Moran, Sampath, Kochunov, et al., 2012; Moran, Sampath, Stein, et al., 2012; Tanabe et al., 2011). The DMN function, which may promote interoceptive feelings of craving and dependence during nicotine abstinence, is perhaps reduced at rest following acute nicotine exposure, and increased ECN may reflect a greater allocation of mental resources driven by the SN (Lerman et al., 2014).

At the same time, acute nicotine exposure might also foster lowered global brain efficiency (perhaps related to metabolic challenges following the introduction of nicotine), such that the RS network changes observed are due to the engagement of compensatory mechanisms, or the result of smoking-related impairments (Wang et al., 2014). However, others report greater local efficiency in limbic and paralimbic regions following nicotine patch administration (Wylie et al., 2012), implicating again the recruitment of the SN, a common feature of both chronic and acute nicotine consumption. Indeed, damage to the insula, part of the SN and DMN, is associated with smoking cessation and reduced craving (Naqvi et al., 2014). Thus, one might suggest that the mechanism of action for acute administration of nicotine is to allocate greater neural resources to cognitive control networks, such as the DMN, ECN, and the FPN that straddles these systems (Vincent et al., 2008) and may be associated with subjective experience of improved attention.

Conversely, the effects of chronic nicotine exposure on RS connectivity are less clear, but appear to broadly and perhaps most consistently alter frontostriatal connectivity. Future RS studies examining the effects of chronic smoking must, however, adopt measures of causality, and control for the longitudinal variances in smoking behavior described in the preceding paragraphs. Yet, to account for the lack of clarity in studies of chronic nicotine use, further information can be gleaned from the few studies that have examined correlations with craving, dependence, and withdrawal effects on RS networks.

Nicotine Craving and Resting-State Networks

Craving scores and severity of nicotine dependence were correlated with measures of functional connectivity in seven studies reviewed here (Cole et al., 2010; Hong et al., 2009; Huang et al., 2014; Janes et al., 2014; Lerman et al., 2014; Lin et al., 2015; Moran, Sampath, Kochunov, et al., 2012; Moran, Sampath, Stein, et al., 2012). The duration of cigarette use, and the severity of nicotine dependence, as measured by the Fagerström score, is associated with greater reduced global efficiency across the whole brain and increased local efficiency in visual regions in chronic smokers compared to nonsmokers (Lin et al., 2015). Such a finding might indicate that smoking heightens visual and attentional bias, perhaps in line with cue-induced addiction behavior (Goldstein & Volkow, 2011). Additionally, when smoking participants experienced more craving during the abstinence or placebo condition in some of the acute studies, there was less inhibition of the DMN (Lerman et al., 2014; Wang et al., 2014), which might be associated with greater interoceptive awareness of homeostatic tension (need for acetylcholine nicotine receptor occupancy), for example. Furthermore, the experience of craving may be associated with reduced RS activity between the insula and striatum during acute nicotine administration (Moran, Sampath, Kochunov, et al., 2012; Moran, Sampath, Stein, et al., 2012), perhaps an indication of the feelings of satiety. Moreover, nicotine abstinence and craving in those who smoke is linked to a reduced ability to concentrate on cognitive tasks (Hughes, 2007; Parrott, Garnham, Wesnes, & Pincock, 1996), and acute nicotine consumption may temporarily improve attention and other executive functions. This lends support to the tendency that people seem to smoke when they are cognitively overloaded and anxious (e.g., worries about work, money, and relationships), and continue to smoke despite knowledge of health risks.

Adherence to and desire for cigarette smoking may coincide with temporary improvements in concentration, with consequent changes to brain circuitry associated with the subjective experience of effective problem-solving and task-focused brain function in the ECN. For example, a dual role regarding the effects of nicotine on the brain has been proposed via the release of dopamine from the ventral tegmental area, to influence the mesolimbic pathway for addiction, and the mesocortical pathway for enhanced cognition (see Jasinska et al., 2014). Clinical studies of those who smoke versus those who do not confirm that acute administration of nicotine improves neuropsychological performance and subjective experience (Caldirola et al., 2013), and might explain why self-medication with cigarettes is observed in up to 80% of people with psychiatric conditions, such as mood disorders and schizophrenia (Drusch et al., 2013). Furthermore, the stimulation of acetylcholine receptors in the ECN by the intake of nicotine, for example, may alleviate some psychiatric symptoms in mental disorders that are associated with deficits in these brain circuits. Thus, nicotine may temporarily reestablish disrupted connectivity between the insula and striatum, enhancing subjective experiences of effective problem solving and modulation of reward responses (Jasinska et al., 2014).

Some limitations of this systematic review deserve consideration. For example, while the Fagerström Test for Nicotine Dependence (FTND) (Fagerström, Heatherton, & Kozlowski, 1990) is a validated measure of smoking addiction severity and is often used in RS studies, with a score of 6 or above indicating moderate to severe nicotine addiction, individual smoking patterns may differ. These differences may alter the chemical yields of nicotine, carbon monoxide levels, and the toxicological effects on brain function. Additionally, caution must be taken when interpreting the effects of acute (e.g., lozenge and patch) versus chronic (e.g., cigarette) nicotine consumption on RS neural activation as these methods of consumption differ dramatically. Also, it is not known to what degree the acute effects of cigarette consumption versus nicotine vehicle differ. Different cigarette products on the market vary in the levels of nicotine and other chemicals, and so the consumer's choice of cigarette can also play a role. Moreover, individual usage of the cigarette, such as the way it is held in the mouth, the length of inhalation and exhalation, and how much of the cigarette smoke is consumed at each setting can alter the amounts of nicotine entering a person's system over a chronic time period. Nicotine-dependence symptoms also vary and may be associated with physiological factors and the incidences of metabolic syndrome X, for example, and this can influence the number of cigarettes consumed, and a compulsion to continue smoking despite knowledge of the health risks. A person who is dependent on nicotine, when attempting to quit, may experience physical withdrawal symptoms, such as increased perspiration and heart rate, raised blood pressure, and general malaise pertaining to differences in brain function.

Additionally, although studies used similar seed regions, they were not identical. Some studies incorporated one seed region, while others used a network as a seed (e.g., the DMN and ECN). The location of the seed can influence the connectivity profile associated with nicotine administration. Nevertheless, most seed regions included in this review were from a similar area of the brain implicated in addiction-related deficits. Furthermore, there was a large variability in the way the networks were classified (e.g., seed based, ICA, graph theory, ReHo, fALFF, GBC, RAI, and SVM), and therefore the inconsistency in methods renders the findings of RS-fMRI studies difficult to generalize. Moreover, only one RS-fMRI study directly examined causality, or the directionality of functional networks, which is a useful avenue for future studies to explore, and can be done with Granger Causality techniques, for example (Ding & Lee, 2013). Additionally, this review included a small number of eligible studies ($n = 19$), though the average sample size was relatively large by neuroimaging standards ($N = 45$). A novelty of this review is that we discussed the intrinsic functional connectivity correlates of acute and chronic smoking in relation to the different methodologies and different study parameters, which is hopefully a progression to the review of Fedota and Stein (2015).

In conclusion, it can be clearly seen that nicotine exposure alters brain function at rest. The effects of chronic nicotine exposure in comparison to those who have never smoked are inconsistent, however, although it appears that the frontostriatal circuitry, encompassing the DMN and ECN, as well as the ACC and insular cortex contributes to nicotine dependence. Inconsistencies in the findings of the effects of chronic smoking on RS networks are likely due to a variety of confounding factors, including differences in smoking behavior and physiological condition of the smoker. Conversely, RS-fMRI studies on the effects of acute nicotine exposure in smokers versus placebo, for example, have fewer confounds, although differences in nicotine vehicle (e.g., lozenge, patch, and cigarette) must be considered. The most consistent findings reported across studies are that acute nicotine exposure may temporarily improve cognitive performance associated with enhanced ECN and SN connectivity, anticorrelating with and reducing connectivity within the DMN. Enhanced cognitive performance following acute nicotine exposure may occur via the effects of nicotine on the ECN that is associated with attention control, and also with the anticorrelated suppression of DMN networks that are associated with internal arousal and homeostatic regulation, perhaps mediated by the activation of the SN. Acute nicotine exposure, if repeated, might promote a conditioned, subjective experience of the beneficial effects of smoking, and provide a heuristic for biomarkers of addiction (e.g., Fedota & Stein, 2015). The ACC, insula, and parietal regions are perhaps most implicated by RS studies to coincide with nicotine addiction severity. Furthermore, these regions are associated with differential resource allocation of the SN during the experience of craving and higher scores of nicotine addiction. Future RS-fMRI studies could examine the reversibility of the effects of nicotine consumption on brain activation at rest, and whether clinical interventions such as cognitive training can decrease symptoms of nicotine dependence by altering activity within these circuits.

Acknowledgments

S.J.B. was funded by the NIH NIDA (R21 DA040492). Thanks are due to Ms. Linsay Blows for her contribution to this manuscript at the beginning of the project, namely in helping to search for studies as part of her MSc project.

Authors' Contributions

S.J.B. conceptualized the project, supervised students in data collection, analysis, and interpretation of the findings, and wrote the manuscript; J.I. contributed to the interpretation and some of the writing of the manuscript (e.g., interpreting resting-state findings in relation to nicotine consumption). D.J.S. helped to write the manuscript.

References

Barkhof, F., Haller, S., & Rombouts, S. A. (2014). Resting-state functional MR imaging: A new window to the brain. *Radiology, 272*(1), 29–49.

Beckmann, C. F., DeLuca, M., Devlin, J. T., & Smith, S. M. (2005). Investigations into resting-state connectivity using independent component analysis. *Philosophical Transactions of the Royal Society of London. Series B, Biological Sciences, 360*, 1001–1013.

Biswal, B., Yetkin, F. Z., Haughton, V. M., & Hyde, J. S. (1995). Functional connectivity in the motor cortex of resting human brain using echo-planar MRI. *Magnetic Resonance Medicine, 34*(4), 537–541.

Breckel, T. P., Thiel, C. M., & Giessing, C. (December 30, 2013). The efficiency of functional brain networks does not differ between smokers and non-smokers. *Psychiatry Research, 214*(3), 349–356.

Brody, A. L., Mandelkern, M. A., Jarvik, M. E., Lee, G. S., Smith, E. C., Huang, J. C., … London, E. D. (2004). Differences between smokers and non-smokers in regional grey matter volumes and densities. *Biological Psychiatry, 55*, 77–84.

Buckner, R. L., Sepulcre, J., Talukdar, T., Krienen, F. M., Liu, H., … Johnson, K. A. (2009). Cortical hubs revealed by intrinsic functional connectivity: Mapping, assessment of stability, and relation to Alzheimer's disease. *The Journal of Neuroscience, 29*, 1860–1873.

Bueichekú, E., Ventura-Campos, N., Palomar-García, M.Á., Miró-Padilla, A., Parcet, M. A., & Ávila, C. (2015). Functional connectivity between superior parietal lobule and primary visual cortex "at rest" predicts visual search efficiency. *Brain Connectivity, 5*(8), 517–526.

Caldirola, D., Dacco, S., Grassi, M., Citterio, A., Menotti, R., Cavedini, P., … Perna, G. (2013). Effects of cigarette smoking on neuropsychological performance in mood disorders: A comparison between smoking and nonsmoking inpatients. *Journal of Clinical Psychiatry, 74*(2), e130–136.

Chu, S., Xiao, D., Wang, S., Peng, P., Xie, T., He, Y., & Wang, C. (2014). Spontaneous brain activity in chronic smokers revealed by fractional amplitude of low frequency fluctuation analysis: A resting state functional magnetic resonance imaging study. *Chinese Medical Journal, 127*(8), 1504–1509, 2014.

Cohen, M. X. (2014). Fluctuations in oscillation frequency control spike timing and coordinate neural networks. *The Journal of Neuroscience, 34*(27), 8988–8998, 2.

Cole, D. M., Beckmann, C. F., Long, C. J., Matthews, P. M., Durcan, M. J., & Beaver, J. D. (2010). Nicotine replacement in abstinent smokers improves cognitive withdrawal symptoms with modulation of resting brain network dynamics. *NeuroImage, 52*(2), 590–599. http://dx.doi.org/10.1016/j.neuroimage.2010.04.251.

Ding, X., & Lee, S. W. (2013). Changes of functional and effective connectivity in smoking replenishment on deprived heavy smokers: A resting-state FMRI study. *PLoS One, 8*(3), e59331.

Drusch, K., Lowe, A., Fisahn, K., Brinkmeyer, J., Musso, F., Mobascher, A., … Wölwer, W. (2013). Effects of nicotine on social cognition, social competence and self-reported stress in schizophrenia patients and healthy controls. *European Archives of Psychiatry and Clinical Neuroscience, 263*(6), 519–527.

Evans, D. E., & Drobes, D. J. (2009). Nicotine self-medication of cognitive-attentional processing. *Addiction Biology, 14*(1), 32–42, 2009.

Fagerström, K. O., Heatherton, T. F., & Kozlowski, L. (1990). Nicotine addiction and its assessment. *Ear, Nose, & Throat Journal, 69*(11), 763–765.

Fedota, J. R., Matous, A. L., Salmeron, B. J., Gu, H., Ross, T. J., & Stein, E. A. (2016). Insula demonstrates a non-linear response to varying demand for cognitive control and weaker resting connectivity with the executive control network in smokers. *Neuropsychopharmacology, 41*(10), 2557–2565 (Epub ahead of print).

Fedota, J. R., & Stein, E. A. (September 2015). Resting-state functional connectivity and nicotine addiction: Prospects for biomarker development. *Annals of the New York Academy of Sciences, 1349*(1), 64–82.

Freund, Y., & Schapire, R. E. (1997). A decision-theoretic generalization of on-line learning and an application to boosting. *Journal of Computer and System Sciences, 55*(1), 119–139.

Gallinat, J., Meisenzahl, E., Jacobsen, L. K., Kalus, P., Bierbrauer, J., Kienast, T., … Staedtgen, M. (2006). Smoking and structural brain deficits: A volumetric MR investigation. *European Journal of Neuroscience, 24*, 1744–1750.

Goldstein, R. Z., & Volkow, N. D. (2011). Dysfunction of the prefrontal cortex in addiction: Neuroimaging findings and clinical implications. *Nature Reviews. Neuroscience, 12*(11), 652–669, 20.

Greicius, M. D., Krasnow, B., Reiss, A. L., & Menon, V. (2003). Functional connectivity in the resting brain: A network analysis of the default mode hypothesis. *Proceedings of the National Academy of Sciences of the United States of America, 100*(1), 253–258.

Heishman, S. J., Kleykamp, B. A., & Singleton, E. G. (2010). Meta-analysis of the acute effects of nicotine and smoking on human performance. *Psychopharmacology, 210*(4), 453–469.

Hong, L. E., Gu, H., Yang, Y., Ross, T. J., Salmeron, B. J., Buchholz, B., … Stein, E. A. (2009). Association of nicotine addiction and nicotine's actions with separate cingulate cortex functional circuits. *Archives of General Psychiatry, 66*(4), 431–441.

Huang, W., King, J. A., Ursprung, W. W., Zheng, S., Zhang, N., Kennedy, D. N., … DiFranza, J. R. (2014). The development and expression of physical nicotine dependence corresponds to structural and functional alterations in the anterior cingulate-precuneus pathway. *Brain and Behavior, 4*(3), 408–417.

Hughes, J. R. (2007). Effects of abstinence from tobacco: Valid symptoms and time course. *Nicotine Tobacco Research, 9*(3), 315–327.

Janes, A. C., Farmer, S., Frederick, B. D., Nickerson, L. D., & Lukas, S. E. (2014). An increase in tobacco craving is associated with enhanced medial prefrontal cortex network coupling. *PLoS One, 9*(2), e88228.

Janes, A. C., Nickerson, L. D., Frederick Bde, B., & Kaufman, M. J. (October 1, 2012). Prefrontal and limbic resting state brain network functional connectivity differs between nicotine-dependent smokers and non-smoking controls. *Drug and Alcohol Dependence, 125*(3), 252–259.

Jasinska, A. J., Zorick, T., Brody, A. L., & Stein, E. A. (September 2014). Dual role of nicotine in addiction and cognition: A review of neuroimaging studies in humans. *Neuropharmacology, 84*, 111–122.

Kiyatkin, E. A. (2014). Critical role of peripheral sensory systems in mediating the neural effects of nicotine following its acute and repeated exposure. *Reviews in the Neurosciences, 25*(2), 207–221.

Kühn, S., Schubert, F., & Gallinat, J. (2010). Reduced thickness of medial orbitofrontal cortex in smokers. *Biological Psychiatry, 68*, 1061–1065.

Lerman, C., Gu, H., Loughead, J., Ruparel, K., Yang, Y., & Stein, E. A. (2014). Large-scale brain network coupling predicts acute nicotine abstinence effects on craving and cognitive function. *JAMA Psychiatry, 71*(5), 523–530.

Lin, F., Wu, G., Zhu, L., & Lei, H. (July 2015). Altered brain functional networks in heavy smokers. *Addiction Biology, 20*(4), 809–819.

Lu, H., & Stein, E. A. (2014). Resting state functional connectivity: Its physiological basis and application in neuropharmacology. *Neuropharmacology, 84*, 79–89.

McClernon, F. J., & Gilbert, D. G. (2004). Human functional neuroimaging in nicotine and tobacco research: Basics, background, and beyond. *Nicotine Tobacco Research, 6*(6), 941–959.

Mel'nikov, M. E., Shtark, M. B., Korostyshevskaya, A. M., Sevelov, A. A., Petrovskii, E. D., Pokrovskii, M. A., … Kosykh, E. P. (2014). Dynamic mapping of the brain in substance-dependent individuals: Functional magnetic resonance imaging. *Bulletin of Experimental Biology and Medicine, 158*(2), 260–263.

Mihov, Y., & Hurlemann, R. (2012). Altered amygdala function in nicotine addiction: Insights from human neuroimaging studies. *Neuropsychologia, 50*(8), 1719–1729.

Moran, L. V., Sampath, H., Kochunov, P., & Hong, L. E. (2012). Brain circuits that link schizophrenia to high risk of cigarette smoking. *Schizophrenia Bulletin, 39*(6), 1373–1381.

Moran, L. V., Sampath, H., Stein, E. A., & Hong, L. E. (2012). Insular and anterior cingulate circuits in smokers with schizophrenia. *Schizophrenia Research, 142*(1–3), 223–229.

Myers, C. S., Taylor, R. C., Moolchan, E. T., & Heishman, S. J. (2008). Dose-related enhancement of mood and cognition in smokers administered nicotine nasal spray. *Neuropsychopharmacology, 33*(3), 588–598.

Naqvi, N. H., Gaznick, N., Tranel, D., & Bechara, A. (2014). The insular: A critical neural substrate for craving and drug seeking under conflict and risk. *Annals of the New York Academy of Sciences, 1316*, 53–70.

Pariyadath, V., Stein, E. A., & Ross, T. J. (2014). Machine learning classification of resting state functional connectivity predicts smoking status. *Frontiers in Human Neuroscience, 8*, 425.

Parrott, A. C., Garnham, N. J., Wesnes, K., & Pincock, C. (1996). Cigarette smoking and abstinence: Comparative effects upon cognitive task performance and mood state over 24 hours. *Human Psychopharmacology and Clinical Experiments, 11*, 391–400.

Piasecki, T. M., Trela, C. J., Hedeker, D., & Mermelstein, R. J. (2014). Smoking antecedents: Separating between- and within-person effects of tobacco dependence in a multiwave ecological momentary assessment investigation of adolescent smoking. *Nicotine & Tobacco Research, 16*(Suppl. 2), S119–S126.

Raichle, M. E., MacLeod, A. M., Snyder, A. Z., Powers, W. J., Gusnard, D. A., & Shulman, G. L. (2001). A default mode of brain function. *Proceedings of the National Academy of Sciences of the United States of America, 98*(2), 676–682.

Song, X. W., Dong, Z. Y., Long, X. Y., Li, S. F., Zuo, X. N., Zhu, C. Z., … Zang, Y. F. (2011). REST: A toolkit for resting-state functional magnetic resonance imaging data processing. *PLoS One, 6*(9), e25031.

Sutherland, M. T., McHugh, M. J., Pariyadath, V., & Stein, E. A. (2012). Resting state functional connectivity in addiction: Lessons learned and a road ahead. *NeuroImage, 62*(4), 2281–2295.

Tanabe, J., Nyberg, E., Martin, L. F., Martin, J., Cordes, D., … Tregellas, J. R. (2011). Nicotine effects on default mode network during resting state. *Psychopharmacology, 216*(2), 287–295.

Tang, J., Liao, Y., Deng, Q., Liu, T., Chen, X., Wang, X., … Hao, W. (2012). Altered spontaneous activity in young chronic cigarette smokers revealed by regional homogeneity. *Behaviour and Brain Function, 8*, 44.

Tomasi, D., & Volkow, N. D. (2010). Functional connectivity density mapping. *Proceedings of the National Academy of Sciences of the United States of America, 107*, 9885–9890.

Tomasi, D., & Volkow, N. D. (2011). Association between functional connectivity hubs and brain networks. *Cerebral Cortex, 21*, 2003–2013.

Tomasi, D., & Volkow, N. D. (2012). Gender differences in brain functional connectivity density. *Human Brain Mapping, 33*, 849–860.

Vincent, J. L., Kahn, I., Snyder, A. Z., Raichle, M. E., & Buckner, R. L. (December, 2008). Evidence for a frontoparietal control system revealed by intrinsic functional connectivity. *Journal of Neurophysiology, 100*(6), 3328–3342, 2008 Dec.

Wang, Z., Faith, M., Patterson, F., Tang, K., Kerrin, K., Wileyto, E. P., … Lerman, C. (2007). Neural substrates of abstinence-induced cigarette cravings in chronic smokers. *The Journal of Neuroscience, 27*(51), 14035–14040.

Wang, K., Yang, J., Zhang, S., Wei, D., Hao, X., Tu, S., & Qiu, J. (2014). The neural mechanisms underlying the acute effect of cigarette smoking on chronic smokers. *PLoS One, 9*(7), e102828.

Weiland, B. J., Sabbineni, A., Calhoun, V. D., Welsh, R. C., & Hutchison, K. E. (2015). Reduced executive and default network functional connectivity in cigarette smokers. *Human Brain Mapping, 36*(3), 872–882.

Wu, G., Yang, S., Zhu, L., & Lin, F. (2015). Altered spontaneous brain activity in heavy smokers revealed by regional homogeneity. *Psychopharmacology, 232*(14), 2481–2489.

Wylie, K. P., Rojas, D. C., Tanabe, J., Martin, L. F., & Tregellas, J. R. (October 15, 2012). Nicotine increases brain functional network efficiency. *Neuroimage, 63*(1), 73–80.

Yu, R., Zhao, L., Tian, J., Qin, W., Wang, W., Yuan, K., … Lu, L. (July 2013). Regional homogeneity changes in heavy male smokers: A resting-state functional magnetic resonance imaging study. *Addiction Biology, 18*(4), 729–731.

Zanetti, F., Giacomello, M., Donati, Y., Carnesecchi, S., Frieden, M., & Barazzone-Argiroffo, C. (November 2014). Nicotine mediates oxidative stress and apoptosis through cross talk between NOX1 and Bcl-2 in lung epithelial cells. *Free Radical Biology and Medicine, 76*, 173–184.

Zhang, X., Salmeron, B. J., Ross, T. J., Gu, H., Geng, X., Yang, Y., & Stein, E. A. (2011). Anatomical differences and network characteristics underlying smoking cue reactivity. *NeuroImage, 54*(1), 131–141.

Zou, Q. H., Zhu, C. Z., Yang, Y., Zuo, X. N., Long, X. Y., Cao, Q. J., … Zang, Y. F. (July 15, 2008). An improved approach to detection of amplitude of low-frequency fluctuation (ALFF) for resting-state fMRI: Fractional ALFF. *Journal of Neuroscience Methods, 172*(1), 137–141.

PART IV

DRUGS OF ABUSE AND BRAIN STRUCTURE AND FUNCTION

CHAPTER

29

Novel Psychoactive Substances: A New Behavioral and Mental Health Threat

M.T. Zanda[1], L. Fattore[2]

[1]University of Cagliari, Cagliari, Italy; [2]Institute of Neuroscience-Cagliari, National Research Council, Cagliari, Italy

INTRODUCTION

Use of novel psychoactive substances (NPS) has dramatically increased since 2000s and represents a serious risk for the public health. Definition of NPS includes any substance that has recently become available and has been designed purposely to replace illegal drugs, although not necessarily of new synthesis. Indeed, the term "new" does not automatically refer to ex novo developed drugs, as several NPS have been synthesized more than 50 years ago. The increasing popularity of NPS is attributable in part to the emergence of the Internet as crucial marketplace where these drugs are sold as "legal highs" or "research chemicals" (EMCDDA, 2015a). To elude controls, regulated NPS are sold in the deep web (or "dark net") networks, which are not accessible by standard search and allow transactions in anonymity thanks to crypto currencies like Bitcoin (EMCDDA, 2015a).

Legislation

The United Nation Office on Drugs and Crime reported 95 member states and territories in which NPS have appeared during December 2014 (UNODC, 2015). Of these, most were located in Europe (39) followed by Asia (27), Africa (14), the Americas (13), and Oceania (2). Cayman Island, Montenegro, Peru, and Seychelles also reported for the first time presence of NPS in 2014 (UNODC, 2015), the most popular being synthetic cannabinoids (39%), phenethylamines (18%), and synthetic cathinones (15%) (EMCDDA, 2015b; UNODC, 2015). It is particularly difficult to monitor NPS use due to their large assortment, transient nature, and the appealing street names that differ among countries, rendering their identification and regulation a puzzling issue (EMCDDA, 2015b). Many NPS detected in 2013 were no longer present on the market in 2014, and the majority are currently limited to few countries and available for a short period of time (e.g., the tryptamine 5-MeO-DPT337). During 2000s, several deaths associated with the use of NPS served as warning bell for a tight regulation of these substances, which started to be placed in Schedules I and II by the United Nations Single Convention on Narcotic Drugs of 1961 and the Convention on Psychotropic Substances of 1971 (UNODC, 2015).

Origin and Production

The first mention to NPS dates back to early 2000s with benzylpiperazine (BZP) and methylone, followed by mephedrone, marketed as amphetamines or new types of ecstasy (EMCDDA, 2015b). They are produced typically in clandestine laboratories when there is low availability of illicit precursor drugs or are derived from accidental synthesis, as in the case of *para*-methoxymethylamphetamine (PMMA) originated from 3,4-methylenedioxymethamphetamine or ecstasy (MDMA) (EMCDDA, 2015b). Manufactured generally in China and India, NPS are then introduced and distributed into European markets (EMCDDA, 2015b).

Availability of NPS

To elude controls, products containing NPS are labeled "not for human use" and sold as "natural" herbal blends like spice drugs (containing synthetic cannabinoids), "bath salts" (containing synthetic cathinones), "research chemicals for scientific purposes," or "food supplements" like adrafinil (claimed to increase energy, memory, and attention) (EMCDDA, 2015b; Fattore & Fratta, 2011; Seely, Lapoint, Moran, & Fattore, 2012). According to the Global Drug Survey 2015, Poland and

Addictive Substances and Neurological Disease
http://dx.doi.org/10.1016/B978-0-12-805373-7.00029-3

Switzerland are the countries with the highest and lowest rate, respectively, of "research chemicals" use in the past 12 months (http://www.globaldrugsurvey.com/the-global-drug-survey-2015-findings/#).

Users and Reason to Use

Although data concerning users of NPS are not entirely reliable, some studies suggest that NPS users are rather assorted and include students, prisoners, people who attend parties, injecting drug users or first-time users, and psychonauts, that is, people who experiment with psychoactive drugs, have detailed technical knowledge, and participate in online discussions sharing experiences and suggestions (Davey, Schifano, Corazza, & Deluca, 2012). Use of NPS among young people is increasing (EMCDDA, 2015a), and consumption of spice drugs, benzylpiperazine and mephedrone, in young people (16–24 years old) is higher than in the adult population (25–59 years old) and prevalent in the male population (Wood, Hunter, Measham, & Dargan, 2012). Reasons to use NPS are different and range from their potent psychoactive effects to the easy availability online, from the affordable cost to the quasilegal status, from their ability to elude drug-screening test to the perception of "safer" and/or "natural" drugs (EMCDDA, 2015a, 2015b; Zawilska & Andrzejczak, 2015).

Classification of NPS

NPS classification is typically based on pharmacological effects or chemical structure. As such, NPS are classified, respectively, into synthetic cannabinoids, stimulant-type drugs, downers/sedative-type drugs, hallucinogenic drugs, and dissociative drugs; or into synthetic cannabinoids, synthetic cathinones, phenethylamines, piperazines, ketamine and phencyclidine-type substances, and tryptamines. However, current NPS classification includes also synthetic cocaine, synthetic opioids, $GABA_A$ and $GABA_B$ receptor agonists, herbs/plants, prescribed products, and performance- and image-enhancing drugs (PIEDs) (Papaseit, Farré, Schifano, & Torrens, 2014; Schifano, Orsolini, Duccio Papanti, & Corkery, 2015) (see Fig. 29.1).

SYNTHETIC CANNABINOIDS

The group of synthetic cannabinoids includes the greatest number of NPS monitored by EMCDDA and comprises compounds that mimic the effects of delta-9-tetrahydrocannabidiol (THC), the psychoactive ingredient of *Cannabis sativa* (Bush & Woodwell, 2014). Synthetic cannabinoids appeared for the first time in the European market in 2004 and soon started spreading worldwide under different brand names such as "Spice," "Spice Diamond," "Spice Gold," "Arctic Spice," "Yucatan Fire", "Silver," "Aroma," "K2," "K3," "Genie," "Scene," "Dream," or "Smoke" (Brents & Prather, 2014; Bush & Woodwell, 2014). They are sold as harmless products such as incense, potpourri, or air fresheners, and are still easily accessible on the Internet and in "head shops" (Fattore & Fratta, 2011; Seely et al., 2012).

Origin and Classification

Although abuse of synthetic cannabinoids is a recent phenomenon, the synthesis of some of them dates back to 1980s. There are currently over 100 different synthetic cannabinoids (Harris & Brown, 2012) that can be

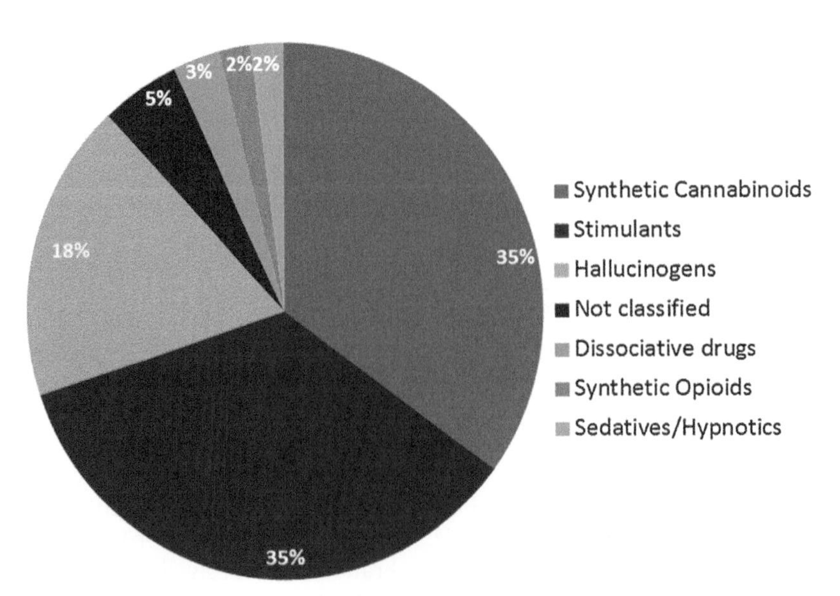

FIGURE 29.1 Relative distribution of novel psychoactive substances in use. *UNODC, Early Warning Advisory on NPS, 2015.*

classified based on their chemical structure. Besides classical cannabinoids, like THC, other constituents of *Cannabis*, and their structurally related synthetic analogs like HU-210 (a tricyclic compound with long duration of action synthesized by Raphael Mechoulam at Hebrew University), there are "nonclassical" cannabinoids such as cyclohexylphenols or 3-arylcyclohexanols like CP-47,497-C8, CP-55,940, and CP-55,244 (Thakur, Nikasa, & Makriyannis, 2005). An important series is represented by aminoalkylindole analogs that show high affinities for the central (CB_1) and peripheral (CB_2) cannabinoid receptors and are more potent than THC. This category includes the so-called "JWH" compounds (e.g., JWH-18, JWH-073, JWH-081, JWH-122, and JWH-200), synthesized by John W. Huffman in 1980s at Clemson University (Huffman, Dai, Martin, & Compton, 1994) and some of the "AM" compounds (e.g., AM-694), synthesized by Alexandros Makriyannis at Northeastern University, while evaluating the therapeutic potential of synthetic cannabinoid agents.

Chemical Structures and Pharmacology

The chemical structures of synthetic cannabinoids change continuously which makes their detection in human fluid samples particularly difficult (Bush & Woodwell, 2014). In some cases, esters and amides of fatty acid have been identified, but often herbal mixtures containing synthetic cannabinoids also contain preservatives, additives, high amount of vitamin E, and β2-adrenergic agonists (Seely et al., 2012). Synthetic cannabinoids show higher affinity for CB_1 and CB_2 cannabinoid receptors and induce long-lasting and more adverse effects as compared to marijuana. Factors contributing to the high incidence of negative/undesired effects following use may include pyrolysis by-products of the vegetable material on which these compounds are sprayed or shared metabolic pathways with frequently used medications that likely lead to adverse drug–drug reactions (Fattore, 2016). Biotransformation in active metabolites may extend and intensify CB_1 receptors activation (Chimalakonda et al., 2012). Moreover, it seems that synthetic cannabinoids interfere with other receptors and may (1) act as inhibitors of the nicotinic acetylcholine receptor and agonists of the ionotropic glutamate (*N*-methyl-D-aspartate, NMDA) receptor, (2) modulate glycine and 5-HT, and (3) inhibit monoamine oxidase activity, which exposes the subject to an enhanced risk of serotonin syndrome (Fattore, 2016; Pertwee, 2010).

Routes of Administration, Users, and Reasons to Use

Synthetic cannabinoids are typically smoked by pipe, rolled in cigarette paper, ingested, or inhaled (Fattore & Fratta, 2011). Users are prevalently young males who enjoy smoking these products for their marijuana-like effects, the legal status, and the affordable cost. Other reasons for using synthetic cannabinoids are their availability on the Internet and their ability to bypass detection screening tests (Bush & Woodwell, 2014; Papanti et al., 2013; Schifano, Papanti, Orsolini, & Corkery, 2016).

Toxic Effects

Desired effects of synthetic cannabinoids are similar to those experienced with *Cannabis* and are accompanied by mild hallucinogenic effects (Fattore & Fratta, 2011). However, undesired effects have also been reported which involve agitation, anxiety, nausea, vomiting, tremors, seizures, tachycardia, hyperglycemia, visual and auditory hallucinations, mydriasis, dyspnea, and tachypnea (Bush & Woodwell, 2014; Schifano et al., 2016). Use of synthetic cannabinoids has been associated with psychiatric (e.g., suicidal ideation, aggressive behavior, panic attacks, psychosis, and delirium), neurological (e.g., cognitive impairment, disorganized thought, and nystagmus) and somatic effects (e.g., hyperemesis, encephalopathy, acute kidney injury, rhabdomyolysis, hyperthermia, serotonin syndrome, kidney toxicity, and liver failure), coma, and death (Fattore, 2016; Patton et al., 2013; Savasman et al., 2014).

Legislation Status

Starting from 2009, different European countries subjected all products containing synthetic cannabinoids to the narcotics laws, preventing their sale in head shops and online stores (Fattore & Fratta, 2011; Seely et al., 2012). In 2011, JWH-018, JWH-073, JWH-200, CP-47,497, and (C8)-CP-47,497 were classified as Schedule I of the Controlled Substances Act, and 177 different synthetic cannabinoids were reported in 2014 to the United Nations Office on drugs and crime early warning advisory from 58 different countries and territories (UNODC, 2015). At present (2016), current legislations are adopting a more general approach by providing a broader definition of regulated compounds to target entire classes of substances derived from a certain chemical structure rather than specific molecules.

SYNTHETIC CATHINONES

Synthetic cathinones have a chemical structure similar to methamphetamine and MDMA. Marketed as bath salts, bath crystals, plant food, herbal incense, or fertilizers, they are commonly known as Blue Silk, Purple Wave, White Knight, White Lightening, Ocean Burst, Ivory Snow, Ivory

Wave, Pure Ivory, Charge+, Snow Leopard, Stardust, and Vanilla Sky (Fass, Fass, & Garcia, 2012; German, Fleckenstein, & Hanson, 2014; Miliano et al., 2016).

Origin and Classification

4-Methylmethcathinone (mephedrone), 3,4-methylenedioxymethcathinone (methylone), and 3,4-methylenedioxyprovalerone (MDPV) are the most common synthetic cathinones, all originated from cathinone, an alkaloid with stimulant properties present in the leaves of *Catha edulis*, and known as khat (German, Fleckenstein, & Hanson, 2014; Valente, Guedes de Pinho, de Lourdes Bastos, Carvalho, & Carvalho, 2014). Synthesis of first cathinones dates back to 1920s. Methcathinone was the first to be synthesized in 1928 (Sanchez, 1929) but started being abused only during 1990s (Emerson & Cisek, 1993). Mephedrone was synthesized in 1929 (Sanchez, 1929) and MDPV in 1967 (G.m.b.H., 1967), while the more recent methylone was synthesized in 1996 (Jacob Peyton III, 1996). Yet, they were used in designer drugs and abused worldwide only from early 2000s (German et al., 2014).

Synthetic cathinones could be classified in four main families: (1) compounds with an N-alkylated group, originally synthesized for therapeutic purposes, like methylone, ethylone, butylone, and pentylone; (2) compounds with a methylenedioxy group added to the benzyl ring, like MDMA, 3,4-methylenedioxy-*N*-ethyl-amphetamine (MDEA), *N*-methyl-1,3-benzodioxolylbutanamine (MBDB), and *N*-methyl-1,3-benzodioxolylpentanamine (MBDP); (3) compounds with a pyrrolidine group added to the nitrogen atom, like α-PPP; and (4) compounds presenting a combination of methylenedioxy and N-pyrrolidine groups, like MDPPP, MDPBP, and MDPV (Valente et al., 2014).

Chemical Structures and Pharmacology

Synthetic cathinones are β-ketoamphetamines structurally related to amphetamine (Valente et al., 2014) and probably with a similar mechanism of action. They interfere with the monoamine neurotransmitter system and increase the level of extracellular monoamines by increasing the presynaptic release or blocking the reuptake (German et al., 2014). In rats, high doses of mephedrone reduce the activity of the dopamine (DAT) and serotonin (SERT) transporter (German et al., 2014).

Route of Administration, Users, and Reason to Use

Synthetic cathinones are sold as crystalline powder, capsules, or tablets (Valente et al., 2014). Commonly snorted, they can also be ingested, administrated rectally, or injected intravenously or intramuscularly (German et al., 2014; Schifano et al., 2015). Users are diversified, but typically are adolescents or young adults (<40 years) with a previous history of drug abuse (Fass et al., 2012). Synthetic cathinones are used in different social contexts, during parties or in nightclubs, consumed alone or with other substances (German et al., 2014). They represent a valid alternative to MDMA and cocaine due to their hallucinogenic and stimulant properties, have low cost, and are perceived as more potent and with less side effects than other stimulants (German et al., 2014; Schifano et al., 2015).

Toxic Effects

Positive and desired effects reported after use of synthetic cathinones include intense euphoria, increased energy and concentration, talkativeness, empathy, and increase of sexual desire (Winstock et al., 2011). Yet, negative effects may also occur, which include memory impairment, panic, paranoia, agitation, anxiety, visual and auditory hallucinations, and aggressiveness. Bruxism, high body temperature, body sweating, tremor, cold extremities, skin rash, blurry vision, shortness of breath, headache, chest pain, and vomiting are quite common following use of synthetic cathinones (Winstock et al., 2011). Severe effects after use are hyperthermia, kidney failure, rhabdomyolysis, and convulsions (Schifano et al., 2015). Repeated use of these substances may lead to dependence and withdrawal symptoms upon interruption of use, for example, tiredness, impaired concentration, insomnia, and nasal congestion (German et al., 2014; Winstock et al., 2011).

Fatal intoxications have been reported after use of mephedrone alone (Morris, 2010) or in combination with other substances such as cocaine, amphetamine, benzodiazepines, codeine, and alcohol, as well as after methylone and butylone use (Loi et al., 2015; Warrick et al., 2012). Deaths are caused frequently by an overstimulation of the dopaminergic and serotoninergic systems, leading to serotonin syndrome and/or multiorgan failure (Loi et al., 2015).

Legislation Status

Mephedrone and MDPV were classified as Schedule I of the Controlled Substances Act of 2012 in the United States, followed by methylone and other 40 substances in 2013. In Europe, mephedrone started being controlled by December 2010, but it was soon replaced by new synthetic cathinones (German et al., 2014).

PHENETHYLAMINES

Phenethylamines are stimulant, entactogenic, and hallucinogenic substances that share similar chemical structures with amphetamine, catecholamines, synthetic cathinones, and other substances (Nelson, Bryant, & Aks, 2014).

Origin and Classification

Phenethylamines include both natural and synthetic substances classified into (1) classical phenethylamines like MDMA, MDEA, and MBDB; (2) psychoactive phenethylamines that include mescaline-derived compounds like TMA, DOM, DOET, DOI, and DOC; and (3) more recent compounds including bromodragonfly, benzofuran, N-benzyl substituted phenethylamines substances, and the "2C-series" (e.g., 2C-I, 2C-E, and 2C-B) (Schifano et al., 2015, 2016).

Chemical Structures and Pharmacology of 2C Molecules

Modification of the aromatic ring in different positions is able to create different compounds with altered neurochemical actions (Nelson et al., 2014). The majority of 2C compounds show affinity for different subtypes of serotonin 5-HT_2 receptor, some interfere with the reuptake of dopamine, serotonin, and noradrenaline, while 2C-B acts as α1-adrenergic receptor agonist (Nelson et al., 2014; Schifano et al., 2016).

Route of Administration, Users, and Reason to Use

2C molecules are typically marketed as tablets or powder and are generally ingested (Nelson et al., 2014). Users are predominantly young males, often with a history of polydrug abuse, that bump into 2C drugs at raves or buy them through the Internet (Nelson et al., 2014).

Toxic Effects

Phenethylamines induce stimulant and hallucinogenic effects but can also cause hallucinations, confusion, depression, dizziness, nausea, vomiting, diarrhea, headaches, and body pains (Nelson et al., 2014). Intoxication includes serotonin syndrome and profound cerebral vasculopathy (Ambrose, Bennett, Lee, & Josephson, 2010). Few cases of death have been reported after use of 2C compounds due to cardiac arrest (Nelson et al., 2014).

Legislation Status

Many 2C compounds are now classified as Schedule I substances. However, as for other NPS, new phenethylamines continue to be synthesized to replace those undergoing law control (Nelson et al., 2014).

PIPERAZINES

Piperazines are products of synthesis originally marketed as antihelminthic substances and proposed later as antidepressant drugs (Nelson et al., 2014).

Origin and Classification

This class of NPS includes benzylpiperazines [BZP, MBZP (1-methyl-4-benzylpiperazine), and DBZP (dibenzylpiperazine)] and phenylpiperazines [TFMPP (3-trifluoromethylphenylpiperazine), CPP (1-(3-chlorophenyl) piperazine), MPP (1-methyl-3-phenylpiperazine), and MeOPP (para-methoxyphenylpiperazine)]. They are claimed by suppliers to be herbal products, but piperazine and its derivatives are all synthetic substances.

Chemical Structures and Pharmacology

Benzylpiperazines have a chemical structure similar to amphetamine but only one-tenth of its potency and are likely involved in both dopamine and noradrenaline release and inhibition of monoamine reuptake (Smith, Sutcliffe, & Banks, 2015; Wikström, Holmgren, & Ahlner, 2004). In contrast, phenylpiperazines primarily enhance the release of serotonin by acting on the serotonin receptor or the reuptake transporter, and have weaker action on dopamine and noradrenaline systems (Nelson et al., 2014).

Route of Administration, Users, and Reason to Use

Piperazines can be found in pills or powder, and consumed alone, as a mixture of two different piperazines, or in combination with different substances (Nelson et al., 2014). They are taken orally, at doses ranging from 50 to 250 mg and are frequently used by prisoners (Nelson et al., 2014; Wikström et al., 2004). In most cases piperazines are taken unconsciously by people convinced to take amphetamine, but they can also be used purposely for their legal status (Wikström et al., 2004).

Toxic Effects

Effects induced by piperazines are similar to amphetamine, but with lower intensity (Schifano et al., 2015). At high doses they induce unpleasant reactions like hallucinations, confusion, paranoia, anxiety, insomnia, tremors, excessive sweating, headaches, dizziness, nausea, palpitation, and shortness of breath (Nelson et al., 2014).

Acute toxicity includes seizures, hyponatremia, prolongation of QT interval, and serotonin syndrome. This clinical picture can be complicated by status epilepticus, hyperthermia, intravascular coagulation, rhabdomyolysis, and kidney failure (Nelson et al., 2014; Schifano et al., 2015). Fatal intoxications have also been reported when used in combination with MDMA and amphetamines (Smith et al., 2015; Wikström et al., 2004).

Legislation Status

Benzylpiperazines were classified as controlled drugs in United States (2002) and Sweden (2003), while in Canada and other European countries they are not regulated yet (Wikström et al., 2004).

KETAMINE AND PHENCYCLIDINE-TYPE SUBSTANCES

This class of NPS includes substances that cause primarily dissociative effects and have chemical structure similar to phencyclidine and ketamine, for example, methoxetamine (MXE), diphenidine (DND), methoxphenidine (MXP), and dextromethorphan (DXM) (Schifano et al., 2015). MXE is marketed as "safer" alternative to ketamine with different street names, such as "Special K," "Mexxy," "M-ket," "MEX," "Kmax," "Special M," "MA," "legal ketamine," "Minx," "Jipper," and "Roflçoptr" (Zawilska, 2014).

Origin and Classification

MXE was first synthesized in United Kingdom by a pharmaceutical scientist for therapeutical purpose, to alleviate phantom limb pain and antidepressant-like effects (Morris, 2011), while DND and MXP are known since 1924 and 1989, respectively, but only recently have started being abused (Helander, Beck, & Bäckberg, 2015; Wink, Michely, Jacobsen-Bauer, Zapp, & Maurer, 2016). DXM was synthesized in 1952 and approved by the Food and Drug Administration (FDA) in 1958 as substitute to codeine for antitussive properties (Morris & Wallach, 2014).

Chemical Structure and Pharmacology

Compared to phencyclidine and ketamine, MXE presents a 3-methoxy group on the phenyl ring and an N-ethylamino group which gives more potency and duration of action but reduce analgesic and anesthetic properties (Adamowicz & Zuba, 2015). DND and MXP are analogs of lefetamine, while DXM is a methylated analog of levorphanol, a nonopioid derivative of morphine with antitussive properties (Schifano et al., 2015). These compounds are supposed to act mainly as noncompetitive NMDA receptor antagonists, although they likewise stimulate dopamine receptors and inhibit the SERT. DND also possesses opioid agonism property (Schifano et al., 2015).

Route of Administration, Users, and Reason to Use

MXE is found as white powder, tablets, capsules, or as liquid, alone or in combination with other drugs, and is used by ingestion, sublingual consumption, nasal insufflations (sniffing or snorting), intramuscular or intravenous injections, or rectal administration, with effects lasting for 2–5 h (Zanda, Fadda, Chiamulera, Fratta, & Fattore, 2016; Zawilska, 2014). DND and MXP are ingested, insufflated, or injected and the effects last for 8–12 h (Schifano et al., 2015). Users are generally males under 30 years who use ketamine-like substances in nightlife settings (Wood et al., 2012). Reasons to use include the easy availability via the Internet, the very affordable cost and, for MXE, the false belief that it is less toxic for bladder than ketamine (Craig & Loeffler, 2014).

Toxic Effects

MXE is abused for its recreational effect such as euphoria, increased empathy, and social interactions, accompanied by pleasant ones like sense of peacefulness and calmness, audiovisual hallucinations, dissociative experiences ("M-Hole"), amelioration of phantom limb pain, brief antidepressant effects, and spiritual and transcendental experiences (Adamowicz & Zuba, 2015; Corazza et al., 2012; Zawilska, 2014). Recent animal studies confirmed the rewarding effects of MXE and its ability to stimulate the mesolimbic dopamine transmission (Chiamulera, Armani, Mutti, & Fattore, 2016; Mutti et al., 2016) which likely underline its abuse potential in humans. However, at high doses neurological/cerebellar and sympathomimetic symptoms typically arise (Craig & Loeffler, 2014). Serious intoxication following high doses of DND, MXP, and DXM abuse lead to serotonin syndrome (Schifano et al., 2015). Numerous fatal intoxications have been reported after use of MXE, DND, and MXP (Helander et al., 2015; Wikström, Thelander, Dahlgren, & Kronstrand, 2013).

Legislative Status

MXE is currently classified as Schedule I substance in Europe, China, Japan, Poland, Russia, Switzerland, Turkey, and United Kingdom (De Paoli, Brandt, Wallach, Archer, & Pounder, 2013). MXP, DND, and DXM are not regulated at present in some US states, although its sales is forbidden to minors (Morris & Wallach, 2014).

TRYPTAMINES

Tryptamines are natural compounds or designer drugs that originate from the decarboxylation of tryptophan (Araujo, Carvalho, Bastos Mde, Guedes de Pinho, & Carvalho, 2015; Tittarelli, Mannocchi, Pantan, & Romolo, 2015). Plants like *Banisteriopsis caapi* and *Psychotria viridis* are the source of tetrahydroharmine (THH) and *N,N*-dimethyltryptamine (DMT), respectively, which are often ingested together to increase their effects; fungi like *Psilocybe cubensis* mushrooms contain psilocybin (4-phosphoryloxy-*N,N*-dimethyltryptamine) and psilocin (4-hydroxy-*N,N*-dimethyltryptamine); animals like *Bufo alvarius* provide bufotenine, 5-hydroxy-*N,N*-dimethyltryptamine (5-OH-DMT), and 5-methoxy-*N,N*-dimethyltryptamine (5-MeO-DMT) (Araujo et al., 2015; Tittarelli et al., 2015). Synthetic tryptamines are derived from modifications of natural tryptamines. LSD, for example, is obtained by the modification of lysergic acid amine (LSA), an analog present in the seeds of *Argyreia nervosa* and *Ipomoea violacea* (Araujo et al., 2015). Recently, abused novel tryptamines are alpha-methyltryptamine (AMT), 5-methoxy-α-methyltryptamine (5-MeO-AMT), 5-methoxy-*N,N*-diisopropyltryptamine (5-MeO-DIPT), known with the streets name of "foxy" or "foxy methoxy," 5-methoxy-*N,N*-diallyltryptamine (5-MeODALT), and *N,N*-diallyltryptamine (DALT) (Araujo et al., 2015; Schifano et al., 2015).

Origin and Classification

Bufotenine was first isolated 1934 by Handovsky, and then synthesized in 1935 by Hoshino (Araujo et al., 2015; Hoshino & Shimodaira, 1935). LSD was synthesized in 1938 by Hofmann, while AMT was commercialized in 1960 as an antidepressant drug with the brand name of Indopan (Araujo et al., 2015). Tryptamines can be divided essentially into (1) simple tryptamines, (2) tryptamines presenting modification on the indole ring in position 4, and (3) tryptamines with modification in position 5. Other conformations reduce the hallucinogenic properties of tryptamines (Tittarelli et al., 2015).

Chemical Structures and Pharmacology

All tryptamines share the basic structure of the amino acid tryptophan, that is, two indole rings with an aminoethyl group at position 3. Tryptamines are derived by its decarboxylation, followed by one of the following steps: (1) an addition of α-alkyl substituent (AMT); (2) a modification on nitrogen atom of the side chain (DALT and DIPT); (3) a modification in position 4 (4-OH-DALT); or (4) in position 5 (5-MeO-DIPT) (Araujo et al., 2015). The indole nucleus is responsible for the hallucinogenic properties, while the hydroxyl and methoxy groups in positions 4 and 5 enhance drug potency (Araujo et al., 2015).

Contrary to other hallucinogenic compounds, many tryptamines display higher affinity than serotonin for other monoamines. Affinity for the serotonin 5-HT_{2A} subtype receptor shown by tryptamines is very low as compared to phenethylamines, with the exception of LSD that displays the highest affinity for the 5-HT_{2A} receptors (Fantegrossi, Murnane, & Reissig, 2008). AMT plays a role in the release and reuptake of dopamine (Tittarelli et al., 2015), while DMT is an agonist of the serotonin 5-HT_{2A} and 5-HT_{2C}, receptors and shows affinity also for α1 and α2-adrenergic receptors, dopamine D1, and sigma-1 receptors (Tittarelli et al., 2015). DPT acts as a strong serotonin receptors reuptake inhibitor and partial agonist to the 5HT_{1A} receptors, while DiPT shows affinity to 5HT_{2A} and acts as partial agonist to 5HT_{1A} receptors (Tittarelli et al., 2015). 5-MeO-AMT is an agonist of 5-HT_{2A} and 5-HT_{1A} receptors and inhibits monoamine reuptake (Tittarelli et al., 2015). 5-MeO-DMT has high affinity for the serotonin 5-HT_{1A} receptor, while 5-MeO-DiPT inhibits serotonin reuptake (Tittarelli et al., 2015).

Route of Administration, Users, and Reasons to Use

Tryptamines are sold as tablets or powder and consumed through different routes of administration (Araujo et al., 2015; Corkery, Durkin, Elliott, Schifano, & Ghodse, 2012). In particular, the *Psilocybe* mushrooms are eaten raw or infused in tea to extract the active principles, while other tryptamines are insufflated (snorting or sniffing), inhaled, injected intravenously, ingested in capsules or wrapped in a cigarette paper, or administered rectally (Araujo et al., 2015; Corkery et al., 2012). Generally, tryptamine users are young males; 5-MeO-DIPT is frequently abused by homosexuals (Araujo et al., 2015). Main reasons to use tryptamines include the curiosity to try a different hallucinogen, its easy accessibility due to legal status, the low cost, and the fact that they are not detectable with routine screening drug tests (Araujo et al., 2015; Winstock, Kaar, & Borschmann, 2014).

Toxic Effects

More common effects after use of tryptamines are hallucinogenic effects accompanied by euphoria, increased body energy, difficulty in walking due to loss of limbs control, and "out of body" experiences (Corkery et al., 2012; Nelson et al., 2014). At high doses vasoconstriction, enhanced blood pressure, rapid heartbeat, headache, sweating, bruxism, dilated pupils, and paranoia (i.e., bad trip) may occur along with anxiety, nausea, strong desire to eat, increased alertness, and agitation (Araujo et al., 2015; Corkery et al., 2012). Fatal intoxications have been reported after use of AMT, 5-MeO-AMT, and 5-MeO-DIPT due to cardiac failure or as a consequence of the hallucinatory effects (Corkery et al., 2012).

Legislation Status

Psilocybin, psilocin, and psilocybin mushrooms are Schedule I substances in the United States. LSD was scheduled in 1970, while AMT and 5-MeO-DIPT are Schedule I substances since 2005. 5-MeO-DALT is regulated in few countries (e.g., Bulgaria, Finland, and Romania), and in Japan it is classified as "designated substance" by the Pharmaceutical Affairs Law, which makes this drug illegal to possess or sell (Araujo et al., 2015; Corkery et al., 2012).

OTHER SUBSTANCES

In this category are included substances that can be either synthesized for clinical purposes (e.g., synthetic opioids, $GABA_A$, and $GABA_B$ receptor agonists) have a natural origin (e.g., plants) or used to improve performance and/or body aspects. Only few of them are scheduled drugs.

Synthetic Cocaine Derivatives

Different synthetic cocaine derivatives sold as "research chemicals" have been identified as potential pharmacological drugs, two of which are under tight observation because of their high abuse potential (EMCDDA, 2015a, 2015b). Among these substances are 3-(*p*-fluorobenzoyloxy)tropane (*p*FBT), with chemical structure very similar to cocaine, and dimethocaine, with chemical structure very similar to procaine, a local anesthetic lacking psychoactive property (EMCDDA, 2015a, 2015b). Both these substances are sold as white powder and used by insufflations (snorted). There is scarce information about pharmacology and toxicology of *p*FBT and dimethocaine, mostly coming from users' online fora. They seem to act as local anesthetic and possess stimulant properties; users reported that *p*FBT induces hypertension, tachycardia, anxiety, and transiently psychosis (EMCDDA, 2015a, 2015b). Currently, both substances are not scheduled drugs in Europe, except for Denmark and Romania where *p*FBT and dimethocaine, respectively, are under law control (EMCDDA, 2015a, 2015b).

Synthetic Opioids

The opioid MT-45 was originally synthesized by a Japanese company, the Dainippon Pharmaceutical Co., during 1970s as substitute of morphine for its analgesic properties (Papsun, Krywanczyk, Vose, Bundock, & Logan, 2016). Recently, MT-45 and other synthetic opioids started being abused for recreational purposes. Being chemically a 1-substituted-4-(1,2-diphenylethyl) piperazine derivative, MT-45 does not possess a chemical structure similar to other opioids; yet, it acts as a μ/δ-opioid receptor agonist (Schifano et al., 2015) with the same potency as that of morphine (Papsun et al., 2016). Like morphine, MT-45 presents analgesic and depressant properties and induces respiratory depression (Papsun et al., 2016). Marketed as di-hydrochloride salt formed in combination with synthetic cannabinoids and synthetic cathinones, it is typically ingested, injected, insufflated, or rectally administered (Papsun et al., 2016). Adverse effects include nausea, itching, bilateral hearing loss, and dissociative-like symptoms. Withdrawal symptoms have also been reported following repeated use, and several fatal intoxications occurred after use of MT-45 alone or in combination with other psychoactive drugs (Papsun et al., 2016).

AH-7921 was first synthesized by Allen and Hanburys Ltd company during 1970s and proposed as analgesic drug; in 2012 it appeared in the market as "legal opioid." In vivo animal studies confirmed its action as μ-opioid receptor agonist and show that it induces analgesia, hyperthermia, respiratory depression, and addictive behavior (Katselou, Papoutsis, Nikolaou, Spiliopoulou, & Athanaselis, 2015), but no information is available on its effects in humans. A number of deaths have been reported in Europe following its use, and now AH-7921 is a controlled drug in Sweden, Poland, Romania, Finland, the Netherlands, and Norway (Katselou et al., 2015).

Other synthetic opioids are nortilidine, W15, W18, and IC-26 on which very little information is available to date (Schifano et al., 2015).

GABA Receptor Agonists

GHB (gamma-hydroxybutyric acid) is a fatty acid present in the human brain, synthesized in 1874 and approved by FDA in 2002 for treatment of narcolepsy (White, 2016). It acts mainly as $GABA_B$ receptor agonist, but it also increases dopamine levels by inhibiting reuptake (White, 2016). GHB started being abused in body-building gyms (to stimulate release of the endogenous growth hormone) and during parties for its desired effect that include increased euphoria and sexual arousal. However, GHB may induce adverse effects like nausea, vomiting, sweating, miosis, hypothermia, ataxia, and depression of the central nervous system including mild somnolence, retrograde amnesia, and coma (White, 2016). GHB possesses rewarding properties and may induce dependence (Fattore, Martellotta, Cossu, & Fratta, 2000); its abuse is associated with withdrawal symptoms like insomnia, muscular cramping, tremor, and anxiety (Schifano et al., 2015) and, if consumed with other substances, it may cause deaths for respiratory depression (White, 2016). GHB is a Schedule I drug in the United States since 2000, and a Schedule IV drug in Europe.

Baclofen is a $GABA_B$ receptor agonist synthesized in 1962, in the light of the promising data obtained in animal studies (Fattore, Cossu, Martellotta, Deiana, & Fratta, 2001; Fattore, Cossu, Martellotta, & Fratta 2002; Fattore et al., 2009; Spano, Fattore, Fratta, & Fadda, 2007), that is used for detoxification from alcohol, nicotine, cocaine, and heroin addiction, although recently it has also been used for the treatment of obesity and binge eating disorder (Gorsane et al., 2012). Desired effects are euphoria, enhanced sociability, and antidepressant like-effects, but acute intoxication is frequent and includes hypotonia, delirium, sedation, respiratory depression, and coma. Baclofen is not a scheduled drug, but several cases of death have been reported after overdose (Schifano et al., 2015).

Phenibut (4-amino-3-phenyl-butyric acid) is approved in Russia for the treatment of anxiety, alcohol withdrawal, and insomnia (Owen, Wood, Archer, & Dargan, 2015). This drug is primarily a $GABA_B$ receptor agonist, but it also acts as $GABA_A$ and dopamine receptor agonist (Owen et al., 2015). Desired effects include euphoria and reduced social anxiety with consequent increased sociability. Withdrawal symptoms are insomnia, hangover, dizziness, nausea, and motor coordination impairment (Owen et al., 2015). Phenibut is not regulated at present.

Prescribed Drugs

Pregabalin is an analog of GABA that was approved in Europe for the treatment of central and peripheral neuropathic pain, generalized anxiety disorder, and adult partial epilepsy (Aldemir, Altintoprak, & Coskunol, 2015; Schifano et al., 2015). Its mechanism of action involves activation of calcium voltage channels that decreases the release of excitatory neurotransmitters such as glutamate, noradrenaline, and substance P (Aldemir et al., 2015; Schifano et al., 2015). Users of pregabalin have typically a history of polydrug abuse and report positive effects like increased energy, reduced anxiety, increased sociability, and visual hallucinations (Aldemir et al., 2015). It likely induces dependence, since sudden interruption of use leads to pessimism, aggression, anxiety, suicidal ideation, fatigue, excessive sleep, tremors, and vomiting (Aldemir et al., 2015).

Phenazepam is a benzodiazepine synthesized in Russia in 1975 and used for treating anxiety, convulsions, and alcohol withdrawal that started being abused in 1999 (Lomas & Maskell, 2015). It is a full $GABA_A$ receptor agonist that can lead to death if used in combination with other substances (Lomas & Maskell, 2015).

Olanzapine and quetiapine are atypical antipsychotics commonly used in the treatment of schizophrenia and bipolar disorder. Recently, these two second-generation antipsychotics became self-prescribed drugs to treat NPS-induced psychosis and are now used as "trip terminator" (olanzapine) and "come off psychedelic trip" (quetiapine) (Schifano et al., 2015; Valeriani et al., 2015).

Herbs/Plants

Salvia divinorum, commonly known as "herb of the gods," is used in Mexico by Mazatec Indians in ritual practices. It acts as agonist of κ-opioid receptors, possesses abuse liability, and is associated with cognitive alteration (González-Trujano et al., 2016; Serra et al., 2015). Commonly consumed orally (infused as tea or chewing the leaves) or by inhalation, *Salvia divinorum* has anxiolytic and sedative properties and induces alteration in perception and psychoactive effects (González-Trujano et al., 2016).

Mitragyna speciosa, better known as "kratom," is a tropical tree cultivated in Southeast Asia (Warner, Kaufman, & Grundmann, 2016). Its leaves contain over 25 alkaloids, of which mitragynine and 7-hydroxymitragynine are considered the most active, and are reduced in powder or chewed to reduce fatigue, in particular among manual labors (Warner et al., 2016). Mitragynine is a μ- and δ-opioid partial agonist 13 times more potent than morphine, while 7-hydroxymitragynine is a μ-opioid agonist, 4 times more potent than mitragynine (Schifano et al., 2015). Kratom induces stimulant and depressant effects according to the dose used: low doses (1–5 g) induce stimulant effects, reduce fatigue, alertness, and sociability and increase sexual desire, as well as produce mild adverse effects like anxiety and agitation; intermediate doses (5–15 g) induce pain and withdrawal relief; high doses (>15 g) cause stupor, sweating, dizziness, diarrhea, nausea, and dysphoria. Prolonged use of high doses is associated with tremor, anorexia, weight loss, convulsions, and psychosis symptoms (Warner et al., 2016).

Piper methysticum (kava) is a shrub of large leaves of South Pacific Island (Ujváry, 2014) used in traditional medicine to treat fever, respiratory problems, convulsion, and urogenital problems. It is sold as food supplements and used as self-medication to treat anxiety, depression, and insomnia (Ujváry, 2014). Ingestion of kava product (drinking) in small quantity induces mild euphoria and increases sociability but high doses induce incoordination, somnolence, and muscle relaxation (Ujváry, 2014). Kava is toxic for liver and kidney and causes gastrointestinal problems and vision alteration. Since 2002, many European countries restricted the sale of kava due to its hepatotoxicity (Ujváry, 2014).

Sceletium tortuosum ("kanna") is a plant of South Africa used by indigenes to improve mood and alleviate sense of hunger and thirst (Ujváry, 2014). Psychoactive effects are caused by its alkaloids, among which is mesembrine, a serotonin 5-HT reuptake inhibitor. Anti-stress, antidepressant, narcotic, anxiolytic, and anti-addictive effects have been associated with use of kanna (Ujváry, 2014).

Ayahuasca is a beverage derived from *Banisteriopsis caapi* used for religious purpose (Callaway & Grob, 1998). Alkaloids contained in ayahuasca inhibit the enzymatic activity of monoamine oxidase (MAO) (Callaway & Grob, 1998). Lethality is observed if ayahuasca is coadministered with an MAO inhibitor of type A or an inhibitor of the reuptake of serotonin (i.e., serotonin syndrome) (Callaway & Grob, 1998).

Tabernanthe iboga is a shrub of West Africa, whose roots are used for stimulatory, hallucinogenic, and sedative properties (Ujváry, 2014). Ibogaine is the main active alkaloid of this plant that possesses stimulant properties at low doses but hallucinogenic activity at high doses. Tremor, ataxia, cardiac toxicity, and lethality may occur after its use (Ujváry, 2014).

Performance- and Image-Enhancing Drugs

PIEDs include drugs used to increase muscles mass, ameliorate body aspect (weight, skin coloration, and hairs), improve sexuality, enhance cognitive functions and mood, and facilitate sociability (Bersani et al., 2015). Internet represents a crucial market for people willing to increase body muscles. A variety of PIEDs are marketed online with different brand names such as "Super Tren-MG," "Tren Bomb," "Super Halo," "Straight Epi," "Protodrol," "Ultra Mass," "11-Sterone," and "P-MAG," labeled as "food supplements" for bodybuilders and declared safer than anabolic steroids (Abbate et al., 2014). Yet, they contain steroids that, when used regularly, lead to serious health consequences. This is particularly worrying when considering that users are teenagers and young men who want to enhance their aspect by increasing their muscles mass (Abbate et al., 2014). Chronic use of methasterone (17α-alkylated compound), for example, induces hepatotoxicity (Abbate et al., 2014).

2,4-Dinitrophenol (DNP) was first used in France during the World War I, and then sold as antiobesity drug due to its capacity to reduce body weight (Zack, Blaas, Goos, Rentsc, & Büttner, 2016). The FDA revoked approval of DNP during 1930s due to its serious side effects. DNP inhibits the production of ATP and induces toxic effects like sweating, nausea, vomiting, agitation, tachycardia, high level of lactate in serum, hyperthermia (45°C), cramps, colic, dyspnea, loss of consciousness, and coma (Zack et al., 2016). Fatal intoxications have been reported even with low doses of DNP (Zack et al., 2016).

Melanotan I and II are synthetic analogs of naturally occurring melanocortin peptide hormone α-melanocyte-stimulating hormone (α-MSH). Used to increase skin pigmentation (i.e., skin tanning), weight loss, or sexual arousal, these substances are easily available on the Internet and widely sold in gym, hairdressers, tanning, and beauty salons, and may induce adverse effects like facial flushing, spontaneous erections, nausea, and vomiting (Breindahl et al., 2014).

Piracetam (2-(2-oxopyrrolidin-1-yl)acetamide) is a cognitive enhancer used to improve study and work performance that is frequently abused for its psychedelic and/or mood-ameliorating effects (Corazza et al., 2014). Its mechanism of action is still unclear. Some users report no cognitive impairment or psychedelic effects but rather unpleasant side effects like psychomotor agitation, dysphoria, tiredness, dizziness, memory loss, headache, and severe diarrhea (Corazza et al., 2014).

CONCLUSIONS

The recent emergence of NPS with their ever-evolving chemical structure and the possibility to distribute in real time through social networks information about their use and effects have dramatically challenged public health and drug policies internationally. NPS has recently attracted great attention, but most are still unregulated and proposed online as legal alternatives to traditional illicit drugs. Unfortunately, this area is still poorly investigated and very limited information is available so far on their nature and potential risks. The phenomenon of NPS requires multinational and multidisciplinary collaborations to improve our knowledge on this changing drug market, to share information, and define good practices at a global level. Political and educational efforts are indispensable to regulate this mutable scenario and to inform the public about health consequences of NPS use. Clinicians and emergency staff should be aware that NPS may cause severe health consequences and unexpected adverse effects, and be informed on how to recognize and treat specific intoxication cases.

References

Abbate, V., Kicman, A. T., Evans-Brown, M., McVeigh, J., Cowan, D. A., Wilson, C., ... Walker, C. J. (2014). Anabolic steroids detected in bodybuilding dietary supplements – A significant risk to public health. *Drug Testing and Analysis, 7*, 609–618.

Adamowicz, P., & Zuba, D. (2015). Fatal intoxication with methoxetamine. *Journal of Forensic Sciences, 60*, S264–S268.

Aldemir, E., Altintoprak, A. E., & Coskunol, H. (2015). Pregabalin dependence: A case report. *Turk Psikiyatri Dergisi, 26*, 217–220.

Ambrose, J. B., Bennett, H. D., Lee, H. S., & Josephson, S. A. (2010). Cerebral vasculopathy after 4-bromo-2,5-dimethoxyphenethylamine ingestion. *The Neurologist, 16*, 199–202.

Araujo, A. M., Carvalho, F., Bastos Mde, L., Guedes de Pinho, P., & Carvalho, M. (2015). The hallucinogenic world of tryptamines: An updated review. *Archives of Toxicology, 89*, 1151–1173.

Bersani, F. S., Coviello, M., Imperatori, C., Francesconi, M., Hough, C. M., Valeriani, G., ... Corazza, O. (2015). Adverse psychiatric effects associated with herbal weight-loss products. *BioMed Research International*, 120679.

Breindahl, T., Evans-Brown, M., Hindersson, P., McVeigh, J., Bellis, M., Stensballe, A., & Kimergård, A. (2014). Identification and characterization by LC-UV-MS/MS of melanotan II skin-tanning products sold illegally on the Internet. *Drug Testing and Analysis, 7*, 164–172.

Brents, L. K., & Prather, P. L. (2014). The K2/Spice phenomenon: Emergence, identification, legislation and metabolic characterization of synthetic cannabinoids in herbal incense products. *Drug Metabolism Reviews, 46*, 72–85.

Bush, D. M., & Woodwell, D. A. (2014). *Update: Drug-related emergency department visits involving synthetic cannabinoids. The CBHSQ report.* Rockville, MD: Substance Abuse and Mental Health Services Administration (US).

Callaway, J. C., & Grob, C. S. (1998). Ayahuasca preparations and serotonin reuptake inhibitors: A potential combination for severe adverse interactions. *Journal of Psychoactive Drugs, 30*, 367–369.

Chiamulera, C., Armani, F., Mutti, A., & Fattore, L. (2016). The ketamine analogue methoxetamine generalizes to ketamine discriminative. *Behavioural Pharmacology, 27*, 204–210.

Chimalakonda, K. C., Seely, K. A., Bratton, S. M., Brents, L. K., Moran, C. L., Endres, G. W., ... Moran, J. H. (2012). Cytochrome P450-mediated oxidative metabolism of abused synthetic cannabinoids found in K2/Spice: Identification of novel cannabinoid receptor ligands. *Drug Metabolism and Disposition: The Biological Fate of Chemicals, 40*, 2174–2184.

Corazza, O., Bersani, F. S., Brunoro, R., Valeriani, G., Martinotti, G., & Schifano, F. (2014). The diffusion of performance and image-enhancing drugs (PIEDs) on the internet: The abuse of the cognitive enhancer piracetam. *Substance Use & Misuse, 49*, 1849–1856.

Corazza, O., Schifano, F., Simonato, P., Fergus, S., Assi, S., Stair, J., ... Scherbaum, N. (2012). Phenomenon of new drugs on the internet: The case of ketamine derivative methoxetamine. *Human Psychopharmacology, 27*, 145–149.

Corkery, J. M., Durkin, E., Elliott, S., Schifano, F., & Ghodse, A. H. (2012). The recreational tryptamine 5-MeO-DALT (*N,N*-diallyl-5-methoxytryptamine): A brief review. *Progress in Neuro-Psychopharmacology & Biological Psychiatry, 39*, 259–262.

Craig, C. L., & Loeffler, G. H. (2014). The ketamine analog methoxetamine: A new designer drug to threaten military readiness. *Military Medicine, 179*, 1149–1157.

Davey, Z., Schifano, F., Corazza, O., & Deluca, P. (2012). e-Psychonauts: Conducting research in online drug forum communities. *Journal of Mental Health, 21*, 386–394.

De Paoli, G., Brandt, S. D., Wallach, J., Archer, R. P., & Pounder, D. J. (2013). From the street to the laboratory: Analytical profiles of methoxetamine, 3-methoxyeticyclidine and 3-methoxyphencyclidine and their determination in three biological matrices. *Journal of Analytical Toxicology, 37*, 277–283.

EMCDDA, European Monitoring Centre for Drugs and Drug Addiction. (2015a). *European drug report 2015: Trends and developments.* Luxembourg: Publications Office of the European Union.

EMCDDA, European Monitoring Centre for Drugs and Drug Addiction (March 2015b). *New psychoactive substances in Europe. An update from the EU early warning system.* Luxembourg: Publications Office of the European Union.

Emerson, T. S., & Cisek, J. E. (1993). Methcathinone: A Russian designer amphetamine infiltrates the rural midwest. *Annals of Emergency Medicine, 22*, 1897–1903.

Fantegrossi, W. E., Murnane, K. S., & Reissig, C. J. (2008). The behavioral pharmacology of hallucinogens. *Biochemical Pharmacology, 75*, 17–33.

Fass, J. A., Fass, A. D., & Garcia, A. S. (2012). Synthetic cathinones (bath salts): Legal status and patterns of abuse. *The Annals of Pharmacotherapy, 46*, 436–441.

Fattore, L. (2016). Synthetic cannabinoids-further evidence supporting the relationship between cannabinoids and psychosis. *Biological Psychiatry, 79*, 539–548.

Fattore, L., Cossu, G., Martellotta, M. C., Deiana, S., & Fratta, W. (2001). Baclofen antagonizes intravenous self-administration of gamma-hydroxybutyric acid in mice. *Neuroreport, 12*, 2243–2246.

Fattore, L., Cossu, G., Martellotta, M. C., & Fratta, W. (2002). Baclofen antagonizes intravenous self-administration of nicotine in mice and rats. *Alcohol and Alcoholism, 37*, 495–498.

Fattore, L., & Fratta, W. (2011). Beyond THC: The new generation of cannabinoid designer drugs. *Frontiers in Behavioral Neuroscience, 5*, 60.

Fattore, L., Martellotta, M. C., Cossu, G., & Fratta, W. (2000). Gamma-hydroxybutyric acid: An evaluation of its rewarding properties in rats and mice. *Alcohol, 20*, 247–256.

Fattore, L., Spano, M. S., Cossu, G., Scherma, M., Fratta, W., & Fadda, P. (2009). Baclofen prevents drug-induced reinstatement of extinguished nicotine-seeking behaviour and nicotine place preference in rodents. *European Neuropshycopharmacology, 19*, 487–498.

G.m.b.H., B. I.. (1967). 1-(3′,4′-methylenedioxy-phenyl)-2-pyrrolidino-alkanones-(1). In *Office USP.* United States.

German, C. L., Fleckenstein, A. E., & Hanson, G. R. (2014). Bath salt sand synthetic cathinones: An emerging designer drug phenomenon. *Life Sciences, 97*, 2–8.

González-Trujano, M. E., Brindis, F., López-Ruiz, E., Ramírez-Salado, I., Martínez, A., & Pellicer, F. (2016). Depressant effects of *Salvia divinorum* involve disruption of physiological sleep. *Phytotherapy Research.* http://dx.doi.org/10.1002/ptr. 5617 (Epub ahead of print).

Gorsane, M. A., Kebir, O., Hache, G., Blecha, L., Aubin, H. J., Reynaud, M., & Benyamina, A. (2012). Is baclofen a revolutionary medication in alcohol addiction management? Review and recent updates. *Substance Abuse, 33*, 336–349.

Harris, C. R., & Brown, A. (2012). Synthetic cannabinoid intoxication: A case series and review. *The Journal of Emergency Medicine, 44*, 360–366.

Helander, A., Beck, O., & Bäckberg, M. (2015). Intoxications by the dissociative new psychoactive substances diphenidine and methoxphenidine. *Clinical Toxicology, 53*, 446–453.

Hoshino, T., & Shimodaira, K. (1935). Synthese des Bufotenins und über 3-Methyl-3-β-oxyäthyl-indolenin. Synthesen in der Indol-Gruppe. XIV. *Justus Liebigs Annalen der Chemie, 520*, 19–30.

Huffman, J. W., Dai, D., Martin, B. R., & Compton, D. R. (1994). Design, synthesis and pharmacology of cannabimimetic indoles. *Bioorganic & Medicinal Chemistry Letters, 4*, 563–566.

Jacob Peyton, A. T. S., III (1996). Novel n-substituted-2-amino-3′,4′-methylene-dioxypropiophenones. In *Office EP*.

Katselou, M., Papoutsis, I., Nikolaou, P., Spiliopoulou, C., & Athanaselis, S. (2015). AH-7921: The list of new psychoactive opioids is expanded. *Forensic Toxicology, 33*, 195–201.

Loi, B., Claridge, H., Goodair, C., Chiappini, S., Gimeno Clemente, C., & Schifano, F. (2015). Deaths of individuals aged 16-24 in the UK after using mephedrone. *Human Psychopharmacology, 30*, 225–232.

Lomas, E. C., & Maskell, P. D. (2015). Phenazepam: More information coming in from the cold. *Journal of Forensic and Legal Medicine, 36*, 61–62.

Miliano, C., Serpelloni, G., Rimondo, C., Mereu, M., Marti, M., & De Luca, M. A. (2016). Neuropharmacology of new psychoactive substances (NPS): Focus on the rewarding and reinforcing properties of cannabimimetics and amphetamine-like stimulants. *Frontiers in Neuroscience, 10*, 153.

Morris, H. (2010). Mephedrone: The phantom menace. *Vice*, 98–100.

Morris, H. (2011). Interview with a ketamine chemist: Or to be more precise, an arylcyclohexylamine chemist. *Vice Magazine*. Retrieved from https://www.vice.com/read/interview-with-ketamine-chemist-704-v18n2.

Morris, H., & Wallach, J. (2014). From PCP to MXE: A comprehensive review of the non-medical use of dissociative drugs. *Drug Testing and Analysis, 6*, 214–232.

Mutti, A., Aroni, S., Fadda, P., Padovani, L., Mancini, L., Collu, R., ... Chiamulera, C. (2016). The ketamine-like compound methoxetamine substitutes for ketamine in the self-administration paradigm and enhances mesolimbic dopaminergic transmission. *Psychopharmacology, 233*, 2241–2251.

Nelson, M. E., Bryant, S. M., & Aks, S. E. (2014). Emerging drugs of abuse. *Disease-a-Month, 60*, 110–132.

Owen, D. R., Wood, D. M., Archer, J. R., & Dargan, P. I. (2015). Phenibut (4-amino-3-phenyl-butyric acid): Availability, prevalence of use, desired effects and acute toxicity. *Drug and Alcohol Review*. http://dx.doi.org/10.1111/dar.12356.

Papanti, D., Schifano, F., Botteon, G., Bertossi, F., Mannix, J., Vidoni, D., ... Bonavigo, T. (2013). 'Spiceophrenia': A systematic overview of 'spice'-related psychopathological issues and a case report. *Human Psychopharmacology, 28*, 379–389.

Papaseit, E., Farré, M., Schifano, F., & Torrens, M. (2014). Emerging drugs in Europe. *Current Opinion in Psychiatry, 27*, 243–250.

Papsun, D., Krywanczyk, A., Vose, J. C., Bundock, E. A., & Logan, B. K. (2016). Analysis of MT-45, a novel synthetic opioid, in human whole blood by LC-MS-MS and its identification in a drug-related death. *Journal of Analytical Toxicology, 40*, 313–317.

Patton, A. L., Chimalakonda, K. C., Moran, C. L., McCain, K. R., Radominska-Pandya, A., James, L. P., ... Moran, J. H. (2013). K2 toxicity: Fatal case of psychiatric complications following AM2201 exposure. *Journal of Forensic Sciences, 58*, 1676–1680.

Pertwee, R. G. (2010). Receptors and channels targeted by synthetic cannabinoid receptor agonists and antagonists. *Current Medicinal Chemistry, 17*, 1360–1381.

Sanchez, S. B. (1929). Sur un homologue de l'ephedrine. *Bulletin de la Société Chimique de France, 45*, 284–286.

Savasman, C.M., Peterson, D.C., Pietak, B.R., Dudley, M.H., Clinton Frazee, C., III, Garg, U. (2014). Two fatalities due to the use of synthetic cannabinoids alone. *Presented at the 66th Annual Meeting of the American Academy of Forensic Sciences, Seattle* (Vol. 316). Denver: Publication Printers Inc.

Schifano, F., Orsolini, L., Duccio Papanti, G., & Corkery, J. M. (2015). Novel psychoactive substances of interest for psychiatry. *World Psychiatry, 14*, 15–26.

Schifano, F., Papanti, G. D., Orsolini, L., & Corkery, J. M. (2016). Novel psychoactive substances: The pharmacology of stimulants and hallucinogens. *Expert Review of Clinical Pharmacology, 4*, 1–12.

Seely, K. A., Lapoint, J., Moran, J. H., & Fattore, L. (2012). Spice drugs are more than harmless herbal blends: A review of the pharmacology and toxicology of synthetic cannabinoids. *Progress in Neuro-Psychopharmacology & Biological Psychiatry, 39*, 234–243.

Serra, V., Fattore, L., Scherma, M., Collu, R., Spano, M. S., Fratta, W., & Fadda, P. (2015). Behavioural and neurochemical assessment of salvinorin A abuse potential in the rat. *Psychopharmacology, 232*, 91–100.

Smith, J. P., Sutcliffe, O. B., & Banks, C. E. (2015). An overview of recent developments in the analytical detection of new psychoactive substances (NPSs). *Analyst, 140*, 4932–4948.

Spano, M. S., Fattore, L., Fratta, W., & Fadda, P. (2007). The $GABA_B$ receptor agonist baclofen prevents heroin-induced reinstatement of heroin-seeking behavior in rats. *Neuropharmacology, 52*, 1555–1562.

Thakur, G. A., Nikasa, S. P., & Makriyannis, A. (2005). CB1 cannabinoid receptor ligands. *Mini Reviews in Medicinal Chemistry, 5*, 631–640.

Tittarelli, R., Mannocchi, G., Pantan, F., & Romolo, F. S. (2015). Recreational use, analysis and toxicity of tryptamines. *Current Neuropharmacology, 13*, 26–46.

Ujváry, I. (2014). Psychoactive natural products: Overview of recent developments. *Annali dell'Istituto superiore di sanita, 50*, 12–27.

United Nations Office on Drugs and Crime. (2015). *World drug report 2015* (United Nations publication, Sales No. E.15.XI.6).

Valente, M. J., Guedes de Pinho, P., de Lourdes Bastos, M., Carvalho, F., & Carvalho, M. (2014). Khat and synthetic cathinones: A review. *Archives of Toxicology, 88*, 15–45.

Valeriani, G., Corazza, O., Bersani, F. S., Melcore, C., Metastasio, A., Bersani, G., & Schifano, F. (2015). Olanzapine as the ideal "trip terminator"? Analysis of online reports relating to antipsychotics' use and misuse following occurrence of novel psychoactive substance-related psychotic symptoms. *Human Psychopharmacology, 30*, 249–254.

Warner, M. L., Kaufman, N. C., & Grundmann, O. (2016). The pharmacology and toxicology of kratom: From traditional herb to drug of abuse. *International Journal of Legal Medicine, 130*, 127–138.

Warrick, B. J., Wilson, J., Hedge, M., Freeman, S., Leonard, K., & Aaron, C. (2012). Lethal serotonin syndrome after methylone and butylone ingestion. *Journal of Medical Toxicology, 8*, 65–68.

White, C. M. (2016). Pharmacologic, pharmacokinetic, and clinical assessment of illicitly used gamma-hydroxybutyrate (GHB). *The Journal of Clinical Pharmacology*. http://dx.doi.org/10.1002/jcph.767.

Wikström, M., Holmgren, P., & Ahlner, J. (2004). A2 (*N*-benzylpiperazine) a new drug of abuse in Sweden. *Journal of Analytical Toxicology, 28*, 67–70.

Wikström, M., Thelander, G., Dahlgren, M., & Kronstrand, R. (2013). An accidental fatal intoxication with methoxetamine. *Journal of Analytical Toxicology, 37*, 43–46.

Wink, C. S., Michely, J. A., Jacobsen-Bauer, A., Zapp, J., & Maurer, H. H. (2016). Diphenidine, a new psychoactive substance: Metabolic fate elucidated with rat urine and human liver preparations and detectability in urine using GC-MS, LC-MSn, and LC-HR-MSn. *Drug Testing and Analysis*. http://dx.doi.org/10.1002/dta. 1946 (Epub ahead of print).

Winstock, A. R., Kaar, S., & Borschmann, R. (2014). Dimethyltryptamine (DMT): Prevalence, user characteristics and abuse liability in a large global sample. *Journal of Psychopharmacology, 28*, 49–54.

Winstock, A., Mitcheson, L., Ramsey, J., Davies, S., Puchnarewicz, M., & Marsden, J. (2011). Mephedrone: Use, subjective effects and health risks. *Addiction, 106*, 1991–1996.

Wood, D. M., Hunter, L., Measham, F., & Dargan, P. I. (2012). Limited use of novel psychoactive substances in South London nightclubs. *The Quarterly Journal of Medicine, 105*, 959–964.

Zack, F., Blaas, V., Goos, C., Rentsc, D., & Büttner, A. (2016). Death within 44 days of 2,4-dinitrophenol intake. *International Journal of Legal Medicine*. http://dx.doi.org/10.1007/s00414-016-1378-4.

Zanda, M. T., Fadda, P., Chiamulera, C., Fratta, W., & Fattore, L. (2016). Methoxetamine, a novel psychoactive substance with important pharmacological effects: A review of case reports and preclinical findings. *Behavioural Pharmacology, 27*, 489–496 (April 28, 2016 Epub ahead of print).

Zawilska, J. B. (2014). Methoxetamine – A novel recreational drug with potent hallucinogenic properties. *Toxicology Letters, 230*, 402–407.

Zawilska, J. B., & Andrzejczak, D. (2015). Next generation of novel psychoactive substances on the horizon – A complex problem to face. *Drug and Alcohol Dependence, 157*, 1–17.

C H A P T E R

30

Cholesterol and Caffeine Modulate Alcohol Actions on Cerebral Arteries and Brain

A.N. Bukiya, A.M. Dopico

The University of Tennessee Health Science Center, Memphis, TN, United States

INTRODUCTION

Alcohol Intake

Alcohol (ethyl alcohol, ethanol: EtOH) drinking is widespread in our society, with episodic alcohol intake being the prevalent form of consumption. Blood alcohol levels (BAL) above 17 mM (≈80 mg/dL) constitute legal intoxication in most US states. These levels of alcohol in the blood are usually reached after consuming three or more (depending on body weight) regular drinks, such as a bottle of beer or a glass of wine. Higher BALs (35–80 mM) are reached during moderate-to-heavy alcohol drinking with >100 mM BAL being lethal to most alcohol-naive humans (Diamond, 1992, pp. 44–47; Wallner, Hanchar, & Olsen, 2006). Excessive alcohol consumption is considered responsible for 1 in 10 deaths among working-age adults (http://www.cdc.gov/alcohol/fact-sheets/alcohol-use.htm). Among the health risks posed by excessive alcohol consumption, the consequences of alcohol use on brain function attract the most attention, as excessive alcohol consumption is linked to violent behaviors, increased risk for accidents, traumatic injury, and sexual aggression (Abbey, Wegner, Woerner, Pegram, & Pierce, 2014; Dinh-Zarr, Goss, Heitman, Roberts, & DiGuiseppi, 2004; Fitterer, Nelson, & Stockwell, 2015). A need for effective intervention against disruption of brain physiology and behavior by alcohol represents the driving force behind studies aimed at understanding the mechanisms of alcohol actions in the brain, including cerebral vasculature and perfusion. This understanding requires identification of relevant targets in neurons, central glia and cerebral arteries that mediate perturbation of normal physiology and behavior by alcohol consumption.

A growing body of clinical and basic research points to cerebrovascular entities as critical factors in the physiology and pathology of the central nervous system. For example, progressive decline in cerebrovascular function underlies age-associated cognitive impairment (Deak, Freeman, Ungvari, Csiszar, & Sonntag, 2016), dementia (including Alzheimer's disease) (Raz, Knoefel, & Bhaskar, 2016; Snyder et al., 2015), and some forms of parkinsonism (Korczyn, 2015). Considering that alcohol severely impairs the cerebral circulation in most species, including humans (Altura & Altura, 1984), alcohol-driven effects on cerebral vessels should be considered as a bona fide component of alcohol effects in the brain.

Caffeine Intake

Caffeine is a methylxanthine alkaloid that belongs to the purine family (Ashihara & Crozier, 1999). A number of plants from South America and East Asia, including the coffee bean of the *Coffea* plant, are the most common source of caffeine. Despite its bitter taste, caffeine became the most widely used psychoactive agent in the world (Meredith, Juliano, Hughes, & Griffiths, 2013; http://www.globaldrugsurvey.com/facts-figures/the-global-drug-survey-2014-findings/). In the USA, about 90% of adults consume caffeine regularly (Frary, Johnson, & Wang, 2005; Juliano, Evatt, Richards, & Griffiths, 2012). The effect of caffeine on health in general and brain function in particular has been a controversial issue. There is some consensus, however, that caffeine intake in moderation (400 mg/day, or up to five cups of coffee daily) is safe for healthy humans (Nehlig, 2016). In adults, this form of caffeine consumption increases alertness and wakefulness by acting on autonomic centers in the pons and medulla (Urry & Landolt, 2015). However, concerns have been raised that early caffeine exposure (prenatal in particular) may increase neuronal excitability and predispose the developing brain to seizures (Tchekalarova, Kubová, & Mareš, 2014). On the other hand, several studies on humans and animal models have shown a neuroprotective

Addictive Substances and Neurological Disease
http://dx.doi.org/10.1016/B978-0-12-805373-7.00030-X

effect of caffeine in Parkinson's disease (Prediger, 2010; Schwarzschild et al., 2003). Furthermore, caffeine has been postulated as a starting point for a rational drug design to treat Parkinson's disease (Petzer & Petzer, 2015). Lastly, a growing number of studies show beneficial effects of caffeine consumption against progression of Alzheimer's disease (Panza et al., 2015).

Cholesterol

Cholesterol belongs to the cholestanol family, that is, steroids with 27 carbon atoms. Cholesterol is a common component of Western diet and a key lipid regulator of cardiovascular health. There are two major sources of cholesterol: dietary intake and de novo synthesis by liver and other organs. It is hard to estimate the contribution of each source to the final pool of cholesterol in the body, as the contributions of these sources are dynamic. Using animal models, however, it has been established that excessive dietary intake of cholesterol downregulates cholesterol biosynthesis and thus, dietary cholesterol prevails in the final cholesterol pool (Gould, 1951; Jones et al., 1996; Siperstein, 1970). In humans, the downregulation of cholesterol biosynthesis by dietary cholesterol has also been reported, although this downregulation is minimal (Jones et al., 1996).

Cholesterol serves as a major structural component for plasma membranes, precursor of steroid hormones, and a critical regulator of protein function (Dopico, Bukiya, & Singh, 2012; Epand, 2006; Gimpl, Burger, & Fahrenholz, 1997; Petrov, Kasimov, & Zefirov, 2016; Pine, 1987). In the brain, cholesterol metabolism is somewhat isolated from the rest of the body due to the blood–brain barrier (Zhang & Liu, 2015). The net level of cholesterol in brain, however, is high: cholesterol constitutes a key component of myelin, which harbors the largest pool of cholesterol in mammals (Saher & Stumpf, 2015). Cholesterol deficiency and excess both contribute to a wide range of neurological disorders, including developmental abnormalities, mental retardation, and several forms of dementia (e.g., Alzheimer's disease) (Nowaczyk & Irons, 2012; Petrov et al., 2016; Polidori, Pientkam, & Mecocci, 2012).

MODULATION OF ALCOHOL ACTION BY CAFFEINE

Remarkably, caffeine consumption often coincides with alcohol intake. The combined use and abuse of alcohol and caffeine has become increasingly popular. This co-consumption is especially prevalent in adolescent and young adult populations, reaching epidemic proportions across US college campuses: more than 50% of US college students reported mixing "energy drinks" (e.g., drinks, in which the main active ingredient is caffeine) with alcohol in the past month (Malinauskas, Aeby, Overton, Carpenter-Aeby, & Barber-Heidal, 2007; McMurtrie, 2014). Thus, caffeine modulation of alcohol-driven behaviors emerges as a research topic of great relevance to human health.

Caffeine is well known for increasing alertness and wakefulness (Urry & Landolt, 2015). Conceivably, one of the factors reported as motivational force behind mixing caffeine with alcohol is a "sobering" (e.g., energizing) effect of the caffeinated drink. Research on both animal models and human subjects shed light on caffeine modulation of alcohol effects. Experiments on Fischer-344 rats chronically receiving caffeine showed that such caffeine intake increased the animal's sensitivity to the stimulating effect of ethanol on locomotor activity in an open-field test (Sudakov, Rusakova, & Medvedeva, 2003). Research on human drinkers has documented that co-consumption of alcohol and caffeine decreased the perceived intensity of alcohol intoxication, and also enhanced stimulation (Marczinski & Fillmore, 2014). When compared with their alcohol-only drinker counterparts, caffeine–alcohol drinkers reported reduced feeling of tiredness (Flotta et al., 2014; Jones, 2011) and increased ability to stay awake and party for longer times (Jones, Barrie, & Berry, 2012). Moreover, combination of caffeine with alcohol reduced alcohol-driven impairment on several cognitive tasks: caffeine reversed alcohol-driven decrease in divided attention, executive function, and motor tasks requiring some executive function, such as hand–eye coordination, visual–motor tracking skills and the digit symbol substitution test (Burns & Moskowitz, 1989; Fillmore, Roach, & Rice, 2002; Fillmore & Vogel-Sprott, 1995; Mackay, Tiplady, & Scholey, 2002; McKetin, Coen, & Kaye, 2015; Rush, Higgins, Hughes, Bickel, & Wiegner, 1993). Mixing of alcohol with high volumes of energy drinks also decreased alcohol-induced impairment of psychomotor function and global information processing (Peacock, Cash, & Bruno, 2015). Moreover, high doses of caffeine (300 mg) have been reported to improve the memory deficits caused by alcohol intoxication (Drake, Roehrs, Turner, Scofield, & Roth, 2003). However, caffeine failed to improve reaction time and simple motor tasks (Howland et al., 2011; McKetin et al., 2015). Considering that overall cognitive and motor performance is an integral of several brain functions and corresponding central neurocircuits, it remains unclear whether mixing caffeine with alcohol improves net (overall) motor and cognitive performance when compared to drinking alcohol alone.

It should be noted that the so-called "sobering" effect of caffeine may mask alcohol effects on the mind, such as judgment impairment, which may lead to alcohol poisoning and increased engagement in risky behaviors (Pennay, Lubman, & Miller, 2011). Caffeine-induced masking of alcohol actions extends to well-known signs of alcohol intoxication, such as dehydration. Moreover, data suggest that the combined use of alcohol and caffeine may increase the incidence of alcohol-related injury (Reissig, Strain, & Griffiths, 2009).

A large body of evidence points at the negative consequences of mixing alcohol with caffeine on the development of an excessive drinking habit itself. In a rat model, intraperitoneal injections of 5 mg/kg caffeine promoted ethanol drinking in a limited access, free choice experimental paradigm. However, lower or higher doses of caffeine did not modify ethanol drinking, the bases of these differences remaining unclear (Kunin, Gaskin, Rogan, Smith, & Amit, 2000a). Analysis of epidemiological data revealed a strong association between daily energy drink consumption and alcohol dependence (Arria et al., 2011), as well as an increased susceptibility to smoking (Azagba & Sharaf, 2014). A study on young adults has shown that combining energy drinks with alcohol increases the urge to drink. Participants of the study also liked caffeinated alcohol more than alcohol alone (McKetin & Coen, 2014). Along similar lines, subjects who co-consume caffeine and alcohol are three times more likely to binge drink than drinkers who do not mix these agents (Fritz, Quoilin, Kasten, Smoker, & Boehm, 2016; Mintel International Group Ltd., 2007). In spite of these epidemiological data and clinical-controlled studies, the molecular mechanisms that are triggered by caffeine–alcohol mixture to promote alcohol consumption remain largely unknown. A study on rats showed that although caffeine or alcohol alone failed to increase circulating levels of the stress hormone corticosterone, the caffeine–alcohol mixture significantly increased plasma corticosterone levels (Kunin, Gaskin, Rogan, Smith, & Amit, 2000b). It has been speculated that this effect of caffeine–alcohol mixtures may trigger a desire to drink more alcohol, as repeated stress has been linked to higher alcohol consumption in both humans and laboratory animals (Engdahl, Dikel, Eberly, & Blank, 1998; Edwards et al., 2013; Pastor et al., 2011).

In contrast to the large number of studies showing behavioral consequences of mixing caffeine with alcohol, not much research has been conducted on how co-consumption of alcohol and caffeine affects brain *perfusion*. Numerous epidemiological studies have documented the deleterious effects of alcohol itself on cerebrovascular health (Altura & Altura, 1984; Patra et al., 2010; Puddey, Rakic, Dimmitt, & Beilin, 1999; Reynolds et al., 2003). Moderate-to-heavy alcohol intake is associated with an increased risk for cerebrovascular ischemia (Puddey et al., 1999), stroke, and death from

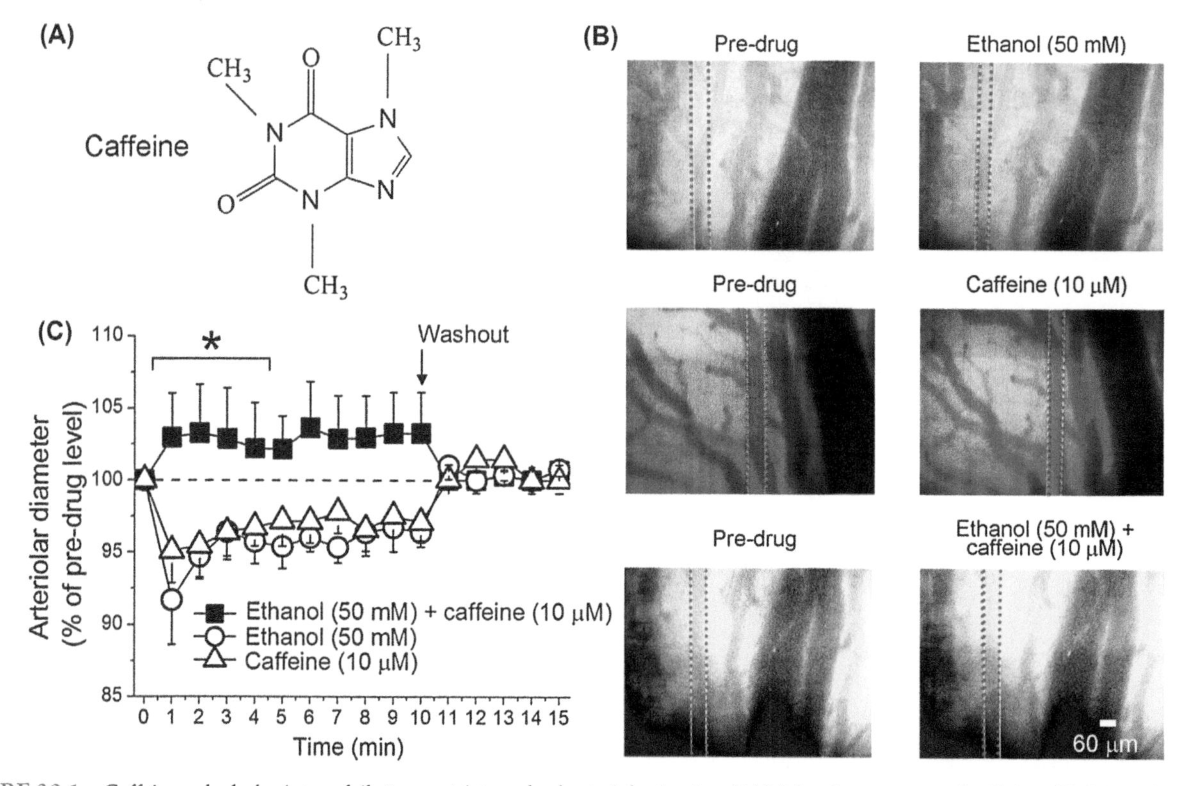

FIGURE 30.1 Caffeine–alcohol mixture fails to constrict cerebral arterioles in vivo. (A) Molecular structure of caffeine. (B) Original snapshots of pial arterioles before and after carotid artery infusion of 50 mM ethanol, 10 μM caffeine, or their combination into carotid artery of anesthetized rat. (C) Averaged pial arteriolar diameter (as percentage of diameter before drug application). Each data point represents an average from no less than three animals. *Different from administration of 50 mM ethanol ($p < .05$). *With slight modifications from Chang, J., Fedinec, A. L., Kuntamallappanavar, G., Leffler, C. W., Bukiya, A. N., & Dopico, A. M. (2016). Endothelial nitric oxide mediates caffeine antagonism of alcohol-induced cerebral artery constriction.* The Journal of Pharmacology and Experimental Therapeutics, 356, *106–115.*

ischemic stroke (Reynolds et al., 2003). In contrast, caffeine may exert protective effects: the Third National Health and Nutrition Examination Survey (NHANES III; 1988–93) reported that high doses of caffeine (equivalent to >3 cups of coffee per day) were associated with a decreased risk for stroke even when smoking or high cholesterol levels in the blood were more frequent among heavy consumers of coffee (Liebeskind, Sanossian, Fu, Wang, & Arab, 2015). Moreover, the simultaneous administration of alcohol and caffeine ("caffeinol") has been proposed as neuroprotection against ischemic stroke in rat models (Aronowski, Strong, Shirzadi, & Grotta, 2003) and humans (Piriyawat, Labiche, Burgin, Aronowski, & Grotta, 2003; Sacco, Chong, Prabhakaran, & Elkind, 2007).

Alcohol- and caffeine-driven changes in cerebral artery diameter may underlie or, at least, contribute to the effect of these substances on cerebrovascular health. BALs (35–80 mM) reached in systemic circulation following moderate-to-heavy alcohol drinking episode constrict cerebral arteries in several species, including humans (Altura & Altura, 1984; Bukiya, Dopico, Leffler, & Fedinec, 2014; Bukiya, Liu, & Dopico, 2009; Liu, Ahmed, Jaggar, & Dopico, 2004). In contrast, the literature on caffeine action on artery diameter is highly heterogeneous, with both dilation and constriction being reported. These variant outcomes have been mainly attributed to differences across vascular nets and species, length of caffeine administration, and differential involvement of

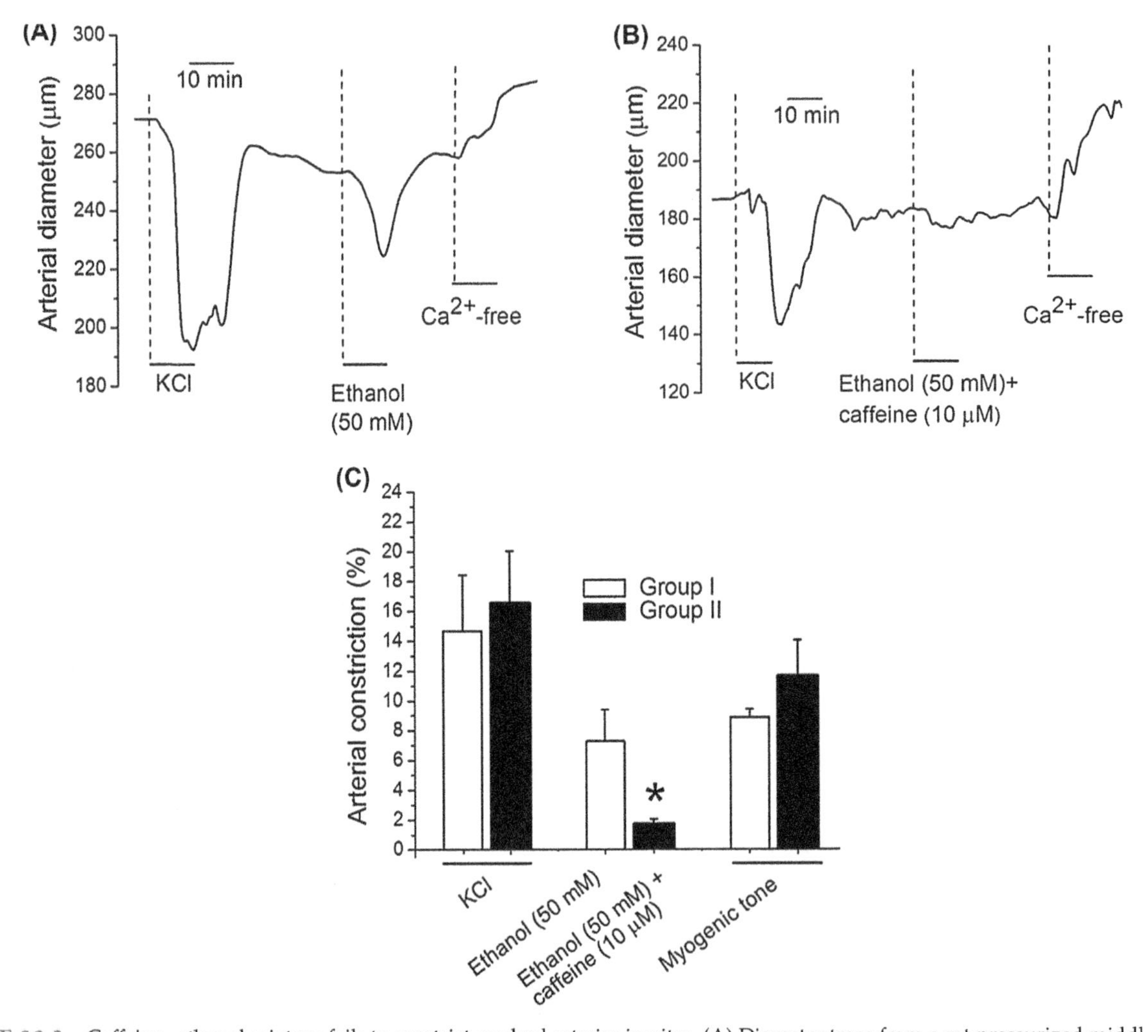

FIGURE 30.2 Caffeine–ethanol mixture fails to constrict cerebral arteries in vitro. (A) Diameter trace from a rat pressurized middle cerebral artery. After development of myogenic tone, artery was probed with 60 mM KCl to test maximal contraction. Following KCl washout, artery chamber was perfused with either 50 mM ethanol (A) or 50 mM ethanol + 10 μM caffeine mixture (B); (C) Average change in artery diameter evoked by KCl, 50 mM ethanol, or 50 mM ethanol + 10 μM caffeine mixture. Last set of bars also depicts averaged percentage of myogenic tone. Hollow bars (group 1) represent data from arteries that were exposed to ethanol only (n = 4); black bars (group 2) reflect data from arteries that were challenged with the ethanol + caffeine mixture (n = 6). *Different from constriction by 50 mM ethanol ($p < .05$). *Figure was originally published in Chang, J., Fedinec, A. L., Kuntamallappanavar, G., Leffler, C. W., Bukiya, A. N., & Dopico, A. M. (2016). Endothelial nitric oxide mediates caffeine antagonism of alcohol-induced cerebral artery constriction.* The Journal of Pharmacology and Experimental Therapeutics, 356, *106–115.*

adenosine receptors, as well as to contribution from endothelium- versus smooth muscle-originated mechanisms (Addicott et al., 2009; Echeverri, Montes, Cabrera, Galán, & Prieto, 2010; Frishman, Del Vecchio, Sanal, & Ismail, 2003; Higgins & Babu, 2013).

Using intravital microscopy on a closed-cranial window and diameter monitoring of isolated, pressurized cerebral arteries of rat, our group studied the effect of alcohol and caffeine co-administration on pial arteriole and cerebral artery diameter. Caffeine at concentrations found in human circulation following ingestion of 1–2 cups of coffee (≈10 μM) antagonized the endothelium-independent constriction of cerebral arteries evoked by ethanol concentrations found in blood during moderate-to-heavy alcohol intoxication (40–70 mM). Caffeine antagonism against alcohol was similar whether evaluated in pial arterioles using a closed cranial window and intravital microscopy (Fig. 30.1) or in pressurized, isolated middle cerebral arteries (Chang et al., 2016). Remarkably, both alcohol and caffeine constricted pial arterioles in vivo and cerebral arteries in vitro when these substances were administered separately (Bukiya, Dopico, et al., 2014; Chang et al., 2016). The ability of caffeine to antagonize ethanol-induced constriction of isolated cerebral arteries when both agents are administered in vitro (Fig. 30.2) indicates that caffeine antagonism of alcohol-induced constriction of cerebral arteries is independent of pharmacokinetic interference, systemic circulating factors, and complex neuronal networks. Incubation of isolated arteries with the caffeine–alcohol mixture, yet not with alcohol alone, triggered a robust release of NO• (Chang et al., 2016). Moreover, caffeine protection against alcohol effect disappeared upon block of NO• synthase. A central role of NO• synthase in the vascular effect of the caffeine–alcohol mixture was determined by using pressurized cerebral arteries of endothelial NO• synthase knockout (eNOS$^{-/-}$) mouse: caffeine protection against alcohol-induced constriction was not observed in these arteries.

Caffeine protection against alcohol was absent in rat arteries with denuded endothelium, this organ being a major source of eNOS-generated NO•. However, incubation of de-endothelialized (e.g., eNOS-lacking) cerebral arteries with the NO• donor sodium nitroprusside (10 μM) fully restored the protective effect of caffeine. Thus, endothelial NO• was identified as the critical mediator that enables protective effect of caffeine against alcohol on (cerebro)vascular targets. Considering that de-endothelialized cerebral artery is largely conformed by smooth muscle cells (Lee, 1995), downstream targets of NO• release in presence of the caffeine–alcohol mixture are likely located within the vascular smooth muscle itself (Fig. 30.3). Identification of these targets represents an ongoing line of investigation in our laboratory.

CHOLESTEROL MODULATION OF ALCOHOL EFFECT

Although alcohol consumption has been recognized to modulate cholesterol homeostasis in the brain (Guizzetti & Costa, 2007), the possibility that cholesterol modulates alcohol effects on the brain remained largely unexplored. Cholesterol concentration in plasma membranes of mammals ranges between 20 and 50 mol% (Gimpl et al., 1997). At concentrations >10 mol%, cholesterol has been shown to reduce the affinity of ethanol for liposomes that had a phospholipid bilayer composition similar to that of the plasma membrane (Trandum, Westh, Jorgensen, & Mouritsen, 2000). At concentrations <10 mol%, cholesterol increased the affinity of ethanol for the phospholipid bilayer (Trandum et al., 2000). Thus, it is expected that variations in plasma membrane cholesterol levels in the brain would modulate alcohol's central actions simply by regulating alcohol's partition in the cell membrane and thus, its eventual distribution into different cell compartments. Many other mechanisms, mostly dealing with perturbation of specific physical properties of the lipid bilayer secondary to cholesterol and ethanol presence, have been proposed for cholesterol–alcohol interactions in lipid membranes; these mechanisms have been extensively discussed elsewhere (Crowley, Treistman, & Dopico, 2003; Peoples, Li, & Weight, 1996; Sun & Sun, 1985). On top of these lipid-driven mechanisms, cholesterol–alcohol interactions in brain cells may result from the participation of proteins that are common targets of cholesterol and ethanol.

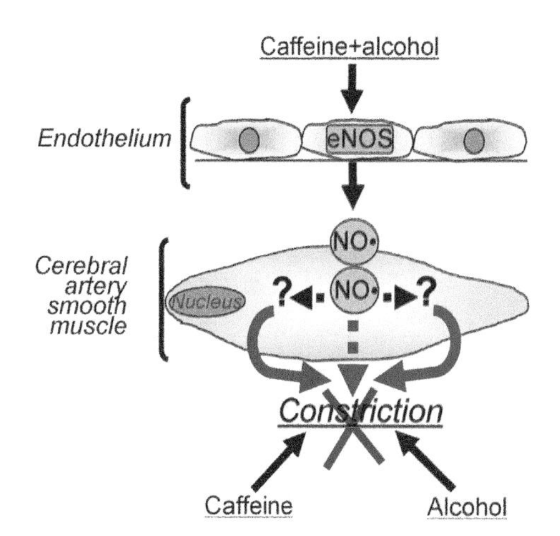

FIGURE 30.3 Schematic depiction of mechanisms that underlie effect of caffeine–alcohol mixture on cerebral artery diameter. Caffeine (10 μM) or alcohol (50 mM) alone constricts cerebral arteries independent of functional endothelium (Chang et al., 2016). Caffeine only protects against alcohol-induced constriction of cerebral arteries in presence of nitric oxide. Smooth muscle molecular targets of nitric oxide release triggered by caffeine–alcohol mixture remain unknown. *Arrows with uninterrupted lines* point at known mechanisms. *Arrows with dotted lines* represent areas of research that require further exploration.

These include ion channel proteins, such as voltage- and calcium-gated potassium channels of large conductance (BK), an ionotropic receptor of critical importance in the function of both central neurons and cerebral artery smooth muscle, which we have studied in particular.

BK channels result from tetrameric association of Slo1 gene products, termed "slo1 channels" or BK α subunits (Orio, Rojas, Ferreira, & Latorre, 2002). They are widely expressed in both central neurons and brain vasculature; and this channel's activity is sensitive to alcohol or cholesterol presence (Dopico, Bukiya, & Martin, 2014; Dopico et al., 2012). In vitro alcohol exposure of slo1 channels usually results in increased ionic current (Dopico et al., 2014). In an elegant work on artificial lipid bilayers made of two phospholipid species where the slo1 channel from human brain (hslo1) was reconstituted into, Crowley et al. (2003) demonstrated that an increase in cholesterol concentration from 0 to 50 mol% of lipid mixture was paralleled by a gradual decrease in alcohol-induced BK current potentiation (Crowley et al., 2003). The molecular bases of cholesterol–alcohol interaction on slo1 channels remain speculative. Remarkably, an alcohol-sensing site and several Cholesterol Recognition Amino Acid Consensus (CRAC) motifs have both been mapped to the cytosolic tail domain of slo1 proteins (Bukiya, Kuntamallappanavar, et al., 2014; Singh et al., 2012). However, these alcohol- and cholesterol-recognition protein areas do not overlap. Thus, it is unlikely that cholesterol and alcohol directly compete for a common binding site and thus, counteract each other's action on slo1 channels. Rather, we hypothesize that each ligand, upon recognition by distinct sites located in the slo1 subunit, induces conformational changes that converge at the channel gate: the changes evoked by cholesterol binding are ultimately counteracting those evoked by ethanol binding.

Cholesterol itself is a critical lipid player in cerebrovascular health, with low circulating levels (<100 mg/dL) considered predictive of poor outcome in critically ill patients (Vyroubal et al., 2008). On the other hand, circulating cholesterol levels above 240 mg/dL are diagnosed as hypercholesterolemia, which favors incidence and poorer prognosis of vascular disease (Ajufo & Rader, 2016). Driven by their relevance to human health, cholesterol–alcohol interactions in cerebral circulation have been extensively studied.

In a rat model of high dietary intake of cholesterol (2% cholesterol diet for 18–23 weeks), alcohol-induced constriction of cerebral arterioles in vivo was reduced when compared to alcohol action on arterioles of rats

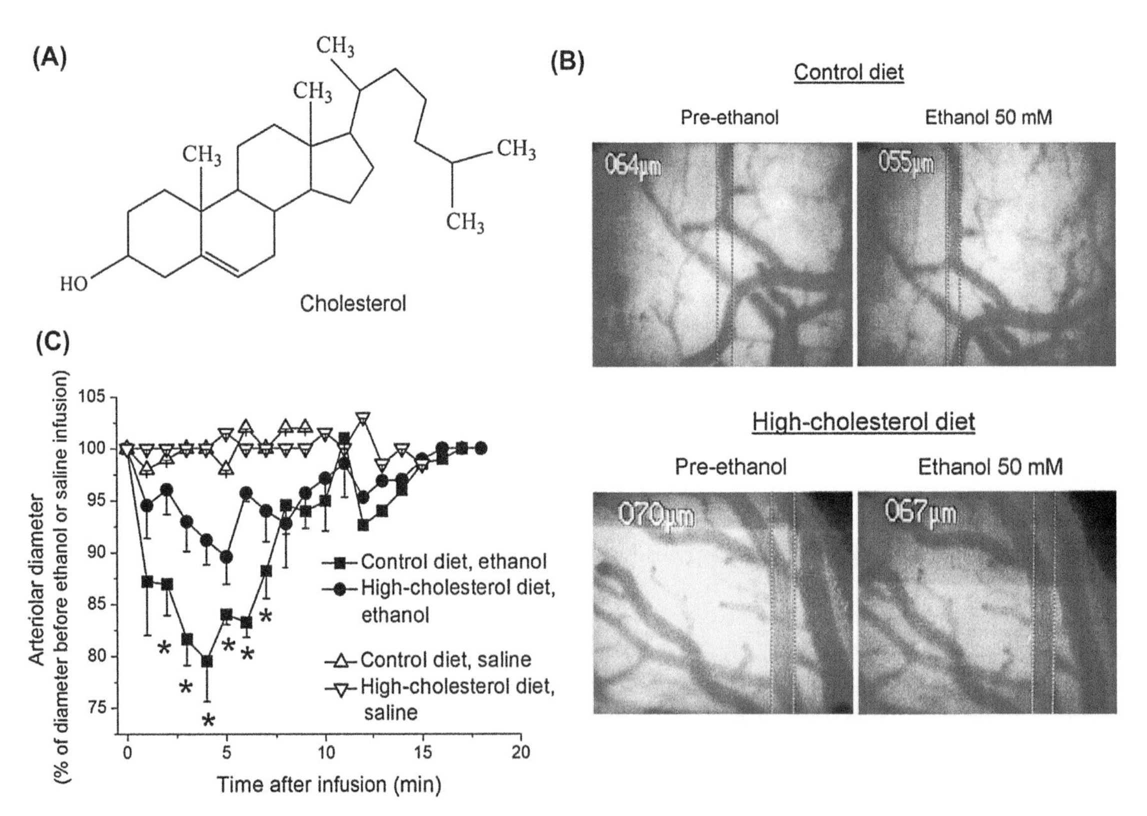

FIGURE 30.4 Dietary cholesterol protects against alcohol-induced pial arteriole in vivo constriction. (A) Molecular structure of cholesterol. (B) Screenshots of brain surface obtained through a closed cranial window. *Vertical dashed lines* highlight external diameter of pial arteriole. Pial arteriole diameter was monitored before (left) and after (right) a 50 mM EtOH infusion into the cerebral circulation via catheter in carotid artery. Washout of EtOH was achieved by infusion of sodium saline. (C) Averaged data show a significant decrease in pial arteriole diameter (PAD) in response to EtOH in rats on high-cholesterol diet (n = 4) compared to control group (n = 4). $^{*}p < .05$, when compared to EtOH-induced arteriole constriction in control group. *With slight modifications from Bukiya, A., Dopico, A. M., Leffler, C. W., & Fedinec, A. (2014). Dietary cholesterol protects against alcohol-induced cerebral artery constriction.* Alcohololism, Clinical and Experimental Research, 38, *1216–1226.*

on control diet (Fig. 30.4) (Bukiya, Dopico, et al., 2014). Further studies into the cholesterol–alcohol interaction in isolated, in vitro pressurized cerebral artery revealed that there was a bell-shaped dependence of alcohol action in cerebral arteries on cholesterol levels in circulation: both low and high cholesterol levels were protective against alcohol-induced constriction (Fig. 30.5A) (Bukiya, Dopico, et al., 2014; Bukiya, Vaithianathan, Kuntamallappanavar, Asuncion-Chin, & Dopico, 2011). Moreover, this bell-shaped dependence was paralleled by cholesterol levels in the arterial wall (Fig. 30.5B). Thus, as reported for the caffeine–ethanol interaction, cholesterol–ethanol interaction on cerebral arteries occurs at the isolated vessel itself. However, unlike caffeine (Fig. 30.3), cholesterol regulation of alcohol effect on cerebral artery diameter does not require the presence of a functional endothelium, as protection could be observed in de-endothelialized arteries (Bukiya et al., 2011; Bukiya, Dopico, et al., 2014).

Alcohol-induced inhibition of BK channels in *vascular smooth muscle* was identified as an underlying cause of alcohol-induced constriction of cerebral arteries (Liu et al., 2004). In vascular smooth muscle, the BK channel complex is composed of the slo1 protein homotetramer accessorized by small, smooth muscle–abundant proteins (termed "β1" and encoded by the *KCNMB1* gene) (Fig. 30.5C) (Brenner et al., 2000). BK β1 subunits regulate several aspects of BK channel gating leading to an increase in the channel's apparent calcium sensitivity. Thus, BK channel activity plays a major role in negatively feedbacking on depolarization-induced calcium entry and smooth muscle contraction (Brenner et al., 2000). In addition, BK β1 subunits drastically modify the channel pharmacology, as alcohol fails to inhibit smooth muscle BK channels and constrict cerebral arteries in *KCNMB1* knockout mice (Bukiya et al., 2009). In rat and mouse cerebral artery myocytes expressing wild-type (e.g., β1 subunit–containing) BK channels, cholesterol depletion resulted in loss of alcohol-induced BK channel inhibition (Bukiya et al., 2011). This effect was replicated in a very simplified experimental system consisting of a two phospholipid species

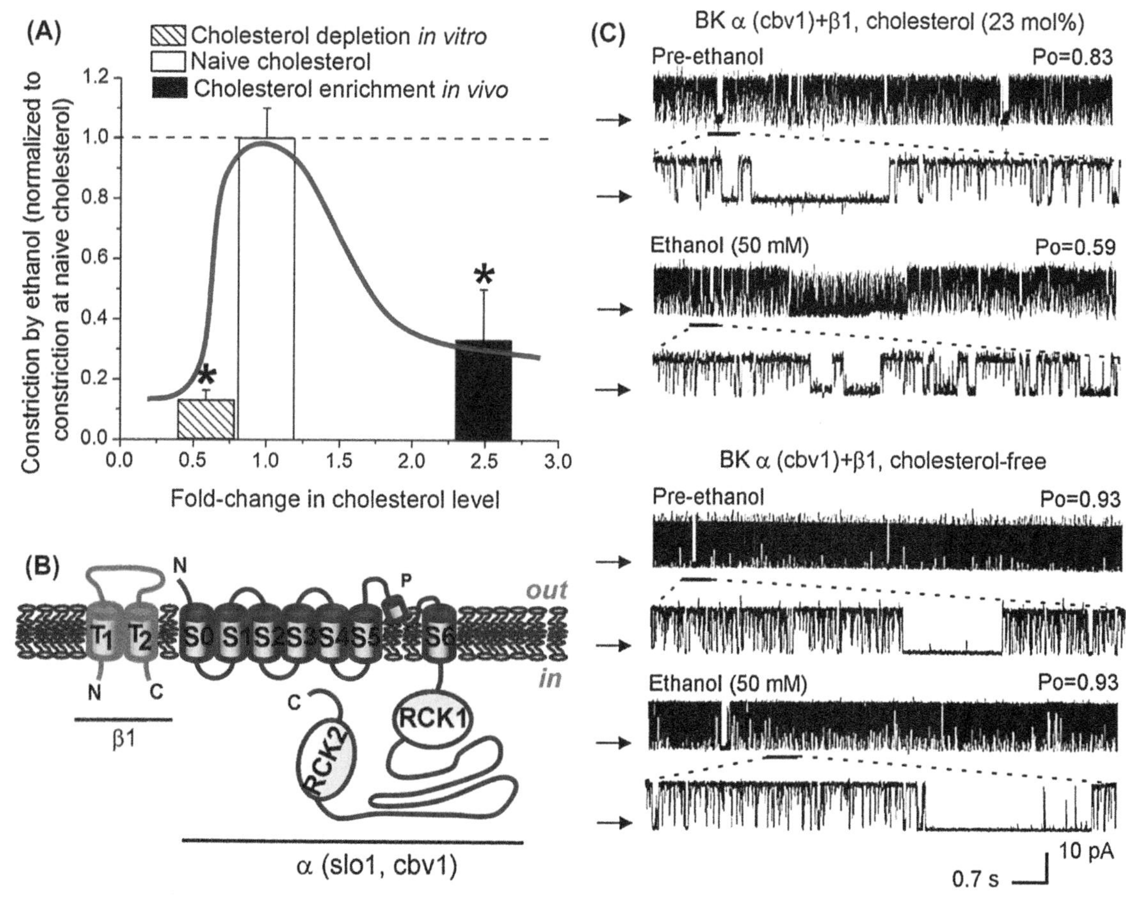

FIGURE 30.5 Current layers of knowledge on cholesterol modulation of alcohol effect in cerebral arteries. (A) Changes in alcohol-induced constriction of cerebral arteries upon cholesterol depletion versus cholesterol enrichment. Bell shape of the changes is highlighted by *red contour line.* (B) Schematic structure of BK channel heteromer in the vascular smooth muscle. (C) Original records of BK (cbv1+β1) channel activity in artificial lipid bilayers showing alcohol-induced BK channel inhibition in bilayers with cholesterol (top set of traces) as opposed to cholesterol-free bilayers (bottom set of traces). *Traces were originally published in Bukiya, A. N., Vaithianathan, T., Kuntamallappanavar, G., Asuncion-Chin, M., & Dopico, A. M. (2011). Smooth muscle cholesterol enables BK β1 subunit-mediated channel inhibition and subsequent vasoconstriction evoked by alcohol.* Arteriosclerosis, Thrombosis and Vascular Biology, *31, 2410–2423.*

planar lipid bilayer where the slo1 and BK β1 subunit proteins were reconstituted (Bukiya et al., 2011). While alcohol-induced BK channel inhibition was observed in bilayers that contained 23 mol% cholesterol, alcohol failed to inhibit the heteromeric BK channel in cholesterol-free bilayers (Fig. 30.5D). Conceivably, the lack of cholesterol decouples BK β1 subunits from their slo1 counterparts leading to an outcome (i.e., loss of ethanol sensitivity) that mimics that of the *KCNMB1* knockout mouse. While the mechanisms and targets underlying blunting of alcohol-induced constriction of cerebral arteries by *decreased* cholesterol levels have been narrowed down to the BK channel protein in vascular myocyte plasmalemma, location of mechanisms and targets involved in blunting of such ethanol action by *excessive* cholesterol levels remains largely unresolved.

CONCLUDING REMARKS

There is little doubt that dietary components such as caffeine and cholesterol play a critical role in shaping alcohol effects on brain and cerebral arteries. While the behavioral consequences of mixing caffeine and alcohol are well studied, little is known about modulation of alcohol-driven behaviors by cholesterol. In cerebral arteries, both caffeine administration and changes in cholesterol levels ablate the cerebral artery constriction evoked by alcohol concentrations reached in blood during moderate-to-heavy drinking episodes. Moreover, the *final effectors* of caffeine- and cholesterol-blunting of alcohol action on cerebral artery diameter seem to reside within the vascular smooth muscle. Whether caffeine and cholesterol may share subcellular pathways in modulating alcohol effect remains unknown. The identification of the entire spectra of mechanisms involved in caffeine or cholesterol modulation of alcohol actions on cerebral artery and brain function represents a major task for the near future. Many intriguing findings are yet to come.

Acknowledgments

This work was supported by NIH awards R37 AA11560 (A.D.) and R01 AA023764 (A.B.)

References

Abbey, A., Wegner, R., Woerner, J., Pegram, S. E., & Pierce, J. (2014). Review of survey and experimental research that examines the relationship between alcohol consumption and men's sexual aggression perpetration. *Trauma, Violence & Abuse, 15*, 265–282.

Addicott, M. A., Yang, L. L., Peiffer, A. M., Burnett, L. R., Burdette, J. H., Chen, M. Y., … Laurienti, P. J. (2009). The effect of daily caffeine use on cerebral blood flow: How much caffeine can we tolerate? *Human Brain Mapping, 30*, 3102–3114.

Ajufo, E., & Rader, D. J. (2016). Recent advances in the pharmacological management of hypercholesterolaemia. *Lancet Diabetes & Endocrinology, 4*, 436–446.

Altura, B. M., & Altura, B. T. (1984). Alcohol, the cerebral circulation and strokes. *Alcohol, 1*, 325–331.

Aronowski, J., Strong, R., Shirzadi, A., & Grotta, J. C. (2003). Ethanol plus caffeine (caffeinol) for treatment of ischemic stroke: Preclinical experience. *Stroke, 34*, 1246–1251.

Arria, A. M., Caldeira, K. M., Kasperski, S. J., Vincent, K. B., Griffiths, R. R., & O'Grady, K. E. (2011). Energy drink consumption and increased risk for alcohol dependence. *Alcoholism, Clinical and Experimental Research, 35*, 365–375.

Ashihara, H., & Crozier, A. (1999). Biosynthesis and metabolism of caffeine and related purine alkaloids in plants. *Advances in Botanical Research, 30*, 117–205.

Azagba, S., & Sharaf, M. F. (2014). Is alcohol mixed with energy drinks consumption associated with susceptibility to smoking? *Preventive Medicine, 61*, 26–28.

Brenner, R., Peréz, G. J., Bonev, A. D., Eckman, D. M., Kosek, J. C., Wiler, S. W., … Aldrich, R. W. (2000). Vasoregulation by the beta1 subunit of the calcium-activated potassium channel. *Nature, 407*, 870–876.

Bukiya, A., Dopico, A. M., Leffler, C. W., & Fedinec, A. (2014). Dietary cholesterol protects against alcohol-induced cerebral artery constriction. *Alcohololism, Clinical and Experimental Research, 38*, 1216–1226.

Bukiya, A. N., Kuntamallappanavar, G., Edwards, J., Singh, A. K., Shivakumar, B., & Dopico, A. M. (2014). An alcohol-sensing site in the calcium- and voltage-gated, large conductance potassium (BK) channel. *Proceedings of the National Academy of Sciences of the United States of America, 111*, 9313–9318.

Bukiya, A. N., Liu, J., & Dopico, A. M. (2009). The BK channel accessory beta1 subunit determines alcohol-induced cerebrovascular constriction. *Federation of European Biochemical Societies Letters, 583*, 2779–2784.

Bukiya, A. N., Vaithianathan, T., Kuntamallappanavar, G., Asuncion-Chin, M., & Dopico, A. M. (2011). Smooth muscle cholesterol enables BK β1 subunit-mediated channel inhibition and subsequent vasoconstriction evoked by alcohol. Arteriosclerosis. *Thrombosis and Vascular Biology, 31*, 2410–2423.

Burns, M., & Moskowitz, H. (1989). Two experiments on alcohol–caffeine interaction. *Alcohol Drugs Driving, 5*, 303–315.

Chang, J., Fedinec, A. L., Kuntamallappanavar, G., Leffler, C. W., Bukiya, A. N., & Dopico, A. M. (2016). Endothelial nitric oxide mediates caffeine antagonism of alcohol-induced cerebral artery constriction. *The Journal of Pharmacology and Experimental Therapeutics, 356*, 106–115.

Crowley, J. J., Treistman, S. N., & Dopico, A. M. (2003). Cholesterol antagonizes ethanol potentiation of human brain BKCa channels reconstituted into phospholipid bilayers. *Molecular Pharmacology, 64*, 365–372.

Deak, F., Freeman, W. M., Ungvari, Z., Csiszar, A., & Sonntag, W. E. (2016). Recent developments in understanding brain aging: Implications for Alzheimer's disease and vascular cognitive impairment. *The Journals of Gerontology. Series A, Biological Sciences and Medical Sciences, 71*, 13–20.

Diamond, I. (1992). *Cecil textbook of medicine*. Philadelphia: W. B. Saunders, Co.

Dinh-Zarr, T., Goss, C., Heitman, E., Roberts, I., & DiGuiseppi, C. (2004). Interventions for preventing injuries in problem drinkers. *The Cochrane Database of Systematic Reviews, 3*, CD001857.

Dopico, A. M., Bukiya, A. N., & Martin, G. E. (2014). Ethanol modulation of mammalian BK channels in excitable tissues: Molecular targets and their possible contribution to alcohol-induced altered behavior. *Frontiers in Physiology, 5*, 466.

Dopico, A. M., Bukiya, A. N., & Singh, A. K. (2012). Large conductance, calcium- and voltage-gated potassium (BK) channels: Regulation by cholesterol. *Pharmacology & Therapeutics, 135*, 133–150.

Drake, C. L., Roehrs, T., Turner, L., Scofield, H. M., & Roth, T. (2003). Caffeine reversal of ethanol effects on the multiple sleep latency test, memory, and psychomotor performance. *Neuropsychopharmacology, 28*, 371–378.

Echeverri, D., Montes, F. R., Cabrera, M., Galán, A., & Prieto, A. (2010). Caffeine's vascular mechanisms of action. *International Journal of Vascular Medicine*, 834060.

Edwards, S., Baynes, B. B., Carmichael, C. Y., Zamora-Martinez, E. R., Barrus, M., Koob, G. F., & Gilpin, N. W. (2013). Traumatic stress reactivity promotes excessive alcohol drinking and alters the balance of prefrontal cortex-amygdala activity. *Translational Psychiatry, 3*, e296.

Engdahl, B., Dikel, T. N., Eberly, R., & Blank, A., Jr. (1998). Comorbidity and course of psychiatric disorders in a community sample of former prisoners of war. *American Journal of Psychiatry, 155*, 1740–1745.

Epand, R. M. (2006). Cholesterol and the interaction of proteins with membrane domains. *Progress in Lipid Research, 45*, 279–294.

Fillmore, M. T., Roach, E. L., & Rice, J. T. (2002). Does caffeine counteract alcohol-induced impairment? The ironic effects of expectancy. *Journal of Studies on Alcohol and Drugs, 63*, 745–754.

Fillmore, M. T., & Vogel-Sprott, M. (1995). Behavioral effects of combining alcohol and caffeine: Contribution of drug-related expectancies. *Experimental and Clinical Psychopharmacology, 3*, 33–38.

Fitterer, J. L., Nelson, T. A., & Stockwell, T. (2015). A review of existing studies reporting the negative effects of alcohol access and positive effects of alcohol control policies on interpersonal violence. *Frontiers in Public Health, 3*, 253.

Flotta, D., Micò, R., Nobile, C., Pileggi, C., Bianco, A., & Pavia, M. (2014). Consumption of energy drinks, alcohol, and alcohol-mixed energy drinks among Italian adolescents. *Alcoholism, Clinical and Experimental Research, 38*, 1654–1661.

Frary, C. D., Johnson, R. K., & Wang, M. Q. (2005). Food sources and intakes of caffeine in the diets of persons in the United States. *Journal of the American Dietetic Association, 105*, 110–113.

Frishman, W. H., Del Vecchio, A., Sanal, S., & Ismail, A. (2003). Cardiovascular manifestations of substance abuse: Part 2: Alcohol, amphetamines, heroin, cannabis, and caffeine. *Heart Disease, 5*, 253–271.

Fritz, B. M., Quoilin, C., Kasten, C. R., Smoker, M., & Boehm, S. L., 2nd (2016). Concomitant caffeine increases binge consumption of ethanol in adolescent and adult mice, but produces additive motor stimulation only in adolescent animals. *Alcoholism, Clinical and Experimental Research, 40*, 1351–1360.

Gimpl, G., Burger, K., & Fahrenholz, F. (1997). Cholesterol as modulator of receptor function. *Biochemistry, 36*, 10959–10974.

Gould, R. G. (1951). Lipid metabolism and atherosclerosis. *American Journal of Medicine, 11*, 209–227.

Guizzetti, M., & Costa, L. G. (2007). Cholesterol homeostasis in the developing brain: A possible new target for ethanol. *Human & Experimental Toxicology, 26*, 355–360.

Higgins, J. P., & Babu, K. M. (2013). Caffeine reduces myocardial blood flow during exercise. *American Journal of Medicine, 126*, 730.e1-8.

Howland, J., Rohsenow, D. J., Arnedt, J. T., Bliss, C. A., Hunt, S. K., Calise, T. V., ... Gottlieb, D. J. (2011). The acute effects of caffeinated versus non-caffeinated alcoholic beverage on driving performance and attention/reaction time. *Addiction, 106*, 335–341.

Jones, S. C. (2011). You wouldn't know it had alcohol in it until you read the can: Adolescents and alcohol-energy drinks. *Australasian Marketing Journal, 19*, 189–195.

Jones, S. C., Barrie, L., & Berry, N. (2012). Why (not) alcohol energy drinks? A qualitative study with Australian university students. *Drug and Alcohol Review, 31*, 281–287.

Jones, P. J., Pappu, A. S., Hatcher, L., Li, Z. C., Illingworth, D. R., & Connor, W. E. (1996). Dietary cholesterol feeding suppresses human cholesterol synthesis measured by deuterium incorporation and urinary mevalonic acid levels. *Arteriosclerosis, Thrombosis, and Vascular Biology, 16*, 1222–1228.

Juliano, L. M., Evatt, D. P., Richards, B. D., & Griffiths, R. R. (2012). Characterization of individuals seeking treatment for caffeine dependence. *Psychology of Addictive Behaviors, 26*, 948–954.

Korczyn, A. D. (2015). Vascular parkinsonism—characteristics, pathogenesis and treatment. *Nature Reviews Neurology, 11*, 319–326.

Kunin, D., Gaskin, S., Rogan, F., Smith, B. R., & Amit, Z. (2000a). Caffeine promotes ethanol drinking in rats. Examination using a limited-access free choice paradigm. *Alcohol, 21*, 271–277.

Kunin, D., Gaskin, S., Rogan, F., Smith, B. R., & Amit, Z. (2000b). Augmentation of corticosterone release by means of a caffeine-ethanol interaction in rats. *Alcohol, 22*, 53–56.

Lee, R. M. (1995). Morphology of cerebral arteries. *Pharmacology & Therapeutics, 66*, 149–173.

Liebeskind, D. S., Sanossian, N., Fu, K. A., Wang, H. J., & Arab, L. (2016). The coffee paradox in stroke: Increased consumption linked with fewer strokes. *Nutritional Neuroscience, 19*, 406–413.

Liu, P., Ahmed, A., Jaggar, J., & Dopico, A. (2004). Essential role for smooth muscle BK channels in alcohol-induced cerebrovascular constriction. *Proceedings of the National Academy of Sciences of the United States of America, 101*, 18217–18222.

Mackay, M., Tiplady, B., & Scholey, A. B. (2002). Interactions between alcohol and caffeine in relation to psychomotor speed and accuracy. *Human Psychopharmacology, 17*, 151–156.

Malinauskas, B. M., Aeby, V. G., Overton, R. F., Carpenter-Aeby, T., & Barber-Heidal, K. (2007). A survey of energy drink consumption patterns among college students. *Nutrition Journal, 31*(6), 35.

Marczinski, C. A., & Fillmore, M. T. (2014). Energy drinks mixed with alcohol: What are the risks? *Nutrition Reviews, 72*, 98–107.

McKetin, R., & Coen, A. (2014). The effect of energy drinks on the urge to drink alcohol in young adults. *Alcoholism, Clinical and Experimental Research, 38*, 2279–2285.

McKetin, R., Coen, A., & Kaye, S. (2015). A comprehensive review of the effects of mixing caffeinated energy drinks with alcohol. *Drug and Alcohol Dependence, 151*, 15–30.

McMurtrie, B. (2014). *Why colleges haven't stopped binge drinking*. Chronicles of Higher Education.

Meredith, S. E., Juliano, L. M., Hughes, J. R., & Griffiths, R. R. (2013). Caffeine use disorder: A comprehensive review and research agenda. *Journal of Caffeine Research, 3*, 114–130.

Mintel International Group Ltd. (2007). *Energy drinks*. Chicago, IL: Mintel International Group Ltd.

Nehlig, A. (2016). Effects of coffee/caffeine on brain health and disease: What should I tell my patients? *Practical Neurology, 16*, 89–95.

Nowaczyk, M. J., & Irons, M. B. (2012). Smith-Lemli-Opitz syndrome: Phenotype, natural history, and epidemiology. *American Journal of Medical Genetics. Part C, Seminars in Medical Genetics, 160C*, 250–262.

Orio, P., Rojas, P., Ferreira, G., & Latorre, R. (2002). New disguises for an old channel: MaxiK channel beta-subunits. *News in Physiological Sciences, 17*, 156–161.

Panza, F., Solfrizzi, V., Barulli, M. R., Bonfiglio, C., Guerra, V., Osella, A., ... Logroscino, G. (2015). Coffee, tea, and caffeine consumption and prevention of late-life cognitive decline and dementia: A systematic review. *The Journal of Nutrition Health and Aging, 19*, 313–328.

Pastor, R., Reed, C., Burkhart-Kasch, S., Li, N., Sharpe, A. L., Coste, S. C., ... Phillips, T. J. (2011). Ethanol concentration-dependent effects and the role of stress on ethanol drinking in corticotropin-releasing factor type 1 and double type 1 and 2 receptor knockout mice. *Psychopharmacology (Berlin), 218*, 169–177.

Patra, J., Taylor, B., Irving, H., Roerecke, M., Baliunas, D., Mohapatra, S., & Rehm, J. (2010). Alcohol consumption and the risk of morbidity and mortality for different stroke types—a systematic review and meta-analysis. *BMC Public Health, 10*, 258.

Peacock, A., Cash, C., & Bruno, R. (2015). Cognitive impairment following consumption of alcohol with and without energy drinks. *Alcoholism, Clinical and Experimental Research, 39*, 733–742.

Pennay, A., Lubman, D., & Miller, P. (2011). Combining energy drinks and alcohol – A recipe for trouble? *Australian Family Physician, 40*, 104–107.

Peoples, R. W., Li, C., & Weight, F. F. (1996). Lipid vs protein theories of alcohol action in the nervous system. *Annual Reviews of Pharmacology and Toxicology, 36*, 185–201.

Petrov, A. M., Kasimov, M. R., & Zefirov, A. L. (2016). Brain cholesterol metabolism and its defects: Linkage to neurodegenerative diseases and synaptic dysfunction. *Acta Naturae, 8*, 58–73.

Petzer, J. P., & Petzer, A. (2015). Caffeine as a lead compound for the design of therapeutic agents for the treatment of Parkinson's disease. *Current Medicinal Chemistry, 22*, 975–988.

Pine, S. (1987). In K. S. Misler, & S. Tenney (Eds.), *Organic chemistry* (5th ed.). New York: McGraw-Hill Company.

Piriyawat, P., Labiche, L. A., Burgin, W. S., Aronowski, J. A., & Grotta, J. C. (2003). Pilot dose-escalation study of caffeine plus ethanol (caffeinol) in acute ischemic stroke. *Stroke, 34*, 1242–1245.

Polidori, M. C., Pientkam, L., & Mecocci, P. (2012). A review of the major vascular risk factors related to Alzheimer's disease. *Journal of Alzheimer's Disease, 32*, 521–530.

Prediger, R. D. (2010). Effects of caffeine in Parkinson's disease: From neuroprotection to the management of motor and non-motor symptoms. *Journal of Alzheimer's Disease, 20*, S205–S220.

Puddey, I. B., Rakic, V., Dimmitt, S. B., & Beilin, L. J. (1999). Influence of pattern of drinking on cardiovascular disease and cardiovascular risk factors—a review. *Addiction, 94*, 649–663.

Raz, L., Knoefel, J., & Bhaskar, K. (2016). The neuropathology and cerebrovascular mechanisms of dementia. *Journal of Cerebral Blood Flow and Metabolism, 36*, 172–186.

Reissig, C. J., Strain, E. C., & Griffiths, R. R. (2009). Caffeinated energy drinks—a growing problem. *Drug and Alcohol Dependence, 99*, 1–10.

Reynolds, K., Lewis, B., Nolen, J. D., Kinney, G. L., Sathya, B., & He, J. (2003). Alcohol consumption and risk of stroke: A meta-analysis. *Journal of American Medical Association, 289*, 579–588.

Rush, C. R., Higgins, S. T., Hughes, J., Bickel, W. K., & Wiegner, M. (1993). Acute behavioral and cardiac effects of alcohol and caffeine, alone and in combination, in humans. *Behavioral Pharmacology, 4*, 562–572.

Sacco, R. L., Chong, J. Y., Prabhakaran, S., & Elkind, M. S. (2007). Experimental treatments for acute ischaemic stroke. *Lancet, 369*, 331–341.

Saher, G., & Stumpf, S. K. (2015). Cholesterol in myelin biogenesis and hypomyelinating disorders. *Biochimica et Biophysica Acta, 1851*, 1083–1094.

Schwarzschild, M. A., Xu, K., Oztas, E., Petzer, J. P., Castagnoli, K., Castagnoli, N., Jr., & Chen, J. F. (2003). Neuroprotection by caffeine and more specific A2A receptor antagonists in animal models of Parkinson's disease. *Neurology, 61*, S55–S61.

Singh, A. K., McMillan, J., Bukiya, A. N., Burton, B., Parrill, A. L., & Dopico, A. M. (2012). Multiple cholesterol recognition/interaction amino acid consensus (CRAC) motifs in cytosolic C tail of Slo1 subunit determine cholesterol sensitivity of Ca^{2+}- and voltage-gated K^{+} (BK) channels. *Journal of Biological Chemistry, 287*, 20509–20521.

Siperstein, M. D. (1970). Regulation of cholesterol biosynthesis in normal and malignant tissues. In B. L. Horecker, & E. R. Stadtman (Eds.), *Current topics in cellular regulation* (pp. 65–100). New York: Academic Press.

Snyder, H. M., Corriveau, R. A., Craft, S., Faber, J. E., Greenberg, S. M., Knopman, D., … Carrillo, M. C. (2015). Vascular contributions to cognitive impairment and dementia including Alzheimer's disease. *Alzheimer's & Dementia, 11*, 710–717.

Sudakov, S. K., Rusakova, I. V., & Medvedeva, O. F. (2003). Effect of chronic caffeine consumption on changes in locomotor activity of WAG/G and Fischer-344 rats induced by nicotine, ethanol, and morphine. *Bulletin of Experimental Biology and Medicine, 136*, 563–565.

Sun, G. Y., & Sun, A. Y. (1985). Ethanol and membrane lipids. *Alcoholism, Clinical and Experimental Research, 9*, 164–180.

Tchekalarova, J. D., Kubová, H., & Mareš, P. (2014). Early caffeine exposure: Transient and long-term consequences on brain excitability. *Brain Research Bulletin, 104*, 27–35.

Trandum, C., Westh, P., Jorgensen, K., & Mouritsen, O. G. (2000). A thermodynamic study of the effects of cholesterol on the interaction between liposomes and ethanol. *Biophysical Journal, 78*, 2486–2492.

Urry, E., & Landolt, H. P. (2015). Adenosine, caffeine, and performance: From cognitive neuroscience of sleep to sleep pharmacogenetics. *Current Topics in Behavioral Neurosciences, 25*, 331–366.

Vyroubal, P., Chiarla, C., Giovannini, I., Hyspler, R., Ticha, A., Hrnciarikova, D., & Zadak, Z. (2008). Hypocholesterolemia in clinically serious conditions—review. *Biomedical Papers of the Medical Faculty of the University Palacky Olomouc, Czechoslovakia, 152*, 181–189.

Wallner, M., Hanchar, H. J., & Olsen, R. W. (2006). Low-dose alcohol actions on alpha4beta3delta $GABA_A$ receptors are reversed by the behavioral alcohol antagonist Ro15-4513. *Proceedings of the National Academy of Sciences of the United States of America, 103*, 8540–8545.

Zhang, J., & Liu, Q. (2015). Cholesterol metabolism and homeostasis in the brain. *Protein & Cell, 6*, 254–264.

CHAPTER

31

Sleep, Caffeine, and Physical Activity in Older Adults

M.A. Schrager
Stetson University, DeLand, FL, United States

INTRODUCTION

In most countries worldwide, rates of physical activity are low and trending lower (Al-Hazzaa, 2004; Brownson, Boehmer, & Luke, 2005; Paffenbarger et al., 1994). This trend continues despite the known benefits of physical activity, and it is particularly deleterious in older persons who require beneficial reductions in bone loss (Howe et al., 2011), sarcopenia (Janssen, Heymsfield, & Ross, 2002), and risk of falls (Hourigan et al., 2008). Potential causes of physical inactivity are numerous, and they include environmental, occupational, social, psychological, and genetic factors (Bauman et al., 2012; Brownson et al., 2005; Sallis et al., 1989; Simonsick, Guralnik, & Fried, 1999). Increasing attention has been paid to the role of the "built environment" on physical inactivity through initiatives such as the US Department of Health and Human Services' *Healthy People 2020* (Koh, 2010) and the American College of Sports Medicine's *Exercise Is Medicine* (Sallis, 2009; www.exerciseismedicine.org). For example, in urban communities where most Americans live, greater access/proximity to walking trails and green spaces, higher population density, and other factors have been shown to increase physical activity across the entire adult age span (Li et al., 2008; Saelens, Sallis, & Frank, 2016; Takano, Nakamura, & Watanabe, 2002).

TWO FACTORS RELATED TO PHYSICAL ACTIVITY: SLEEP AND CAFFEINE

Another factor that has received greater attention of late in obesity and physical inactivity research is the age-associated changes in sleep experienced by older persons (Kaufmann et al., 2016). Therefore, this chapter focuses on the physiological interrelationships between sleep and physical activity. It also focuses on direct and indirect effects of caffeine on sleep and physical activity. It is hoped that this chapter will help initiate future controlled studies in the older persons to specifically address the effects of caffeine on physical activity generally in the older population (Ancoli-Israel, 2009) and in various potentially more vulnerable subpopulations, for example, older Hispanic (Kaufmann et al., 2016) and African-American (Ancoli-Israel et al., 1995; Surani, 2013) individuals.

PHYSIOLOGICAL CONTRIBUTORS TO PHYSICAL ACTIVITY LEVEL

Research into age-related physiological causes of physical inactivity suggests that a complex network of physiological factors, including sleep, is in play. Although these factors are not fully understood, they likely include age-related declines in maximal aerobic capacity (Tanaka et al., 1997) and increases in fatigue (Simonsick et al., 1999). A relatively new hypothesis that attempts to connect the physiology of energy metabolism with physical activity has its basis in a new bioenergetics paradigm generated using data from the Baltimore Longitudinal Study of Aging. The primary hypothesis of this paradigm asserts that when energy to perform activities becomes deficient with aging, adaptive behaviors (e.g., sedentary behavior) develop in order to conserve energy. This information was supported by a study by Schrager, Schrack, Simonsick, and Ferrucci (2014), which indicated that energy usable for physical activity, or "available energy"—operationally defined as the difference between oxygen consumption during peak walking and resting oxygen consumption (Schrack, Simonsick, & Ferrucci, 2010)—

Addictive Substances and Neurological Disease
http://dx.doi.org/10.1016/B978-0-12-805373-7.00031-1

had significant, independent, and positive relationships to total and vigorous physical activity levels in health, of independently mobile persons aged 45–96. Physical activity requires considerable levels of energy expenditure and, particularly in lower-fitness individuals, "available energy" may be an important limiting determinant of physical inactivity in older persons. Preservation of this physiologically based factor through a lifestyle focused on physical activity (both aerobic and resistance training) may therefore be critical in preventing older persons from falling below critical thresholds in parameters such as "available energy," in order to maintain levels in physical activity, and reap its established health benefits (Morris, Clayton, Everitt, Semmence, & Burgess, 1990; Morris, Kagan, Pattison, & Gardner, 1966; Paffenbarger et al., 1994).

BENEFITS OF PHYSICAL ACTIVITY ON OBSTRUCTIVE SLEEP APNEA

From the perspective of "Exercise is Medicine," the aforementioned known health benefits of exercise and physical activity have been studied in clinical sleep disordered breathing conditions, most notably obstructive sleep apnea (OSA). Age is a considered a major risk factor in the development of OSA, with men 45–64 and women over 65 having the highest reported prevalence of OSA (Carter & Watenpaugh, 2008). Characterized by repeated episodes of complete or partial blockage of the upper airway during sleep, OSA causes sleep fragmentation and oxygen desaturation and is associated with obesity, depression, and coronary disease (Carter & Watenpaugh, 2008; Pamidi, Knutson, Ghods, & Mokhlesi, 2011).

A meta-analysis that included three randomized, controlled trials indicated that independent of changes in body mass index (BMI), primarily aerobic exercise training reduces OSA severity (i.e., the number of apnea or hypopnea events per hours of sleep) by 30–40% relative to controls (Iftikhar, Kline, & Youngstedt, 2014). For example, a small, 12-week randomized, controlled trial by Kline et al. (2011) found that in the exercise group—which performed 150 min/week of moderate-intensity aerobic exercise (and two sessions of resistance training per week)—both the apnea–hypopnea and the oxygen desaturation indices were significantly improved compared to the control (stretching) group (Kline et al., 2011).

Studies consistently support the existence of a vicious cycle that includes weight gain/obesity, type II diabetes, and OSA. Furthermore, these studies also indicate that classes of drugs used to treat type II diabetes and depression, which often accompanies OSA, can both accelerate weight gain in OSA patients (Mohammad & Ahmad, 2016; Schwartz, Nihalani, Jindal, Virk, & Jones, 2004). The effectiveness of physical activity as a (non-pharmacological) treatment for OSA is not surprising because it is the most important factor in the maintenance of weight loss, and it improves blood glucose control in type 2 diabetics (Foreyt & Poston, 1999). As noted by Carter III and Watenpaugh (2008), "...medications may emerge to treat obesity, OSA, and their sequelae with minimal side effects. However, there are effective ways to approach these problems now without waiting for the magic pill." The most effective current approach to OSA appears to be to simply increase physical activity levels. Future randomized, controlled studies in older OSA patients should be larger and longer term with a focus on populations older than the 55 years. They should also endeavor to determine appropriate/optimal/minimal doses, intensities, and modalities of exercise, and assess whether short-duration/high-intensity interval training may be effective. One related practical issue that has been minimally explored is that of timing of exercise with respect to sleep. In young persons, sleep was surprisingly improved with late-night, intense exercise (Brand et al., 2014). However, the exercise timing issue remains unexplored in older persons with OSA. Finally, given the increasing prevalence of OSA in younger populations (de la Eva, Baur, Donaghue, & Waters, 2002; Narang et al., 2016), future research should also include this portion of the age span to have a more progressive, far-reaching impact on health.

BENEFITS OF PHYSICAL ACTIVITY ON SLEEP IN SEDENTARY OLDER PERSONS

As mentioned, benefits of the "Exercise is Medicine" perspective also extend to subclinical forms of disordered sleep in older persons. Age is associated with a trend toward greater rates of insomnia, with as many as 50% of older Americans experiencing disrupted or inefficient sleep (Ancoli-Israel, Poceta, Stepnowsky, Martin, & Gehrman, 1997; Ohayon, Carskadon, Guilleminault, & Vitiello, 2004); however, it appears that this trend may be less an effect of age per se than a reflection of the age-associated increase in prevalence of comorbidities, several of which are contributed independently by obesity (Ancoli-Israel, 2009). Chief among these age- and obesity-related comorbidities are cardiovascular disease, osteoarthritis, and depression, each of which is known to benefit from increased physical activity (Karlsson, Johnell, Sigström, Sjöberg, & Fratiglioni, 2016; Lakatta & Levy, 2003; Palazzo, Nguyen, Lefevre-Colau, Rannou, & Poiraudeau, 2016).

Actigraphic data have allowed for longer-term, more objective data collection in the home environment, and these data indicate that sleep's effects on obesity are mediated not only by sleep duration but also by (1) irregularity of sleep duration and (2) daytime napping; that is, the greater the variability in sleep duration and daytime napping, the greater the likelihood of obesity (Patel et al., 2014). Thus, it is likely that by treating—through exercise and other approaches—obesity and its linked comorbidities, age-associated trends toward these and other indicators of insomnia, will also improve in older populations.

The salient benefits of reducing the consequences of disordered sleep through increased physical activity encompass decreased risk of cardiovascular disease, metabolic disease, falls, diminished physical and cognitive function, and mortality. In the context of functional mobility and physical activity, these are vitally important to independence and the quality of life (Ancoli-Israel, 2009).

The Lifestyle Interventions and Independence for Elder (LIFE) study is a randomized, controlled trial designed to assess a physical activity program's effects on measures of functional mobility and mobility disability. To further investigate whether physical activity positively affects sleep, objective accelerometry–based physical activity data has recently been evaluated in 1635 sedentary, mobility-limited, community-dwelling older persons during the baseline assessment of individuals enrolled in the LIFE Study. While no significant associations between mildly severe sleep–wake disturbances and physical inactivity were reported in this cross-sectional study, the authors stated that this may have been due to a lack of severity and variability in the sleep disturbances found across participants (Vaz Fragoso et al., 2014). Follow-up, longitudinal analyses comprising 2.7 years of exercise training in the same LIFE Study cohort indicated that an intervention of moderate-intensity, structured physical activity—about 150 min/week of (primarily) walking with some strength, flexibility, and balance training—resulted in no significant benefits in persons who entered the study with prevalent sleep difficulty (Vaz Fragoso et al., 2015). This lack of significant benefit was in contrast to the improvements found in a younger (i.e., middle-aged) cohort who completed 16 weeks of higher-intensity (60–75% of heart rate reserve) exercise training (King, Oman, Brassington, Bliwise, & Haskell, 1997). However, Vaz Fragoso et al. (2015) did demonstrate in their intervention LIFE study–based trial a preventive effect such that those in the exercise intervention group had a lower likelihood of developing poor sleep quality over the about 2-year intervention period compared with individuals in a group receiving only health education over this period. This preventive effect of physical activity on disordered sleep may have critical public health implications in a growing population of community-dwelling, sedentary persons, as well as other, younger segments of the population.

EFFECTS OF SLEEP ON PHYSICAL ACTIVITY IN OSA PATIENTS

The presence of a bidirectional relationship between sleep and physical activity would seem logical, as daytime sleepiness leading to physical inactivity may potentially occur simultaneously with physical inactivity leading to disrupted circadian regulation of the sleep–wake cycle (Vaz Fragoso et al., 2015). In OSA patients treated using continuous positive airway pressure (CPAP) treatment, evidence is lacking for increased physical activity or energy-expenditure levels due to resultant improved sleep (Diamanti et al., 2013; West, Kohler, Nicoll, & Stradling, 2009), suggesting that CPAP treatment alone was inadequate to increase physical activity in OSA patients. A multifaceted intervention comprising CPAP and education on eating behavior and physical activity did not improve levels of physical activity, either; however, the physical activity component of the intervention in the study by Igelström et al. (2014) was only educational and was not structured. Future studies should assess whether CPAP in combination with a more structured exercise would encourage habitual physical activity in OSA patients.

EFFECTS OF SLEEP ON PHYSICAL ACTIVITY IN HEALTHY OLDER PERSONS

In healthy older persons, as in OSA patients, there is limited evidence that through an intervention focused on improving sleep, a greater level of physical activity will result. More direct study of this question is needed, though, to clarify the question of a bidirectional relationship between improving sleep and improving physical activity. As noted by Kline (2014): "Thus, additional research is needed to overcome the current conundrum—although exercise may be an important behavioral treatment for improving poor and/or disordered sleep, poor sleep may be a key impediment to initiating and/or maintaining a physically active lifestyle."

However, there is substantial evidence that poor sleep has acute detrimental effects in healthy individuals on key mobility-related measures that can influence physical activity such as risk of falling (Hill, Cumming, Lewis, Carrington, & Couteur, 2007), pain (Edwards, Almeida, Klick, Haythornthwaite, & Smith, 2008), time to exhaustion, and mood

(Meney, Waterhouse, Atkinson, Reilly, & Davenne, 1998). Poor sleep in older women was also linked to low heart rate variability, which is in turn associated with increased cardiac morbidity and overall mortality risk (Virtanen, Kalleinen, Urrila, Leppänen, & Polo-Kantola, 2015). Poor sleep quality has also been shown in 426 community-dwelling older adults to be predictive of lower levels of physical inactivity 2–7 years later (Holfeld & Ruthig, 2014). More directly, longer-duration (16 days) actigraphic data showed that higher sleep efficiency predicted greater next-day physical activity levels in healthy, older women (Lambiase, Gabriel, Kuller, & Matthews, 2013).

Pertaining to the relationship between insufficient sleep and obesity, research indicates that inadequate sleep affects metabolism and physical activity, and it is a key independent risk factor for weight gain. Markwald et al. (2013) studied 16 young adults over about 2 weeks to assess the effects of 5 days of poor sleep. In women, total energy expenditure increased by about 5%, but energy intake, primarily after dinner, was greater than the level needed to maintain energy balance, which resulted in positive energy balance and weight gain. The authors concluded that this increased energy intake during periods of insufficient sleep is a physiological, adaptive response geared to provide adequate energy for additional time spent awake; when readily available, intake of food surpasses the amount required in this scenario, likely due to disrupted circadian systems (Markwald et al., 2013). Similarly, fragmented sleep was associated with sleepiness as well as increased carbohydrate oxidation and decreased fat oxidation, which likely contributed to weight gain, and, surprisingly, greater physical activity level (Hursel, Rutters, Gonnissen, Martens, & Westerterp-Plantenga, 2011). This increase in physical activity may have been due to the required act of turning off set clock alarms 7 times per night, and it contradicts findings from Schmid et al. (2009) showing a highly significant decrease in accelerometry-assessed physical activity linked to fragmented sleep under free-living conditions. These studies demonstrate the important influences of sleep on energy balance, metabolism, and physical activity. However, limited related research exists on older persons.

EFFECTS OF CAFFEINE ON SLEEP, METABOLISM, AND PHYSICAL ACTIVITY

Due to the potentially strong detrimental influence of poor sleep on metabolism, weight, and physical activity, it is important to consider the indirect effects caffeine has on these parameters. Caffeine blocks the action of adenosine, which is a neuromodulator involved in regulation of the sleep–wake cycle. Considered the world's most popular drug, caffeine is habitually consumed by as many as 85–90% of the adult population (Nawrot et al., 2003), and is most commonly consumed in people in their sixth and seventh decades of life. Additionally, this demographic group was more sensitive than young individuals to caffeine's dose-dependent effects on sleep, with greater shortening, fragmentation, and shallowing of sleep than found in younger persons (Robillard, Bouchard, Cartier, Nicolau, & Carrier, 2015). Other studies have shown that caffeine can increase sleep latency and reduce sleep duration, especially if large amounts of caffeine are ingested close to usual bedtime of the individual, and especially in individuals not accustomed to high caffeine consumption (Zwyghuizen-Doorenbos, Roehrs, Lipschutz, Timms, & Roth, 1990). Given these findings related to greater sensitivity to—especially high levels of—caffeine consumption in older persons, it would seem advisable to minimize habitual caffeine consumption in order to optimize sleep and its aforementioned metabolic effects, including those on weight maintenance and fat metabolism, as well as its downstream effects on daytime sleepiness, risk of falling, mood, and next-day physical activity.

More direct effects of caffeine on thermogenesis have been considered to be a means to prevent positive energy balance, weight gain (Bérubé-Parent, Pelletier, Doré, & Tremblay, 2005), and increase 24-h energy expenditure and fat oxidation (Dulloo et al., 1999) through products such as green tea and/or Guarana extract. Moreover, caffeine is a known ergogenic substance with stimulatory effects on the central nervous system, a property which should have direct effects on physical activity. High doses of caffeine have also been shown to increase sympathetic nervous activity and blood pressure in nonhabituated coffee drinkers; this increase in blood pressure should also be considered as a safety issue during exercise in older persons (Corti et al., 2002). In the only study to assess longer-term metabolic caffeine intake under free-living conditions, Júdice et al. (2013) studied young males and found no significant differences either acutely or long term (i.e., over the 4-day study) in terms of low and moderate–high physical activity (Júdice et al., 2013). In summary, there is limited evidence for caffeine's ability to promote weight loss and physical activity. Along with potentially risky increases in blood pressure and negative effects on sleep—especially in older persons—that appear to promote weight gain, fatigue, and diminished physical activity, the properties of caffeine make it difficult at this point to recommend as a means to increase physical activity in older persons. However, more research is needed in this specific demographic to better address this question.

References

Al-Hazzaa, H. M. (2004). Prevalence of physical inactivity in Saudi Arabia: A brief review. *Eastern Mediterranean Health Journal = La Revue de Santé de La Méditerranée Orientale = Al-Majallah Al-Ṣiḥḥīyah Li-Sharq Al-Mutawassiṭ, 10*(4–5), 663–670.

Ancoli-Israel, S. (2009). Sleep and its disorders in aging populations. *Sleep Medicine, 10*(Suppl. 1), S7–S11. http://doi.org/10.1016/j.sleep.2009.07.004.

Ancoli-Israel, S., Klauber, M. R., Stepnowsky, C., Estline, E., Chinn, A., & Fell, R. (1995). Sleep-disordered breathing in African-American elderly. *American Journal of Respiratory and Critical Care Medicine, 152*(6 Pt 1), 1946–1949. http://doi.org/10.1164/ajrccm.152.6.8520760.

Ancoli-Israel, S., Poceta, J. S., Stepnowsky, C., Martin, J., & Gehrman, P. (1997). Identification and treatment of sleep problems in the elderly. *Sleep Medicine Reviews, 1*(1), 3–17.

Bauman, A. E., Reis, R. S., Sallis, J. F., Wells, J. C., Loos, R. J. F., & Martin, B. W. (2012). Correlates of physical activity: Why are some people physically active and others not? *Lancet, 380*(9838), 258–271. http://doi.org/10.1016/S0140-6736(12)60735-1.

Bérubé-Parent, S., Pelletier, C., Doré, J., & Tremblay, A. (2005). Effects of encapsulated green tea and Guarana extracts containing a mixture of epigallocatechin-3-gallate and caffeine on 24 h energy expenditure and fat oxidation in men. *The British Journal of Nutrition, 94*(3), 432–436.

Brand, S., Kalak, N., Gerber, M., Kirov, R., Pühse, U., & Holsboer-Trachsler, E. (2014). High self-perceived exercise exertion before bedtime is associated with greater objectively assessed sleep efficiency. *Sleep Medicine, 15*(9), 1031–1036. http://doi.org/10.1016/j.sleep.2014.05.016.

Brownson, R. C., Boehmer, T. K., & Luke, D. A. (2005). Declining rates of physical activity in the United States: What are the contributors? *Annual Review of Public Health, 26*(1), 421–443. http://doi.org/10.1146/annurev.publhealth.26.021304.144437.

Carter, R., III, & Watenpaugh, D. E. (2008). Obesity and obstructive sleep apnea: Or is it OSA and obesity? *Pathophysiology, 15*(2), 71–77. http://doi.org/10.1016/j.pathophys.2008.04.009.

Corti, R., Binggeli, C., Sudano, I., Spieker, L., Hänseler, E., Ruschitzka, F., ... Noll, G. (2002). Coffee acutely increases sympathetic nerve activity and blood pressure independently of caffeine content role of habitual versus nonhabitual drinking. *Circulation, 106*(23), 2935–2940. http://doi.org/10.1161/01.CIR.0000046228.97025.3A.

Diamanti, C., Manali, E., Ginieri-Coccossis, M., Vougas, K., Cholidou, K., Markozannes, E., ... Alchanatis, M. (2013). Depression, physical activity, energy consumption, and quality of life in OSA patients before and after CPAP treatment. *Sleep & Breathing = Schlaf & Atmung, 17*(4), 1159–1168. http://doi.org/10.1007/s11325-013-0815-6.

Dulloo, A. G., Duret, C., Rohrer, D., Girardier, L., Mensi, N., Fathi, M., ... Vandermander, J. (1999). Efficacy of a green tea extract rich in catechin polyphenols and caffeine in increasing 24-h energy expenditure and fat oxidation in humans. *The American Journal of Clinical Nutrition, 70*(6), 1040–1045.

Edwards, R. R., Almeida, D. M., Klick, B., Haythornthwaite, J. A., & Smith, M. T. (2008). Duration of sleep contributes to next-day pain report in the general population. *Pain, 137*(1), 202–207. http://doi.org/10.1016/j.pain.2008.01.025.

de la Eva, R. C., Baur, L. A., Donaghue, K. C., & Waters, K. A. (2002). Metabolic correlates with obstructive sleep apnea in obese subjects. *The Journal of Pediatrics, 140*(6), 654–659. http://doi.org/10.1067/mpd.2002.123765.

Foreyt, J. P., & Poston, W. S. (1999). The challenge of diet, exercise and lifestyle modification in the management of the obese diabetic patient. *International Journal of Obesity and Related Metabolic Disorders: Journal of the International Association for the Study of Obesity, 23*(Suppl. 7), S5–S11.

Hill, E. L., Cumming, R. G., Lewis, R., Carrington, S., & Couteur, D. G. L. (2007). Sleep disturbances and falls in older people. *The Journals of Gerontology Series A: Biological Sciences and Medical Sciences, 62*(1), 62–66.

Holfeld, B., & Ruthig, J. C. (2014). A longitudinal examination of sleep quality and physical activity in older adults. *Journal of Applied Gerontology: The Official Journal of the Southern Gerontological Society, 33*(7), 791–807. http://doi.org/10.1177/0733464812455097.

Hourigan, S. R., Nitz, J. C., Brauer, S. G., O'Neill, S., Wong, J., & Richardson, C. A. (2008). Positive effects of exercise on falls and fracture risk in osteopenic women. *Osteoporosis International: A Journal Established as Result of Cooperation between the European Foundation for Osteoporosis and the National Osteoporosis Foundation of the USA, 19*(7), 1077–1086. http://doi.org/10.1007/s00198-007-0541-7.

Howe, T. E., Shea, B., Dawson, L. J., Downie, F., Murray, A., Ross, C., ... Creed, G. (2011). Exercise for preventing and treating osteoporosis in postmenopausal women. *Cochrane Database of Systematic Reviews (Online), 7*, CD000333. http://doi.org/10.1002/14651858.CD000333.pub2.

Hursel, R., Rutters, F., Gonnissen, H. K., Martens, E. A., & Westerterp-Plantenga, M. S. (2011). Effects of sleep fragmentation in healthy men on energy expenditure, substrate oxidation, physical activity, and exhaustion measured over 48 h in a respiratory chamber. *The American Journal of Clinical Nutrition, 94*(3), 804–808. http://doi.org/10.3945/ajcn.111.017632.

Iftikhar, I. H., Kline, C. E., & Youngstedt, S. D. (2014). Effects of exercise training on sleep apnea: A meta-analysis. *Lung, 192*(1), 175–184. http://doi.org/10.1007/s00408-013-9511-3.

Igelström, H., Helena, I., Emtner, M., Margareta, E., Lindberg, E., Eva, L., ... Pernilla, Å. (2014). Tailored behavioral medicine intervention for enhanced physical activity and healthy eating in patients with obstructive sleep apnea syndrome and overweight. *Sleep & Breathing = Schlaf & Atmung, 18*(3), 655–668. http://doi.org/10.1007/s11325-013-0929-x.

Janssen, I., Heymsfield, S. B., & Ross, R. (2002). Low relative skeletal muscle mass (sarcopenia) in older persons is associated with functional impairment and physical disability. *Journal of the American Geriatrics Society, 50*(5), 889–896.

Júdice, P. B., Magalhães, J. P., Santos, D. A., Matias, C. N., Carita, A. I., Armada-Da-Silva, P. A. S., ... Silva, A. M. (2013). A moderate dose of caffeine ingestion does not change energy expenditure but decreases sleep time in physically active males: A double-blind randomized controlled trial. *Applied Physiology, Nutrition, and Metabolism, 38*(1), 49–56. http://doi.org/10.1139/apnm-2012-0145.

Karlsson, B., Johnell, K., Sigström, R., Sjöberg, L., & Fratiglioni, L. (2016). Depression and depression treatment in a population-based study of individuals over 60 years old without dementia. *The American Journal of Geriatric Psychiatry: Official Journal of the American Association for Geriatric Psychiatry.* http://doi.org/10.1016/j.jagp.2016.03.009.

Kaufmann, C. N., Mojtabai, R., Hock, R. S., Thorpe, R. J., Canham, S. L., Chen, L.-Y., ... Spira, A. P. (2016). Racial/Ethnic differences in insomnia trajectories among U.S. Older adults. *The American Journal of Geriatric Psychiatry: Official Journal of the American Association for Geriatric Psychiatry, 24*(7), 575–584. http://doi.org/10.1016/j.jagp.2016.02.049.

King, A. C., Oman, R. F., Brassington, G. S., Bliwise, D. L., & Haskell, W. L. (1997). Moderate-intensity exercise and self-rated quality of sleep in older adults. A randomized controlled trial. *JAMA, 277*(1), 32–37.

Kline, C. E., Crowley, E. P., Ewing, G. B., Burch, J. B., Blair, S. N., Durstine, J. L., ... Youngstedt, S. D. (2011). The effect of exercise training on obstructive sleep apnea and sleep quality: A randomized controlled trial. *Sleep, 34*(12), 1631–1640. http://doi.org/10.5665/sleep.1422.

Koh, H. K. (2010). A 2020 vision for healthy people. *New England Journal of Medicine, 362*(18), 1653–1656. http://doi.org/10.1056/NEJMp1001601.

Lakatta, E. G., & Levy, D. (2003). Arterial and cardiac aging: Major shareholders in cardiovascular disease enterprises. *Circulation, 107*(1), 139–146. http://doi.org/10.1161/01.CIR.0000048892.83521.58.

Lambiase, M. J., Gabriel, K. P., Kuller, L. H., & Matthews, K. A. (2013). Temporal relationships between physical activity and sleep in older women. *Medicine and Science in Sports and Exercise, 45*(12), 2362–2368. http://doi.org/10.1249/MSS.0b013e31829e4cea.

Li, F., Harmer, P. A., Cardinal, B. J., Bosworth, M., Acock, A., Johnson-Shelton, D., & Moore, J. M. (2008). Built environment, adiposity, and physical activity in adults aged 50–75. *American Journal of Preventive Medicine, 35*(1), 38–46. http://doi.org/10.1016/j.amepre.2008.03.021.

Markwald, R. R., Melanson, E. L., Smith, M. R., Higgins, J., Perreault, L., Eckel, R. H., & Wright, K. P. (2013). Impact of insufficient sleep on total daily energy expenditure, food intake, and weight gain. *Proceedings of the National Academy of Sciences of the United States of America, 110*(14), 5695–5700. http://doi.org/10.1073/pnas.1216951110.

Meney, I., Waterhouse, J., Atkinson, G., Reilly, T., & Davenne, D. (1998). The effect of one night's sleep deprivation on temperature, mood, and physical performance in subjects with different amounts of habitual physical activity. *Chronobiology International, 15*(4), 349–363.

Mohammad, S., & Ahmad, J. (2016). Management of obesity in patients with type 2 diabetes mellitus in primary care. *Diabetes & Metabolic Syndrome: Clinical Research & Reviews, 10*(3), 171–181. http://doi.org/10.1016/j.dsx.2016.01.017.

Morris, J. N., Clayton, D. G., Everitt, M. G., Semmence, A. M., & Burgess, E. H. (1990). Exercise in leisure time: Coronary attack and death rates. *British Heart Journal, 63*(6), 325–334.

Morris, J. N., Kagan, A., Pattison, D. C., & Gardner, M. J. (1966). Incidence and prediction of ischaemic heart-disease in London busmen. *Lancet, 2*(7463), 553–559.

Narang, I., McCrindle, B., Al-Saleh, S., Manlhiot, C., Slorach, C., Mertens, L., … Hamilton, J. (2016). Obstructive sleep apnea and vascular stiffness in obese adolescents. In *A27. Advances in Pediatric Sleep* (Vols. 1–301, p. A1221). American Thoracic Society. Retrieved from http://www.atsjournals.org/doi/abs/10.1164/ajrccm-conference.2016.193.1_MeetingAbstracts.A1221.

Nawrot, P., Jordan, S., Eastwood, J., Rotstein, J., Hugenholtz, A., & Feeley, M. (2003). Effects of caffeine on human health. *Food Additives & Contaminants, 20*(1), 1–30. http://doi.org/10.1080/0265203021000007840.

Ohayon, M. M., Carskadon, M. A., Guilleminault, C., & Vitiello, M. V. (2004). Meta-analysis of quantitative sleep parameters from childhood to old age in healthy individuals: Developing normative sleep values across the human lifespan. *Sleep, 27*(7), 1255–1273.

Paffenbarger, R. S., Jr., Kampert, J. B., Lee, I. M., Hyde, R. T., Leung, R. W., & Wing, A. L. (1994). Changes in physical activity and other lifeway patterns influencing longevity. *Medicine and Science in Sports and Exercise, 26*(7), 857–865.

Palazzo, C., Nguyen, C., Lefevre-Colau, M.-M., Rannou, F., & Poiraudeau, S. (2016). Risk factors and burden of osteoarthritis. *Annals of Physical and Rehabilitation Medicine, 59*(3), 134–138. http://doi.org/10.1016/j.rehab.2016.01.006.

Pamidi, S., Knutson, K. L., Ghods, F., & Mokhlesi, B. (2011). Depressive symptoms and obesity as predictors of sleepiness and quality of life in patients with REM-related obstructive sleep apnea: Cross-sectional analysis of a large clinical population. *Sleep Medicine, 12*(9), 827–831. http://doi.org/10.1016/j.sleep.2011.08.003.

Patel, S. R., Hayes, A. L., Blackwell, T., Evans, D. S., Ancoli-Israel, S., Wing, Y. K., … Study of Osteoporotic Fractures (SOF) Research Groups. (2014). The association between sleep patterns and obesity in older adults. *International Journal of Obesity (2005), 38*(9), 1159–1164. http://doi.org/10.1038/ijo.2014.13.

Robillard, R., Bouchard, M., Cartier, A., Nicolau, L., & Carrier, J. (2015). Sleep is more sensitive to high doses of caffeine in the middle years of life. *Journal of Psychopharmacology (Oxford, England), 29*(6), 688–697. http://doi.org/10.1177/0269881115575535.

Saelens, B. E., Sallis, J. F., & Frank, L. D. (2016). Environmental correlates of walking and cycling: Findings from the transportation, urban design, and planning literature. *Annals of Behavioral Medicine, 25*(2), 80–91. http://doi.org/10.1207/S15324796ABM2502_03.

Sallis, R. E. (2009). Exercise is medicine and physicians need to prescribe it! *British Journal of Sports Medicine, 43*(1), 3–4. http://doi.org/10.1136/bjsm.2008.054825.

Sallis, J. F., Hovell, M. F., Hofstetter, R. C., Faucher, P., Elder, J. P., Blanchard, J., … Christenson, G. M. (1989). A multivariate study of determinants of vigorous exercise in a community sample. *Preventive Medicine, 18*(1), 20–34. http://doi.org/10.1016/0091-7435(89)90051-0.

Schmid, S. M., Hallschmid, M., Jauch-Chara, K., Wilms, B., Benedict, C., Lehnert, H., … Schultes, B. (2009). Short-term sleep loss decreases physical activity under free-living conditions but does not increase food intake under time-deprived laboratory conditions in healthy men. *The American Journal of Clinical Nutrition, 90*(6), 1476–1482. http://doi.org/10.3945/ajcn.2009.27984.

Schrack, J. A., Simonsick, E. M., & Ferrucci, L. (2010). The energetic pathway to mobility loss: An emerging new framework for longitudinal studies on aging. *Journal of the American Geriatrics Society, 58*(Suppl. 2), S329–S336. http://doi.org/10.1111/j.1532-5415.2010.02913.x.

Schrager, M. A., Schrack, J. A., Simonsick, E. M., & Ferrucci, L. (2014). Association between energy availability and physical activity in older adults. *American Journal of Physical Medicine & Rehabilitation/Association of Academic Physiatrists, 93*(10), 876–883. http://doi.org/10.1097/PHM.0000000000000108.

Schwartz, T. L., Nihalani, N., Jindal, S., Virk, S., & Jones, N. (2004). Psychiatric medication-induced obesity: A review. *Obesity Reviews: An Official Journal of the International Association for the Study of Obesity, 5*(2), 115–121. http://doi.org/10.1111/j.1467-789X.2004.00139.x.

Simonsick, E. M., Guralnik, J. M., & Fried, L. P. (1999). Who walks? Factors associated with walking behavior in disabled older women with and without self-reported walking difficulty. *Journal of the American Geriatrics Society, 47*(6), 672–680.

Surani, S. (2013). Are diabetic patients being screened for sleep related breathing disorder? *World Journal of Diabetes, 4*(5), 162–164.

Takano, T., Nakamura, K., & Watanabe, M. (2002). Urban residential environments and senior citizens' longevity in megacity areas: The importance of walkable green spaces. *Journal of Epidemiology and Community Health, 56*(12), 913–918. http://doi.org/10.1136/jech.56.12.913.

Tanaka, H., Desouza, C. A., Jones, P. P., Stevenson, E. T., Davy, K. P., & Seals, D. R. (1997). Greater rate of decline in maximal aerobic capacity with age in physically active vs. sedentary healthy women. *Journal of Applied Physiology, 83*(6), 1947–1953.

Vaz Fragoso, C. A., Miller, M. E., Fielding, R. A., King, A. C., Kritchevsky, S. B., McDermott, M. M., … Lifestyle Interventions and Independence in Elder Study Group. (2014). Sleep–wake disturbances in sedentary community-dwelling elderly adults with functional limitations. *Journal of the American Geriatrics Society, 62*(6), 1064–1072. http://doi.org/10.1111/jgs.12845.

Vaz Fragoso, C. A., Miller, M. E., King, A. C., Kritchevsky, S. B., Liu, C. K., Myers, V. H., … Lifestyle Interventions and Independence for Elders Study Group. (2015). Effect of structured physical activity on sleep–wake behaviors in sedentary elderly adults with mobility limitations. *Journal of the American Geriatrics Society, 63*(7), 1381–1390. http://doi.org/10.1111/jgs.13509.

Virtanen, I., Kalleinen, N., Urrila, A. S., Leppänen, C., & Polo-Kantola, P. (2015). Cardiac autonomic changes after 40 hours of total sleep deprivation in women. *Sleep Medicine, 16*(2), 250–257. http://doi.org/10.1016/j.sleep.2014.10.012.

West, S. D., Kohler, M., Nicoll, D. J., & Stradling, J. R. (2009). The effect of continuous positive airway pressure treatment on physical activity in patients with obstructive sleep apnoea: A randomised controlled trial. *Sleep Medicine, 10*(9), 1056–1058. http://doi.org/10.1016/j.sleep.2008.11.007.

Zwyghuizen-Doorenbos, A., Roehrs, T. A., Lipschutz, L., Timms, V., & Roth, T. (1990). Effects of caffeine on alertness. *Psychopharmacology, 100*(1), 36–39. http://doi.org/10.1007/BF02245786.

CHAPTER

32

Ketamine: Neurotoxicity and Neurobehavioral Disorders

S.C. Cartágenes[1], *L.M.P. Fernandes*[1], *E. Fontes de Andrade, Jr.* [1], *R.D. Prediger*[2], *C.S.F. Maia*[1]

[1]Federal University of Pará, Belém, Brazil; [2]Federal University of Santa Catarina, Florianópolis, Brazil

INTRODUCTION

Ketamine is a dissociative anesthetic synthesized during early 1960s as an alternative to phencyclidine (Morgan & Curran, 2012; Niesters, Martini, & Dahan, 2014). When it was first introduced for clinical use, ketamine was regarded as an ideal and complete anesthetic drug, since it provides all the required components of surgical anesthesia: pain relief, immobility, amnesia, and loss of consciousness (Annetta, Iemma, Garisto, Tafani, & Proietti, 2005). However, the identification of psychedelic effects (euphoria, sensory distortions, and delirium) related to this drug has limited its clinical use (Dillon, Copeland, & Jansen, 2003; Kalsi, Wood, & Dargan, 2011). Although hallucinogenic effects limit the ketamine use in clinical practice, this effect has been the main reason for ketamine popularity as a drug of misuse (Niesters et al., 2014).

The initial reports of recreational use of ketamine were registered in 1960s. However, during 1990s the notifications have become more significant, primarily in the United States and the United Kingdom (Corazza, Assi, & Schifano, 2013; Morgan & Curran, 2012; Trujillo, 2011). Epidemiological data about the consumption of illicit ketamine are scarce. Isolated studies suggest lower rates of ketamine use, ranging from 0.1% to 0.4% among the population. According to the questionnaire of Psychoactive Substance review for 35th Expert Committee on Drug Dependence (ECDD 35) of the World Health Organization (WHO), 16 of the 64 countries studied have reported the recreational use of this substance (Report, 2012).

Ketamine or "Special K" is frequently used by adolescent and young adults at bars, nightclubs, concerts, and festivals such as "raves" in order to maintain energy levels for the dance or attain an altered consciousness state (Koesters, Rogers, & Rajasingham, 2002; Wu, Schlenger, & Galvin, 2006). According to users, low doses of the substance are able to promote stimulation, excitement, euphoria, sensory distortions, and lucid intoxication that is intensified by feelings of empathy (Bonta, 2004; Corazza et al., 2013; Dillon et al., 2003), while high doses can produce hallucinations as a "K-hole" intense dissociative experience that includes views and distortion of time, direction and identity—sometimes outside the body, or experiences of near death or rebirth (Dillon et al., 2003; Jansen & Darracot-Cankovic, 2011).

A single dose of ketamine promotes cognitive impairment, exacerbates psychotic symptoms in schizophrenic patients, and induces positive and negative symptoms similar to disease in healthy subjects (Adler et al., 1999; Trujillo, 2011). These effects observed in humans and animal models of schizophrenia are related to glutamatergic dysfunction in this psychiatric disorder (Krystal et al., 2005; Trujillo, 2011).

The main effect in the acute use of ketamine consists of the decrease of immediate awareness of the ambient, thus exposing the user to a physical injury. Therefore, reduced consciousness includes a sense of depersonalization, de-realization, reduced perception of pain, and potentially unconsciousness (Dillon et al., 2003; Morgan & Curran, 2012; Niesters et al., 2014). These effects are accompanied by the lack of coordination, temporary paralysis, inability to move, blurred vision, and inability to speak (Dillon et al., 2003). Thus, users put themselves at risk of significant injury by jump from height, traffic accidents, drowning, and hypothermia (Jansen, 2000).

The neurotoxic effects of ketamine are demonstrated in animal studies that revealed apoptotic neuronal death

Addictive Substances and Neurological Disease
http://dx.doi.org/10.1016/B978-0-12-805373-7.00032-3

in brain development in rodents (Liu, Paule, Ali, & Wang, 2011). The neuronal damage is displayed through the loss of inhibitory pathways which leads to increased excitatory neuronal activity (Jevtovic-Todorovic & Carter, 2005; Slikker et al., 2007). In adults, the toxic effect of ketamine was observed by cerebral volumes of ketamine addicts who showed a decrease of gray and white matter in the bilateral frontal cortex, as well as degeneration of the left temporoparietal cortex white matter (Liao et al., 2011, 2010). These changes were related to memory deficits in healthy volunteers (working, episodic, and semantic memories) and in schizotypal symptoms (Liao et al., 2011).

An important observation in the recreational ketamine users relies on the absence of urological symptoms in clinical patients. Frequently, the ketamine abusers may present ulcerative cystitis with symptoms of high urgency and frequency of urination, dysuria, urge incontinence, and hematuria (Li et al., 2011; Morgan & Curran, 2012).

The etiology of ulcerative cystitis induced by ketamine it is unclear, but this occurrence is directly associated with the frequency of the abuse. Generally, the urological symptoms may recover after cessation of ketamine use; however, in a long time–misuse scenario, urological symptoms may persist for long periods after cessation of the drug use (Bergman, 1999; Liao et al., 2011; Niesters et al., 2014).

In this context, this chapter discusses the chemistry and kinetics, the plausible neurotoxic mechanisms, and behavioral effects resulted from ketamine recreational use.

CHEMISTRY AND KINETICS

Ketamine is an arylcyclohexylamine, which includes several psychoactive drugs, such as phencyclidine and methoxetamine, that have been used in addiction therapy, but sometimes are targets of abuse. Chemically known as 2-O-chlorophenyl-2-(methylamino) cyclohexanone, ketamine has a molecular weight of 238 g/mol and p*Ka* of 7.5. It is widely soluble in water and lipids (lipid solubility is about 10 times greater than that of thiopental), which facilitates an easy body-wide distribution (Reich & Silvay, 1989).

The asymmetry of the second carbon of cyclohexanone ring lends chirality to the ketamine molecule (Fig. 32.1). Thus, ketamine is typically available as a racemic mixture of equal proportions of the enantiomers S(−) and R(+)-ketamine. Besides the chemical peculiarities, this feature reveals important influence on the pharmacology, since the S(−)-ketamine enantiomer has anesthetic potency twice of the racemic mixture and four times greater than that of the R(+)-ketamine isomer (Mion & Villevieille, 2013; Reich & Silvay, 1989).

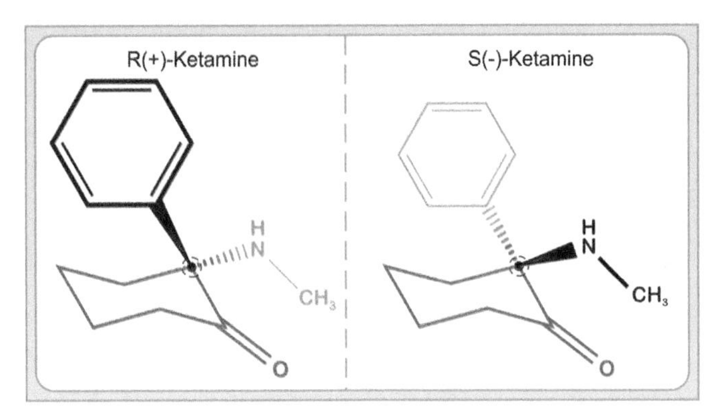

FIGURE 32.1 The enantiomers of R(+)-ketamine and S(−)-ketamine.

Routes of Administration and Absorption

Ketamine can be administered by several routes, including intrathecal, intraperitoneal, intranasal, and rectal. However, commonly the intravenous and intramuscular routes are the most utilized. The relationship of the drug with the oral route is quite erratic. Despite the facility of crossing the biomembranes, the drug reaches a very limited bioavailability (17–24%) due to the intense hepatic first-pass metabolism (Fanta, Kinnunen, Backman, & Kalso, 2015; Kharasch & Labroo, 1992). Intriguingly, in the oral pathway the effects may be prolonged for 4–6 h (Quibell, Prommer, Mihalyo, Twycross, & Wilcock, 2011).

The intravenous route produces maximum bioavailability and rapid onset of effects (less than 1 min). In the intramuscular administration, a rapid absorption occurs, reaching a bioavailability of up to 93%, with a peak in about 5–15 min from the administration (Clements, Nimmo, & Grant, 1982; Mion & Villevieille, 2013).

Nasal insufflation is a form closely linked to the abuse of this drug. It is to be explained, at least in part, by the rapid absorption, with bioavailability of about 45%. Brain effects are achieved in about 5 min, associated with short half-life, which contribute to the stimulation of repeated administrations (Yanagihara et al., 2003).

Distribution

The easy dissolution of ketamine in aqueous and lipid content allied to the low plasma protein binding (10–30%) contribute to its rapid distribution throughout the body, with high concentrations in the central nervous system (CNS). The volume of distribution is about 3–5 L/kg following a two-compartment model (Dayton, Stiller, Cook, & Perel, 1983; Hijazi & Boulieu, 2002). On the other hand, these kinetic features induce a rapid redistribution and biotransformation, that provides relatively short time of action (intramuscular: 30 min to 2 h; oral route: 4–6 h) (Potter & Choudhury, 2014).

Elimination

Excretion of ketamine in its unaltered form through the feces and urine corresponds to only 10% of the bioavailability (Quibell et al., 2011). In fact, the main ketamine elimination process consists of the hepatic biotransformation, although lung, intestine, and kidney metabolism also exists (Edwards & Mather, 2001; Park, Manara, Mendel, & Bateman, 1987).

N-demethylation is the major route of ketamine metabolism that involves the cytochrome P450 (CYP) enzymes and is responsible for about 80% of the biotransformation (Fanta et al., 2015). Among the enzymes involved in this process, the isoform CYP3A4 is considered the main pathway of N-demethylation. In addition, CYP2C9 and CYP2B6 have also been implicated in the ketamine biotransformation (Hijazi & Boulieu, 2002; Yanagihara et al., 2001). The ketamine N-demethylation product norketamine consists of a metabolite that retains about one-fifth to one-third of the precursor activity (Potter & Choudhury, 2014; Reich & Silvay, 1989). This feature has been the subject of studies for use in therapy, but no conclusive results have been reached (Fanta et al., 2015). In a second metabolic step norketamine undergoes hydroxylation whose product, mainly 6-hydroxy-norketamine, is conjugated to glucuronic acid (third step) and excreted through the bile and kidney (Mion & Villevieille, 2013).

MECHANISMS OF NEUROTOXICITY AND NEUROBEHAVIORAL EFFECTS

The neurotoxicity of ketamine is accompanied by the neuropsychiatric symptoms, characterized by a marked change in cognitive function and psychological well-being (Morgan & Curran, 2012). These neurological disruptions derive from the effects on the dense population of glutamate NMDA receptors located throughout the cerebral cortex and hippocampus, as well as on the transmission of modulatory monoamines such as dopamine (DA) and serotonin (5-HT) in the striatum and cortex (Aalto et al., 2002; Adler et al., 1999; Morgan, Muetzelfeldt, & Curran, 2010) (Fig. 32.2).

Several symptoms induced by neuropsychiatric drugs are transient, reversible, and influenced by weather conditions, dosage, and pathway of administration (Keilhoff, Becker, Grecksch, Wolf, & Bernstein, 2004). The acute consumption of ketamine reduces perception of pain and loss of sense of environment and of depersonalization. This is compounded by addicts who report incoordination, temporary paralysis, inability to move, blurred vision, and inability to speak (Dillon et al., 2003; Kalsi et al., 2011).

The long-term recreational use of ketamine promotes persistent neuropsychiatric symptoms, generally characterized as schizophrenia-like symptoms, as well as the impairment of the working and episodic memories and semantic processing (Morgan & Curran, 2012; Morgan et al., 2010). In fact, prolonged exposure to ketamine, which blocks NMDA receptors continuously, causes cell death in the developing brain by a mechanism that involves a compensatory up-regulation of NMDA receptor subunits. This up-regulation could be associated with toxic accumulation of intracellular calcium, increased oxidative stress, and activation of nuclear factor kappa B (NF-κB) signaling pathway, which makes neurons more vulnerable even after ketamine withdrawal (Liao et al., 2011; Shibuta, Morita, Kosaka, Kamibayashi, & Fujino, 2015; Wang, Fu, Wilson, & Ma, 2006).

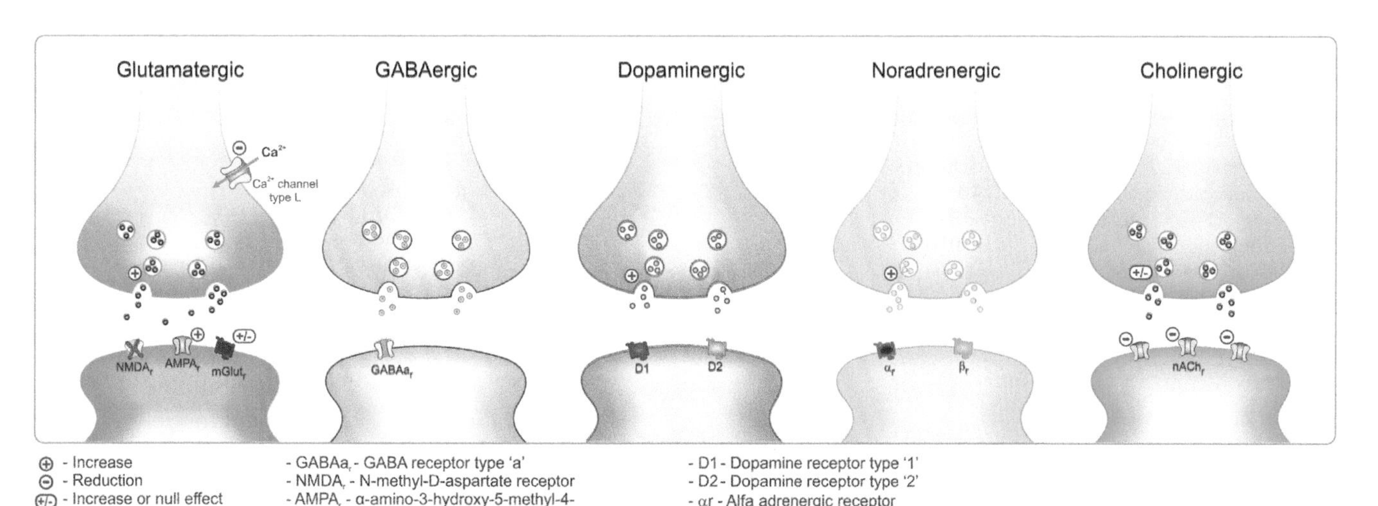

FIGURE 32.2 Effects on neurotransmission of ketamine.

Effects on Neurotransmission

As previously mentioned, ketamine modulates neurotransmission at postsynaptic receptors via glutamate NMDA receptors and gamma-aminobutyric acid (GABA) receptors (Durieux, 1995; Kapur & Seeman, 2002; Morgan & Curran, 2012; Rabiner, 2007), as well as exhibits strong affinity to DA receptors in the prefrontal cortex (PFC) (Tan, Rudd, & Yew, 2011).

It is noteworthy that the hippocampus and PFC present dense populations of NMDA receptors (Wang, Ramakrishnan, Fletcher, Prochownik, & Genetics, 2015). Thus, both acute and chronic exposure to ketamine increases the glutamate levels in the synaptic cleft. This effect triggers excessive cell input of calcium ions (Ca^{2+}) that initiate a set of cytoplasmic and nuclear processes, which leads to neurodegeneration and psychosis (Lipton, 2004). According to Xu and Lipsky (2015), the up-regulation of NMDA receptors in the frontal cortex underlies the mechanism of ketamine-induced persistent psychosis in human, as well the semantic-processing deficits in the hippocampus.

On the other hand, GABA is the major inhibitory neurotransmitter in the CNS and the $GABA_A$ receptors belong to a family of transmembrane ionic channel activated by ligands. As described earlier, ketamine increases the levels of extracellular glutamate in the PFC, possibly via the blockade of NMDA receptors on GABAergic interneurons, that results in disinhibition of glutamate neurotransmission (Duman & Li, 2012) (Fig. 32.3).

As the PFC has been described to be involved in the central executive control of cognitive processing, alterations of interconnections among neurons in this area promotes failure in the integrate information followed by a subsequent decline of cognitive function (Baddeley, 2003; Tan, Lam, Wai, Yu, & Yew, 2012).

Studies of Kapur and Seeman (2002) have demonstrated that ketamine modulates the dopaminergic system, indicating its strong affinity with dopamine D_2 receptors in the nucleus accumbens (Hamida et al., 2008; Jin et al., 2013; Li et al., 2015), which may be associated with the development of drug dependence (Li et al., 2015). Several studies have demonstrated that a single subanesthetic ketamine dose rapidly increased DA release in the PFC cortex of rats. Besides, in the repeated ketamine administration such increase in the DA levels was supported by tyrosine hydroxylase (TH) up-regulation, suggesting a longer duration effect after repeated ketamine administration (Tan et al., 2012).

In the ketamine dependence mechanism, the primary process involves the activation of the cerebral reward system, namely the mesencephalon–limbic DA system. In this system, dopaminergic neurons in the brain stem ventral tegmental area (VTA) project to the nucleus accumbens septum (NAs), the main areas receiving projects from the VTA. This VTA–NAs dopamine pathway is the core site mediating the rewarding effects of drug abuse (Lindefors, Barati, & O'Connor, 1997; Tan et al., 2012).

Thus, chronic ketamine administration potentiates the discharge of VTA dopaminergic neurons, that leads to increased release of DA into the NAs, modulation of dopaminergic signaling, and the rewarding effects of

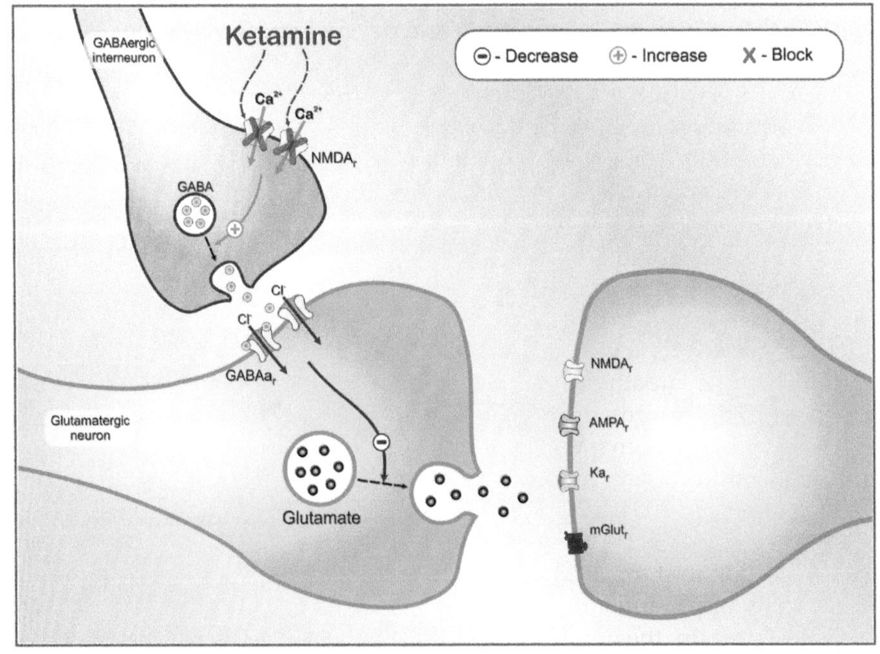

FIGURE 32.3 Mechanism on glutamate release increase by ketamine. The blockade of NMDA receptors on GABAergic interneurons leads to GABA release reduction, which diminishes its negative modulation on the glutamatergic neurons. As a result there will be an increase in glutamatergic neurotransmission. *AMPA$_r$*, α-amino-3-hydroxy-5-methyl-4-isoxazolepropionic acid receptor; *GABA*, gamma-aminobutyric acid; *GABAa$_r$*, GABA a-type receptor; *Ka$_r$*, kainate receptor; *mGlut$_r$*, metabotropic glutamate receptors; *NMDA$_r$*, N-methyl-D-aspartate receptor.

drug abuse (Tan et al., 2012). Additionally, DA receptors in the ventral midbrain fire in distinct tonic and phasic patterns (Grace & Bunney, 1984; Soden et al., 2013), which provides essential signals to cortical and striatal circuits responsible for several forms of motivation, learning, salience processing, and attention (Bromberg-Martin, Matsumoto, & Hikosaka, 2010; Schultz, 2007; Soden et al., 2013).

Neuroinflammation

Fanta et al. (2015) reported for the first time that reduction of serum levels of tumor necrosis factor alpha (TNF-α) and an increase of interleukin (IL)-6 and IL-18 may cause the psychotic-like symptoms in ketamine chronic users. In rats, repeated administration of ketamine display neurodegenerative changes in brain regions similarly as schizophrenia disease (Keilhoff et al., 2004).

TNF-α is a prototypical inflammatory cytokine widely involved in several biological or pathological processes, including inflammation, the immune response, apoptosis and necrosis, and neurodegenerative diseases (McCoy, Runh, Blesch, & Tansey, 2011; Zhou, Zheng, & Xia, 2015). Moreover, TNF-α was involved in dopaminergic processes and plays a key role in the pathogenesis of dopaminergic neurodegeneration. Thus, this cytokine has been associated to schizophrenia-like psychiatric symptoms (Nakajima, 2004; Niwa et al., 2007). According to the clinical study by Fan et al. (2015), the effects of TNF-α on brain functions and psychotic symptoms may be different under acute versus chronic conditions. The authors suggest that a bidirectional pathway of TNF-α may occur under different circumstances, in which the serum levels of cytokines such as TNF-α may be beneficial rather than producing harmful effects.

On the other hand, IL-6 consists of a proinflammatory cytokine that promotes immune response to infection and inflammation by migration of leukocytes to inflammatory sites and/or by activating inflammatory cells (Erta, Quintana, & Hidalgo, 2012). Although the role of IL-6 in brain development and CNS disease is not well understood, several studies have demonstrated that production of IL-6 may affect neuronal functionality. For instance, this cytokine induces the cholinergic phenotype of sympathetic neurons and has been implicated in the impairment of working memory caused by peripheral inflammation and in the amount of stored information during in vivo long-term potentiation (LTP) (Balschun et al., 2004; Erta et al., 2012; Tancredi et al., 2000).

In addition to IL-6, IL-18 consists of a member of the IL-1 family of pro-inflammatory cytokines, which plays a critical role in the immune responses, since it is a pleiotropic cytokine modulator (Alboni, 2012; Tanaka, Shintani, Fujii, Yagi, & Asai, 2000). In fact, studies have showed a variety of these cytokines in different brain regions including the hippocampus, hypothalamus, and cerebral cortex. Besides, IL-18 mediates two distinct immunological regulatory pathways of cytotoxic and inflammatory responses under neuropathological conditions (Felderhoff-Mueser, Schmidt, Oberholzer, Bührer, & Stahel, 2005), which establishes the link between the immune system and CNS, mediates neuroinflammation and neurodegenerative processes, as well as influences homeostasis and behavior (Alboni, 2012).

Oxidative Stress

Reactive oxygen species (ROS) have important role in the pathogenesis of many diseases, especially in neurological and psychiatric diseases (Oliveira et al., 2009; Takuma, Baba, & Matsuda, 2004). High levels of ROS can lead to cellular and DNA damage, and oxidative stress that can elicit cell survival or apoptosis mechanisms depending on severity and duration of exposure (Hou, Janczuk, & Wang, 1999).

Oxidative stress is manifested as an increase of lipid peroxidation end products, DNA (and often RNA) base oxidation products, and oxidative protein damage (Halliwell, 2006). It is an important event that has been related to the pathogenesis of many diseases affecting the CNS, such as neurodegenerative disorders, Parkinson and Alzheimer's diseases (Fendri et al., 2006). Studies have reported that subanesthetic doses of ketamine at 4 and 10 mg/kg increases thiobarbituric acid reactive substances (TBARS), which indicates an increase in lipid peroxidation (Arvindakshan et al., 2003; Mahadik & Mukherjee, 1996; Petronijević, Mićić, Duričić, Marinković, & Paunović, 2003; Srivastava et al., 2001).

In mice, subchronic administration of ketamine induced an increase in the nicotinamide adenine dinucleotide phosphate (NADPH) oxidase activity and oxidative stress in the PFC, hippocampus, and thalamus, that leads to a loss of parvalbumin interneurons, similarly to what occurs in schizophrenia (Behrens & Sejnowski, 2009; Powell, Sejnowski, & Behrens, 2012).

Moreover, the studies of Oliveira et al. (2009) have revealed an increase in the carbonyl content allied to a decrease in the sulfhydryl content of protein, two parameters commonly used to evaluate protein damage, in the cerebellum, hippocampus, striatum, and cortex of schizophrenic-like rat model, after a single injection of intraperitoneal subanesthetic doses of ketamine (4, 10, or 30 mg/kg). In this work, the protein oxidation was intensified similarly as occurs in the pathogenesis of schizophrenia, which may account for the negative and positive symptoms in ketamine addict. In addition to lipid peroxidation and protein oxidation, the authors reported that a reduction in the antioxidant enzymatic system may be present.

BEHAVIORAL EFFECTS

Recreational ketamine use displays hallucinations, euphoria, loss of capacity of judgment, amnesia, anxiety-related behaviors, and other behavioral alterations (Hayase, Yamamoto, & Yamamoto, 2006). In rodents and monkeys, acute doses and subanesthetic doses of NMDA receptor antagonists produce behavioral alterations as hyperlocomotion, increased stereotypic behavior, cognitive and sensorimotor deficits, and social interaction avoidance (Silva et al., 2010).

Anxiety

Babar et al. (2001) have shown anxiogenic-like effect in a ketamine subanesthetic protocol. However, the recreational ketamine exposure may produce anxiety behavior. The exact mechanism between recreational ketamine exposure and vulnerability to anxiety disorders are not well established, although some studies suggest the occurrence of alterations in neural structures related to fear/anxiety, such as activation of 5-HT transmission within the dorsal raphe system (Maier, Grahn, & Watkins, 1995; Maswood, Barter, Watkins, & Maier, 1998).

The brain serotonin system consists of a small group of neurons located in the raphe nuclei of the midbrain that are unique since they express the brain-specific isoform of the rate-limiting enzyme for serotonin synthesis, the tryptophan hydroxylase-2 (TPH2) (Lenicov, Lemonde, Czesak, Mosher, & Albert, 2007; Walther et al., 2003). 5-HT neurons of the rostral raphe nuclei, including the dorsal and median raphe nuclei, project widely throughout the brain to innervate key brain regions involved in anxiety and depression (Albert, Vahid-Ansari, & Luckhart, 2014).

5-HT released in the raphe activates the 5-HT1A autoreceptor, which negatively regulates the firing of serotonin system. The release of 5-HT at target neurons activates serotoninergic heteroreceptors (i.e., 5-HT1A), which is abundantly expressed in the hippocampus, septum, amygdala, and PFC, that in turns mediates 5-HT actions on fear, anxiety, stress, and cognitive function (Albert, Zhou, Van Tol, Bunzow, & Civelli, 1990).

Depression

Clinical and preclinical studies reveal the beneficial effects of ketamine low-dose intravenous infusions to attenuate depressive symptoms within hours in subjects with treatment-resistant depression (Berman et al., 2000; Owolabi, Akanmu, & Adeyemi, 2014; Zunszain, Horowitz, Cattaneo, Lupi, & Pariante, 2013).

Ketamine is able to enhance extracellular glutamate levels in the PFC by NMDA receptor currents block on GABAergic interneurons, that in turn promotes glutamate transmission disinhibition (Homayoun & Moghaddam, 2007). This mechanism is associated with the antidepressant-like activity of ketamine in rodents following a single administration (0.5 mg/kg, i.v.) observed in behavioral tests of depression after 4 h of administration, which is long-lasting, for 3 days (Owolabi et al., 2014).

In contrast, our group observed that the administration of ketamine (10 mg/kg per day, i.p.) on 3 consecutive days followed by abstinence of 3 h in adolescent rodents induced increased immobility time in the Porsolt test (unpublished data). The mechanism that underlies such harmful effect in the recreational use of ketamine is still unclear.

Schizophrenia

As previously mentioned, the use of ketamine induces positive and negative symptoms similar to those associated with schizophrenia (Keilhoff et al., 2004; Pietraszek, 2003), both in humans (Bressan & Pilowsky, 2003; Keilhoff et al., 2004) and in animal models (Becker & Grecksch, 2004).

Studies suggest that the ketamine-induced psychotic symptoms result of a dopaminergic hyperfunction that causes the imbalance between cortical and subcortical dopaminergic systems (Trujillo, 2011). The hypofunction of the dopaminergic system in the PFC may be responsible for the development of negative symptoms, while positive symptoms could be attributed to increased dopaminergic activity in the limbic system (Davis, Kahn, Ko, & Davidson, 1991; Knable & Weinberger, 1997).

Learning and Memory

Hartvig et al. (1995) have shown in a double-blind randomized crossover study with five healthy volunteers that short-term memory could be impaired dose dependently by ketamine administration at doses of 0.1 and 0.2 mg/kg (i.v), as assessed by a word recall test. Amann et al. have shown that a single dose of ketamine induces dose-dependent impairment in working and episodic memory which would impact profoundly on user ability (Amann et al., 2009). In the same study, the authors have reported significant impairments in working memory, verbal recognition memory, and planning in frequent ketamine users in comparison to infrequent ketamine users, abstinent ketamine users, and individuals with no history of ketamine use (Amann et al., 2009).

The effect of ketamine on cognition is associated, at least in part, with the blockage of NMDA receptors in the hippocampal region (Morris, Anderson, Lynch, & Baudry, 1986; Wang et al., 2006). The hippocampus is a critical structure responsible for spatial memory. The NMDA receptors are distributed broadly in the brain and densely within the hippocampus, in which these receptors play a vital role on the LTP process and learning and memory processes (Cotman & Monaghan, 1988; Moosavi, Yadollahi Khales, Rastegar, & Zarifkar, 2012).

Other hypothesis related to cognition deficits induced by ketamine misuse relies on the alterations in the neurotrophins-mediated signaling. The NMDA receptors interact with the brain-derived neurotrophic factor (BDNF)/tropomyosin receptor kinase B (TrkB) pathway to promote synaptic plasticity (Keilhoff et al., 2004; Morgan, Riccelli, Maitland, & Curran, 2004). Subchronic ketamine administration has been shown to alter the expression of mRNAs for neurotrophins, including BDNF, in the rat brain (Becker & Grecksch, 2004). Thus, alterations in BDNF levels might be involved in cognitive deficits induced by ketamine (Goulart et al., 2010), since within the hippocampus, BDNF plays a pivotal role on neural plasticity and memory formation. In fact, animal and human studies have suggested that hippocampal BDNF might play a major role in influencing recognition memory (Heldt, Stanek, Chhatwal, & Ressler, 2007).

PROSPECTS AND FINAL COMMENTS

The recreational use of ketamine is poorly discussed. Recently, ketamine has been the target of several studies more for its pharmacological benefits than its dependence/misuse risk. However, despite the scarce studies and as discussed in this chapter, ketamine recreational use displays neurobehavioral alterations, such as depression, anxiety, cognition deficits, and schizophrenia. The duration of these behavioral harmful effects, as well as the exact mechanism of the neuronal damage are unclear. In this context, further studies involving ketamine abuse protocol are necessary, in order to identify the physiopathology and the neuronal damage that may reflect cognitive and behavioral impairment to support clinical practices.

References

Aalto, S., Hirvonen, J., Kajander, J., Scheinin, H., Någren, K., Vilkman, H., & Hietala, J. (2002). Ketamine does not decrease striatal dopamine D2 receptor binding in man. *Psychopharmacology, 164*(4), 401–406.

Adler, C. M., Malhotra, A. K., Elman, I., Goldberg, T., Egan, M., Pickar, D., & Breier, A. (1999). Comparison of ketamine-induced thought disorder in healthy volunteers and thought disorder in schizophrenia. *American Journal of Psychiatry, 156*(10), 1646–1649.

Albert, P. R., Vahid-Ansari, F., & Luckhart, C. (June 2014). Serotonin-prefrontal cortical circuitry in anxiety and depression phenotypes: Pivotal role of pre- and post-synaptic 5-HT1A receptor expression. *Frontiers in Behavioral Neuroscience, 8*, 199.

Albert, P. R., Zhou, Q. Y., Van Tol, H. H., Bunzow, J. R., & Civelli, O. (1990). Cloning, functional expression, and mRNA tissue distribution of the rat 5-hydroxytryptamine1A receptor gene. *The Journal of Biological Chemistry, 265*(10), 5825–5832.

Alboni, S. (2012). Alboni 2012 interleukin 18 in the CNS. *Interleukin 18 in the CNS*, 1–12.

Amann, L. C., Halene, T. B., Ehrlichman, R. S., Luminais, S. N., Ma, N., Abel, T., & Siegel, S. J. (2009). Chronic ketamine impairs fear conditioning and produces long-lasting reductions in auditory evoked potentials. *Neurobiology of Disease, 35*, 311–317.

Annetta, M. G., Iemma, D., Garisto, C., Tafani, C., & Proietti, R. (2005). Ketamine: New indications for an old drug. *Current Drug Targets, 6*(7), 789–794.

Arvindakshan, M., Sitasawad, S., Debsikdar, V., Ghate, M., Evans, D., Horrobin, D. F., & Mahadik, S. P. (2003). Essential polyunsaturated fatty acid and lipid peroxide levels in never-medicated and medicated schizophrenia patients. *Biological Psychiatry, 53*(1), 56–64.

Babar, E., Ozgunen, T., Melik, E., Polat, S., & Akman, H. (2001). Effects of ketamine on different types of anxiety/fear and related memory in rats with lesions of the median raphe nucleus. *European Journal of Pharmacology, 431*, 315–320.

Baddeley, A. (2003). Working memory: Looking back and looking forward. *Nature Reviews Neuroscience, 4*(10), 829–839.

Balschun, D., Wetzel, W., Del Rey, A., Pitossi, F., Schneider, H., Zuschratter, W., & Besedovsky, H. O. (2004). Interleukin-6: A cytokine to forget. *FASEB Journal: Official Publication of the Federation of American Societies for Experimental Biology, 18*(14), 1788–1790.

Becker, A., & Grecksch, G. (2004). Ketamine-induced changes in rat behaviour: A possible animal model of schizophrenia. Test of predictive validity. *Progress in Neuro-Psychopharmacology and Biological Psychiatry, 28*(8), 1267–1277.

Behrens, M. M., & Sejnowski, T. J. (2009). Does schizophrenia arise from oxidative dysregulation of parvalbumin-interneurons in the developing cortex? *Neuropharmacology, 57*(3), 193–200.

Bergman, S. A. (1999). Ketamine: Review of its pharmacology and its use in pediatric anesthesia. *Anesthesia Progress, 46*(1), 10–20.

Berman, R. M., Cappiello, A., Anand, A., Oren, D. A., Heninger, G. R., Charney, D. S., & Krystal, J. H. (2000). Antidepressant effects of ketamine in depressed patients. *Society of Biological Psychiatry, 47*(4), 351–354.

Bonta, I. L. (2004). Schizophrenia, dissociative anaesthesia and near-death experience; three events meeting at the NMDA receptor. *Medical Hypotheses, 62*(1), 23–28.

Bressan, R. A., & Pilowsky, L. S. (2003). Hipótese glutamatérgica da esquizofrenia. *Revista Brasileira de Psiquiatria, 25*, 177–183.

Bromberg-Martin, E. S., Matsumoto, M., & Hikosaka, O. (2010). Dopamine in motivational control: Rewarding, aversive, and alerting. *Neuron, 68*(5), 815–834.

Clements, J. A., Nimmo, W. S., & Grant, I. S. (1982). Bioavailability, pharmacokinetics, and analgesic activity of ketamine in humans. *Journal of Pharmaceutical Sciences, 71*(5), 539–542.

Corazza, O., Assi, S., & Schifano, F. (2013). From "special K" to "special M": The evolution of the recreational use of ketamine and methoxetamine. *CNS Neuroscience & Therapeutics, 19*(6), 454–460.

Cotman, C. W., & Monaghan, D. T. (1988). Excitatory amino acid neurotransmission: NMDA receptors and Hebb-type synaptic plasticity. *Annual Reviews of Neuroscience*, 61–80.

Davis, K. L., Kahn, R. S., Ko, G., & Davidson, M. (1991). Dopamine in schizophrenia: A review and reconceptualization. *American Journal of Psychiatry, 148*(11), 1474–1486.

Dayton, P. G., Stiller, R. L., Cook, D. R., & Perel, J. M. (1983). The binding of ketamine to plasma proteins: Emphasis on human plasma. *European Journal of Clinical Pharmacology, 24*, 825–831.

Dillon, P., Copeland, J., & Jansen, K. (2003). Patterns of use and harms associated with non-medical ketamine use. *Drug and Alcohol Dependence, 69*(1), 23–28.

Duman, R. S., & Li, N. (2012). A neurotrophic hypothesis of depression: Role of synaptogenesis in the actions of NMDA receptor antagonists. *Philosophical Transactions of the Royal Society B: Biological Sciences, 367*(1601), 2475–2484.

Durieux, M. E. (1995). Inhibition by ketamine of muscarinic acetylcholine receptor function. *Anesthesia Analgesia, 81*(1), 57–62.

Edwards, S. R., & Mather, L. E. (2001). Tissue uptake of ketamine and norketamine enantiomers in the rat indirect evidence for extrahepatic metabolic inversion. *Life Sciences, 69*(17), 2051–2066.

Erta, M., Quintana, A., & Hidalgo, J. (2012). Interleukin-6, a major cytokine in the central nervous system. *International Journal of Biological Sciences, 8*(9), 1254–1266.

Fan, N., Luo, Y., Xu, K., Zhang, M., Ke, X., Huang, X., ... He, H. (2015). Relationship of serum levels of TNF-α, IL-6 and IL-18 and schizophrenia-like symptoms in chronic ketamine abusers. *Schizophrenia Research, 169*, 10–15.

Fanta, S., Kinnunen, M., Backman, J. T., & Kalso, E. (2015). Population pharmacokinetics of S-ketamine and norketamine in healthy volunteers after intravenous and oral dosing. *European Journal of Clinical Pharmacology, 71*(4), 441–447.

Felderhoff-Mueser, U., Schmidt, O. I., Oberholzer, A., Bührer, C., & Stahel, P. F. (2005). IL-18: A key player in neuroinflammation and neurodegeneration? *Trends in Neurosciences, 28*(9), 487–493.

Fendri, C., Mechri, A., Khiari, G., Othman, A., Kerkeni, A., & Gaha, L. (2006). Oxidative stress involvement in schizophrenia pathophysiology: A review. *L'Encephale, 32*(1), 244–252.

Goulart, B. K., de Lima, M. N. M., de Farias, C. B., Reolon, G. K., Almeida, V. R., Quevedo, J., ... Roesler, R. (2010). Ketamine impairs recognition memory consolidation and prevents learning-induced increase in hippocampal brain-derived neurotrophic factor levels. *Neuroscience, 167*(4), 969–973.

Grace, A. A., & Bunney, B. S. (1984). The control of firing pattern in nigral dopamine neurons: Burst firing. *Journal of Neuroscience, 4*(11), 2877–2890.

Halliwell, B. (2006). Reactive species and antioxidants. Redox biology is a fundamental theme of aerobic life. *Plant Physiology, 141*(2), 312–322.

Hamida, S. B., Plute, E., Cosquer, B., Kelche, C., Jones, B. C., & Cassel, J. C. (2008). Interactions between ethanol and cocaine, amphetamine, or MDMA in the rat: Thermoregulatory and locomotor effects. *Psychopharmacology, 197*(1), 67–82.

Hartvig, P., Valtysson, J., Lindner, K. J., Kristensen, J., Karlsten, R., Gustafsson, L. L., ... Antoni, G. (1995). Central nervous system effects of subdissociative doses of (S)-ketamine are related to plasma and brain concentrations measured with positron emission tomography in healthy volunteers. *Clinical Pharmacology and Therapeutics, 58*(2), 165–173.

Hayase, T., Yamamoto, Y., & Yamamoto, K. (2006). Behavioral effects of ketamine and toxic interactions with psychostimulants. *BMC Neuroscience, 7*, 25.

Heldt, S. A., Stanek, L., Chhatwal, J. P., & Ressler, K. J. (2007). Hippocampus-specific deletion of BDNF in adult mice impairs spatial memory and extinction of aversive memories. *Molecular Psychiatry, 12*(7), 656–670.

Hijazi, Y., & Boulieu, R. (2002). Contribution of CYP3A4, CYP2B6, and CYP2C9 isoforms to N-demethylation of ketamine in human liver microsomes. *Drug Metabolism and Disposition, 30*(7), 853–858.

Homayoun, H., & Moghaddam, B. (2007). NMDA receptor hypofunction produces opposite effects on prefrontal cortex interneurons and pyramidal neurons. *The Journal of Neuroscience: The Official Journal of the Society for Neuroscience, 27*(43), 11496–11500.

Hou, Y. C., Janczuk, A., & Wang, P. G. (1999). Current trends in the development of nitric oxide donors. *Current Pharmaceutical Design, 5*(6), 417–441.

Jansen, K. L. R. (2000). A review of the nonmedical use of ketamine: Use, users and consequences. *Journal of Psychoactive Drugs, 32*(4), 419–433.

Jansen, K. L., & Darracot-Cankovic, R. (2011). The nonmedical use of ketamine, part two: A review of problem use and dependence. *Journal of Psychoactive Drugs, 33*(2), 151–158.

Jevtovic-Todorovic, V., & Carter, L. B. (2005). The anesthetics nitrous oxide and ketamine are more neurotoxic to old than to young rat brain. *Neurobiology of Aging, 26*(6), 947–956.

Jin, J., Gong, k., Zou, X., Wang, R., Lin, Q., & Chen, J. (2013). The blockade of NMDA receptor ion channels by ketamine is enhanced in developing rat cortical neurons. *Neuroscience Letters, 539*, 11–15.

Kalsi, S. S., Wood, D. M., & Dargan, P. I. (2011). The epidemiology and patterns of acute and chronic toxicity associated with recreational ketamine use. *Emerging Health Threats Journal, 4*(1), 1–10.

Kapur, S., & Seeman, P. (2002). NMDA receptor antagonists ketamine and PCP have direct effects on the dopamine D(2) and serotonin 5-HT(2)receptors-implications for models of schizophrenia. *Molecular Psychiatry, 7*(8), 837–844.

Keilhoff, G., Becker, A., Grecksch, G., Wolf, G., & Bernstein, H. G. (2004). Repeated application of ketamine to rats induces changes in the hippocampal expression of parvalbumin, neuronal nitric oxide synthase and cFOS similar to those found in human schizophrenia. *Neuroscience, 126*(3), 591–598.

Kharasch, E. D., & Labroo, R. (1992). Metabolism of ketamine stereoisomers by human liver microsomes. *Anesthesiology, 77*(6), 1201–1207.

Knable, M. B., & Weinberger, D. R. (1997). Dopamine, the prefrontal cortex and schizophrenia. *Journal of Psychopharmacology, 11*(2), 123–131.

Koesters, S. C., Rogers, P. D., & Rajasingham, C. R. (2002). MDMA ('ecstasy') and other "club drugs": The new epidemic. *Pediatric Clinics of North America, 49*(2), 415–433.

Krystal, J. H., Perry, E. B., Jr., Gueorguieva, R., Belger, A., Madonick, S. H., Abi-Dargham, A., ... D'Souza, D. C. (2005). Comparative and interactive human psychopharmacologic effects of ketamine and amphetamine: Implications for glutamatergic and dopaminergic model psychoses and cognitive function. *Archives of General Psychiatry, 62*(9), 985–994.

Lenicov, F. R., Lemonde, S., Czesak, M., Mosher, T. M., & Albert, P. R. (2007). Cell-type specific induction of tryptophan hydroxylase-2 transcription by calcium mobilization. *Journal of Neurochemistry, 103*(5), 2047–2057.

Li, B., Liu, M., Wu, X., Jia, J., Cao, J., Wei, Z., & Wang, Y. (2015). Effects of ketamine exposure on dopamine concentrations and dopamine type 2 receptor mRNA expression in rat brain tissue. *8*(7), 11181–11187.

Li, J.-H., Vicknasingam, B., Cheung, Y.-W., Zhou, W., Nurhidayat, A. W., Des Jarlais, D. C., & Schottenfeld, R. (2011). To use or not to use: An update on licit and illicit ketamine use. *Substance Abuse and Rehabilitation, 2*, 11–20.

Liao, Y., Tang, J., Corlett, P. R., Wang, X., Yang, M., Chen, H., & Fletcher, P. C. (2011). Reduced dorsal prefrontal gray matter after chronic ketamine use. *Biological Psychiatry, 69*(1), 42–48.

Liao, Y., Tang, J., Ma, M., Wu, Z., Yang, M., Wang, X., & Hao, W. (2010). Frontal white matter abnormalities following chronic ketamine use: A diffusion tensor imaging study. *Brain, 133*(7), 2115–2122.

Lindefors, N., Barati, S., & O'Connor, W. T. (1997). Differential effects of single and repeated ketamine administration on dopamine, serotonin and GABA transmission in rat medial prefrontal cortex. *Brain Research, 759*(2), 205–212.

Lipton, S. A. (January 2004). Failures and successes of NMDA receptor antagonists: Molecular basis for the use of open-channel blockers like memantine in the treatment of acute and chronic neurologic insults. *1*, 101–110.

Liu, F., Paule, M. G., Ali, S., & Wang, C. (2011). Ketamine-induced neurotoxicity and changes in gene expression in the developing rat brain. *Current Neuropharmacology, 9*(1), 256–261.

Mahadik, S. P., & Mukherjee, S. (1996). Free radical pathology and antioxidant defense in schizophrenia: A review. *Schizophrenia Research, 19*(1), 1–17.

Maier, S. F., Grahn, R. E., & Watkins, L. R. (1995). 8-OH-DPAT microinjected in the region of the dorsal raphe nucleus blocks and reverses the enhancement of fear conditioning and interference with escape produced by exposure to inescapable shock. *Behavioral Neuroscience, 109*(3), 404–412.

Maswood, S., Barter, J. E., Watkins, L. R., & Maier, S. F. (1998). Exposure to inescapable but not escapable shock increases extracellular levels of 5-HT in the dorsal raphe nucleus of the rat. *Brain Research, 783*(1), 115–120.

McCoy, M. K., Ruhn, K. A., Blesch, A., & Tansey, M. G. (2011). TNF: A key neuroinflammatory mediator of neurotoxicity and neurodegeneration in models of Parkinson's disease. *Advances in Experimental Medicine and Biology, 691*, 539–540.

Mion, G., & Villevieille, T. (2013). Ketamine pharmacology: An update (pharmacodynamics and molecular aspects, recent findings). *CNS Neuroscience & Therapeutics, 19*(6), 370–380.

Moosavi, M., Yadollahi Khales, G., Rastegar, K., & Zarifkar, A. (2012). The effect of sub-anesthetic and anesthetic ketamine on water maze memory acquisition, consolidation and retrieval. *European Journal of Pharmacology, 677*(1–3), 107–110.

Morgan, C. J. A., & Curran, H. V. (2012). Ketamine use: A review. *Addiction, 107*(1), 27–38.

Morgan, C. J. A., Muetzelfeldt, L., & Curran, H. V. (2010). Consequences of chronic ketamine self-administration upon neurocognitive function and psychological wellbeing: A 1-year longitudinal study. *Addiction, 105*(1), 121–133.

Morgan, C. J. A., Riccelli, M., Maitland, C. H., & Curran, H. V. (2004). Long-term effects of ketamine: Evidence for a persisting impairment of source memory in recreational users. *Drug and Alcohol Dependence, 75*(3), 301–308.

Morris, R. G. M., Anderson, E., Lynch, G. S., & Baudry, M. (1986). Selective impairment of learning and blockade of long-term potentiation by an N-methyl-D-aspartate receptor antagonist, AP5. *Nature, 319*(6056), 774–776.

Nakajima, A. (2004). Role of tumor necrosis factor- in methamphetamine-induced drug dependence and neurotoxicity. *Journal of Neuroscience, 24*(9), 2212–2225.

Niesters, M., Martini, C., & Dahan, A. (2014). Ketamine for chronic pain: Risks and benefits. *British Journal of Clinical Pharmacology, 77*(2), 357–367.

Niwa, M., Nitta, A., Yamada, Y., Nakajima, A., Saito, K., Seishima, M., ... Nabeshima, T. (2007). Tumor necrosis factor-α and its inducer inhibit morphine-induced rewarding effects and sensitization. *Biological Psychiatry, 62*(6), 658–668.

Oliveira, L., Cecilia, C. M., Bortolin, T., Canever, L., Petronilho, F., Gonçalves Mina, F., & Zugno, A. I. (2009). Different subanesthetic doses of ketamine increase oxidative stress in the brain of rats. *Progress in Neuro-Psychopharmacology and Biological Psychiatry, 33*(6), 1003–1008.

Owolabi, R. A., Akanmu, M. A., & Adeyemi, O. I. (2014). Effects of ketamine and *N*-methyl-d-aspartate on fluoxetine-induced antidepressant-related behavior using the forced swimming test. *Neuroscience Letters, 566*, 172–176.

Park, G. R., Manara, A. R., Mendel, L., & Bateman, P. E. (1987). Ketamine infusion. Its use as a sedative, inotrope and bronchodilator in a critically ill patient. *Anaesthesia, 42*(9), 980–983.

Petronijević, N. D., Mićić, D. V., Duričić, B., Marinković, D., & Paunović, V. R. (2003). Substrate kinetics of erythrocyte membrane Na,K-ATPase and lipid peroxides in schizophrenia. *Progress in Neuro-Psychopharmacology and Biological Psychiatry, 27*(3), 431–440.

Pietraszek, M. (2003). Significance of dysfunctional glutamatergic transmission for the development of psychotic symptoms. *Polish Journal of Pharmacology, 55*(3), 133–154.

Potter, D. E., & Choudhury, M. (2014). Ketamine: Repurposing and redefining a multifaceted drug. *Drug Discovery Today, 19*(12), 1848–1854.

Powell, S. B., Sejnowski, T. J., & Behrens, M. M. (2012). Behavioral and neurochemical consequences of cortical oxidative stress on parvalbumin-interneuron maturation in rodent models of schizophrenia. *Neuropharmacology, 62*(3), 1322–1331.

Quibell, R., Prommer, E. E., Mihalyo, M., Twycross, R., & Wilcock, A. (2011). Ketamine*. *Journal of Pain and Symptom Management, 41*(3), 640–649.

Rabiner, E. A. (2007). Imaging of striatal dopamine release elicited with NMDA antagonists: Is there anything there to be seen? *Journal of Psychopharmacology (Oxford, England), 21*(3), 253–258.

Reich, D. L., & Silvay, G. (1989). Ketamine: An update on the first twenty-five years of clinical experience. *Canadian Journal of Anaesthesia/Journal Canadien D'anesthésie, 36*(2), 186–197.

Report, T. (2012). WHO expert committee on drug dependence. *World Health Organization Technical Report Series, 973*, 1–26.

Schultz, W. (2007). Multiple dopamine functions at different time courses. *Annual Review of Neuroscience, 30*, 259–288.

Shibuta, S., Morita, T., Kosaka, J., Kamibayashi, T., & Fujino, Y. (2015). Only extra-high dose of ketamine affects L-glutamate-induced intracellular Ca^{2+} elevation and neurotoxicity. *Neuroscience Research, 98*, 9–16.

Silva, F. C. C., Dantas, R. T., Citó, M. D. C. D. O., Silva, M. I. G., De Vasconcelos, S. M. M., Fonteles, M. M. D. F., ... De Sousa, F. C. F. (2010). Ketamina, da anestesia ao uso abusivo: Artigo de revisão. *Revista Neurociencias, 18*(2), 227–237.

Slikker, W., Zou, X., Hotchkiss, C. E., Divine, R. L., Sadovova, N., Twaddle, N. C., ... Wang, C. (2007). Ketamine-induced neuronal cell death in the perinatal rhesus monkey. *Toxicological Sciences, 98*(1), 145–158.

Soden, M. E., Jones, G. L., Sanford, C. A., Chung, A. S., Güler, A. D., Chavkin, C., ... Zweifel, L. S. (2013). Disruption of dopamine neuron activity pattern regulation through selective expression of a human KCNN3 mutation. *Neuron, 80*(4), 997–1009.

Srivastava, N., Barthwal, M. K., Dalal, P. K., Agarwal, A. K., Nag, D., Srimal, R. C., ... Dikshit, M. (2001). Nitrite content and antioxidant enzyme levels in the blood of schizophrenia patients. *Psychopharmacology, 158*(2), 140–145.

Takuma, K., Baba, A., & Matsuda, T. (2004). Astrocyte apoptosis: Implications for neuroprotection. *Progress in Neurobiology, 72*(2), 111–127.

Tan, S., Lam, W. P., Wai, M. S. M., Yu, W. H. A., & Yew, D. T. (2012). Chronic ketamine administration modulates midbrain dopamine system in mice. *PLoS One, 7*(8).

Tan, S., Rudd, J. A., & Yew, D. T. (2011). Gene expression changes in $GABA_A$ receptors and cognition following chronic ketamine administration in mice. *PLoS One, 6*(6), 1–8.

Tanaka, K. F., Shintani, F., Fujii, Y., Yagi, G., & Asai, M. (2000). Serum interleukin-18 levels are elevated in schizophrenia. *Psychiatry Research, 96*(1), 75–80.

Tancredi, V., D'Antuono, M., Cafe, C., Giovedi, S., Bue, M. C., D'Arcangelo, G., ... Benfenati, F. (2000). The inhibitory effects of Interleukin-6 on synaptic plasticity in the rat hippocampus are associated with an inhibition of mitogen-activated protein kinase ERK. *Journal of Neurochemistry, 75*(2), 634–643.

Trujillo, K. (2011). The neurobehavioral pharmacology of ketamine: Implications for drug abuse, addition, and psychiatric disorders.

ILAR Journal/National Research Council, Institute of Laboratory Animal Resources, 52(3), 366–378.

Walther, D. J., Peter, J.-U., Bashammakh, S., Hörtnagl, H., Voits, M., Fink, H., & Bader, M. (2003). Synthesis of serotonin by a second tryptophan hydroxylase isoform. *Science (New York, N.Y.), 299*(5603), 76.

Wang, J. H., Fu, Y., Wilson, F. A. W., & Ma, Y. Y. (2006). Ketamine affects memory consolidation: Differential effects in T-maze and passive avoidance paradigms in mice. *Neuroscience, 140*(3), 993–1002.

Wang, H., Ramakrishnan, A., Fletcher, S., Prochownik, E. V., & Genetics, M. (2015). *HHS Public Access, 2*(2), 145–155.

Wu, L. T., Schlenger, W. E., & Galvin, D. M. (2006). Concurrent use of methamphetamine, MDMA, LSD, ketamine, GHB, and flunitrazepam among American youths. *Drug and Alcohol Dependence, 84*(1), 102–113.

Xu, K., & Lipsky, R. H. (2015). Repeated ketamine administration alters N-methyl-d-aspartic acid receptor subunit gene expression: Implication of genetic vulnerability for ketamine abuse and ketamine psychosis in humans. *Experimental Biology and Medicine, 240*(2), 145–155.

Yanagihara, Y., Kariya, S., Ohtani, M., Uchino, K., Aoyama, T., Yamamura, Y., & Iga, T. (2001). Involvement of CYP2B6 in *N*-demethylation of ketamine in human liver microsomes. *Drug Metabolism and Disposition, 29*(6), 887–890.

Yanagihara, Y., Ohtani, M., Kariya, S., Uchino, K., Hiraishi, T., Ashizawa, N., ... Iga, T. (2003). Plasma concentration profiles of ketamine and norketamine after administration of various ketamine preparations to healthy Japanese volunteers. *Biopharmaceutics & Drug Disposition, 24*(1), 37–43.

Zhou, J., Zheng, X., & Xia, Y. (2015). The role of TNF-α in regulating ketamine-induced hippocampal neurotoxicity. *Archives of Medical Science, 11*(6), 1296–1302.

Zunszain, P. A., Horowitz, M. A., Cattaneo, A., Lupi, M. M., & Pariante, C. M. (2013). Ketamine: Synaptogenesis, immunomodulation and glycogen synthase kinase-3 as underlying mechanisms of its antidepressant properties. *Molecular Psychiatry, 18*(12), 1236–1241.

CHAPTER

33

Left/Right Hemispheric "Unbalance" Model in Addiction

R. Finocchiaro, M. Balconi

Catholic University of the Sacred Heart, Milan, Italy

BIAS OF REWARD MECHANISMS IN ADDICTION

Brain Dysfunctions of Reward and Inhibition Systems

Neuroscience studies have identified addiction as a chronic brain disease with genetic, neurobiological, and environment components that lead to changes in whole-brain functioning and long-lasting impairments to specific brain structures involved in attention, working memory, decision-making processes, judgment, and gratification, with negative consequence on cognition performances, emotion regulation, and social adaptation (Baler & Volkow, 2006; Bechara & Damasio, 2002; Bechara & Martin, 2004; Li & Sinha, 2008; Li et al., 2013; Yan et al., 2014). The principal neural circuits that seem to be involved in the "addicted brain" are the mesostriatocortical system and the frontocortical area (Volkow, Wang, Tomasi, & Baler, 2013). Indeed, repeated drug administration triggers neuroplastic modifications with a modified dopamine (DA) activity in the mesocorticolimbic circuit, an alteration of glutamate neurotransmission, and a cortical excitability modulation, which influence cognition, emotion, and behavior (Volkow & Baler, 2014). Decrease in DA's stimulation in the nucleus accumbens (NAcc), which is a major component of the ventral striatum and a key structure involved in mediating motivational and emotional processes, creates a strong consolidation on the motivational system to take more substance. It enhances the brain's reactivity to drug cues and reduces the sensitivity to nondrug rewards, as consequences of weakening self-regulation and increasing the sensitivity to stressful stimuli and dysphoria (Volkow & Li, 2004; Volkow & Morales, 2015). Moreover, neuroimaging studies using functional magnetic resonance imaging (fMRI) or positron emission tomography (PET) suggest reduction in DA (D2) receptors and a decrease in the release of DA in the ventral striatum (Volkow, Fowler, & Wang, 2003), which contribute to reduce the sensitivity to natural reinforcements in addict population. Another study showed an overactivity of the orbitofrontal cortex (OFC) connected to the limbic system (Yamamoto et al., 2014). Specifically, literature suggests that OFC is involved in decoding, representing, learning, and reversing associations of stimuli to the reinforcers and, also, in controlling reward-related adjustment and punishment-related behavior (Rolls, 2004), thus increasing of OFC activity in addict population is probably linked to extreme focus on drug-related rewards.

Behavioral Inhibition System and Behavioral Activation System Components

Several studies showed that behavioral addictions such as pathological gambling (PG) or Internet addiction disorder (IAD) share the same dysfunction in reward mechanisms and cognitive control as found in substance-addiction disorders (Wareham & Potenza, 2010; Yuan et al., 2011). Specifically, a reduction in the activity in the mesolimbic reward system in PG (Reuter et al., 2005) and structural abnormalities in gray and white matter volume in left posterior limbic and dorsolateral prefrontal cortex (DLPFC), which are linked to functional impairments in cognitive control in IAD, were found (Yuan et al., 2011). Thus, altered prefrontal activity with enhanced striatal responses to addicted drug- or addicted behavior-related salient stimuli perpetuates habitual drug- or behavioral object-seeking despite the negative consequences.

Addictive Substances and Neurological Disease
http://dx.doi.org/10.1016/B978-0-12-805373-7.00033-5

In our study we tested reward sensitivity in cocaine addiction population (Balconi & Finocchiaro, 2015). We focused on the behavioral motivational responses that are crucial to the generation of emotions relevant to approach (reward) and withdrawal (inhibition) in the decisional process (Gray, 1981). Carver and White (1994), developed the behavioral inhibition system/behavioral activation system (BIS/BAS) scales, a self-report measure composed of 24 items; also BAS scale includes three subscales (Reward, five items; Drive, four items; and Fun Seeking, four items). The BAS seems to activate behavior in response to conditioned, rewarding, and nonpunishment stimuli, and is supposed to be mediated by dopaminergic pathways from the ventral tegmental area (VTA) to the NAcc and VS (Fowles, 1994). Normal level of BAS functionally affects positive emotional attitude, but extreme levels of BAS have been linked to impulsivity disorders such as attention-deficit and hyperactivity disorder (ADHD), or addictive diseases, risk, and antisocial behavior. Instead, the BIS appears to be preferentially activated by stimuli conditioned as being aversive, thus the BIS is responsive to nonreward stimuli, preventing individuals from negative or painful outcomes. A dysfunction in the direction of hyperactivity of this system could generate pathological disorders such as generalized anxiety disorder (GAD) or obsessive–compulsive disorder (OCD). Several studies showed a strong correlation between BIS/BAS systems and the cortical brain activity. Specifically a greater left frontal activity seems to characterize individuals with higher BAS and lower BIS scores (Harmon-Jones & Allen, 1997; Sutton & Davidson, 1997). On the other hand, an increase in right frontal activity seems to be related with higher BIS and lower BAS sensitivity (Balconi & Mazza, 2009, 2010). In different studies we related the BIS/BAS scale to the Iowa Gambling Task (IGT), which is a typical risky decision-making test developed by Bechara, Damasio, Damasio, and Anderson (1994) to experimentally capture the decision-making deficits of patients with ventral medial prefrontal cortex (VmPFC) damage. The IGT is a sensitive tool that is able to discriminate people with frontal lobe dysfunctions, in addition to adults with VmPFC damage (Balconi & Finocchiaro, 2015; Balconi, Finocchiaro, & Campanella, 2014; Balconi, Finocchiaro, & Canavasio, 2014a; Bechara & Martin, 2004). There are other groups who perform poorly on the task, including people who report being high in risk-taking behaviors and people who abuse substances, such as drugs and alcohol (Bechara et al., 1994; Brevers, Bechara, Cleeremans, & Noël, 2013). The IGT requires continuous selections to be made from decks of cards with varying rewards and punishments. Some decks have high initial rewards but result in high punishments over time and thus are disadvantageous in the long run. Other decks have lower initial rewards but also lower punishments over time, making them advantageous in the long run. We assumed the BAS is a predictive marker of dysfunctional behavior in IGT, and also we focused on self-reported metacognitive measures concerning the decisional process (Balconi, Finocchiaro, & Campanella, 2014). We found that an increase in the reward sensitivity (higher BAS and BAS reward) explained a poorer performance on the IGT and a dysfunctional metacognition ability (unrealistic representation) in cocaine addicts (CA) group compared with the control group. Generally, high level in the BAS reward responsiveness may be considered a predictive measure of risk-taking and dysfunctional behavior, not only in pathological (CA) individuals, but also in subclinical individuals (Balconi, Finocchiaro, & Canavesio, 2014b; Balconi, Finocchiaro, Canavesio, & Messina, 2014). Individuals with CA often evidence poor cognitive control: in a study Worhunsky et al. (2013) used fMRI method to investigate frontocingular connectivity network which was supposed to underlie cognitive control processes, in CA patients, who were asked to perform a Stroop task for testing selective attention and inhibition of control. A reduced connection of a "top-down" frontocingular network contributing to conflict monitoring correlated with better treatment retention was found. However, greater involvement of two "bottom-up" subcortical and ventral prefrontal networks related to cue-elicited motivational processing correlated with abstinence during treatment. Authors argued that these brain networks (frontocingular, subcortical, and ventral prefrontal) linked to cocaine abstinence and treatment retention could represent important targets in novel treatment for CA. Moreover, another study demonstrated that cocaine users had difficulties to inhibit their own behaviors, particularly when working memory demands during cue-induced craving for the drug, and they showed reduced activity in anterior cingulate and right prefrontal cortex, which are thought to be critical for cognitive control (Hester & Garavan, 2004).

Decision-Making Processes

Authors suggest that cognitive functions may play an important role in prolonging abuse or predisposing cocaine abusers toward relapse (Hester & Garavan, 2004). Thus, it is accepted that some specific cognitive processes seem to be affected in substance-use disorder (SUD). In our study we demonstrated that the SUD group showed a strong lateralization effect in DLPFC, which is involved during the decisional process: the SUD group revealed an increase in left hemisphere activation in response to immediate reward choices, and this cortical unbalance effect seems to be related to the lower performance in IGT (Balconi & Finocchiaro, 2015). It seems that there are some structural effects of substance on neural systems mediating cognition and motivation in decision making. For example, Makris et al. (2008) found a correlation between thinner prefrontal cortex and reduced performance during judgment and decision making in addicts. It was suggested that brain-structure abnormalities in addicts could be related not only to drug use but also to predisposition of development of addiction disease. Furthermore, anomalous brain activity was found in behavioral addiction like PG, and it seems that the same brain pathways are affected both in substance or nonsubstance addiction disorders. Potenza et al. (2014) investigated impulse control behavior using fMRI: the PG group performed a Stroop task to test attention and response inhibition during the presentation of congruent and incongruent stimuli, and the authors found that in response to infrequent incongruent stimuli, PG group showed a decreasing activity in the left VmPFC compared to control group. PET studies indicate that substance-dependent individuals show altered prefrontal activity on the IGT task. Specifically, reductions in right prefronal activity during decision making may reflect impaired working memory, stimulus reward valuation, or cue reactivity in substance-dependent individuals (Tanabe et al., 2007). In our studies we investigated the motivational traits considering the approach or withdrawal tendencies in CA and subclinical individuals. We focused on the BAS reward trait that seems to characterize addict's personality (Balconi & Finocchiaro, 2015; Balconi, Finocchiaro, & Campanella 2014; Balconi et al., 2014a, 2014b; Balconi, Finocchiaro, Canavesio, et al., 2014). We considered the hypothesis that individuals with high-activation system in motivational dimension (high-BAS) could show similarity with CA profile (Finocchiaro & Balconi, 2015). We postulated that high-BAS individuals (HBIs) having a similar dysfunction mechanisms in decision-making process (lower performance in IGT) related to a higher left hemisphere activation could be more vulnerable to develop an addiction even if they were not clinical population. We consider this "cortical unbalance effect" as a critical marker of dysfunctional decision making in high-risk population, and a factor able to explain the tendency to opt in favor of more reward-related conditions (Finocchiaro & Balconi, 2015). Literature shows that deficits in cognitive performance are correlated with altered brain activity also in Internet Gaming Disorder (IGD) (Ko et al., 2014). A 2014 fMRI study focused on response inhibition using a Go/No-go paradigm in a population of IGD showed higher brain activation in IGD while they were processing response inhibition over the left orbital frontal lobe and bilateral caudate nucleus in comparison of control group; moreover, the activation over the right insula was lower in the individuals with IGD (Ko et al., 2014). Thus, authors suggest that frontostriatal network involved in response inhibition which contributes to error processing could be damaged in individuals with IGD; for this reason they could have impaired insular function in error processing, and lower abilities to maintain their response inhibition performance.

Psychological Traits

Although scientific evidences support a neurobiological basis for SUD, the link between the mechanisms underlying dysfunctional behaviors and biological systems are still unclear. A 2014 study found that individual differences in personality traits such as extraversion, neuroticism, and constraint, which implies intentional and volitional motor control, are related to genetic profile that could lead high-risk individuals to develop an addiction disease (Belcher, Volkow, Moeller, & Ferré, 2014). Indeed, authors postulated that some personality traits underlying dimension of sensitivity to signals of punishment or reward, moderated by genes, interact dynamically with the environment to determine the degree of vulnerability or resilience to the development of SUD (Belcher et al., 2014). Personality traits of impulsivity and sensation seeking are highly prevalent in SUD individuals (De Wit, 2009). In particular, sensation seeking has been linked with the onset of substance abuse, and impulsivity has been associated with the development and maintenance of dependence (Belin, Mar, Dalley, Robbins, & Everitt, 2008). Impulsivity seems to be a pathological trait marker of addiction, for example, impulsive choice in SUD individuals correlates with impaired function of prefrontal cortical areas, such as the OFC. In a study with heroin-dependent (HD) patients, higher impulsivity scores (Barratt Impulsiveness Scale 11, BIS-11) were related to significantly enhanced intrinsic amygdala functional connectivity (iAFC) (Xie et al., 2011). Thus, an altered iAFC network connectivity in HD patients may contribute to the loss of impulsive control. Therefore, changes in neurocircuitry involved in impulse control has significant implications for understanding addiction vulnerability. Literature report that the onset of addictive disorders is mainly concentrated in adolescence and young adulthood phases (Chambers, Taylor, & Potenza, 2003). Thus, neuromaturational changes in frontal cortical and subcortical systems in adolescence may promote learning for adaptation to social roles but may also confer on them greater vulnerability to the addictive behaviors.

NEUROPHYSIOLOGICAL CORRELATES IN ADDICTION

Cortical Oscillations

Studies with frequency-band analysis focused on modification in cortical oscillations during cognitive tasks in different types of addiction behaviors. Specifically, several studies used electroencephalographic (EEG) method to analyze the asymmetry between PFC activity in the left and right hemispheres and the associations to affective and motivational behavior, and clinical outcomes (Coan & Allen, 2003; Davidson, 2004; Sutton & Davidson, 1997). Coan and Allen (2003) found that PFC asymmetry index may be considered as an indicator of risk for individual's propensity, and it could be useful in prognoses and treatment interventions. Regarding the association between cortical activity and advantageous/disadvantageous choices in IGT, in our research we focused on the hypothesis that left hemisphere dominance should index greater approach–attitude tendency, maybe reinforced by and related to the positive experience of immediate reward, which are higher in the disadvantageous decks. Furthermore, left dominant individuals should indicate less sensitivity to punishment than right dominant individuals. Thus, in our recent study we aimed to investigate the decision-making process and the effect of the reward sensitivity, considering the BAS-Reward construct, on the IGT performance. We considered the impact of the BAS motivational system to the frontal left and right cortical activity on individuals' decisions. More specifically, we hypothesized a specific lateralization effect, which is supposed to be related to the increased activation of the left (BAS-Reward-related) hemisphere, in delta, theta, alpha, and beta cortical bands for HBIs. Also, behavioral responses (gain/loss options), metacognition dimensions (self-knowledge, strategic planning, flexibility, and efficacy) were investigated. Thirty participants were divided into high-BAS and low-BAS groups. In comparison with low-BAS, the high-BAS group showed an increased tendency to opt in favor of the immediate reward (losing strategy) instead of the long-term option (winning strategy). Moreover, the high-BAS group was more impaired in metacognitive monitoring of their strategies and showed an increased left hemisphere activation when they responded to losing choices. A "reward bias" effect was confirmed to act for high BAS, based on a left-hemisphere hyperactivation (Balconi et al., 2014a). In another study we considered addict population and tested specifically the activity of alpha (α) band modulation during an IGT performance. Activity in the α band oscillations is used as an inverse index of cortical activity, which assumes that a brain region producing α rhythms is in a state of cortical loafing. Thus, the more α appearing in the EEG track of a brain region the less active or engaged it is. We found that the SUD group increased the tendency to opt in favor of the immediate reward, which is a losing strategy more than the long-term option, which is a winning strategy, compared to the control group. Moreover, higher reward-subscale scores were observed in SUD. Finally, SUD showed an increase in left-hemisphere activation in response to immediate rewarding choices. We conducted regression analysis for BAS subscales, and found that higher BAS traits could explain this imbalanced left hemispheric effect related to the main behavioral deficits (Balconi et al., 2014a). Moreover, we found the same cortical lateralization effects in a sample of high-risk individuals with high scores in BAS scale, only for α band analysis. Thus a "reward-bias" effect was supposed to explain both the bad strategy and the unbalanced hemispheric activation for high-BAS and more risk-taking subjects. These findings could have relevance for prevention in high-risk population (Balconi et al., 2014b; Balconi, Finocchiaro, & Canavesio 2015; Finocchiaro & Balconi, 2015).

Event-Related Potential Measures

Event-related potential (ERP) analysis is widely used to investigate cortical brain response to cognitive and emotional task (Harper, Malone, & Bernat, 2014). Specifically, the amplitude of P300 component in addiction research was largely investigated. P300 was identified as a centroparietal positive voltage peak in the ERP waveform, and it occurs between 300 and 500 ms after stimulus that occurs when an individual attends or responds to an infrequent but task-relevant stimulus presentation. Literature showed a reduction on amplitude in P300 in alcoholics and in high-risk individuals, and this could be considered as a biological marker of cognitive deficits in addiction (Porjesz et al., 2005). Literature suggested that drug-addicted individuals have a diminished cortical response to nondrug rewards. An ERP study of Parvaz et al. (2012) investigated the P300 amplitude using an attention task with gradual increase of monetary incentives in a sample of CA. Authors found that the P300 amplitude was not modulated by the varying amounts of money, but the group that was negative to the substance showed lower sensitivity to money and poorer task accuracy with the lowest P300 amplitudes compared with the positive for the substance group and control group. Authors inferred that, when the levels of impairment are most severe, cocaine users seem to self-medicate with substance to avoid or compensate for underlying cognitive and emotional difficulties despite the long-term damaging consequences on sensitivity to nondrug reward.

Another ERP component called feedback-related negativity (FRN) was commonly used as an index of cognitive dysfunction in addiction diseases (Kamarajan et al., 2010; Torres et al., 2013). It is a frontal–central deflection thought to reflect anterior cingulate cortex (ACC) processing, with an onset of 250/300 ms after feedback stimulus is presented, and it is typically largest for unexpected aversive outcomes, reflecting a binary evaluation of good versus bad feedback (Hajcak, Moser, Holroyd, & Simons, 2006). In our ERP study we analyzed the FRN and P300 effects as predictive markers of IGT performance (Balconi et al., 2015). The BAS-reward measure was applied to distinguish between high-BAS and low-BAS group. It was found that higher-BAS group opted in favor of the immediate reward (disadvantageous decks), with a concomitant dysfunctional metacognition of their strategy (unrealistic representation). Finally, a consistent "reward bias" affected the high-BAS performance reducing the P300 and FRN in response to unexpected (loss) events. In addition, the source cortical localization (standardized low-resolution brain electromagnetic tomography, sLORETA) of ERPs showed the contribution by distinct PFC areas and posterior areas.

Brain Connectivity

It is widely accepted that addictive drug use is related to abnormal functional network in addict's brain. In the past years several neuroscientific studies aimed to identify this type of abnormality within the brain networks implicated in addiction, often by measuring resting-state functional connectivity, which offers a direct measure of functional interactions between the brain areas (Kelly et al., 2011). Ma et al. (2010) found that chronic heroin users showed increased functional connectivity between NAcc and ventral/rostral ACC, between NAcc and OFC, and between amygdala and OFC, but reduced functional connectivity among PFC, OFC, and ACC. Authors argue that findings may provide additional evidence supporting the theory of addiction that emphasizes enhanced salience value of the substance and ineffective cognitive control of cues-related condition, which could have a major role in the maintenance of the addictive behavior (Ma et al., 2010). Individuals with CA showed greater connectivity of the right insula cortex with the dorsomedial prefrontal cortex, inferior frontal gyrus, and bilateral DLPFC (Cisler et al., 2013). These data confirm the hypothesis that CA is related to altered functional interactions of the insular cortex with prefrontal networks, thus this could have negative influence in cognitive control and decision-making processes. Moreover, literature confirms the hypothesis that Internet Gaming Addiction (IGA) shares similar neurobiological abnormalities with substance addictive disorders. In a recent resting-state fMRI study it was found that IGA group showed increased functional connectivity in posterior lobe of the bilateral cerebellum and middle temporal gyrus, in spite of decreased connectivity in the bilateral inferior parietal lobule and right inferior temporal gyrus, and that these different patterns of brain activity in IGA group were correlated with the severity of IA and impulsivity (Ding et al., 2013). Often addiction models emphasize the role of disrupted frontal circuitry supporting cognitive control processes. However, it is useful to consider addiction-related alterations in functional interactions among brain regions, especially between the cerebral hemispheres that are only occasionally analyzed. Kelly et al. (2011) observed reduced prefrontal interhemispheric connectivity in CA. Specifically, they demonstrated a severe cocaine-dependence–related reduction in interhemispheric connectivity among nodes of the dorsal attention network (frontal and parietal areas), which were associated with self-reported attentional deficits. Their findings confirmed a link between chronic abuse to cocaine and disruptions in brain circuitry supporting cognitive control (Kelly et al., 2011). Another study focused on IA investigated interhemispheric functional and structural connectivity in adolescents (Bi et al., 2015). Authors showed decreased activity of DLPFC which was negatively correlated with the duration of IA, and also a lower integrity of white matter and lower connectivity in corpus callosum. Moreover, in a Go/No-go study, adolescents with IA fail to recruit the indirect frontal–basal ganglia pathway which was engaged by response inhibition in healthy subjects (Li et al., 2014). All these evidences indicate that addictive disorders (with or without substance) have similarities in the neural basis of poor impulse control, and this fact is important for understanding the neurobiological mechanisms of addiction.

AN INTEGRATIVE CORTICAL UNBALANCE MODEL

As we underlined, the behavioral interhemispheric balance, which is reflected in a functional decision-making process to adopt the most effective strategies for the individuals' well-being, has a fundamental importance in order to avoid styles of behavior at risk that can lead to addiction. Considering all the evidences described in the foregoing, we propose a model, the Integrative Cortical Unbalance Model (ICUM) that includes the relationship between the motivational system

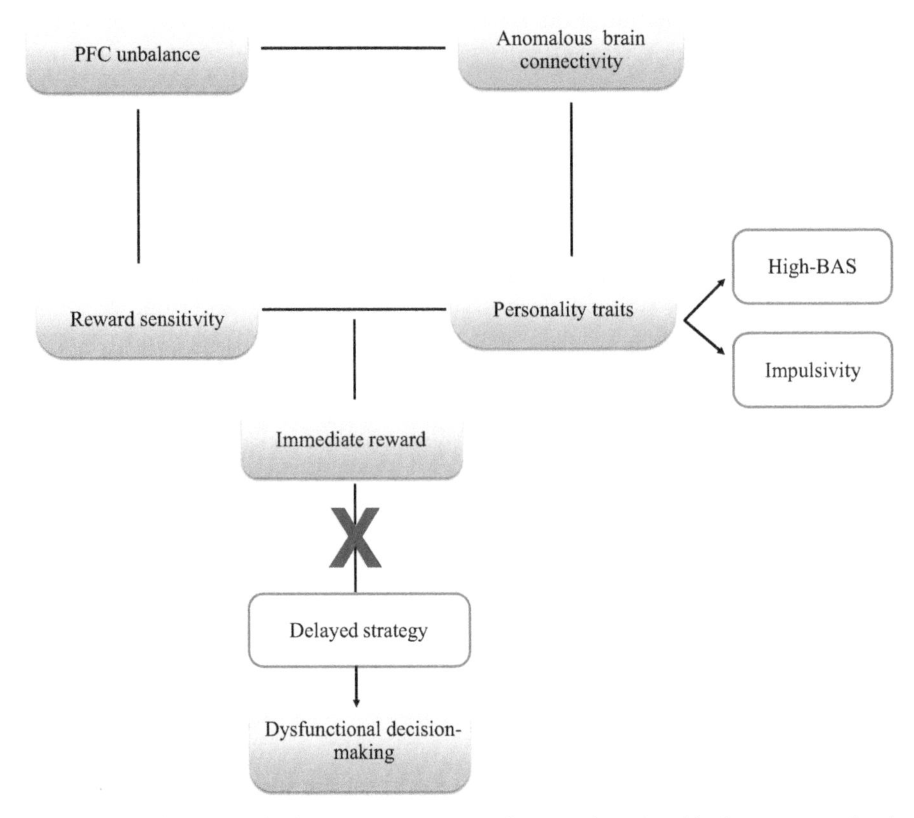

FIGURE 33.1 Higher activity in the left PFC and a decrease in connectivity between frontal and limbic systems related to dysfunctional reward mechanisms, in combination with personality traits, such as higher levels of BAS and impulsivity, may determine attitude toward more rewarding conditions, loss of control, and reinforcing of compulsive behavior in addiction disease. *BAS*, behavioral activation system; *PFC*, prefrontal cortex.

and reward sensitivity related to the cortical unbalance effect, and altered functional connectivity on brain oscillations (Fig. 33.1). To resume, in our studies we explored the impact of cortical frontal asymmetry (left-lateralization effect) and BAS on SUD in decision-making processes using the IGT, and we found that in clinical and subclinical individuals (high-risk individuals) the lower performance in IGT and an unbalanced left hemispheric effect were related to higher BAS trait. Furthermore, metacognitive deficits seem to accompany this type of vulnerability profile to risky behaviors. Therefore, it is possible to assume that there is a biological vulnerability expressed by higher left hemisphere reactivity by SUD and HBIs, and this general interhemispheric unbalance may determine the cognitive bias toward more appetitive and rewarding conditions. Thus, frontal mechanisms could have a relevant and consistent effect on the goal-directed behavior, producing a general dysfunctional strategy based on a reward-bias effect, which could maintain the maladaptive behavior and lead to relapses in addiction individuals and to increase the vulnerability to develop addiction disease in HBIs. Moreover, we suggest to integrate this first model of addiction with another biological component concerning the structurally and functionally reduced connectivity between PFC and subcortical regions. This reduced connectivity may reflect an enhanced salience value of cues-related addiction, with an ineffective cognitive control and maintenance of the addictive behavior. Additionally, it could be suggested that the tendency for an "immediate-reward" trait, related to alteration in the brain activity, may issue a new model of potential sensitivity to addictive behaviors, because high reward sensitivity and high-risky profiles share the reward vulnerability with SUDs but without the concomitant injurious effects of substance abusing.

CONCLUSIONS AND FUTURE PERSPECTIVES

The current state of knowledge from neuroscientific studies suggests that there may exist a similar pathological pathway between SUD and nonsubstance-related disorder (e.g., gambling, food, sex, or Internet addiction), involving dysfunctional reward mechanisms and deficit in cognitive control and decision-making processes. From a neurobiological point of view, there is a reduction in the DA (D2) receptor on compulsive feeding (Wang,

Volkow, & Fowler, 2002) and gambling related to deficits of the frontal cortex in pathological gambling (Potenza, 2008). It is widely accepted that addiction disease results from both vulnerability and/or post-alterations (drug effects) on corticostriatal circuits that are essential for cognitive control, motivation, reward, dependent learning, and emotional processing. Thus, dysfunctions in corticostriatal circuits are thought to relate to the core features of addiction, which include compulsive drug use or behavior, loss of the ability to control drug intake or maladaptive behaviors, and the emergence of negative emotional states (Koob & Volkow, 2010).

In the last decades, thanks to the use of noninvasive and low-cost neuromodulation and given the fact that pharmacological strategies are of limited effectiveness, interest has grown in the application of brain stimulation techniques as treatment of addiction. These experimental techniques as treatment of addiction could be applied also for individuals at high risk of dependence who could have a vulnerability in the development of dysfunctional reward mechanisms and unbalance of PFC activity. Specifically, brain stimulation is induced by electric or magnetic energy to improve brain function. It is applied for both research and treatment of psychiatric and neurological disorders, which do not always fully respond to conventional treatments (pharmacological or psychotherapeutic). Stimulation with electrical or magnetic energy interacts with the neurons of the cortex, causing the release of neurotransmitters that can inhibit or excite specific cortical networks (Amiaz, Levy, Vainiger, Grunhaus, & Zangen, 2009) and modulate the cortical activity. In particular, two techniques have been mainly tested in neuroscientific studies, namely the transcranial magnetic stimulation (TMS) and the transcranial direct current stimulation (tDCS). The TMS modulates the activity of the brain with magnetic pulses focused on a limited portion of the scalp by a coil (high frequency: excitatory; low frequency: inhibitory). Repeated sessions of TMS over the DLPFC were observed to reduce drug craving, drug seeking, and drug consumption, and relapse in nicotine addicts and CA (Amiaz et al., 2009; Camprodon, Martínez-Raga, Alonso-Alonso, Shih, & Pascual-Leone, 2007); and to improve cognitive abilities in alcohol dependence (Del Felice et al., 2016). The tDCS can induce functional changes in the cerebral cortex. It consists in applying on the scalp two electrodes, one anode (excitatory) and one cathode (inhibitory), delivering a continuous current of low intensity that is not perceptible by the individuals, crossing the scalp and influencing neuronal functions. Literature shows that tDCS anodal stimulation over the right DLPFC induces a reduction in risky behavior in CA (Gorini, Lucchiari, Russell-Edu, & Pravettoni, 2014), induces decreased ACC activity after visualization of drug cues (Conti & Nakamura-Palacios, 2014), and reduces nicotine, cocaine, and alcohol craving (Batista, Klauss, Fregni, Nitsche, & Nakamura-Palacios, 2015; Boggio et al., 2008; Fregni et al., 2008). These findings support the hypothesis that excessive risk propensity in addict patients might be due to a hypoactivity of the right DLPFC and hyperactivity of the left DLPFC, as was found in previous studies (Balconi & Finocchiaro, 2015; Balconi et al., 2014a).

Thus, all these data highlight the importance of an integrative model of addiction that takes into account the reward-bias system related to anomalous lateralized response in cortical activity (unbalance effect) and, at the same time, the possibility to induce, by neuromodulation or neurostimulation, an improvement of the symptomatology, through a balancing of the cortical activity.

References

Amiaz, R., Levy, D., Vainiger, D., Grunhaus, L., & Zangen, A. (2009). Repeated high-frequency transcranial magnetic stimulation over the dorsolateral prefrontal cortex reduces cigarette craving and consumption. *Addiction, 104*, 653–660.

Balconi, M., & Finocchiaro, R. (2015). Decisional impairments in cocaine addiction, reward bias, and cortical oscillation "unbalance". *Neuropsychiatric Disease and Treatment, 11*, 777–786.

Balconi, M., Finocchiaro, R., & Campanella, S. (2014). Reward sensitivity, decisional bias, and metacognitive deficits in cocaine drug addiction. *Journal of Addiction Medicine, 8*, 399–406.

Balconi, M., Finocchiaro, R., & Canavesio, Y. (2014a). Reward-system effect (BAS rating), left hemispheric "unbalance"(alpha band oscillations) and decisional impairments in drug addiction. *Addictive Behaviors, 39*, 1026–1032.

Balconi, M., Finocchiaro, R., & Canavesio, Y. (2014b). Left hemispheric imbalance and reward mechanisms affect gambling behavior the contribution of the metacognition and cortical brain oscillations. *Clinical EEG and Neuroscience, 46*, 197–207.

Balconi, M., Finocchiaro, R., & Canavesio, Y. (2015). Reward sensitivity (behavioral activation system), cognitive, and metacognitive control in gambling behavior: Evidences from behavioral, feedback-related negativity, and P300 effect. *The Journal of Neuropsychiatry and Clinical Neurosciences, 27*, 219–227.

Balconi, M., Finocchiaro, R., Canavesio, Y., & Messina, R. (2014). Reward bias and lateralization in gambling behavior: Behavioral activation system and alpha band analysis. *Psychiatry Research, 219*, 570–576.

Balconi, M., & Mazza, G. (2009). Brain oscillations and BIS/BAS (behavioral inhibition/activation system) effects on processing masked emotional cues.: ERS/ERD and coherence measures of alpha band. *International Journal of Psychophysiology, 74*, 158–165.

Balconi, M., & Mazza, G. (2010). Lateralisation effect in comprehension of emotional facial expression: A comparison between EEG alpha band power and behavioural inhibition (BIS) and activation (BAS) systems. *Laterality: Asymmetries of Body, Brain and Cognition, 15*, 361–384.

Baler, R. D., & Volkow, N. D. (2006). Drug addiction: The neurobiology of disrupted self-control. *Trends in Molecular Medicine, 12*, 559–566.

Batista, E. K., Klauss, J., Fregni, F., Nitsche, M. A., & Nakamura-Palacios, E. M. (2015). A randomized placebo-controlled trial of targeted prefrontal cortex modulation with bilateral tDCS in patients with crack-cocaine dependence. *International Journal of Neuropsychopharmacology, 18*, 1–11.

Bechara, A., & Damasio, H. (2002). Decision-making and addiction (part I): Impaired activation of somatic states in substance dependent individuals when pondering decisions with negative future consequences. *Neuropsychologia, 40*, 1675–1689.

Bechara, A., Damasio, A. R., Damasio, H., & Anderson, S. W. (1994). Insensitivity to future consequences following damage to human prefrontal cortex. *Cognition, 50*, 7–15.

Bechara, A., & Martin, E. M. (2004). Impaired decision making related to working memory deficits in individuals with substance addictions. *Neuropsychology, 18*, 152–156.

Belcher, A. M., Volkow, N. D., Moeller, F. G., & Ferré, S. (2014). Personality traits and vulnerability or resilience to substance use disorders. *Trends in Cognitive Sciences, 18*, 211–217.

Belin, D., Mar, A. C., Dalley, J. W., Robbins, T. W., & Everitt, B. J. (2008). High impulsivity predicts the switch to compulsive cocaine-taking. *Science, 320*, 1352–1355.

Bi, Y., Yuan, K., Feng, D., Xing, L., Li, Y., Wang, H., ... Tian, J. (2015). Disrupted inter-hemispheric functional and structural coupling in internet addiction adolescents. *Psychiatry Research: Neuroimaging, 234*, 157–163.

Boggio, P. S., Sultani, N., Fecteau, S., Merabet, L., Mecca, T., Pascual-Leone, A., & Fregni, F. (2008). Prefrontal cortex modulation using transcranial DC stimulation reduces alcohol craving: A double-blind, sham-controlled study. *Drug and Alcohol Dependence, 92*, 55–60.

Brevers, D., Bechara, A., Cleeremans, A., & Noël, X. (2013). Iowa Gambling Task (IGT): Twenty years after-gambling disorder and IGT. *Frontiers in Psychology, 4*(665), 1–14.

Camprodon, J. A., Martínez-Raga, J., Alonso-Alonso, M., Shih, M. C., & Pascual-Leone, A. (2007). One session of high frequency repetitive transcranial magnetic stimulation (rTMS) to the right prefrontal cortex transiently reduces cocaine craving. *Drug and Alcohol Dependence, 86*, 91–94.

Carver, C. S., & White, T. L. (1994). Behavioral inhibition, behavioral activation, and affective responses to impending reward and punishment: The BIS/BAS scales. *Journal of Personality and Social Psychology, 67*, 319–333.

Chambers, R. A., Taylor, J. R., & Potenza, M. N. (2003). Developmental neurocircuitry of motivation in adolescence: A critical period of addiction vulnerability. *American Journal of Psychiatry, 160*, 1041–1052.

Cisler, J. M., Elton, A., Kennedy, A. P., Young, J., Smitherman, S., James, G. A., & Kilts, C. D. (2013). Altered functional connectivity of the insular cortex across prefrontal networks in cocaine addiction. *Psychiatry Research: Neuroimaging, 213*, 39–46.

Coan, J. A., & Allen, J. J. (2003). State and trait of frontal EEG asymmetry in emotion. In K. Hugdahl, & R. J. Davidson (Eds.), *The asymmetrical brain* (pp. 566–615). Cambridge: MIT Press.

Conti, C. L., & Nakamura-Palacios, E. M. (2014). Bilateral transcranial direct current stimulation over dorsolateral prefrontal cortex changes the drug-cued reactivity in the anterior cingulate cortex of crack-cocaine addicts. *Brain Stimulation, 7*, 130–132.

Davidson, R. J. (2004). What does the prefrontal cortex "do" in affect: Perspectives on frontal EEG asymmetry research. *Biological Psychology, 67*, 219–234.

De Wit, H. (2009). Impulsivity as a determinant and consequence of drug use: A review of underlying processes. *Addiction Biology, 14*, 22–31.

Del Felice, A., Bellamoli, E., Formaggio, E., Manganotti, P., Masiero, S., Cuoghi, G., ... Serpelloni, G. (2016). Neurophysiological, psychological and behavioural correlates of rTMS treatment in alcohol dependence. *Drug and Alcohol Dependence, 158*, 147–153.

Ding, W. N., Sun, J. H., Sun, Y. W., Zhou, Y., Li, L., Xu, J. R., & Du, Y. S. (2013). Altered default network resting-state functional connectivity in adolescents with Internet gaming addiction. *PLoS One, 8*, 1–8.

Finocchiaro, R., & Balconi, M. (2015). Reward-system effect and "left hemispheric unbalance": A comparison between drug addiction and high-BAS healthy subjects on gambling behavior. *Neuropsychological Trends, 17*, 37–45.

Fowles, D. C. (1994). A motivational theory of psychopathology. *Nebraska Symposium on Motivation, 41*, 181–238.

Fregni, F., Liquori, P., Fecteau, S., Nitsche, M. A., Pascual-Leone, A., & Boggio, P. S. (2008). Cortical stimulation of the prefrontal cortex with transcranial direct current stimulation reduces cue-provoked smoking craving: A randomized, sham-controlled study. *The Journal of Clinical Psychiatry, 69*, 32–40.

Gorini, A., Lucchiari, C., Russell-Edu, W., & Pravettoni, G. (2014). Modulation of risky choices in recently abstinent dependent cocaine users: A transcranial direct-current stimulation study. *Frontiers in Human Neuroscience, 8*(661), 1–9.

Gray, J. A. (1981). A critique of Eysenck's theory of personality. In H. J. Eysenck (Ed.), *A model for personality* (pp. 246–276). Heidelberg: Springer-Verlag Berlin.

Hajcak, G., Moser, J. S., Holroyd, C. B., & Simons, R. F. (2006). The feedback-related negativity reflects the binary evaluation of good versus bad outcomes. *Biological Psychology, 71*, 148–154.

Harmon-Jones, E., & Allen, J. J. (1997). Behavioral activation sensitivity and resting frontal EEG asymmetry: Covariation of putative indicators related to risk for mood disorders. *Journal of Abnormal Psychology, 106*, 159–163.

Harper, J., Malone, S. M., & Bernat, E. M. (2014). Theta and delta band activity explain N2 and P3 ERP component activity in a go/no-go task. *Clinical Neurophysiology, 125*, 124–132.

Hester, R., & Garavan, H. (2004). Executive dysfunction in cocaine addiction: Evidence for discordant frontal, cingulate, and cerebellar activity. *The Journal of Neuroscience, 24*, 11017–11022.

Kamarajan, C., Rangaswamy, M., Tang, Y., Chorlian, D. B., Pandey, A. K., Roopesh, B. N., ... Porjesz, B. (2010). Dysfunctional reward processing in male alcoholics: An ERP study during a gambling task. *Journal of Psychiatric Research, 44*, 576–590.

Kelly, C., Zuo, X. N., Gotimer, K., Cox, C. L., Lynch, L., Brock, D., ... Milham, M. P. (2011). Reduced interhemispheric resting state functional connectivity in cocaine addiction. *Biological Psychiatry, 69*, 684–692.

Ko, C. H., Hsieh, T. J., Chen, C. Y., Yen, C. F., Chen, C. S., Yen, J. Y., ... Liu, G. C. (2014). Altered brain activation during response inhibition and error processing in subjects with Internet gaming disorder: A functional magnetic imaging study. *European Archives of Psychiatry and Clinical Neuroscience, 264*, 661–672.

Koob, G. F., & Volkow, N. D. (2010). Neurocircuitry of addiction. *Neuropsychopharmacology, 35*, 217–238.

Li, B., Friston, K. J., Liu, J., Liu, Y., Zhang, G., Cao, F., ... Hu, D. (2014). Impaired frontal-basal ganglia connectivity in adolescents with internet addiction. *Scientific Reports, 4*(5027), 1–8.

Li, C. S. R., & Sinha, R. (2008). Inhibitory control and emotional stress regulation: Neuroimaging evidence for frontal–limbic dysfunction in psycho-stimulant addiction. *Neuroscience & Biobehavioral Reviews, 32*, 581–597.

Li, X., Zhang, F., Zhou, Y., Zhang, M., Wang, X., & Shen, M. (2013). Decision-making deficits are still present in heroin abusers after short-to long-term abstinence. *Drug and Alcohol Dependence, 130*, 61–67.

Makris, N., Gasic, G. P., Kennedy, D. N., Hodge, S. M., Kaiser, J. R., Lee, M. J., ... Breiter, H. C. (2008). Cortical thickness abnormalities in cocaine addiction—A reflection of both drug use and a pre-existing disposition to drug abuse? *Neuron, 60*, 174–188.

Ma, N., Liu, Y., Li, N., Wang, C. X., Zhang, H., Jiang, X. F., ... Zhang, D. R. (2010). Addiction related alteration in resting-state brain connectivity. *NeuroImage, 49*, 738–744.

Parvaz, M. A., Maloney, T., Moeller, S. J., Woicik, P. A., Alia-Klein, N., Telang, F., ... Goldstein, R. Z. (2012). Sensitivity to monetary reward is most severely compromised in recently abstaining cocaine addicted individuals: A cross-sectional ERP study. *Psychiatry Research: Neuroimaging, 203*, 75–82.

Porjesz, B., Rangaswamy, M., Kamarajan, C., Jones, K. A., Padmanabhapillai, A., & Begleiter, H. (2005). The utility of neurophysiological markers in the study of alcoholism. *Clinical Neurophysiology, 116*, 993–1018.

Potenza, M. N. (2008). The neurobiology of pathological gambling and drug addiction: An overview and new findings. *Philosophical Transactions of the Royal B Society, 363*, 3181–3189.

Potenza, M. N., Leung, H. C., Blumberg, H. P., Peterson, B. S., Fulbright, R. K., Lacadie, C. M., ... Gore, J. C. (2014). An FMRI Stroop task study of ventromedial prefrontal cortical function in pathological gamblers. *American Journal of Psychiatry, 160*, 1990–1994.

Reuter, J., Raedler, T., Rose, M., Hand, I., Gläscher, J., & Büchel, C. (2005). Pathological gambling is linked to reduced activation of the mesolimbic reward system. *Nature Neuroscience, 8*, 147–148.

Rolls, E. T. (2004). The functions of the orbitofrontal cortex. *Brain and Cognition, 55*, 11–29.

Sutton, S. K., & Davidson, R. J. (1997). Prefrontal brain asymmetry: A biological substrate of the behavioral approach and inhibition systems. *Psychological Science, 8*, 204–210.

Tanabe, J., Thompson, L., Claus, E., Dalwani, M., Hutchison, K., & Banich, M. T. (2007). Prefrontal cortex activity is reduced in gambling and nongambling substance users during decision-making. *Human Brain Mapping, 28*, 1276–1286.

Torres, A., Catena, A., Cándido, A., Maldonado, A., Megías, A., & Perales, J. C. (2013). Cocaine dependent individuals and gamblers present different associative learning anomalies in feedback-driven decision making: A behavioral and ERP study. *Frontiers in Psychology, 4*(122), 1–14.

Volkow, N. D., & Baler, R. D. (2014). Addiction science: Uncovering neurobiological complexity. *Neuropharmacology, 76*, 235–249.

Volkow, N. D., Fowler, J. S., & Wang, G. J. (2003). The addicted human brain: Insights from imaging studies. *The Journal of Clinical Investigation, 111*, 1444–1451.

Volkow, N. D., & Li, T. K. (2004). Drug addiction: The neurobiology of behaviour gone awry. *Nature Reviews Neuroscience, 5*, 963–970.

Volkow, N. D., & Morales, M. (2015). The brain on drugs: From reward to addiction. *Cell, 162*, 712–725.

Volkow, N. D., Wang, G. J., Tomasi, D., & Baler, R. D. (2013). Unbalanced neuronal circuits in addiction. *Current Opinion in Neurobiology, 23*, 639–648.

Wang, G. J., Volkow, N. D., & Fowler, J. S. (2002). The role of dopamine in motivation for food in humans: Implications for obesity. *Expert Opinion on Therapeutic Targets, 6*, 601–609.

Wareham, J. D., & Potenza, M. N. (2010). Pathological gambling and substance use disorders. *The American Journal of Drug and Alcohol Abuse, 36*, 242–247.

Worhunsky, P. D., Stevens, M. C., Carroll, K. M., Rounsaville, B. J., Calhoun, V. D., Pearlson, G. D., & Potenza, M. N. (2013). Functional brain networks associated with cognitive control, cocaine dependence, and treatment outcome. *Psychology of Addictive Behaviors, 27*, 477–488.

Xie, C., Li, S. J., Shao, Y., Fu, L., Goveas, J., Ye, E., ... Yang, Z. (2011). Identification of hyperactive intrinsic amygdala network connectivity associated with impulsivity in abstinent heroin addicts. *Behavioural Brain Research, 216*, 639–646.

Yamamoto, D. J., Reynolds, J., Krmpotich, T., Banich, M. T., Thompson, L., & Tanabe, J. (2014). Temporal profile of fronto-striatal-limbic activity during implicit decisions in drug dependence. *Drug and Alcohol Dependence, 136*, 108–114.

Yan, W. S., Li, Y. H., Xiao, L., Zhu, N., Bechara, A., & Sui, N. (2014). Working memory and affective decision-making in addiction: A neurocognitive comparison between heroin addicts, pathological gamblers and healthy controls. *Drug and Alcohol Dependence, 134*, 194–200.

Yuan, K., Qin, W., Wang, G., Zeng, F., Zhao, L., Yang, X., ... Tian, J. (2011). Microstructure abnormalities in adolescents with internet addiction disorder. *PLoS One, 6*, e20708, 1-8.

Further Reading

Kravitz, A. V., Tomasi, D., LeBlanc, K. H., Baler, R., Volkow, N. D., Bonci, A., & Ferré, S. (2015). Cortico-striatal circuits: Novel therapeutic targets for substance use disorders. *Brain Research, 1628*, 186–198.

Volkow, N. D., & Baler, R. D. (2015). NOW vs LATER brain circuits: Implications for obesity and addiction. *Trends in Neurosciences, 38*, 345–352.

Index

'*Note*: Page numbers followed by "f" indicate figures and "t" indicate tables.'

Printed and bound by CPI Group (UK) Ltd, Croydon, CR0 4YY
08/06/2025

01896880-0001